2004 NSTI Nanotechnology Conference and

NSTI Nanotech 2004
Volume 2

Boston, March 7-11 2004

The Nanotechnology Conference and Trade Show

An Interdisciplinary Integrative Forum on
Nanotechnology, Biotechnology and Microtechnology

March 7-11, 2004
Boston Sheraton Hotel
Boston, Massachusetts, USA
www.nsti.org

Nano Science and Technology Institute
Boston • Geneva • San Francisco

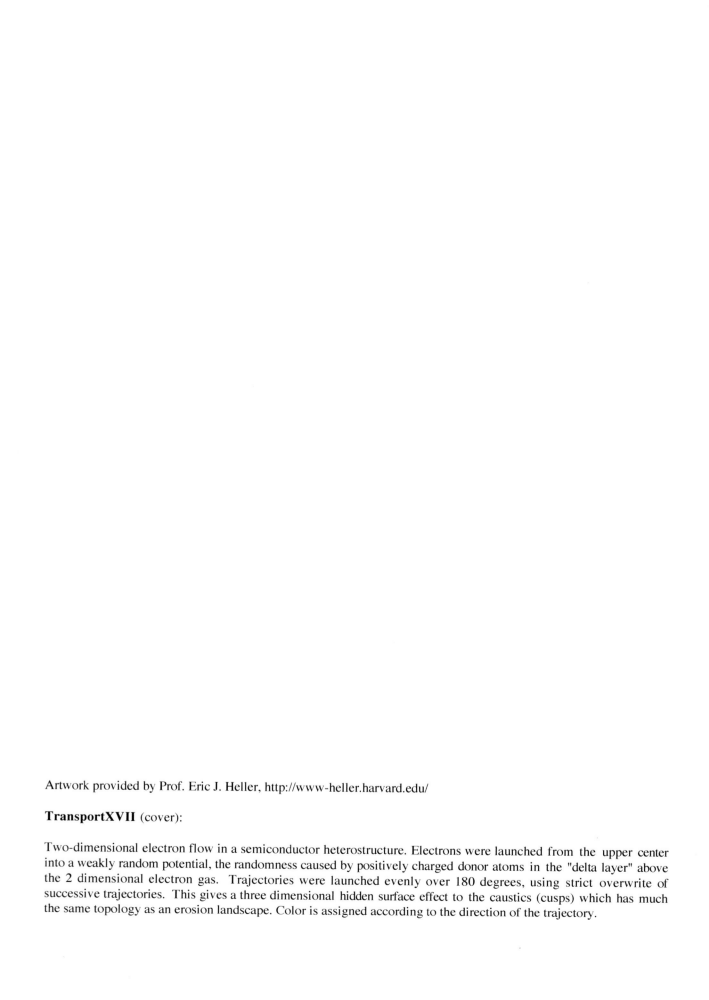

TransportXVII (cover):

Two-dimensional electron flow in a semiconductor heterostructure. Electrons were launched from the upper center into a weakly random potential, the randomness caused by positively charged donor atoms in the "delta layer" above the 2 dimensional electron gas. Trajectories were launched evenly over 180 degrees, using strict overwrite of successive trajectories. This gives a three dimensional hidden surface effect to the caustics (cusps) which has much the same topology as an erosion landscape. Color is assigned according to the direction of the trajectory.

2004 NSTI Nanotechnology Conference and Trade Show

NSTI Nanotech 2004

Volume 2

An Interdisciplinary Integrative Forum on
Nanotechnology, Biotechnology and Microtechnology

March 7-11, 2004
Boston, Massachusetts, USA
http://www.nsti.org

NSTI Nanotech 2004 Joint Meeting

The 2004 NSTI Nanotechnology Conference and Trade Show includes:
7th International Conference on Modeling and Simulation of Microsystems, MSM 2004
4th International Conference on Computational Nanoscience and Technology, ICCN 2004
2004 Workshop on Compact Modeling, WCM 2004

NSTI Nanotech 2004 Proceedings Editors:

Matthew Laudon
mlaudon@nsti.org

Bart Romanowicz
bfr@nsti.org

Nano Science and Technology Institute
Boston • Geneva • San Francisco

**Nano Science and Technology Institute
One Kendall Square, PMB 308
Cambridge, MA 02139
U.S.A.**

The papers in this book comprise the proceedings of the 2004 NSTI Nanotechnology Conference and Trade Show, Nanotech 2004, Boston, Massachusetts, March 7-11 2004. They reflect the authors' opinions and, in the interests of timely dissemination, are published as presented and without change. Their inclusion in this publication does not necessarily constitute endorsement by the editors, Nano Science and Technology Institute, or the sponsors. Dedicated to our family and friends that have supported and suffered all of our late nights. And to Avinash and Nina, study nicely – Ed.

Nano Science and Technology Institute Order Number PCP04020392
ISBN 0-9728422-8-4
ISBN 0-9728422-5-X (www)
ISBN 0-9728422-6-8 (CD-ROM, Vol. 1-3)

Additional copies may be ordered from:

Nano Science and Technology Institute
Publishing Office
One Kendall Sq., PMB 308
Cambridge, MA 02139, USA
http://www.nsti.org

Printed in the United States of America

**Nano Science and Technology Institute
Boston • Geneva • San Francisco**

Sponsors

Nano Science and Technology Institute

Intel Corporation

Motorola

Texas Instruments

Ciphergen Biosystems, Inc.

iMediasoft Group

Frontier Carbon Corporation

Veeco Instruments

Hitachi High Technologies America, Inc.

Racepoint Group, Inc.

International SEMATECH

FEI Company

Keithley

Saint-Gobain High Performance Materials

Zyvex Corporation

Accelrys, Inc.

ANSYS, Inc.

Atomistix

PolyInsight

Georgia Department of Industry & Trade

Swiss Business Hub USA

Australian Government - Invest Australia

State of Bavaria, Germany - The United States Office For Economic Development

m+w zander

Engis Corporation

COMSOL

Engelhard Corporation

Nanonex, Inc.

nanoTITAN

Tegal Corporation

Umech Technologies

MEMSCAP

Swiss House for Advanced Research and Education SHARE

Basel Area Business Development

Greater Zürich Area

Development Economic Western Switzerland

Location:Switzerland

Nanoworld AG

Nanosensors

Nanofair

Burns, Doane, Swecker & Mathis, L.L.P.

Mintz Levin Cohn Ferris Glovsky and Popeo, PC

Jackson Walker L.L.P.

Supporting Organizations

American Physical Society

IEEE Electron Devices Society

IEEE Boston Section

The American Ceramic Society

ASME Nanotechnology Institute

NANOPOLIS

The Global Emerging Technology Institute, Ltd.

Nanotechnology Researchers Network Center of Japan

TIMA-CMP Laboratory, France

Squire, Sanders & Dempsey

Boston University

Massachusetts Technology Collaborative

CIMIT (Center for Integration of Medicine & Innovative Technology)

Draper Laboratory

International Scientific Communications, Inc

DARPA (Defense Advanced Research Projects Agency)

Swiss Federal Institute of Technology

Ibero-American Science and Technology Education Consortium

Applied Computational Research Society

Media Sponsors

Technology Review

iMediasoft Group

Taylor & Francis

NanoApex

Institute of Physics

nanotechweb.org

Photonics Spectra

R&D Magazine and Micro/Nano

Earth and Sky Radio Series

Nano Nordic Network

Table of Contents

Compact Modeling

Circuits

System Level Modeling

MEMS Modeling

Modeling Fundamental Phenamena in MEMS

Computational Methods and Numerics

NSTI Nanotech 2004 Program Committee

TECHNICAL PROGRAM CO-CHAIRS

Matthew Laudon	*Nano Science and Technology Institute, USA*
Bart Romanowicz	*Nano Science and Technology Institute, USA*

TOPICAL AND REGIONAL SCIENTIFIC ADVISORS AND CHAIRS

Nanotechnology

Matthew Laudon	*Nano Science and Technology Institute, USA*
Philippe Renaud	*Swiss Federal Institute of Technology, Switzerland*
Mihail Roco	*National Science Foundation, USA*
Wolfgang Windl	*Ohio State University, USA*

Biotechnology

Robert S. Eisenberg	*Rush Medical Center, Chicago, USA*
Srinivas Iyer	*Los Alamos National Laboratory, USA*

Pharmaceutical

Kurt Krause	*University of Houston, USA*

Microtechnology

Narayan R. Aluru	*University of Illinois Urbana-Champaign, USA*
Bernard Courtois	*TIMA-CMP, France*
Anantha Krishnan	*Defense Advanced Research Projects Agency, USA*
Bart Romanowicz	*Nano Science and Technology Institute, USA*

Semiconductors

David K. Ferry	*Arizona State University, USA*
Andreas Wild	*Motorola, Germany*

NANOTECHNOLOGY CONFERENCE COMMITTEE

M.P. Anantram	*NASA Ames Research Center, USA*
Phaedon Avouris	*IBM, USA*
Xavier J.R. Avula	*Washington University, USA*
Wolfgang S. Bacsa	*Université Paul Sabatier, France*
Roberto Car	*Princeton University, USA*
Franco Cerrina	*University of Wisconsin - Madison, USA*
Murray S. Daw	*Clemson University, USA*
Alex Demkov	*Motorola, USA*
Toshio Fukuda	*Nagoya University, Japan*
David K. Ferry	*Arizona State University, USA*
Sharon Glotzer	*University of Michigan, USA*
William Goddard	*CalTech, USA*
Gerhard Goldbeck-Wood	*Accelrys, Inc., UK*
Niels Gronbech-Jensen	*UC Davis and Berkeley Laboratory, USA*
Karl Hess	*University of Illinois at Urbana-Champaign, USA*
Charles H. Hsu	*Maximem Limited, Taiwan*
Hannes Jonsson	*University of Washington, USA*
Anantha Krishnan	*Defense Advanced Research Projects Agency, USA*
Alex Liddle	*Lawrence Berkeley National Laboratory, USA*
Chris Menzel	*Nano Science and Technology Institute, USA*
Stephen Paddison	*Los Alamos National Laboratory, USA*
Sokrates Pantelides	*Vanderbilt University, USA*
Philip Pincus	*University of California at Santa Barbara, USA*
Joachim Piprek	*University of California at Santa Barbara, USA*
Serge Prudhomme	*University of Texas at Austin, USA*
Nick Quirke	*Imperial College, London, UK*
PVM Rao	*Indian Institute of Technology, Delhi, India*
Philippe Renaud	*Swiss Federal Institute of Technology of Lausanne, Switzerland*
Robert Rudd	*Lawrence Livermore National Laboratory, USA*
Douglas Smith	*University of San Diego, USA*
Clayton Teague	*National Nanotechnology Coordination Office, USA*
Dragica Vasilesca	*Arizona State University, USA*
Arthur Voter	*Los Alamos National Laboratory, USA*
Phillip R. Westmoreland	*University of Massachusetts Amherst, USA*
Wolfgang Windl	*Ohio State University, USA*

| Gloria Yueh | *Midwestern University, USA* |
| Xiaoguang Zhang | *Oakridge National Laboratory, USA* |

BIOTECHNOLOGY CONFERENCE COMMITTEE

Dirk Bussiere	*Chiron Corporation, USA*
Amos Bairoch	*Swiss Institute of Bioinformatics, Switzerland*
Stephen H. Bryant	*National Institute of Health, USA*
Fred Cohen	*University of California, San Francisco, USA*
Daniel Davison	*Bristol Myers Squibb, USA*
Andreas Hieke	*Ciphergen Biosystems, Inc., USA*
Leroy Hood	*Institute for Systems Biology, USA*
Sorin Istrail	*Celera Genomics, USA*
Srinivas Iyer	*Los Alamos National Laboratory, USA*
Brian Korgel	*University of Texas-Austin, USA*
Kurt Krause	*University of Houston, USA*
Daniel Lacks	*Tulane University, USA*
Mike Masquelier	*Motorola, Los Alamos NM, USA*
Atul Parikh	*University of California, Davis, USA*
Andrzej Przekwas	*CFD Research Corporation, USA*
George Robillard	*BioMade Corporation, Netherlands*
Jonathan Rosen	*Center for Integration of Medicine and Innovative Technology, USA*
Tom Terwilliger	*Los Alamos National Laboratory, USA*
Michael S. Waterman	*University of Southern California, USA*

MICROSYSTEMS CONFERENCE COMMITTEE

Narayan R. Aluru	*University of Illinois Urbana-Champaign, USA*
Xavier J. R. Avula	*University of Missouri-Rolla, USA*
Stephen F. Bart	*Nano Science and Technology Institute, USA*
Bum-Kyoo Choi	*Sogang University, Korea*
Bernard Courtois	*TIMA-CMP, France*
Robert W. Dutton	*Stanford University, USA*
Gary K. Fedder	*Carnegie Mellon University, USA*
Elena Gaura	*Coventry University, USA*
Steffen Hardt	*Institute of Microtechnology Mainz, Germany*
Lee W. Ho	*Corning Intellisense, USA*
Eberhard P. Hofer	*University of Ulm, Germany*
Michael Judy	*Analog Devices, USA*
Yozo Kanda	*Toyo University, Japan*
Jan G. Korvink	*University of Freiburg, Germany*
Anantha Krishnan	*Defense Advanced Research Projects Agency, USA*
Mark E. Law	*University of Florida, USA*
Mary-Ann Maher	*MemsCap, France*
Kazunori Matsuda	*Naruto University of Education, Japan*
Tamal Mukherjee	*Carnegie Mellon University, USA*
Andrzej Napieralski	*Technical University of Lodz, Poland*
Ruth Pachter	*Air Force Research Laboratory, USA*
Michael G. Pecht	*University of Maryland, USA*
Marcel D. Profirescu	*Technical University of Bucharest, Romania*
Marta Rencz	*Technical University of Budapest, Hungary*
Siegried Selberherr	*Technical University of Vienna, Austria*
Sudhama Shastri	*ON Semiconductor, USA*
Armin Sulzmann	*Daimler-Chrysler, Germany*
Mathew Varghese	*The Charles Stark Draper Laboratory, Inc., USA*
Dragica Vasilesca	*Arizona State University, USA*
Gerhard Wachutka	*Technical University of Münich, Germany*
Jacob White	*Massachusetts Institute of Technology, USA*
Thomas Wiegele	*BF Goodrich Aerospace, USA*
Wenjing Ye	*Georgia Institute of Technology, USA*
Sung-Kie Youn	*Korea Advanced Institute of Science and Technology, Korea*
Xing Zhou	*Nanyang Technological University, Singapore*

CONFERENCE OPERATIONS MANAGER

| Sarah Wenning | *Nano Science and Technology Institute, USA* |

NSTI Nanotech 2004 Proceedings Topics

2004 NSTI Nanotechnology Conference and Trade Show

NSTI Nanotech 2004, Vol. 1, ISBN 0-9728422-7-6

- Bio Nano Systems and Chemistry
- Bio Nano Analysis and Characterization
- Bio Molecular Motors
- Ion Channels
- Bio Nano Computational Methods and Applications
- Bio Micro Sensors and Systems
- Micro Fluidics and Nanoscale Transport
- MEMS Design and Application
- Smart MEMS and Sensor Systems
- Micro and Nano Structuring and Assembly
- Wafer and MEMS Processing

NSTI Nanotech 2004, Vol. 2, ISBN 0-9728422-8-4

- Advanced Semiconductors
- Nano Scale Device Modeling
- Compact Modeling
- Circuits
- System Level Modeling
- MEMS Modeling
- Modeling Fundamental Phenomena in MEMS
- Computational Methods and Numerics

NSTI Nanotech 2004, Vol. 3, ISBN 0-9728422-9-2

- Nano Photonics, Optoelectronics and Imaging
- Nanoscale Electronics and Quantum Devices
- Atomic and Mesoscale Modelling of Nanoscale Phenomena
- Nano Devices and Systems
- Carbon Nano Structures and Devices
- Nano Composites
- Surfaces and Films
- Nano Particles and Molecules
- Characterization and Parameter Extraction
- Micro and Nano Structuring and Assembly
- Commercial Tools, Processes and Materials

Message from the Program Committee

The Nano Science and Technology Institute is proud to present the 2004 Nanotechnology Conference and Trade Show (Nanotech 2004). The charter of the Nanotech conference, and its numerous sub-conferences, remains the same since its original conception in 1997. The Nanotech provides for a single interdisciplinary integrative community, allowing for core scientific advancements to disseminate into a multitude of industrial sectors and across the breadth of traditional science and technology domains converging under Nanotechnology, Biotechnology and Microtechnology.

The Nanotech Program Committee makes every effort to provide a scientifically outstanding environment, through its review and ranking process. The Nanotech received a total of 656 abstracts for the 2004 event. All abstracts submitted into the Nanotech are reviewed and scored by a minimum of three (3) expert reviewers. Of the 656 submitted abstracts, 25% were accepted as oral presentations, 40% were accepted as poster presentations resulting in a 35% rejection rate. In addition to the regular program there are a number of invited sessions, panels, overview presentations and tutorials provided for completeness. The Nanotech program committee thanks the hard work of all of this year's reviewers (116 active reviewers in total). This grassroots self-review process by the nano, bio and micro technology communities is a source of pride for the Nanotech conference. This process is in place to provide for a yearly evolution and technical validation of the Nanotech conference content. We thank the authors for submitting their latest work, making a meeting of this caliber possible.

We hope the reader will find the papers assembled in these proceedings rewarding to read, and that the conference continues to foster further advances in this fascinating and multi-disciplinary field. Although the Nanotech conference makes every effort to be as comprehensive as possible, due to the rapid advancements in science and industry, there will inevitably be under-represented sections of this event. We look to you, as the true science and business leaders of small-technologies, to assist us in identifying the needs of our participants so that we can continue to grow the content of this event to best serve this community.

We would like to take this opportunity to thank the many individuals who have worked so hard to make this meeting happen, and to welcome the new members of the program committee. Conferences of this scope are possible only because of the continuing interest and support of the community, expressed both by their submission of papers of high quality and by their attendance. The Nanotech 2004 program committee is grateful to all keynote speakers, authors and session chairs for contributing to the success of the event. We are also indebted to the foundations, agencies and companies whose financial contributions made this meeting possible. We encourage your feedback and participation. Additionally, if you have an interest in conference or session organizational assistance, please contact the conference manager and we will attempt to accommodate. Information concerning next year's conference is posted online at URL: http://www.nsti.org. We look forward to seeing you again next year, and thank you for your continuing support and participation.

Technical Program Co-chairs, Nanotech 2004

Matthew Laudon, Nano Science and Technology Institute, USA

Bart Romanowicz, Nano Science and Technology Institute, USA

An electrothermal solution of the heat equation for MMICs based on the 2-D Fourier series

Agostino Giorgio, Anna Gina Perri

Laboratorio di Dispositivi Elettronici, Dipartimento di Elettrotecnica ed Elettronica, Politecnico di Bari
Via E. Orabona 4, 70125, Bari, Italy
Phone: +39 - 80 - 5963427/5963314/5963239 Fax: +39 - 80- 5963410 E-mail:perri@poliba.it

Abstract
In this paper a 2-D Fourier transform-based analytical method for the thermal and electrical solution of multilayer structure electronic devices is proposed.
It takes into account the dependence of the thermal conductivity of all the layers on the temperature; the heat equation is coupled with the device current-voltage relation in order to give physical consistence to the experimental evidence that the temperature increase causes a degradation of the electrical performances and that the electrical power is not uniformly distributed.

I. INTRODUCTION

The general evolution of electronic devices for high frequency applications and the recent interest in integrating power devices on Microwave Monolithic Integrated Circuits (MMICs) emphasize the growing importance of the thermal problem in the design process.

Particularly in GaAs technology, one of the main problems to overcome is the low thermal conductivity of the semiconductor, which focuses the designer's interest on the design thermal optimization when good reliability is to be achieved.

An electrothermal solution of the heat equation for MMICs based on the 2-D Fourier series was presented by Johnson *et al.* [1]. Unfortunately, in this case the ipothesis of uniform channel temperature was assumed, which, on one hand, greatly reduces the computational effort with respect to a fine discretization of the heat source but, on the other hand, neglects the non-uniform power dissipation under the gate, which is a well-known phenomenon. Furthermore, in case of multifinger devices or thermal coupling between contiguous devices, the inaccuracy introduced by the ipothesis of uniform channel temperature can be relevant.

This paper presents an analytical procedure for the thermal and electrical solution for multilayer structure integrated devices. In particular, it addresses the solution to the 3-D steady-state heat equation with temperature-dependent thermal conductivity for a single electronic device or a given configuration of two or more devices.

Section II outlines the proposed mathematical model, whereas in Section III an example of application to a GaAs MESFET is shown and the numerical results of the simulation are discussed and compared with experimental data.

II. THE PROPOSED METHOD

In order to calculate the I-V characteristics of a GaAs MESFET, the temperature-dependent physical properties of the device, such as low-field mobility, saturation velocity, built-in voltage, energy gap, parasitic source-to-gate and gate-to-drain resistance, have been evaluated at the actual channel temperature.

The temperature distribution has been computed using the 3-D solution of the heat equation in steady-state condition for a multiple layer structure with an embedded thermal source [2-5]:

$$\overline{\nabla} \cdot \left[k_{TH}(T) \overline{\nabla} T(x,y,z) \right] = -Q_V(x,y,z) \qquad (1)$$

where $T(x,y,z)$ is the temperature field, k_{TH} is the temperature-dependent thermal conductivity and $Q_V(x,y,z)$ is the dissipated power density.
The basic assumptions of the model are:
- the device and package structure can be represented as a set of superimposed homogeneous layers
- the thickness of each layer is constant and the extension of the layers in the x and y direction is infinite
- the contact thermal resistance is neglected
- the thermal source is modeled as a 2-D geometrical shape $Q_S(x,y)$, located at the interface between two contiguous layers, say the k-th and the $(k+1)$-th.

Eqn. (1) can be solved considering the following:

$$\overline{\nabla} \cdot \left[k_{TH}(T) \overline{\nabla} T(x,y,z) \right] = 0 \qquad (2)$$

and accounting for the heat source in the Boundary Conditions (BCs).
Dirichlet and Neumann BCs can be expressed as:

$$T_i(x,y,z) = T_{i+1}(x,y,z) \quad i = 1...m-1 \qquad (3a)$$

$$k_{TH1}(T_1) \frac{\partial T_1(x,y,z)}{\partial z} = 0 \qquad (3b)$$

$$T_m(x, y, z) = T_\infty \qquad (3c)$$

$$\Psi_i(\alpha, \beta, z_{i+1}) = \Psi_{i+1}(\alpha, \beta, z_{i+1}) \qquad (8a)$$

$$k_{THi}(T_i)\frac{\partial T_i(x, y, z)}{\partial z} = k_{THi+1}(T_{i+1})\frac{\partial T_{i+1}(x, y, z)}{\partial z} \qquad (3d)$$

$$k_{R1}\frac{d\Psi_1(\alpha, \beta, z_1)}{dz} = 0 \qquad (8b)$$

$$k_{THk}(T_k)\frac{\partial T_k(x, y, z)}{\partial z} - k_{THk+1}(T_{k+1})\frac{\partial T_{k+1}(x, y, z)}{\partial z} = Q_S(x, y) \qquad (3e)$$

$$\Psi_m(\alpha, \beta, z_m) = 0 \qquad (8c)$$

Eqn. (3b) and (3c) refer to an adiabatic top surface and to an isothermal bottom surface with reference room temperature T_∞ respectively, whereas Eqn. (3e) refers to the interface containing the heat source. Eqns. (3a) and (3d) impose the temperature and heat flux continuity across the interfaces.

In order to linearize Eqn. (2), the Kirchhoff transformation can be applied to each layer in the following form:

$$\theta_i(x, y, z) = \frac{1}{k_{Ri}}\int_{T_\infty}^{T_i(x,y,z)} k_{THi}(\tau)d\tau \quad i = 1...m \qquad (4)$$

In Eqn. (4) $\theta_i(x,y,z)$ is the transformed temperature or the so-called "pseudo-temperature" of the i-th layer, $T_i(x,y,z)$ is the actual temperature, k_{Ri} is the temperature-dependent thermal conductivity $k_{THi}(T)$ evaluated at $T = T_\infty$.
Hence, Eqn. (2) is transformed into the well-known Laplace equation:

$$\nabla^2\theta_i(x, y, z) = 0 \qquad (5)$$

Unfortunately, applying the Kirchhoff transformation to Eqns. (3a) ÷ (3e), the non-linearity of the problem equation is shifted to the boundary conditions. Thus, if a first order Taylor expansion for the inverse transform is considered, i.e. $T \approx \theta + T_\infty$, the resulting problem is linear in each layer.

Furthermore, to simplify the mathematical steps, the 2-D Fourier transform can be applied to (5) in the following form:

$$\Psi_i(\alpha, \beta, z) = \int_{-\infty}^{+\infty}\int_{-\infty}^{+\infty}\theta_i(x, y, z)e^{-j\alpha x}e^{-j\beta y}dxdy$$

This leads to the one-dimensional ordinary differential equation:

$$\frac{d^2\Psi_i}{dz^2} - (\alpha^2 + \beta^2)\Psi_i = 0 \qquad (6)$$

with the following solution in each i-th layer:

$$\Psi_i(\alpha, \beta, z) = C_i'e^{-\gamma z} + C_i''e^{\gamma z} \qquad (7)$$

and with the transformed BCs:

$$k_{Ri}\frac{d\Psi_i(\alpha, \beta, z_{i+1})}{dz} = k_{Ri+1}\frac{d\Psi_{i+1}(\alpha, \beta, z_{i+1})}{dz} \qquad (8d)$$

$$k_{Rk}\frac{d\Psi_k(\alpha, \beta, z_{k+1})}{dz} - k_{Rk+1}\frac{d\Psi_{k+1}(\alpha, \beta, z_{k+1})}{dz} = \Im_{\alpha\beta}(Q_s(x, y)) \qquad (8e)$$

In Eqn. (8e), the right-hand side is the Fourier transform of the 2-D heat source, which can be easily calculated once the geometrical shape has suitably been described. The heat source can be modeled as a rectangle located at the active layer-to-substrate interface but, to account for the non-uniform power dissipation, it is more convenient to approximate the 2-D shape as a set of elementary point sources. In order to link the power dissipation to the device current, each point source has been associated with a part of the device dissipation region assuming that the whole current can be expressed as the sum of contributions, corresponding to elementary devices. In this way the original problem results split in elementary problems in which an unit hot spot is associated with an unit device. Since the problem (6) with the BCs (8a) – (8e) is linear, it is possible to solve the elementary problem and then reconstructing the overall solution by applying the superposition of effects.

The right-hand side of (8e) for a point heat source is:

$$\Im_{\alpha\beta}(Q_s(x, y)) = \int_{-\infty}^{+\infty}\int_{-\infty}^{+\infty}Q_0\delta(x - a)\delta(y - b)e^{-j\alpha x}e^{-j\beta y}dxdy = Q_0$$

where $\delta(x-x_0)$ is the Dirac function centered in x_0 and Q_0 is the power dissipated by the unit device. The electrothermal feedback can be implemented by evaluating the current of each elementary device at the actual channel temperature, which is approximated with the hot spot temperature.

The solution to (6) can be calculated by substituting (7) into (8a) - (8e), which leads to the linear system:

$$M(\gamma)C(\alpha, \beta) = U(\alpha, \beta) \qquad (9)$$

where $\gamma^2 = \alpha^2 + \beta^2$, $M(\gamma)$ is the $2m \times 2m$ coefficient matrix, C is the integration constants vector containing the unknowns C_1', C_1'', .., C_m', C_m'', U is the column vector containing the Fourier transform of the heat source and having only the $(k+1)$-th non-zero entry. It has to be remarked that M is a function of γ and not of α and β separately, while this is not generally true for C and U but results in the case of point heat source.

Unfortunately, the solution to Eqn. (9) is not a trivial problem since $M(\gamma)$ is not a numeric matrix but contains the Fourier frequencies α and β as parameters. It could be possible to give a closed-form expression of (7) after solving (9) by applying the Cramer rule and substituting C_1', C_1'', .., C_m', C_m'' into (7), but just for a limited number of layers, e.g. five. However, it would be a very tedious and almost impossible operation to carry out for a large number of layers. Furthermore, Eqn. (7) has to be back-transformed involving a double integration in a large domain of a very complicated expression. In this work the Discrete Fourier Transform (DFT) has been applied in order to show that the linear system (9) can be solved for any m and the simultaneous solution of the pseudo-temperature θ_i of all layers can be obtained.

The proposed technique consists of sampling (9) i.e.:

$$\forall \alpha = \alpha_p, \beta = \beta_q \Rightarrow M(\gamma_{pq})C(\alpha_p,\beta_q) = U(\alpha_p,\beta_q) \quad (10)$$

which can be easily solved since it is a numeric system:

$$C(\alpha_p,\beta_q) = M^{-1}(\gamma_{pq})U(\alpha_p,\beta_q) \quad (11)$$

Thus, after substituting (11) into (7), the samples of the 2-D Fourier transform are:

$$\Psi_i(\alpha_p,\beta_q,z) = C_i'(\alpha_p,\beta_q)e^{-\gamma_{pq}z} + C_i''(\alpha_p,\beta_q)e^{\gamma_{pq}z}$$

where index i refers to the i-th layer.

In order to perform the 2-D inverse DFT, which is a computationally advantageous approach, it is useful to evaluate (12) on specific surfaces, e.g. on the interfaces between contiguous layers, so as to obtain samples of the 2-D function $\theta_i(x,y,z_i)$. It can be easily shown that if the point thermal source is normalized to unit, Eqn. (9) can be solved just once and the inverse transform of (12), referred to the $(k+1)$-th, represents the normalized unit thermal profile on the source surface. It can be used to calculate the whole thermal field by multiplying it by the dissipated power of a specific elementary device and by shifting the resulting function to the device location. Updating the elementary dissipated powers and solving iteratively, the device current results consistent with the actual channel temperature.

III. NUMERICAL RESULTS AND EXPERIMENTAL DATA

The proposed technique has been applied to the multi-gate GaAs MESFET, the TC252MB produced by Alcatel.

A one-dimensional I-V MESFET equation [2] has been implemented in order to consider the feedback between the device current and the active layer temperature distribution. The most widely accepted empiric relations between FET physical parameters and temperature have also been taken into account [4-5].

In Fig. 1 the cross section of the overall thermal profile for the given device is shown. In this case the dissipated power is $P = 1.18$ W. The solid line refers to the temperature profile along the y-axis after the Kirchhoff transform whereas the dash-dotted line refers to the same profile before the transformation. As one can clearly see, the difference becomes relevant in the device area, which confirms that the non-linear dependence of the GaAs thermal conductivity cannot be neglected. Furthermore, as the peak temperature rise above the reference temperature of 300 K is about 140 K and the mean temperature rise in the active area is about 110 K, the linear approximation of the boundary condition leads to a 6% error, which is acceptable.

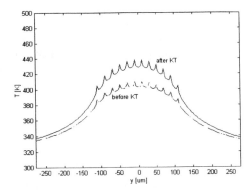

Fig. 1 – Cross section of the thermal profile on the source surface along the y direction: comparison between the results after the Kirchoff transform (solid line) and before the transformation (dash-dotted line).

Finally, in Fig. 2 the overall thermal profile on the heat source surface is shown. The plot has been derived for dissipated power $P = 1.18$ W at $V_{GS} = -1$ V and $V_{DS} = 5$ V.

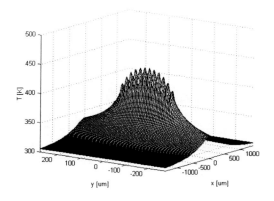

Fig. 2 – Overall thermal field on the surface containing the heat source for P = 1.18 W, V_{GS} = -1 V and V_{DS} = 5 V.

The calculation time for the given example can be quantified in less than 10 minutes for the main part of the algorithm, that is the temperature field of the elementary source and the current of the elementary device, and in a few minutes for the graphic routines on a 400MHz compatible PC. It is worthwhile to remark that the software, by which all the calculations have been carried out, has been implemented just for academic non-commercial purpose.

IV. CONCLUSION

A mathematical model for the solution to the heat flow equation in a multilayer structure has been presented. An example of application to a multifinger GaAs MESFET has also been given. The FET drain current has been related to the actual channel temperature coupling a 2-D electric model with a 3-D thermal model. The non-linearity of the problem has been handled by the Kirchhoff transform and the transformed heat flux equation has been solved by the 2-D Fourier transform. The method is general and can be applied to wide range of integrated devices provided their structure can be approximated with a multiple layer structure. The limitation in the number of layers allowed by previous presented methods has also been overcome.

V. REFERENCES

[1] R. G. Johnson, W. Batty, A. J. Panks, C. M. Snowden "Fully physical coupled electro-thermal simulations and measurements of power FETs", *IEEE MTT-S*, vol. 1, pp. 461-464, Boston, 2000.

[2] A. Giorgio, A.G. Perri, "Dispositivi Elettronici: fisica tecnologia e modelli", Ed. Biblios, Bari (Italy), 1996.

[3] R. Anholt "Electrical and thermal characterization of MESFETs, HEMTs and HBTs", Artech House Inc., 1995.

[4] J. S. Blakemore " Semiconducting and other major properties of GaAs", *J. Appl. Phys.*, vol. **53**, pp. R123-R181, October 1982.

[5] J. C. Brice "Properties of gallium arsenide", EMIS Datareviews Series No. 2, INSPEC, London and New York, 1990.

Spectral Analysis of Channel Noise in Nanoscale MOSFETS

Giorgio Casinovi

School of Electrical and Computer Engineering
Georgia Institute of Technology
Atlanta, GA 30332-0250

ABSTRACT

This paper describes an algorithm for numerical computation of the power spectral density (PSD) of channel noise in nanoscale MOSFETs. Noise generation phenomena inside the channel are modeled as random processes, represented by distributed sources that are added to the equations describing charge transport in the channel. The resulting set of differential equations is then solved using a frequency-domain simulation algorithm, yielding the total noise PSD at the device terminals. Comparisons between simulated and measured values of the noise PSD can then be used to test the validity of noise models for nanoscale MOSFETs. Simulation results obtained on a MOSFET under various DC bias conditions are presented.

Keywords: Device modeling; simulation; noise.

1 INTRODUCTION

Noise generation in semiconductor devices has been studied for over fifty years. The best known model of channel noise in MOSFETs is due to van der Ziel [1]. While this model is sufficiently accurate for long-channel devices, discrepancies with experimental data in submicron devices were already observed more than fifteen years ago [2]. In fact, the van der Ziel model underestimates the thermal noise generated in short-channel devices; the origin of this excess noise has been the object of intense investigation for several years.

Some researchers have tried to obtain more accurate models by resorting to approaches based on the so-called impedance field method [3], which leads to modeling the channel as a linear, one-dimensional transmission line in which noise generation phenomena are represented by Langevin sources distributed throughout the length of the channel [4], [5]. Because the line parameters vary along the channel, better accuracy can be obtained by dividing the channel into multiple segments, where each segment is modeled as a uniform transmission line [6].

Based on the results of simulations performed using the approaches described above, it has been suggested that the excess noise can be explained in terms of nonlocal effects on the carrier velocity that are captured only by higher-order transport models, such as the hydrodynamic model [6], [7].

This work was supported in part by the National Science Foundation under grant CCR-0306343.

Other researchers, however, have offered different explanations of the same phenomenon, based on effects such as carrier velocity saturation and carrier heating. In short channel devices, carrier velocity saturates near the drain end of the channel due to the presence of high lateral electrical fields [8], [9]. This affects the local value of the channel conductance, which is normally expressed as: $G = \mu(x)q$, where $\mu(x)$ is the carrier mobility and q is the inversion layer charge density. Since the values of the local noise sources that model thermal noise in the channel depend on G, variations in carrier mobility along the channel must be accounted for. Moreover, the assumption that thermal equilibrium exists in the channel is no longer necessarily valid in the presence of high electric fields [9]. In order to account for the deviation of carrier density from its equilibrium value, a field-dependent equivalent temperature is introduced. Since the exact relationship between the equivalent temperature and the electric field cannot be obtained analytically, empirical formulae expressing a linear [9] or quadratic [10], [11] dependence on the electric field are used instead. Several models explaining the excess channel noise as due to carrier velocity saturation and carrier heating have been published in the literature [9]–[12].

This paper describes an algorithm to compute the noise PSD at the drain and source terminals of a MOSFET from frequency-domain simulations of the ordinary or partial differential equations used to model the device's behavior (e.g. the drift-diffusion equations). Specifically, noise generation inside the device is modeled as a stationary random process, and a relationship between the PSD of the noise sources and that of the resulting noise voltages or currents at the device terminals is obtained by linearizing the charge transport equation around the bias point of the device. This makes it possible to simulate a variety of models for noise generation phenomena inside MOSFETS and to compare the simulation results with experimental measurements to verify the validity of those models. A detailed description of the algorithm is given in the next section, and simulation results are presented in Sec. 3.

2 NOISE SPECTRAL ANALYSIS

The cross-section of a MOSFET device is shown in Fig. 1. For the purpose of noise analysis, a very simple model for the channel charge will be used, as described by the equa-

Figure 1: MOSFET cross-section

tion:

$$Q(x) = -C_{ox}[V(x) + V_t]$$

where Q is the charge in the channel, V the local voltage in the channel with respect to the *gate*, C_{ox} the gate oxide capacitance, and V_t the MOSFET threshold voltage. Based on this model, the voltage and current in the device channel must satisfy the following pair of partial differential equations:

$$\begin{aligned}\frac{\partial V}{\partial x} &= \frac{I}{\mu C_{ox}(V + V_t)} \\ \frac{\partial I}{\partial x} &= -C_{ox}\frac{\partial}{\partial t}(V + V_t)\end{aligned} \quad (1)$$

Noise generation mechanisms inside the channel can be modeled by the introduction of noise sources in the above equations:

$$\begin{aligned}\frac{\partial V}{\partial x} &= \frac{I}{\mu C_{ox}(V + V_t)} + v_n(x,t) \\ \frac{\partial I}{\partial x} &= -C_{ox}\frac{\partial}{\partial t}(V + V_t) + i_n(x,t)\end{aligned}$$

where $v_n(x,t)$ and $i_n(x,t)$ model noise generation phenomena inside the channel.

The noise sources cause perturbations v and i in the channel voltage and current, respectively. Assuming that those perturbations are small, they can be computed by linearizing the equations above around the DC solution:

$$\begin{aligned}\frac{\partial v}{\partial x} &= -\frac{I}{\mu C_{ox}(V + V_t)^2}v + \frac{1}{\mu C_{ox}(V + V_t)}i \\ &\quad + v_n(x,t) \qquad\qquad\qquad (2) \\ \frac{\partial i}{\partial x} &= -C_{ox}\frac{\partial v}{\partial t} + i_n(x,t)\end{aligned}$$

This set of differential equations can be rewritten in matrix and vector notation as:

$$\frac{\partial \mathbf{f}}{\partial x} = \mathbf{G}(x)\mathbf{f} - \mathbf{C}\frac{\partial \mathbf{f}}{\partial t} + \mathbf{u}(x,t) \qquad (3)$$

where: $\mathbf{f} = [\, v \; i \,]^T$ and

$$\begin{aligned}\mathbf{G}(x) &= \frac{1}{\mu C_{ox}}\begin{bmatrix} \frac{I}{[V(x)+V_t]^2} & \frac{-1}{[V(x)+V_t]} \\ 0 & 0 \end{bmatrix} \\ \mathbf{C} &= \begin{bmatrix} 0 & 0 \\ C_{ox} & 0 \end{bmatrix} \\ \mathbf{u}(x,t) &= \begin{bmatrix} v_n(x,t) \\ i_n(x,t) \end{bmatrix}\end{aligned}$$

Equation (3) can be discretized with respect to x by selecting points $x_0 = 0 < x_1 < \ldots < x_N = L$ and using an appropriate numerical integration algorithm, such as the trapezoidal method:

$$\begin{aligned}\mathbf{f}_{n+1} &= \mathbf{f}_n + \frac{h_n}{2}\left[\mathbf{G}_{n+1}\mathbf{f}_{n+1} - \mathbf{C}\frac{d\mathbf{f}_{n+1}}{dt}\right. \\ &\quad \left. + \mathbf{u}_{n+1}(t) + \mathbf{G}_n\mathbf{f}_n - \mathbf{C}\frac{d\mathbf{f}_n}{dt} + \mathbf{u}_n(t)\right] \quad (4)\end{aligned}$$

where $h_n = x_{n+1} - x_n$ and for the sake of brevity the following notation has been used:

$$\begin{aligned}\mathbf{f}_n &= \mathbf{f}(x_n, t) \\ \mathbf{G}_n &= \mathbf{G}(x_n) \\ \mathbf{u}_n(t) &= \mathbf{u}(x_n, t)\end{aligned}$$

Equation (4) can be rewritten as:

$$\begin{aligned}\frac{h_n}{2}\mathbf{C}\frac{d\mathbf{f}_n}{dt} + \frac{h_n}{2}\mathbf{C}\frac{d\mathbf{f}_{n+1}}{dt} &= (\mathbf{I} + \frac{h_n}{2}\mathbf{G}_n)\mathbf{f}_n \\ &\quad - (\mathbf{I} - \frac{h_n}{2}\mathbf{G}_{n+1})\mathbf{f}_{n+1} + \frac{h_n}{2}[\mathbf{u}_{n+1}(t) + \mathbf{u}_n(t)]\end{aligned}$$

This system of differential equations can be written in matrix form as:

$$\mathcal{C}\frac{d\mathbf{F}}{dt} + \mathcal{G}\mathbf{F} = \mathbf{U}(t) \qquad (5)$$

where:

$$\mathbf{F} = \begin{bmatrix} \mathbf{f}_0 \\ \vdots \\ \mathbf{f}_N \end{bmatrix}, \quad \mathbf{U}(t) = \begin{bmatrix} \frac{h_0}{2}[\mathbf{u}_0(t) + \mathbf{u}_1(t)] \\ \vdots \\ \frac{h_{N-1}}{2}[\mathbf{u}_{N-1}(t) + \mathbf{u}_N(t)] \end{bmatrix}$$

and block matrices $\mathcal{C}$ and $\mathcal{G}$ are shown in Fig. 2.

It will be assumed that $\mathbf{U}(t)$ is a realization of a zero-mean stationary stochastic process. Let $\mathbf{R}_U(t_1, t_2)$ be its autocorrelation function:

$$\mathbf{R}_U(t_1, t_2) = E\left\{\mathbf{U}(t_1)\mathbf{U}^T(t_2)\right\}$$

Since the process is stationary, $\mathbf{R}_U(t_1, t_2)$ depends only on $\tau = t_1 - t_2$: $\mathbf{R}_U(t_1, t_2) = \mathbf{R}_U(t_1 - t_2) = \mathbf{R}_U(\tau)$, and the process's PSD is given by the Fourier transform of $\mathbf{R}_U(\tau)$ [13]:

$$\mathbf{S}_U(j\omega) = \int_{-\infty}^{+\infty} \mathbf{R}_U(\tau)e^{-j\omega\tau}\,d\tau$$

The autocorrelation function of $\mathbf{F}(t)$ can be computed by multiplying (5) by its transposed and taking the expected value of both sides of the resulting equation:

$$\begin{aligned}&E\left\{[\mathcal{C}\dot{\mathbf{F}}(t_1) + \mathcal{G}\mathbf{F}(t_1)][\mathcal{C}\dot{\mathbf{F}}(t_2) + \mathcal{G}\mathbf{F}(t_2)]^T\right\} \\ &= E\left\{\mathbf{U}(t_1)\mathbf{U}^T(t_2)\right\} \qquad\qquad\qquad (6)\end{aligned}$$

$$
\mathcal{C} = \begin{bmatrix} h_0\mathbf{C}/2 & h_0\mathbf{C}/2 & & \\ & h_1\mathbf{C}/2 & h_1\mathbf{C}/2 & \\ & & \ddots & \\ & & h_N\mathbf{C}/2 & h_N\mathbf{C}/2 \end{bmatrix}
$$

$$
\mathcal{G} = \begin{bmatrix} -\mathbf{I} - h_0\mathbf{G}_0/2 & \mathbf{I} - h_0\mathbf{G}_1/2 & & \\ & -\mathbf{I} - h_1\mathbf{G}_1/2 & \mathbf{I} - h_1\mathbf{G}_2/2 & \\ & & \ddots & \\ & & -\mathbf{I} - h_N\mathbf{G}_{N-1}/2 & \mathbf{I} + h_N\mathbf{G}_N/2 \end{bmatrix}
$$

Figure 2: Block matrices $\mathcal{C}$ and $\mathcal{G}$

It can be shown [13] that:

$$
\begin{aligned}
E\left\{\dot{\mathbf{F}}(t_1)\mathbf{F}^T(t_2)\right\} &= \frac{\partial \mathbf{R}_F}{\partial t_1} \\
E\left\{\mathbf{F}(t_1)\dot{\mathbf{F}}^T(t_2)\right\} &= \frac{\partial \mathbf{R}_F}{\partial t_2} \\
E\left\{\dot{\mathbf{F}}(t_1)\dot{\mathbf{F}}^T(t_2)\right\} &= \frac{\partial^2 \mathbf{R}_F}{\partial t_1 \partial t_2}
\end{aligned}
$$

Therefore (6) can be rewritten as:

$$
\begin{aligned}
\mathcal{C}\frac{\partial^2 \mathbf{R}_F}{\partial t_1 \partial t_2}\mathcal{C}^T + \mathcal{C}\frac{\partial \mathbf{R}_F}{\partial t_1}\mathcal{G}^T & \\
+ \mathcal{G}\frac{\partial \mathbf{R}_F}{\partial t_2}\mathcal{C}^T + \mathcal{G}\mathbf{R}_F\mathcal{G}^T &= \mathbf{R}_U(t_1, t_2)
\end{aligned} \tag{7}
$$

Since $\mathbf{F}(t)$ is also a realization of a stationary process, its autocorrelation function depends only on $t_1 - t_2$: $\mathbf{R}_F(t_1, t_2) = \mathbf{R}_F(t_1 - t_2) = \mathbf{R}_F(\tau)$. Consequently:

$$
\begin{aligned}
\frac{\partial \mathbf{R}_F}{\partial t_1} &= \mathbf{R}'_F(\tau) \\
\frac{\partial \mathbf{R}_F}{\partial t_2} &= -\mathbf{R}'_F(\tau) \\
\frac{\partial^2 \mathbf{R}_F}{\partial t_1 \partial t_2} &= -\mathbf{R}''_F(\tau)
\end{aligned}
$$

Thus taking the Fourier transform of (7) yields:

$$
\begin{aligned}
\omega^2 \mathcal{C}\mathbf{S}_F(j\omega)\mathcal{C}^T + j\omega \mathcal{C}\mathbf{S}_F(j\omega)\mathcal{G}^T & \\
- j\omega \mathcal{G}\mathbf{S}_F(j\omega)\mathcal{C}^T + \mathcal{G}\mathbf{S}_F(j\omega)\mathcal{G}^T &= \mathbf{S}_U(j\omega)
\end{aligned}
$$

or equivalently:

$$
\mathcal{H}(j\omega)\mathbf{S}_F(j\omega)\mathcal{H}^*(j\omega) = \mathbf{S}_U(j\omega) \tag{8}
$$

where $\mathbf{S}_F(j\omega)$ is the PSD of $\mathbf{F}$, $\mathcal{H}(j\omega) = j\omega\mathcal{C} + \mathcal{G}$ and $\mathcal{H}^*$ denotes the adjoint (conjugate transposed) of $\mathcal{H}$.

This equation can be solved by performing an LU decomposition of $\mathcal{H}$: $\mathcal{H}(j\omega) = \mathcal{L}(j\omega)\mathcal{U}(j\omega)$. In principle, it is possible to solve for $\mathbf{S}_F(j\omega)$ directly:

$$
\begin{aligned}
&\mathbf{S}_F(j\omega) \\
&= \mathcal{U}^{-1}(j\omega)\mathcal{L}^{-1}(j\omega)\mathbf{S}_U(j\omega)[\mathcal{L}^*(j\omega)]^{-1}[\mathcal{U}^*(j\omega)]^{-1}
\end{aligned}
$$

Numerical round-off error introduced in the computation of $\mathbf{S}_U(j\omega)$, however, could yield a solution that is not a positive-definite self-adjoint matrix, a non-physical result. For this reason it is preferable to compute $\mathbf{S}_F(j\omega)$ in factored form through a Cholesky decomposition of $\mathbf{S}_U(j\omega)$:

$$
\begin{aligned}
\mathbf{S}_U(j\omega) &= \mathbf{L}_U(j\omega)\mathbf{L}_U^*(j\omega) \\
\mathbf{W}_F(j\omega) &= \mathcal{U}^{-1}(j\omega)\mathcal{L}^{-1}(j\omega)\mathbf{L}_U(j\omega) \\
\mathbf{S}_F(j\omega) &= \mathbf{W}_F(j\omega)\mathbf{W}_F^*(j\omega)
\end{aligned}
$$

This ensures that the resulting $\mathbf{S}_F(j\omega)$ is a positive-definite self-adjoint matrix.

3 NUMERICAL RESULTS

The algorithm described in the previous section was used to compute the noise PSD at the source and drain terminals of a MOSFET under various bias conditions. It was assumed that the noise sources represented by $i_n(x,t)$ and $v_n(x,t)$ in (2) where white, uniformly distributed in the device channel, and uncorrelated. Under these assumptions, $\mathbf{S}_U(j\omega)$ is block-diagonal and independent of ω. Figures 3 and 4 show the simulated noise PSD at the drain (solid line) and source (dashed line) ends of the MOSFET for two different values of V_{ds}.

This approach can be applied to more complex and accurate models of carrier transport and noise generation mechanisms inside the channel. For example, (1) can be replaced by the drift-diffusion equations:

$$
\begin{aligned}
\varepsilon\nabla^2 V &= -q(N_D - N_A + p - n) \\
\frac{1}{q}\nabla \cdot \mathbf{J}_n &= \frac{\partial n}{\partial t} - (G - R) \\
\frac{1}{q}\nabla \cdot \mathbf{J}_p &= -\frac{\partial p}{\partial t} + (G - R) \\
\mathbf{J}_n &= -q\mu_n n\nabla V + qD_n\nabla n \\
\mathbf{J}_p &= -q\mu_p p\nabla V - qD_p\nabla p
\end{aligned}
$$

If a one-dimensional model of the device is used, only one spatial variable is retained, and this set of partial differential equations can be discretized following the approach de-

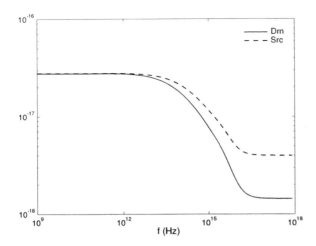

Figure 3: Noise power spectral density, $V_{ds} = 0.2\text{V}$

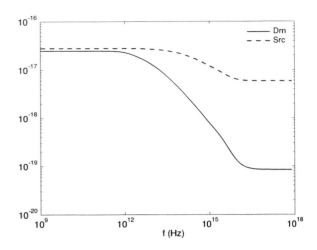

Figure 4: Noise power spectral density, $V_{ds} = 0.44\text{V}$

scribed in Section 2. For better accuracy, two- or three-dimensional device geometries can be used, in which case more complex discretization schemes become necessary.

Similarly, noise generation phenomena inside the device can be modeled more accurately. For example, experimental data and theoretical analysis indicate that the noise sources inside the device are not uncorrelated [14]. Correlation among noise sources can be accounted for by introducing off-diagonal terms in $\mathbf{S}_U(j\omega)$. Regardless of the complexity of the geometry of the device and of the models, computation the noise PSD at the device terminals can be performed following the approach described earlier. Comparisons of simulation results with experimental data, can provide information useful in developing noise models for nanoscale MOSFET devices.

REFERENCES

[1] A. van der Ziel, *Noise in Solid State Devices and Circuits.* New York, NY: Wiley, 1986.

[2] A. A. Abidi, "High-frequency noise measurements on FET's with small dimensions," *IEEE Trans. Electron Devices*, vol. ED-33, no. 11, pp. 1801–1805, November 1986.

[3] W. Shockley, J. A. Copeland, and R. P. James, "The impedance field method of noise calculation in active semiconductor devices," in *Quantum Theory of Atoms, Molecules, and the Solid State*, P. O. Lowdin, Ed. New York, NY: Academic Press, 1966, p. 537.

[4] A. Cappy and W. Heinrich, "High-frequency FET noise performance: A new approach," *IEEE Trans. Electron Devices*, vol. 36, no. 2, pp. 403–409, February 1989.

[5] F. Danneville, H. Happy, G. Dambrine, J.-M. Belquin, and A. Cappy, "Microscopic noise modeling and macroscopic noise models: How good a connection?" *IEEE Trans. Electron Devices*, vol. 41, no. 5, pp. 779–786, May 1994.

[6] J.-S. Goo, C.-H. Choi, F. Danneville, E. Morifuji, H. S. Momose, Z. Yu, H. Iwai, T. H. Lee, and R. W. Dutton, "An accurate and efficient high frequency noise simulation technique for deep submicron MOSFETs," *IEEE Trans. Electron Devices*, vol. 47, no. 12, pp. 2410–2419, December 2000.

[7] J.-S. Goo, C.-H. Choi, A. Abramo, J.-G. Ahn, Z. Yu, T. H. Lee, and R. W. Dutton, "Physical origin of the excess thermal noise in short channel MOSFETs," *IEEE Electron Dev. Lett.*, vol. 22, no. 2, pp. 101–103, February 2001.

[8] B. Wang, J. R. Hellums, and C. G. Sodini, "MOSFET thermal noise modeling for analog integrated circuits," *IEEE Jour. Solid-State Circuits*, vol. 29, no. 7, pp. 833–835, July 1994.

[9] P. Klein, "An analytical thermal noise model of deep submicron MOSFET's," *IEEE Electron Dev. Lett.*, vol. 20, no. 8, pp. 399–401, August 1999.

[10] D. P. Triantis, A. N. Birbas, and D. Kondis, "Thermal noise modeling for short-channel MOSFET's," *IEEE Trans. Electron Devices*, vol. 43, no. 11, pp. 1950–1955, November 1996.

[11] G. Knoblinger, P. Klein, and M. Tiebout, "A new model for thermal channel noise of deep submicron MOSFET'S and its application in RF-CMOS design," in *Proc. 2000 Symp. VLSI Circuits*, 2000, pp. 150–153.

[12] C. H. Chen and M. J. Deen, "High frequency noise of MOSFETs. I: Modeling," *Solid-State Electronics*, vol. 42, no. 11, pp. 2069–2081, November 1998.

[13] A. Papoulis, *Probability, Random Variables, and Stochastic Processes.* New York, NY: McGraw-Hill Book Co., 1965.

[14] J.-P. Nougier, "Fluctuations and noise of hot carriers in semiconductor materials and devices," *IEEE Trans. Electron Devices*, vol. 41, no. 11, pp. 2034–2049, Nov. 1994.

Gate Length Scaling Effects in ESD Protection Ultrathin Body SOI Devices

Jam-Wem Lee[1], Yiming Li[1,2,*] and S. M. Sze[3,4]

[1]Department of Nano Device Technology, National Nano Device Laboratories, Hsinchu 300, TAIWAN
[2]Mircoelectronics and Information Systems Research Center, National Chiao Tung University, Hsinchu 300, TAIWAN
[3]National Nano Device Laboratories, Hsinchu 300, TAIWAN
[4]Institute of Electronics, National Chiao Tung University, Hsinchu 300, TAIWAN
[*]P.O. Box 25-178, Hsinchu 300, TAIWAN, ymli@faculty.nctu.edu.tw

ABSTRACT

In this paper we experimentally explore the gate length scaling effects that related to the abrupt degradation of electrostatic discharge (ESD) robustness for ultra-thin body silicon on insulator (SOI) devices and integrated circuits (ICs). It is found that, for the ultra-thin body SOI, the ESD protection devices fail when the gate length of protection devices is smaller than the 0.18 micron meter (um). Taking the effects into consideration, it is believed that optimizations among the device profiles, geometries, and the protection efficiency should be done simultaneously for high performance VLSI circuit design, in particular for modern system-on-chip (SoC). This observation is very useful in both the nano-scale CMOS fabrication technology and VLSI circuit design.

Keywords: SOI, ESD, SoC, VLSI, optimization, Ultrathin body SOI, short channel

1 INTRODUCTION

ESD protection for device and circuit has been of important reliability concerns with the rapid progress in the miniaturization and process technology recently [1-4]. It is known that ESD robustness reveals the major demerit for the SOI technology and application [2]. Diverse works have been reported in the investigation of ESD robustness for SOI devices recently [3]; the improvement is very limited, especially for the ultrathin body film SOI devices. Therefore, a constructive investigation of ESD robustness for ultra-thin body SOI devices should at the same time take the SOI circuit into consideration.

In this paper, we investgate the ESD robustness of devices which fabricated on two kinds of SOI wafers. The first wafer has its silicon thickness 140 nm and the other one is 90 nm. For all gate lengths, without considering the turn on voltage and turn on resistance, the thicker body film has a stable robustness (~ 3 mA/μm). On the other hand, an abruptly decreasing of ESD robustness is found in the thinner one. The abrupt robustness reduction in short gate-length devices (gate length < 0.18 μm) with thin silicon film is examined that a narrow conduction path occurs when devices turn on. The narrow path is caused from a lower doping concentration regime near the interface of silicon/buried oxide [4]. For thin silicon film devices, the high n-type concentration regime of source/drain extensions are extending all over the silicon thin film that connecting to the low boron-doping region. Therefore, a punch-through path is generated and burn out this narrow trace. Our investigation demonstrates that the optimization is necessary and has to be achieved by considering the device profiles, geometries, and efficiency.

2 DEVICE FABRICATION AND MEASUREMENT

Two kinds of SOI wafers are used in fabrication test devices. The first one has its silicon thickness of 140 nm and the other one is 90 nm. Figure 1 shows the test structure of all devices that a 90 nm CMOS technology with the gate oxide thickness of 1.2 nm is used. ESD robustness is characterized by transmission line pulse (TLP) with its diagram shown in figure. 2, moreover, in well agreement with the ESD robustness measurement from HBM, the pulse width is setup to 100 nsec.

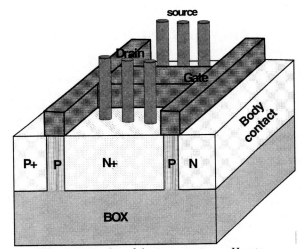

Figure 1: A plot of the test structure – H-gate.

One loop of the TLP measurement includes following steps; firstly the switch will be turned to node 1, at the meanwhile, transmission line will be charged to a designed voltage by a high voltage source. Secondly, the charged transmission line will discharge to the device under test (DUT) whenever the switch closes on the node 2. During the discharge process, a digital waveform scope is used to automatically calculate transient IV characteristics of the DUT.

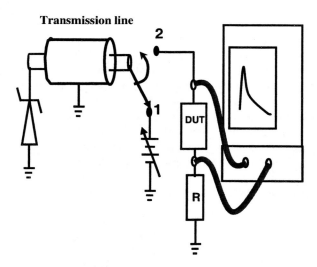

Figure 2: An illustration of the TLP measurement configuration in the study.

Soon after the discharge step finished, a quasi-static IV measurement is done to evaluate the health of DUT. If a healthy status is decided, a new TLP loop will be taken to a higher charge voltage. Otherwise, the measurement will be ended that the TLP IV measurement is finished. The quasi-static IV characteristics are measured by using the HP4156B semiconductor analyzer.

3   RESULTS AND DISCUSSION

Figure 3 shows the TLP IV characteristics of the thicker silicon film devices. It could be easily found that a longer gate length will result in a higher turn voltage, moreover, a snap back Phenomena is also observed in the longer channel device. Those results indicate different turn on mechanisms existed in the thick body devices. A more detail mechanism could be drawn that turn on behaviors of parasitic bipolar junction transistor (BJT) dominates IV characteristics of longer channel devices. Consequently the drain junction breakdown initiated parasitic BJT turn on and thus got a higher turn on voltage and a snap back behavior. On the other hand, drain induced barrier height lowering (DIBL) at drain side plays the major role in short devices; therefore we can have a lower turn on voltage IV curve without having a snap back characteristic.

Though the turn on characteristics is strongly related to the channel length of the SOI devices; the ESD robustness keeps almost the same for all the thick body devices. The result is mainly caused from the fact that the narrow silicon film allows very limited current flows. Owing to the very finite current path, a slight higher current flow can melt silicon thin film. This is the major challenge in design the SOI circuits.

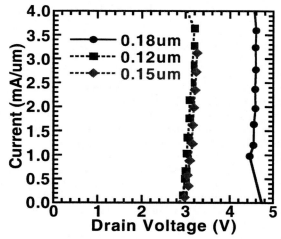

Figure 3: TLP IV characteristics of thick silicon sample with various gate lengths.

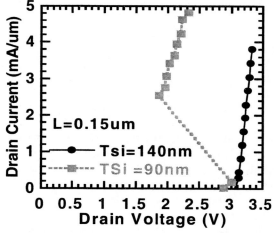

Figure 4: TLP IV characteristics of both thin and thick silicon samples

Figure 4 compares TLP IV characteristics difference between the thin and thick body SOI devices. It could be clearly explored that the thin body device will be burned out whenever the device turned on. This observation is not only true for 0.15 um lengthen device, but also real for any thin body device with its channel length shorter than 0.15 um. More details could be investigated form the Figure 5 that ESD robustness sustains constant when the channel length larger than 0.15um. On the other hand, the ESD robustness is abruptly degrading to nearly zero while the channel length of device is scaled down to less than 0.15um. This result is very important in designing the ESD protections on the SOI circuits. Moreover, mechanisms that dominate the ESD characteristic must be also discussed to have a better achievement in SOI circuit.

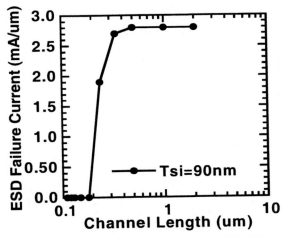

Figure 5: ESD failure current vs. gate length of the thin silicon sample

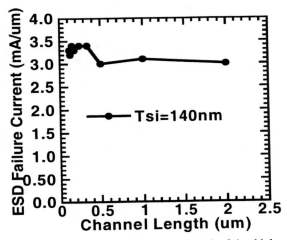

Figure 6: ESD failure current vs. gate length of the thick silicon sample

Unlike the thin body SOI device, the thick body device has a nearly channel length independent ESD robustness. This result could be referred to Figure 6 that thick body SOI device has ESD robustness about 3mA/um. In comparing with the long channel (Lg > 0.15) thin body SOI devices, the thick body device has a better ESD robustness. The result is very reasonable that a wider current path will be existed in the thicker body device; therefore, sustains a high current flow. The abruptly degradation phenomena occurred in the thin body SOI device could be explained by the following two figures.

Figure 7 shows the cross section view of the thick body SOI device. Owing to the segregation effects, the boron doping near the buried oxide (BOX) will greatly diffuse into the BOX; therefore, boron-doping concentration near the BOX will be largely lowered. It could be simplified that owing to the segregation effect, there exists a low boron doping concentration region as shown in the figure 7. The low boron doping concentration region is a relatively lower barrier height area, thus existing an easier punch through path. Fortunately, the easier punch-through region dose not connect to the source/drain extensions directly; the device would not turn on through this low boron concentration. The turn on path of the thick body device would flow through the source/drain extensions instead of the narrow low boron-doping path.

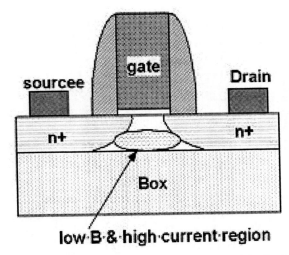

Figure 7: Cross section view of the current path in the thick silicon devices

On the contrary, thinning the thickness of the silicon film will make the low boron-doping region distributed much nearer to the source drain extensions. Therefore, as the channel length becoming shorter and shorter, the punch through path will take place of the normally turn on path.

Owing to the very limited punch through path, the path will be burned out soon after the device turned on. This effect will greatly limit the scaling ability of the SOI device. The effect will also limit the fabrication ability of fully depleted SOI devices. It is worse that the newly developed FinFET structure, the silicon film is thinning down to lower than 50nm. In those fully depleted thin silicon films, a similar phenomenon could be occurred. Accordingly, this result explores a serious challenge for the future proposed FinFET liked devices.

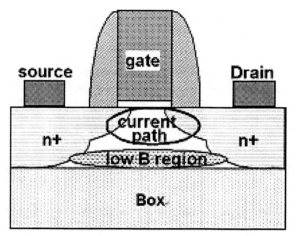

Figure 8: Cross section view of the current path in the thick silicon devices

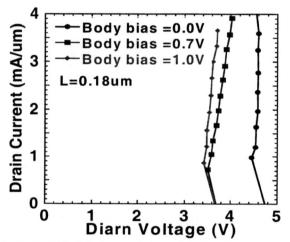

Figure 9: TLP IV characteristics of thick silicon sample with various body bias, where L = 0.18 um.

Substrate biased (body biased) technology is also performed on the thick body SOI device. Figure 9 and Figure 10 demonstrate that body bias did lower the turn on voltage of the long channel thick SOI device, but do not affect turn characteristic of the short channel devices. Those results are greatly consisted with our previous ratiocination that parasitic BJT dominates turn on behavior of the long channel thick body SOI device. Though a lower turn on voltage, the ESD robustness does not be affected by body bias (owing to a limited current path).

4   CONCLUSIONS

We have studied an abruptly degradation of ESD robustness owing to gate length reduction. Two kinds of SOI wafers have been used in the fabrication of test devices. The first one is with its silicon thickness 140 nm and the other is with 90 nm. We have observed that the robustness abruptly decreasing in the thinner one when the gate length is smaller than 0.18 um. For all gate lengths, the SOI with thicker body films has a stable robustness (~ 3 mA/um). Our investigation demonstrates that the optimization is necessary and has to be achieved by considering the device profiles, geometries, and efficiency.

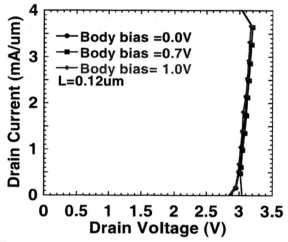

Figure 10: TLP IV characteristics of thick silicon sample with various body bias, where L = 0.18 um.

5   ACKNOWLEDGEMENTS

This work is supported in part by the TAIWAN NSC grants: NSC – 92 – 2112 – M – 429 - 001 and NSC – 92 - 815 – C – 492 – 001 - E. It is also supported in part by the grant of the Ministry of Economic Affairs, Taiwan under contract No. 91 – EC – 17 – A – 07 - S1 - 0011.

REFERENCES

[1] Yiming Li, Jam-Wem Lee, and S. M. Sze, "Optimization of the Anti-punch-through Implant for ESD Protection Circuit Design," Japanese Journal of Applied Physics, 42, 2152, 2003.

[2] P. Raha, S. Ramaswamy, and E. Rosenbaum, "Heat flow analysis for EOS/ESD protection device design in SOI technology," IEEE Transactions on Electron Devices 44, 464, 1997.

[3] M. Chan, J.C. King, P. K. Ko, and C. Hu, "High performance bulk MOSFET fabricated on SOI substrate for ESD protection and circuit applications," Proceedings of IEEE International SOI Conference 61, 1994.

[4] H. Park, E. C. Jones, P. Ronsheim, C. Jr. Cabral, C. D'Emic, G. M. Cohen, R. Young, and W. Rausch, "Dopant redistribution in SOI during RTA: a study on doping in scaled-down Si layers," Technical Digest of International Electron Devices Meeting (IEDM) 337, 1999.

A Unified Mobility Model for Excimer Laser Annealed Complementary Thin Film Transistors Simulation

Hsiao-Yi Lin[1], Yiming Li[2,3,*], Jam-Wem Lee[2], Chaung-Ming Chiu[1], and S.M. Sze[4,5]

[1]Toppoly Optoelectronic Corp., Miao-Li 350, TAIWAN
[2]Department of Nano Device Technology, National Nano Device Laboratories, Hsinchu 300, TAIWAN
[3]Mircoelectronics and Information Systems Research Center, National Chiao Tung University, Hsinchu 300, TAIWAN
[4]National Nano Device Laboratories, Hsinchu 300, TAIWAN
[5]Institute of Electronics, National Chiao Tung University, Hsinchu 300, TAIWAN
[*]P.O. Box 25-178, Hsinchu 300, TAIWAN, ymli@faculty.nctu.edu.tw

ABSTRACT

In this paper, a unified SPICE mobility model for excimer leaser annealed lower temperature polycrystalline silicon (LPTS) complementary thin film transistors (TFTs) is proposed. In comparing with the conventional RPI model, the proposed mobility model exhibits very good property in representing the vertical electrical field induced mobility degradation. Our model improves the correctness of the circuit simulation without increasing any complexity and having no any convergence issue. Comparisons among the conventional RPI TFT mobility model, and measurement data, this new mobility model demonstrated very good accuracy in the simulation of laser annealed LTPS TFTs. This TFT model is very useful in precisely modeling the circuit characteristics for the complementary system on panel (SOP) circuit.

Keywords: excimer laser annealed, LTPS, complementary TFT mode, SPICE mobility model, gate induced mobility degradation.

1 INTRODUCTION

Excimer laser annealing technique has recently been proposed in the fabrication of low temperature polycrystalline silicon; in particular for the applications to active-matrix liquid crystal display (AMLCD) [1] and achieving system on panel. In comparing with the traditional thin film transistors, the most attractive property of the laser annealed polycrystalline silicon is due to its relatively larger grain size and higher electron-hole mobility. Therefore, the embedded driving circuit could be easily achieved for replacing the additional driving integration circuits (IC's) in LCD's. It is known that circuit models play important role in the design of the embedded driving circuit using laser annealed LTPS TFTs. Unfortunately, most of mobility models are valid only for some conventional TFTs and can not be applied to the simulation of the laser annealed LTPS TFTs [2-3] accurately.

In this work, we propose a unified mobility model which is suitable for the simulation of both the n- and p-type laser annealed LTPS TFTs. This unified mobility model for complementary TFT devices is successfully developed, verified and implemented in the SPICE circuit simulator. By considering the channel mobility degraded by the vertical electric field, this empirical model is similar to the BISM4-liked MOSFET mobility model. With the well-known RPI TFT model, this mobility model can be directly incorporated into SPICE circuit simulator without any convergence problems. To test the accuracy of this new RPI TFT model, we firstly apply genetic algorithm to optimize it with the measured I-V data. This intelligent model parameter optimization methodology has been successfully developed by us in HBT characterization recently [4]. Therefore, a set of characterized model parameters can be extracted precisely. Comparison with the conventional model is also performed on different TFTs [5-7]. Simulation and measurement results for the n- and p-type laser annealed LTPS TFTs are reported and discussed in this work. It is found that simulation with our model demonstrates very good agreement with the measurement. The improvement is unified for both the n- and p-type laser annealed LTPS TFTs. The driving circuit designed with the complementary TFTs is known to have a lower leakage current, and then extend the battery lifetime of the personal data assistant. Thus, this result is very promising and useful in designing the driving circuit by complementary TFTs.

Section 2 discusses the fabrication and measurement procedure for the test devices. In section 3, we state the conventional RPI TFT model. Section 4 reports the proposed TFT mobility model. Comparison results are discussed in this section. Section 5 is focus on the model parameter extraction and accuracy issues. In section 6, we apply the model to simulation n- and p-channel TFT. Section 7 draws the conclusions and suggests the future works.

2   DEVICE FABRICATION AND MEASUREMENT

The complementary TFTs are fabricated on the glass substrate by using the novel excimer laser annealing process. A 50 nm thicken amorphous silicon film is firstly deposited by using a PECVD cluster tool at 300℃. The silicon film was then taken into a novel excimer laser system to transform the amorphous silicon into the large grained poly crystalline silicon. The key process will improve the mobility of both electrons and holes.

After the source/drain formatted by using the high power plasma implantation, a 100 nm thick oxide was deposited by using the TEOS source in PECVD clusters. Gates, contacts, and interconnections were finally deposited and etched to have a good electrical connection between the designed TFTs. The IV characteristics are measured by using the HP4156B semiconductor analyzer.

3   THE CONVENTIONAL RPI MODEL

The RPI model is a SPICE compact TFTs model developed on the single crystalline MOSFET model [6]. This model contents following features:

1. Field effect mobility that becomes a function of gate bias
2. Effective mobility that accounts for trap states.
3. Reverse bias drain current function of electric field near drain and temperature.
4. A design independent of channel length.
5. A unified DC model that includes all four regimes for channel lengths down to 4 um; those four regions are: leakage (thermionic emission), subthreshold (diffusion-like model), above threshold (c-Si-like, with mFet), and Kink effect (impact ionization with feedback).
6. An AC model that accurately reproduces C_{gc} frequency dispersion.
7. An automatic scaling of model parameters that accurately model a wide range of device geometries

Though the vertical electrical field induced mobility degradation effect has been introduced in the RPI model, the expression could not reflect mobility behavior well. The mobility model used in RPI is represented as following function:

$$\frac{1}{\mu_{FET}} = \frac{1}{MU0} + \frac{1}{Tmul \cdot \left(\dfrac{2 \cdot V_{gte}}{V_{sth}} \right)^{MMU}} \qquad (1)$$

It is found from the n-channel TFT optimization result shown in Fig. 1 that one cannot model Gm characteristics well by using the conventional mobility model. In comparing with the measured date, when the measured Gm decreasing at the high gate bias region, the simulated Gm will sustain flat.

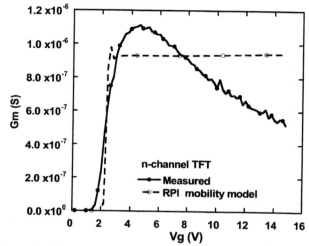

Figure 1: Comparison of the measured and simulated Gm characteristics for an n-channel TFT (width / length (W/L) = 20/4 [um]).

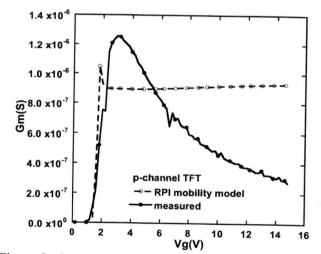

Figure 2: Comparison of the measured and modeled Gm characteristic for a p-channel TFT (width / length (W/L) = 20/4 [um]).

Similar result is also found in the Fig. 2 that the p-channel TFT has a Gm decrement for the high gate bias condition; but the RPI model could not handle this characteristic well. The error of Gm should cause an accumulation of inaccuracy and finally results a wrong circuit function. The problem will be more critical for analog circuit design.

4   THE PROPOSED MODEL

It is known that BISM 4 model is one of the most popular SPICE compact models for the simulation of deep sub-micron CMOS devices. The physical based mobility model is one of the most attractive features in the BISM 4 model. The model we considered here incorporates the following effects into the electron/hole mobility, they are: high field, substrate bias, and temperature effects. Owing to the fact that no body bias is usually applied on the TFT devices; therefore the substrate effects should not be considered. Accordingly, the proposed mobility model has the following form:

$$\frac{\mu_0}{1+(U_a)\left(\dfrac{V_G+V_T}{T_{ox}}\right)+U_b\left(\dfrac{V_G+V_T}{T_{ox}}\right)^2} \tag{2}$$

Gm characteristics simulated with the modified mobility model are shown in the following two figures. It is found that the new mobility model greatly improves the high field Gm behavior for both the n-channel and p-channel TFTs. Those characteristics express the gate voltage reduced mobility effect successfully, owing to the better agreement of the Gm characteristics. It is expected that the simulation result of the TFT circuit will have higher correctness, especially for the AC simulations.

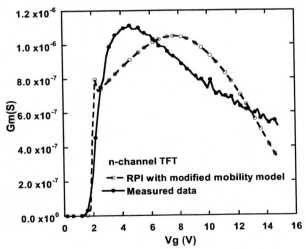

Figure 3: Comparison of the measurement and our new model for an n-channel TFT.

5   MODEL EXTRACTION

An auto-expiation tool that developed by our research team has been applied to perform the parameters extraction.

Tab. 1 shows the error calculated from measured data and RPI model. It should be noticed that RMS error at the high gate biased region would sustain above 3 percents, this is not acceptable for the most applications; however, could not make any improvement under the conventional mobility structure. The similar optimization process is also performed on our proposed mobility modified model. It could be found from Tab. 2 that our correctness will greatly enhance the agreement between the model and measured data. By using our mobility modification, the error can be easily reduced to less than 1 percent. In comparing with the conventional RPI TFT model result, the improvement is significant and easy to achieve.

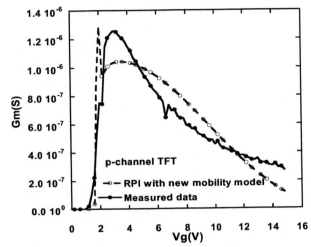

Figure 4: Comparison of the measurement and our new model for a p-channel TFT.

Table 1: Errors of the conventional RPI TFT model.

Measured Vth	2.0505 V
Extracted Vth	2.0500 V
RMS Error of I_D-V_D	3.73%
RMS Error of I_D-V_D in Sec.1	1.30%
RMS Error of I_D-V_D in Sec.2	3.19%
RMS Error of I_D-V_D in Sec.3	4.66%
RMS Error of I_D-V_G	3.33%
RMS Error of I_D-V_G in Sec.1	0.02%
RMS Error of I_D-V_G in Sec.2	1.35%
RMS Error of I_D-V_G in Sec.3	7.01%

Table 2: Errors of the proposed model.

Measured Vth	2.0505 V
Extracted Vth	2.0381 V
RMS Error of I_D-V_D	0.94%
RMS Error of I_D-V_D in Sec.1	0.45%
RMS Error of I_D-V_D in Sec.2	0.48%
RMS Error of I_D-V_D in Sec.3	1.32%
RMS Error of I_D-V_G	0.60%
RMS Error of I_D-V_G in Sec.1	0.01%
RMS Error of I_D-V_G in Sec.2	0.65%
RMS Error of I_D-V_G in Sec.3	1.11%

6 IV CHARACTERISTICS SIMULATION

Output characteristics have been simulated in this work and shown in the following two figures. It is found that the high field mobility effects dominate output characteristics at high gate biased situations. This point is the most important way when designing the driving circuit by using the complementary TFT devices. This is causing from the fact that driving capability will be strongly affected by the output resistance. Therefore, a precisely modeling of high field mobility is not only helpful in device fabrication but also essential in TFT circuit design. Observing from the modeling results, a better agreement will be found, especially for the p-channel TFT. Those superior characteristics are mainly based on the correctness of the proposed mobility model.

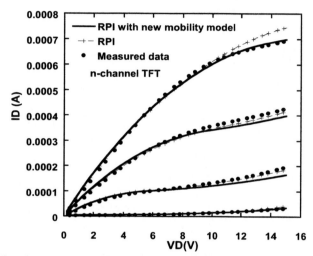

Figure 5: Comparison of the measured n-channel TFT output characteristics between the RPT, and new mobility models.

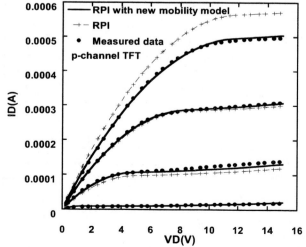

Figure 6: Comparison of the measured p-channel TFT output characteristics between the RPT, and new mobility models.

7 CONCLUSIONS

It has been demonstrated that the mobility modified RPI model will greatly improve the agreement between the experimental device and models. With the modification, the optimization process could be easily achieved; moreover, no additional complexity is added. The convergence and efficiency tests have been also done on the SPICE circuit simulator, it is proven that no convergence and efficiency problems are found.

To verify the usability of the new model, the convergence test should be done on SPICE circuit Simulator. After a series tests, it have been found that no any convergence problem existed in our model. Computationally it has shown very good efficiency in different testing cases.

8 ACKNOWLEDGEMENTS

This work is supported in part by the TAIWAN NSC grants: NSC – 92 – 2112 – M – 429 - 001 and NSC – 92 - 815 – C – 492 – 001 - E. It is also supported in part by the grant of Toppoly Optoelectronic Corp., Miao-Li 350, TAIWAN and the grant of the Ministry of Economic Affairs, Taiwan under contract No. 91 – EC – 17 – A – 07 - S1 - 0011.

REFERENCES

[1] S. Jagar *et al.*, "A SPICE model for thin-film transistors fabricated on grain-enhanced polysilicon film," IEEE Trans. Electron Devices 50, 1103, 2003.

[2] A. Wang and K. C. Saraswat, "A strategy for modeling of variations due to grain size in polycrystalline thin-film transistors," IEEE Trans. Electron Devices 47, 1035, 2000.

[3] F. V. Farmakis *et al.*, "On-current modeling of large-grain polycrystalline silicon thin-film transistors," IEEE Trans. Electron Devices 48, 701, 2001.

[4] Y. Li *et al.*, "A Genetic Algorithm Approach to InGaP/GaAs HBT Parameters Extraction and RF Characterization," Jpn. J. Appl. Phys. 42, 2371, 2003.

[5] G.-Y. Yang *et al.*, :S-TFT: an analytical model of polysilicon thin-film transistors for circuit simulation," Proceedings of the IEEE Custom Integrated Circuits Conference, 213, 2000.

[6] M. S. Shur *et al.*, "Modeling and scaling of a-Si:H and Poly-Si Thin Film Transistors," Material Research Society Proceeding, Amorphous and Microcrystalline Silicon Technology, 467, 1997.

[7] G. A. Armstrong *et al.*, "Modeling of Laser-Annealed Polysilicon TFT Characteristics," IEEE Electron Device Letters 18, 315, 1997.

Impact of quantum mechanical tunneling on off-leakage current in double-gate MOSFET using a quantum drift-diffusion model

M.-A. Jaud[1], S. Barraud and G. Le Carval

CEA-LETI, 17 rue des Martyrs
F-38054 Grenoble Cedex 9, France, marie-anne.jaud@cea.fr

ABSTRACT

With the growing use of wireless electronic systems, off-state leakage current in MOSFETs appears as one of the major physical limitations. Measurements of quantum tunnel current between source-drain (S-D) have recently shown that it will become detrim[1]ental in bulk MOSFET architecture for channel lengths around 5nm and at low temperature ($\leq$100K) [1]. In this paper we investigate, using a 2D quantum drift-diffusion model, the influence of source-to-drain tunneling on off-state-leakage current in double-gate MOSFETs.

It is shown that in double-gate MOSFET architecture (contrary to bulk architecture) quantum tunnel current component will be a non negligible part of the off-state leakage current for ultra-thin film thicknesses, even at room temperature.

Keywords: Simulation, density-gradient, Schrödinger, source-to-drain tunneling, off-leakage current, double-gate MOSFET.

1 INTRODUCTION

With the downscaling of MOSFETs into the nanometer regime, the subthreshold electrical characteristics are mainly affected by quantum confinement effects, tunneling current through insulators and source-to-drain tunneling current [1-2]. A measurement of quantum tunnel current between source-drain has recently shown that it will become detrimental in bulk MOSFET architecture for channel length around 5nm and at low temperature ($\leq$100K) [1]. So, to pursue towards low power and high performance circuits, the subthreshold leakage current should be well controlled.

The architecture used in this study is a double-gate MOSFET transistor. This architecture is expected to be a good candidate for ultimate device shrinking, due to its low sensitivity to short-channel-effects and to the improvement of carrier transport properties which is expected from low channel doping [3]. After a careful calibration of quantum drift-diffusion model (QDDM) [4], we present in this paper a systematic study on the impact of source-drain direct tunneling on off-state leakage current in double-gate

MOSFET architecture describe in Fig. 1. A large range of temperatures (100K < T < 300K), channel lengths (10nm < L_c < 80nm) and Si-film thicknesses (5nm < t_{si} < 15nm) are used. We show that in double-gate MOSFET architecture (contrary to bulk architecture) the quantum tunnel current component will be a non negligible part of the off-state leakage current for ultra-thin film thicknesses, even at room temperature.

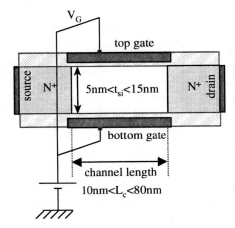

Figure 1: Schematic cross-section of the simulated double-gate nMOSFET (S/D: 10^{20}cm^{-3}, channel: 10^{16}cm^{-3}).

This paper is organized as follow : first of all, we demonstrate the ability to model quantum confinement effects with a quantum drift-diffusion model. This model is calibrated with C-V curve simulations in long MOS capacitor in respect of quantum simulations based on 1D Poisson/Schrödinger equations solver (PS). Secondly, Id-Vg characteristics of double-gate MOSFET are simulated using both quantum (QDDM) and semi-classical drift-diffusion (SC) approaches to estimate the ability to model quantum mechanical tunneling with QDDM and quantify its impact on the subthreshold current.

2 MODEL CALIBRATION

In nanoscale devices, the transport of carriers is expected to be dominated by quantum effects throughout the channel. However, the coupling of quantum effects to transport equations is not well established within a consistent conceptual framework [5]. If a self-consistent Poisson-Schrödinger solver is an acknowledged way to

---
[1] Present employer : STMicroelectronics, F-38926 Crolles

model quantum confinement effects at the SiO_2/Si interface, it is not well suited to a complete treatment of quantum transport because it is too computationally demanding to be used for everyday engineering analysis. Since many years, the Density-Gradient formalism (e.i. quantum drift-diffusion model) appears as an accurate way to include quantum effects to the transport equations [6-7]. The quantum drift-diffusion model introduces lowest-order-effects of non-locality of quantum mechanics through a quantum correction potential added to the classical potential. The electron drift-diffusion current density may be expressed as [4]:

$$J_n = qD_n \nabla n - q\mu_n n \nabla \Psi^Q \qquad (1)$$

with

$$\Psi^Q = \Psi^{CL} + \frac{\gamma \hbar^2}{6m_n}\left(\frac{\nabla^2 \sqrt{n}}{\sqrt{n}}\right) \qquad (2)$$

where γ is a fit factor which comes from the fact that QDDM is based on a lowest-order-development, m_n is the carrier effective mass, n the electron density, and $\hbar$ the reduced Planck constant. It can be noted that when $\hbar \to 0$, Eq. (1) leads to the classical electron current density. The implementation of this formalism makes this approach simple and attractive for an engineering tool.

A carefully calibration of the QDDM approach has been performed against the results of a Poisson/Schrödinger equations solver (PS). C-V curves have been simulated on a long MOS capacitor with 10Å gate oxide thickness with a channel doping varying from $10^{16}cm^{-3}$ to $10^{18}cm^{-3}$. The effects of confinement on the C-V characteristics are shown in Fig. 2. We plot C-V curves using semi-classical drift-diffusion model (SC) and quantum drift-diffusion model (QDDM) and compare them with results obtained by resolution of Poisson/Schrödinger equations (PS). An excellent agreement is demonstrated between quantum drift-diffusion model (QDDM) and Poisson-Schrödinger solver (PS) with $\gamma = 3.2$ used in all the range of channel doping. The results show higher threshold voltage and lower gate-channel capacitance in accumulation and inversion regimes when quantization effects are included (cf. Fig. 2). The quantum confinement effects below the gate oxide are well-reproduced by QDDM and practically introduced by adding the quantum correction potential (cf. Eq. 2).

To extend the validation of QDDM, it is now applied to very thin silicon films. The carrier concentration has been calculated in long double-gate MOSFETs using both QDDM and PS. A channel length $L_c = 800nm$ has been used to only considered 1D quantum confinement effects with a film thickness t_{Si}, ranging from 50nm to 5nm. The channel doping $N_A = 10^{16}cm^{-3}$ is chosen to be the same that the one used in the next section. The resulting carrier density profiles calculated from 1D Poisson/Schrödinger solver (PS) and quantum drift-diffusion model (QDDM) are shown in Fig. 3. Using a single value for γ ($\gamma = 3.2$), the curves exhibit the quantum repulsion of carrier density

at the Si/SiO_2 interface due to the wave nature of electrons. A very good agreement is obtained between QDDM and PS approaches. The validation of QDDM has been performed even in a thin-film double-gate architecture (down to 5nm).

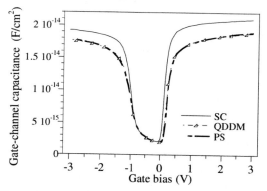

Figure 2: Simulated C-V curves using semi-classical drift-diffusion model (SC), quantum drift-diffusion model (QDDM) and Poisson/Schrödinger equations solver (PS) with a channel doping $N_A = 10^{18}cm^{-3}$.

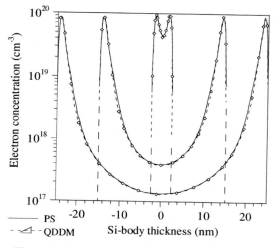

Figure 3: Inversion layer charge distribution calculated from Poisson/Schrödinger solver (PS) and quantum drift-diffusion model (QDDM) with $N_A = 10^{16}cm^{-3}$. $V_{GS} = 1V$.

3 SIMULATION RESULTS

The schematic device structure and device parameters used in the simulations are shown in Fig. 1. The source/drain regions and the substrate are doped to $N_D = 10^{20}cm^{-3}$ and $N_A = 10^{16}cm^{-3}$, respectively. To investigate the impact of source-to-drain tunneling on the electrical behaviour of the device, a channel length ranging from 80nm to 10nm and a Si-film thickness ranging from 15nm to 5nm are used. The calculated Id-Vg characteristics are obtained at low drain voltage, e.i Vds=10mV using both quantum (QDDM) and semi-classical drift-diffusion (SC) approaches.

To estimate the ability to model quantum mechanical tunneling with QDDM, we show here the results for a Si-film thickness equal to 5nm. Subthreshold characteristics of double-gate MOSFETs having channel lengths from 80nm to 10nm are shown in Fig. 4. Using semi-classical drift-diffusion approach the subthreshold slope is degraded with the shrinking of the channel length. This degradation can be attributed to short-channel-effects. Using quantum approach (QDDM), results clearly show an increase of leakage current in comparison to SC results as the channel length is decreased (Fig. 4). This electrical behaviour in subthreshold regime is confirmed by results obtained from a full 2D numerical code of quantum ballistic transport including both thermoionic emission and quantum tunneling of carriers [2].

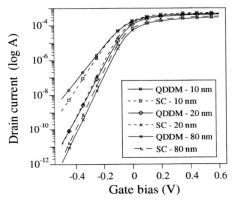

Figure 4: Id-Vg characteristics for a double-gate MOSFET. Comparison is made between semi-classical drift-diffusion model (SC) and quantum drift-diffusion model (QDDM). Silicon film thickness t_{si}=5nm and V_{DS}=10mV.

Indeed, the carriers which come from the source feel steep gradients from the potential barrier. Inclusion of the wave-functions penetration into the barrier is efficiently described with QDDM and induces an additional charge inside the barrier. This charge is self-consistently taken into account in the model and induces a new component to thermal current as observed in Fig. 4. These results show the ability to qualitatively describe the source-to-drain tunneling with a quantum drift-diffusion model.

To consolidate the idea that source-to-drain tunneling is included in the quantum drift-diffusion model, the dependence of subthreshold current on temperature has been investigated. The temperature dependence on Id-Vg characteristics using semi-classical and quantum simulations is shown in Fig. 5 and Fig. 6, respectively.

In comparison to semi-classical drift-diffusion simulations, the subthreshold slope shows a weaker dependence on temperature, which is one of the main properties of measured [1] and calculated [8-10] source-to-drain tunneling current.

Contrary to electrical transport at room temperature, the source-to-drain tunneling plays an important role when the temperature decreases and becomes predominant over the thermoionic current component. These results are in concordance with experimental data [1].

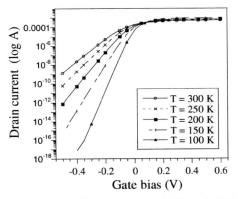

Figure 5: Simulated Id-Vg curves for a double-gate MOSFET with varying temperature using semi-classical drift-diffusion model (SC). Silicon film thickness t_{si}=5nm, channel length L_C=10nm and V_{DS}=10mV.

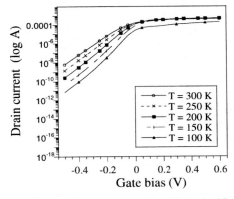

Figure 6: Simulated Id-Vg curves for a double-gate MOSFET with varying temperature using quantum drift-diffusion model (QDDM). The silicon film thickness t_{si}=5nm, channel length L_C=10nm and V_{DS}=10mV.

The relative deviation of subthreshold current calculated at room temperature between QDDM and SC is shown in Fig. 7 for three Si-film thicknesses (5nm, 10nm and 15nm). Two regions may be identified. In the first region 'R1' ($L_c \gtrsim$ 30nm), the relative deviation of subthreshold current is essentially due to 1D quantum confinement effects which shift the threshold voltage. In the second region 'R2' ($L_c \lesssim$ 30nm) 2D quantum effects characterized by quantum confinement effects and source-to-drain tunneling are present. We show that if quantum tunnelling is not predominant over thermal current for large Si-film thicknesses, a noticeable increase of S-D tunnel current component appears when the Si-film thickness decreases, even at room temperature.

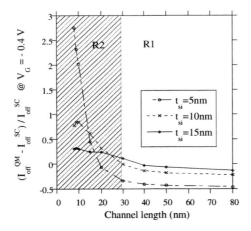

Figure 7: I_{off} relative deviation calculated with quantum drift-diffusion model (QDDM) and semi-classical drift-diffusion model (SC). Regions 'R1' and 'R2' characterize respectively vertical quantum confinement effects and 2D quantum effects. T=300K.

An evaluation of the percentage of source-to-drain tunnel current calculated for different Si-film thicknesses and two temperatures is reported in Fig. 8 with gate lengths ranging from 80nm to 10nm. The calculation is performed for $V_{GS}-V_T = -0.2V$. As experimentally observed at room temperature, the calculation predicts a thermal current dominant for large Si-film thickness [1]. Nevertheless, it is shown that in double-gate MOSFET architecture quantum tunnel current component will be a non negligible part of off-state leakage current for ultra-thin film thicknesses, at low temperature but also at room temperature.

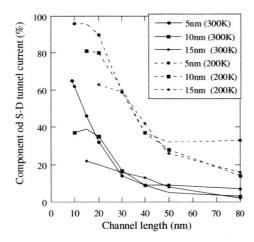

Figure 8: Component of source-drain quantum tunnel current on off-state leakage current (%) for various channel lengths and Si-film thicknesses.

4   CONCLUSION

First of all, we show that a 2D quantum drift-diffusion model (QDDM), based on the Density-Gradient approach, is well-suited to device modelling for which quantum confinement effects significantly affect the electrical behaviour of devices.

Subthreshold electrical characteristics of double-gate MOSFETs with channel lengths ranging from 80nm to 10nm and Si-film thicknesses ranging from 15nm to 5nm have been performed to estimate the ability to model quantum mechanical tunneling with QDDM. For ultra-short channel lengths, a comparison between classical and quantum simulations shows a degradation of the subthreshold slope with inclusion of quantum effects. The additional current induced by QDDM has a similar behaviour to source-to-drain tunnel current (subthreshold slope nearly independent of the temperature) and leads to extend the validation of QDDM to reproduce the source-to-drain tunneling.

We show that at room temperature, the thermoionic current is the main mechanism of electrical transport for large Si-film thicknesses even for short channel lengths ($t_{Si} \geq 15nm$ and $L_c \approx 10nm$). Nevertheless, at low temperature or/and for ultra-thin Si-film thicknesses the influence of tunnel current component has been clearly shown.

REFERENCES

[1] H. Kawaura, T. Sakamoto, IEICE Trans. Electron., vol. E84-C, pp. 1037-1042, 2001.

[2] D. Munteanu, J.-L. Autran, Proc. of ULIS 2002, Munich, Germany, pp. 119-122, 2002.

[3] S.-H. Oh et al., IEEE Electron Device Lett., vol. 21, n°9, pp. 397-399, 2000.

[4] Atlas user's manual, Silvaco, vol. 2, pp.12-4, 2002.

[5] R. Akis, L. Shifren, D.-K. Ferry, D. Vasileska, Phys. Stat. Sol. (b), vol. 226, pp. 1-8, 2001.

[6] A. Wettstein, A. Schenk, W. Fichtner, IEEE Trans. Electron Devices, vol. 48, pp. 279, 2000.

[7] A. Asenov, G. Slavcheva, A.-R. Brown, J.-H. Davies, and S. Saini, IEEE Trans. Electron Devices, vol. 48, pp. 722-729, 2001.

[8] J.-R. Walting, A.-R. Brown, A. Asenov, A. Svizhenko, M.-P. Anantran, pp. 267-270, Proc. of SISPAD, 2002.

[9] A. Schenk, A. Wettstein, Proc. of MSM, pp. 552-555, 2002.

[10] V.-A. Sverdlov, T.-J. Walls, K. Likharev, IEEE Trans. Electron. Dev., vol. 50, pp. 1926-1933, 2003.

Methodology for Prediction of Ultra Shallow Junction Resistivities Considering Uncertainties with a Genetic Algorithm Optimization

C. Renard[*], P. Scheiblin[**], F. de Crécy[**], A. Ferron[*], E. Guichard[*], P. Holliger[**], and C. Laviron[**]

[*]SILVACO Data Systems, Montbonnot St-Martin, France, cyril.renard@cea.fr
[**]CEA-LETI, Grenoble, France, pascal.scheiblin@cea.fr

ABSTRACT

The accurate prediction of arsenic activation after spike annealing is mandatory for Ultra Shallow Junction (USJ) sheet resistance optimization for advanced NMOS transistors engineering. For the first time, we propose a fast and efficient methodology which consists in both predicting coefficients which model the arsenic activation, and in calibrating a physically-based mobility model from experimental data. Calibration was obtained by a genetic algorithm optimization of a criterion taking into account the difference between simulation and measurement, and both experimental and modelling uncertainties.

Keywords: arsenic activation, modelling, calibration, DoE, optimization, genetic algorithm, analysis of variance

1 INTRODUCTION

As device downscaling continues, the sheet resistance of source/drain diffusion areas (S/D) becomes the major limiting factor of the deep submicron device performance [1]. To obtain low resistance USJ as required for nanoscale devices, spike anneals are usually performed. These anneals (a few seconds at 1050°C) allow to achieve high doping level activation with little diffusion. In the case of arsenic it is known that for high doping concentration, a fraction of dopants remains electrically inactive after the anneal. It is accepted that the inactive unprecipitated dopant is in a clustered form, in a mass action equilibrium with the ionized As [2]. The rigorous prediction of inactive fraction would require dynamic models that describe the formation and the kinetic evolution of each population of a great number of different arsenic-vacancies defects. However, such a model has not been reported in literature until now.

In this work, we propose a new methodology to determine, from an empirical modelling, the electrically active arsenic distribution after a 1050°C spike anneal for a wide range of ion implantation conditions.

2 EXPERIMENTAL DATA

In order to ensure the statistical confidence on the results, experiments have been performed following two adjacent standard 3^2 Design of Experiments (DoE) on the two implantation factors : energy and dose. The center of each design was replicated 3 times on different wafers to estimate the experimental dispersion. 8" P-type substrates were implanted with arsenic through a 2 nm screen oxide, with an EATON NV8200P implanter. The details of the ion implantation conditions are given in table 1. Finally dopants activation was performed with a 1050°C spike annealing.

	Energy (keV)	Dose (cm^{-2})	Rsheet measured values ($\Omega/\square$)	σ (%)	Rsheet optimized values ($\Omega/\square$)
P1	3	10^{13}	14500	14	14214
P2	3	$2.5 \cdot 10^{14}$	1120	9.1	1091
P3	3	$5 \cdot 10^{14}$	750	4.7	749
P4	3	$1.75 \cdot 10^{15}$	320	4	333
P5	3	$3 \cdot 10^{15}$	290	5.2	289
P6	9	10^{13}	5700	12.5	5423
P7	9	$2.5 \cdot 10^{14}$	555	4.5	546
P8	9	$2.5 \cdot 10^{14}$	580	5.2	583
P9	9	$2.5 \cdot 10^{14}$	545	4.6	541
P10	9	$2.5 \cdot 10^{14}$	560	3.6	577
P11	9	$5 \cdot 10^{14}$	374	4	385
P12	9	$1.75 \cdot 10^{15}$	220	4.6	225
P13	9	$1.75 \cdot 10^{15}$	220	3.2	218
P14	9	$1.75 \cdot 10^{15}$	205	6	200
P15	9	$1.75 \cdot 10^{15}$	218	3.7	212
P16	9	$3 \cdot 10^{15}$	193	3.6	185
P17	15	10^{13}	4500	6.7	4557
P18	15	$2.5 \cdot 10^{14}$	430	4.7	417
P19	15	$5 \cdot 10^{14}$	273	4.5	267
P20	15	$1.75 \cdot 10^{15}$	161	4.3	162
P21	15	$3 \cdot 10^{15}$	152	3.8	159

Table 1: Experimental ranges for arsenic implantation following two adjacent standard 3^2 DoE, R$_{sheet}$ measurements with its standard deviation σ, and R$_{sheet}$ values predicted after the optimization stage.

The As chemical profile is obtained from SIMS measurements. The MCs$_2^+$ technique was used [3] on a Cameca IMS-5f instrument with a primary (Cs$^+$ beam) impact energy and incidence angle of 1 keV and 50° respectively, in order to reduce ion beam mixing and equilibration depths.

Four-point probe measurements were also performed to get the sheet resistance (see table 1).

3 ELECTRICAL RESPONSE MODELLING

To determine the As active distribution we adopted an inverse modelling methodology. Actually we look for the profile which modelled R_{sheet} value is as close as possible to the corresponding experimental value.

The sheet resistance is modelled by :

$$R_{sheet} = \left(\int q\, N_{active}(x)\, \mu(N_{active}(x), N_{inactive}(x))dx \right)^{-1} \quad (1)$$

$N_{active}(x)$ and $N_{inactive}(x)$ are the active and inactive arsenic distributions respectively. μ is the mobility model as explained in section 3.2.

3.1 Active arsenic distribution

The active arsenic profile $N_{active}(x)$ is obtained by truncating the chemical profile $C_{As}(x)$ at a maximum concentration value C_lim representing the electrical solubility threshold of arsenic into silicon. C_lim is usually set to 2×10^{20} cm^{-3} [2].

As the calculation of R_{sheet} with a constant C_lim leads to big discrepancies with the experimental values (see "Literature data" in figure 5), we have improved the modelling of the active profile. Indeed, on the one hand we assumed that C_lim is a function of the implantation conditions, and on the other hand the As active distribution is assumed to have a constant drop at the interface SiO₂/Si due to pile-up (figure 1).

$$N_{active}(x) = \frac{x}{C_1} C\_lim \quad \text{for } x < C_1 \quad (2)$$

where C_1 is the characteristic length of the pile-up decay.

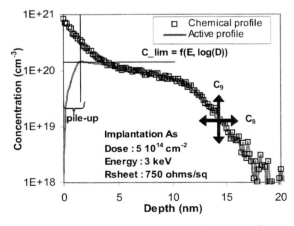

Figure 1: Example of the chemical and active As profile obtained after optimization

Then C_lim was extracted for each experiment of the DoE while keeping the mobility model parameters to their literature value.

Finally a quadratic model of the response log(C_lim) as a function of the factors log(dose) and energy was generated with the software ECHIP [4].

$$\begin{aligned} \log(C\_lim) = &\, C_2 + C_3 \log(dose) + \\ &\, C_4 \log(dose) \cdot energy + C_5 (\log(dose))^2 + \\ &\, C_6\, energy^2 \end{aligned} \quad (3)$$

The linear term in energy of that empirical model has been removed since its effect is negligible according to the Pareto effects graph in figure 2.

Pareto effects graph for response 'log(C_lim)'

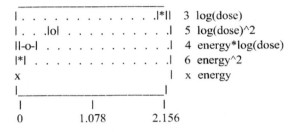

Figure 2: Pareto effects graph giving the classified effects of the factors scaled to units of the response log(C_lim), with their confidence intervals.

With an adjusted R^2 of 0.99, the quality of the RSM model is very satisfactory. Moreover figure 2 shows small confidence intervals, then the accuracy of the terms of the model may be quite good. The coefficients generated by ECHIP are displayed in table 2.

In figure 5 we compare results obtained with literature data, and those using the empirical model of C_lim and including the pile-up effect (called "First fit"). It appears that the new modelling of the active profile gives better predicted R_{sheet} values than before.

3.2 Mobility model

The mobility model μ from (1) is a Mathiessen's combination of two mobility models μ_{active} and $\mu_{inactive}$ depending on the active and inactive As profile respectively.

$$\frac{1}{\mu} = \frac{1}{\mu_{active}} + \frac{1}{\mu_{inactive}} \quad (4)$$

μ_{active} corresponds to the Masetti's model with parameters set to their initial value [5].

$$\mu_{active} = \mu_0 + \frac{\mu_{max} - \mu_0}{1 + \left(\dfrac{N_{active}(x)}{C_r}\right)^{\alpha}} - \frac{\mu_1}{1 + \left(\dfrac{C_s}{N_{active}(x)}\right)^{\beta}} \quad (5)$$

$\mu_{inactive}$ is a term added by Rousseau et al. [6] to take into account carrier interaction with neutral inactive defects.

$$\mu_{inactive} = \frac{K\,q}{m^* \, N_{inactive}(x)} \quad (6)$$

where q is electron charge, m* is its effective mass (we used m*=0.26 of electron mass). K is the Rousseau's fitting parameter for neutral scattering ($K=4.59\times10^7$ cm^{-3}s).

4 OPTIMIZATION CONSIDERING UNCERTAINTIES

To determine more rigorously the value of the parameters of (1) and particularly those used to obtain $N_{active}(x)$, we have to choose a criterion to be minimized. According to the Maximum Likelihood theory [7], the most probable parameters should minimize the following expression :

$$S = \sum_{j=1}^{21} \frac{\left(h_{measured\,j} - h_{predicted\,j}(C_1, ..., C_{13})\right)^2}{\sigma_j^2 + \sum_{i=1}^{13}\left(\dfrac{\partial h_{predicted\,j}(C_1, ..., C_{13})}{\partial C_i}\right)^2 . \sigma_{C_i}^2} \quad (7)$$

where $h=R_{sheet}^{-1}$ is a transformation which linearize S to make the numerical optimization easier.

The parameters C_1, ..., C_{13} are to be optimized and their initial value is given in table 2. C_1 corresponds to the characteristic length of the pile-up drop. C_2, ..., C_6 correspond to the coefficients of the quadratic model (3). C_7 is a parameter that makes the transition between the chemical profile and C_lim smooth. The SIMS error on the profile depth and the concentration [3] are also taken into account with parameters C_8 and C_9 respectively. Their values are close to the unity.

$$C_{As}(x, C_8, C_9) = C_9 . SIMS(C_8 . x) \quad (8)$$

where SIMS(x) is the experimental chemical profile of arsenic.

Finally C_{10}, ..., C_{13} are the mobility parameters μ_1, β, Cs and K chosen to be optimized as done in [6].

The coefficient σ_j in the expression (7) is the standard deviation of each R_{sheet}^{-1} measurement. σ_{Ci} is the standard deviation of each model parameter obtained from analysis of results given by ECHIP for the coefficients of the quadratic model (3) and from literature data ([2], [3], [5]

and [6]) for the other parameters. The search range of the parameters is taken to the nominal value $\pm \sigma_{Ci}$.

	Lit. data value	First fit	FEP optim.
C_1 (nm) $\in$ [0.1,3]	--	1.5	1.4
$C_2 \in$ [19,21]	20.3	19.6945	19.655
$C_3 \in$ [0.8,0.9]	--	0.85	0.89
$C_4 \in$ [-0.0081,-0.0075]	--	-0.0791	-0.00786
$C_5 \in$ [-0.4,-0.3]	--	-0.3561	-0.4
$C_6 \in$ [0.0006,0.00064]	--	0.000625	0.00061
$C_7 \in$ [0.7,0.99]	--	0.9	0.76
$C_8 \in$ [0.975,1.025]	1	1	0.975
$C_9 \in$ [0.95,1.05]	1	1	0.95
$C_{10}=\mu_1$ (cm^2/Vs) $\in$ [30,50]	43.4	43.4	36.4
$C_{11}=\beta \in$ [1.8,3]	2	2	3
$C_{12}=Cs$ (cm^{-3}) $\in$ [2.5-4 10^{20}]	3.43 10^{20}	3.43 10^{20}	2.61 10^{20}
$C_{13}=K$ (cm^{-3}s) $\in$ [3-6 10^7]	4.59 10^7	4.59 10^7	3 10^7
S	178	16.5	6.8

Table 2: R_{sheet} model parameters before and after optimization, with their search range estimated from the confidence intervals of the Pareto effects graph (Figure 2).

Facing this high dimensional (13 parameters) optimization task, we have implemented in the software MATHCAD [8] a Fast Evolutionnary Program (FEP) based on Yao work [9]. This global optimization method is a particular case of Genetic Algorithms, known to find the global optimum in most cases without getting trapped by local minima. The flowchart of the algorithm used is detailed in figure 3.

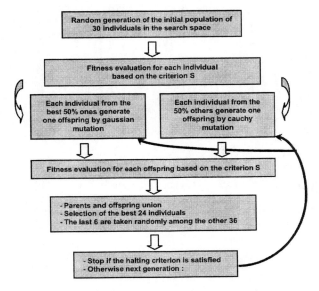

Figure 3: Flowchart of the Evolutionnary Programming algorithm used in the parameters optimization

5 RESULTS

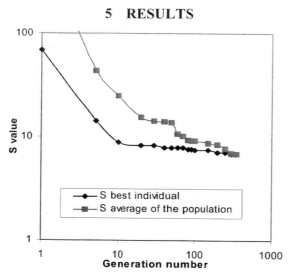

Figure 4: Convergence of the FEP.

Figure 4 shows the evolution of the value of the S criterion when the number of generations is increased. It indicates that after 350 generations, the algorithm has probably converged on the global minimum since most individuals of the population are very close to the best individual. In addition the best individual of the population changes only a little after the generation 100.

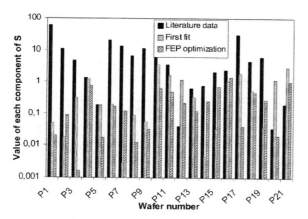

Figure 5: Comparison of each component value of the S criterion obtained with literature data parameters, with values from the empirical model of C_lim and considering pile-up effect, and with parameters given by the genetic optimization.

The values of the thirteen parameters obtained after the genetic optimization are given in table 2. One can see that the optimized set of coefficients gives the lowest criterion value of the three trials. The values of R_{sheet} predicted by (1) after the FEP optimization are displayed in table 1. These

values are quite close to the experimental ones. Indeed figure 5 shows that the components of the S criterion have been, on the whole, improved after the FEP optimization compared with previous work.

6 CONCLUSION

An efficient and statistically rigorous methodology has been developped to calibrate a high number of model parameters required to simulate the USJ resistivity. The method accurately predicts the arsenic active distribution for a wide range of ion implantation conditions and a 1050°C spike anneal. The method also optimizes, with a genetic algorithm, the relevant set of model parameters, considering the experimental and modelling uncertainties.

ACKNOWLEDGEMENT

The authors would like to thank Professor Brini from the Institute of Microelectronics, Electromagnetism and Photonics (IMEP France), as well as Dr. Roger [10] from SILVACO Data Systems, for helpful discussions.

REFERENCES

[1] H. Iwai, IEEE Journal of Solid-State Circuits, Vol. 34, pp. 357-366, 1999.
[2] P. B. Griffin, ESSDERC Proceedings, pp. 60-63, 2000.
[3] P. Holliger, F. Laugier and J. C. Dupuy, Surface and Interface Analysis, Vol. 34, pp. 472-476, 2002.
[4] ECHIP 6.4 user's guide.
[5] G. Masetti, M. Severi and S. Solmi, IEEE Transactions on Electron. Devices, Vol. ED-30, pp. 764-769, 1983.
[6] P.M. Rousseau, P. B. Griffin, S. Luning and J. D. Plummer, IEEE Transactions on Electron. Devices, Vol. 43, pp. 2025-2027, 1996.
[7] Y. Cortial, Bulletin de l'Union des Physiciens, Vol. 725, pp. 769-791, 1990.
[8] MATHCAD 8 user's guide.
[9] X. Yao, Y. Liu and G. Lin, IEEE Transactions on Evolutionary Computation, Vol.3, pp. 82-102, 1999.
[10] F. Roger, Thèse de Doctorat, 2002.

Full-band Particle-based Simulation of Germanium-On-Insulator FETs

S. Beysserie[*], J. Branland[*], S. Aboud[*,***], S. Goodnick[**], T. Thornton[**], and M. Saraniti[*]

[*]Department of Electrical and Computer Engineering,
Illinois Institute of Technology, 3301 South Dearborn, Chicago, IL 60616-3793, beysseb@iit.edu
[**]Department of Electrical Engineering and Center for Solid State Electronics Research, Arizona State
University, Tempe, AZ, USA
[***]Molecular Biophysics Department, Rush University, Chicago, IL

ABSTRACT

We model and simulate novel fully depleted (FD) sub-50nm gate lengths MOSFET structures using a full-band particle simulator based on the Cellular Monte Carlo (CMC) method that provides an accurate transport model at the high electric fields present in these nanoscale devices. Simulations of Germanium- and Silicon-On-Insulator devices (GOI and SOI, respectively) are performed to quantitatively investigate the predicted increase of performance of GOI technology. A comparison of static and dynamic properties of similar GOI and SOI devices is performed for 50 nm and 35 nm gate lengths.

Keywords: full-band simulation, SOI MOSFET, frequency analysis, high-κ dielectric, Germanium

1 INTRODUCTION

The general trend of microelectronic technology towards higher circuit integration has driven the size of semiconductor devices into the sub-micron regime; to this end FD Silicon-On-Insulator technology [1] has generated a tremendous interest due to the operating speed, lower voltage and power-consumption in comparison with traditional bulk devices. However, Silicon (Si) inversion layers exhibit asymmetric low-field electron and hole mobilities, resulting in a degradation of the performance of n-channel devices over their p-channel counterpart. This asymmetric behavior is less pronounced in Germanium (Ge), which makes it an attractive material for extending CMOS to sub-micron devices. The higher electron and hole mobilities of Ge also translate into larger saturation drain currents, enhanced transconductance and higher cut-off frequencies [2].

In this work, we model and simulate charge transport in several novel sub-50nm GOI and SOI MOSFETs using a full-band particle-based simulator based on the Cellular Monte Carlo (CMC) method [3, 4]. Simulation of a realistic representation of the layout of such devices is particularly challenging due to the fact that the high doping gradients used in the layout of these down-scaled devices requires extremely fine discretization schemes resulting in a large computational burden. In order to investigate the predicted increase of performance of GOI over SOI structures we

compare the transport properties of several GOI and SOI MOSFETs. A preliminary frequency characterization of these devices is also performed through the analysis of their transient response [5].

In the following section, the CMC algorithmic approach is summarized. In Section 3 we present the different simulated GOI and SOI structures. Finally, we compare and discuss the simulation results in Section 4.

2 CMC FULL-BAND SOLVER

The use of a full-band solver is of critical importance to realistically model the devices considered in this work; indeed, analytical approximations of the band structure fail to accurately account for the pronounced warping of the valence bands in Ge and the transport properties at high electric fields that are present in the small structures of interest. The full-band particle-based solver used in this work is based on the CMC approach, which was developed to address the need for computational resources associated with the Ensemble Monte Carlo (EMC) method [6]. This efficient simulation code allows the full characterization of a nanoscale MOSFET structure within a realistic amount of time. This quality is particularly advantageous for the simulation of p-channel GOI structures for which the low effective mass of holes results in the requirement of a time resolution that can be one order of magnitude larger than that used for simulating SOI devices [2]. Within the CMC framework, the electronic structure is computed with the empirical pseudo-potential method [7] inclusive of the spin-orbit interaction, while the full phonon structure is computed with the valence shell model [8]. Both longitudinal acoustic and optical phonon modes are considered in the computation of the deformation potential scattering [6], and impact ionization is modeled using an energy-dependent analytical fit of the momentum-dependent anisotropic ionization rates, as done in [9]. Steady-state field-dependent velocity and energy characteristics are calibrated for different materials with scattering parameters reported in literature [6, 10]. The band structure of Si and Ge as used within the simulation code are shown in Fig.1 (a) and (b), respectively.

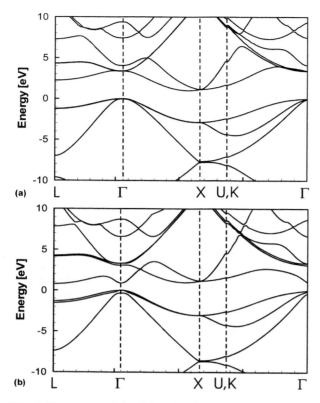

Fig. 1. Representation of the electronic band structure along the L, Γ, X and U,K directions for Si (a) and Ge (b).

3 DEVICE DESIGN AND SIMULATION

To investigate the behavior of GOI FETs, several novel-structure devices have been simulated in this work, including a 50 nm gate, n-channel SOI together with a p-channel SOI and a GOI device with 50 nm and 35 nm gate lengths. For all these devices, the source and drain are separated by 30 nm from the gate ends. A 1.5 nm equivalent oxide thickness layer of high-κ dielectric is used as an insulator between the channel and the gate contact, whereas SiO_2 is used for the 200 nm-thick buried oxide. To ensure full depletion at threshold, a 15 nm doped layer has been simulated in the channel region. A Gaussian doping profile with a peak concentration of 10^{20} cm^{-3} is used in the longitudinal direction, extending 11 nm underneath both sides of the gate. This results in an effective channel length of 28 nm and 13 nm, respectively, for the 50 nm and 35 nm gate structures. The actual layout and doping profile of these devices are shown in Fig. 2 for the 50 nm gate p-MOSFET structure. All devices have been mapped to a 2-dimensional inhomogeneous 256 by 65 grid. Following the standard approach used by particle-based simulation methods [6], the fixed "free flight" time step was set to 0.2 fs whereas the time between successive solutions of Poisson's equation ranged from 10 fs to 0.2 fs depending on the semiconductor material and the doping concentration. Furthermore the use of a very short Poisson time step

greatly reduces unphysical oscillations of the output current related to the high doping concentrations in the simulated structures, and facilitates the analysis of the device transient response.

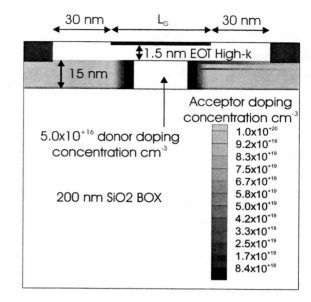

Fig. 2. Layout and doping profile of the simulated devices with L_G = 50 nm

4 RESULTS

4.1 Static Analysis

In Si-based CMOS technology, p-channel transistors are the performance bottleneck due to the lower mobility and saturation velocity of holes.

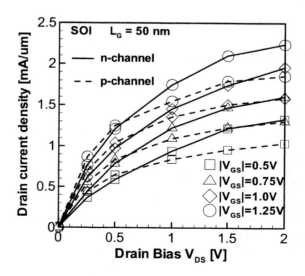

Fig. 3. Comparison of the current density versus drain voltage of 50 nm gate n and p-type SOI MOSFETs.

The resulting degradation of the saturation current in p-type devices is illustrated in Fig. 3, showing the simulated current-voltage characteristics of both p-channel and n-channel SOI MOSFETs with 50 nm gates. The saturation current is 25% higher for the n-channel transistors. In order to study the impact of using similar Ge devices, we compared both GOI and SOI p-channel MOSFETs. Simulation results of the drain current density versus drain voltage for different gate biases ranging from –0.5V to –1.25V are depicted on Fig. 4 (a) and (b) for the 50 nm and 35 nm gate devices, respectively.

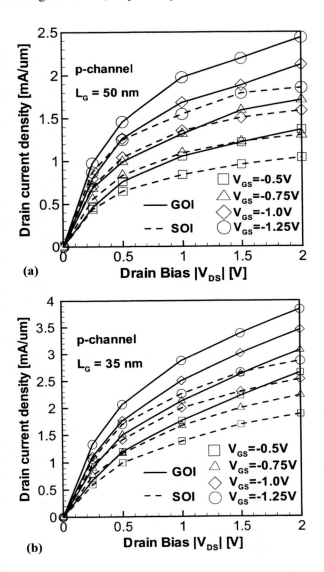

(a)

(b)

Fig. 4. Current density versus drain voltage comparison of p-channel SOI and GOI MOSFETs for (a) 50 nm and (b) 35 nm gate lengths.

For a drain bias of –2.0V, the observed increase in drain current ranges from 30% to 40% for GOI devices in comparison with SOI technology, which is in agreement with the increased peak velocity of holes inside the channel as shown in Fig.5 for the 35 nm gate p-channel MOSFETs. In both Si- and Ge-based devices, quasi-ballistic transport is achieved in the high field region underneath the gate where the peak velocity of holes is 1.7 times higher than the bulk saturation velocity. In the case of the 35 nm device, the simulated peak velocity of holes is 1.55×10^5 m/s versus 1.15×10^5 m/s, which is a 35% increase as shown by the current-voltage characteristics.

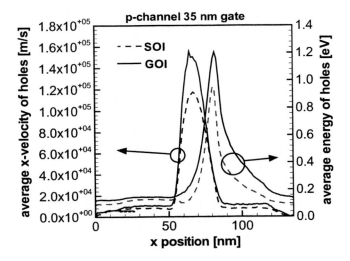

Fig. 5. Average time hole energy and longitudinal velocity along the channel for p-type 35 nm SOI and GOI devices.

The static, DC simulations confirm the expected increase in performance that Ge can provide for CMOS technology. Indeed, the simulated p-channel GOI MOSFETs deliver as much current as n-channel SOI MOSFETs for the same biases, and alleviate the performance differences between p- and n-type Si devices.

4.2 Dynamic Analysis

The frequency analysis of these devices is crucial to study the impact of using Ge on the voltage gain. The method used to investigate the frequency behavior of these devices is based on the Fourier decomposition of the current transients in response to a step voltage perturbation on the drain and gate electrodes. Each device was simulated for 25 ps using a 0.2 fs Poisson time step and 100,000 particles to represent the total hole population, simulation results are typically obtained within 60 hours when performed on a Pentium IV Xeon 2.4 GHz processor. The first 10 ps of the simulation are used to let the device reach steady state about the operating point, then a step voltage on either the gate or drain electrode is applied during the remaining 15 ps. Initial, low resolution results of the voltage gain as a function of frequency are depicted on Fig.6 for the 50nm gate SOI p-channel, n-channel, and the p-channel GOI.

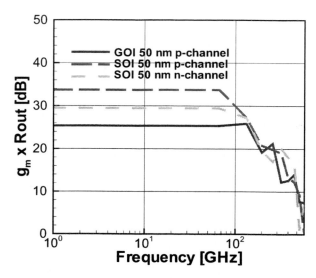

Fig. 6. Voltage gain in dB versus frequency of 50 nm gate GOI and SOI devices.

Within the constant gain bandwidth, the SOI p-MOSFET delivers a relatively higher amplification than the GOI device, nevertheless the –3 dB frequency of the GOI MOSFET is significantly higher than for the p-type SOI device. Longer simulation times will be employed in order to increase the spectral resolution of the present analysis and quantitatively investigate the behavior of these devices within the 30 to 130 GHz frequency range.

CONCLUSION

We have successfully performed a first-hand particle-based investigation of the performance provided by Ge devices used in CMOS technology. The higher mobility observed in Ge with respect to Si translates into higher saturation currents that place p-type GOI MOSFETs on the same performance level as n-type SOI MOSFETs. Similarly promising results have also been obtained with the dynamic analysis. Additional investigation should be performed in sub-threshold regime to further validate the advantages of Ge over Si and confirm GOI as a suitable solution to reduce the performance gap between p- and n-MOSFETs.

ACKNOWLEDGEMENTS

This work has been partially supported by the award numbers ECS-9976484 and ECS-0115548 of the National Science Foundation.

REFERENCES

[1] J.P. Colinge, "Silicon-On-Insulator Technology: Materials to VLSI, 2nd Edition," Kluwer Academic Publishers, 2000.

[2] S.M. Sze, "Physics of Semiconductor Devices, 2nd Edition," Wiley Interscience, 1981.

[3] M. Saraniti, and S.M. Goodnick, "Hybrid Full-band Cellular Automaton Monte Carlo Approach for fast Simulation of Charge Transport in Semiconductors," IEEE Transaction on Electron Devices, vol. 47, pp. 1909-1915, (2000).

[4] M. Saraniti, J. Tang, S.M Goodnick, and S.J.Wigger, "Numerical challenges in particle-based approaches for the simulation of semiconductor devices," Mathematical and Computer in Simulations, 2002.

[5] J. Branlard, S. Aboud, S. Goodnick, and M. Saraniti, "Frequency analysis of 3D GaAs MESFET structure using full-band particle-based simulations," Proc. of the 9th Int. Workshop on Computational Electronics (IWCE-9), Rome, Italy, June 2003.

[6] M.V. Fischetti and S.E. Laux, "Monte Carlo analysis of electron transport in small semiconductor devices including band-structure and space-charge effects", Phys. Rev. B, vol. 38, pp 9721-9745, Nov. 1988.

[7] J.R. Chelikowsky and M.L Cohen, "Nonlocal pseudo potential calculations for the electronic structure of eleven diamond and zincblend semiconductors", Phys. Rev. B, vol 14, pp. 556-582, July 1976.

[8] K. Kunc and O.H. Nielsen, "Lattice dynamics of Zincblende structure compounds-II: Shell model", Compu. Phys. Commun., vol. 17, pp. 413-422, July/Aug. 1979.

[9] M.V. Fischetti, N. Sano, S.E. Laux, and K. Natori, "Fullband-structure theory of high-field transport and impact ionization of electrons and holes in Ge., Si., and GaAs", in Proceedings 1996 Int. Conf. Semiconductor Processes and Devices, Tokyo, Japan, 1996.

[10] W. Pötz and P. Vogl, "Theory of optical-phonon deformation potentials in tetrahedral semiconductors", Phys. Rev. B, vol. 24, pp. 2025-2037, Aug. 1981.

A Technology-Independent Model for Nanoscale Logic Devices

[*]University of Florida, Depts. of CISE and ECE
CSE Bldg., Room 301, Gainesville, FL, mpf@cise.ufl.edu

ABSTRACT

We discuss how to model nanoscale logic devices without making any assumptions about what type of physical mechanism (electrical, mechanical, optical, etc.) they are based on. Starting from core facts of quantum field theory, we review how generic physical quantities such as entropy and energy relate to computational concepts such as capacity and performance. We advocate partitioning our model into subsystems playing certain generic roles. Finally, we illustrate how our device-independent perspective lets us infer strong, general facts about any future nanocomputing technology. *E.g.*, standard irreversible logic can *never* perform more than ~10^{22} bit-ops/sec per 100 Watts of power. Furthermore, achieving logic frequencies above about 9 THz (in ~20 years) will *require* performing controlled manipulations of logical bits at generalized temperatures well above room temperature, which would also allow reversible computing to achieve sub-kT dissipation per bit-operation despite ambient thermal noise and decoherence.

Keywords: nanocomputing, devices, compact models, fundamental limits, reversible computing, decoherence

1 INTRODUCTION

At this time, a wide variety of different mechanisms for performing digital information processing at the nanoscale have been proposed. In the literature, one finds proposals based variously on electronic, mechanical, chemical, and optical principles, and various combinations of these. Even if we narrow our attention to the all-electronic technologies, we encounter a broad range of proposed devices, based variously on semiconductors, conductors, or superconductors; field-effect transistors, resonant tunneling transistors, or Josephson junctions; quantum dots or wires; metal, silicon crystals, carbon nanotubes, and organic molecules. For encoding information, electron position, voltage, current, or spin states could be used, or even atomic and nuclear configurations, motions, and spin states. This is not to mention all the possible permutations that also utilize photonic & electromagnetic phenomena, chemical transitions, *etc.*

We would like to provide a theoretical foundation for nanocomputing that will allow us to characterize the limits of nanocomputing, as well as to analyze, compare and optimize different candidate nanocomputer architectures. But, how are we to do this, when there is such a broad range of wildly differing technologies that have been proposed, with

no clear long-term winner among them? Can we bring some order to this chaos?

One approach to this problem is to develop *technology-independent* theoretical models of nanocomputing, based not on the particular design constraints of any specific technology, but on more generic physical considerations that must apply to *any* physically possible technology.

We contend is that this technology-independent modeling effort is both feasible and useful. It is feasible because all nanotechnologies are ultimately subject to the same underlying laws of physics. At the nanoscale, the relevant "gold standard" theory is quantum electrodynamics (QED), which has stood for more than 40 years now as an extremely precise underlying model for all experimentally accessible, non-gravitational phenomena involving only photons, electrons, and stable nuclei. It thus subsumes virtually all of chemical, electrical, optical, and materials science. Within the scope of its domain of applicability, QED's predictions have been empirically confirmed to as many as 11 decimal places of precision, and no clear contradictions between the theory and experiment have been found. For processes involving very high-energy interactions, and more exotic, unstable particles, QED has been successfully extended to yield the Standard Model of particle physics, which has reigned supreme for about 30 years now as the basis of all known physical phenomena (except gravity).

Modern theories such as QED and its relatives assure us that all physical systems and processes, regardless of their makeup, can ultimately be characterized in terms of a few universal, domain-independent physical concepts, such as entropy, energy, heat, temperature, and momentum.

Meanwhile, in computer engineering, we also ultimately care only about a range of other universal, technology-independent concepts, such as operating frequency (clock speed), energy dissipation, information propagation speed, information bandwidth and bandwidth density, heat flux, throughput, latency, performance, cost, and so forth.

In the end, any particular device technology (whether it involves carbon nanotube transistors, superconducting junctions, or spintronic valves) can be viewed as just being an interfacial "glue" layer, which executes a mapping (though possibly a complex one) between one essentially technology-independent domain (that of fundamental physics) and another one (that of computer engineering).

Thus, we ought to be able to model devices in *all* nanocomputing technologies generically, by abstractly characterizing how they carry out this mapping.

What are the advantages of this unified approach, as opposed to using a different, specific model for each different

NSTI-Nanotech 2004, www.nsti.org, ISBN 0-9728422-8-4   Vol. 2, 2004   29 device technology? (One disadvantage is that a technology-specific model might well be more accurate.)

The advantages of a generic model are that:

(1) We are not forced to make a guess (which would probably be wrong anyway) concerning which specific nanocomputing technolog(y/ies) will be commercially viable 30 years from now, and thus are worth the time of developing a detailed theory for modeling them;

(2) The model can be easily adapted to quantitatively fit whatever specific nanocomputing technology does eventually become dominant.

(3) Barring an (extremely unlikely) discovery of a huge flaw in modern fundamental physics that has eluded detection by large swarms of researchers for many decades, general qualitative results obtained from our generic model *cannot become obsolete* as new device technology concepts are developed. At most, certain quantitative predictions will need to be further refined.

(4) The model provides a framework that device physicists can use to translate from the low-level characteristics of their specific technology to system-level figures of merit (*e.g.*, performance per unit cost) that will apply to a complete, large-scale digital system design that is based on those devices. This will help technology designers to steer their efforts towards the most useful technologies.

(5) The model provides a basis for nanocomputer architecture that is by and large independent of the nanocomputing technology that is used.

The general effort to develop and explore models of computing that are based soundly on universal physical principles I call *physical computing theory*. In this document, we include a brief outline of a particular theoretical physical model of computing that we are currently developing, which we call *CORP* (Computing with Optimal, Realistic Physics). We previously described CORP in a bit more detail in [1]. Later, we will discuss some results obtained from the CORP model.

One technique that has been useful in building CORP is to start by first reinterpreting fundamental physics itself in computational terms, which allows us to identify the key physical concepts that impact computation.

2 PHYSICS AS COMPUTATION

As we previously discussed in [1], all of the received quantum field theories, such as QED and the Standard Model, can be approximated to any desired accuracy using the Q3M (quantum 3d mesh) model [2], a type of parallel quantum computer consisting of a regular 3-dimensional array of cells having only a finite number of qubits per cell (representing, *e.g.*, the number of fundamental particle quanta of each type in that cell). Each cell continuously interacts locally with its nearest neighbors, exchanging particles by means of a Hamiltonian derivable from the Schrödinger equation, while simultaneously updating its internal state according to another Hamiltonian describing the interac-

tions between fundamental particles. Such a mesh, at a fine level of granularity, can apparently accurately capture all of Standard Model physics.

In such computational models of physics, various important physical quantities can be given precise, well-defined computational meanings. There is not space to justify and detail all of these here, so we merely summarize the most important results:

(1) The physical *entropy* in any subsystem is just the amount of incompressible (non-decomputable) information in that subsystem. Information can be effectively incompressible either when it is unknown, or when it is known but random, or even (effectively) when it is known and non-random, if its underlying pattern of order is effectively inaccessible (such as an encrypted file, when the decryption key is lost). The *maximum* entropy of a system is its total physical information content, the logarithm of its number of distinguishable states. The non-entropy physical information in a system can be called *extropy*.

(2) The physical *energy* in any subsystem is the *rate of quantum physical computation* in that subsystem. (This can be given a precise meaning based on the maximum rate of rotation of quantum state vectors in Hilbert space.) The rate of useful bit-operations is $R=2E/h$ by the Margolus-Levitin theorem [3]; *e.g.* 1 eV is 484 Tbops (trillion bit-ops per second). *Heat* is then just the energy in that part of the information that is entropy—*i.e.* it is the rate at which the random bits of physical information are changing. For example, 1 BTU (British Thermal Unit) turns out to be a rate of 3×10^{36} random bit-flips per second.

(3) The thermodynamic *temperature* of a subsystem is then, roughly speaking, the heat per bit of entropy, that is, the "clock speed" for updating of random information [4]. *E.g.*, each degree Kelvin is a frequency $f \approx 2K/(hk \ln 2) = 28.9$ GHz of bit-updating. We can generalize temperature to *generalized temperature*, which is the total energy per bit of information, even for those bits that are not entropy. A system's generalized temperature (overall physical clock speed) can be higher than its thermal temperature, although non-thermal energy tends to degrade into heat, unless the system's high-energy extropic degrees of freedom are very well-isolated from parasitic interactions that will leech off their energy into entropic (thermal) degrees of freedom.

The above observations serve as a basis for our generic technology-independent model of computation which respects all the fundamental laws of physics.

3 CORP DEVICE MODEL

CORP (Computing with Optimal, Realistic Physics) is the theoretical physical model of computation that we are developing. CORP's device model is essentially a "lumped element model" of the underlying computational model of physics described in section 2. That is, each device is a compound subsystem that may include a large number of underlying quantum bits of state. However, as a lumped system, it still has an energy, an entropy, and thermody-

Device				
Coding Subsystem		Non-coding Subsystem		
Logical Subsystem	Redundancy Subsystem	Structural Subsystem	Power Subsystem	Thermal Subsystem

Figure 1. Conceptual hierarchy of subsystems in a CORP device. If desired, a separate *timing subsystem* can also be split out from the coding or non-coding subsystem.

namic and generalized temperatures. Further, we can still conceptualize its dynamics as decomposing into Hamiltonians for its self-interaction and its interactions with neighboring systems.

Now, not all of the degrees of freedom in a real physical device are actually used for computation. So, we break each device down into subsystems that play different roles. Figure 1 shows the conceptual structure of a device in the CORP model. The *coding subsystem* is the part of the state that is varied in a controlled way to store and manipulate logical information as part of the computation of interest. We can further divide it into the *logical subsystem*, the bits being represented, and the *redundancy subsystem*, other redundant physical bits that are included for purposes such as noise immunity and error-correction.

The rest of the device's state is its *non-coding subsystem*. We can break this into the *structural subsystem*—the part of the state that must remain unchanged if the device is to continue to operate properly—the *power subsystem*—which supplies low-entropy energy, and the *thermal subsystem*—the part of the state that is allowed to vary randomly and provides a pathway for removal of waste heat.

Each of these subsystems can be itself characterized by its energy, total physical information content (entropy plus extropy), and thermal and generalized temperatures. In addition, between each pair of subsystems within a device—as well as between it and its neighbors—there is an interaction energy and a generalized *interaction temperature* that characterize the rate at which bits of one subsystem are changed due to interactions with the other.

Ideally, all of the device's entropy is kept isolated within the thermal subsystem, although it may tend to creep into the coding subsystem (noise) or the structural subsystem (degradation) due to unwanted parasitic interactions between the thermal subsystem and the other subsystems. In general, active error correction and structural repair mechanisms (both of which are forms of refrigeration) must be used in order to keep the coding and structural subsystems clear of entropy indefinitely.

Next, we define the device's spatial geometry (region of space occupied), and identify which physical degrees of freedom within that region make up its state. This allows us to determine what assemblages of devices can exist without overlapping. Portions of the device's coding state are identified as I/O channels for communication with neighboring devices.

Finally, we summarize the device's overall computational behavior with a quantum transition function, which is

a unitary map U from its input+internal logical state to its internal+output state a moment later. Such a map is necessarily invertible, so logically irreversible operations can only be implemented by extending the part of the internal state that is involved in the transformation to include not only the logical subsystem, but also the thermal subsystem. All bits discarded from the coding subsystem thus end up as entropy in the thermal subsystem.

One way to summarize the device, for technology-independent computer-engineering purposes is to specify, in addition to its transition function, also its length ℓ, area A, volume V, information capacity I, information I/O bandwidth B, maximum operation frequency f_{max}, standby power consumption P_{leak} and rate of standby entropy generation S_t, its energy and entropy coefficients E_f and S_f when doing reversible operations, its energy dissipation and entropy generation E_i and S_i for logically irreversible operations, and the maximum energy and entropy flows P_{max} and $S_{t,max}$ sustainable in its power and thermal subsystems.

In previous work [2], we have used such models to show that architectures that are predominantly logically reversible are asymptotically faster and more cost-efficient than traditional irreversible architectures, for the broadest class of applications, whenever there is a fixed limit on either total system power, or on per-area entropy flux S_{At}. Irreversible machines have a fundamental limit on their performance within room-temperature environments of $(100 \text{ W})/(k\, 300\, \text{K} \ln 2) = 3 \times 10^{22}$ bit-operations per second, per 100 Watts of power consumption. This is only about 5 orders of magnitude beyond today's technology. Reversible computing is the only possible way to exceed this limit.

Note that these results all hold true completely independently of which domain of device technology is used (electrical, optical, mechanical, chemical). This illustrates the power of technology-independent modeling.

In the next section, we illustrate how our unified physical-computational perspective can also be applied to a more technology-specific device-level problem, of estimating the minimum entropy generation per op in field-effect devices.

4 MINIMUM ENTROPY/OP FOR FETS

For purposes of this analysis, let a device be characterized by the following independent parameters: T_g – Average generalized temperature for operations in the entire coding subsystem, including timing signals. E_{lb} – Energy per amount of coding-state information representing one logical bit. t_{tr} – The elapsed time of one useful logical bit-operation (transition between distinguishable states of a logical bit). t_d – The average time between local decoherence events for each bit within the coding subsystem. P_{lk} – Leakage power per stored logical bit. S_t – Rate of standby entropy generation per logical bit due to parasitic thermally activated transitions.

From these, we can derive the following dependent parameters: $I_{lb} = E_{lb}/T_g$ – Physical information per logical bit. The dimensionless ratio $r = I_{lb}/\text{bit}$ is called the *redun-*

dancy factor. E.g., in a voltage-coded logic circuit node, r is the number of electron states between Fermi levels at high and low voltage states. Energy per physical bit: $E_{pb} = E_{lb}/r = T_g \cdot b = kT_g \ln 2$. Rate of physical computation per logical bit: $C_{lb} = I_b \cdot \text{step}/t_{tr} = I_b \cdot (\text{op/bit})/t_{tr} = r \cdot \text{op}/t_{tr} = (E_b/T_g) t_{tr}(\text{op/bit})$. Rate of energy transfer (power transfer) involved in switching each bit: $P_{tr} = E_b / t_{tr}$.

We are also subject to the following constraints: $I_b \geq 1$ bit, since a bit of logical information obviously cannot be encoded in less than 1 bit of physical information. The Margolus-Levitin relation [3] tells us that the time to change each physical bit is lower-bounded by its energy, so $t_{tr} \geq h/2E_{pb} = h/2bT_g$. (The logical bit cannot change faster than its redundant physical bits can.) Thus, for example, if the generalized temperature of the coding subsystem is only 300K, then at least 0.115 ps are required to change a bit.

In a field-effect device switched over voltage V, $P_{tr}/P_{lk} = i_{tr}V/i_{lk}V = i_{tr}/i_{lk}$, where i_{tr} and i_{lk} are the currents during desired and leakage transitions. Now, the on/off ratio $i_{tr}/i_{lk} \leq \exp(V/kT) \approx^* \exp(E_{lb}/rkT) = \exp[E_{lb}/(E_{lb}/T_g \cdot \text{bit})kT] = \exp[(\ln(2)k/k)(T_g/T)] = 2^c$ where $c = T_g/T$, the ratio of generalized to thermal temperature in the coding subsystem.

Decoherence will mean that that $S_t \geq I_{lb}/t_d$, and leakage will mean that $S_t \geq P_{lk}/T$. Actually we can represent the total S_t as a sum of these factors, $S_t = I_{lb}/t_d + P_{lk}/T$. Then, the total entropy generated over a bit-cycle is $\Delta S_{lbc} = S_t t_{tr} = I_{lb}(t_{tr}/t_d) + P_{lk}t_{tr}/T$. However, from earlier we have that $P_{lk} \geq P_{tr}/2^c$, so $\Delta S_{lbc} \geq I_{lb}(t_{tr}/t_d) + P_{tr}t_{tr}/2^cT$. But now $P_{tr} = E_b/t_{tr}$, so $\Delta S_{lbc} \geq I_{lb}(t_{tr}/t_d) + E_{lb}/2^cT$. Since $E_b = I_{lb}T_g$, and $T_g/T = c$, we get:

$$\Delta S_{lbc} \geq I_{lb}\left(\frac{t_{tr}}{t_d} + \frac{c}{2^c}\right) \geq 1 \text{ bit} \cdot \left(\frac{1}{q} + \frac{c}{2^c}\right) \qquad (1)$$

where $q = t_d/t_{tr}$ is the quantum *quality factor* and c is the *coding speedup.* The value of q can also be expressed as T_g/T_d where T_d is the *decoherence temperature*, which is the decoherence rate per bit, or in other words the interaction temperature between coding and non-coding subsystems.

Note that this expression for entropy generation can take on arbitrarily small values, but that this requires that both the $1/q$ and $c/2^c$ terms be made comparably small. Fortunately, both terms can be made small simultaneously by making T_g—the generalized temperature of the coding system—large relative to both T and T_d.

In other words, perhaps counter-intuitively, in order to minimize the entropy generated per logical bit-operation in field-effect devices, the energy per physical bit in the coding system should be made large, relative to both thermal and decoherence temperatures in the system.

Intuitively, the coding temperature T_g needs to be larger than the decoherence temperature T_d so that decoherence events don't have time to happen over the course of a logic operation. Meanwhile, it also needs to be larger than the

thermal temperature, in order to suppress thermally-activated leakage of electrons over the potential energy barriers set up in the field effect devices.

We argue that a closely analogous scaling analysis ought to still hold true in any digital switching technology, since this will always involve the raising and lowering of potential energy barriers between states, activated by transitions occurring between states in other similar devices.

5 CONCLUSIONS

The laws of physics can ultimately be understood from a computational perspective. Similarly, computation can be fundamentally described in terms of physical concepts. By using universal physical concepts such as entropy, energy, and temperature, we can compose theoretical models of nanocomputing devices in a way that is independent of the particular nanotechnology that is being used, and obtain results that will apply to all future nanotechnologies.

In this document, we briefly outlined two types of results that have already been obtained using these methods. The first was a high-level computer systems engineering analysis showing that in the long run, reversible computing, if possible, is more cost-effective than irreversible computing. The second was a demonstration that reversible computing with arbitrarily little entropy generation per operation is indeed possible in a readily generalizable model of switched FET-like devices, even when accounting for decoherence effects and thermally-activated leakage, so long as the generalized temperature in the coding system—the maximum rate of transitions per bit—can be made large relative to ambient rates of decoherence and thermal transitions. At room temperature (300 K), thermal bits change at a rate of (300 K)(1 bit)/($h/2$) = 8.7 THz. If we (reasonably) assume that decoherence temperatures are also at around this level, then present-day GHz-speed computers are still more than 3 orders of magnitude away from the point where sub-kT computing becomes manifestly possible according to this analysis. Historically, a factor of 1,000 in frequency takes only about 20 years to achieve. But, whenever we reach this point, reversible computing principles will be absolutely vital in order to make any further progress in nanocomputer power-performance beyond it, regardless of our choice of device technology.

REFERENCES

[1] Michael P. Frank, "Nanocomputer Systems Engineering," Nanoengineering World Forum, International Engineering Consortium, Marlborough, MA, June 23-25, 2003.

[2] Michael P. Frank, "Reversibility for Efficient Computing," Ph.D. thesis, MIT, June 1999.

[3] Norman H. Margolus and Lev B. Levitin, "The maximum speed of dynamical evolution," Physica D 120, 188-195, 1998.

[4] Seth Lloyd, "Ultimate physical limits to computation," Nature 406, 1047-1054, 2000.

---

* $V \approx E_{lb}/r$ because the density of states increases with energy, so the majority of the r electron states will have energy closer to the voltage-V Fermi level than to the ground level.

Hierarchical Simulation Approaches for the Design of Ultra-Fast Amplifier Circuits

J. Desai[1], S. Aboud[1,2], P. Chiney[1], P. Osuch[1], J. Branlard[1], S. Goodnick[3], and M. Saraniti[1]

[1]Electrical and Computer Engineering Department,
Illinois Institute of Technology, Chicago, IL, USA, desajai@iit.edu
[2]Molecular Biophysics Department, Rush University, Chicago, IL, USA
[3]Department of Electrical Engineering, Arizona State University, Tempe, AZ, USA

ABSTRACT

The silicon-on-insulator (SOI) technology is one of the most promising technologies as the semiconductor industry shifts to 0.13µm and smaller devices. Fully depleted (FD) SOI transistors offer a nearly ideal behavior for application in analog circuits, particularly in high frequency and low power operation. In this work, the design and development of a highly efficient power amplifier circuit is investigated for SOI technology. A fully depleted NMOS SOI transistor is built and characterized, which exhibits TeraHertz cutoff frequencies. The device parameters of this transistor are then extracted to build a compact circuit model for use in the PSPICE circuit simulator. Finally, a low power and high frequency class E power amplifier is designed based on the SOI transistor, and a full analysis of the performance compared to bulk Si technology is performed.

Keywords: FDSOI, DST, class E power amplifier, power-added-efficiency, drift diffusion

1 INTRODUCTION

The inherent characteristics of SOI based structures improve device speed by 15% to 35% over that exhibited by bulk CMOS technology [1]. Previous works demonstrate the advantages of SOI in low power digital baseband circuits such as microcontroller CPUs, SRAM, DRAM and ALU [2]. More recently, results of receiver functions such as low noise amplifiers, mixers and VCOs implemented in SOI have been reported. Lack of successful demonstration of a power amplifier has been one element preventing implementation of a complete SOI RF transceiver [3]. In battery-operated devices like cell phones, the talk time directly depends on the efficiency of the power amplifier (PA) in the transmitter. Input power consumption of other blocks in the transceiver (DSP/baseband circuitry, oscillators, mixers, filters, LNA etc.) is often negligible to that of the PA and much research is focused on how to improve the efficiency of PA circuits.

Currently, bulk Si MOS technology is the dominating technology in the semiconductor and integrated circuit industry. The main goal in this work is to investigate the performance of the class E PA circuit built on SOI technology. The n-channel SOI transistor designed here is based on the DST transistor [4] fabricated by Intel. Model parameters are extracted from the simulation data of the device in order to build a compact circuit model of the SOI transistor for use in the class E PA circuit.

2 DEVICE CHARACTERIZATION

The circuit design process begins with the modeling and characterization of a 70nm fully depleted SOI NMOS transistor. Figure 1 and Fig. 2 show the schematic layout of the simulated device and the corresponding electric potential profile, respectively. Where available, the internal device parameters correspond to the values of the DST transistor fabricated by the Intel group [4]. The sub-70nm transistor consists of a thin silicon body 30nm thick fabricated on top of a 200nm thick buried oxide. The physical gate oxide thickness is equal to 1.5nm.

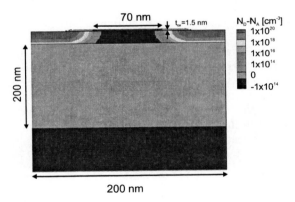

Figure 1: Schematic layout of FDSOI device simulated with the DESSIS

The device is simulated with both a fullband particle-based simulation tool [5] and with the ISE-TCAD drift-diffusion simulation tool, DESSIS [6]. Figure 3 shows the corresponding I-V curves of the device simulated with the drift diffusion transport model. The on current is found to be 0.75mA/um at $V_{dd} = 1.3V$ compared with 1.18 mA/um in the DST device. This is due to the fact that the experimental device has a raised source and drain contact region that improves the on current by 20% [3]. The computed saturation current is also higher than the experimental values, and this discrepancy can be attributed to the lack of experimental information available about the doping profile and the source and drain contact resistances.

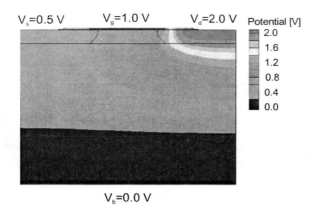

V_s=0.5 V   V_g=1.0 V   V_d=2.0 V

$V_s=0.5$ V $V_g=1.0$ V $V_d=2.0$ V

V_b=0.0 V

$V_b=0.0$ V

Figure 2: Potential profile for $V_{gs} = 1.0$V and $V_{ds} = 2.0$V

The DST transistor is reported to have an operational frequency in the TeraHertz range [3]. The frequency analysis of the device was conducted with both the particle-based simulation tool and DESSIS to further calibrate the proposed FDSOI structure. Within the particle-based approach, a constant bias was applied to the gate electrode while a step voltage was applied at the drain. The drain current was recorded and the complex impedance was obtained as a function of frequency, following the approach of [7]. The results show a cutoff frequency of approximately 1 THz. To further verify the frequency behavior, ISExtract [6] was used to extract the small-signal model parameters of the FDSOI device. The parameters were then used to build the equivalent circuit in PSPICE and the frequency response was then recalculated. Figure 4 shows the plot of the frequency response of the device obtained from the PSPICE simulation. It can be seen that the 3 dB frequency is approximately 1 Thz.

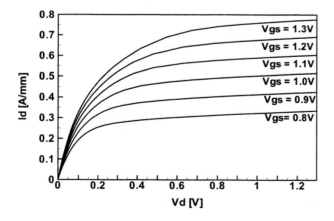

Figure 3: Output characteristics of the FDSOI transistor presented in the above figures

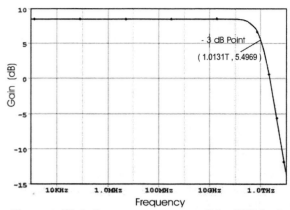

Figure 4: High frequency response of the SOI device

3   CLASS E POWER AMPLIFIER

The high frequency operation of the simulted FDSOI transistor makes it particularly well suited for analog applications. The FDOSI device is used to build a class E power amplifier to show the performance improvement over bulk Si technology. Class E power amplifiers were introduced by Sokal and Sokal in 1975 [8]. These types of amplifiers belong to the switching type of PAs rather than the conventional PAs, where transistors operate as voltage controlled current sources. The final circuit of class E PA is shown in Fig. 6. The output stage consists of a switch, which is the FDSOI transistor, a grounded capacitor C, and a high order reactive network composed of a series combination of inductors and capacitors. The radio frequency chock (RFC) simply provides a DC path to the supply and approximates an open circuit at RF. The capacitor C is placed in such a way that it can absorb any device output capacitance. The reactive network reduces the switch loss by forcing a zero voltage and a zero slope at the turn-on of the switch [9]. The idea behind the class E PA is to employ non-overlapping voltage and current waveforms to improve the power efficiency. Hence, the transistor in this configuration is treated as a switching device. In the case of negligible switching loses the circuit efficiency approaches 100%. However, the switching operation makes the PA highly non-linear and it requires extra circuitry to linearize its operation when the input waveform does not have a constant envelop [10]. The non-linearity is a trade-off with high power efficiency, a relative simple design, a high tolerance to circuit variations and a small number of required components.

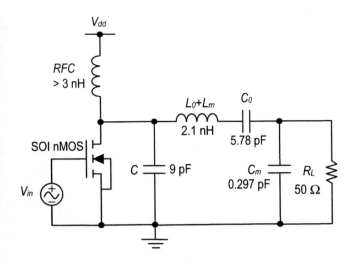

Figure 6: Class E Power Amplifier Circuit

A unique feature of the class E circuit is known as "soft switching", which describes the non-overlapping current and voltage waveform, and results in an efficiency of 100% in the ideal case [10]. The load resistance required for "soft switching" is [9],

$$R_{load} \approx 0.577 \cdot \frac{V_{DD}^2}{P_o} \quad , \qquad (1)$$

where V_{DD} is the supply voltage and P_o is the desired output power. It should be noted that R_{load} given by the above equation does not actually match the actual load resistance at small V_{DD} and a transformation network is employed to effectively transform the load resistance to the equivalent resistance. In Fig. 6, the inductor L_m and capacitor C_m represent a simple *L-match* network for the downward transformation of the load resistance.

The proposed power amplifier (PA) is designed for operation at 1.95GHz Universal Mobile Telecommunication System (UMTS) transmission frequency with the target output power of 500mW. Here, the frequency 1.95GHz corresponds to the central frequency of the UMTS uplink band 1920 MHz -1980 MHz [10]. Transistor sizing is an important issue for hand-held devices, and a initial step is the PA design. In the simulated circuit design, the device was sized to obtain the desired output power and to avoid unrealistic high current densities that could give reliability problems. This means that the FDSOI transistor must be fairly large to provide low on-resistance, and to minimize the losses during the on state. The performance of the class E PA is measured by two quantities, namely, drain or output efficiency and power-added-efficiency (PAE). The drain efficiency of a power amplifier is defined as,

$$\eta = \frac{P_{out}}{P_{dc}} \quad , \qquad (2)$$

where P_{out} is the output RF power and P_{dc} is the supply power. While PAE is given by,

$$PAE = \frac{P_{out} - P_{in}}{P_{dc}} \quad . \qquad (3)$$

PAE takes power gain into account and simply replaces the RF output power with the difference between the output and input power in the drain efficiency equation. Here, P_{in} is the input power for the power amplifier. In this design the input terminal of the transistor is driven by a sinusoidal voltage source, although the ideal driving signal for a class E PA is a squarewave or trapezoidal voltage [10]. At GHz frequencies it is difficult to efficiently generate such pulses and hence a sinusoidal driving signal provides a good approximation and a realistic option. A similar circuit design is found in literature [10] with three different technologies. Table 1 compares the performance of the class E power amplifier built with the FDSOI with that of BJT, HBT and CMOS technologies.

Table 1: Comparison of results for four Class E Pas

Parameters	Technology			
	Results from [Mil03]			Results from This Work
	BJT	HBT	CMOS	SOI
Frequency (GHz)	1.95	1.95	1.95	1.95
Supply Voltage (V)	3	3	1.2	1.2
Output Power (mW)	498	482	410	492
Output Efficiency, η (%)	79	89.6	79.6	97.8
PAE (%)	75	86	-	88.1
Power Gain (dB)	13.5	14.3	-	10.04

It can be seen that the SOI technology achieves the highest drain efficiency and PAE. The drop in the efficiency from the theoretical 100% value is caused by several factors. A transistor can only approximate the switching action and will exhibit finite turn-on and turn-off transitions. Thus, there will be a certain overlapping of non-zero voltage and current that will introduce losses. Due to the excellent speed performance characteristics of SOI transistor, the turn-on and turn-off transition times are greatly reduced, which in turn contribute to higher drain efficiency and PAE.

Studies were also conducted on a 3D tri-gate FDSOI structure, which are based on the 25nm Omega FET [11]. These devices represent an attractive alternative to the FDSOI because they show superior scalability, both from a reduction in short-channel effects and an improvement in the gate control over the channel [11]. The simulated tri-gate structure and the corresponding potential profile are shown in Fig. 7 (a) and (b), respectively. Further simulations are being run to characterize the tri-gate FET and extract the small signal parameters for use in the amplifier circuit.

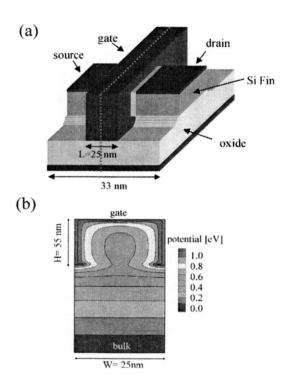

Figure 7: (a) Schematic layout of 25 nm CMOS Tri-gate FET and (b) corresponding potential profile for a gate and drain bias of both 1.0 V.

4   CONCLUSION

In this paper he performance of class E PA operation based on SOI technology is studied. A fully depleted 70nm SOI NMOS transistor is characterized and simulated in order to obtain the compact circuit model of the transistor. The class E PA circuit is then designed based on the SOI device and simulated. Simulation results show that PA has 97.8% drain efficiency and 88% PAE. However, the results should be adopted with caution as ideal passive components have been used. A similar design can be found in [10] for BJT, HBT and bulk Si CMOS technologies and the comparison of results shows that SOI technology significantly improves the performance of the PA.

Acknowledgements

This work has been partially supported by the NFS grant #ECS-0115548.

REFERENCES

[1] Shahidi, G.G., "SOI Technology for the GHz era," *VLSI Technology, Systems, and Applications, Proceedings of Technical Papers, International Symposium on*, pp. 11-14, April 2001.

[2] Colinge, J-P., *SILICON-ON-INSULATOR TECHNOLOGY: Materials to VLSI*, 2nd Edition, Kluwer Academic Publishers, Boston, 1997.

[3] Ngo, D., Huang, W.M., Ford, J.M., Spooner, D., "Power Amplifiers on Thin-Film-Silicon-on-Insulator (TFSOI) Technology," *IEEE International SOI Conference*, Oct. 1999.

[4] Chau, R., Kavalieros, J., Doyle, B., Murthy, A., Paulsen, N., Lionberger, D., Barlage, D., Arghavani, R., Roberds, B., and Doczy, M., "A 50nm Depleted-Substrate CMOS Transistor (DST)," *Electron Devices Meeting 2001, IEDM Technical Digest*, pp. 29.1.1-29.1.1, December 2001.

[5] M. Saraniti, S.J. Wigger, and S.M. Goodnick in *Proceedings of 2nd International Conf. on Modeling and Simulation of Microsystems*, April 1999.

[6] ISE TCAD Manuals, Version 8.0, www.ise.com

[7] R. W. Hockney and J. W. Eastwood, *Computer Simulation Using Particles*, Adam Hilger, Bristol, 1988.

[8] Sokal, N.O., and Sokal, A.D., "Class E-A new class of high efficiency tuned single-ended switching power amplifier," *IEEE J. Solid-State Circuits*, vol. SC-10, pp. 168-176, June 1975.

[9] Lee, T. H., *"The Design of CMOS Radio-Frequency Integrated Circuits,"* Cambridge University Press, New York, 1998.

[10] Milosevic, D., Van Der Tang, J., Van Roermund, A., "On the feasibility of application of class E RF power amplifiers in UMTS," *Circuits and Systems, ISCAS '03. Proceedings of the 2003 International Symposium on*, Volume: 1, pp. 149-152, May 2003.

[11] Yang, F.L. *et al. "*25 nm CMOS Omega FET*"*, *IEDM Technical Digest*, pp. 255-258, 2002

Principles of Metallic Field Effect Transistor (METFET)

Slava V. Rotkin, Karl Hess

Beckman Institute for Advanced Science and Technology,
University of Illinois at Urbana–Champaign,
405 N.Mathews, Urbana, IL 61801, USA;
Email: rotkin@uiuc.edu
Fax: (217) 244–4333

ABSTRACT

Novel type of a field effect transistor (FET) is described. A metallic channel of a metallic nanotube FET is proposed to be switched ON/OFF by applying electric fields of a local gate. Very inhomogeneous electric fields may lower the nanotube symmetry and open a band gap, as shown by tight–binding calculations.

Keywords: metal FET, nanotube, molecular devices, theory, symmetry breaking

1 INTRODUCTION

Field effect transistors in current use are mostly semiconductor devices [1]. While the scaling trend to ever smaller dimensions calls for ever higher doping and channel conductance of these devices [2], the doping level is limited by the nature of a metallic state. A high conductivity in bulk metallic systems inevitably prevents the penetration of the electric field except for extremely short distances, too short to achieve device function.

In this paper we put forward a concept for a novel transistor with a metallic channel. We model a new mechanism to control electronic transport in metallic one dimensional (1D) systems by use of the inhomogeneous electric field induced by a highly localized gate such as an STM tip or a nanotube tip [3]. Also, ultra narrow leads [4], fabricated closely to the nanotube channel, special chemical function groups [5] at the tube sidewalls or inside the tube may be used as a local gate. We notice that use of a dual gate (both local and backgate) may be beneficial because of the uniform backgate controls the Fermi level (charge density) while the local gate controls the conductivity and switches OFF/ON the channel. The weak screening of electric potentials in 1D channels enhances effect of the electric fields. In a subsequent article [6] we discuss that the highly localized gate induces a high electric field in a narrow region and impedes the electron transport.

The key effect which is employed for metallic field effect transistors (METFET's) is the symmetry breaking. The electronic structure of an armchair metallic single wall nanotube (M–SWNT) is robust to perturbations that have a *long range* potential. In contrast, inhomogeneous electric fields have *short range* and may lead to the opening of a band gap in carbon nanotubes due to lowering of the symmetry of the band structure. The possibility to open a band gap by use of inhomogeneous electric fields and the high (ballistic) conductance of the armchair M–SWNTs make them a suitable material to design METFET's.

In this paper we focus on the mechanisms for the band gap opening and discuss the physics of this effect. (For more details on applications of the effect in a METFET we refer a reader to Ref. [6].) We use a tight-binding approximation (TBA) to study the band-structure of the metallic nanotube. It is well known that for the SWNT which has a moderate curvature the one–band TBA captures most of features of the band structure. Thus we include in the calculation only one orbit per carbon atom and use one–parameter TB Hamiltonian with a hopping integral $\gamma \simeq 2.7$ eV.

We propose two main mechanisms for the local band gap opening in the SWNT: (i) the non-linear Stark effect and (ii) the band-structure modulation by a very non-uniform (multipole) electric field. We notice that other possible mechanisms are thought which may include a combination of an electric field and other perturbation potential, such as molecular/atomic environment (tube wrapping or decoration, for example), deformation or NT–lattice effects. The latter has been already demonstrated to be an effective mechanism for the gap opening in the ropes of armchair SWNTs [7].

The action of the transistor device is based on the gap opening in the following way: the gated region of the channel (with the semiconductor band gap at non zero gate voltage) represents a barrier for charge carriers. The barrier height and length can be modulated by the gate voltage. The transport through the gated region is due to (1) tunnelling and (2) thermionic emission at non zero temperature. We consider examples of typical armchair tubes: [10,10] SWNT of 1.4 nm diameter and [5,5] SWNT of 0.7 nm diameter. The gap is larger for smaller tubes as we will show below.

2 ARMCHAIR NANOTUBES

Use of metallic SWNT's to realize the possibility of 1D METFET's is appealing not only because of the extremely small size of the tube, but also for other rea-

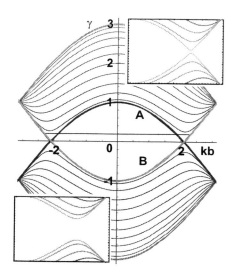

Figure 1: Bandstructure of armchair [10,10] M–SWNT. Upper right Inset shows zoom view of the Fermi point with opening of a gap due to *high–multipole* perturbation as described in the text. Lower left Inset shows how this gap grows linearly with applied potential.

sons, for example high (ballistic) conductance. IBM researchers have enunciated the vision of combining metallic and semiconducting nanotubes (M–SWNT's and S–SWNT's) in circuits [8–11]. The S–SWNT's would serve as active devices and the M–SWNT's as interconnects (also proposed by NASA group for conventional devices [12]) . Here we add a possibility to also use the *metallic tubes* as transistors which would be advantageous for various reasons including their high current capacity and conductance. The METFET will add to the predominantly p-type S–SWNT's an ambipolar device without using extra doping and/or complicated control on a contact work function. The METFET could be combined to circuits in analogy to CMOS.

Armchair M–SWNT's have also the advantage of a low phonon scattering-rate [13] and an equally low rate of scattering by impurities with a long range potential [14,15]. This follows from a special symmetry (selection rules) applicable to the transitions between the closest subbands of the armchair SWNT's.

The wave-function of the electron in a conduction band can be written as a product of an envelope function and an amplitude in a unit cell. Each unit cell consists of two atoms (of different sublattices). Thus, the amplitude has two components, which are often referred as components of a pseudospinor in the nanotube literature. The pseudospin has the opposite sign, ± 1, for states of the valence and conduction band. The corresponding TB wave function is given by:

$$\langle \vec{r} | \psi_{m,k,\zeta} \rangle = \frac{e^{ikz} e^{im\alpha}}{\sqrt{2\pi L}} \frac{\langle \vec{r} - \vec{R}_I | \varphi \rangle + \zeta c_{mk} \langle \vec{r} - \vec{R}_{II} | \varphi \rangle}{\sqrt{2}},$$

(1)

here c_{mk} is to be found as an eigenvector of the TB Hamiltonian; the index m stays for an angular momentum of the electron and labels orbital subbands; k is a longitudinal momentum; $\zeta = \pm 1$ is the pseudospin. The components of the pseudospinor vector are atomic-like wave functions, defined in the unit cell, $\varphi_{I/II}$. The electron is considered at the surface of a cylinder. Its coordinates ($\vec{r}$) are z, and α, along the tube axis and along the tube circumference respectively. Here R is the nanotube radius and L is the length of the tube.

The two (closest) crossing subbands are shown in Fig.1 as bold green and blue curves. Other subbands of orbital quantization are shown in color from red to blue (from top to bottom in conduction band) and correspond to $m = 0, 1 \ldots 9$. Two closest subbands, A and B (blue and green) have $m = 10$, but the pseudospin has opposite sign. Hence, the pseudospinor overlap integrals of the states of these subbands are zero [16].

A common approximation of solid state theory is, that the spatial variation of any long range external potential is small at the scale of a unit cell (effective mass theorem). Thus, the matrix element of the potential is a product of the unit cell overlap integral and the Fourier component of the potential: $\langle \psi | U | \psi \rangle = \langle \varphi | \varphi \rangle U_q$. It is evident that this approximation leads to a zero mixing/scattering between two closest subbands of the armchair tube as $\langle \varphi | \varphi \rangle \equiv 0$ and is suggestive for ballistic transport in the armchair SWNT channel [17]. Thus, the ideal armchair SWNT is a great candidate for a 1D METFET as its conductance in the ON state has to be about $2G_o = 2e^2/h$, 4 times of the conductance quantum (for 2 spin and 2 space channels), maximum conductance for a circuit with macroscopic leads [18].

Group theory proves that the direct matrix element is non–zero if the Fourier transform of the potential $U_{q,\delta m}$ is a full scalar with respect to the rotations of the symmetry group of the nanotube [16]. This is fulfilled for $\delta m = s \cdot n$, where $s = 2, 4, \ldots$ is an positive even integer and n is the integer appearing in the notation [n,n] for armchair SWNT's [19]. Then, the energy dispersion of two new subbands $|A \pm B\rangle$ in vicinity of the Fermi point reads as: $E_\pm = \pm \sqrt{E_A(q)^2 + U_{q,\delta m}^2}$. Here $E_A(q) = -E_B(q) = E_B(-q)$ is the dispersion in A/B subband at zero gate voltage. At the Fermi point $E_A(0) = 0$ and, thus, a gap opens:

$$E_g^{(sn)} = 2|U_{q,sn}| \qquad (2)$$

This is a *direct mixing* of the subbands A and B. The band gap is linear in the applied gate potential (see Inset in Figure 1). There is no upper limit for the magnitude of the opened gap in this case (except for a natural condition that the field must not cause an electric breakdown).

The direct mixing of the crossing subbands is tricky as one needs to break the mirror reflection symmetry

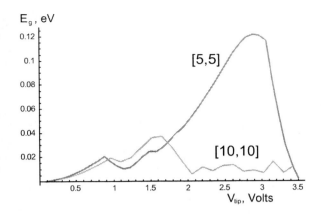

Figure 2: Band gaps in [10,10] and [5,5] SWNTs for *dipole + quadrupole* potential as a function of applied voltage.

along the tube axis *and* keep the full symmetry of the applied potential which requires very high multipole moment of the potential $\delta m = 2n$. Although such potential may be possibly created by applying the gate voltage to a chemically modified/decorated surface of the SWNT (to be discussed elsewhere), this is a challenging problem for future technology.

However, if only first condition is fulfilled (mirror symmetry breaking), the gap opens by *indirect* mixing of the subbands in a higher order of applied potential. In particular, by applying a uniform field (dipole component of the potential) across the armchair SWNT *together* with a higher order (quadrupole) component we obtained the gap which scales as a third power of the gate potential. Perturbation theory for nanotubes (similar to what was used in [20]) predicts a maximum gap which depends only on the size of the tube and scales as R^{-2} or n^{-2} (see below).

To prove the gap opening due to this mechanism, the non–linear Stark mechanism, we consider a number of armchair SWNTs in the fields with two components: the dipole and the quadrupole. A simplest realization of such potential can be given by two tips at the top and the bottom of the SWNT, which are kept at the potentials $+2V_{tip}$ and $-V_{tip}$. (Other potentials were simulated as well and led to similar results.) This potential is also mimic to the potential of a local gate over the SWNT on the insulating substrate with a dielectric permittivity ϵ. The image charge in the substrate is $(\epsilon - 1)/(\epsilon + 1)$ fraction of the tip charge and has the opposite sign. The results of the calculation for [10,10] and [5,5] SWNTs are presented in Fig.2.

3   SWITCHING OFF A QUASI-METALLIC NANOTUBE

While opening the energy gap in the metallic armchair SWNT requires a special symmetry of the gate potential, the quasi–metallic tubes are best candidates for band gap engineering. As we already noticed in Ref. [20], the uniform external electric field applied across a zigzag quasi–metallic nanotube opens the gap. This is just due to the Stark effect and, therefore, the gap is proportional to an interband transition overlap integral $\langle C|V \rangle$ and to the square of the potential, $|V|$ (cf. the third power for non–linear Stark effect in armchair tubes). Though, similarly to the case of the armchair tube, this gap will close eventually with increasing external potential. Thus, there is a maximum gap opening which is defined solely by the tube radius $E_g^{(c)} \sim R^{-2}$.

The closing of the opened gap happens with increasing the gate potential when its value is about the distance between other subbands. Thus, the critical potential is $\sim \hbar v_F / R$, where $v_F = 3\gamma b / 2\hbar$ is the Fermi velocity in a M–SWNT, a bond length is $b \simeq 0.14$ nm.

For the armchair SWNT we go in a high order of the perturbation theory and at the critical potential the matrix element of the perturbation cancels with the energy level separation. As a result the maximum gap opening is a universal function for armchair and zigzag nanotubes:

$$E_g^{(c)} \sim \langle C|V \rangle |V^{(c)}| \sim \frac{\hbar v_F b}{2R^2} \propto \frac{\gamma}{n^2}, \qquad (3)$$

and the maximum gap $E_g^{(c)}$ is proportional to the square of the nanotube curvature $(1/R)^2$ (Fig.2) as we also confirmed by a numerical calculation in [20].

Even though the maximum gap due to the Stark effect is small for a quasi–metallic nanotube and thus can be observed only at low temperature, this gap opening might be easier to achieve experimentally, *e.g.*, by use of the inhomogeneous electric field of a tunnelling tip, because of no special symmetry between closest subbands as compared to the case of direct subband mixing by a high–multipole potential. However, the non–armchair SWNT must also have a lower ON conductance due to the same argument. A stronger scattering of carriers happens in this tube, similarly to a S–SWNT [15].

4   FIELD EFFECT MODULATION OF METALLIC CONDUCTANCE

Any opening of the gate induced semiconductor gap along the metallic tube will create a potential barrier for the electrons and, therefore, modulate the conduction. Figure 3 shows typical IV curves (IVC) for a METFET using an armchair metallic nanotube of the radius 0.7 nm ([10,10] SWNT) at T =4K, gate width $W = 15$ nm. The upper (red) IVC corresponds to a zero gate voltage (no gap). The channel is fully open and the ON current is determined by injection from a contact and thus by the quantum conductance $4G_o$ [18]. With increasing gate voltage one observes a substantial decrease of the

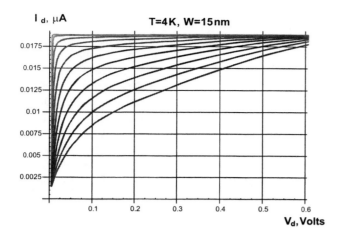

Figure 3: Current of a METFET made of an armchair [10,10] SWNT versus the drain potential at different gate voltages (at different gaps). The gate width is 15 nm. $T = 4$K. Curves from top to bottom correspond to increasing gap from 0 to 0.4 eV (in color: from red to blue).

current due to the opening of the gap and a depletion of electrons in this region. The IVC's correspond to gaps up to 0.4 eV (blue curve).

In conclusion, we put forward a novel type of electronic switching device based on carbon nanotubes: a one–dimensional metallic field effect transistor (1D MET-FET). We have described band gap opening mechanisms in the armchair and zigzag metallic SWNT's. Our calculations demonstrate, at least principle, the possibility to open a band gap by application of inhomogeneous electric fields that may be created by special local gates to break the SWNT symmetry. Two mechanisms of the symmetry breaking and subsequent gap opening in an armchair SWNT are outlined: a subband mixing by a high–multipole potential and a non–linear Stark effect. A quasi–metallic zigzag SWNT can be also used for the METFET. In that case the proposed gap opening mechanism is the Stark effect (in a low order of a perturbation theory). Assuming virtually ballistic transport, the IV curves for the METFET's are obtained.

ACKNOWLEDGMENTS

Authors acknowledge Ms. Yan Li for help in numerical computations. This work was supported by NSF Grant No. 9809520, by the Office of Naval Research grant NO0014-98-1-0604 and the Army Research Office grant DAAG55-09-1-0306. S.V.R. acknowledges partial support of DoE Grant No. DE-FG02-01ER45932, and NSF grant No. ECS–0210495.

REFERENCES

[1] Advanced Theory Of Semiconductor Devices// Karl Hess. New York: IEEE Press, 2000.

[2] International Technology Roadmap for Semiconductors (Semiconductor Industry Association, San Jose, CA, 2001); http://public.itrs.net/.

[3] http://www.itg.uiuc.edu/exhibits/iotw/2003-09-09/

[4] J.-F. Lin, J.P. Bird, L. Rotkina, P.A. Bennett, Appl. Phys. Lett. **82** (5), 802, 2003.

[5] S.V. Rotkin, I.Zharov, Int. Journal of Nanoscience **1** (3/4), 347–356, 2002.

[6] S.V.Rotkin, K.Hess, Appl.Phys.Lett. (submitted).

[7] P. Delaney, H.J.Choi, J.Ihm, S.G.Louie, M.L.Cohen, Nature, **391**, 466, 1998

[8] J. Appenzeller, Device Research Conference, Utah, June 26, 2003

[9] J. Appenzeller, R. Martel, V. Derycke, M. Radosavljevic, S. Wind, D. Neumayer, and Ph. Avouris, Microelectronic Eng., **64**, 391, 2002.

[10] Ph.G. Collins, Ph. Avouris, Sci.Am., 12, 62, 2002. Ph. Avouris, Chemical Physics, **281**, 429, 2002.

[11] S. Heinze, J. Tersoff, R. Martel, V. Derycke, J. Appenzeller, Ph. Avouris, Phys.Rev.Lett. **89**, 106801, 2002.

[12] J. Li, Q. Ye, A. Cassell, H. T. Ng, R. Stevens, J. Han, and M. Meyyappan, Appl.Phys.Lett. 82, 2491 (2003).

[13] H. Suzuura and T. Ando, Phys. Rev. **B 65**, 235412, 2002.

[14] T. Ando and T. Nakanishi, J. Phys. Soc. Jpn. **67**, 1704, 1998. T. Ando, T. Nakanishi and R. Saito, J. Phys. Soc. Japan **67**, 2857, 1998.

[15] A.G.Petrov, S.V.Rotkin, submitted.

[16] T. Vukovic, I. Milosevic, M. Damnjanovic, Phys.Rev. B 65(04), 5418, 2002.

[17] Although, there is a possibility of scattering to the other valley of the same subband, this scattering amplitude is small due to a large momentum transfer $q \simeq 2\pi/\sqrt{3}b$ (see thin line between left and right Fermi points in Fig.1), where $b \simeq 0.14$ nm is a carbon–carbon bond length.

[18] For a circuit with macroscopic leads to the M–SWNT channel the total conductance will be about $4G_o = 2e^2/h$. This gives a minimum resistance, a contact resistance, of the SWNT device $\sim 6.5k\Omega$. The lower resistance can be expected in the case of entirely nanotube circuit [8]. The quantum contact resistance will not limit anymore the ON current in this case. That device can fully exploit all advantages of the METFET.

[19] Y.Li, S.V.Rotkin, U.Ravaioli, and K.Hess, unpublished.

[20] Y.Li, S.V.Rotkin and U.Ravaioli, Nano Letters, vol. 3(2), 183–187, 2003.

Ab Initio Simulation on Mechanical and Electronic Properties of Nanostructures under Deformation

Y. Umeno* and T. Kitamura*

* Dept. of Engineering Physics and Mechanics,
Graduate School of Engineering, Kyoto University
Yoshida-hommachi, Sakyo-ku, Kyoto 606-8501, Japan, umeno@kues.kyoto-u.ac.jp

ABSTRACT

Nanostructures have been attracting attention because of their prominent properties, and their applications for novel devices with advanced functions have been attemted. Large stress and strain occur in local regions in materials with nanostructures owing to their complex structures. In this study, we conduct *ab initio* simulations to elucidate the mechanical and electronic properties of materials, which are essential for designing functional nanomaterials. Firstly, we investigate in detail the deformation of crystals, such as silicon and nickel, and the change in their electronic structures under deformation. We find that the strain largely changes the electronic structures of the crystals, resulting in the changes in the electric conductivity of silicon and the magnetic property of nickel. Secondly, we conduct simulations on the deformation of carbon nanotubes, which are among greatly notable nanomaterials. We find that the electric conductivity of the carbon nanotubes shows transitions from metallic to semiconducting and vice versa under axial tension. We also find that the semiconducting nanotubes become metallic under radial compression, while the metallic nanotubes do not show the transition.

Keywords: ab initio, deformation, mechanical property, electronic structure, nanostructure

1 Introduction

With the advent of the nano-processing technology, microscopic devices, which include complex structures in nanometer scale, have been intensively developed and their size is continuously reducing. The complex structure may cause local strain condition owing to the mismatch of bimaterials, and this leads to the change in the electronic structure in the local region. As the size of the device is reduced, the effect of the local strain becomes beyond negligible degree, which results in the deterioration of the functions of the device. Conversely, such effect can be utilized for designing novel devices with preferable functions as various properties may be controlled by making use of local strain. Therefore, it is of great importance to clarify the correlation between the electronic properties and the deformation behavior of materials.

We have conducted simulations on the problem by using the *ab initio* DFT calculation based on the pseudopotential plane wave method, by which the mechanical and electronic properties are easily evaluated. In this paper, simulation results about the effect of deformation on the electronic properties in the cases of single crystals and carbon nanotubes are presented.

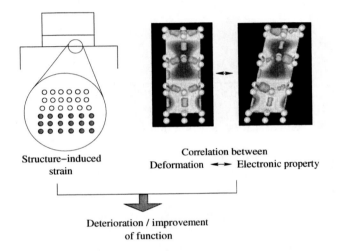

Figure 1: Schematic of local strain caused by nanostructure.

2 Mechanical and Electronic Properties under Deformation

2.1 Single crystal

The ideal deformation of the silicon single crystal, which is commonly used for the substrate of electronic devices, is simulated by the *ab initio* pseudopotential method[1] based on the LDA. The simulation cell shown in Fig. 2(a) is used and the shear strain, γ_{zx}, which corresponds to $[\bar{1}01]$, is applied. All the stress components except τ_{zx} are controlled to be zero by adjusting the cell size, and the atom configuration is relaxed during the shear deformation. Figure 2(b) shows the stress-strain relation. The curve shows the peak, $\tau_{zx} = 10\text{GPa}$

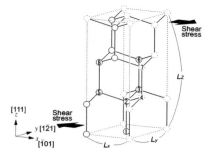

(a) Simulation cell.

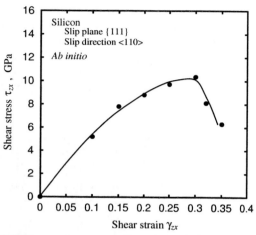

(b) Shear stress as a function of shear strain.

Figure 2: Ideal shear deformation of silicon single crystal.

at $\gamma_{zx} = 0.3$, which indicates the ideal shear strength. The deformation in the region up to $\gamma_{zx} = 0.3$ is elastic without the bond breaking. The crystal becomes back to the initial state when unloaded. On the other hand, the rapid decrease in the shear stress over $\gamma_{zx} = 0.3$ corresponds to the bond breaking, which we can denote the "mechanical fracture" under this deformation mode.

Figure 3 shows the change in the density of states (DOS) during the shear. The initial state has the band gap energy and it is eliminated at $\gamma_{zx} = 0.2$, which means that the crystal loses the electronic property as the semiconductor under the shear deformation. These results suggest that the electrical deterioration may take place before the mechanical fracture.

The deformation of the nickel crystal, which is one of typical ferromagnetic metals used for magnetic devices, is investigated by *ab initio* calculations based on the LSDA, where the partial core correction[2] is taken into account. Figure 4 shows the magnetic moment of the nickel single crystal, μ, as a function of the isotropic strain, ε. The magnetic moment, μ, increases with the isotropic strain. On the other hand, μ is reduced under

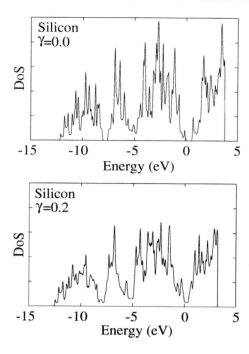

Figure 3: Change in density of states under shear deformation.

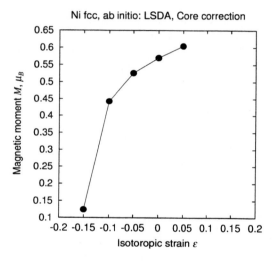

Figure 4: Magnetic moment as a function of isotropic strain in nickel single crystal.

contraction and becomes almost zero at $\varepsilon = -0.15$, which means the crystal is no longer ferromagnetic. This corresponds to the similar calculation conducted by Černý *et al.* [3]. The result can be applied for the device designing, for example, if the tensile strain condition is created by the component structure, the magnetic property of the device can be improved.

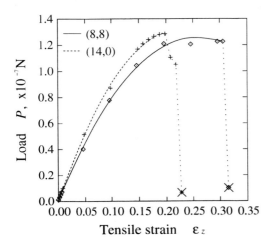

Figure 5: Load-strain relation of CNT under axial tension.

2.2 Carbon nanotube

Carbon nanotubes (CNTs) have been attracting attention because of their prominent mechanical and electronic properties. The deformation of single-walled carbon nanotubes (SWCNTs) is investigated by tight-binding calculations and the change in the band gap energy is examined [4]. The focus is placed on the dependency of the mechanical and electronic properties on the CNT structure, which is denoted by the chiral vector, (m, n). The validity of the tight-binding calculations is verified by *ab initio* DFT calculations based on the pseudopotential method as preliminaly calculations. Two deformation modes, axial tension and radial compression, are considered.

The axial load as a function of the axial strain for the (14,0) and the (8,8) tubes is shown in Fig. 5. This result points out that CNTs can show large elastic elongation in the axial tension and that the elongation depends on the chiral structure. This suggests that it is meaningful to investigate the change in the electronic properties of CNTs under high axial strain. Figure 6 shows the change in the band gap energy, E_{gap} during the axial tension. CNTs are classified into three groups; A:$m - n = 3q$, B:$m - n = 3q + 1$, and C:$m - n = 3q + 2$. The CNTs show different patterns of transition of the electric conductivity under the deformation. Type A shows M→S→M, where M and S denote "metallic" and "semiconducting", respectively. On the other hand, The transition of Types B and C is S→M→S.

Figure 7 shows the relationship between the applied load, P, and the normalized displacement, δ, under the radial compression. The curves show the difference in the deformation behavior because of the difference in the diameters of the tubes. In the case of the (8,0) tube, the diameter of which is 0.626nm, the load gradually

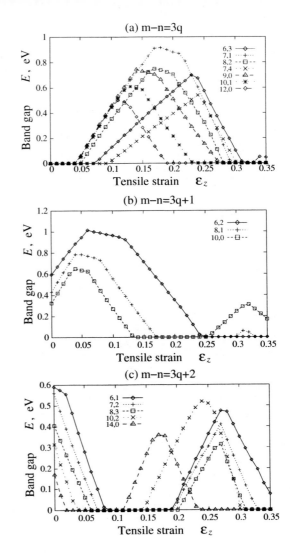

Figure 6: Change in band gap energy during axial tension.

increases with the displacement. The inclination, which corresponds to the resistance against the deformation, also shows monotonic increase. This is becase the CNT wall has elasticity so that it has the resistance against the curvature change. On the other hand, in the case of the (14,0) tube, which has a larger diameter, 1.10nm, the inclination of the curve is relatively low up to $\delta = 0.7$ and then it increases rapidly. The rapid increase of the load at $\delta > 0.7$ indicates the tube becomes stiffer, which is explained by the repulsion between the top and the bottom walls of the CNT because the distance between the upper and the lower wall when the curve shows the rapid increase is about 0.33nm, which is a little smaller than the interlayer distance in graphite, 0.335nm.

It should be noted that the shape of the tubes becomes back to the initial state when unloaded.

Figure 8 shows the change in the band gap energy of

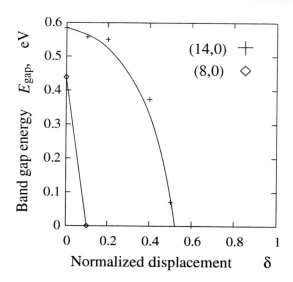

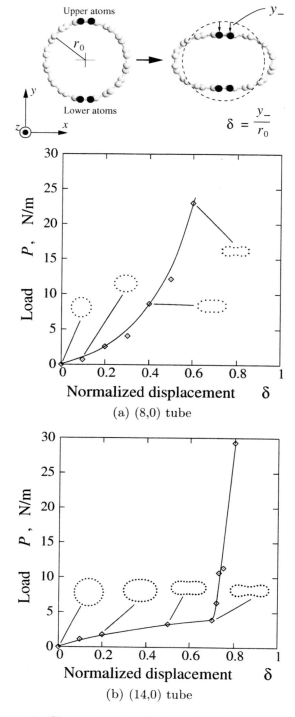

Figure 8: Change in band gap energy during radial compression of CNT.

to eliminate the band gap than the (14,0) tube ($\delta = 0.5$ and $P = 4.1\text{N/m}$). The result indicates that it is possible to control the sensitivity of the band gap energy to the deformation by the diameter.

3 Summary

We presented simulation results on the mechanical and electronic properties under deformation of crystals and CNTs. We found these materials have the close correlation in those properties. *Ab initio* and semi-empirical band calculations are suitable for evaluating the correlation, which is essential for designing functional nanodevices.

Acknowledgment

This study was partly supported by Hattori Houk-oukai Foundation, Asahi Glass Foundation, and Industrial Technology Research Grant Program from New Energy and Industrial Technology Development Organization (NEDO).

REFERENCES

[1] Umeno,Y. and Kitamura,T., *Mater. Sci. Eng.*, B, **88**(2002), 79–84.

[2] Louie,S.G., Froyen,S. and Cohen,M.L., *Phys. Rev.*, B, **26**(1982), 1738–1742.

[3] Černý,M., Pokluda,J., Šob,M., Friák,M. and Šandera,P., *Phys. Rev.*, B, **67**(2003), 035116.

[4] Umeno,Y., Kitamura,T. and Kushima,A., *Comp. Mater. Sci.*, submitted.

(a) (8,0) tube

(b) (14,0) tube

Figure 7: Change in load during radial compression of CNT.

the (8,0) and the (14,0) tubes under the radial compression. Both tubes have the band gap energy at the initial state, which decreases with the displacement and finally becomes zero. However, the (8,0) tube needs much smaller displacement ($\delta = 0.1$) and load ($P = 0.9\text{N/m}$)

Atomistic Process and Simulation in the Regime of sub-50nm Gate Length

Ohseob Kwon, Kidong Kim, Jihyun Seo, and Taeyoung Won

Department of Electrical Engineering, School of Engineering, Inha University
253 Yonghyun-dong, Nam-gu, Incheon, Korea 402-751, E-mail: kos@hsel.inha.ac.kr

ABSTRACT

In this paper, we report an atomistic simulation approach for sub-50nm gate length FETs. The proposed atomistic approach consists of the coupling the molecular dynamics (MD) simulations of the collision cascades for ion implantation process and Kinetic Monte Carlo (KMC) simulations for the subsequent diffusion process. The impurity profiles from the MD and KMC calculations were interfaced with the quantum-mechanical device simulations. The device performance of FinFET with 20nm physical gate length is discussed in this paper.

Keywords: molecular dynamics, kinetic monte carlo, ion implantation, diffusion, device modeling, FinFET

1 INDRODUCTION

Recently, a great deal of attractions have been made on the development of novel nano-scale devices for sub-50nm gate length, such as double-gate (DG) MOSFET, vertical replacement-gate (VRG) MOSFET, ultra thin body (UTB) MOSFET, and FinFET [1-4]. These devices demand new modeling methodology for the simulations, i.e. an atomistic and quantum mechanical (QM) scheme.

In this paper, we report a simulation method based on atomistic approach for sub-50nm gate length. Molecular dynamics (MD) is implemented for the ion implantation process to form ultra-shallow junctions [5-6]. Thereafter, the diffusion process is simulated by using Kinetic Monte Carlo (KMC) with the damages and impurities distribution from the ion implantation profile in MD [7]. A device simulation is performed by using profiles from the results of KMC. As an exemplary case, we chose a FinFET structure with 20nm physical gate length. Let us start with a discussion of the simulation model (Section 2), followed by presentation of the results and discussion (Section 3).

2 SIMULATION MODEL

The concentration distribution of dopants during the ion implantation is calculated from the MD module. The MD approach accurately calculates the concentration distribution of dopants in the ion implantation using the recoil ion approximation (RIA). MD simulations can calculate range profiles of different ion species implanted into crystalline Si for nano-scale devices in ultra low energy

regime. In our work, MD with a damage model has been employed. In order to model the ultra shallow junction, we used a modified RIA for dose-dependent damage. The Ziegler-Biersack-Littmark (ZBL) potential model, Eq. (1), has been used for the interaction among atoms. In order to model the electronic stopping power, the density functional theory by Echenique et al [8] was implemented in this work. Furthermore, the Firsov model was employed in order to model the energy loss during the inelastic collisions [9].

$$E_{ZBL} = V(r_{ij}) = \frac{Z_1 Z_2 e^2}{R_{ij}} \sum_{k=1}^{4} c_k \exp(-d_k \frac{r_{ij}}{a}), \qquad (1)$$

For an accurate atomic-scale model for diffusion of intrinsic point defects (I, V) and impurities (B) in ion-implanted silicon, we used a Kinetic Monte Carlo (KMC) method. In this type of KMC model, point defects and dopants are treated at an atomic scale while they are considered to diffuse in accordance with the reaction rates, which are given as input parameters [10]. These input parameters can be obtained from either first-principles calculations, classical MD simulation, or experimental data. Especially, the formation of clusters and extended defects, which usually control the annealing kinetics after ion implantation, should be minimized in the range of low dose in an effort to create dilute concentrations of I and V. Therefore, a simple kick-out mechanism has been tested and a good agreement with the experimental data [11] was verified in this condition. However, a more recent model, interstitialcy mechanism, is preferred when compared to the traditional kick-out mechanism by *ab initio* molecular dynamics [12].

We propose an atomistic diffusion mechanism involving fast-migrating intermediate species of the form. The reactions $X_s + I \Leftrightarrow X_m$ and $I + V \Leftrightarrow 0$ are essentially diffusion limited, with capture radius of second neighbor atomic length (3.84Å) and direction of particle migration is limited to six neighbor sites. Here X_s is the immobile substitution impurity, which through reaction with a self-interstitial (I) forms a fast-migrating species X_m, which diffuse at a rate D_m. Ordinary, the diffusion rate is form in Arrhenius type ($D_0 \exp(-E_{act}/KT)$). KMC simulations are performed by using the damage profiles and defects distribution from the MD simulations. The KMC is an event-driven technique, i.e., simulate events at random with probabilities according to the corresponding event rates. In this way, it self-adjusts the reasonable time step as the simulation proceeds. In our work, the only event that a point defect (I,V) can perform is a jump. Their jump rate is given by Eq. (2).

$$J_{rate} = 6 \times D_0 \times exp(-R_m / kT) / \lambda^2, \qquad (2)$$

The change of the transition probability density, $f(t)$ over some short time interval, dt is proportional to the uniform transition probability, r, dt, and f because f gives the probability density that the object still remains at time t. The solution is given by Eq. (4) with boundary conditions. There, the simulation time is updated with $t' = t + \Delta t$ according to event hopping rate, Eq. (5).

$$df(t) = -rf(t)dt, \qquad (3)$$

$$f(t) = re^{-rt}, f(0) = r, \qquad (4)$$

$$\Delta t = -\frac{\log u}{R}, \qquad (5)$$

For the device of sub-50nm gate length, the nonlinear Poisson, Eq. (6) and Schödinger equation, Eq. (7) solved by using a finite different method (FDM) with Gunter's grid in self-consistent manner, followed by solving the continuity equation, Eq. (8) for current density. For QM solution, an iterative procedure is employed.

3 RESULTS AND DISCUSSION

In MD simulation, all simulations were performed on a Si {100} target at 300K. Dopants and damage profiles were simulated for B and As ions with the dose 1×10^{14} ions/cm^2. Fig.1 and Fig. 2 show the simulation results with the energy of 0.5, 1, 2, 4, 8, and 16 keV B implant into Si, respectively. As boron ion dose increases, local damage accumulation affects dopant distribution in the cases of ultra-low energy ion implantation. In other words, the channeling tail drops very steeply with dose increase. In case of As ion, the channeling tail drops very steeply with depth due to large atomic mass of the arsenic ion. Fig. 3 shows 3D distribution with the energy of 5 keV B implant into Si after 4,000 B ion. In this simulation, the mean range is 741.31 ?, and the sputtered target atoms are 13.

As the implantation energy increases, the electric characteristics of semiconductor devices exhibited noticeable improvement in the wide distribution of dopants. It was, however, difficult to obtain the shallow junction depth. The small and light ions like B caused less lattice damage near the surface even with high implantation energy. However, the nuclear stopping power is more pronounced than the electronic stopping power at the depth of the target material when ion energy is gradually decreased. The concentration distribution difference in the deeper region increases because ions were scattered to sides due to the collisions with lattice damages. On the other hand, heavy ions like As and P cause much lattice damage near the surface.

Fig. 4(a) shows the defect distribution of the initial profile from the result of MD simulation. Fig. 4(b) shows the profile after 10^9 fs. In this simulation, the length recombination is 4 . After ion implantation at 2 x 10^4 ions/cm^2 dose rate, the defects hopping distance is 2.35? . After 10^9 fs, many defects are recombined and vanish. The rate of recombination is dependant to annealing temperature. In high temperature (1500K), the defects (self-interstitials and vacancies) are recombined rapidly as shown Fig.5.

Fig. 6 shows boron and Si damage profiles after 50keV B ion implantation with the dose of 1×10^{11}/cm^2 using MARLOWE [13]. The dashed lines represent I and V damage distributions. For the low dose implantation, the split between interstitial and vacancy distributions is little. The solid lines represent the distribution of boron after thermal annealing for 15 min at 450℃. A significant fraction of B atoms in the shallow marker are displaced and migrate a considerable distance ($\sim 10^2$ nm) from their initial positions. Fig. 7 shows 3D distribution of boron after thermal annealing. Table 1 shows simulation parameters. D_I and D_V are the I and V diffusivities, respectively, and E's are the potential energies of simulation particles.

Fig. 8 shows a schematic diagram illustrating the structure of FinFET. FinFET has characteristics such as the ultra thin Si fin for suppression of short-channel effects, two gates which are self-aligned to each other and to the S/D, the raised S/D to reduce parasitic resistance, a short (50nm) Si fin for quasi-planar topography, and the compatibility with low-T, high-k gate dielectrics. Fig. 9 shows a schematic view illustrating the Gunter gird implemented for FDM. Fig. 10 shows the electron density with Vg = 1.5V, 0.5V, and -1.5V. As the voltage ranges 1.5V to -1.5V, the electrons move from the middle of the Si-fin to source and drain.

4 CONCLUSIONS

In this paper, we present our simulation results with basis on an atomistic and quantum mechanical approach. In this work, we employed an MD method for ion implantation process and KMC method for diffusion process, followed by the device simulation. As an exemplary device, the electron density of FinFET with 20nm physical gate length has been investigated.

ACKNOLEDGEMENT

This work was supported partly by the Ministry of Information & Communication (MIC) of Korea through Support Project of University Information Technology Research Center (ITRC) Program supervised by IITA (Institute of Information Technology Assessment), and the author(Ohseob Kwon) would like to express thanks to Mr. Stefan of WSI(Walter Schottky Institute) for fruitful discussion.

REFERENCES

[1] D.Hisamoto, *et al*, IEEE Trans. ED, 47, p.2320 (2000).

[2] L.Chang, *et al*, Tech.Dig.IEDM, p.715 (2001).

[3] B. Yu *et al*, Tech.Dig.IEDM, p.251 (2002).

[4] L.Chang, *et al*, IEEE Trans. ED, 49, p.2288 (2002).

[5] J.Nord, *et al*, Phys. Rev. B (2002).

[6] J. Peltola, *et al.*, Nucl. Instr. Meth. Phys. Res. B, (2002).

[7] M. Jaraiz, *et al.*, MRS Symp. Proc. 532, 43 (1998).

[8] P. M. Echenique *et al.*, Appl. Phys. A 71, 503 (2000).

[9] P. Keblinski *et al*, Phys. Rev. B 66, 4104 (2002).

[10] Martin Jaraiz, Atomistic simulations in Materials Processing.

[11] N.E.B. Cowern *et al*, Phys Rev Lett Vol 69 (1992).

[12] Paola Alippi *et al*, Phys Rev B, Vol 64 (2001).

[13] M.T. Robinson *et al*, Appl. Phys. Lett. 56, 1787 (1990).

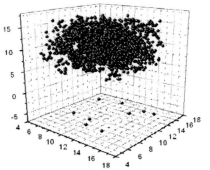

Fig. 3   A plot showing the simulation result with the energy of 5 keV B implant into Si after 4,000 B ions.

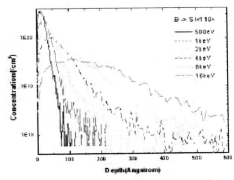

Fig. 1   A plot showing the simulation results with the energy of 0.5, 1, 2, 4, 8, and 16 keV and the dose 1×10^{14}ions/cm^2 B implant.

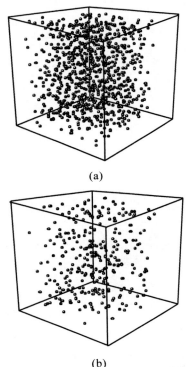

(a)

(b)

Fig.4   Plots showing the defect distributions of (a) the initial profile and (b) after 10^9 fs.

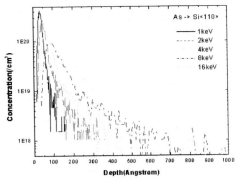

Fig. 2   A plot showing the simulation results with the energy of 1, 2, 4, 8, and 16 keV and the dose 1×10^{14}ions/cm^2 As.

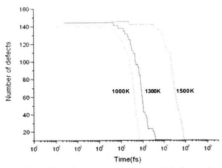

Fig. 5 A Plot showing the number of defects dependent temperature.

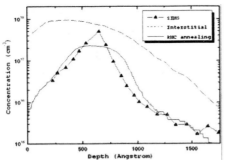

Fig. 6 A plot showing B dopant profile modified by 50keV Si ion implantation to dose of $1 \times 10^{11}/cm^2$.

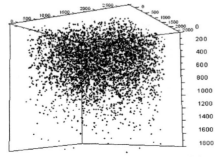

Fig. 7 A plot showing 3D distribution of boron after thermal annealing.

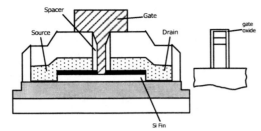

Fig. 8 A Schematic view illustrating FinFET structure.

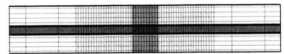

Fig. 9 A Schematic view illustrating the Gunter gird implemented for FDM.

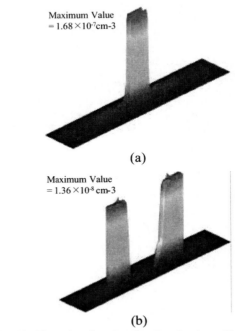

Maximum Value $= 1.68 \times 10^{-7} cm^{-3}$

(a)

Maximum Value $= 1.36 \times 10^{-8} cm^{-3}$

(b)

Fig. 10 Plots showing the electron density with (a) Vg = 1.5V and (b) Vg = -1.5V.

Table 1 Parameters for KMC diffusion.

Diffusivity (cm2/s)		Potential Energy (eV)	
(Single Point Defects : migration)			
D_{m}_V	1e-3	E_{m}_V	0.43
D_{m}_I	5.0	E_{m}_I	1.0
D_{m}_B	0.0	E_{m}_B	5.0
(Interstitial – Vacancy Pair)			
D_{m}_IV	0.0	E_{m}_IV	5.0
D_{b}_IV	0.0	Eb_IV	5.0
D_{z}_IV	0.0	E_{z}_IV	0.0
(Interstitial - Boron Pair)			
Dm_IB	1.0	E_{m}_IB	4.0
D_{b}_IB	15.0	E_{b}_IB	1.0
D_{z}_IB	1e-3	E_{z}_IB	1.05
(Bi atom)			
D_{m}_Bi	1.5e-3	E_{m}_Bi	0.3
D_{b}_Bi	0.0	E_{b}_Bi	5

Sub-Threshold Electron Mobility in SOI MESFET

T. Khan, D. Vasileska and T. J. Thornton

Arizona State University, Tempe, AZ, 85287-5706, USA

Fax: (480) 965-8058, Phone: (480) 727-7522

ktk@asu.edu, vasileska@asu.edu, t.thornton@asu.edu

ABSTRACT

Micropower circuits use subthreshold MOSFETs that consume minimal power resulting from the combination of ultra-low drain currents ($10^{-11} < I_d < 10^{-5}$ A/µm) and small drain voltages required for saturation (V_d^{sat} ~150-200mV). Unfortunately, sub-threshold CMOS is a slow technology, with micropower circuits limited to operating frequencies below ~ 1MHz. To achieve higher sub-threshold operating frequencies, we have proposed a Schottky Junction Transistor (SJT) [1] as an alternative to sub-threshold MOSFET devices. It adopts a MESFET architecture and exhibits higher electron mobility. The enhanced MESFET mobility leads to a corresponding increase in the cut-off frequency f_T compared to a similar gate length MOSFET carrying the same amount of current

Keywords: Schottky Junction Transistor (SJT), subthreshold conduction, micropower.

1. INTRODUCTION

Micropower circuits based on sub-threshold MOSFETs are used in a variety of applications ranging from digital watches to medical implants. The principal advantage of sub-threshold transistor operation is the minimal power consumption that results from the combination of ultra-low drain currents ($10^{-11} < I_d < 10^{-5}$ A/µm) and small drain voltages required for saturation (V_d^{sat} ~ 150-200 mV). However, the sub-threshold CMOS is a slow technology, with micropower circuits limited to operating frequencies below ~ 1MHz due to its low cut-off frequency $f_T = \mu V_T/2\pi L_g^2$, where μ is the carrier mobility, $V_T = kT/e$ is the thermal voltage and L_g is the gate length. In the sub-threshold regime, it is impractical to increase f_T by reducing the gate length because of difficulties with transistor matching. The only remaining option to increase f_T is to increase the carrier mobility. In a MOSFET, the inversion electron mobility is typically 600-700 cm²/Vs but falls to only 100-200 cm²/Vs in weak inversion, and we, therefore, expect a cut-off frequency in the range 40-80 MHz for a sub-threshold MOSFET with $L_g = 1$ µm. To achieve higher sub-threshold operating frequencies we have proposed a Schottky Junction Transistor (SJT) as an alternative sub-threshold device [1] that adopts MESFET architecture. To confirm that the carriers exhibit higher mobility in this device structure, we have performed Monte Carlo simulations of similar geometry MOSFET and MESFET devices (see Fig. 1).

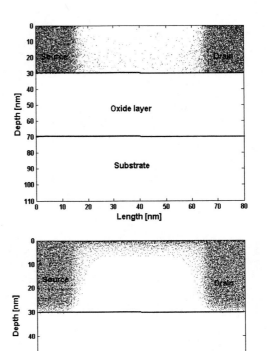

Fig. 1: The geometry of the MESFET (upper figure) and MOSFET (lower figure) used in the simulations. Also shown is the electron distribution across the channel.

2. MOBILITY CALCULATION

Electron mobility is an important parameter as it describes the ease with which carriers respond to applied electric fields. For bulk material systems with low impurity concentration, the mobility is limited by phonon scattering. Coulomb scattering plays an important role if the doping density is in excess of 10^{17} cm⁻³. In the device structures from Fig. 1, surface-roughness scattering plays significant role at high transverse fields. Thus, for accurate simulation of the output current, modeling of the surface mobility is very important. At low electric fields, the mobility can be found as a ratio of the average particle drift-velocity and the electrical field. Instead of the cell based calculation, we calculate the electron mobility for each particle in the channel using $\mu(x) = v(x)/\overline{E}_x(x)$. An average over time (in the steady state) and space is taken to calculate the average mobility. The method is noisy, but it allows us to investigate the role played by interface-roughness on the electron mobility for devices with different channel thickness.

3. DESCRIPTION OF THE 2D DEVICE SIMULATOR

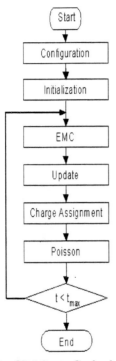

Figure 2: 2D Monte Carlo device simulator.

The 2D ensemble Monte Carlo method, coupled with a 2D Poisson solver, is used for the device simulation. It is based on the usual *Si* band structure for three dimensional electrons, in a set of non-parabolic Δ-valleys with energy dependent effective masses. The six conduction band valleys, inherent to the silicon band structure, are included through three pairs. The longitudinal and the transverse masses are very important in this case and they have been included in the model using the Herring-Vogt transformation [2]. It is well known that intra-valley scattering is limited to acoustic phonons, but for inter-valley scattering we have included both *g* and *f*–phonon processes. Coulomb and surface roughness scattering are included in this model as they strongly affect the carrier mobility at low and high electric fields, respectively.

The successive over relaxation (SOR) method has been used for the solution of the 2D Poisson equation and the Monte Carlo simulation has been used to obtain the charge

distribution in the device. Within a Monte Carlo scheme, the charges are distributed within a continuum mesh instead of discrete grid points. The particle mesh coupling is used to perform the switch between the continuum in a cell and discrete grid points at the corners of the cell. The charge assignment has been carried out using the Nearest-Element-Center (NEC) method [3,4]. This method leads to zero self-force for non-uniform meshes and spatially varying dielectric constants. The flow chart of the Monte Carlo device simulator is shown in Fig. 2.

4. RESULTS AND DISCUSSION

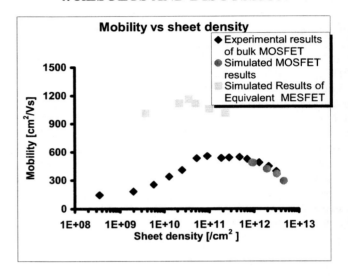

Figure 3: Mobility vs. sheet density.

In Fig. 3 we show the variation of SOI MESFET mobility and the MOSFET mobility with sheet density. It is seen that at high sheet density the mobility degrades quickly due to surface-roughness scattering. For the case of a MOSFET device, there exists a peak mobility value of ~ 600 cm^2/Vs at a sheet density of $N_s \sim 10^{12}$ cm^2/V-s but this quickly falls off as N_s is reduced below 10^{10} cm^{-2}, corresponding to the sub-threshold regime of weak inversion. In contrast, the electron mobility in the MESFET channel has a higher peak value of 1200cm^2/V-s and does fall down with decreasing N_s, but its magnitude is larger than the MOSFET mobility. The same is true for higher sheet charge densities. This behavior can be explained by the different role played by surface-roughness scattering. In the case of a MOSFET, the stronger electric field pulls the carriers closer to the surface. In MESFETs, it pulls carriers away from the interface. As seen from Fig. 1, the current flow in the MESFET device is away from the upper interface when compared to the MOSFET device. As a result of this, fewer current carrying electrons interact with the rough interface and the average mobility is increased. Of course, as the thickness of the MESFET channel is reduced, a greater fraction of the electrons are forced to interact with the back interface and the average mobility will degrade. This behavior is shown in Fig. 4. Also note, by comparing the results shown in Fig. 3, that the MOSFET mobility follows the experimental values [5] and the MESFET mobility is 5 to 10 times higher than that of the MOSFET device. The enhanced MESFET mobility will lead to a corresponding increase in the cut-off frequency f_T compared to a similar gate length MOSFET carrying the same amount of current.

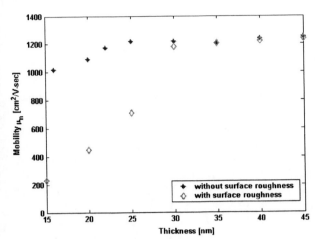

Figure 4: Mobility vs. silicon thickness with and without surface roughness scattering.

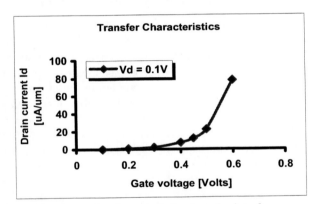

Figure 5: Transfer characteristics to calculate f_T

From the transfer characteristics of the MESFET device shown in Fig. 5 for drain voltage $V_d = 0.1V$, we calculate the transconductance g_m, which is given by

$$g_m = \left(\frac{\partial I_D}{\partial V_g}\right)_{V_d = cons\tan t} = 40\mu S/\mu m.$$

The gate capacitance C_{gs} equals to

$$C_{gs} = \left(\frac{\partial Q}{\partial V_g}\right)_{V_d = cons\tan t} = 1.47 \times 10^{-13} \text{ F/cm},$$

thus leading to a cut off frequency of

$$f_T = \frac{1}{2\pi}\frac{g_m}{C_{gs}} = 433 \text{ GHz.}$$

CONCLUSIONS

In this work we have utilized 2D Monte Carlo device simulator to successfully simulate Silicon MESFETs operating in the subthreshold regime that exhibit high cutoff frequency. The enhancement in the cut-off frequency is due to increased electron mobility and decreased capacitance of the structure, making it suitable for r.f. micropower circuit applications.

ACKNOWLEDGEMENTS

The authors would like to thank Prof. Stephen M. Goodnick and David K. Ferry for valuable discussions during the preparation of this manuscript. The financial support from the Office of Naval Research, under contract number N00014-02-1-0783 is also acknowledged.

REFERENCE

[1]   T. J. Thornton "Physics and Applications of the Schottky Junction Transistor" *IEEE Trans. Electron Devices* 48, 2421 (2001)

[2]   C. Herring and E. Vogt, "Transport and Deformation-Potential Theory for Many-Valley Semiconductors with Anisotropic Scattering", *Phys. Rev.* 101, 944 (1956).

[3]   C. Jacoboni and L. Reggiani, "The Monte Carlo Method for the Solution of Charge Transport in Semiconductors with Applications to Covalent Materials", *Rev. Modern Phys.* 55, 645 (1983).

[4]   R. W. Hockney and J. W. Eastwood, *Computer Simulation Using Particles*. Maidenhead, U.K. McGraw-Hill, 1981

[5]   J. T. C. Chen and R. S. Muller, "Carrier mobilities at weakly inverted silicon surfaces," *J. App. Phys.* 45, 828 (1974).

Technology Limits and compact model for SiGe Scaled FETs

Robert W. Dutton and Chang-Hoon Choi
Center for Integrated Systems
Stanford University, Stanford, CA, USA, rdutton@stanford.edu

ABSTRACT

Stress relaxation in strained-Si MOSFETs can be significant in the presence of compressive stress imposed by trench isolation, especially for highly scaled active regions. Stress of the strained region is reduced by ~2/3 when the active region is scaled from L_{active}=0.4 μm to 0.1 μm. Mobility can be lower by 50 % for narrow active widths resulting from the strain relaxation. The strain relaxation may restrict the use of strained-Si MOSFETs for technology nodes beyond 25 nm. Electrical and thermal characteristics of strained-Si devices are investigated and a compact junction capacitance model for strained-Si MOSFET suitable for circuit simulation is proposed.

Keywords: Strained-Si, STI, stress, ESD, capacitance.

1 Introduction

Strained-Si MOSFET becomes a promising device for CMOS beyond the 50 nm node to overcome scaling limitations, since the strained-Si layer grown on relaxed SiGe increases carrier mobility. However, recent studies show that the mobility of strained Si NMOS degrades as the gate length becomes shorter [6], which raises scalability issues in strained-Si MOSFETs. Also, the thermal conductivity of $Si_{0.8}Ge_{0.2}$ is 15 times lower than that of bulk-Si, such that self-heating problems similar to that for SOI can cause performance degradation in strained-Si devices [6]. This article investigates technology limits of of strained-Si devices in terms of electrical and thermal characteristics.

2 STI Effects on Strained Si

Shallow Trench Isolation (STI)-induced compressive stress reduces carrier mobility and drive current in bulk NMOS transistors as design rules continue to shrink [2]. The compressive stress in silicon from trench isolation makes the NMOS drive current highly sensitive to the transistor layout. Thus, problems arise when the process-induced stress can relax the tensile stress in the strained layer for scaled active regions. In this work, stress analysis is performed based on lattice and thermal mismatch models for SiGe and silicon depositions, as well as STI formation by using 2 & 3D process/device simulations [3].

Fig. 1 shows the cross-section of a strained Si layer, which is grown on the relaxed SiGe seed layer with a Ge mole fraction of 0.2. The lattice constant of SiGe is based on Vegard's law and the lattice mismatch strain is calculated when Si is grown on the SiGe layer. Stress relaxation is calculated for a structure where the trench is etched out and oxide filling is then used.

Fig. 2 shows two-dimensional stress analysis results of a strained active region with STI for different lengths of active layers (L_{active}). The initial stress (S_{xx}) in the strained Si layer is ~1400 MPa for L_{active} = 0.4 μm (L_G = 0.1 μm), which is reduced to 1033 Mpa after the STI process. The tensile stress in the strained layer is reduced down to 600 MPa and 350 MPa as L_{active} is scaled to L_{active} = 0.2 μm (L_G = 50 nm) and L_{active} = 0.1 μm (L_G = 25 nm), respectively.

3 Stress Relaxation and Mobility

Strain relaxation is dependent not only on the device geometry, but also on thermal annealing. Three-dimensional device simulations are performed based on the stress-induced bandgap change model, in which the bandgap varies with regard to mechanical stress and strain tensors in silicon region [4]. In addition, the stress-induced mobility model combined with phonon and surface scattering mobility models are utilized to account for mobility corrections due to the mechanical stress; electron mobility becomes anisotropic for stress-induced band shifts and changes in effective mass. The longitudinal and transverse effective electron masses are based on the $k-p$ theory for Si grown on the $Si_{0.8}Ge_{0.2}$ surface [5]. Fig. 3 shows simulated $I_D - V_G$ and $g_m - V_G$ based on stress calculations for different annealing conditions after STI process - 1s, 800°C and 10s, 850°C. This implies that mobility enhancement degrades for the higher temperature annealing condition that enhances strain relaxation. Fig. 4(a) shows S_{xx} contours for different active widths; stress relaxation in the strained layer is enhanced for the narrower active width due to impact of the surrounding STI. As a result, mobility enhancement of the narrow width device (i.e. W = 0.2 μm) is reduced by more than 50 % compared to that of the wider device (i.e. W = 1.7 μm), as shown in Fig. 4(b).

4 Modeling for High Current Operation

The thermal conductivity of $Si_{0.8}Ge_{0.2}$ is 15 times lower than that of bulk-Si, such that self-heating problems similar to that for SOI can cause performance degradation in strained-Si devices [6]. Fig. 5(a) shows device structure of a bulk NMOS and a strained-Si NMOS. In order to determine energy band structures of the strained-Si device, material parameters are used based on Ref. [7]. Fig 5(b) are simulated energy band diagrams for a strained- and a bulk-Si devices by using MEDICI. $E_{n,\parallel}$ is the electric field into current flow direction. In order to determine the phonon mean free path for electrons (λ_n) of strained-Si, the energy relaxation times are calculated, as the mean free path is closely related to the energy relaxation times in the scattering process. Fig. 6 represents calculated energy relaxation times of electrons for strained-Si and bulk-Si with respect to electric field and electron temperature, based on full-band Monte Carlo (FBMC) device simulation [8]. In the Ref. [6] the energy relaxation time was increased by roughly a factor of two from the bulk-Si device. The FBMC simulation results in Fig. 6 show that the increase of energy relaxation time of the strained-Si relative to bulk-Si is reduced as the electric field increases (i.e. at high electron temperature). In this work, λ_n of the strained-Si on $Si_{0.75}Ge_{0.25}$ layer has been assumed to be increased by 20% from that for the bulk-Si ($\lambda_n = 10$ nm for bulk-Si).

5 Electro-Thermal Simulation

Fig. 7(a) shows simulated I_D-V_D curves for strained- and bulk-Si devices ($V_G = 0$ V). The snapback voltage of the strained-Si device is lower than that for bulk-Si device. The hold voltage (V_h), that is the minimum voltage required for the bipolar operation, is also lower for the strained-Si device. The second breakdown triggering current (I_{t2}) for strained-Si device is higher relative to bulk-Si device. As a result, the power density and peak lattice temperature during the parasitic bipolar operation are lower for the strained-Si device than that for bulk-Si device. The main current path is the strained-Si layer during normal operation, while the main current path during the parasitic bipolar action is the buried SiGe layer. Fig. 7(b) shows peak temperature versus drain current (I_D) for bulk- and strained-Si devices. There is an abrupt increase of lattice temperature at ~0.008 A/μm of drain current for the bulk-Si device. As a result, thermal failure occurs for the bulk-Si device ($T > T_c \sim 1650°$K) when the drain current is ~0.01 A/μm (I_{t2}) due to its high power density. This implies that even though the thermal dissipation of strained-Si is worse compared to silicon due to the 30 times lower thermal conductivity of SiGe layer, the local temperature overheating can effectively be suppressed in strained-Si devices owing to the higher bipolar gain (β) and current uniformity.

6 Compact Junction Capacitance Model

Higher permittivity and smaller bandgap of SiGe result in higher junction capacitance in strained-Si devices, which degrades circuit performance of S-Si MOS circuits. Fig. 8 shows drain-to-bulk junction capacitance obtained from device simulations, for strained-Si/$Si_{0.75}Ge_{0.25}$ and unstrained (bulk)-Si NMOS devices. The increase of junction capacitance in strained-Si is about 16% compared to bulk-Si NMOS

at the zero-bias, while the increase is 8% for $V_{DB} = 2.0$ V. The sharper decrease of bias-dependent junction capacitance in strained-Si device can be attributed to the reduced junction built-in potential (ϕ_{BA} and ϕ_{BSW}) for strained-Si, which should be considered in a compact model. Fig. 9 represents junction capacitance components for strained-Si NMOS. It should be noted that sidewall junction capacitance (C_{JSW}) consists of two capacitance components, $C_{JSW,SSi}$ and $C_{JSW,SiGe}$, to take into account the different material parameters– permittivity, bandgap and built-in potential. As a result, the sidewall junction capacitance of S-Si device can be expressed as shown in the equation in the figure.

Fig. 9 shows calaculated drain-to-bulk junction capacitance (C_{DB}) versus V_{DB} of strained-Si NMOS based on the compact model; curves are shown for various Ge mole fractions, $x = 0$ (unstrained-Si), 0.15, 0.25, 0.35, while other parameters remain same. The good agreemetns between the compact model and the numerical model can be attributed to considerations of the two different junction sidewall capacitance components for the strained-Si layer and the SiGe layer.

7 Summary

Stress relaxation in strained-Si layers can be significant in the presence of compressive stress imposed by trench isolation, especially for highly scaled active regions. Stress of the strained region is reduced by ~2/3 when the active region is scaled from 0.4 μm to 0.1 μm. The strain relaxation may restrict the use of strained-Si MOSFETs for technology nodes beyond 25 nm. However, the strained-Si NMOS has higher bipolar current gain and impact ionization rate due to its narrower energy band gap and longer phonon mean-free-path. Thus, contrary to SOI devices, the strained-Si NMOS can be used as an effective ESD protection device, despite its self-heating problems caused by the lower thermal conductivity of SiGe layer. A compact junction capacitance model for strained-Si MOSFET has been developed, which shows good agreement with device simulation results due to considerations of junction sidewall capacitances for both strained-Si and SiGe layers.

Acknowledgment

This project is supported by the MARCO/DARPA Focus Center on Materials, Structures, and Devices (MSD).

REFERENCES

[1] K. Rim, *IEDM Tech. Dig.*, p.43, 2002.
[2] G. Scott, *IEDM Tech. Dig.*, p.827, 1999.
[3] *TAURUS: Process & Device.* Synopsys, Inc., 2002.
[4] J.L. Egley, *Solid-State Elec.*, p.1653, 1993.
[5] N. Cavassilas, *Physics Rev. B64*, p.115207, 2001
[6] K. Rim, *IEEE T. ED*, vol. 47, p.1406, July 2000.
[7] J-S. Goo, *IEEE ED-L*, p.568, Sep. 2003.
[8] C. Jungemann, *IEICE T. Electronics*, no.6, p.870, 1999.

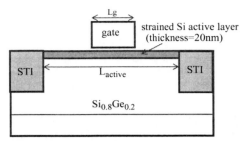

Fig. 1. A cross-section of strained Si active layer on $Si_{0.8}Ge_{0.2}$ layer.

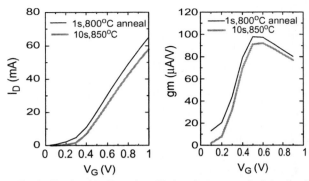

Fig. 3. Simulated I_D-V_G and gm-V_G based on stress calculation for different annealing conditions after STI process - 1s, 800°C and 10s, 850°C.

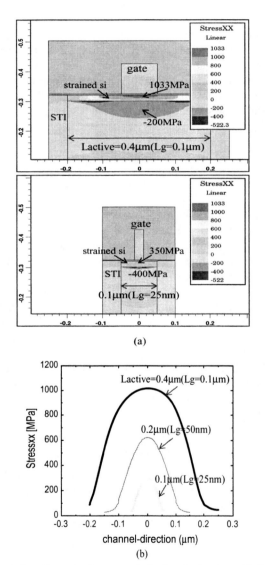

(a)

(b)

Fig. 2. Two-dimensional stress analysis of strained actives with STI effects (a) stress (Sxx) contours for Lactive=0.4μm (Lg=0.1μm) and Lactive=0.1μm (Lg=25nm) (b) comparison of tensile stresses in strained layer along the channel direction for different active lengths (Lactive) .

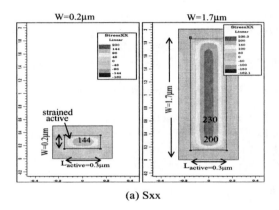

(a) Sxx

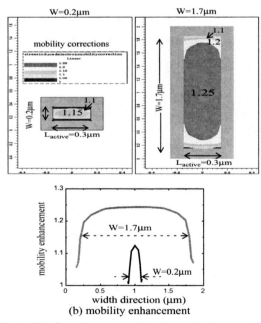

(b) mobility enhancement

Fig. 4. STI effects for narrow width active strained region, (a) Sxx contours for W/L = 0.2μm/0.1μm and 1.7μm/0.1μm (b) Szz (c) stress-induced mobility enhancements for W=0.2 and 1.7μm.

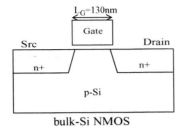

bulk-Si NMOS

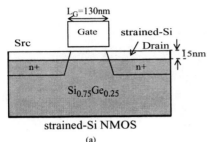

strained-Si NMOS

(a)

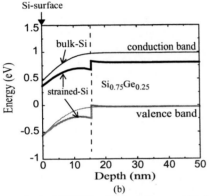

(b)

Fig. 5.. Strained Si NMOS and its band diagrams, (a) bulk-Si NMOS and strained-Si NMOS with 15nm strained layer thickness on $Si_{0.75}Ge_{0.25}$, (b) band-diagrams of the device at $V_G=0$ V into the depth direction at the position denoted in (a) with an arrow.

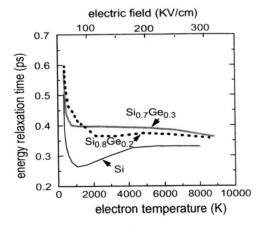

Fig. 6. Electron energy relaxation times vs. electron temperatures for strained-Si and bulk-Si devices from full-band Monte Carlo simulation.

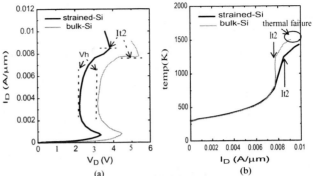

Fig. 7. I_D/V_D vs. temperatures for strained-Si NMOS and bulk-Si NMOS, (a) V_D vs. lattice temperature (b) I_D vs. lattice temperature.

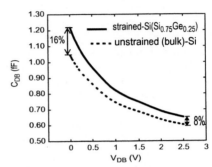

Fig. 8. Simulated drain-to-bulk junction capacitance for strained-Si and bulk (unstrained) Si MOSFETs.

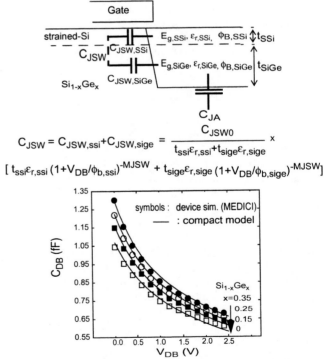

$$C_{JSW} = C_{JSW,ssi}+C_{JSW,sige} = \frac{C_{JSW0}}{t_{ssi}\varepsilon_{r,ssi}+t_{sige}\varepsilon_{r,sige}} \times$$

$$[\ t_{ssi}\varepsilon_{r,ssi}\ (1+V_{DB}/\phi_{b,ssi})^{-MJSW} + t_{sige}\varepsilon_{r,sige}\ (1+V_{DB}/\phi_{b,sige})^{-MJSW}]$$

Fig. 9. Comparisons of junction capacitances betwwen the obtained from the compact model and device simulation.

Ballistic MOS Model (BMM) Considering Full 2D Quantum Effects

Zhiping Yu, Dawei Zhang, and Lilin Tian

Institute of Microelectronics, Tsinghua University
Beijing 100084, China, yuzhip@tsinghua.edu.cn

ABSTRACT

As the channel length of MOSFETs is shrunk to below 50 nm, 2D quantum mechanical (QM) effects becomes profound on both the carrier confinement in the transverse direction to the channel and the carrier transport, mainly ballistic and tunneling, along the channel. A compact MOS model which incorporates the physics-based correction of 2D QM effects on the surface inversion charge density and ballistic transport is developed. The model has been applied to various MOS devices with gate length ranging from 45 nm to 14 nm and excellent agreement with published measurement data is achieved. It is proposed for the first time that WKB theory can be explored to model the subband lowering in the confined dimension because of the open boundary on the other dimension. An empirical formula for 2D-QM-corrected threshold voltage is provided. Mobility dependence on gate bias is investigated and modeled.

Keywords: compact MOS model, 2D quantum effects, ballistic transport, mobility modeling

1 INTRODUCTION

With the gate length of MOSFETs further scaled down into the sub-50nm regime, two non-classical physical mechanisms are generally recognized as playing a critical role in determination of device characteristics: ballistic transport and QM effects.

So far the efforts in modeling MOS with ballistic transport focus mainly on the regular device structures, such as the uniformly doped substrate in [1] for bulk CMOS and undoped silicon thin film as the channel region for double gate MOS in [2]. Practical devices, however, have highly nonuniform channel doping in order to optimize the device performance. In the meanwhile, with lateral (channel) dimension not much bigger than the size of carrier confinement in the transverse dimension, full 2D, not just 1D, QM effects have considerable impact on such parameters as the threshold voltage. Previous research was either based on 1D QM correction [3] to existing model formulation, or relies on 2D/quasi-2D [4] numerical simulation, the results of which are difficult to explore in developing physical insight of compact modeling.

In this work, a compact circuit model for sub-50 nm bulk silicon MOSFETs applicable to realistic channel doping is developed. The two aforementioned physical mechanisms, namely, the ballistic transport and the full 2D QM effects have been incorporated in the model. To emphasize the role of ballistic transport, the model is named as the Ballistic MOS Model, or BMM. An ingenious method based on the quantum mechanical WKB theory is proposed to transform the open boundary condition and the detailed potential profile along the channel to the subband lowering in the carrier confinement dimension. Because of the 2D effects, it is concluded that the threshold voltage increase in a deep-scaled down MOSFET due to the QM effects is less severe than a strict 1D treatment predicts.

The number of parameters in BMM is small compared to the BRIM models and since many parameters are physics based, the extraction process for model parameters is straightforward.

2 MODELING APPROACH

The analytical formula of the ballistic theory is discussed first as the basis for the complete model. The surface charge density of the inversion layer at the peak of the potential barrier along the channel, $Q(0)$, is then modeled using the proposed 2D QM correction scheme.

2.1 Modeling of the Ballistic Theory

The expression for the drain current considering the ballistic transport is described in [2],

$$\frac{I_D}{W} = Q(0)\frac{1-r}{1+r}\left[v_T\frac{\mathcal{F}_{1/2}(\eta)}{\mathcal{F}_0(\eta)}\right]\frac{1 - \frac{\mathcal{F}_{1/2}(\eta-U_D)}{\mathcal{F}_{1/2}(\eta)}}{1 + \frac{1-r}{1+r}\frac{\mathcal{F}_0(\eta-U_D)}{\mathcal{F}_0(\eta)}} \quad (1)$$

where the back scattering coefficient

$$r = \frac{l}{l+\lambda}, \, l = L\left(V_T\frac{\beta}{V_{DS}}\right)^\alpha, \, \lambda = V_T\frac{2\mu}{v_T}\frac{\mathcal{F}_0^2(\eta)}{\mathcal{F}_{-1}(\eta)\mathcal{F}_{1/2}(\eta)}$$

and l is the scattering critical length, λ the mean free path, $V_T = k_B T/q$ the thermal voltage, L and W are the channel length and gate width, respectively, μ the mobility, v_T the thermal velocity. $\mathcal{F}_j(x)$ is the j-th order Fermi integral, V_{DS} the drain bias, $U_D = V_{DS}/V_T$, and

α and β are fitting parameters. $\eta = (E_{FS} - E_{\max})/k_B T$ where E_{FS} is the Fermi energy at the source contact and $E_{\max}$ is the energy for the bottom of the first subband at the channel location, $x_{\max}$, where the position-dependent (along the channel) subband bottom energy reaches the maximum. The value of η is implicitly determined by the gate bias, V_{GS} and we'll find it through $Q(0)$.

The key modeling issue in (1) is $Q(0)$, the surface inversion layer charge density at $x_{\max}$ and it is determined by the applied terminal bias. With 1D QM correction, it is obtained in [3] that

$$Q(0) = C_{ox} V_{g,\text{eff}} \qquad (2)$$

where $C_{ox} = \epsilon_{ox}/t_{ox}$ is the gate-oxide capacitance per unit surface area and $V_{g,\text{eff}}$ is the effective gate voltage,

$$V_{g,\text{eff}} = \frac{2n V_T \ln\left[1 + \exp\left(\frac{V_{od}}{2n V_T}\right)\right]}{1 + 2n C_{ox}\sqrt{\frac{2\phi_S}{q\epsilon_{Si} N_{\text{sub}}}} \exp\left[\frac{V_{od} - 2(V_{GS} - V_{th} - V_{\text{off}})}{2n V_T}\right]}$$
$$(3)$$

where $V_{od} = V_{GS} - V_{FB} - \phi_S - Q_{\text{dep}}/C_{ox}$ In the above, n is the subthreshold swing parameter, V_{off} is a parameter related to the threshold voltage change in different operation regions [3]. ϕ_S is the surface potential when the gate bias equals the threshold voltage, Q_{dep} and V_{th} are the depletion region charge and threshold voltage, respectively. For details, see [3].

On the other hand, according to Natori [1], η is linked to $Q(0)$ through the following relation:

$$Q(0) = \frac{qkT}{2\pi\hbar^2} \sum_{\text{valley}} \sum_i \sqrt{m_x m_y} \left[\ln\left(1 + e^\eta\right) + \ln\left(1 + e^{\eta - U_D}\right)\right]$$

$$\approx 2.5 \frac{qkT}{2\pi\hbar^2} \sqrt{m_t m_l} \left[\ln\left(1 + e^\eta\right) + \ln\left(1 + e^{\eta - U_D}\right)\right] \quad (4)$$

where "valley" refers to counting all degenerate valleys, i counting all subbands, m_x effective mass in the transport direction and m_y in the direction of the device's width. For the substrate on (100) plane and consider only the lowest subband, an approximation can be obtained as in (4) [1], where m_t and m_l are transverse and longitudinal effective masses, respectively. Thus, knowing $Q(0)$, η can be solved for and the drain current can be evaluated in (1) from V_{GS} and V_{DS}.

2.2 Modeling of 2D QM Effects

It is long realized that the open boundary condition along the channel will affect the 1D solution to the Schrödinger equation in the transverse direction. Since the threshold voltage is largely determined by the ground energy level in the surface quantum well in this transverse direction, the proper assessment of the effect of potential distribution along the channel is crucial to the accurate calculation of the threshold voltage considering the 2D QM effects.

Even though the numerical 2D QM simulation can provide the information of the surface charge density at the x_{max}, which is close to the source end of the channel, it can hardly be used to develop a compact model. We propose a simple correction scheme to the subband edge obtained from the 1D Schrödinger equation using the solution of 1D Schrödinger equation along the channel. The solution in the channel direction is provided by the WKB theory for an arbitrary potential profile. Following Natori's coordinate system [1] of $x-$axis along the channel, $y-$axis on channel width, $z-$axis into the depth of the substrate, one can write the solution to the 3D Schrödinger equation using the separation of variables as follows.

$$\Psi(x,y,z) = \underbrace{\frac{A}{\sqrt{p(x)}} \exp\left(\frac{i}{\hbar} \int^x p(\xi) d\xi\right)}_{X(x)} \cdot \underbrace{\sqrt{\frac{2}{W}} \sin\left(\frac{n_y \pi}{W} y\right)}_{Y(y)} \cdot \underbrace{\varphi_{n_z}(z)}_{Z(z)} \quad (5)$$

In deriving the above wavefunction, we have assumed that the along the width of the channel, there is a flat quantum well with infinitely high barrier at $y = 0$ and $y = W$. $\varphi_{n_z}(z)$ is to be solved numerically as is done in 1D Schrödinger/Poisson equation solver. The key for the above solution in terms of our modeling need is the wavefunction along the x-direction. With the WKB theory,

$$p(x) = \sqrt{2m_x[E - U(x,y,z)]} \qquad (6)$$

which may be an imaginary number if $E < U$. If U is a constant, then the solution is reduced to a plane wave and the energy E can never be below U and the kinetic energy $\hbar^2 k^2 / 2m_x$ is always positive. However, the potential distribution along the channel is not uniform and peaks at $x_{\max}$.

We define the subband edge as the energy when the wavefunction in the x-direction becomes localized, which means that we extend the solution range to include the tunneling regime as well. Since when $E < U$, the quantity of $(i/\hbar) \int^x p(\xi) d\xi$ becomes negative, indicating the wave is decaying within the potential barrier. Thus one can define the "locality" as that the exponential term in $X(x)$ becomes negligibly small while integrating from the left classical turning point ($x = a$) to the right ($x = b$) of the barrier. The energy level which demarcates the boundary between the nonlocal and local states can be found by solving the following nonlinear equation numerically.

$$\frac{i}{\hbar} \int_a^b \sqrt{2m_x[E_d - U(x)]} dx = -4 \qquad (7)$$

for $e^{-4} = 1.83\%$ can be considered as negligibly small. From the solved E_d one can further define $\Delta V_{\text{2D-QM}} = (U_0 - E_d)/q$ where U_0 is the peak value of the potential $U(x)$. This ΔV will be used as the 2D QM correction term in the 1D QM formulation for the threshold voltage in the existing model.

2.3 Model for Threshold Voltage

The V_{th} in (3) will be modified using the corrections due to the full 2D QM effects, the short-channel-effects (SCEs) and the drain-induced-barrier-lowering (DIBL) simply as

$$V_{th,\text{eff}} = V_{th} - \Delta V_{\text{2D-QM}} - \Delta V_{\text{SCE}} - \Delta V_{\text{DIBL}} \quad (8)$$

where ΔV_{SCE} and ΔV_{DIBL} are taken from BSIM 4. This update of V_{th} completes the modeling of $Q(0)$ and hence I_D.

2.4 Mobility Model

With varying channel length in the sub-50nm range and nonuniform channel doping, the accurate mobility model is critical to the overall model accuracy. The gate bias dependence of the mobility is investigated and modeled. The following formula is used for mobility degradation

$$\mu(V_{GS}) = \frac{\mu_0}{1 + \theta_1 V_{GS} + \theta_2 V_{GS}^2} \quad (9)$$

where μ_0, θ_1, and θ_2 are fitting parameters.

3 PARAMETER EXTRACTION

Overall, there are seven model parameters which need to be determined to evaluate $I_D(V_{GS}, V_{DS})$. They are t_{ox} and N_{sub} for $V_{g,\text{eff}}$ (hence for $Q(0)$ as well) in (3), α and β in (1), and μ_0, θ_1, and θ_2 in (9). To determine these parameters for a specific MOSFET, at least one transfer curve and three output curves from experimental data are needed. In the process of parameter extraction, first, t_{ox} and N_{sub} are adjusted using the effective oxide thickness (EOT) and the average channel doping as the initial values and setting the other five parameters to default values (α and β taken from [1], and parameters for mobility from MEDICI of Synopsys) to fit the transfer curve. As long as the error is smaller than 30%, the values of t_{ox} and N_{sub} are fixed temporarily. Second, to determine α and β, at least one output curve is needed. The values of α and β are sensitive to the shape of I-V curve and can be determined relatively easily from the measured data. Finally, three parameters for μ are extracted from three output curves with different gate biases. The above steps may need iterations to get the overall error within desired level. Our experience shows that error between analytical results and experimental data can be as low as 5%.

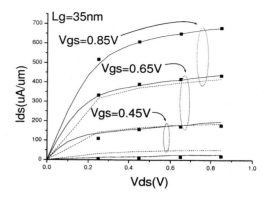

Figure 1: Comparison between simulation (lines) and measurement (symbols) data for Toshiba 35nm CMOS [6]. Dashed line is without 2D QM correction to V_{th}.

4 RESULTS

The compact model is used to calculate the current of the bulk silicon devices in [5]-[7], having respective gate lengths of 15nm, 35nm, and 42nm. The devices are all made either in industry or in the laboratory. The fitted parameters for these three devices are listed in Table 1. The values of N_{sub} are physically reasonable compared to the real doping in [5]-[7] and the values of t_{ox} are also almost the same as their EOTs. β is constant across the devices while α monotonically decreases with the increase of the gate length.

In Fig. 1, the comparison of the output characteristics of the 35nm device in [6] between experimental data and analytical results is presented. The analytical results without modeling ΔV_{th} are also compared to the experimental data. It is clear from Fig. 1 that without considering the QM effects along the channel, the analytical results grossly underestimate the real data, indicating that the strict 1D QM correction in the substrate exaggerates the increase of the threshold voltage due to the QM effects. In Fig. 2, the comparison of the transfer characteristics of 15nm device in [5] between experimental data and analytical results is presented. From Figs. 1-2, it can be seen that the analytical results agree well with the experimental data, proving that this compact model is applicable to those devices with a sub-50nm feature size. Our preliminary investigation even shows that the smaller the device size, the better the fitting accuracy of the BMM model.

5 CONCLUSIONS

A compact model for sub-50 nm MOSFETs has been developed. Advanced physics such as the ballistic transport and 2D quantum mechanical effects are incorporated in the model. The model is applicable to the devices with nonuniform channel doping. The model

Table 1: Parameters extracted for three bulk CMOS devices with gate length of 15, 35, and 42nm.

L_g (nm)	t_{ox} (nm)	N_{sub} (cm^{-3})	α	β	μ_0 (cm^2/V·s)	θ_1 (V^{-1})	θ_2 (V^{-2})
15	1	9×10^{18}	0.3	1.08	60.976	-6.240	11.266
35	1.2	2.3×10^{18}	0.2	1.08	167.504	-5.551	11.568
42	2	1×10^{18}	0.18	1.08	88.81	-3.352	5.711

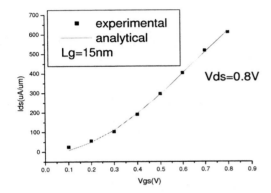

Figure 2: Transfer curve from modeling (solid line) and measurement (symbols) data for AMD 15nm CMOS [5].

has been tested using several published device structures with gate length ranging from 45 nm to 14 nm and the comparison between the simulated and measured data shows the model has good accuracy and predictability.

6 ACKNOWLEDGEMENT

This work is supported by a research grant (973) from Ministry of Science and Technology in China.

REFERENCES

[1] K. Natori, "Ballistic metal-oxide-semiconductor field effect transistor," *J. Appl. Phys.*, Vol. 76, 4879-4889, 1994. K. Natori, "Scaling limit of the MOS transistor – A ballistic MOSFET," *IEICE Trans. Electron.*, Vol. E84-C, 1029-1036, Aug. 2001.

[2] A. Rahman and S. Lundstrom, "A compact scattering model for the nanoscale double-gate MOSFET," *IEEE Trans. Electron Devices*, Vol. 49, 481-489, March 2002.

[3] Yutao Ma, Litian Liu, Lilin Tian, Zhiping Yu, and Zhijian Li, "Analytical charge control and I-V model for submicron and deep-submicron MOSFET's fully comprising quantum mechanical effects," *IEEE Trans. Computer-Aided-Design of Integrated Circuits and Systems*, Vol. 20, No. 4, 495-502, April 2001.

[4] A. Pirovano, A. L. Lacaita, and A. S. Spinelli, "Two-Dimensional quantum effects in nanoscale MOSFETs," *IEEE Trans. Electron Devices*, Vol. 49, 25-31, Jan. 2002.

[5] Bin Yu, Haihong Wang, Amol Joshi, Qi Xiang, Effiong Ibok, and Mingren Lin, "15nm gate length planar CMOS transistor," *IEDM Tech. Dig.*, 11.7.1-11.7.3, Dec. 2001.

[6] S. Inaba, *et al.*, "High performance 35nm gate length CMOS with NO oxynitride gate dielectric and Ni salicide," *IEDM Tech. Dig.*, 29.6.1-29.6.4, Dec. 2001.

[7] Qiuxia Xu, He Qian, Zhengsheng Han, Ming Liu, Ruibing Hou, Baoqing Chen, Haojie Jiang, Yuyin Zhao, and Dexin Wu, "High performance 42nm gate length CMOS device," *Chinese Journal of Semiconductors*, Vol. 24, Supplement, 153-160, April 2003.

Recent enhancements of MOS Model 11

R. van Langevelde, A.J. Scholten and D.B.M. Klaassen

Philips Research Laboratories
Prof. Holstlaan 4, 5656 AA Eindhoven, The Netherlands, ronald.van.langevelde@philips.com

ABSTRACT

MOS Model 11 (MM11) is a surface-potential-based compact MOSFET model, which was introduced in 2001 (level 1100). An update of MM11, level 1101, was introduced in 2002. At the moment a second update of MM11, level 1102, has been completed and is under test. It includes: *i*) an iterative solution of the surface potential; *ii*) an improved description of the velocity saturation yielding a better modelling of the transconductance in saturation; and *iii*) a better description of noise, especially the induced gate current noise. In this paper we describe these improvements and show the resulting improved modelling of transistor performance.

Keywords: MOS Model 11, compact MOSFET modelling, surface potential, thermal noise, induced gate noise

1   INTRODUCTION

There are three categories of compact MOS models: *i*) threshold-voltage-based models, *ii*) inversion-charge-based models, and *iii*) surface-potential-based models. Since the latter category of models give a physics-based and accurate description in all operation regions, experts judge them to be the most successful in describing deep-submicron CMOS technologies for demanding analogue and RF applications [1]. As a result, most new MOS models in the public domain are surface-potential-based [1]–[3].

MOS Model 11 (MM11) is a surface-potential-based compact MOSFET model, which has been developed with special emphasis on the preservation of source-drain symmetry and on an accurate description of distortion behaviour. MM11 was introduced in 2001 (level 1100, indicated by MM1100) and contains all major physical effects such as poly-depletion for the gate, gate tunnelling current and quantum-mechanical quantisation effects. During 2002, an update of MM11, level 1101 (MM1101), has been introduced [4]. The main enhancements of MM1101 over MM1100 are: *i*) the modelling of gate-induced drain leakage, *ii*) a rigorous separation of geometrical and temperature scaling rules for the model parameters, and *iii)* a separate set of geometrical scaling rules for the model parameters, especially designed

for binning. In MM1101, there is a choice to use the physics-based geometrical scaling rules **or** the binning geometrical scaling rules for the model parameters.

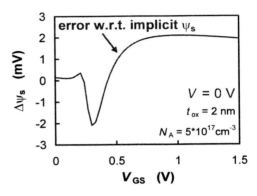

Figure 1: Absolute error in the approximate surface potential (as used in MM1101) with respect to the exact solution of the implicit equation.

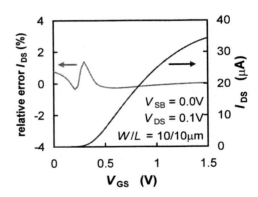

Figure 2: Relative error in the current using the approximated and exact solution of the surface potential corresponding to Fig. 1.

As mentioned before, surface-potential-based models give a physics-based and accurate description in all operation regions. The main drawback of these models, however, is the fact that only an implicit equation for the surface potential can be derived. The solution to this implicit equation can be either approximated an-

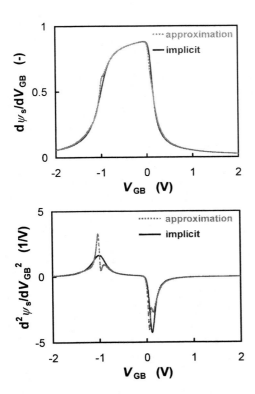

Figure 3: First-order derivative (**top**) and second-order derivative (**bottom**) of the surface potential ψ_s with respect to the gate-bulk voltage. Both the result of the implicit relation for surface potential (*solid line*) and the explicit approximation for surface potential as used in MM1101 (*dashed line*) are shown.

alytically or found exactly by iteration. MM1101 uses an analytical approximated solution of the surface potential, which is computationally very efficient [5]. This approximated solution shows only small deviations with respect to the exact solution of the implicit equation for the surface potential, see Figure 1. The error in the current caused by this deviation is negligible, see Figure 2. When, however, the first-order and, especially, the second-order derivatives of the surface potential with respect to the gate-bulk voltage are inspected, larger deviations show up, see Figure 3. This leads to an error in the input capacitance C_GG, and may be a problem when modelling the distortion of MOS varactors.

More accurate analytical approximations for the surface potential have been published [1], [6]. These approximations are, however, also computationally inefficient compared to our approximation. Therefore we have adopted an iterative solution of the surface potential in MM1102. In addition to this we have also improved the description of the velocity saturation and added a better description of noise, especially the induced gate current noise. These three improvements

will be described in the next Sections.

2 ITERATIVE SOLUTION OF THE SURFACE POTENTIAL

The surface potential ψ_s is defined as the electrostatic potential at the gate oxide/substrate interface with respect to the neutral bulk. For an n-type MOSFET with uniform substrate doping N_A and oxide capacitance C_ox, ψ_s can be calculated from the following implicit relation [7]:

$$\left(\frac{V_\mathrm{GB}^* - \psi_\mathrm{p} - \psi_\mathrm{s}}{k_0}\right)^2 = \psi_\mathrm{s} + \phi_\mathrm{T} \cdot \left[\exp\left(-\frac{\psi_\mathrm{s}}{\phi_\mathrm{T}}\right) - 1\right] \quad (1)$$

$$+ \phi_\mathrm{T} \cdot \exp\left(-\frac{\phi_\mathrm{B} + V}{\phi_\mathrm{T}}\right) \cdot \left[\exp\left(\frac{\psi_\mathrm{s}}{\phi_\mathrm{T}}\right) - \frac{\psi_\mathrm{s}}{\phi_\mathrm{T}} - 1\right]$$

where V_GB^* is the effective gate bias ($V_\mathrm{GB}^* = V_\mathrm{GB} - V_\mathrm{FB}$), V_FB is the flat-band voltage, ψ_p is the potential drop in the polysilicon gate due to poly-depletion, k_0 is the body factor of the silicon substrate ($k_0 = \sqrt{2 \cdot q \cdot \epsilon_\mathrm{Si} \cdot N_\mathrm{A}}/C_\mathrm{ox}$), ϕ_T is the thermal voltage ($\phi_\mathrm{T} = k \cdot T/q$), and ϕ_B is twice the intrinsic Fermi-potential ($\phi_\mathrm{B} = 2 \cdot \phi_\mathrm{T} \cdot \ln(N_\mathrm{A}/n_\mathrm{i})$). The above implicit relation for ψ_s can be solved iteratively using the Newton-Raphson method [8]. Using a simple yet accurate zero-order estimate of ψ_s and applying a first-order Newton-Raphson scheme to the implicit equation for the surface potential, a maximum of four iterations is needed to reach an accuracy of $10^{-10} \cdot \phi_\mathrm{T}$, where ϕ_T is the thermal voltage, see Figure 4. In an attempt to reduce the number of iterations needed to reach this accuracy, a second-order Newton-Raphson scheme has been investigated, see Figure 5. From this figure it can be seen that a maximum of only three iterations is needed to reach the same accuracy. Although the computational effort per iteration is slightly larger for the second-order Newton-Raphson scheme compared to the first-order Newton-Raphson scheme, the overall computation time is smaller for the second-order Newton-Raphson scheme. Hence we adopted the second-order Newton-Raphson scheme for MM1102. In Figure 6 the third-order derivative of the surface potential is shown as a function of gate-bulk voltage. From this figure it can be seen that with MM1102, the distortion modelling of varactors is greatly improved.

3 IMPROVED DESCRIPTION OF VELOCITY SATURATION

Although MM1101 gives an accurate description of the output I_D-V_DS characteristics of short-channel devices, it nevertheless gives a slightly inaccurate description of transconductance g_m in the saturation region, see Figure 7 (top). Close examination shows that this

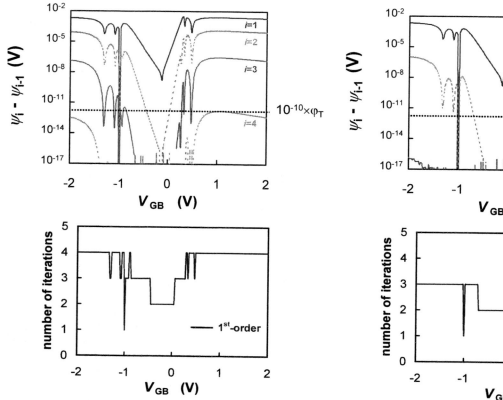

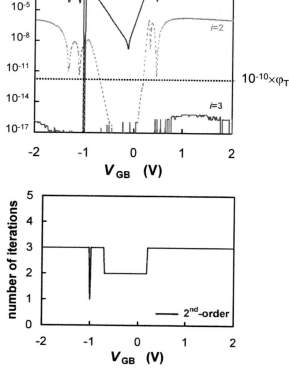

Figure 4: **Top:** Absolute error in surface potential (as a function of gate-bulk voltage) between two successive iterations, where the number of 1st-order Newton-Raphson iterations is indicated by i.
Bottom: Number of iterations that is needed to obtain an absolute error in surface potential between two successive iterations smaller than 10^{-10} times the thermal voltage.

Figure 5: **Top:** Absolute error in surface potential (as a function of gate-bulk voltage) between two successive iterations, where the number of 2nd-order Newton-Raphson iterations is indicated by i.
Bottom: Number of iterations that is needed to obtain an absolute error in surface potential between two successive iterations smaller than 10^{-10} times the thermal voltage.

inaccuracy is caused by the velocity saturation expression used in MM1101. In MM1102, therefore, a more physical and accurate expression for velocity saturation is introduced, see Figure 7 (bottom). In this Section, we will briefly describe the derivation of this new velocity saturation expression.

With an increase in lateral electric field, carriers gain sufficient energy to be scattered by optical phonons, resulting in a decrease of mobility and eventually resulting in the saturation of drift velocity. This is often referred to as velocity saturation. For electrons, an accurate description of velocity saturation is given by [9]–[11]:

$$v = \frac{\mu \cdot \frac{\partial \psi_s}{\partial x}}{\sqrt{1 + \left(\frac{\mu}{v_{\text{sat}}} \cdot \frac{\partial \psi_s}{\partial x}\right)^2}} \qquad (2)$$

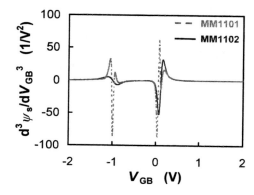

Figure 6: Third-order derivative of the surface potential with respect to the gate-bulk voltage for both level 1101 and level 1102 of MOS Model 11.

where μ is the mobility and v_{sat} is the saturation velocity. This expression can be incorporated in the drain-source channel current I_{DS} by:

$$I_{\text{DS}} = - \frac{\mu \cdot W \cdot Q_{\text{inv}}^*}{\sqrt{1 + \left(\frac{\mu}{v_{\text{sat}}} \cdot \frac{\partial \psi_{\text{s}}}{\partial x}\right)^2}} \cdot \frac{\partial \psi_{\text{s}}}{\partial x} \qquad (3)$$

where Q_{inv} is the inversion-layer charge density, and Q_{inv}^* is given by:

$$Q_{\text{inv}}^* = Q_{\text{inv}} - \phi_{\text{T}} \cdot \frac{\partial Q_{\text{inv}}}{\partial \psi_{\text{s}}} \qquad (4)$$

The above differential equation (3) for I_{DS} is complicated, and is generally approximated by:

$$I_{\text{DS}} = - \frac{\mu \cdot W}{G_{\text{vsat}} \cdot L} \cdot \int_{\psi_{\text{s}_0}}^{\psi_{\text{s}_\text{L}}} Q_{\text{inv}}^* \cdot \mathrm{d}\psi_{\text{s}} \qquad (5)$$

where:

$$G_{\text{vsat}} = \frac{1}{L} \cdot \int_0^L \sqrt{1 + \left(\frac{\mu}{v_{\text{sat}}} \cdot \frac{\partial \psi_{\text{s}}}{\partial x}\right)^2} \cdot \mathrm{d}x \qquad (6)$$

A first-order approximation of the integral in (6) can be obtained by assuming that the lateral field $-\partial \psi_{\text{s}}/\partial x$ is constant along the channel and equal to $\Delta \psi / L$ [10], [11]. In MM1101, a slightly better approximation of (6) is obtained by assuming that the lateral field $-\partial \psi_{\text{s}}/\partial x$ increases linearly along the channel from 0 at the source to $2 \cdot \Delta \psi / L$ at the drain. In this case, G_{vsat} is given by:

$$G_{\text{vsat}} = \frac{1}{2} \cdot \left[\sqrt{1 + \Gamma^2} + \frac{\ln\left(\Gamma + \sqrt{1 + \Gamma^2}\right)}{\Gamma} \right] \qquad (7)$$

where:

$$\Gamma = 2 \cdot \frac{\mu}{v_{\text{sat}}} \cdot \frac{\Delta \psi}{L} \qquad (8)$$

As mentioned before, the above expression for velocity saturation results in an accurate description of the output I_{D}-V_{DS} characteristics, but it nevertheless gives a slightly inaccurate description of transconductance g_{m} in the saturation region, see Figure 7 (top).

In MM1102, velocity saturation is incorporated in a more physical and consequently more accurate way. The differential equation (3) for I_{DS} is rewritten to:

$$I_{\text{DS}} = \sqrt{\left(\mu \cdot W \cdot Q_{\text{inv}}^*\right)^2 - \left(\frac{\mu}{v_{\text{sat}}} \cdot I_{\text{DS}}\right)^2} \cdot \frac{\partial \psi_{\text{s}}}{\partial x} \qquad (9)$$

Under the correct linearisation of Q_{inv}^* with respect to ψ_{s}, the above expression can be integrated along the channel from source to drain, resulting in an implicit relation for I_{DS}. Next, the resulting relation is approximated by a third-order Taylor expansion around

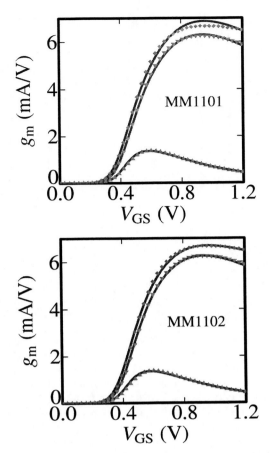

Figure 7: Transconductance of an n-channel MOS-FET with W/L=10μm/0.13μm as a function of V_{GS} at $V_{\text{DS}} = 0.05$, 0.6, and 1.2V and $V_{\text{SB}} = 0$V. Symbols indicate measurements; lines indicate model calculations with MM1101 (**top**), and MM1102 (**bottom**).

$I_{\text{DS}} = 0$, which again leads to (5) where in this case G_{vsat} is given by:

$$G_{\text{vsat}} = \frac{1}{2} \cdot \left[1 + \sqrt{1 + \frac{\Gamma^2}{2}} \right] \qquad (10)$$

The above G_{vsat}-expression is simpler than the MM1101-expression, nevertheless it gives a more accurate description of g_{m} in the saturation region, see Figure 7 (bottom). As a result, MM1102 gives a more accurate description of distortion than MM1101, see Figure 8.

4 IMPROVED NOISE DESCRIPTION

4.1 Introduction

Accurate compact modeling of noise is a prerequisite for RF CMOS circuit design. At the RF-frequencies used, besides the drain current thermal noise also the

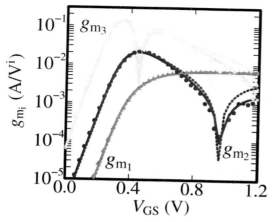

Figure 8: First- (g_{m_1}), second- (g_{m_2}), and third-order (g_{m_3}) derivatives of the drain current with respect to the gate voltage as a function of V_{GS} at $V_{DS} = 1.2$V for the device of Figure 7. Symbols indicate measurements; solid lines and dashed lines indicate model calculations with MM1102 and MM1101, respectively.

induced gate noise plays an important role. Over the years, several compact models for induced gate noise have been introduced [12]–[15]. Nevertheless, all of these models are based on the same Van der Ziel model [12], which is only valid in saturation. In addition, the Van der Ziel model is based on the Klaassen-Prins (KP) equation [16], which does not accurately account for velocity saturation [17]. As a result, the above models are inaccurate for short-channel devices, and more accurate modeling for induced gate noise is needed.

4.2 Correct Incorporation of Velocity Saturation in Noise Model

We use an improved KP-approach [18], where fluctuations in the velocity saturation term are also taken into account. Under the appropriate linearisations, the improved approach can be solved in the ψ_s-framework. The above noise model has been implemented in MM1102, where the impact of series resistance and short-channel effects (CLM, DIBL, etc.) have been incorporated as well. To be able to use the compact model for RF-applications, we furthermore extend the model with a gate resistor and a bulk resistance network.

4.3 Experimental results

Noise measurements were performed on 0.18μm and 0.12μm RF CMOS technologies. Compact model parameters were extracted from DC-measurements, and gate and bulk resistance values were determined from Y-parameter measurements. In Figure 9, noise results are shown for a $L = 0.5\mu$m MOSFET. MM1102 based on the improved KP-approach is in excellent agreement with the data without any additional noise parameter

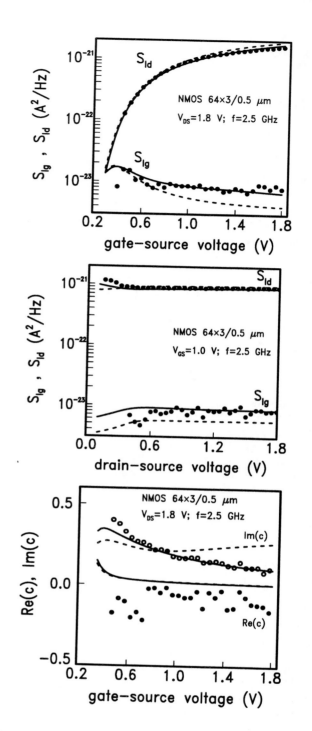

Figure 9: Drain and gate current noise versus gate-source bias (**top**) and versus drain-source bias (**middle**), and correlation coefficient c versus gate-source bias (**bottom**) for an $L = 0.5\mu$m n-channel device in 0.18μm CMOS technology. Symbols denote measurements, dashed lines denote the Van der Ziel model, and solid lines denote MM1102 based on the improved KP-approach.

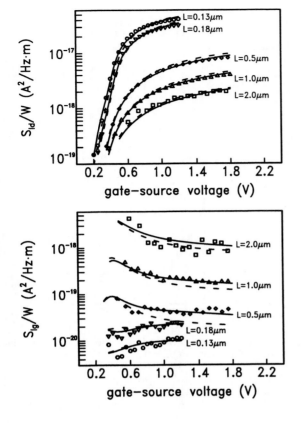

description of the velocity saturation yielding a better modelling of the transconductance in saturation; and *iii*) a better description of noise, especially the induced gate current noise.

The iterative solution results in a more accurate description of surface potential. It has as additional advantage that new physical effects are easily added to the iterative scheme, while for an analytical approximation tedious algebra and often additional approximations for the individual physical effects have to be made. This also implies that: *i*) code maintenance is easier for an iterative solution scheme, and *ii*) the increase in simulation time is very limited. In fact, during test runs with MM1102 on circuits containing several hundreds of transistors an increase in simulation time of only 2 to 3 percent has been observed, while retaining the excellent convergence behaviour of MM1101.

More details and the code of MOS Model 11, level 1102, can be found on our website [19].

REFERENCES

[1] G. Gildenblat *et al.*,
    Proc. of IEEE CICC, pp. 233-240, 2003.
[2] R. van Langevelde *et al.*, *ESSDERC 2001*, p. 81.
[3] M. Miura-Mattausch *et al.*, *IEDM 2002*, p. 109.
[4] R. van Langevelde *et al.*,
    "MOS Model 11, Level 1101," *NL-UR 2002/802*,
    Philips Electronics N.V., 2002, see website of [19].
[5] R. van Langevelde and F.M. Klaassen,
    Solid-State Electron., Vol. 44, pp. 409-418, 2000.
[6] G. Gildenblat and T.-L. Chen,
    Solid-State Electron., Vol. 45, pp. 335-341, 2001.
[7] C.C. McAndrew and J.J. Victory, *IEEE TED*, Vol. ED-49, No. 1, pp. 72-81, 2002.
[8] A.R. Boothroyd *et al.*, *IEEE Trans. Computer-Aided Des.*, Vol. CAD-10, pp. 1512-1529, 1991.
[9] D.L. Scharfetter and H.K. Gummel,
    IEEE TED, Vol. ED-16, No. 1, pp. 64-77, 1969.
[10] R. van Langevelde and F.M. Klaassen,
    IEDM Tech. Dig., pp. 313-316, 1997.
[11] K. Joardar *et al.*, *IEEE TED*, **45**, p. 134, 1998.
[12] A. van der Ziel, *Noise in Solid State Devices and Circuits*, pp. 88-92, 1986.
[13] BSIM4: www-device.eecs.berkeley.edu
[14] C.H. Chen and M.J. Deen,
    Solid-State Elec., **42**, pp. 2069-2081, 1998.
[15] G. Knoblinger, *ESSDERC 2001*, pp. 331-334, 2001
[16] F.M. Klaassen and J. Prins,
    Philips Res. Repts., **22**, pp. 505-514, 1967.
[17] F.M. Klaassen, *IEEE TED*, **17**, pp. 858-862, 1971.
[18] R. van Langevelde *et al.*,
    to be presented at IEDM, Washington D.C., 2003.
[19] Documentation and source code of MOS Model 11,
    levels 1100, 1101 and 1102, can be found at:
    www.semiconductors.philips.com/Philips_Models

Figure 10: Drain current noise (**top**) and gate current noise (**bottom**) versus gate-source bias for *n*-channel devices with different channel length *L*. Symbols denote measurements, dashed lines denote Van der Ziel model, and solid lines denote MM1102 based on the improved KP-approach. ($f = 5$GHz and $V_{DS} = 1.0$V for $L \leq 0.18\mu$m, $f = 2.5$GHz and $V_{DS} = 1.8$V for $L \geq 0.5\mu$m)

fitting. The Van der Ziel model, on the other hand, leads to a considerable error in the gate noise. The measured real part of *c* is slightly negative due to non-quasi-static effects, which are not taken into account in the modelling. Noise results for various channel lengths *L* are shown in Figure 10. MM1102 gives accurate results over all bias conditions and channel lengths. Using the Van der Ziel model leads to an underestimation of gate noise (up to 40%), particularly for intermediate *L* where velocity saturation is important. For short-channel devices, the differences between the two models are obscured due to the dominant impact of gate resistance (see e.g. [18]).

5 SUMMARY

The newest update of MOS Model 11, level 1102 (MM1102), which is now under test, includes: *i*) an iterative solution of the surface potential; *ii*) an improved

Noise Modeling with HiSIM Based on Self-Consistent Surface-Potential Description

M. Miura-Mattausch, S. Hosokawa, D. Navarro, S. Matsumoto, H. Ueno, H. J. Mattausch
T. Ohguro*, T. Iizuka*, M. Taguchi*, T. Kage*, and S. Miyamoto*

Grad. School of Adv. Sc. of Matter, Hiroshima Univ.,
739–8526, Japan, mmm@hiroshima-u.ac.jp
* Semiconductor Technology Academic Research Center, Kanagawa, 222-0033, Japan

ABSTRACT

Accurate prediction of noise characteristics is a prerequisite for RF-circuit simulation. We demonstrate here that the $1/f$ noise is modeled only with the trap density as a model parameter and the thermal drain noise is determined only by the $I\text{-}V$ characteristics even for short-channel MOSFETs. Good agreement with measurements is the justification of the model implemented in HiSIM, the circuit simulation model based on a full surface-potential description. The thermal drain-noise coefficient γ of short-channel MOSFETs increases from 2/3 under the saturation condition. The origin is explained by the potential increase along the channel.

Keywords: MOSFET Model, Noise Measurement, Surface Potential, HiSIM

1 INTRODUCTION

As MOSFET size reduces, noise increase is becoming a serious problem for RF applications [1]. Fig. 1 shows the noise characteristic of a 100nm technology, where the $1/f$ noise and the thermal drain noise are observed, which originate from different physical mechanisms. However, measured low-frequency noise of recent advanced technologies shows no clear $1/f$ characteristics, and the thermal noise is hard to measure directly due to its relatively small magnitude and considerable required experimental effort [2-4]. Here we report that the deviation from the $1/f$ characteristics is due to the inhomogenuity of trap density in the oxide, and thus the averaged measured $1/f$ noise characteristics can be exploited for circuit simulation [5]. Only with the trap density as a model parameter, all measured $1/f$ noise characteristics for any gate lengths are successfully modeled. The thermal drain noise requires no fitting parameter, but inclusion of the potential distribution along the channel is essential for short-channel cases [6].

2 MODEL CONSISTENCY OF HiSIM

HiSIM, a MOSFET circuit-simulation model released as a free software [7], introduces two approximations: the charge-sheet approximation and the gradual-channel approximation. The approximations allow to derive analytical formulations for all device characteristics as a function of surface potentials at the source side ϕ_{S0} and the darin side ϕ_{SL} (for example [8]). These surface potentials are obtained by solving the Poisson equation iteratively. In spite of the iteration the calculation time is not longer than with BSIM3v3 [9]. It has to be emphasized that ϕ_{S0} and ϕ_{SL} are very sensitive to device parameters such as the bulk impurity concentration.

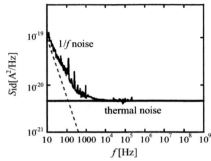

Fig. 1: Noise characteristics as a function of frequency f. Measured values are for L_g=0.11μm in a 100nm technology.

The gradual-channel approximation is only valid for the non-saturation region. Beyond ϕ_{SL} the potential increases steeply, forming the pinch-off condition. The model beyond the pinch-off point is depicted schematically in Fig. 2 [10]. This potential distribution is the origin of all phenomena, and the distribution is determined by device construction. HiSIM considers the complete potential distribution explicitly.

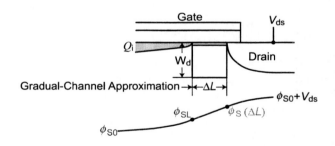

Fig. 2: Schematics depicting correlations among physical quantities in the pinch-off region.

Advanced MOSFET technologies undertake intensive channel engineering. The pocket implantation is one example. We assume a linearly reducing pocket profile along the channel as shown in Fig. 3 [11] to derive analytical formulations. Length of the extension into the channel (L_p) and maximum concentration N_{subp} are extracted from the measured threshold voltage (V_{th}) dependence on the gate length (L_g). Calculated V_{th} values are compared with measurement in Fig. 4. It has to be emphasized again that the impurity concentration is the model parameter, which influences all device characteristics such as the mobility through the surface potential values.

The low-field carrier mobility (μ) is described with three independent contributions [12]

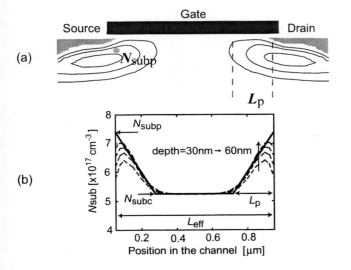

(a)

(b)

Fig. 3: (a) 2D-pocket profile obtained by the 2D-process simulator to reproduce measured V_{th}-L_g characteristics, and (b) its projection. L_{eff} is the channel length and N_{subc} is the substrate concentration.

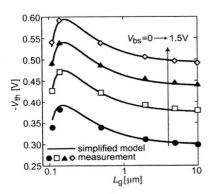

Fig. 4: Pocket implanted threshold voltage V_{th}-L_g characteristics. Calculated V_{th} with HiSIM in comparison to measurements.

$$\frac{1}{\mu} = \frac{1}{\mu_{Clmb}} + \frac{1}{\mu_{ph}} + \frac{1}{\mu_{sr}} \quad (1)$$

$$\mu_{Clmb} = MUECB0 + MUECB1\frac{Q_i}{q \times 10^{11}} \quad (2)$$

$$\mu_{ph} = \frac{MUEPH1}{(T/300K)^{MUETMP} \times E_{eff}^{MUEPH0}} \quad (3)$$

$$\mu_{sr} = \frac{MUESR1}{E_{eff}^{MUESR0}} \quad (4)$$

where μ_{Clmb}, μ_{ph}, and μ_{sr} are the mobility degradation due to the Coulomb scattering, the phonon scattering, and the surface roughness scattering, respectively. $MUECB0$, $MUECB1$, $MUEPH0$, $MUEPH1$, $MUESR0$, and $MUESR1$ are model parameters. The effective electric field E_{eff} is described as [13]

$$E_{eff} = \frac{1}{\epsilon_{Si}}(Q_b + \eta Q_i) \quad (5)$$

where η is $1/2$ for electrons and $1/3$ for holes for normal MOSFETs. $MUEPH0=0.3$ and $MUESR0=2.0$ are known as the mobility universality [14]. Preservation of the universality is the proof of correct calculation of the charges Q_b and Q_i [15]. Fig. 5 demonstrates an extracted low field mobility with HiSIM. The model-parameter set for HiSIM was determined by a conventional parameter extraction based on the I-V characteristics. HiSIM includes around 70 model parameters for describing all MOSFET characteristics [16].

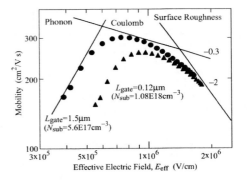

Fig. 5: Extracted low field mobility by HiSIM as a function of effective electric field.

3 1/f NOISE

Origin of the conventional $1/f$ noise in MOSFETs has been understood theoretically as the fluctuation in the number of channel carriers due to trapping/detrapping processes at the gate-oxide interface, as well as by the mobility fluctuation caused by scattering fluctuations due to the charges in the oxide (see for example [17]). Fig. 6 shows measured noise spectrum density (S_{id}) of 30 different chips on a wafer for $L_g = 0.46\mu m$ [5]. As the noise source, the intrinsic part of the device is still dominant and the extrinsic noise source such as diffusion regions generate negligible contribution. Additionally, we confirmed that the gate leakage current of the MOSFETs used for the measurement is negligibly small. The non-$1/f$ noise spectrum of 100nm MOSFETs is caused by the inhomogenuity of the trap density and energy distribution in the depth-direction of the gate oxide. By averaging over many noise spectra of identical devices on a wafer, the non-$1/f$ noise spectra reduce to a conventional $1/f$ noise spectra as shown by a thick line. Thus as a circuit-simulation model it is a subject to describe only this averaged $1/f$ noise characteristics with two boundaries as the worst and the best case. Fig. 7 summarizes the measured gate-voltage (V_{gs}) dependence of the obtained S_{id} for L_g=1.0μm, 0.46μm, and 0.12μm at $f = 100$Hz. All measured points are averaged values over 30 chips on a wafer. The final model equation based on the drift-diffusion approximation is

$$S_{I_{ds}} = \frac{I_{ds}^2 NFTRP}{\beta f L_{eff} W_{eff}} \left\{ \frac{1}{(N_0 + N^*)(N_L + N^*)} \right.$$
$$\left. + \frac{2NFALP \cdot v}{N_L - N_0} \log\left(\frac{N_L + N^*}{N_0 + N^*}\right) + (NFALP \cdot v)^2 \right\} \quad (6)$$

where L_{eff} and W_{eff} are channel length and width, and

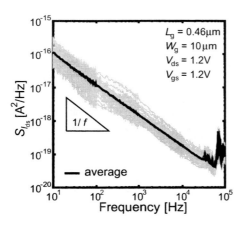

Fig. 6: Calculated $1/f$ noise characteristics in comparison with measurements. Only the trap density $NFTRP$ is the model parameter, valid for all gate lengths [5].

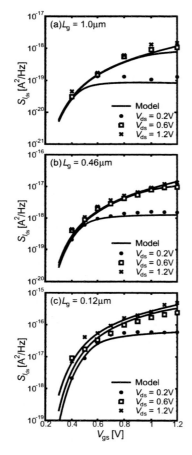

Fig. 7: Calculated $1/f$ noise characteristics in comparison with measurements. Only the trap density $NFTRP$ is the model parameter, valid for all gate lengths [5].

$NFTRP$, $NFALP$, and CIT are model parameters. N_0, N_L and v are carrier concentrations at the source side, the drain side, and carrier velocity, respectively. Due to the drift-diffusion approximation, a single equation is valid for any bias conditions. We have extracted that the contribution of the mobility fluctuation is small

and CIT can be fixed to zero. Thus a channel-length independent single model parameter $NFTRP$ is the only model parameter. A good agreement for any applied voltages and L_g proves consistency of the measurements as well as the model. Additionally we found that the extracted trap density values are quite similar for different technologies, if the technologies are mature.

4 THERMAL DRAIN NOISE

The Nyquist theorem describes the spectral density of the thermal drain-noise current at temperature T by integration along the channel direction y [18]

$$S_{id} = \frac{4kT}{L_{eff}} \int g_{ds}(y)dy = 4kTg_{ds0}\gamma \quad (7)$$

Here k, I_{ds}, $g_{ds}(y)$, g_{ds0}, γ are Boltzmann's constant, drain current, position-dependent channel conductance, channel conductance at $V_{ds} = 0$, and drain-noise coefficient, respectively. The Nyquist theorem predicts that γ decreases from 1 to $2/3$ as a function of V_{ds} for long-channel transistors. For short-channel transistors γ has been reported to be up to an order of magnitude larger [2]. Modeling and explanation of the measured γ-characteristic are based on different reasons; based on the hot electron, the channel length modulation, and also the velocity saturation [2-4]. However, we could not confirm these explanations. Instead we found that the distribution of device quantities along the channel is the major concern for the γ behavior.

Our theoretical approach is based on solving directly the integral along the channel direction in the Nyquist theorem by including the channel-position dependence of the concerning device quantities. Origin of all position dependent device quantities is the potential distribution in the channel. Since HiSIM treats the surface potential ϕ_s in the channel self-consistently as the measure, our integration is done with ϕ_s

$$S_{id} = \frac{4kT}{L_{eff}^2 I_{ds}} \int g_{ds}^2(\phi_s)d\phi_s \quad (8)$$

$$g_{ds}(\phi_s) = W_{eff}\frac{d(\mu(\phi_s)f(\phi_s))}{d\phi_s} \quad (9)$$

Here $f(\phi_s)$ is a characteristic function of HiSIM related to the carrier concentration [16]. The final equation for S_{id} and γ in our compact-modeling approach after solving the integral of Eq. (9) becomes a function of the self-consistent surface potentials as well as surface-potential derivatives at source and drain [6]. Fig. 8 shows the V_{ds}- and L_g-dependence of this S_{id}-model in comparison to measured results for a 100nm pocket-implant technology. The measurement is performed with the noise figure method. Fig. 9 shows the noise coefficient γ. The γ-increase for short L_g is much less drastic than reported in [2] but resembles the reported increase in [4]. γ-values in the linear region are very similar and show the same behavior for long as well as short L_g. The different behavior and the γ-increase for short L_g starts when V_{ds} enters the saturation condition. Under the condition the surface-potential increase becomes obvious. For shorter L_g even the γ-minimum at the end of the linear region becomes larger than $2/3$.

We have tested the apparent characteristic of the γ-V_{ds} relationship by another technology. As Fig. 9 shows, the same characteristic is indeed obtained.

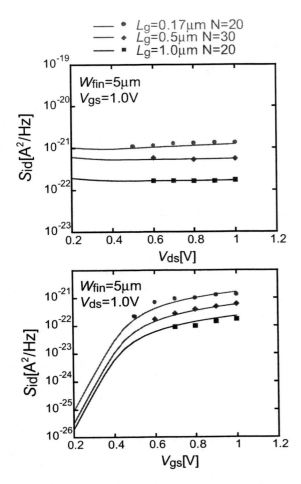

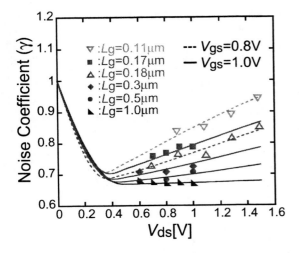

Fig. 8: Calculated thermal noise in comparison with measurements. For the calculation model parameters extracted only from measured *I-V* characteristics are used [6].

Fig. 9: Calculated thermal noise coefficients γ and measurements as a function of the drain voltage V_{ds} for various gate lengths L_g [6].

5 CONCLUSION

Both the $1/f$ and the thermal noise are successfully modeled based on the surface potential in HiSIM. The $1/f$ noise requires practically only the trap density as a L_g independent model parameter. The thermal noise requires no model parameter, and the increase of the noise coefficient γ for short-channel transistors is not so drastical as previously considered.

REFERENCES

[1] T. A. M. Kevenaar, E. J. W. ter Maten, Proc. of SIS-PAD, p.7, 1999.
[2] G. Knoblinger, P. Klein, M. Tiebout, IEEE J. Solid-State Circuits, **36**, No.5, p.831, (2001).
[3] M. Jamal Deen and C.-H. Chen, Proc. of MSM, p.694, (2002).
[4] A. J. Scholten, L. F. Tiemeijer, R. van Langevelde, R. J. Havens, V. C. Venezia, A. T. A. Zegers-van Duijnhoven, B. Neinhüs, C. Jungemann, and D. B. M. Klassen, Tech. Dig. IEDM, p.129, (2002).
[5] S. Matsumoto, H. Ueno, S. Hosokawa, T. Kitamaru, M. Miura-Mattausch, H. J. Mattausch, S. Kumashiro, K. Yamaguchi, K. Yamashita, and N. Nakayama, submitted for publication.
[6] S. Hosokawa, Y. Shiraga, H. Ueno, M. Miura-Mattausch, H. J. Mattausch, T. Ohguro, S. Kumashiro, M. Taguchi, H. Masuda, and S. Miyamoto, Ext. Abs. SSDM, p.20, 2003.
[7] http://www.starc.or.jp/kaihatu/pdgr/hisim/index.html
[8] M. Miura-Mattausch, U. Feldmann, A. Rahm, M. Bollu, and D. Savignac, IEEE Trans. CAD/ICAS, **15**, No.1 p.1, (1996).
[9] BSIM3v3 Manual, University of California, Berkeley, 1996.
[10] D. Navarro, T. Mizoguchi, M. Suetake, S. Ooshiro, K. Hisamitsu, H. Ueno, M. Miura-Mattausch, H. J. Mattausch, S. Kumashiro, K. Yamaguchi, S. Odanaka, and N. Nakayama, submitted for publication.
[11] H. Ueno, D. Kitamaru, K. Morikawa, M. Tanaka, M. Miura-Mattausch, H. J. Mattausch, S. Kumashiro, T. Yamaguchi, K. Yamashita, and N. Nakayama, IEEE Trans. Electron Devices, **49**, p.1783, (2002).
[12] T. Ando, A. B. Fowler, and F. Stern, Rev. Mod. Phys., **54**, p.437, (1982).
[13] Y. Matsumoto and Y. Uemura, Jpn. J. Appl. Phys., Suppl. **2**, p.367, (1974).
[14] A. G. Sabnis, and J. T. Clements, Tech. Dig. IEDM, p.18, (1979).
[15] S. Matsumoto, K. Hisamitsu, M. Tanaka, H. Ueno, M. Miura-Mattausch, H. J. Mattausch, S. Kumashiro, K. Yamaguchi, S. Odanaka, and N. Nakayama, J. Appl. Phys., **92**, No.9, p.5228 (2002).
[16] M. Miura-Mattausch, H. Ueno, H. J. Mattausch, K. Morikawa, S. Itoh, A. Kobayashi, and H. Masuda, IEICE Trans. Electron., **E86-C**, No.6, p.1009, (2003).
[17] K. K. Hung, P. K. Ko, C. Hu, and Y. C. Cheng, IEEE Trans. Electron Dev., **37**, p.1323, (1990).
[18] H. Nyquist, Phys. Rev., **32**, p.110 (1928).

The Development of the Next Generation BSIM for Sub-100nm Mixed-Signal Circuit Simulation

Xuemei (Jane) Xi, Jin He, Mohan Dunga, Chung-Hsun Lin, Babak Heydari, Hui Wan,
Mansun Chan, Ali M. Niknejad, Chenming Hu

Department of Electric Engineering and Computer Sciences, University of California at Berkeley
CA, 94720, janexi@eecs.berkeley.edu

ABSTRACT

This paper describes the next generation BSIM model for aggressively scaled CMOS technology. New features in the model include more accurate non-charge-sheet based physics, completely continuous current and derivatives, and extendibility to non-traditional CMOS based devices including SOI and double-gate MOSFETs.

Keywords: MOSFETs, compact modeling, surface-potential-plus model, small dimensional effects.

1 INTRODUCTION

Device models play a very important role in the advancement of CMOS technology and they appear ubiquitously from fabrication process development to IC design and manufacturing. The standardization procedure setup by EIA Compact Model Council (CMC)[1] helps the semiconductor industry to reduce the effort in evaluating models, and also provides directions for modeling development. Since the standardization of BSIM3V3[2], several high quality models like BSIM4 [3], SP2000 [4], HiSIM [5], EKV[6], ACM[7], USIM[8] models have emerged to meet the needs of advanced process and applications.

With the continuous scaling of CMOS technology following the ITRS roadmap, a number of the assumptions in the development of these traditional models become less valid, specifically, the charge-sheet approach ignoring the vertical carrier distribution in the inversion layer. In addition, a stronger interaction between design and modeling exists as a result of layout dependent characteristics that have been previously ignored. This is particularly true at RF and microwave frequencies where device performance such as f_{max} is a strong function of layout.

To further cope with the reduced time in the adoption of a new technology, there is need for a flexible model architecture to allow circuit simulation together with technology development, even before all physical phenomena are fully understood. To address the needs in modeling nano-CMOS, a new physical core and architecture is proposed for the next generation BSIM model. The model results in consistent, smooth and unified I-V and C-V equations that are critical in predicting distortion in analog and RF circuits. The non-charge-sheet formulation also makes the model easily extendible to non-classical CMOS devices like double-gate (DG) MOSFETs.

2 I-V MODEL FORMULATION

Accurate analytical modeling of devices in deep submicron CMOS technologies requires the physical incorporation of short-channel effects and poly-depletion, quantum effect, halo doping, retrograde doping, etc. Based on our previous work[9], SPP(Surface-Potential-Plus) is now extending to take these effects into account as well as consistent IV and CV models in quasi-2D Poisson equation solution. SPP introduces new concepts with the direct charge calculation in terms of terminal voltage, which are essential to the definition of normalized variables useful for hand calculation. By adopting a physical description of the inversion charge density and the particular structure of the model, with the derivation of continuous expressions for the MOS charges valid in all modes of operation, new deep-submicron CMOS technology effects can be simply integrated into this unified model framework, still keeping a small number of parameters. SPP core model chooses charge density rather than surface-potential as the state variable, on the other hand, maintains good computation efficiency and flexibility.

The band structure at the surface of a MOSFET is shown in Fig. 1 indicating the surface potential and the quasi-fermi potential. The relationship between charge, surface potential and quasi-fermi potential results in drift diffusion current being function of quasi-fermi potential:

$$I_{ds} = \mu_{eff} W Q_i \frac{dV_{ch}}{dy} \tag{1}$$

Following Pao-Sah's gradual channel approach and quasi-fermi potential, potential gradient equation can be derived:

$$\frac{dQ_i}{Q_i} = \frac{d\phi_s - dV_{ch}}{V_t} \tag{2}$$

And applying Gauss Law at Si-SiO₂ interface:

$$Q_i + Q_b = C_{ox}\left(V_{gb} - V_{fb} - \phi_s\right) \tag{3}$$

After eliminating the surface potential, inversion charge area density can be expressed as ($q_I = Q_I / V_t C_{ox}$, $v = V/V_t$):

$$\frac{dq_I}{q_I} + \frac{dq_I}{n_0} = \frac{dv_{GB}}{n_0} - dv_{ch} \tag{4a}$$

and

$$\ln q_I + \frac{q_I}{n_0} = \frac{v_{GB} - v_{FB}}{n_0} - \phi_B - v_{ch} \tag{4b}$$

Integrating (1) from source to drain we can get the drain current:

$$I_{ds0} = \frac{W\mu_{eff}}{L_{eff}}\left[\frac{Q_s^2 - Q_d^2}{2n_0 C_{ox}} + V_t(Q_s - Q_d)\right] \quad (5)$$

Note that the drain current is a result of the superposition of the drift current (the Q^2 terms) and the diffusion current (the Q terms) as shown in Fig. 2.

Short channel effects are incorporated from the quasi-2D solution to Poisson equation yielding:

$$[n_0 + 2f]\left[\frac{dq_1}{q_1} + dv_{ch}\right] + dq_1 = dv_{GB} + d(v_{DB} + v_{SB} + 2v_{bi})f \quad (6a)$$

$$\ln q_1 + \frac{q_1}{n} = \frac{v_{GB} - v_{FB}}{n} + \frac{(v_{DB} + v_{SB} + 2v_{bi})f}{n} - (\phi_B + v_{ch}) \quad (6b)$$

Where: $n = n_0 + 2f$ and $f = \frac{1}{L}\left[1 - \exp\left(-\frac{L}{l}\right)\right]$

$$n_0 = 1 + \frac{C_d}{C_{ox}} = 1 + q_B'(\phi_s) = 1 + \frac{r}{2\sqrt{\phi_s + v_t \exp(\phi_s - 2\phi_f - v_{ch})}}$$

which can be approximated by

$$n_0 \approx 1 + \frac{r}{2\sqrt{\phi_{sa} + v_t \exp(\phi_{sa} - 2\phi_f - v_{ch})}}$$

where $\phi_{sa} = \left(\sqrt{VGB - VFB + \frac{r^2}{4}} - \frac{r}{2}\right)^2$

The source and drain charge can be expressed explicitly in terms of the W-Lambert equation.

Poly-depletion effects and quantum effects can be easily handled by using the n-factor and ϕ_B correction.

$$n = 1 + \frac{C_d}{C_{ox}} + \frac{N_{sub}}{N_{gate}} + \lambda_q \phi_{sa}^{-1/3} \quad (7a)$$

$$\phi_B = \phi_{B0}\left(1 + \frac{N_{sub}}{N_{gate}}\right) + \lambda_q \phi_{s0}^{2/3} \quad (7b)$$

The characteristics of the channel charge distribution including polysilicon depletion and quantum effect are shown in Fig. 3, indicating good agreement with 2-D and quantum numerical simulation results.

Velocity saturation, velocity overshoot and ballistic transports are handled in a unified way using the saturation charge concept with the expression:

$$Q_{dsat} = \frac{Q_s}{1 + C_{ox}E_c L/(Q_s + 2C_{ox}V_t/n)} \quad (8)$$

The saturation drain current is given by the Ids0 expression with Q_d replaced by Q_{dsat}. With velocity overshoot based on the Price's hydrodynamic model (HD) and source-end velocity limited ballistic transport (BT), the unified drain current expression become:

$$I_{ds} = \frac{I_{ds0}}{\left[1 + \frac{Q_s - Q_d}{C_{ox}L_{eff}E_{sat}^{OV}}\right]\left[1 + (v_{sHD}/v_{sBT})^{2\xi}\right]^{1/2\xi}} \quad (9)$$

where v_{sHD} is the velocity from the hydrodynamic model. The model has also been extensively verified by experimental data with gate lengths down to 0.125mm and some of the results are shown in Fig. 4. The physical core also gives correct behavior for analog sensitive parameters such as gm/Id as shown in Fig. 5 with C-∞ continuity facilitating higher order harmonic analysis.

3 CHARGE AND C-V MODEL FORMULATION

The C-V equations can be directly derived from the charges calculated in the previous section resulting in a consistent I-V and C-V model with all major physical effects including poly-depletion, quantum effect, SCE, retrograde doping implemented. The C-V model is symmetric at V_{DS}=0V. The charge equation for current continuity is given by:

$$Q_i(y) = C_{ox}V_t\left(\sqrt{\left(\frac{Q_s}{C_{ox}V_t} + n\right)^2 - \frac{y}{L}\left(\frac{Q_s}{C_{ox}V_t} + n\right)^2 - \left(\frac{Q_d}{C_{ox}V_t} + n\right)^2} - n\right) \quad (10)$$

The source charge and drain charge are then given by:

$$Q_s = W\int_0^L\left[1 - \frac{x}{L}\right]Q_i(x)dx \text{ and } Q_d = W\int_0^L\frac{x}{L}q(x)dx$$

The capacitances can then calculated directly by differentiating the charges:

$$C_{XY} = \delta \cdot \frac{\partial Q_X}{\partial V_Y} = \delta \cdot \left[\frac{\partial Q_X}{\partial Q_d}\frac{\partial Q_d}{\partial V_Y} + \frac{\partial Q_X}{\partial Q_s}\frac{\partial Q_s}{\partial V_Y}\right] \quad (11)$$

with δ =1 for X=Y and −1 otherwise. The behaviors of the terminal charges and capacitances are shown in Fig. 6, together with the result from a quantum simulator.

4 CONCLUSION

In this paper, the formulation of a modular BSIM model is described. The model exhibits smooth derivaties and is thus well suited for analog and RF applications. It also utilizes a new modular architecture with many physical effects decoupled. Advance features including layout dependent mobility, consistent non-quasi-static effects can also be incorporated. The new approach is expected to be able to accelerate the adoption of the rapid evolving CMOS technology and help to develop models together with the technology progress.

REFERENCES

[1] CMC, http://www.eigroup.org/cmc/.
[2] Y. Cheng, etal., IEEE Trans. on Electron Devices , Vol. 44, No. 2, pp. 277-287, February 1997.
[3] BSIM,http://www.device.eecs.berkeley.edu/~bsim3.
[4] G. Gildenblat, etal., pp233-240, CICC2003.
[5] M. Miura-Mattausch, etal., pp258-261, WCM2003
[6] M. Bucher, etal, pp. 670 - 673 WCM2002.
[7] C. Galup-Montoro, etal., pp. 254-257, WCM2003.
[8]H. Gummel and K.Singhal. IEEE Trans on Electron Devices, vol. 48, 1585-1593, 2001.
[9] J. He, etal, pp.262-265, WCM2003

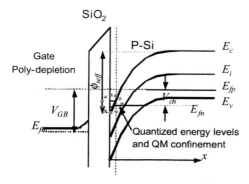

Fig. 1: The detail band diagram of a MOSFETs under strong inversion, that used to formulation equation (1)-(3)

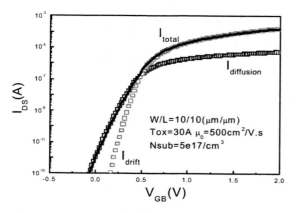

Fig. 2: Verification of drift current in strong inversion and diffusion current in subthreshold region as predicted by the new core model

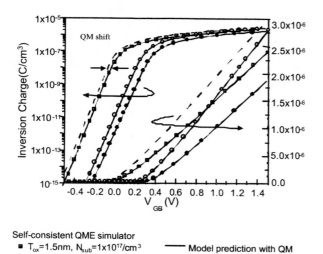

Self-consistent QME simulator
- ■ T_{ox}=1.5nm, N_{sub}=1x10^{17}/cm^3 — Model prediction with QM
- ● T_{ox}=1.5nm, N_{sub}=1x10^{18}/cm^3 -- Model prediction without QM
- ○ T_{ox}=1.0nm, N_{sub}=1x10^{18}/cm^3

Fig. 3: Inversion charge concentration with quantum mechanical effects included as predicted by the model and comparison with self-consistent quantum mechanical simulator

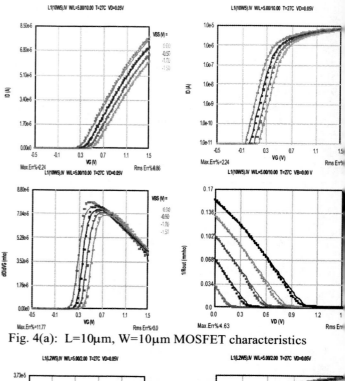

Fig. 4(a): L=10μm, W=10μm MOSFET characteristics

Fig. 4(b): L=2μm, W=10μm MOSFET characteristics

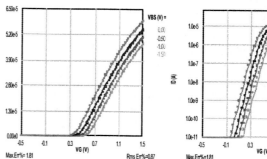

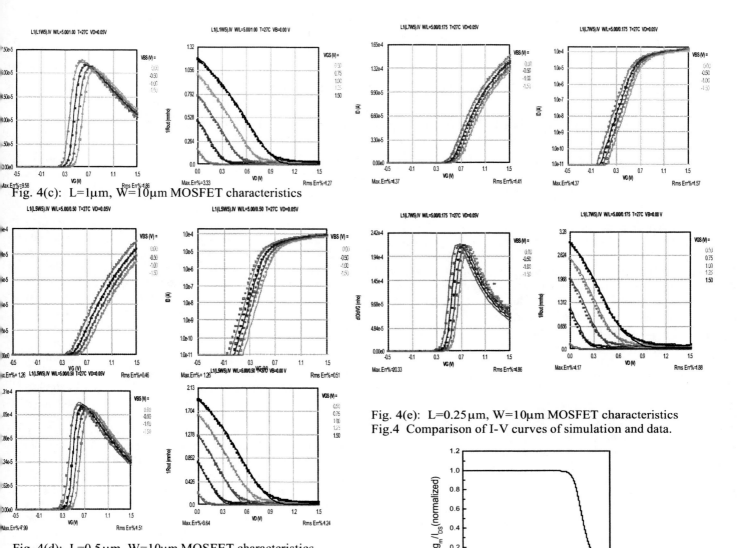

Fig. 4(c): L=1μm, W=10μm MOSFET characteristics

Fig. 4(d): L=0.5μm, W=10μm MOSFET characteristics

Fig. 4(e): L=0.25μm, W=10μm MOSFET characteristics
Fig.4 Comparison of I-V curves of simulation and data.

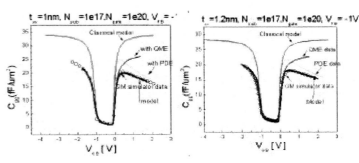

Fig. 5: The behavior of the normalized gm/IDS plot for a long channel device

Fig.6 Comparison of Gds curves of simulation and data.

Fig.4(e): L=0.25um, W=10um MOSFET characteristics

Unified Regional Approach to Consistent and Symmetric DC/AC Modeling of Deep-Submicron MOSFETs

Xing Zhou[*], Siau Ben Chiah[*], Karthik Chandrasekaran[*],
Khee Yong Lim[**], Lap Chan[**], and Sanford Chu[**]

[*]School of Electrical & Electronic Engineering, Nanyang Technological University
Nanyang Avenue, Singapore 639798, exzhou@ntu.edu.sg
[**]Chartered Semiconductor Manufacturing Ltd, 60 Woodlands Industrial Park D, St. 2, Singapore 738406

ABSTRACT

This paper presents our new developments of Xsim, a unified regional threshold-voltage-based model for deep-submicron MOSFETs. New features include complete reformulation with bulk reference, including transverse electric field for effective mobility resulting in source–drain symmetry, charge-based AC model fully consistent with DC without the need for $C–V$ data, and inclusion of poly-depletion effect for both DC and AC models.

Keywords: Compact model, symmetry, charge-based model, deep-submicron MOSFET, poly depletion, Xsim.

1 INTRODUCTION

MOSFET compact models (CMs), which have played a central role in the chip design and fabrication, have been "spiced" up recently as evidenced by the collection of feature articles appeared in the *IEEE Circuits & Devices Magazine* in the recent years [1]. This reflects the call for new CM development to keep up with the next generation CMOS technologies as a result of technology advancement being always ahead of model development. The importance of CMs can never be over-emphasized since it is the cornerstone for bridging the design and the given technology. The challenge in developing next generation CMs lies in the increasing demand in the tradeoff in model accuracy and simplicity, as well as model scalability and predictability for new technology generations.

Regional formulations of MOSFET (charge and current) equations have been used for circuit simulation in the past decades, starting with the Meyer's model [2] through generations of BSIM models. One of the factors of their popularity and success is due to the physically-based regional approach, which is flexible and relatively easy to add new features to keep up with the technology advancements. On the other hand, keep adding fitting parameters to the model (including "binning") makes it more and more empirical and unmanageable. Together with the "intrinsic" problems in model symmetry (due to source referencing) and continuity (due to use of smoothing functions), these inadequacies have become "showstoppers for design." In the mean time, surface-potential and conductance-based models are becoming popular in CM

development [3]. From the history of MOSFET models, surface-potential-based (Ψ_s-based) formulation [4] was developed in 1966, much earlier than charge-based models in 1971 [2]. The fact that an earlier invented (surface-potential-based) model only becomes popular *after* the later developed (charge-based) model facing limitations tells that the charge-based (unified regional) approach has some unique advantages, which can (and should) be extended provided its shortcomings can be overcome. One of the major advantages of the unified regional approach is threshold-voltage (V_t) based (as opposed to Ψ_s-based), in which the model can be characterized to a given technology of varying geometry down to the deep V_t roll-off regime [5]. This aspect of model property (scalability) is becoming equally (if not more) important as others (such as accuracy and continuity over bias) for "scaled" deep-submicron MOSFETs, since geometry variation (in gate length, L_g) has an equivalent effect to bias variation (in V_{ds}) for the effective lateral field, V_{ds}/L_g. This will be traded off with accuracy in the moderate-inversion region due to "pinned" surface potential at strong inversion as well as the use of smoothing functions. However, as will be shown in this paper, with physical modeling of the respective linear/saturation and subthreshold/strong-inversion characteristics of charge and current as well as good smoothing functions, a unified V_t-based regional model can be symmetric, consistent in DC and AC, and continuous in higher-order derivatives.

This paper summarizes the recent development of our unified regional model, Xsim [5], which has been re-derived with bulk as the reference in the drain-current and threshold-voltage formulations, together with a consistent charge model for intrinsic capacitances, as well as inclusion of poly-depletion effect. The new model demonstrates Gummel symmetry while maintaining a small number of fitting parameters that requires a *one-iteration* extraction procedure using minimum measured $I–V$ data (without the need for $C–V$ measurement).

2 MODEL DESCRIPTION

Our source-referenced V_t-based model [5] has shown continuity in third-order derivatives and improved symmetry and scalabe over entire geometry. In this section, we highlight new development in bulk-referenced regional charge models combining our developed smoothing/interpolation functions for consistent current/capacitance modeling. The DC/AC model has also been extended to include poly-depletion effect derived from bulk-referenced charge model and expressed as a simple modification in the flat-band voltage and bulk-charge factor.

2.1 Symmetric DC Model

One source of MOSFET model asymmetry is due to source referencing as a result of bulk-charge linearization [6]. This can be avoided if I_{ds} is derived from charge-sheet model with bulk as the reference [7], which results in the same I_{ds} equation as source-referenced model but different expressions for the threshold voltage

$$V_t = V_{t0} - (A_b - 1)V_{bs} \tag{1a}$$

$$V_{t0} = V_{FB} + \phi_s + \gamma_{eff}\sqrt{\phi_{s0}} \tag{1b}$$

and the bulk-charge factor

$$A_b = 1 + \gamma/(2\sqrt{\phi_s}) \tag{1c}$$

where $\phi_s = 2\phi_F - \Delta\phi_s$ is the (pinned) surface potential at strong inversion ($\Delta\phi_s$ is the barrier lowering for short-channel devices [8], charge-sharing and drain-induced barrier lowering are contained in γ_{eff} [8]). These are well-established regional models, which are readily applicable to combine with our developed smoothing/interpolation functions for the unified model.

Another source of asymmetry is in the effective transverse field (E_{eff}) used in the effective mobility model. If the drain-bias effect is ignored, as proposed in the universal mobility model [9], asymmetry will occur (the same is true for Ψ_s-based model if low-field approximation is assumed). In our new model, E_{eff} is derived from the average transverse field based on the bulk-referenced inversion and bulk charges, given by

$$E_{eff} = \frac{C_{ox}}{2\varepsilon_{Si}}\left[\left(V_{geff} - \frac{A_b V_{deff}}{2}\right) + 2\chi_b\gamma\sqrt{\phi_s}\left(1 + \frac{V_{deff} - 2V_{bs}}{4\phi_s}\right)\right] \tag{2}$$

for nMOSFETs, where V_{geff} and V_{deff} are the interpolation/smoothing functions [8], and χ_b is a fitting parameter.

A third improvement in symmetry is to introduce a "δ_0 factor" for velocity saturation in the lateral-field effective mobility

$$\mu_{eff} = \frac{\mu_0}{1 + \delta_0 V_{deff}/(L_{eff}E_{sat})} \tag{3}$$

in which δ_0 has an empirical form: $\delta_0 = \delta_L(V_{deff}/V_{dsat})$, where V_{dsat} is the saturation voltage, V_{deff} is the smoothing function between linear and saturation regions, and δ_L is a fitting parameter.

2.2 Consistent AC Model

One of the major problems with the source-referenced regional charge model is asymmetry in intrinsic capacitances ($C_{gs} \neq C_{gd}$ and $C_{bs} \neq C_{bd}$ at $V_{ds} = 0$). However, if terminal charges are derived with bulk reference, there is no reason why regional model must suffer asymmetry. When we first start to extend our Xsim DC model to AC (capacitances), we take note of symmetry at the very beginning. Since the well-established charge models [10] in strong inversion based on the Ward and Dutton partition [11] are consistent with the corresponding current model, with the bulk-referenced expressions, we simply used the same smoothing functions (V_{geff}, V_{deff}) as well as similar ideas to the effective gate/drain voltage product (V_{gg}) in our I_{ds} model to come up with unified source/drain terminal-charge expressions such that, in the linear strong-inversion limit ($V_{geff} \rightarrow V_{gs} - V_t$, $V_{deff} \rightarrow V_{ds}$) and subthreshold ($V_{gs} \ll V_t$) region, they approach the well-known regional expressions [10]. The final derived Q_s and Q_d equations are summarized below:

$$Q_s = 0.5C_{ox}(V_{gc,s} + A_s) \tag{4a}$$

$$Q_d = 0.5C_{ox}(V_{gc,d} + A_d) \tag{4b}$$

where

$$V_{gc,s} = \frac{2nv_{th}\ln\left(1 + e^{(V_{gs}-V_t)/(2nv_{th})}\right)V_{dc,s}}{1 + \frac{6nV_{dc,s}(C_{ox}/C_d)e^{-(V_{gs}-V_t-2V_{off})/(2nv_{th})}}{v_{th}(2 + e^{-V_{ds}/v_{th}})}} \tag{5a}$$

$$V_{gc,d} = \frac{2nv_{th}\ln\left(1 + e^{(V_{gs}-V_t)/(2nv_{th})}\right)V_{dc,d}}{1 + \frac{6nV_{dc,d}(C_{ox}/C_d)e^{-(V_{gs}-V_t-2V_{off})/(2nv_{th})}}{v_{th}(1 + 2e^{-V_{ds}/v_{th}})}} \tag{5b}$$

$$V_{dc,s} = 1 - \frac{A_b V_{deff}}{3V_{geff}} \tag{6a}$$

$$V_{dc,d} = 1 - \frac{2A_b V_{deff}}{3V_{geff}} \tag{6b}$$

with the unified auxiliary functions A_d and A_s

$$A_s = 2\mathcal{A}(1 - \mathcal{B}) \tag{7a}$$

$$A_d = 2\mathcal{A}\mathcal{B} \tag{7b}$$

$$\mathcal{A} = \frac{A_b^2 V_{deff}^2}{12(V_{geff} - \frac{1}{2}A_b V_{deff})} \tag{8a}$$

$$\mathcal{B} = \frac{5V_{geff} - 2A_b V_{deff}}{10(V_{geff} - \frac{1}{2}A_b V_{deff})}. \tag{8b}$$

The above equations are fully consistent with $Q_i = Q_s + Q_d$ used in the I_{ds} equation [12].

The final unified bulk charge is expressed as

$$Q_b = C_{ox}V_{teff} \tag{9}$$

with a smoothing function (and fixed smoothing parameter $\delta_b = 0.0001$)

$$V_{teff} = V_{tstr} - \frac{1}{2}\left[V_{tstr} - V_{tsub} - \delta_b + \sqrt{(V_{tstr} - V_{tsub} - \delta_b)^2 + 4\delta_b V_{tstr}}\right] \tag{10}$$

joining strong-inversion and subthreshold regions, which approaches

$$V_{tstr} = \left[\gamma_{eff} \sqrt{\phi_{s0}} - \zeta_b (A_b - 1) V_{bs} \right] + \zeta_b (A_b - 1) V_{deff} \mathcal{D} \quad (11)$$

in strong inversion, where ζ_b is a fitting parameter and $\mathcal{D}$ is an auxiliary function

$$\mathcal{D} = \frac{3V_{geff} - 2A_b V_{deff}}{6 \left(V_{geff} - \frac{1}{2} A_b V_{deff} \right)}, \quad (12)$$

and approaches

$$V_{tsub} = \gamma \sqrt{\phi_{sub}} \quad (13)$$

in subthreshold with a gate-voltage-dependent surface potential ϕ_{sub} [12].

2.3 Poly-Depletion Model

When poly-depletion effect is considered, the inversion charge along the channel is given by [13]

$$Q_i(y) = C_{ox} \left[V_{gb} - V_{FB} - \Psi_p(y) - \Psi_s(y) \right] - Q_b(y) \quad (14)$$

in a bulk-referenced model, where the potential drop across the poly-depletion region can be derived, similar to [13]

$$\Psi_p(y) = V_{gb} - V_{FB} - \Psi_s(y)$$
$$- A_v \left[\sqrt{1 + \frac{2}{A_v} \left[V_{gb} - V_{FB} - \Psi_s(y) \right]} - 1 \right] \quad (15)$$

or re-written, similar to [14], as

$$\Psi_p(y) = A_v \left(\frac{b+1}{2} - \sqrt{b} \right) + K_v \Psi_s(y) \quad (16a)$$

$$b = 1 + \frac{2}{A_v} \left(V_{gb} - V_{FB} \right) \quad (16b)$$

$$A_v = \frac{2\varepsilon_{Si} N_{gate}}{C_{ox}^2} \quad (16c)$$

$$K_v = \frac{\theta_p}{\sqrt{b}} - 1 \quad (16d)$$

where N_{gate} is the poly-gate doping concentration and θ_p is a fitting parameter due to Taylor's series expansion from (15). Substituting (16a) into (14) and integrating across the channel from source ($y = 0$) to drain ($y = L_{eff}$), with the boundary conditions $\Psi_{s0} = 2\phi_F + V_{sb}$ and $\Psi_{sL} = 2\phi_F + V_{db}$, it can be shown that the drain current can be expressed in the same conventional form:

$$I_{ds} = \mu_{eff} C_{ox} \frac{W_{eff}}{L_{eff}} \left[\left(V_{gs} - V_t \right) V_{ds} - \frac{1}{2} A_b V_{ds}^2 \right], \quad (17)$$

with a modification of the flat-band voltage

$$V_{t0} = V_{fb} + \phi_s + \gamma_{eff} \sqrt{\phi_{s0}} \quad (18a)$$

$$V_{fb} = V_{FB} + \Delta V_p \quad (18b)$$

$$V_{FB} = \phi_M - \left(\chi + \frac{1}{2} E_g + \frac{1}{2} \phi_{s0} \right) - q N_{ss} / C_{ox} \quad (18c)$$

and bulk-charge factor

$$A_b = 1 + \gamma / \left(2\sqrt{\phi_s} \right) + K_v . \quad (19)$$

ΔV_p in (18b) represents a "flat-band voltage shift" as a result of poly-depletion effect, given by

$$\Delta V_p = A_v \left[(b+1)/2 - \sqrt{b} \right] - K_v \left(\phi_s - V_{bs} \right), \quad (20)$$

which is gate-bias dependent. The above modifications in V_{fb} and A_b, although derived from integration of inversion charge for the drain current, are also extended to the charge models in Sec. 2.2, which provides consistent modeling of the poly-depletion effect in the intrinsic capacitances.

As our regional drain-current model is V_t based (pinned surface potential in strong inversion), similar idea is applied to ΔV_p in strong inversion: ΔV_p is "pinned" to the value evaluated at $V_{gs} = V_t$: $\Delta V_{p,th} = \Delta V_p |_{V_{gs} = V_t}$, for which a closed-form expression can be solved. This $\Delta V_{p,th}$ is used in the unified drain-current model.

3 RESULTS AND DISCUSSION

The new bulk-referenced DC/AC model has been validated with measurement [12] and Medici numerical device data of a 90-nm technology, as will be shown in this section. Model calibration (parameter extraction) follows the similar one-iteration procedure as developed before [5]. V_t model is first calibrated to the $V_t - L_g$ data extracted from $I_{ds} - V_{gs}$ curves for high poly-gate doping ($N_{gate} = 10^{22}$ cm^{-3}) when poly-depletion effect can be ignored (in principle, V_t data can be obtained from a given technology ET data without $I-V$ measurements). The calibrated $V_t - L_g$ curves are shown in Fig. 1.

To calibrate the model for poly-depletion effect, the flat-band voltage [or workfunction, ϕ_M in (18c)] together with the long-channel doping (N_{ch}) is calibrated to the long-channel ($L_g = 20$ μm) $V_t - V_{bs}$ data (at low V_{ds}) for different poly-gate doping N_{gate} (same as Medici values). The fitted results are shown in Fig. 2(a), and the *predicted* results at short channel ($L_g = 90$ nm) are shown in Fig. 2(b).

After V_t calibration, only **two** "corner" devices ($L_g = 20$ μm and 90 nm; $N_{gate} = 10^{22}$ cm^{-3}) $I_{ds} - V_{gs}$ (at high/low V_{ds} and V_{bs}) are used for I_{ds} model calibration to extract the mobility, bulk charge, series resistance, and subthreshold parameters [5], and $I_{ds} - V_{ds}$ (at high V_{gs} and low V_{bs}) data for **each** gate-length device are used to extract the smoothing parameter (δ_s) in V_{deff}. For each poly-gate device, **one** additional $I_{ds} - V_{gs}$ (at low V_{ds} and high V_{bs}) data from the long-channel (20-μm) device is used to determine the optimum value of θ_p in (16d). **No other device I–V and C–V data have been used in parameter extraction**.

The fitted $I_{ds} - V_{ds}$ curves for long- and short-channel devices are shown in Fig. 3, together with the second-order derivatives (g_{ds}'), which follow the behavior of the data with continuity across $V_{ds} = 0$. While g_{ds}' does not reflect true model symmetry, Gummel symmetry test can be verified as shown in Fig. 4 for the 90-nm (and 20-μm [12]) device. Model symmetry in intrinsic capacitances is shown in Fig. 5, where the capacitances are obtained by the respective partial derivatives of the unified charge equations consistently calibrated to the drain-current data.

The *predicted* capacitances, transconductances, output conductances, and drain currents at various poly-gate doping concentrations are shown in Figs. 6–9. Excellent prediction over the entire range of bias, geometry, and poly doping verifies the physics built into the model. Fig. 9 further demonstrates the model's ability in "extrapolating" to deeply-scaled (in the V_t roll-off region) device, which uses only one $I_{ds} - V_{ds}$ plus the V_t data for the 65-nm device.

4 CONCLUSION

In summary, we have demonstrated that a regional V_t-based model can be symmetric as long as the drain-current and charge equations are formulated based on bulk reference, even though the threshold voltage is source-extrapolated. Symmetric and bias-dependent intrinsic capacitances can be obtained from the unified regional charge equations consistent with the current model without the need for C–V data. Poly-depletion effect is modeled through a modification in the (long-channel) flat-band voltage and bulk-charge factor, which can be extended to all geometry/bias conditions for both DC and AC models. Compared with popular Ψ_s-based models, the regional V_t-based model may not be as accurate in the moderate-inversion region due to the use of smoothing functions. However, with technology characterization of the V_t model, the unified regional approach has the advantage of modeling deeply-scaled (V_t roll-off) devices, a region most Ψ_s-based models do not (or cannot) model accurately. With the small number of parameters and minimum data requirement, together with the one-iteration parameter extraction for the model, our approach to modeling deep-submicron MOSFETs will prove to have a major contribution in predicting next-generation devices and technologies. The significance of our model development lies in providing a choice for next-generation MOSFET models with scalability, symmetry, accuracy, continuity, simplicity, as well as familiarity to the existing modeling infrastructure for circuit simulation and design communities.

REFERENCES

[1] http://www.ntu.edu.sg/home/exzhou/WCM/link.htm# Article

[2] J. E. Meyer, *RCA Rev.*, vol. 32, pp. 42–63, 1971.

[3] M. Chan, *et al.*, *Microelectronics Reliability*, vol. 43, no. 3, pp. 399–404, 2003.

[4] H. C. Pao and C. T. Sah, *Solid-State Electron.*, vol. 9, pp. 927–937, 1966.

[5] X. Zhou, S. B. Chiah, and K. Y. Lim, *Proc. Nanotech 2003*, Feb. 2003, vol. 2, pp. 266–269.

[6] T.-L. Chen and G. Gildenblat, *Electron. Lett.*, vol. 37, pp. 791–793, 2001.

[7] Y. Tsividis, *Operation and Modeling of the MOS Transistor*, McGraw-Hill, 2nd ed., pp. 179–181, 1999.

[8] X. Zhou and K. Y. Lim, *IEEE Trans. Electron Devices*, vol. 48, no. 5, pp. 887–896, 2001.

[9] A. G. Sabnis and J. T. Clemens, *1979 IEDM Tech. Dig.*, pp. 18–21.

[10] N. Arora, *MOSFET Models for VLSI Circuit Simulation—Theory and Practice*, Wien, New York: Springer-Verlag, 1993.

[11] D. E. Ward and R. W. Dutton, *IEEE J. Solid-State Circuits,* vol. SC-13, p. 703–707, Oct. 1978.

[12] S. B. Chiah, X. Zhou, K. Chandrasekaran, K. Y. Lim, L. Chan, and S. Chu, "Threshold-voltage-based regional modeling of MOSFETs with symmetry and continuity," *Proc. Nanotech 2004*, Mar. 2004.

[13] N. D. Arora, R. Rios, and C.-L. Huang, *IEEE Trans. Electron Devices*, vol. 42, no. 5, pp. 935–943, 1995.

[14] R. Rios, N. D. Arora, and C.-L. Huang, *IEEE Electron Device Lett.*, vol. 15, no. 4, pp. 129–131, 1994.

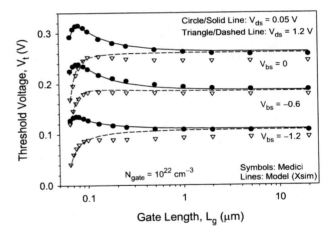

Figure 1: Threshold voltage vs. gate length at high poly-gate doping used to calibrate the V_t model.

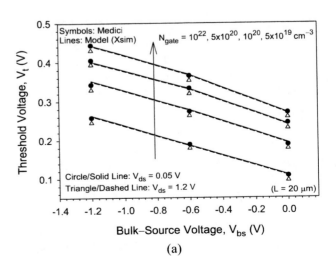

(a)

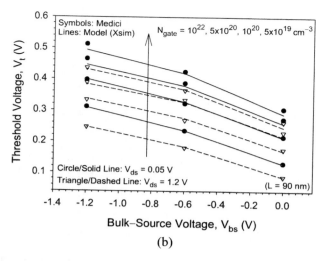

(b)

Figure 2: Threshold voltage vs. bulk voltage for (a) long-channel (linear V_t) used to calibrate the flat-band and channel doping for poly-depletion effect, and (b) short-channel showing the model prediction.

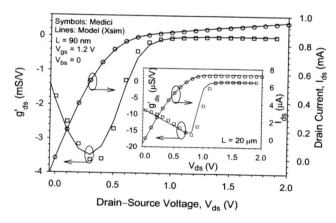

Figure 3: Drain current and its second derivative vs. drain voltage for the short- and long-channel (inset) devices.

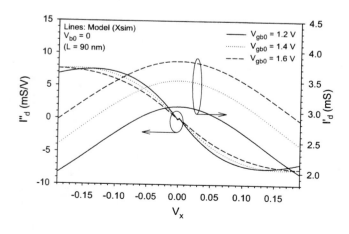

Figure 4: Gummel symmetry test for the 90-nm device.

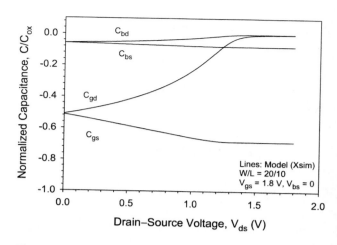

Figure 5: Intrinsic capacitances vs. drain voltage, showing model symmetry ($C_{gs} = C_{gd}$ and $C_{bs} = C_{bd}$ at $V_{ds} = 0$).

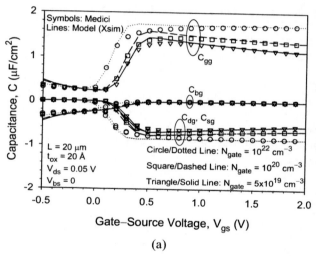

(a)

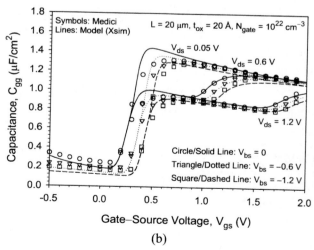

(b)

Figure 6: Predicted intrinsic capacitances vs. gate voltage for (a) various poly-gate dopings at low V_{ds} and V_{bs}, and (b) various V_{ds} and V_{bs} at high poly-gate doping.

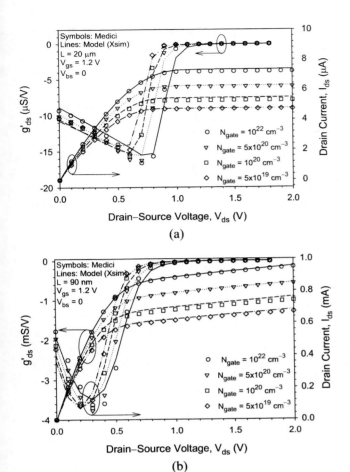

(a)

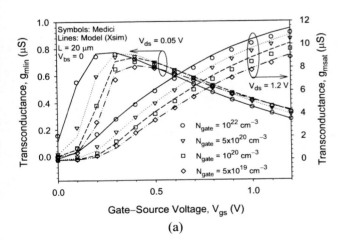

(b)

Figure 7: Predicted drain current and its second derivative vs. drain voltage at various poly-gate dopings for (a) long and (b) short channel devices.

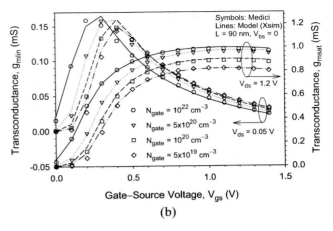

(a)

Figure 8: Predicted linear and saturation transconductance vs. gate voltage at various poly-gate dopings for (a) long and (b) short channel devices.

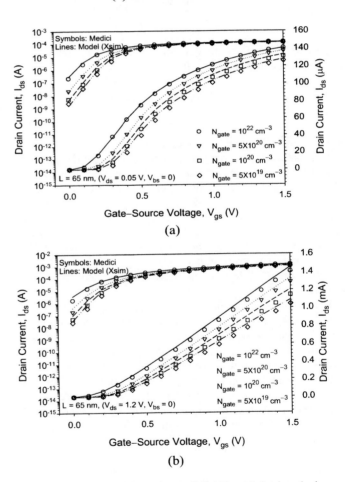

(a)

(b)

Figure 9: Predicted deeply-scaled (65-nm) device drain current vs. gate voltage at various poly-gate dopings in (a) linear and (b) saturation regions of operation.

Modeling and Characterization of On-Chip Inductance for High Speed VLSI Design

Narain D. Arora and Li Song

Cadence Design Systems, Inc., San Jose, CA, 95134, USA

ABSTRACT

Ever increasing circuit density, operating speed, faster on-chip rise times, use of low resistance Copper (Cu) interconnects, and longer wire lengths due to high level of integration in VLSI chip designs, have necessitated the need for modeling of wire inductive (L) effects which were ignored in the past. In this paper we will review different approaches of modeling the on-chip wire inductance, and discuss practical methods of assessing the inductance with special reference to return path in an IC chip. This will be followed by discussion on impact of inductance on performance of high speed VLIS. We then cover methods of validating the models using test chip approach.

Keywords: On-chip inductance, inductance modeling, inductance characterization, inductance test structures.

1 INTRODUCTION

As operational frequency of IC chips with copper (Cu) interconnects increases towards multi-gigahertz (GHz) range, the on-chip inductance effects can no longer be ignored for the state of the art VLSI design. The inductive effect can cause ringing and overshoot problems in clock lines, and reflections of signals due to impedance mismatch. In addition, the switching noise due to inductive voltage drops is an issue for the power distribution network. Thus it is important to model inductive effects accurately for high speed VLSI design.

Inductance of a wire is more complicated to model compared to resistance or capacitance because the inductance describes the magnetic flux being generated by current flowing in a loop. In a VLSI chip, calculation of wire inductance will therefore require knowledge of the current return path. However often the return path is not easily identified as it is not necessarily through the silicon substrate. Calculation is further complicated by the fact that not all the return currents follow DC paths as some are in the form of displacement current through the interconnect capacitances. Furthermore, unlike capacitive coupling, inductive coupling has long range effects because the magnetic field strength decreases much slowly compared to the electric field strength. As such localized windowing (nearest neighbor approximation) commonly used for capacitance calculation may not be valid for inductance calculation [1].

In section 2, we review different approaches of modeling the wire inductance, followed by discussions in section 3 on practical methods of assessing the inductance with special reference to return path in an IC chip. Section

4 covers methods of characterizing on-chip inductance and validating the models using test chip approach.

2. EFFECT OF INDUCTANCE ON IC DESIGN

Unlike line capacitance that is directly proportional to the length of the interconnect, inductance effect in an IC chip is length dependent. The length range where on-chip inductance is important is given by the following bound [2, 3]:

$$\frac{t_r}{2\sqrt{LC}} < l < \frac{2}{R}\sqrt{\frac{L}{C}} \qquad (1)$$

As technology advances, with faster on-chip rise time, higher clock frequencies and the use of Cu wires as interconnect, the range of wire length where inductance is important becomes larger [4] (Fig. 1). For wires of a VLSI chip in this range, the use of RCL models as a distributed network becomes a necessity. Even more critically, a RCLK model, in which not only self inductance L, but the mutual coupling inductance K is considered, is needed. In fact, the inclusion of K becomes more important for technology node of 90nm and below. Figure 2 shows the results of simulation using 3 different interconnect delay modes [5]. The importance of mutual inductive coupling (RLCK model) is evident.

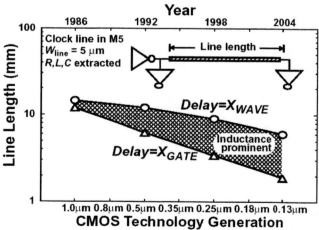

Fig. 1 Inductive effects vs. technology scaling. After Qi. et al. [4].

Traditionally, the inductance extraction and molding problem has been suppressed by design strategies, which minimize the formation of significant long-range inductance on chip. One straightforward approach is that wiring layers containing lines with high current density are sandwiched in between isolating metal planes above and below [6]. With ground planes sufficiently close, the magnetic field is blocked and the couplings to adjacent layers and crosstalk are effectively suppressed [7]. By knowing current return paths, calculation of the loop inductance is easy. However, the isolating layers add significantly to the capacitive coupling of the surrounding lines which adversely impacts the speed and power dissipation of the circuitry.

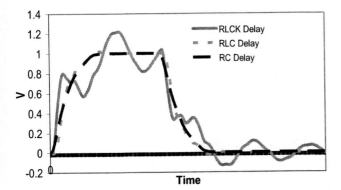

Fig. 2 Signal bus inductive noise simulation using various delay models

Another selective method of reducing inductance between high current wires is to introduce additional inverters in the lines at regular intervals [6]. This causes the current to change direction in each adjacent interval, thus the self inductance of the entire line is reduced significantly. At such point, the RC propagation dominates and inductance models are no longer required. However this approach can be costly in terms of IC area, performance and power consumption [8].

Scaling scheme has been proposed [9] to extend the current design tools and methodologies to cope with inductive effects of future technologies. The scheme includes the optimization of driver size, buffer area, and interconnects line width and length. The critical interconnect length l_{crit} for which the interconnect is neither over-damped nor under-damped can be obtained for a uniform distributed RLC line as a function of interconnect geometry and input vector. For each technology node, while keeping the aspect ratio and interconnect pitch as depicted by ITRS, global line widths are varied until l_{crit} vs. l plot coincides with the 180nm technology case, ensuring inductive effects remain constant across technology nodes. However the increase of the minimum wire width as a result of this scaling scheme creates wireability problem and may increase metal layer numbers and chip area.

Despite problems such as unwanted ringing and overshooting, reflection of signals and switching noise caused by inductive effect, the inductive behavior of the wire can be exploited for new concepts that benefit high-speed circuit design [10].

3. INDUCTANCE MODELING

Inductance, by definition, is for a loop of a wire, wider the current loop, higher the inductance,

$$L_{i,j} = \frac{\psi_{ij}}{I_j} \qquad (2)$$

where Ψ_{ij} represents the magnetic flux in a loop i due to a current I_j in loop j The re are different methods of modeling inductance effect of a wire. The most accurate one is based on Maxwell equations, the so called field solver approach, but is not suitable for circuit level inductance calculation. At the chip level different approaches are used and discussed below.

3.1 Field Solver

The modeling and calculation of inductance of a wire in an IC chip requires knowledge of the return path(s). Three dimensional (3-D) field solvers are used to extract inductance using following current integral equation without the need to know the actual return paths,

$$\frac{J(r)}{\sigma} + \frac{j\omega\mu_0}{2} \int_V \frac{J(r')}{|r - r'|} dr' = -\nabla\Psi(r) \qquad (3)$$

where V is the volume of the conductors, r' is the source point vector, ω is the frequency and μ_0 is the free space permeability.

Finite difference or finite element methods are applied to the governing Maxwell equations in differential form to calculate magnetic flux Ψ and current density J, hence L using Eq. 2. The Maxwell equation approach generates a global 3-D mesh that makes it impractical to extract complex 3-D structures such as the one encountered in a VLSI chip.

Sometimes inductance matrix [L] is derived directly from the capacitance matrix [C] such that $[L] = [C]^{-1}/v_0^2$, where v_0^2 is the phase velocity of the medium. Because the relation holds only for ideal transmission lines, with perfect conductor and perfect dielectric, it will result in underestimation of the on chip wire inductance [3].

3.2 Partial Inductance/PEEC Method

At IC chip level, it is not clear which conductor forms the loop, and the return path is not easily identified. An effective way of analyzing inductive effect of complex wire structures is the partial inductance (PI) method, in which the current is assumed to return at infinity [11]. The infinite loop cancels out when two line segments are subtracted and the determination of the returning path for each wire is not required. Based on PI approach, the self, mutual inductances of two parallel lines of length l, width w, and thickness t, separated by distance d are:

$$L_{self} = \frac{\mu_0}{2\pi}\left[l.\ln\left(\frac{2l}{w+t}\right) + \frac{l}{2} + 0.2235(w+t)\right] \quad (4)$$

$$M = \frac{\mu_0 l}{2\pi}\left[\ln\left(\frac{2l}{d}\right) - 1 + \left(\frac{d}{l}\right)\right] \quad (5)$$

Extending the PI approach to on-chip inductance calculation, the so called Partial Element Equivalent Circuit (PEEC) method [11-13], determines the inductance associated with each wire by breaking it into segments and then calculating self-inductance associated with each wire segment and mutual inductance associated with each pair of wire segments. Because of the long range influence of magnetic field, the inductance matrix is usually very dense and simulation of this scale may not be practical for every wire at chip level. Therefore, this method is often used to extract wire inductance for critical nets only.

Based on PEEC model, a frequency dependent inductance and resistance extractor (FastHenry) was developed [14]. The method divides conductors into filaments and solves for impedance matrix using accelerated multi-pole approach. Figure 3 is a simulation of the inductance vs. frequency of a signal line with multiple coplanar return paths, here w = 0.5μm, s = 0.5μm, h = 0.8μm and l = 1000μm. In this figure, the low frequency inductance value holds up to 1GHz before it starts to decay to coplanar two line return (CTR) model as shown in the figure. In the presence of 32 return paths, the actual inductance is more than double that of the CTR model [15] for frequency up to 1GHz.

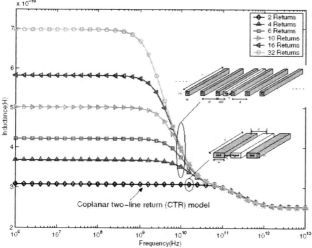

Fig. 3 Inductance vs. frequency using different number of return paths [14].

3.3 Loop Inductance

The formidable dense inductance matrix has prompted the returning of loop inductance approach to extract L for today's VLSI designs. The loop inductance model can be used to find the inductance effect, if the return path is known or can be estimated. The model is suitable for well designed IC structures, such as clock nets shielded by power/grounding wires next to them. For a structure of a single line over power grid, inductive interaction within every segment surrounding a signal line can be incorporated into an effective loop inductance L_{eff} [16]. Excellent linearity is observed for high frequency inductance because return current is mostly confined to the nearest power grid. The inclusion of up to the second nearest pair of grid is sufficient for less than 5% error in L_{eff}. The L_{eff} of diagonal wires over the power grid is insensitive to the relative location of the wires due to the fact that magnetic flux between the diagonal wires and the nearest power grid does not change much by changing the location of the diagonal wires [16].

Even with exact return path unknown, the Min/Max loop inductance estimation can provide an insight into inductance effects of a VLSI design [17]. The loop inductance of two wires is given as:

$$L_{loop} = L_1 + L_2 - 2M \quad (6)$$

where L_1 and L_2 are the partial self inductances and M is the partial mutual inductance. Note that the minimum loop inductance occurs when the mutual inductance is at its maximum, or the returning paths are provided at the nearest possible location. Assuming a constant layer thickness, the optimal spacing and wire width values can be calculated. As expected, the minimum loop inductance occurs when the spacing between the signal and ground wires is minimum. The optimal width of the ground wire is a function of the signal wire width, thickness and spacing [17]. An absolute upper bound for the loop inductance is obtained by assuming a return path at infinity with infinite width. In this case, the loop inductance is equivalent to the partial self inductance of the wire (see Fig. 4).

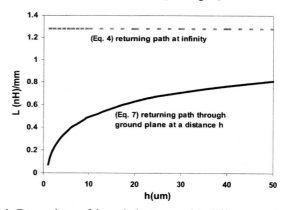

Fig. 4 Comparison of loop inductance with different returning paths (assuming W = 5μm, t = 0.5μm).

Inductance estimation can be obtained from formulations of the characteristic impedance for transmission lines and their relationship with distributed inductances. It has been shown that the loop inductance of a wire over a ground plane and the mutual inductance of two wires on a ground plane are [18]

$$L = 0.2 \ln\left(\frac{2\left(h + {t}/{2}\right)\pi}{w + t}\right) \text{ nH/mm} \qquad (7)$$

$$M = 0.2 \ln\left(1 + \left(\frac{2\left(h + {t}/{2}\right)}{D}\right)^2\right) \text{ nH/mm} \qquad (8)$$

Where h is the wire height from the bottom of the wire to the ground plane. W and t are the width and thickness of metal wire, and D is the center to center wire separation.

Note that in Eq. 7 when the wire height increases, the loop inductance increases and approaches the upper bound which is defined by the self-inductance calculated using PEEC method (see Fig. 4).

3.4 Frequency Dependent Inductance Modeling

Inductance is frequency dependent as evident from Fig. 2. Ideally the frequency dependent parameters of general RLCG model in Fig. 5(a) should be used in the circuit simulation. However the frequency dependent parameters are not supported in many circuit simulators such as SPICE. A parallel RLC model (Fig. 5(b)) that consists of low-resistance, high-inductance path at low frequency, and high-resistance, low-inductance path at high frequency has been proposed [19] to incorporate the frequency-dependent behavior of an interconnect. Multiple parallel branches (Fig. 5(c)) are used to model inductance reduction at high frequency due to capacitive coupling of the signal line to random lines and power grids [20]. By introducing separate "slow-wave factor" (SWF) for parallel and crossing lines, and low-frequency (DC), moderate-frequency (MF) and high frequency (HF) parameters, the high frequency interconnect behavior can be modeled more accurately [20].

$$R_{DC} = R_1 \| R_2 \| R_3$$

$$L_{DC} = L_1 + \left(\frac{R_1}{R_1 + R_2 \| R_3}\right)^2 \left(L_2 + \left(\frac{R_2}{R_2 + R_3}\right)^2 L_3\right)$$

$$R_{MF} = R_1 \| R_2; \qquad L_{MF} = L_1 + \left(\frac{R_1}{R_1 + R_2}\right)^2 L_2$$

$$R_{HF} = R_1; \qquad L_{HF} = L_1. \qquad (9)$$

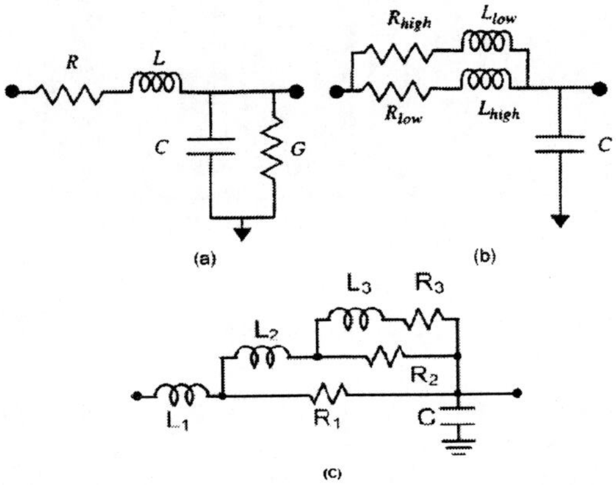

Fig. 5 (a) Telegrapher's equation RLGC model (b) Parallel RCL model [19], (c) Frequency dependent RCL model [20]

4. INDUCTANCE CHARACTERIZATION USING TEST STRUCTURES

The only way to validate inductance extraction, and investigate the inductive effects accurately is to use test chip. Investigations of noise and delay values as functions of interconnect length, width and spacing, under lumped or distributed gate loading, and inductive effects in the presence of various returning paths including substrate, coplanar structures, under layers, power grid, ground bus, and random structures can be performed using test structures. Either direct time-domain approach such as using ring oscillators or frequency-domain measurement such as S-parameter characterization can be used to analyze on-chip inductance.

A ring oscillator in which two legs of the oscillator traverse a reasonably long interconnect path was used to study the performance of a given system of conductors [21]. The conductor configuration consists of aggressor wires and a victim wire. Different cases were studied as 1) no nearby return paths exists, 2) effects of a reference plane, and 3) various signal to return ratios with various return widths. The circuit enables the effects of inductance on delay and overshoots to be measured on the aggressor wires and induced noise to be measured on the victim wire. The loop frequency and noise level measurement showed excellent correlation to the simulation results. The results demonstrated significant overshoot if no nearby return paths were provided.

Ring oscillators were also used for characterizing the ratio between effective driver resistance and characteristic impedance of the wire [19]. Two different configurations of ring (loop) are used to study this behavior as shown in Fig. 6. The mismatch between the wire and the driver will result in overshoot and ringing.

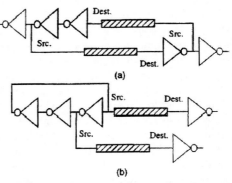

Fig. 6 Ring oscillators with connections at (a) destination (b) source [22].

Inductance-Oscillator whose oscillation frequency is determined by the wire inductance has also been developed to quantitatively evaluate inductance impact on on-chip interconnect delay [22]. It consists of a ring oscillator, long signal lines and on-chip frequency counter. The long signal lines have equal parasitic capacitance and resistance but different loop inductance. Each structure includes different signal power supply and ground wire configuration, and each long signal line employs a coplanar wave guide configuration. The equivalent distance to the nearest ground grid, which serves as the current return path is varied to control wire inductance. The measurement results suggest that if an insufficient ground were used, unpredictable timing difference due to inductance may arise even for coplanar structures.

Frequency domain S parameter characterization is carried out by launching waves from both ends of a signal line and measuring the reflected (S_{11}) and transmitted (S_{21}) network parameters using a network analyzer over the frequency range up to 50-100GHz. Dummy open structures are required to de-embed the pad parasitics. Telegrapher's parameters (RLGC) or various frequency-dependent RLC models parameters (such as R1, L1, R2, L2, etc. as shown in Fig. 4) can be extracted from S parameters results [19-20, 23-24].

Interconnect transmission line effects due to the characteristics of a silicon substrate, ground planes, coplanar lines and power/ground grids were studied and characterized using S parameter measurement [4, 19, and 25]. Measurement results and FastHenry simulation show that at low frequency, most current returns through the minimum resistance path, the current uses as many return paths as possible, and often through the nearest ground wires. The inductance increases for increasing spacing between the wire and the nearest ground wires of the co-planar structure because more flux is enclosed between the signal and the return ground. At higher frequency, current return path is to minimize loop inductance. The inductance is decreased due to the proximity effects between the wires as well as between the wire and the substrate. Time-varying magnetic fields coupled to the substrate generate eddy currents that offer a portion of the current return paths. With eddy current, the substrate, even power grid and random lines can reduce inductance.

For VLSI chip, power and ground are usually distributed through grid structure in order to minimize IR drop and reduce ground bounce. Due to proximity effects, the nearest power and ground wires provide the return paths for most of the signal current through coupling. With substrate excluded, multiple parallel (with signal line) ground wires in ground grids reduce signal wire inductance by providing multiple current return paths. Floating wires do not affect the signal wire inductance if these wires are smaller than 20μm due to negligible eddy currents. However dense grid and ground planes can generate eddy current due to the existence of time-varying magnetic field. Whether the grid structure is floating or grounded does not make much difference in the wire inductance.

Using a test structure that consists of a set of pseudo-random signal lines that are connected to on-chip drivers to mimic a real chip (Fig. 7), the inductance effects were studied and compared to the conventional co-planar wave guide structures [18-19]. The results showed that random

signal lines and power grids reduce high-frequency inductance significantly by providing closer return paths through capacitive coupling. The signal lines could couple inductively to the parallel random lines which effectively reduces the line inductance. They can also couple capacitively to the orthogonal random lines that in turn connect or couple to parallel random lines. Whether the drivers of the random lines are on or off does not significantly affect the inductance because the nature of capacitive couplings to the power grid and signal lines.

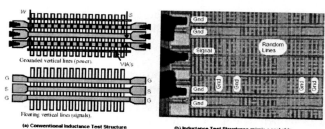

Fig. 7 Inductance test structures [18] (a) conventional (b) mimic a real chip

Due to its low resistance value, the inductive effects of Cu wires are expected to be different from those of Al wires. Figure 8 is the micrograph of inductance test structures designed to study 90nm Cu inductive effects for various current return paths such as co-planar structures, ground plane, power grid, as well as random structures. The results will be reported after the availability of the data.

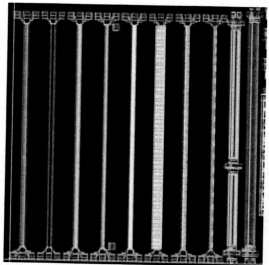

Fig. 8 Micrograph of inductance test structures designed for 90nm Cu process

5. CONCLUSION

The inductance causes unwanted effects such as ringing, overshoot, cross-talk etc. in chip designs. The inductive effects are often suppressed during design using techniques and structures such as micro-strip line and co-planer strip line. Inductance can be modeled using Field solvers, PEEC and loop inductance approaches. Due to the large inductance matrix associated with PEEC, the use of loop inductance approach has been revisited for today's VLSI design. The concept of effective loop inductance is compact and could be applied to extract the inductance of signal lines in the presence of random structures where return paths are difficult to be determined. Ultimately, inductance test structures are the only means of validating inductance models accurately. Using well designed test structures, one can characterize inductance of a wire in the presence of co-planar structures, ground planes, power grids, and random structures etc.

REFERENCES

1. N. D. Arora, "Challenges of Modeling VLIS interconnects in the DSM Era", *Proc. Modeling and Simulation of Microsystems,* pp.645-648, 2002.

2. Y. Ismail, E. G. Friedman and J. L. Neves, "Figures of merit to characterize the importance of on-chip inductance", *IEEE Tran. VLSI System,* Vol. 7, pp. 442-449, Dec. 1999.

3. Y. Ismail, and E. Friedman, "On-chip Inductance in High Speed Integrated-circuits", Kluwer Academic Publishers, 2001.

4. XQ B. Kleveland, G. Wang, ZYu, S. S. Wong, R. W. Dutton, and T. Furusawa, "High Frequency Characterization and Modeling of VLSI On-Chip Interconnects ", *Proc. of the Workshop on Synthesis and System Integration of Mixed Technologies (SASIMI),* Japan, Oct. 2001.

5. N. D. Arora, "Modeling and Characterization of Copper Interconnect for SoC Design", *Proc. SISPAD,* pp.1-6, 2003.

6. M. W. Beattie and L. T. Pileggi, "IC Analyses Including Extracted Inductance Models", *Proc. IEEE DAC,* pp. 915-920, 1999.

7. J. A. Davis, R. Venkatesan, A. Kaloyeros, M. Beylansky, S. J. Souri, K. Banerjee, K. C. Saraswat, A. Rahman, R. Reif, and J. D. Meindl, "Interconnect limits on gigascale integration (GSI) in the 21st century", *Proc. of IEEE ,* Vol. 89, pp. 305-324, March 2001.

8. P. Kapur, G. Chandra, J. P. McVittie, and K. C. Saraswat, "Technology and Reliability Constrained Future Copper Interconnects - Part II: Performance Implication", *IEEE Tran. On Electron Dev.* Vol. 49, pp.598-603, April, 2002.

9. K. Banerjee and A. Mehrotra, "Inductance Aware Interconnect Scaling", *Proc. IEEE ISQED,* pp. 43-47, 2002.

10. S. S. Wong, P. Yue, R. Chang, S. Kim, B. Kleveland and F. O'Mahony, "On-Chip Interconnect Inductance –Friend or Foe", *Proc. IEEE ISQED,* pp. 389-394, 2003

11. A. E. Ruehli, "Inductance calculations in a complex integrated circuit environment," *IBM J. Res. & Dev.,* Vol. 16, pp. 470, 1972.

12. P. A. Brennan, N. Raver and A. Ruehli, "Three dimensional inductance computations with partial element equivalent circuits", *IBM J. Res. & Dev.,* vol. 23, pp.661-668, Nov. 1979.

13. P. Restle, A. Ruehli and S. G. Walker, "Dealing with Inductance in High Speed Chip Design", *Proc. IEEE DAC,* pp. 904-909, 1999.

14. M. Kamon, M. Tsuk, and J. White, FastHenry: A Multi-pole Accelerated 3-D Inductance Extractions Program*", IEEE Trans. Microwave Theory and Techniques,* Vol. 42, No. 9 pp. 1750-1758, 1994.

15. S. Kim, Y. Massoud, and S. S. Wong, "On the Accuracy of Return Path Assumption for Loop Inductance Extraction for 0.1um Technology and Beyond", *Proc. IEEE ISQED,* pp. 401-404, 2003.

16. S. Sim, C. Chao, S. Krishnan, D. M. Petranovic, N. D. Arora, K. Lee and C. Y. Yang, "An Effective Loop Inductance Model for General Non-Orthogonal Interconnect with Random Capacitive Coupling", *Proc. IEEE IEDM,* pp. 315-318, 2002.

17. Y. Lu, M. Celik, T. Young, and L. Pileggi, "Min/Max On-Chip Inductance Models and Delay Metrics", *Proc. IEEE DAC,* pp. 341-346, June, 2001.

18. B. Kleveland, XQ L. Madden, R. W. Dutton, and S. S. Wong, "Line Inductance Extraction and Modeling in a Real Chip with Power Grid" *Proc. IEEE IEDM,* pp.901-904, Dec. 1999.

19. B. Kleveland, XQ L. Madden, T. Furusawa, R. Dutton, M. A. Horowitz, and S. S. Wong, "High-frequency characterization of on-chip digital interconnects", *IEEE ISSC,* vol. 37, pp.716-725, June, 2002.

20. S. Sim, K. Lee, and C. Y. Yang, "High-Frequency On-Chip Inductance Model", *IEEE EDL,* Vol. 23, pp.740-742, December, 2002.

21. S. V. Morton, "On-chip Inductance in Multiconductor Systems", *Proc. IEEE DAC,* pp. 921-926, 1999.

22. T. Sato, and H. Masuda, "Design and Measurement of an Inductance-Oscillator for Analyzing Inductance Impact on On-chip Interconnect Delay", *Proc. IEEE ISQED,* pp. 395-400, 2003.

23. A. Deutsch, P. W. Coteus, G. V. Kopcsay, H. H. Smith, C. W. Surovic, B. L. Krauter, D. C. Edelstein, and P. J. Restle, "On-chip wiring design challenges for gigahertz operation", *Proc. IEEE,* Vol. 89, pp.529-555, April, 2001.

24. B. Krauter and S. Mehrotra, "Layout-based frequency dependent inductance and resistance extraction for on-chip interconnect timing analysis", *Proc. IEEE DAC,* pp. 303-308, 1998.

25. XQ B. Kleveland, ZYu, S. S. Wong, R. Dutton, and T. Young, "On-Chip Inductance Modeling of VLSI Interconnects", *IEEE ISSCC,* pp.172-173, Feb. 2000.

R3, an Accurate JFET and 3-Terminal Diffused Resistor Model

Colin C. McAndrew

Motorola, Tempe, AZ, Colin.McAndrew@motorola.com

ABSTRACT

This paper presents an improved compact model for diffused resistors and JFETs, valid over geometry, bias, and temperature. The model includes a physically based junction depletion model, a new and accurate velocity saturation model derived from data, and self-heating, which is important for low sheet resistance devices.

Keywords: JFET, diffused resistor, SPICE model, compact model, velocity saturation.

1 INTRODUCTION

As we all know from introductory classes in physics and engineering, Ohm's law tells us that $V = IR$ for a resistor. So for IC simulation and design we can take that as given, and concentrate on modeling "real" devices like MOSFETs and BJTs, right? Wrong!

Real resistors do not have linear $I(V)$ characteristics, but deviate from this because of depletion pinching (for diffused resistors, which are JFETs), velocity saturation, self-heating, and Schottky effects and self-heating at contacts. For example, the effective mobility degradation with bias due to self-heating is [1]

$$R/R_0 = \mu_{red} = 1 + \frac{R_{THA}T_{C1}}{\rho_s}\left(\frac{V}{L}\right)^2 \qquad (1)$$

where R_{THA} is the thermal resistance per unit area, and the other terms have their usual meaning. Self-heating is thus important for short resistors in low sheet resistance layers.

The details of resistor nonlinearities can be important for whether a circuit meets or fails specifications, especially where harmonic distortion is a key performance metric, and this depends on the linearity (or otherwise) of resistors. So accurate modeling of the nonlinear $I(V)$ characteristics is critical for some analog and mixed-signal applications.

This paper presents a physically-based compact model for diffused resistors and JFETs. The standard SPICE JFET model does not model depletion pinching accurately, and does not include velocity saturation or self-heating. A significantly improved model was presented in [2]; however, this model does not include self-heating, has only a simple velocity saturation model that does not accurately represent measured data, and has only an empirical approach for pinch-off voltage calculation.

This paper presents a diffused resistor and JFET model based on [2] that is significantly more accurate as it overcomes the deficiencies listed above. It also simplifies the basic depletion pinching formulation of [2], yet retains the same accuracy over geometry. The two section resistance model and 1/6:2/3:1/6 capacitance partitioning of [2] has proved accurate to date, so the formulation presented here is for a single section DC model and assumes implementation is a sectional model like [2].

2 DEPLETION MODEL

Fig. 1 shows a cross section of a diffused resistor.

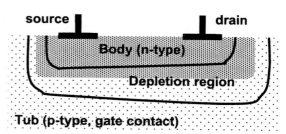

Fig. 1 Diffused Resistor Cross Section

The analysis of [2] gives a resistor current in the presence of depletion pinching, but not velocity saturation, as

$$I = \frac{q\mu_0 N x_0 W}{L}\left(1 - a\sqrt{\psi_a + \overline{V}}\right)\left(1 - \frac{b}{W}\sqrt{\psi_p + \overline{V}}\right)V_{ds} \qquad (2)$$

where W and L are the effective width and length, N and x_0 are the doping and depth of the resistor body, ψ_a and ψ_p are the built-in potentials of the bottom (area) and side-wall (perimeter) junctions, a and b are depletion width factors for the bottom and side-wall, and the average body to tub (gate) bias is $\overline{V} = 0.5(V_{dg} + V_{sg})$. For resistors commonly used in IC processes the bias dependence of the last two terms in (2) is relatively small, and $\psi_a \approx \psi_p$, so expanding the product of the depletion pinching terms in (2) and dropping small terms gives a good approximation

$$I_{depl} = G_F\left(1 - D_F\sqrt{P + \overline{V}}\right)V_{ds} \qquad (3)$$

where $D_F = D_a + D_p/W$ is an effective depletion factor, $P = 0.5(\psi_a + \psi_p)$, and the effective conductance factor is $G_F = 1/(R_0(1 - D\sqrt{P}))$ where R_0 is the zero bias resistance.

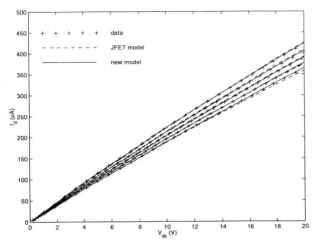

Fig. 2 $I_d(V_{ds}, V_{sg})$ for a wide, long resistor.

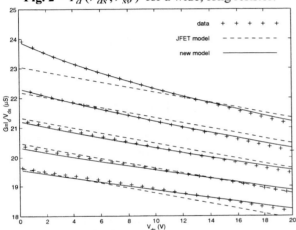

Fig. 3 $G(V_{ds}, V_{sg})$ for a wide, long resistor.

Fig. 2 and Fig. 3 show current and conductance $G = I_d/V_{ds}$ (for a long resistor, so velocity saturation and self-heating are negligible) from data, the model (3), and the SPICE JFET model $I_d = \beta V_{ds}(2(V_{gs} - V_{TO}) - V_{ds})$. The model (3) agrees well with the measured data, the SPICE JFET model does not exhibit even the correct qualitative nonlinear behavior with bias. The nonlinear "depletion" nature of the bias dependence is clear in the data of Fig. 3 and in the model (3), whereas the bias dependence of G is linear with bias for the JFET model. The RMS and maximum errors for the JFET model are 0.89% and 3.7%, respectively, and for the model (3) are 0.25% and 0.59%, respectively.

In practice, to allow greater flexibility in modeling the geometry dependence of the depletion pinching effect, the effective depletion pinching factor is modeling as

$$D_F = D_{F\infty} + D_{FW}/W + D_{FL}/L + D_{FWL}/WL \qquad (4)$$

and by use of a selector the geometries used to calculate the depletion factor can be either design dimensions or effective electrical dimensions. The latter can be more

accurate, but the former can be simpler to characterize. The effective width models used are those of [3].

3 VELOCITY SATURATION

Besides modeling the depletion pinching effect, accurate modeling of the nonlinearity of highly linear resistors requires accurate modeling of velocity saturation. Velocity saturation is thought of as only being important to model short devices. However, its effect on nonlinearity is important even for "long" devices.

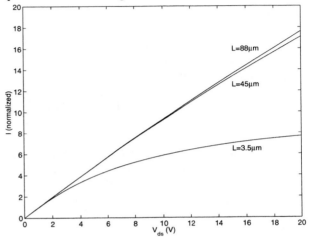

Fig. 4 Normalized $I(V)$ data with velocity saturation.

Fig. 4 shows data from high sheet resistance (so self-heating is negligible) p-type resistors, normalized by low low-field conductance. From (3) the depletion pinching is independent of L so the differences between the curves are primarily from velocity saturation. The effect is noticeable at $L=45\mu m$, and although it looks to be small, it is the difference with respect to a perfectly linear $I(V)$ relation that is important for distortion, and for this purpose the effect is significant.

Rather than analyzing velocity saturation as a velocity versus electric field relation, $v(E)$, a simple and useful way to study the effect is in a reciprocal sense, as an effective mobility reduction factor,

$$\mu_{red}(E) = \mu_0/\mu(E). \qquad (5)$$

The two commonest models for μ_{red} are [4], [5]

$$\mu_{red}(V) = 1 + \frac{\mu_0}{v_{sat}}\frac{|V|}{L}, \quad \mu_{red}(V) = \sqrt{1 + \left(\frac{\mu_0}{v_{sat}}\frac{|V|}{L}\right)^2} \qquad (6)$$

where the symbols have their conventional meaning. The first form is often used for analytic simplicity, the second for improved accuracy and lack of a singularity at $V = 0$. Although both forms exhibit qualitatively the correct behavior, they are not accurate for distortion modeling.

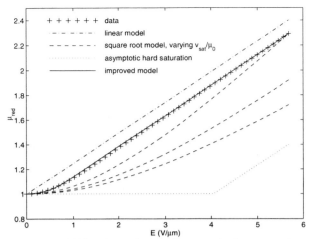

Fig. 5 Mobility degradation due to velocity saturation.

Fig. 5 shows μ_{red} calculated from measured data, with the models (6). The square root model has the critical field $E_{cr} = v_{sat}/\mu_0$ calculated in 3 ways: to match the value of μ_{red} at the highest measured field; to match the slope of $\mu_{red}(E)$ at the highest measured field; and to match the asymptotic slope as $E \to \infty$. The inaccuracy in the models (6) is clear (the linear model is used in [2]).

Note that for small fields the square root model is

$$\mu_{red}(V) \approx 1 + \frac{\mu_0^2}{2v_{sat}^2}\left(\frac{V}{L}\right)^2 \qquad (7)$$

which has the same form as (1). The effects of velocity saturation and self-heating are similar for low bias DC operation (they are qualitatively different for high bias DC operation and AC operation).

The measured μ_{red} data have the following characteristics. Value 1 at $E = 0$, symmetry with respect to E, asymptotically approaches $1 + (E - E_{co})/E_{cr}$ for large E where E_{co} is a corner field, and value $1 + \delta$ at E_{co}. An empirical model that has these characteristics is

$$\mu_{red} = 1 + \sqrt{\left(\frac{E - E_{ce}}{2E_{cr}}\right)^2 + U} + \sqrt{\left(\frac{E + E_{ce}}{2E_{cr}}\right)^2 + U} - \frac{E_{co}}{E_{cr}} \quad (8)$$

where $U = d\mu E_{ce}/E_{cr}$, $d\mu$ is a fitting parameter for the "hardness" of the corner at E_{co} ($d\mu \approx \delta^2 E_{cr}/E_{ce}$), and an effective corner field

$$E_{ce} = \sqrt{E_{co}^2 + (2d\mu E_{cr})^2} - 2d\mu E_{cr} \qquad (9)$$

is used to ensure that the asymptote for large E is correct. Fig. 6 shows how the model is parameterized, initial parameter values are calculated from $\mu_{red}(E)$ characteristics and then refined using optimization.

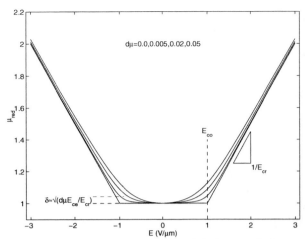

Fig. 6 Velocity saturation model characteristics.

The current in the improved model is

$$I_{ds} = I_{depl}/\mu_{red} \qquad (10)$$

where I_{depl} is from (2) and μ_{red} is from (8).

4 SATURATION VOLTAGE

The drain end of a resistor pinches off when the mobile carrier density there is zero, i.e. when V_{dg} is

$$V_P = 1/D_F^2 - P. \qquad (11)$$

This was the basic definition used in [2]. However, because of approximations made during derivation of (1) the small-signal conductance $g_d = \partial I_d/\partial V_d$ is negative at V_P. A better definition of saturation is when $g_d = 0$. For (2)

$$V_{sat} = \frac{1 - 3D_F^2(P + V_{sg}) + \sqrt{1 + 3D_F^2(P + V_{sg})}}{9D_F^2/4}. \qquad (12)$$

However, this does not take velocity saturation into account. For a general velocity saturation model applied on top of an existing model, as in (10), it is not possible to guarantee that a closed form solution for the V_{ds} at which $g_d = 0$. Indeed, there is no closed form solution for $g_d = 0$ for the models (2) and (8). The potential problem with an approximate solution is that if this value is greater than the actual value at which $g_d = 0$ then the model exhibits a negative output conductance glitch, which is highly undesirable.

Fortuitously, and somewhat counter-intuitively, if a μ_{red} model that is less than the true value is used (actually that has a greater $(1/\mu_{red})\partial\mu_{red}/\partial E$), then this gives a calculated saturation voltage that is guaranteed to be less than the actual value where $g_d = 0$, hence guaranteeing that there is not negative output conductance glitch.

V_{sat} is therefore calculated as the value of V_{ds} at which the derivative of

$$I = \frac{G_F\left(1 - D_F\sqrt{P + \overline{V}}\right)V_{ds}}{1 + (V_{ds}/L - E_{co})/E_{cr}} \tag{13}$$

is zero. This gives a fourth order equation that has an analytic solution. For low field velocity saturation is not important and the underestimation of V_{sat} is small. For high field the simple model $1 + (E - E_{co})/E_{cr}$ is very close to the more accurate model (8) and so the accuracy is again good. V_{ds} is limited to V_{sat} via the limiting function

$$V = \frac{2V_{ds}V_{sat}}{\sqrt{(V_{ds} - V_{sat})^2 + 4A^2} + \sqrt{(V_{ds} + V_{sat})^2 + 4A^2}} \tag{14}$$

which maintains symmetry.

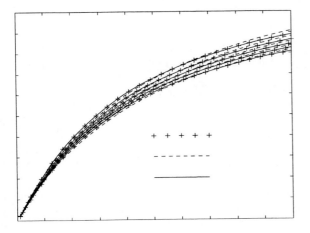

Fig. 7 $I_d(V_{ds}, V_{sg})$ for a wide, short resistor.

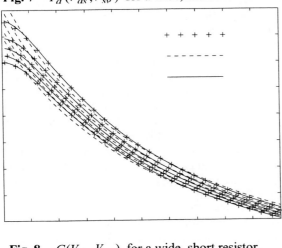

Fig. 8 $G(V_{ds}, V_{sg})$ for a wide, short resistor.

Fig. 7 and Fig. 8 show current and $G = I_d/V_{ds}$ for a wide, short resistor. Also shown are the model (13) and the depletion pinching model (2) with the linear model for mobility reduction due to velocity saturation in (6). The improved accuracy of the velocity saturation model (8) is clear. In particular, the simple linear velocity saturation model of (6), apart from having a singularity at zero, is inaccurate and qualitatively incorrect near zero.

5 CONCLUSIONS

An improved model for diffused resistors and JFETs has been presented. The model improves on that of [2] by having a simpler formulation, having a more accurate velocity saturation model, and having an accurate and analytic saturation voltage calculation. V_{sg} is limited to then pinch-off voltage of (11), with a log-exp form, and this gives a current in subthreshold operation that varies exponentially with gate bias (lowly doped well resistors, that have a low JFET threshold voltage, in high voltage processes can be biased below threshold).

Although not detailed here, the model also includes self-heating modeling, and a separate contact resistance model. The contact resistance model includes a separate, implicit self-heating model based on (1), which allows the anomalous decrease in resistance with increasing current, seen in contacts to p-type material [7], to be modeled.

REFERENCES

[1] C. C. McAndrew, "Predictive Technology Characterization, Missing Links Between TCAD and Compact Modeling," *Proc. IEEE SISPAD*, pp. 12-17, 2000.

[2] R. V. H. Booth and C. C. McAndrew, "A 3-Terminal Model for Diffused and Ion-Implanted Resistors," *IEEE Trans. ED*, vol. 44, no. 5, pp. 809-814, May 1997.

[3] C. C. McAndrew, S. Sekine, A. Cassagnes, and Z. Wu, "Physically-Based Effective Width Modeling of MOSFETs and Diffused Resistors," *Proc. IEEE ICMTS*, pp. 169-174, 2000.

[4] F. N. Trofimenkoff, "Field-Dependent Mobility Analysis of the Field-Effect Transistor," *Proc. IEEE*, vol. 53, no. 11, pp. 1765-1766, Nov. 1965.

[5] D. M. Caughey and R. E. Thomas, "Carrier Mobilities in Silicon Empirically Related to Doping and Fields," *Proc. IEEE*, vol. 55, no. 12, pp. 2192-2193, Dec. 1967.

[6] C. C. McAndrew, "Useful Numerical techniques for Compact Modeling," *Proc. IEEE ICMTS*, pp. 121-126, 2002.

[7] K. Banerjee, A. Amerasekera, G. Dixit and C. Hu, "Temperature and Current Effects on Small-Geometry-Contact Resistance," *Proc. IEEE IEDM*, pp. 115-118, 1997.

Advanced MOSFET Modeling for RF IC Design

Yuhua Cheng

Skyworks Solutions, 5221 California Ave. Irvine, CA 92612

ABSTRACT

In this paper, advanced MOSFET modeling for radio-frequency (RF) integrated-circuit (IC) design is discussed. An introduction of the basics of RF modeling of MOSFET is given first. A simple sub-circuit model is then presented with comparisons of the data for both y parameter and f_T characteristics. The high frequency (HF) noise and distortion modeling issues are also discussed by showing the validation results against measured data. The developed RF MOSFET model can be the basis of a predictive and statistical modeling approach for RF applications.

Keywords: mosfet modeling, rf modeling, hf mosfet modeling, rf ic design, rf cmos.

1   INTRODUCTION

The down-scaling of CMOS technology has resulted in a significant improvement of RF performance of MOS devices [1]. Also, CMOS technology offers other advantages such as low power consumption, high integration and lower cost than III-V and SiGe RF technologies. With the fast growth of radio frequency (RF) wireless communications market, RF designers have begun to explore the use of CMOS in RF circuits. Accurate and efficient RF MOSFET models are required. Compared with the MOSFET modeling at low frequency, compact RF models are more complex to develop. Recently, work has been reported to model the RF performance of submicron MOS devices [2-8]. Basically, they are all developed with the subcircuit approach by adding parasitic components to a core intrinsic MOSFET model. They have demonstrated good accuracy up to 10GHz. However, there are still a lot of issues to be studied, and some examples are listed as follows: (1) The added parasitic components should be physics-based and linked to process and geometry information to ensure the scalability and prediction capability of the model; (2) Clear and efficient parameter extraction methodologies should be adopted to enable statistical modeling; (3) Efficient models to account for NQS effects are required; (4) Capability in predicting DC, AC, HF noise and distortion behavior should be ensured; and   (5) Better model implementation to improve the simulation efficiency and convergence.

In this paper, we discuss some issues that must be properly accounted for in modeling a MOSFET at RF and present some model prediction results by using a sub-circuit MOSFET model as an example. The model are verified by measured small signal AC, HF noise and distortion data over biases and frequency, while some important effects such as Non-quasi-static (NQS) and induced gate noise are reviewed.

2   AC SMALL SIGNAL MODELING

As shown in Fig. 1, a four terminal MOSFET contains many parasitic components. They will influence significantly the device performance at high frequency.

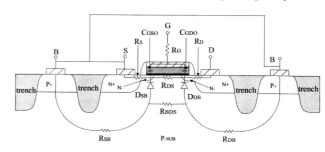

Fig. 1 A MOSFET cross-section with parasitic components.

2.1   Modeling of Gate Resistance

The gate resistance consists mainly of the poly-silicon sheet resistance. Signal delay at the gate due to the distributed transmission line effect at high frequency has been studied. A factor of 1/3 or 1/12 is introduced, depending on the layout structures of the gate connection, to account for the distributed RC effects when calculating the gate resistance at RF [9, 10]. This effect will become more severe as the gate width becomes wider and the operation frequency becomes higher. So multi-finger devices are used in the circuit design with narrow gate widths for each finger to reduce the influence of this effect. Complex numerical models for the gate delay have been proposed. However, a simple gate resistance model with the factor of 1/3 or 1/12 for the distributed effect has been found accurate up to ½ f_T [8, 10], even though additional bias dependence of gate resistance may need to be included to account for the non-quasi-static effect (NQS) effect.

The NQS effect or the distributed RC effect of the channel is another effect that should be accounted for in modeling the HF behavior of a MOSFET. It has been proposed that an additional component is added to the gate resistance to represent the channel distributed RC effect as shown in Fig. 2 [10].

When a MOSFET operates at high frequency, the contribution to the effective gate resistance is not only from the physical gate electrode resistance but also from the distributed channel resistance, which can be "seen" by the signal applied to the gate. Thus, the effective gate resistance

R_g consists of two parts: the distributed gate electrode resistance (R_{geltd}) and the distributed channel resistance seen from the gate (R_{gch}), which is a function of biases [10]. This bias dependent R_g model is one of the approaches to account for the NQS effect, as we will discuss again later.

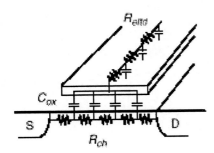

Fig. 2 Illustration of gate electrode resistance R_{geltd}, channel resistance R_{ch}, and gate capacitance C_{ox} [10].

2.2 Modeling of Source/Drain Resistances

The source and drain resistances consist of several parts in a MOSFET, such as the via resistance, the salicide resistance, the salicide-to-salicide contact resistance, and the sheet resistance in LDD region, etc. However, the total resistance is usually dominated by the contact and LDD sheet resistances.

It has been known that the source/drain resistances are bias dependent. In some compact models such as BSIM3v3 [11, 12], these bias dependencies are included. However, these parasitic resistances are treated only as virtual components in the I-V expressions of BSIM3 to account for the DC voltage drop across these resistances and therefore they are invisible by signal in ac simulation. External components for these series resistances need to be added outside the intrinsic model to accurately describe noise characteristics and the input AC impedance of the device.

2.3 Modeling of Substrate Resistance

The influence of the substrate resistance can be ignored in the compact model for digital and analog circuit simulation at low frequency. However, at high frequencies, the signal at the drain couples to the source and bulk terminals through the source/drain junction capacitance and the substrate resistance. The substrate resistance influences mainly the output characteristics, and can contribute as much as 20% or more of the total output admittance [4]. Recently, work on the modeling of substrate components are reported. Several different substrate networks have been proposed to account for the influence of substrate resistance at RF [2-8]. An equivalent circuit (EC) for the substrate network is proposed to describe the HF substrate-coupling-effect (SCE), as shown in Fig. 3, which has been used in RF modeling with good accuracy up to 10GHz [7, 8].

Analytical model equations can be found for the substrate resistance components R_{sb}, R_{dsb}, and R_{db} respectively, which are functions of process and layout parameters such as substrate doping concentration, the space and depth of field (or trench) isolation, etc.

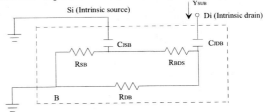

Fig. 3 Proposed equivalent circuit for substrate network. C_{jsb} and C_{jdb} are capacitances of source/bulk and drain/bulk junctions [8].

2.4 Modeling of Parasitic Capacitances

The parasitic capacitances in a MOSFET can be divided into five components as shown in Fig. 4: 1) the outer fringing capacitance between the polysilicon gate and the source/ drain, C_{FO}; 2) the inner fringing capacitance between the polysilicon gate and the source/drain, C_{FI}; 3) the overlap capacitances between the gate and the heavily doped S/D regions (and the bulk region), C_{GSO} & C_{GDO} (C_{GBO}), which are relatively insensitive to terminal voltages; 4) the overlap capacitances between the gate and lightly doped S/D region, C_{GSOL} & C_{GDOL}, which changes with biases; and 5) the source/drain junction capacitances, C_{JD} & C_{JS}. Most of them have been modeled for digital/analog circuit simulation. It would be preferred that these capacitance models are still applicable to RF simulation. For that purpose, an efficient and correct parameter extraction methodology considering the cases for both low frequency and RF is needed. However, additional parasitic capacitance models may have to be developed if the present models cannot meet the requirements at RF [8].

2.5 Modeling of NQS Effects

Most MOSFET models available in circuit simulators use the quasi-static (QS) approximation. In a QS model, the channel charge is assumed to be a unique function of the instantaneous biases: i.e. the charge has to respond a change in voltages with infinite speed. Thus, the finite charging time of the carriers in the inversion layer is ignored. In reality, the carriers in the channel do not respond to the signal immediately, and thus, the channel charge is not a unique function of the instantaneous terminal voltages (quasi-static) but a function of the history of the voltages (non-quasi-static). This problem may become pronounced in RF applications, where the input signals may have rise or fall times comparable to, or even smaller than, the channel transit time. For long channel devices, the channel transit time is roughly inversely proportional to $(V_{gs}-V_{th})$ and proportional to L^2 [9]. Because the carriers in these devices cannot follow the changes of the applied signal, the QS models may give inaccurate or anomalous simulation results that cannot be used to guide circuit design.

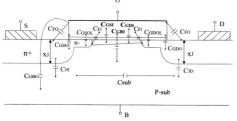

Fig. 4 Capacitance components in a MOSFET.

The NQS effect can be modeled with different approaches for RF applications: (a) R_g approach in which a bias-dependent gate resistance is introduced to account for the distributed effects from the channel resistance as discussed earlier [10], (b) R_i approach in which a resistance R_i (well-used in modeling a MESFET) is introduced to account for the NQS effect [13], (c) transadmittance approach in which a voltage-control-current-source (VCCS) is connected in parallel to the intrinsic capacitances and transconductances to model the NQS effect [7], and (d) core model approach in which the NQS effect can be modeled in the core intrinsic model [8]. It should be pointed out that all of these approaches would have to deal with a complex implementation.

Both R_g and R_i approaches will introduce additional resistance besides the existing physical gate and channel resistance, so the noise characteristics of the model using either R_g or R_i approach need to be examined. Ideally, the NQS effect should be included in the core intrinsic model if it can predict both NQS and noise characteristics without a large penalty in the model implementation and simulation efficiency.

2.6 A Subcircuit RF Model

Based on the above analysis, a complete subcircuit model for RF MOSFETs is given in Fig. 5. The core intrinsic model can be any MOSFET model that is used for analog applications, and here it is BSIM3v3 [11], which has included a bias-dependent overlap capacitance.

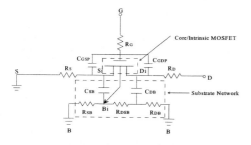

Fig.5 A subcircuit RF model [8].

The model has been examined at different bias conditions, and shows satisfactory agreement to experiments. As an example, Fig. 6 shows the comparison of the y-parameter characteristics between measurements and the model for a three finger device with $W_f/L_f=12/0.36$ at $V_g=V_d=1.5V$. Good match between the model and data shows the simple EC model can work up to 10GHz, which is about half of f_T for the given device. Fig.7 gives the comparison of f_T-I_D characteristics between the model and measurements for different devices.

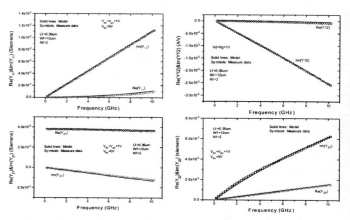

Fig. 6 Comparison of the Y parameters for a two finger device with W/L=12/0.36 at Vg=Vd=1.0V [18].

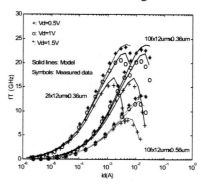

Fig. 7 Comparison of f_T-I_D characteristics between the model and measurements for different devices [8].

3 HF NOISE MODELING

Different noise sources, associated with terminal resistances, and channel resistance, exist in a MOSFET. In this paper, we will validate the subcircuit model discussed above with the measured HF noise data. Also, we address the issue of induced-gate noise that has become one of the interesting topics in RF modeling.

3.1 Thermal Noise Modeling

The HF noise sources in a MOS transistor include the contributions from the terminal resistances at the gate, source and drain, the channel resistance, and the substrate resistances [14]. Fig. 8 shows a complete equivalent noise circuit model for a MOSFET operated at RF.

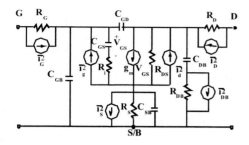

Fig. 8 Equivalent noise circuit model for a MOSFET [8].

With the extracted parameters from the measured data for a 0.25um RF CMOS technology [14], we verify the noise characteristics of the RF model discussed above. The four noise parameters calculated by the correlation matrix technique (CMT) [15] from the simulated noise characteristics are given in Fig. 9 against the measured data for a 0.36um device at different bias conditions. In Fig. 9, the solid lines represent the simulation results of the RF model, and the symbols with solid squares and open circles are the measured data for V_{GS}=1V and V_{DS}=1V and for V_{GS}=2V and V_{DS}=1V, respectively. While the RF model with extracted parameters fits accurately the measured y-parameters data as shown in previous section, it can also predict the HF noise characteristics of the device as given in Fig. 9. It has been found that the transconductance and trancapacitances are the key components determining the HF noise characteristics besides the resistive components. For a model to predict well the HF noise characteristics, the accuracy in both DC and AC fittings has to be ensured while the noise model itself is developed with the inclusion of important physical effects such as velocity saturation (VS) and hot carrier effects (HCE). In this RF model, the influence of the VS effect has been included in the core model; however, the contribution of the HCE to thermal noise is not considered even though the influence of impact ionization (and hence HCE) to the channel conductance has been incorporated in the DC model [12]. In Fig. 9, a discrepancy in R_n characteristics between the model and the measured data at V_{GS}=2V has been found. The inaccuracy in either DC or AC models can result in this discrepancy and here we believe this was caused by the inaccurate capacitance prediction of the model at that bias condition since an obvious disagreement in the simulated and measured imaginary part of y_{12} was found and the contribution from capacitive components to R_n becomes comparable to that from the transconductance at HF.

The noise characteristics of several noise models including the subcircuit RF model discussed above are verified with the extracted channel thermal noise with the discussed methodology to explore the physical nature and accuracy of the models. Fig. 10 shows the curves of the channel thermal noise vs. bias current, from the measured data, and simulations of the RF model and several other noise models. It shows that the calculated channel thermal noise based on the equation $i_d^2 = 8kTG_m/3$, $i_d^2 = 8kTG_{ds}/3$, and $i_d^2 = 8kT(G_m + G_{ds})/3$, where G_{ds} is the channel

conductance, cannot predict the channel thermal noise extracted from measured data. The noise prediction of this subcircuit RF model has much better accuracy at the given bias conditions.

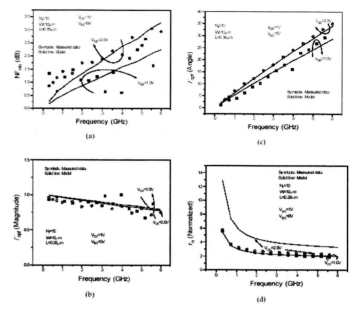

Fig. 9 Comparisons of measured data for minimum noise figure, NF_{min}, the magnitude of the optimized source reflection coefficient, G_{opt}, the phase of the optimized source reflection coefficient, Γ_{opt}, the noise resistance normalized to 50Ω, r_n, with simulations at different bias conditions [8].

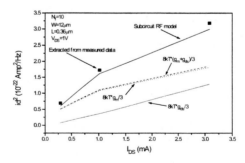

Fig. 10 Power spectral densities of channel thermal noise vs. bias current of a 0.36 μm n-channel MOSFET. They are extracted from the measured data and calculated from different channel thermal noise models [8].

3.2 Induced-Gate Noise

The concept of the induced-gate noise has been introduced for three decades [15]. But it is still an issue that many researchers are debating the existence of this noise source. At high-frequencies, it is believed that the local channel voltage fluctuations due to thermal noise couple to the gate through the oxide capacitance and cause an induced gate noise current to flow [16]. This noise current can be modeled by a noisy current source connected in parallel to the intrinsic gate-to-source capacitance C_{gsi}. Since the physical origin of the induced gate noise is the same as for the channel thermal noise at the drain, the two noise sources are partially correlated with a correlation factor [7].

As shown in Fig. 11, in a 0.18um technology, the devices with shorter channel length has negligible induced gate noise compared with the channel noise of the same devices shown in Fig. 12 [17]. However, the induced gate noise becomes comparable for devices with longer channel length at higher frequency.

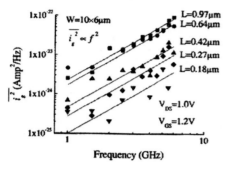

Fig. 11 Measured IGN of devices in a 0.18um process [17].

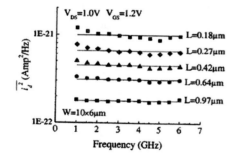

Fig. 12 Measured channel noise of devices in Fig. 11 [17].

Currently, the induced gate noise and moreover its correlation to the thermal noise at the drain are not implemented completely in compact models yet. A further detailed investigation is needed to understand the induced-gate noise issue and model it correctly.

4 LINEARITY MODELING

For a MOSEFT transistor, its distortion is primarily caused by the non-linearity of I_d. As a first-order, the output power at the fundamental tune of the device is basically correlated to its transconductance G_m, while the second and the third harmonic tunes are correlated to the first- and second-order derivatives of G_m. Therefore, to ensure the model can predict the distortion behavior, the voltage

dependences of threshold voltage V_{th}, mobility and I_d (and hence G_m) have to be precisely captured in the extraction.

Besides, as for a properly tuned circuit operating at high power, the output voltage swing can be very large. Therefore, the voltage conductance G_{out} should also be precisely modeled. Furthermore, to be able to predict the distortion at RF, the capacitance and charge components, which in turn determine the transit times of the device, of the model have to be properly extracted.

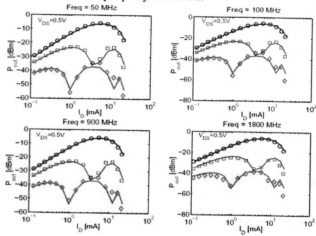

Fig. 13 Modeled (solid lines) and measured (symbols) P_{out}'s vs drain current I_{ds} for the 10x12x0.36 device at different frequencies when V_{ds}=0.5V and P_{in}=-10dBm; (a) f_{in}=50M, (b) f_{in}=100M, (c) f_{in}=900MHz, and (d) f_{in}=1.8GHz [18].

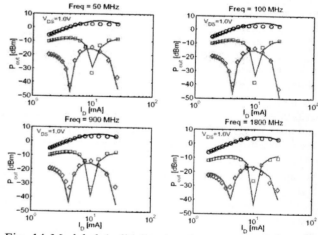

Fig. 14 Modeled (solid lines) and measured (symbols) P_{out}'s vs. I_{ds} for the 10x12x0.36 device at different frequencies when V_{ds}=1.0V and P_{in}=0dBm; (a) f_{in}=50M, (b) f_{in}=100M, (c) f_{in}=900MHz, and (d) f_{in}=1.8GHz [18].

Fig. 13 shows the fitting of the distortion simulations against measurements for a transistor operating at linear region, while Fig. 14 shows the fitting of the model for a transistor operating at saturation region [18]. The poor fitting in the region when P_{out}>0dBm results from the limited dynamic range of the measurement system. The model can accurately describe the fundamental, second and third harmonics over a wide range of bias conditions and frequencies. The model is generated based on DC and AC measurements without any extra tuning with regards to the large-signal distortion. This demonstrates that a model with good fittings in both DC (both current and the derivatives) and AC (both capacitance and s-parameters) characteristics can predict the distortion behavior of a transistor up to a operating frequency at which the device is still capable of delivering decent power gain.

The modeling methodology presented is based on the quasi-static (QS) condition, which assumes that the small-signal AC characteristics of a device can be derived from its DC/static characteristics. We have demonstrated that such an approach can yield an accurate model for distortion prediction up to a few GHz, although none of the non-quasi-static (NQS) effects are accounted.

5 SUMMARY

In this paper, we have discussed some important issues in RF MOSFET modeling. The modeling of parasitic components in MOSFETs is necessary to describe the HF behavior of MOS devices at GHz frequency. An accurate RF MOSFET model with a simple substrate network is presented. The model has been verified by high frequency measurements for AC small signal, noise and distortion. Good model accuracy at different bias conditions has been found for devices with different channel length, width and fingers. The developed RF MOSFET model can be the basis of a predictive and statistical modeling approach for RF applications.

The modeling approaches of NQS effects have been analyzed. A RF model including the NQS effect is desirable without introducing complex implementation and simulation time penalty.

REFERENCES

[1] H. Iwai, "Current status and future of advanced CMOS technologies," ASDAM'98, pp. 1-10, 1998.

[2] D. R. Pehlke, M. Schroter, A. Burstein, M. Matloubian and M. F. Chang, "High-Frequency Application of MOS Compact Models and their Development for Scalable RF MOS Libraries," Proc. IEEE Custom Integrated Circuits Conference, pp. 219-222, May 1998.

[3] J.-J. Ou, X. Jin, I. Ma, C. Hu and P. Gray, "CMOS RF Modeling for GHz Communication IC's," Proc. of the VLSI Symposium on Technology, pp. 94-95, June 1998.

[4] S. H. Jen, C. Enz, D. R. Pehlke, M. Schroter, B. J. Sheu, "Accurate MOS Transistor Modeling and Parameter Extraction Valid up to 10-GHz," Proc. of the European Solid-State Device Research Conference, Bordeaux, pp. 484-487, Sept. 1998.

[5] Y. Cheng, M. Schroter, C. Enz, M. Matloubian and D. Pehlke, "RF Modeling Issues of Deep-submicron MOS-FETs for Circuit Design," Proc. of the IEEE International Conference on Solid-State and Integrated Circuit Technology, pp. 416-419, Oct. 1998.

[6] W. Liu, R. Gharpurey, M. C. Chang, U. Erdogan, R. Aggarwal and J. P. Mattia, "R.F.MOSFET Modeling Ac-counting for Distributed Substrate and Channel Resistances with Emphasis on the BSIM3v3 SPICE Model," Technical Digest of International Electron Devices Meeting, pp. 309-312, Dec. 1997.

[7] C. Enz and Y. Cheng, "MOS Transistor Modeling for RF IC Design," IEEE Journal of Solid-State Circuits, Vol. 35, no. 2, pp. 186-201, 2000.

[8] Y. Cheng et al, "High Frequency Small-signal AC and Noise Modeling of MOSFETs for RF IC Design," IEEE Trans. on Electron Devices, Volume: 49 Issue: 3 , pp. 400-408, 2002.

[9] Y Tsividis, Operation and Modeling of the MOS Transistor, 2nd Edition, Mc-Graw_Hill, 1999

[10] X. Jin, J. -J. Ou, C.-H. Chen, W. Liu, M. J. Deen, P. R. Green and C. Hu, "An Effective Gate Resistance Model for CMOS RF and Noise Modeling," Technical Digest of International Electron Devices Meeting, Dec., p. 981, 1998.

[11] Y Cheng, M. Chan, K. Hui, M. Jeng, Z. Liu, J. Huang, K. Chen, P. Ko, and C. Hu, BSIM3v3.1 user's manual, Memorandum No. UCB/ERL M97/2, 1997

[12] Y. Cheng and C. Hu, MOSFET Modeling & BSIM3 User's Guide, Kluwer Academic publishers, 1999.

[13] C. H. Chen and M. J. Deen, "High Frequency Noise of MOSFETs I: Modeling," Solid-State Electronics, vol. 42, pp. 2069-2081, Nov. 1998.

[14] C. H. Chen, M. J. Deen, M Matloubian, and Y Cheng, "Extraction of Channel Thermal Noise of MOSFETs," International Conference on Microelectrics Test Structures, pp. 42-47, 2000.

[15] H. E. Halladay and A. Van der Ziel, "On the High Frequency Excess Noise and Equivalent Circuit Representation of the MOSFET with n-type Channel," Solid-State Electronics, vol. 12, pp. 161-176, 1969.

[16] D. P. Triantis, A. N. Birbas and S. E. Plevridis, "Induced Gate Noise in MOSFETs Revisited: The Submicron Case," Solid-State Electronics, vol. 41, No. 12, pp. 1937-1942, 1997.

[17] C. H. Chen, M. J. Deen, M. Matloubian and Y. Cheng, "Extraction of the induced gate noise, channel thermal noise and their correlation in sub-micron MOSFETs from RF noise measurements," Proceedings of IEEE international conference on microelectronic test structures, Kobe, Japan, March, 2001.

[18] T. Lee and Y. Cheng, "MOSFET HF distortion behavior and modeling for RFIC design," pp. 138-142, CICC 2003.

RF Noise Models of MOSFETs- A Review

S. Asgaran, and M. Jamal Deen

Electrical and Computer Engineering Department, CRL 226
McMaster University, Hamilton, ON L8S 4K1, Canada, jamal@mcmaster.ca

ABSTRACT

A thorough study of high frequency MOSFET noise compact modeling with emphasis on channel thermal noise modeling is presented. Although the modeling of MOSFET noise dates back to many years ago, the enhanced noise generated in short channel MOSFETs has made researchers revisit the problem to develop better models, especially in recent years. In this review, a detailed discussion of the most recent models published in the literature is provided. Each model is investigated in terms of physical and analytical aspects of the model, and some comments are made in each case. The induced gate noise, which is important in today's high frequency applications, is also briefly reviewed at the end of this paper.

Keywords: MOSFET, compact modeling, thermal noise

1 INTRODUCTION

The ever decreasing channel length of MOSFET has enabled circuit designers to design circuits that can operate at very high frequencies. An important application of these circuits can be found in wireless communications where most often the signal received by the receiver is so small that only a limited amount of noise can be tolerated in the system. Therefore, it is very important for circuit designers to be able to predict the noise of MOS devices with reasonable accuracy and also to understand the noise dependence on the geometrical and biasing conditions of the device. Thus, a simple and yet accurate compact noise model is called for.

Two types of noise are present in MOSFET in strong inversion at moderate frequencies: Flicker noise and thermal noise. Although, the effect of flicker noise at high frequencies is negligible, another type of noise, which is generated due to the capacitive coupling between the MOSFET channel and its gate, arises at very high frequencies and is called induced gate noise. Fig. 1 shows the contributions of different types of noise in MOSFET versus frequency [1].

The noise sources that generate thermal noise in MOSFET are the channel, S_{iD}, the actual resistances associated with the device electrodes, i.e. R_G, R_D, R_S, resistances between the substrate and source/drain, i.e. R_{DB}, R_{SB}, R_{DSB}, and the induced gate noise, S_{iG}.
Fig. 2 shows the equivalent compact circuit MOSFET model, including the noise sources, which is suitable for high frequency applications [2].

In this paper we present a comprehensive review of major channel thermal noise models reported hitherto in the literature, followed by a discussion on the induced gate noise and other noise sources in MOSFET.

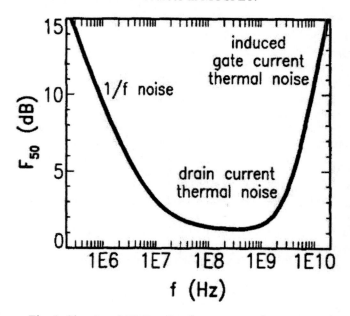

Fig. 1: Simulated 50-Ω noise figure versus frequency, for a 1000/0.25 (μm/μm) n-channel MOSFET [1].

2 MOSFET CHANNEL THERMAL NOISE

Modeling of the thermal noise generated in the MOSFET channel goes back to a few decades ago. The well-known Klaassen-Prins [3] model gives the MOSFET channel noise as

$$S_{iD} = \frac{4kT}{L^2 I_D} \int_0^{V_{DS}} g^2(V_0)dV_0 . \qquad (1)$$

Here, k is the Boltzman constant, T is the device temperature, L is the MOSFET channel length, I_D is the drain current of the device, V_0 is the dc potential at point x along the channel with reference to the source electrode, and $g(V_0)$ is the channel conductance at point x. This model is obtained by calculating the current noise of small sections of the MOSFET channel and integrating the noise currents all over the channel. Van der Ziel [4] includes the so-called hot electron effects in the model by replacing the lattice temperature with carrier temperature, $T_e(x)$, and modified the model to

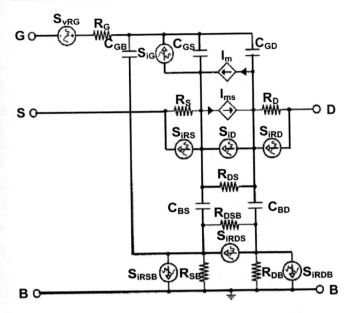

Fig.2: The equivalent compact circuit model of the MOSFET, which is suitable for high frequency applications [2].

$$S_{iD} = \frac{4kT}{L^2 I_D} \int_0^{V_{DS}} \frac{T_e(x)}{T} g^2(V_0) dV_0 \,. \qquad (2)$$

It is assumed that the relation between these two temperatures can be described with a quadratic function as [4], [5]

$$T_e(x) = T\left\{1 + \delta\left[\frac{E(x)}{E_C}\right]^2\right\}, \qquad (3)$$

where δ is an empirical constant, $E(x)$ is the longitudinal electric field along the channel, and E_C is the critical channel field beyond which the carriers are velocity saturated. It is experimentally shown in [5] that this relation is inadequate for the channel fields beyond E_C, and $T_e(x)$ dependence on $E(x)$ is better described by an exponential relation.

Using a similar approach, a rather simpler model is developed by Tsividis [6] as

$$S_{iD} = \frac{4kT\mu}{L^2} Q_{inv} \,, \qquad (4)$$

where μ is the mobility of carriers in the channel and Q_{inv} is the total inversion layer charge.

The models presented in (1) and (4) are developed for long channel devices, where the carriers are in thermal equilibrium with the lattice. However, as the MOSFET channel becomes shorter, significant increase in the drain current noise is observed [7], as shown in Fig. 3

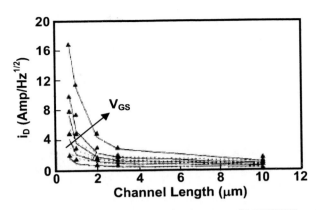

Fig.3: Channel thermal noise current of MOSFETs with different channel lengths, and different operating points (V_{GS}) [7].

In the past few years, several models have been developed to explain this enhanced noise. In these models, the MOSFET channel is divided into two sections as depicted in Fig. 4. The first section is called the linear region, where the gradual channel approximation (GCA) holds, i.e. the channel field is much smaller than the critical field. The second part is where the carriers travel at their saturation velocity. The difference between different models is the approach taken to model the noise of each of the two sections of the channel. The following is a review of major channel thermal noise models published recently.

3 CHANNEL THERMAL NOISE IN SHORT CHANNEL MOSFETS: A REVIEW

3.1 Model of Park et al.

The drain current noise spectral density of MOSFET is given in [8] as

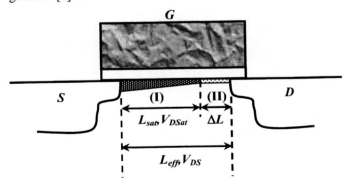

Fig. 4: Cross section of the channel of a MOSFET, divided into two regions: (I) gradual and (II) velocity saturation regions. L_{sat} is the point in the channel where the channel field $E(x)$ is equal to the critical field, E_c, and the velocity of carriers is equal to their saturation velocity, v_{sat}. ΔL is the length of the pinch-off or velocity saturated region.

$$S_{iD} = g_{DS}^2 \left\{ \frac{4kTV_c}{I_D} \left[\frac{\frac{\alpha^2}{3} V_C^2 - \alpha V_{GT} V_C + V_{GT}^2}{(V_{GT} - \alpha V_C)^2} + \delta \right] \times \right.$$
$$\left. \cosh^2(L_{sat}/l) + \frac{4qDI_D}{3\varepsilon_s^2 W^2 x_j^2 v_s^3} L_{sat}^3 \right\}. \quad (5)$$

Here, g_{DS} is the channel conductance, V_C is the channel potential at the point where $E(x)=E_C$, α is the bulk-charge effect, $V_{GT}=V_{GS}-V_{th}$ where V_{th} is the MOSFET threshold voltage, δ is the same as δ in (3), L_{sat} is the length of the pinch-off region of the channel in saturation and equal to $L_{sat}=l\sinh^{-1}((V_{DS}-V_C)/E_Cl)$, where $l=(x_jt_{ox}\varepsilon_s/\varepsilon_{ox})^{1/2}$, D is the high field diffusion constant, W is the width of the device, x_j is the drain junction depth, q is the electron charge, and v_s is the saturated velocity of the dipole layer.

In this model, the voltage noise in the linear part of the channel is calculated using the simple drain current model provided by BSIM [9]. However, the mobility degradation effect due to the longitudinal electric field in this part of the channel is not taken into account, which in turn could result in reduced accuracy of the model, especially in very short channel devices. The carrier temperature, i.e. T_e in (3), is used in the linear part of the channel to account for hot electrons.

In this model, the noise in the velocity saturation part of the channel is believed to come from intrinsic diffusion noise sources in that region. The total drain current noise of MOSFET is the product of the channel conductance and the sum of noise voltages generated in the two sections of the channel. This way of calculating the total drain current noise is not correct simply because the MOSFET channel conductance is a function of the position along the channel and is not constant.

Moreover, this model has been verified experimentally by using the measured data presented by Abidi in [10], which may not be as reliable as recent measured noise data. This is because his noise measurements were carried out under very high V_{DS}, i.e. $V_{DS}=4$ V, and hence a significant part of the noise comes from avalanche noise, which is caused by drain to bulk current [11].

3.2 Model of Triantis et al. [12]

Similar to the model presented in [8], this model calculates the *voltage* noise generated in the two sections of the channel first and the total drain current noise is the product of the channel conductance and the total voltage noise of the channel. Carrier temperature is used in both the linear and velocity saturated parts of the channel, with different models used to model the temperature of the carriers.

In this model, it is assumed that the noise generated in the second part of the channel is thermal noise generated from small sections of Δx, located at point x, with resistance Δr_D, given by

$$\Delta r_D = \frac{E_{II}(x)\Delta x}{I_D} \quad (6)$$

where $E_{II}(x)$ is the longitudinal electric field at point x in the second part of the channel and I_D is the drain current. This way of modeling the noise is questionable because of the fact that Δr_D is an ac resistance and does not contribute any thermal noise.

To calculate the voltage fluctuations at the drain side of the channel, it is assumed that the electric field in the linear part of the channel is simply the voltage across this region divided by the length of the region. This assumption is not correct, especially in very short channel devices since in the linear part of the channel, the electric field is a nonlinear function of the position [13]. To experimentally verify the model, the measured data presented in [10] is used in [12].

3.3 Model of Knoblinger et al.

The channel noise of MOSFET is given in [14] in a closed-form as

$$S_{iD} = \frac{4kT}{L^2} \mu_{eff} Q_{inv} + \delta \frac{4kTI_D}{L^2 E_C^2} V_{DS}$$
$$+ \frac{4kTI_D}{L^2 E_C} \frac{2}{\alpha} \left\{ \arctan\left[\exp(\alpha\Delta L)\right] - \arctan(1) \right\} + \quad (7)$$
$$\delta \frac{4kTI_D}{L^2 E_C} \frac{1}{\alpha} \sinh(\alpha\Delta L) \cdot$$

Here, μ_{eff} is the effective carrier mobility, ΔL is the length of the pinch-off region, α is the inverse of l, which is defined before, and the rest of the parameters are the same as before.

On the contrary to the previously mentioned models, this model calculates the *current* noise generated from each part of the channel by integrating the elemental thermal noise currents generated from small sections of the channel.

The hot electron effect is included in the model by using the carrier temperature, (3), in both parts of the channel with different models for the longitudinal electric field, $E(x)$. It is interesting to note that in this model, the carrier mobility is defined as $\mu(x)=v(x)/E(x)$ in both linear and saturation parts of the channel. This is of course not true because of the fact that the carriers are velocity saturated in the second part and using this relationship is not very meaningful.

3.4 Model of Scholten et al.

Based on the previously mentioned Klaassen-Prins model, the current noise spectral density of MOSFET is given in [1] as

$$S_{iD} = \frac{1}{L^2 I_D} \int_0^{V_{DSat}} 4kT_e(x)g^2(V)dV \qquad (8)$$

where T_e is the temperature of the carriers, and $g(V)$ is given by

$$g(V) = W\mu Q_{inv}(V) \qquad (9)$$

where V is the quasi-Fermi potential. The effect of velocity saturation on the carrier mobility is modeled by (10) as

$$\mu = \frac{\mu_{eff}}{\left[1 + \left(\dfrac{E}{E_C}\right)^p\right]^{1/p}} \qquad (10)$$

where μ_{eff} is the effective mobility and $p=1$ and 2 for p-channel and n-channel MOSFETs, respectively [1].

The relation between T and T_e in this model is given as

$$T_e = T\left(1 + \frac{E}{E_C}\right)^n \qquad (11)$$

where n is between 0 and 2. A comparison between measured and modeled channel thermal noise currents using the model in (8) is given in [1] and is shown in Fig. 5. It can be seen from Fig. 5, that the measured values are best modeled when there is no carrier heating, i.e. $n=0$. Therefore, introducing the carrier temperature [8], [12], [14], to model the channel thermal noise is questionable.

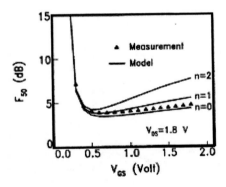

Fig. 5: Measured (symbols) and modeled (lines) 50-Ω noise figure of an NMOSFET with channel length L=0.18 μm versus the gate to source voltage, V_{GS} [1].

3.5 Model of Chen et al.

A simple model similar to the model presented in [6] is developed in [15] as

$$S_{iD} = \frac{4kT\mu_{eff}}{L_{elec}^2}Q_{inv} + \delta\frac{4kTI_D}{L_{elec}^2 E_C^2}V_{DSat}. \qquad (12)$$

This model is based on the model in (4), with L replaced by L_{elec}, which is the electrical channel length of the

MOSFET defined as $L_{elec}=L_{eff}-\Delta L$, where L_{eff} is the effective channel length and ΔL is the length of the pinch-off region. The second term in (12) is added to account for the carrier heating effect. However, in experimental verification of (12), δ is set to 0, and very good agreement with the measured data is achieved [15]. This is in agreement with [1] that no carrier heating is needed to model the channel thermal noise.

To develop (12), the channel thermal noise generated from the first section of the channel is obtained by integrating the noise currents of small sections of the channel from the source side of the channel to L_{elec}. It is argued in [15] that the noise generated from the velocity saturated part of the channel is negligible. This is because in this part of the channel, the carriers do not respond to the local electric field fluctuations, caused by the voltage noise. This argument is experimentally verified in [15], where it is shown that the noise dependence on the drain to source voltage is negligible, as shown in Fig. 6.

3.6 Model of Scholten et al.

A revised version of the model given in (8) is presented in [16]. However, in agreement with [15] the electrical channel length is used instead of L, to include the channel length modulation effect. The velocity saturation effect is accounted for via the local channel conductance, $g(V)$ [16]. The noise contribution from the second part of the channel is neglected in this model following the argument made in [15].

A closed-form expression, based on (8), is given in [14] for the drain current noise spectral density, (eq. (24) in [16]).

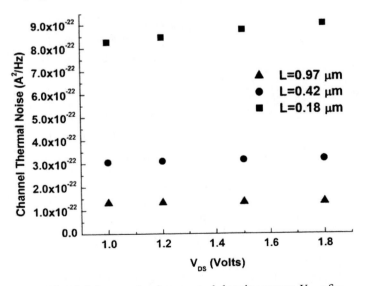

Fig. 6: Measured noise spectral density versus V_{DS} of n-type MOSFETs with channel width W=60 μm and channel lengths L=0.97, 0.42 and 0.18 μm, biased at V_{GS}=1 V. Measured noise data are from [15].

However, since this expression is a function of the channel surface potential, it is not suitable for fast calculation and does not show the explicit dependence of the noise current on the biasing conditions of the MOSFET.

In addition, the experimental data given in [16] show that the model presented in [6] cannot accurately predict the noise of deep-submicron devices at relatively high gate to source voltages, as shown in Fig. 7. This may be due to the approximations made to calculate the channel conductance, $g(V)$.

4 INDUCED GATE NOISE

The MOSFET channel acts as a distributed RC network at high frequencies [4]. The capacitive coupling to the gate represents the distributed capacitance and the MOSFET channel represents the distributed resistance. Therefore, the voltage fluctuation in the MOSFET's channel is coupled to its gate, causing the induced gate noise current to flow. This noise is modeled by a current source connected in parallel with C_{gs} (see Fig. 2). Unlike channel thermal noise, the power spectral density of induced gate noise is dependent on the frequency of operation an is given as [2]

$$S_{iG} = 4kT\delta \frac{\omega^2 C_{gs}}{5g_{ms}} \qquad (13)$$

where g_{ms} is the source transconductance, and δ is a bias dependent factor equal to 4/3 in long channel devices.

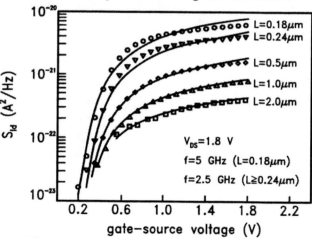

Fig. 7: Measured (symbols) and modeled (lines) drain current noise spectral density of MOS devices with different channel lengths versus V_{GS}. The drain to source voltage is fixed at 1.8 V [16].

Since the channel thermal noise and induced gate noise have the same origin, they are correlated with a correlation factor c, which is equal to $j0.395$ in long channel devices, and slightly smaller (between $j0.35$ to $j0.3$) in short channel devices [17].

The induced gate noise, S_{iG}, and the correlation noise, $\overline{i_g i_d^*}$, of NMOSFETs with different channel length versus frequency are shown in Figs. 8 and 9, respectively [18]. It has been found that the induced gate noise and the correlation noise are proportional to f^2 and f, respectively.

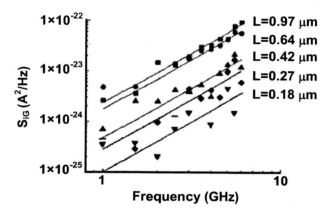

Fig. 8: Induced gate noise of NMOSFETs with different channel length biased at V_{DS}=1 and V_{GS}=1.2 V, versus frequency [18].

It can also be seen from Figs. 8 and 9 that both noises decrease with decreasing the channel length because of the reduction in the gate to source capacitance, C_{gs}.

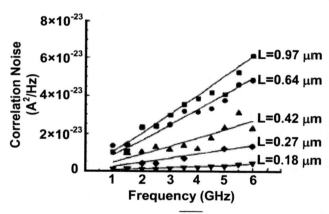

Fig. 9: Correlation noise, $\overline{i_g i_d^*}$, of NMOSFETs with different channel length biased at V_{DS}=1 and V_{GS}=1.2 V, versus frequency [18].

The dependence of the induced gate noise on V_{GS} and V_{DS}, is shown in Figs. 10 and 11, respectively [16].

5 THERMAL NOISE FROM PARASITIC RESISTANCES

As mentioned before, four types of resistance exist in a MOSFET: gate resistance, source and drain resistance, and the resistance between substrate and source/drain electrodes (see Fig.2). The gate resistance has a strong impact on the maximum oscillation frequency of the MOSFET, and the source/drain resistances have an impact on the current drive capability of the MOSFET [19]. The thermal noise associated with these resistances can be easily expressed as

$$S_{iD} = \frac{4kT}{R} \qquad (14)$$

where R is the value of each resistance, which depends on the extraction method used. It should be noted that different extraction techniques yield slightly different values [19].

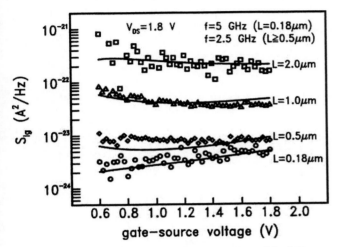

Fig.10: Induced gate noise of MOSFETs with different channel length versus the gate to source voltage V_{GS} [16].

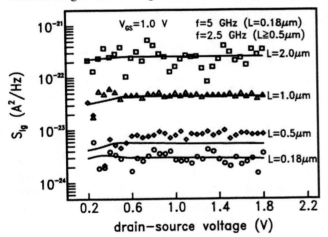

Fig.11: Induced gate noise of MOSFETs with different channel length versus the drain to source voltage V_{DS} [16].

6   CONCLUSIONS

A review of the MOSFET high frequency thermal noise modeling, with emphasis on channel thermal noise, is presented. Several channel noise models are reviewed and discussed in detail. The models are common in the fact that they divide the MOSFET channel into two sections of linear and velocity saturation. However, each model has its own approach of modeling the noise in each part of the channel. Most of these models use a carrier temperature, which is a function of the lattice temperature to model the hot electron effects. However, it is shown that there is no need to include the carrier temperature to model the noise. In addition, the induced gate noise modeling, which is important at high frequencies, is briefly reviewed together with the noise generated from parasitic resistances of MOSFET.

Since the demand for low power circuit design is increasing everyday, the operating region of MOSFET is being pushed to weak and moderate inversion. However, there has been little published data on MOSFET noise modeling in these regions of operation especially at high frequencies. Therefore, in the near future, analytical noise models that are valid in moderate and weak inversion regions of operation will be required.

REFERENCES

[1] A. J. Scholten et al. *IEDM* 1999, pp. 155-158.

[2] C. Enz, *IEEE Trans. Microwave Theory Tech.*, vol. 50, no. 1, Jan. 2002.

[3] F. M. Klaassen, and J. Prins, "Thermal noise of MOS transistors," *Philips Res. Reps.*, vol. 22, pp. 504-514, 1967.

[4] A. van der Ziel, *Noise in Solid State Devices and Circuits*. New York: Wiley, 1986.

[5] D. Gasquet, Ph.D. thesis, Universite de Montpellier II. P. Glinelli et al., "Noise temperature and hot carrier thermal conductivity in semiconductors," in second *ELEN Workshop*, Grenoble, France, 1995.

[6] Y. P. Tsividis, *Operation and Modeling of the MOS transistor*. New York: Wiley, 1987.

[7] P. Klein, *IEEE Electron Dev. Letters*, vol. 20, no. 8, pp. 399-401, Aug. 1998.

[8] C. H. Park, and Y. J. Park, *Solid State Elctronics*, vol. 44, pp. 2053-2057, 2000.

[9] B. J. Sheu, D. L. Scharfetter, P-K. Ko, and M-C. Jeng, *IEEE J. Solid-State Circuits*, vol. sc-22, pp.558-566, Aug. 1987.

[10] A. Abidi, *IEEE Trans. Electron Devices*, ED-33, no. 11, NOV. 1986.

[11] A. J. Scholten et al. *IEDM* 1999, pp. 129-132.

[12] D. P. Triantis, A. N. Birbas, and D. Kondis, *IEEE Trans. Electron Devices*, vol. 43, no. 11, pp.1950-1955, Nov. 1996.

[13] K. Takeuchi, and M. Fukuma, *IEEE J. Solid-State Circuits*, vol. 41, pp. 1623-1627, Sep. 1994.

[14] G. Knoblinger, P. Klein, and M. Tiebout, *IEEE J. Solid-State Circuits*, vol. 36, pp. 831-837, May 2001.

[15] C. H. Chen and M. J. Deen, *IEEE Trans. Electron Devices*, vol. 49, pp. 1484-1487, Aug. 2002.

[16] A. J. Scholten, L. F. Tiemeijer, R. van Langevelde, R. J. Havens, A. T. A. Zegeres-van Duijnhoven, and V. C. Venezia, *IEEE Trans. Electron Devices*, vol. 50, pp. 1-15, Mar. 2003.

[17] T. Manku, *IEEE J.Solid-State Circuits*, vol. 34, pp. 277–285, Mar. 1999.

[18] M. J. Deen, and C. H. Chen, *IEEE Custom Integrated Circuits Conference*, pp. 201-208, May 2002.

[19] C. H. Chen, and M. J. Deen, *Solid-State Electronics*, vol. 42, no. 11, pp. 2069-2081, 1998.

Modeling of charge and collector field in Si-based bipolar transistors

H. Tran and M. Schroter

Chair for Electron Devices and Integrated Circuits, Dresden University of Technology
Mommsenstr. 13, 01062 Dresden, Germany, mschroter@ieee.org

ABSTRACT

An analytical formulation for the voltage and current dependent electric field in the collector of a bipolar transistor is presented. The new field expression is then employed for calculating the base-collector depletion capacitance and the field related transit time components. Comparison to device simulation results show good agreement.

1 Introduction

Rapidly increasing mask cost and reduced design cycles are putting increasing pressure on EDA tool capabilities, including compact models. Furthermore, Si/SiGe bipolar technology development has accelerated tremendously over the past couple of years, leading to an increasing boost for advanced compact bipolar transistor models and modeling methodologies. For instance, the capability for predictive and statistical modeling (and design) has become an important requirement from design houses for many foundries. Such a capability in turn requires physics-based compact models and extraction strategies.

The electrical behavior of bipolar transistors, regardless whether they include a heterojunction at the collector or not, is strongly determined by the conditions in the collector region, especially at medium and high current densities, i.e. at peak f_T and beyond. One of the key variables here is the electric field in the collector, in particular at the base-collector (BC) junction, which is linked to the BC capacitance, minority carrier density in base and collector, and the avalanche breakdown. In III-V HBTs, and possibly also in future SiGe HBTs, velocity overshoot occurs in the collector region, causing the conventional model formulations for the above mentioned quantities and effects to become inaccurate. In addition, certain parameter extraction methods such as determining the transit time from $1/(2\pi f_T)$ vs $1/I_C$ cannot be easily applied anymore [1].

In this work, a first version of a bias dependent analytical model for the electric field in the collector is presented. Based on this formulation, the current dependent BC depletion capacitance and transit time are described. The resulting model equations are compared to 1D device simulations in order to verify the fundamental suitability and accuracy, and to possibly identify areas of improvement.

2 Investigated technology

The SiGe HBT under investigation contains a "conventional" doping profile as shown in Fig. 1, with a high emitter and moderate base concentration. The collector

doping corresponds to that of a "power" or high-voltage transistor type in such processes. The peak transit frequency of this transistor is about 35GHz at $V_{BC}=0V$.

Since the investigations in this paper are based on 1D device simulations, currents, charges and capacitances are normalized to the unit area $A_E=1\mu m^2$.

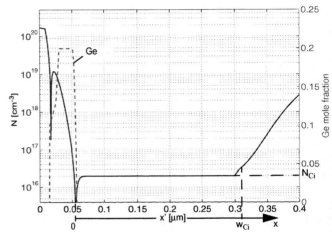

Fig. 1 Doping profile of the investigated transistor. The dashed line indicates the Ge profile.

3 Modeling the electric field

According to Gauss' law, the charge Q_{BC} in the collector region is associated with the electric field E_{jc} at the BC junction (x=0 in Fig. 1) via the following relation,

$$Q_{BC}(V_{BC}, I_{Tf}) = \varepsilon E_{jc}(V_{BC}, I_{Tf}) \ . \qquad (1)$$

Since it is more appropriate from a modeling and (circuit) application point of view, the (internal) BC terminal voltage V_{BC} and the quasi-static forward transfer current I_{Tf} have been selected as independent variables rather than V_{BC} and V_{BE}. For quasi-static operation, Q_{BC} can be obtained from a path independent integration over the independent variables. Thus, one can write for the incremental change

$$dQ_{BC}(V_{BC}, I_{Tf}) = C_{jCi} \, dV_{BC} + \tau_{BC} \, dI_{Tf} \qquad (2)$$

where the variables

$$C_{jCi}(V_{BC}, I_{Tf}) = \left. \frac{\partial Q_{BC}}{\partial V_{BC}} \right|_{I_{Tf}} \text{ and } \tau_{BC}(V_{BC}, I_{Tf}) = \left. \frac{\partial Q_{BC}}{\partial I_{Tf}} \right|_{V_{BC}} \qquad (3)$$

can be defined as bias dependent (internal) BC depletion capacitance and BC transit time, respectively.

According to (1), the above elements can be calculated as a

function of bias, once the electric field is known. This also holds for other elements in a transistor, such as the base portion of the transit time and the avalanche current, which are strongly current dependent and have been difficult to describe sufficiently physics-based. In addition, "non-stationary" transport effects can be included in a compact model to first order. In practice, the difficulty though has been to find a sufficiently *simple* and *continuously differentiable* formulation for E_{jc} that at the same time ensures *accurate* derivatives.

Generally, the electric field in the collector can be obtained from solving Poisson's equation. However, suitable compact analytical formulations require certain simplifying assumptions, mainly

- a spatially independent doping concentration N_{Ci} within the (lightly doped) collector region w_{Ci}, and
- an abrupt transition at w_{Ci} to the much larger buried layer doping concentration.

The consequence of these assumptions is that two mathematically different solutions are obtained, depending on the bias point (V_{BC}, I_{Tf}), which are repeated here for further discussion. For a partially depleted collector holds (note that E_{jc} is defined as positive here)

$$E_{jc} = E_{wc} + \sqrt{\frac{2qN_{Ci}}{\varepsilon}} \sqrt{\left(1 - \frac{I_{Tf}}{I_{lim}}\right)(v_{ceff} - E_{wc}w_{Ci})} \quad (4)$$

with the field at the buried layer side at $x = w_{Ci}$,

$$E_{wc} = \frac{I_{Tf}}{qA_E\mu_{nCi}(E_{wc})N_{Ci}} \quad . \quad (5)$$

Here, μ_{nCi} is the field dependent mobility in the collector; $v_{ceff} = V_{DCi} - V_{BC}$ is the effective voltage across the collector region [0, w_{Ci}] with V_{DCi} as built-in voltage; finally, $I_{lim} = qN_{Ci}v_{sn}$ with v_{sn} as electron saturation velocity.

In contrast, for a fully depleted collector holds

$$E_{jc} = \frac{v_{ceff} + V_{PT0}(1 - I_T/I_{lim})}{w_{Ci}} \quad (6)$$

with the low-current punch-through voltage

$$V_{PT0} = \frac{qN_{Ci}}{2\varepsilon}w_{Ci}^2 \quad . \quad (7)$$

The above solutions were employed in, e.g., [2][3] for modeling the base transit time. However, there are several issues with the above formulations regarding their use in compact models. First of all, the square root dependence (4) on V_{BC} is not suitable for describing the BC depletion capacitance accurately enough for realistic applications. This will be discussed later on more detail. Second, the equations contain various numerical instabilities (i.e. conditions causing arithmetic errors) that have to be taken care of properly to arrive at a *reliable* compact model formulation. For non-zero current, there is neither a continuous first derivative of E_{jc} with respect to current or

voltage nor a smooth transition from high to low voltages. Third, the equations are not valid in the high-current region (i.e. close to peak f_T and beyond). Some of these issues were addressed in [2], but a satisfying reliable formulation was not obtained; also, the square-root dependence was maintained, so that modeling of C_{jCi} could not be addressed.

Fig. 2 and Fig. 3 show the field and carrier distributions in the drift-type SiGe HBT investigated here. While the distributions in the collector are similar to that of the low-emitter concentration transistor in [2], they are quite different in the base region, where both the doping and Ge gradient cause a large field that influences E_{jc} at high current densities. Fig. 2 also contains a comparison between the electrostatic field, E_ψ, and the field E_φ that is defined by the gradient of the electron quasi-fermi potential. The latter is the actual driving force for the current in the BC region, and thus should be used in the model.

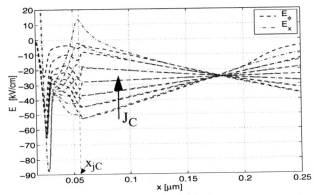

Fig. 2 Electric field distribution in base and collector region for selected bias points.
J_C/[mA/µm^2] = 0.074, 0.155, 0.225, 0.3, 0.37, 0.45, 0.51.

The consequences of the high electric field in the base are a dip in the electron density as shown in Fig. 3. Thus, the simple diffusion-type model applied in [2] to the transit time has to be generalized to make a model applicable to a wide range of process technologies.

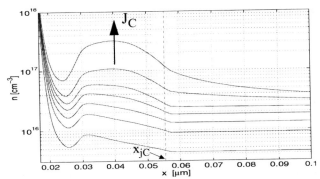

Fig. 3 Electron distribution in the base and BC region for selected bias points (cf. Fig. 2).

At this point, it is instructional to take a look at the current dependence of E_{jc}, which is shown in Fig. 4 for the transistor defined in Fig. 1. As expected, the field decreases with current. At low voltages, the square-root like dependence on current can be observed while in the punch-through region the dependence is quite linear as expected. At high current densities, the *electrostatic* field becomes negative in a SiGe (and any other) DHBT, while $|E_{\varphi}|$ reaches its minimum given by $E_{\infty} = 2V_T/w_C$ which can be derived from simple theory. Furthermore, it is interesting to note that the transition from the almost linear decrease of $|E_{\varphi}|$ to E_{∞} can be well described as a function of voltage by the critical current I_{CK} in HICUM [4], which is indicated by arrows in Fig. 4.

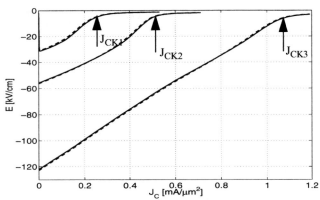

Fig. 4   Bias dependent electric field at the BC junction calculated from electron quasi-fermi potential (solid lines) and electrostatic potential (dashed lines) at $V_{BC}/V = 0.5, 0, -2$. The arrows indicate the critical current I_{CK}.

For a compact model, the analytical description of E_{jc} has to be kept as simple as possible mainly in order (i) to be able to guarantee the numerical stability of the overall formulation and (ii) to minimize the arithmetic operations. Thus, the bias dependent electric field is described here by

$$E_{jc} = E_{\infty} + f_e E_{lim} \qquad (8)$$

with the field $E_{lim} = v_{sn}/\mu_{nCi}(E=0) = V_{lim}/w_{Ci}$. The smooth transition from medium to high current densities is accomplished by the function

$$f_e = \frac{e_j + \sqrt{e_j^2 + g_{jc}\dfrac{E_{CK}}{E_{lim}}}}{2} \qquad (9)$$

with the (model) parameter g_{jc} and the argument [8]

$$e_j = \frac{(E_{jC0} - E_{\infty}) - (E_{jC0} - E_{CK})\dfrac{I_T}{I_{CK}}}{E_{lim}} . \qquad (10)$$

It depends on the critical current I_{CK} and the bias dependent critical field

$$E_{CK} = E_{lim}\frac{v_{ceff}/V_{lim}}{\sqrt{1 + (v_{ceff}/V_{lim})^2}} , \qquad (11)$$

which takes into account that at very low voltages the onset of high current effects occurs when the field curve becomes horizontal. Furthermore, E_{jc0} in (10) is the field at $I_{Tf}=0$ which, according to (1), can be described continuously differentiable as a function of v_{ceff} by using the depletion charge expression for Q_{jCi} already available in HICUM [4]. This allows to adjust the absolute value of the field based on measurements and eliminates the inaccurate square dependence on v_{ceff}. In order to capture the ohmic voltage drop, that occurs in the partial depletion case at the end of the collector, v_{BC} has been replaced by $v_{BC}+\Delta v_{pd}$ with

$$\Delta v_{pd} = V_{lim}\frac{I_{Tf}}{I_{lim}}\left(1 + \frac{I_{Tf}}{I_{lim}}\right) . \qquad (12)$$

Above expression follows from (5) after converting the numerically unstable term $1/(1-I_{Tf}/I_{lim})$ to the numerically stable expression $(1+I_{Tf}/I_{lim})$.

Fig. 5 shows a comparison between the analytical field model and the results obtained from device simulation over a wide voltage range V_{BC}. The agreement is fairly good.

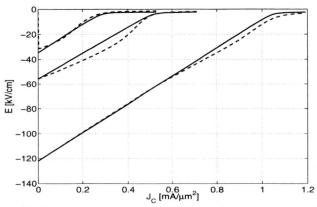

Fig. 5   Bias dependent electric field at the BC junction: comparison between analytical model (solid lines) and device simulation (dashed lines); $V_{BC}/V = 0.5, 0, -2$.

4 Base-collector depletion capacitance

The capacitance is calculated analytically from E_{jc} applying (3). In addition, the voltage dependent formulation for the critical current I_{CK} and its derivative have been included. Fig. 6 shows the depletion capacitance vs. current and voltage over a wide bias range. The peak in the current dependence at low voltages is caused by the ohmic voltage drop Δv_{pd} in the partial depletion case, which increases the forward biasing of the voltage across the junction. At higher voltages and medium current densities, the capacitance becomes flat when the punch-through case occurs. Once the electric field at the junction starts to collapse at I_{CK}, the capacitance decreases and then disappears at high current densities.

Overall, the analytical voltage *and* current dependent BC depletion capacitance agrees fairly well with the device simulation. The accuracy of the corresponding charge is equivalent to that of the electric field (cf. Fig. 5).

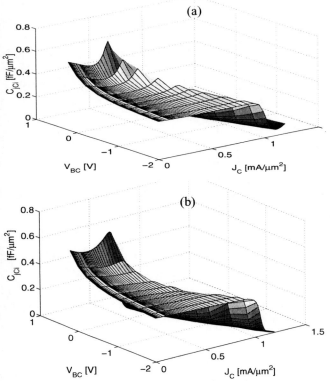

Fig. 6 Voltage and current dependence of the BC depletion capacitance: (a) device simulation, (b) analytical equation.

5 Transit time

The forward transit time τ_f represents the minority charge storage and determines the dynamic transistor behavior especially at medium and high current densities. τ_f can be partitioned into its components associated with the neutral and space-charge regions (SCRs),

$$\tau_f = \tau_{pE} + \tau_{BE} + \tau_{Bf} + \tau_{BC} + \tau_{pC}, \qquad (13)$$

which is shown in Fig. 7 for the selected transistor. Here, τ_{pE}, τ_{Bf}, τ_{pC} represent the charge storage in the neutral regions, and τ_{BE}, τ_{BC} are associated with the SCRs. Such a partitioning is very useful for compact modeling since it allows to separate the various effects, determine their relative importance, and derive adequate model expressions.

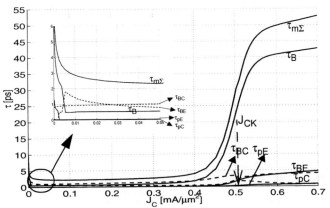

Fig. 7 Transit time components in a SiGe DHBT as a function of collector current density. $V_{BC} = 0V$.

Various partitioning methods have been proposed in literature. The classic regional approach (e.g. [5]) considers d.c. carrier concentrations and, thus, does generally not reproduce the small-signal transit time. The often referenced small-signal based method described in [6] only works for bipolar transistors with unrealistically high collector doping, but not for practical profiles and at high current densities. Furthermore, in advanced BJTs and HBTs the large doping gradients and intentional bandgap changes can produce "spikes" in the (small-signal) carrier densities, that turned out to make also other published partitioning methods (e.g. [7]) known to the authors unsuitable. As a consequence, the small-signal based carrier (and charge) partitioning definition described in [9] had to be extended in order to obtain clearly defined regions and a smooth bias dependence.

From Fig. 7, it can be observed that at low current densities (i.e. below I_{CK}) the contributions of neutral base and emitter as well as the BC SCR are very similar, while the other two components are negligible. At I_{CK} the base transit time increases rapidly due to the collapse of the electric field E_{jc} and the associated formation of the barrier at the location where the Ge decreases into the collector. In contrast to homo-junction transistors, the increase of the collector transit time is negligible and even the saturation value very small. The latter results from the relatively thin collector width, which gives $\tau_{pCs}=5.6ps$ according to theory [4][10], and can be considerably larger for power transistors with thicker epi. At very small current densities, an increase of the BE component, which is associated with the so-called neutral charge in the BE SCR, can be observed, but is irrelevant for circuit applications due to the very small absolute value of the charge in that bias region.

An analytical model for describing the transit time τ_f and its components was presented in [10] for Si BJTs. Although the formulation is flexible enough to allow any partitioning of the base and collector transit time at high current densities, it does not explicitly include the impact of the bias dependent electric field E_{jc} on the base component. Therefore, the existing theory for the base component has been extended to yield

$$\Delta\tau_{bfv} = \tau_{bfvl}f_u\left[1-\frac{I_{Tf}}{E_{lim}}\frac{1}{u}\left(1-\left(\frac{v_n}{v_{sn}}\right)^{\gamma_u}\right)\frac{dE_{jC}}{dI_{Tf}}\right]\exp\left(-b_{hc}\frac{u}{2}\right) \quad (14)$$

with [10] τ_{bfvl} as base transit time at negligible collector current, the normalized field variables

$$u = \frac{E_{jc}(V_{BC}, I_{Tf})}{E_{lim}}, \quad u_0 = \frac{E_{jc}(V_{BC}, 0)}{E_{lim}} = \frac{E_{jc0}}{E_{lim}}, \quad (15)$$

$v_n(E_{jc})$ as field dependent electron velocity ($\gamma_n = 2$) and the associated saturation velocity v_{sn}, and the function

$$f_u = \frac{(1+u^{\gamma_u})^{1/\gamma_u}}{(1+u_0^{\gamma_u})^{1/\gamma_u}}\frac{u_0}{u}\underbrace{\frac{C_{jCi}(V_{BC}, I_{Tf})}{C_{jCi}(0, I_{Tf})}}_{1/c}. \quad (16)$$

Furthermore, the last term $\exp(-b_{hc}u/2)$, which has already been given in [2], represents the barrier effect with the (new) model parameter b_{hc}.

Fig. 8 shows a comparison between the analytical model with the new description of the base transit time and the device simulation. Good agreement is obtained over a wide voltage and current density range.

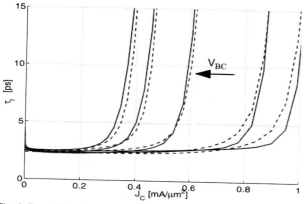

Fig. 8 Transit time vs. collector current density: comparison between device simulation (dashed lines) and new analytical equation (solid lines). $V_{BC}/V = 0.2, 0, -0.5, -1.5, -2$.

6 Modeling velocity overshoot

The analytical expression for the electric field as a function of bias can be used to model the effect of velocity overshoot observed in certain III-V HBTs. In this paper, the goal is only to evaluate the suitability of the new formulation for compact modeling of the transit time in AlGaAs HBTs. For this, the standard velocity-field expression for electrons in GaAs

$$v_n = v_{sn}\frac{(v_{max}/v_{sn})(E_{jc}/E_{lim}) + (E_{jc}/E_{lim})^4}{1 + (E_{jc}/E_{lim})^4} \quad (17)$$

is inserted into the transit time formulation. The component most impacted by the different velocity-field relation is now the base-collector delay time τ_{BC}.

Fig. 9 contains a representation that is typically used to determine the transit time, and which can be used to display any velocity overshoot (e.g. [1]). Except for the transit time, all other parameters for calculating the transit frequency f_T were taken directly from device simulation, so that the difference is solely due to the different velocity-field expressions. The result in Fig. 9 clearly indicates for the GaAs expression a significant drop of $1/(2\pi f_T)$ before it starts to increase rapidly in the high-current region. Thus, it is expected that the new transit time formulation will be applicable to III-V HBTs, although the present model is based on an a-priori known bias dependence of the electric field; i.e. the influence of the now modified carrier distribution on the field is neglected. This would require an iterative solution which does not seem suitable for a compact model.

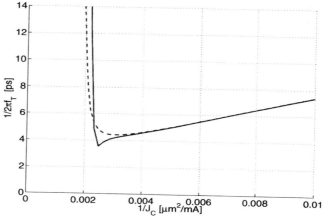

Fig. 9 Reciprocal of the transit frequency vs. reciprocal collector current with different velocity-field models: Si (dashed line) and GaAs (solid line).

7 Conclusion

A first version of a bias dependent description of the electric field in the collector of a bipolar transistor has been presented, which is suitable for compact modeling. Due to the complexity of the problem, caused especially by the various numerical instabilities in the fundamental (classical) solutions, the goal was to keep the new formulation as simple as possible.

The focus of the present work has been on transistors realized in advanced SiGe process technologies. However, as already indicated by Fig. 9, the formulation is also assumed to be applicable to III-V HBTs, since the

NSTI-Nanotech 2004, www.nsti.org, ISBN 0-9728422-8-4   Vol. 2, 2004 availability of an analytical solution for the electric field in the collector allows to include velocity overshoot to first order. Further work will attempt to

- improve the accuracy in certain regions of operation,
- validate the formulations for a larger variety of device designs, including III-V HBTs and experimental data, and
- possibly simplify certain expressions further.

Acknowledgments: This project has been partially supported by the German Ministry for Research and Technology within the project SFB 358 and by Atmel Germany, Heilbronn, within the project HG-DAT.

References

[1] P. Zampardi et al., "III-V HBT modeling, scaling and parameter extraction using TRADICA and HICUM", Power Amplifier Workshop, San Diego, Sept. 2003.

[2] S. Wilms and H.-M. Rein,"Analytical high-current model for the transit time of SiGe HBTs", Proc. Bipolar/ BiCMOS Circuits and Technology Meeting, Minneapolis, pp. 199-202, 1998.

[3] M. Friedrich and H.-M. Rein, "A novel transistor model for simulating avalanche-breakdown effects in Si bipolar circuits ", IEEE J. Solid-State Circuits, Vol. 37, 2002, pp. 1184-1197.

[4] M. Schroter, "HICUM, a Scalable Physics-based Compact Bipolar Transistor Model", http://www.iee.et.tu-dresden.de/iee/eb, December 2001.

[5] R. Schilling, "A Regional Approach for Computer-Aided Transistor Design", IEEE Trans. Education, Vol. E-12, pp. 152-161, 1969.

[6] Van den Biesen, "A Simple Regional Analysis of Transit Times in Bipolar Transistors", Solid-State Elec., Vol. 29, pp. 529-534 , 1986.

[7] Palestri et al. "A Drift-Diffusion/Monte Carlo Simulation Methodology for $Si_{1-x}Ge_x$ HBT Design", IEEE. Trans. Electron Dev., Vol. 49, pp. 1242-1249, 2002.

[8] M. Schroter, SFB Project F2, Report for period 2000-2001.

[9] M. Schroter, "Physical models for fast silicon bipolar transistors - A review", Habilitation thesis (in German), Ruhr-University Bochum, Bochum, Germany, 1993.

[10] M. Schröter and T.-Y. Lee, „A physics-based minority charge and transit time model for bipolar transistors", IEEE Trans. Electron Dev., vol. 46, pp. 288-300, 1999.

[11] M. Schroter, "A compact physical large-signal model for high-speed bipolar transistors with special regard to high current densities and two-dimensional effects", Ph. D. thesis (in German), Ruhr-University Bochum, Bochum, Germany, 1988.

Quasi-2D Compact Modeling for Double-Gate MOSFET

Mansun Chan[1], Tze Yin Man[1], Jin He[2], Xuemei Xi[2], Chung-Hsun Lin[2], Xinnan Lin[1], Ping K. Ko[1], Ali M. Niknejad[2] and Chenming Hu[3]

Department of Electrical and Electronic Engineering,
Hong Kong University of Science & Technology, Clear Water Bay, Hong Kong
Tel: +(852) 2358-8519; Fax: +(852) 2358-1485; E-mail: mchan@ee.ust.hk

[2]Department of Electrical Engineering and Computer Science,
University of California at Berkeley, Berkeley, CA 94720-1770, USA

[3]Taiwan Semiconductor Manufacturing Company, Taiwan, ROC

ABSTRACT

This paper presents an approach to model the characteristics of undoped Double-Gate MOSFETs without relying on the charge-sheet approximation. Due to the extremely thin silicon film used, the inversion charge thickness becomes comparable to the silicon film thickness and cannot be ignored. Together with volume inversion and quantum effect, the carriers are distributed along the vertical direction perpendicular to the direction of current flow. Therefore, a 2-D modeling approach considering vertical current distribution and lateral carrier transport is required. To simplify the 2-D problem, the quasi-Fermi potential has been taken as a reference to develop a quasi 2-D DG MOSFET model.

Keywords: CMOS Device, Double-gate MOSFET, Circuit Simulation, Device model, SPICE, BSIM

1 INTRODUCTION

Double-gate (DG) MOSFETs have been a topic of interest recently due to its scalability beyond the limit of convention technology in the sub-50nm gate length [1]. The short-channel effect (SCE) of DG MOSFET is controlled by the thin silicon film that enables strong gate coupling to the channel and elimination of sub-surface leakage paths. The physics that govern the operation of DG MOSFETs, however, also become more complicated. A number of effects that are considered as secondary become dominant in DG MOSFETs. Traditional approaches in MOSFET modeling usually based on charge-sheet approximation [2] to simplify the current expression in different bias regions. This approach ignores the vertical carrier distribution, which maybe valid in bulk MOSFETs, cannot fully account for the behaviors observed in DG MOSFETs such as volume inversion and quantum confinement by the dielectric barriers. A more rigorous 2-D approach that includes the vertical carrier distribution and lateral transports is required to predict the characteristics of DG MOSFETs.

As a full 2-D modeling of DG MOSFET is extremely difficult, we proposed a quasi-2D framework to model the performance of DG MOSFETs. The model is based on the exaction solution of Poisson's equation along the vertical direction. Quantum effects are included based on the result of 2-D quantum simulation using a self-consistence Schrodinger and Poisson solver taking wavefunction penetration into account. The model has been verified with extensive 2-D/3-D device simulations and limited device data available.

2 MODELING STRATEGY

A flexible DG MOSFET model should be able to handle different mode of operations, including symmetric DG (SDG) MOSFETs and asymmetric DG (ADG) MOSFETs. The ADG mode also includes a wide range of devices such as ground plate MOSFETs and ultra-thin-body (UTB) MOSFETs. To simplify the 2-D modeling, we assumed that the quasi-Fermi level along the vertical direction (or x) remains constant. This assumption implies that the current only flows in the lateral (or y) direction and no carrier exchange in the x direction. That is the current density at any location (x, y) is given by.

$$J(x,y) = J(x) \tag{1}$$

Under this assumption, a generic expression for current can be obtained [3] as given below:

$$I_{DS} = \frac{\mu_{eff} W_{eff}}{L_{eff}} C_{ox} \left(\frac{kT}{q}\right)^2 \left[\frac{Q_S^2 - Q_D^2}{2} + (Q_S - Q_D) \right] \tag{2}$$

Expression in (2) is a very powerful expression because it implies that as long as the source charge and drain charge are know, the I-V expression can be obtained easily. The same expression applied to SDG, ADG and ground plane MOSFETs. The next task in the model formulation is then to find the source/drain charge, and the mobility.

3 GENERIC CHARGE FORMULATION

The formulation of DG MOSFETs starts from undoped body as it is more meaningful in the regime that requires DG MOSFETs. The detail formulation of source and drain charges for undoped SDG MOSFETs has been derived in detail by another paper in the same conference [4]. However, a more generic expression is required from a modeling perspective to account ADG and ground-plane MOSFET. The cost for going for more generic solution is the more complicated boundary conditions resulting in more complex solution. The generic solution can be obtained with a similar method as in [4], but using the minimum potential point in the channel rather than the middle of the channel as a boundary condition. Starting from

$$\frac{d^2\psi}{dx^2} = \frac{q}{\varepsilon_{si}} n_i e^{\frac{q\psi}{kT}} \tag{3}$$

we perform integration with the reference to the location of the minimum potential at $x=x_0$ where $\psi(x_0) = \psi_{min}$. We consider a few special cases here to simplify the discussion. In SDG case, $x_0=T_{Si}/2$. In UTB case, $x_0=T_{Si}$. In ADG case, if we assume the device has n+/p+ gates with workfunction difference of 1V, and V_{DD} is less than the workfunction difference in ultra-small devices, then $x_0=T_{Si}$ in all operation regions similar to the UTB case. By integrating (3) up to x_0, a generic solution is obtained:

$$\frac{d\psi}{dx} = -\sqrt{\frac{2kTn_i}{\varepsilon_{si}}\left(e^{\frac{q\psi}{kT}} - e^{\frac{q\psi_{min}}{kT}}\right) + \left(\frac{d\psi}{dx}\Big|_{x=x_0}\right)^2} \tag{4}$$

Letting

$$A = \frac{2kTn_i}{\varepsilon_{si}} \quad \text{and} \quad B = A\exp\left(\frac{q\psi_{min}}{kT}\right) - \left(\frac{d\psi}{dx}\Big|_{x=x_0}\right)^2 \tag{5}$$

and integrating (4) again, we obtain the potential as a function of spatial coordinate in the x direction.

$$\psi(x) = \frac{2kT}{q}\ln\left[\sqrt{\frac{A}{B}}\cos\left(\frac{q\sqrt{B}}{2kT}(x_0-x) + \cos^{-1}\left(\sqrt{\frac{B}{A}}e^{-\frac{q\psi_{min}}{kT}}\right)\right)\right] \tag{6}$$

relating electron density and potential using (6) we got

$$n(x) = \frac{n_i\left(\dfrac{B}{A}\right)}{\cos^2\left[\dfrac{q\sqrt{B}}{2kT}(x_0-x) + \cos^{-1}\left(\sqrt{\dfrac{B}{A}}e^{-\frac{q\psi_{min}}{kT}}\right)\right]} \tag{7}$$

The total inversion charge is then obtained by integrating (7) from $x=0$ to $x=x_0$ giving

$$Q_{inv} = \frac{2kTn_i\left(\dfrac{B}{A}\right)}{\sqrt{B}}\left[\tan\left(\frac{q\sqrt{B}}{2kT}(x_0-x) + \cos^{-1}\left(\sqrt{\frac{B}{A}}e^{-\frac{q\psi_{min}}{2kT}}\right)\right) - \tan\left(\cos^{-1}\left(\sqrt{\frac{B}{A}}e^{-\frac{q\psi_{min}}{2kT}}\right)\right)\right] \tag{8}$$

By using this inversion charge term, and following similar derivation as in [4], the I-V equation can be derived. One difficulty in this formulation is the dependent of the solution in term of the boundary condition of x_0 and ψ_{min} which the analytical expression in term of device structure and bias condition is not yet formulated besides the special cases as described before. The general solution with arbitrary device structures and bias conditions requires some fitting function or extrapolation to find x_0 and ψ_{min}.

4. EFFECTIVE MOBILITY FORMULATION

The mobility formulation follows the charge formulation for consistency. For devices with relatively thick silicon film,

some experimental results show that the electron effective mobility (μ_{eff}) agrees well with the universal mobility when the inversion layers at both sides of the silicon film weakly interact. In this case, the current transport can be represented by two channels in parallel [5]. When the T_{si} is reduced, the inversion electrons are confined by the two oxide barriers and the form factor increases. Therefore, the phonon limited electron mobility decreases with reducing T_{si} [6-8]. As the T_{si} is scaled below 5nm, most of the electrons are populated in the unprimed subband, featuring a higher μ_{eff} due to its lower conductivity mass [6-8].

While many mechanisms that affect the mobility has been proposed, the most of them can be modeled by the constant term μ_0 in the universal mobility model. A more important part of the model, which is not as widely addressed, is the bias dependent of the mobility. The impact of bias towards the mobility is mainly through the effective vertical field (E_{eff}). Recent experimental results show that the bias dependent of the mobility still more or less follows the universal mobility model. Even if not, it is likely that the direct link between mobility to E_{eff} still exist. With these assumptions in mind, an analytical expression of E_{eff} as a function of device geometry and bias is the most important bridge to model the bias dependence of carrier mobility in DG MOSFETs.

Following the classical approach, E_{eff} is defined as [9-10]

$$E_{eff} = \left[\int_0^{x_0} |E(x)| \cdot n(x)dx\right] / \left[\int_0^{x_0} n(x)dx\right] \tag{9}$$

It can be interpreted as the average electric field experienced by the carriers in the inversion layer towards the surface of the channel. As the E_{eff} increases, the carriers in the inversion layer have a larger chances to interact with the Si/SiO$_2$ interface, which degrades the carrier mobility according to the universal mobility model [11].

$$\mu_{eff} = \frac{\mu_o}{1 + \left(E_{eff}/E_o\right)^v} \tag{10}$$

To calculate the E_{eff}, the Poisson's Equation is solved along the vertical direction in the silicon channel, considering only the mobile charge (electron) density as the body is undoped. From equation (7) and Guass Law, the electric field as a function of position can be found and the expression is

$$E(x) = \frac{2kTn_i\left(\dfrac{B}{A}\right)}{\varepsilon_{si}\sqrt{B}}\left[\tan\left(\begin{array}{c}\dfrac{q\sqrt{B}}{2kT}(x_0-x)\\[2mm] + \cos^{-1}\left(\sqrt{\dfrac{B}{A}}e^{-\frac{q\psi_{min}}{2kT}}\right)\end{array}\right) - \tan\left(\cos^{-1}\left(\sqrt{\dfrac{B}{A}}e^{-\frac{q\psi_{min}}{2kT}}\right)\right)\right] - \frac{d\psi}{dx}\Big|_{x=x_0} \tag{11}$$

By substituting equation (11) and (7) into (9) and using equation (8) to simplify the expression, we obtain a simple generic expressions of the E_{eff}

$$E_{eff} = \frac{Q_{inv}}{2\varepsilon_{si}} - \frac{d\psi}{dx}\Big|_{x=x_0} \tag{12}$$

It should be noted that the expression for SDG with symmetrical boundaries eliminate the second term in equation

(12) and double the inversion charge with the second channel. It results in an even simpler expression of

$$E_{eff} = \frac{Q_{inv}}{4\varepsilon_{si}} \qquad (13)$$

To verify the validity of the formulation of E_{eff}, extensive numerical simulations with and without quantum effect have been performed. Fig. 1 illustrates the E_{eff} against the Q_{inv} with different T_{Si} (without quantum effect) for SDG and UTB MOSFETs. The ADG case is very similar to that of UTB except for the different $-\frac{d\psi}{dx}\Big|_{x=T_{Si}}$ term used in equation (12). It shows good agreement between the analytical solution and the simulation result. Also, it is interesting to notice that the $1/4\varepsilon_{si}$ relationship in SDG and the $1/2\varepsilon_{si}$ relationship in UTB device are independent of the T_{Si}.

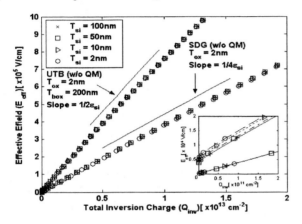

Fig. 1: The plots of the E_{eff} against the Q_{inv} with different T_{Si} for SDG and UTB MOSFETs without quantum effect. The insert magnifies the low inversion charge region showing the offset due to $d\psi/dx$

The plots of E_{eff} versus Q_{inv} including quantum effect for SDG and UTB devices with different T_{si} are shown in Fig. 2. It shows that the functional dependence of $1/4\varepsilon_{si}$ and $1/2\varepsilon_{si}$ with Q_{inv} in both SDG and UTB is not influenced by the quantum effect. We can, in general model the quantum effect only through a correction to the zero-field mobility (μ_o) rather than the E_{eff} term.

Fig. 3 shows the μ_{eff} against the Q_{inv} for a SDG and an UTB MOSFETs with same T_{si}. When the SDG and the UTB device have the same amount of Q_{inv}, the SDG case has a much smaller E_{eff} than the UTB or ADG devices according to equation (12) and (13), which allows it to have a significantly higher μ_{eff}. Therefore, SDG is a more optimal structure for small devices when current drive and speed are considered.

Recently, experimental results show that the mobility in SDG is higher than that in UTB device when the Q_{inv} of UTB device is half of the Q_{inv} of SDG MOSFETs especially when the amount of inversion charge is relatively low. This effect is not fully explained in previous studies [9]. According to equation (12) and (13), when the Q_{inv} of UTB devices is half of

the Q_{inv} MOSFETs, the E_{eff} of UTB devices is larger than the that of SDG MOSFETs due to the finite electric field ($-\frac{d\psi}{dx}\Big|_{x=T_{Si}}$) at the backside Si/SiO$_2$ interface of UTB (and also ADG) MOSFETs. This also explained the higher μ_{eff} in SDG MOSFET at low inversion charge as shown in Fig. 4.

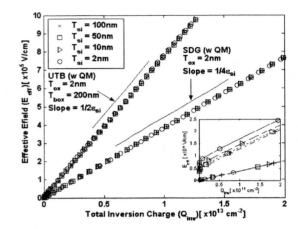

Fig. 2: The plots of the E_{eff} against the Q_{inv} for SDG and ADG MOSFETs with different T_{si} including quantum effect.

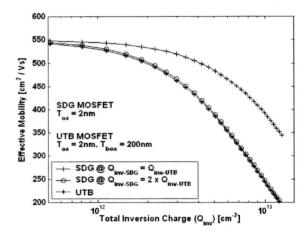

Fig. 3: The plot of the μ_{eff} versus the Q_{inv} for SDG and UTB MOSFETs with $T_{si} = 5nm$. Less mobility degradation in SDG MOSFET at the low inversion charge region when Q_{inv_UTB} is half of the Q_{inv_SDG}.

Fig. 4 illustrates the electric field at the backside Si/SiO$_2$ interface ($-\frac{d\psi}{dx}\Big|_{x=T_{Si}}$) of UTB MOSFETs as a function of Q_{inv} with different T_{si}. It shows that the backside electric field is relatively constant with respect to Q_{inv}. As the amount of inversion charge increases, the contribution of the $-\frac{d\psi}{dx}\Big|_{x=T_{Si}}$ term in the UTB device becomes smaller, which makes the E_{eff} of UTB devices behave more like E_{eff} of SDG MOSFETs under the $Q_{inv\_UTB} = Q_{inv\_SDG} / 2$ condition. Therefore, as the Q_{inv} is increased, the μ_{eff} becomes essentially the same in both SDG and UTB device (Fig. 4) when the $Q_{inv\ UTB}$ is half of the $Q_{inv\ SDG}$.

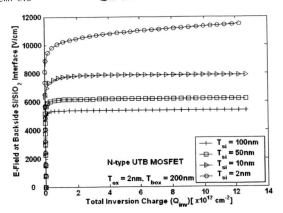

Fig. 4: Simulated electric field at the backside Si/SiO$_2$ interface of UTB MOSFETs as a function of Q_{inv} with different T_{si}.

5 MODELING CHANNEL ELECTRIC FIELD OF SDG MOSFET IN THE SATURATION REGION

The modeling of DG MOSFETs so far has been focused on the linear region. The modeling of the saturation characteristics is through the introduction of the saturation charge Q_{Dsat} at the end of the gradual channel region, which is independent of the detail physics in the velocity saturation region (VSR). In most cases, the approach is sufficient in giving reasonable I-V characteristics of a DG MOSFET. However, to have more detail understand in the operation inside the VSR such as channel length modulation, a more detail analysis in the velocity saturation region is required. As an initial trial to model the electric field in the velocity saturation region, we first focus on SDG MOSFET due to its simpler boundary condition. The model will be extended to the ADG case in the future.

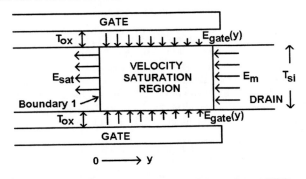

Fig. 5: The velocity saturation region of the undoped SDG MOSFET where Gauss's Law is applied.

For the device operating in saturation regime, the channel of the DG MOSFET can be divided into gradual channel approximation (GCA) region and VSR. A schematic cross-section of the VSR of a SDG MOSFET with undoped body is shown in Fig. 5. "Boundary 1" in the figure indicates the location in the channel where the velocity of carriers just saturates.

The model is developed based on the pseudo two-dimensional approach proposed by Elmansy [12] and Ko [13]. When Gauss's law is applied to the VSR (Fig. 6), we have

$$\varepsilon_{si} \int_{-\frac{T_{Si}}{2}}^{\frac{T_{Si}}{2}} E_{sat} \cdot dx - \varepsilon_{si} \int_{-\frac{T_{Si}}{2}}^{\frac{T_{Si}}{2}} \frac{\partial V}{\partial y} \cdot dx - 2\varepsilon_{gate} \int_{0}^{y} E_{gate}(k) \cdot dk$$

$$= -q \int_{-\frac{T_{Si}}{2}}^{\frac{T_{Si}}{2}} \int_{0}^{y} Q_{inv}(x,k) \cdot dk dx \qquad (14)$$

where ε_{gate} are the dielectric coefficient of the gate insulator. The integral can be performed with surface potential, but channel voltage (which differs with the surface potential only by a constant) is used in this case to link the result directly to the external voltage. To account for the non-uniform distribution of the lateral electric field and allow the problem to be solved analytically, it is assumed that the ratio of average lateral electric field to the surface lateral electric field in the VSR is independent of position y and is equal to the value at boundary 1 [12]. We can then write

$$\int_{-\frac{T_{Si}}{2}}^{\frac{T_{Si}}{2}} \frac{\partial V}{\partial y} \cdot dx = A \cdot T_{Si} \cdot \frac{dV_{sur}}{dy} \qquad (15)$$

where V_{sur} is the voltage at the silicon/oxide interface. The value of A at boundary 1 can be calculated using the vertical potential profile proposed by Taur [14]

$$V(x) = -\frac{2kT}{q} \ln[\cos(B \cdot x)] + V_o \qquad (16)$$

where V_o is the potential at the center of the film and $B = \sqrt{\frac{q^2 n_i}{2\varepsilon_{si}kT} \cdot e^{\frac{qV_o}{2kT}}}$. For simplicity, the potential profile is approximated by a parabolic function with a parameter K

$$V(x) \approx K \cdot x^2 + V_o \qquad (17)$$

Since the parameter K will be eliminated in the calculation of A, the exact value of K is not shown in the following calculations. The value of A is calculated by substituting (17) into (15), giving

$$A = \frac{1}{3} + \frac{2}{3} \cdot \frac{\partial V_o}{\partial V_{sur}} \qquad (18)$$

The partial derivative in (18) is calculated using equation (16) giving

$$A = \frac{1}{3} + \frac{2}{3\left(1 + \frac{B \cdot T_{Si}}{2} \cdot \tan\left(\frac{B \cdot T_{Si}}{2}\right)\right)} \qquad (19)$$

Substituting (15) into (14) and differentiating both sides with respect to y, we get

$$\varepsilon_{si} \cdot A \cdot T_{Si} \cdot \frac{d^2 V_{sur}}{dy^2} = q \int_{-\frac{T_{Si}}{2}}^{\frac{T_{Si}}{2}} Q_{inv}(x,y)dx - 2\varepsilon_{gate} \cdot E_{gate}(y) \quad (20)$$

Using similar approach as [13], the integrated mobile charge term in (20) can be replaced by the gate electric field at boundary 1 with the assumption that GCA still holds at that location. Equation (20) becomes

$$\varepsilon_{si} \cdot A \cdot T_{Si} \cdot \frac{d^2 V_{sur}}{dy^2} = 2\varepsilon_{gate} \cdot \left[E_{gate}(0) - E_{gate}(y) \right] \quad (21)$$

where the gate electric field $E_{gate}(y) = \dfrac{V_G - V_{FB} - V_{sur}(y)}{T_{gate}}$, V_G

and V_{FB} is the gate and flat band voltage, respectively. T_{gate} is the gate insulator thickness. (T_{ox} is not used as high-k material are allowed in the formulation) Equation (21) is rewritten to

$$\frac{d^2 V_{sur}}{dy^2} = \frac{\left[V_{sur}(y) - V_{sur}(0)\right]}{\lambda^2} \quad (22)$$

where $\lambda = \sqrt{\dfrac{A \cdot \varepsilon_{si}}{2\varepsilon_{gate}}} \cdot \left(T_{Si}\right)^{\frac{1}{m}} \cdot \left(T_{gate}\right)^{\frac{1}{n}} \quad (23)$

with $m = n = 2$. Using the conditions at boundary 1 that $V(0) = V_{Dsat}$ and $\dfrac{dV_{sur}(y)}{dy}\Big|_{y=0} = E_{sat}$, the solution of (22) is given by

$$E_{sur}(y) = E_{sat} \cdot \cosh(\frac{y}{\lambda}) \quad (24)$$

and

$$V_{sur}(y) = V_{Dsat} + \lambda \cdot E_{sat} \cdot \sinh(\frac{y}{\lambda}) \quad (25)$$

It can be shown that the maximum lateral electric field (E_m) at the end of the channel is

$$E_m = \sqrt{\frac{(V_D - V_{Dsat})^2}{\lambda^2} + E_{sat}^2} \quad (26)$$

where V_D is voltage at the end of the channel.

To verify the result, a 50nm SDG NMOSFET with mid-gap work function gate-electrode simulated by MEDICI and compared with the model. 2-D simulation results show that the parameter m in equation (23) should be linearly from 2.03 to 1.97 as the T_{Si} increasing from 5nm to 20nm. Fig. 6 compares the results calculated from the model with the results obtained from MEDICI for the SDG device with T_{ox} =2nm, T_{si} = 10nm and E_{sat} = 3.5x10^4 V/cm [15]. Good agreement, is obtained between the model and the 2D simulation result especially near the peak E-field region.

Fig. 7 shows E_m as a function of T_{Si} with different T_{ox} under the same biasing condition. The results from the model, again, show good agreement with the simulation results. As predicted by the model, the maximum electrical field increases with the reduction of T_{si} and T_{ox}. The impact of geometry scaling is included through λ which does not have an obvious scaling trend. However, there is a minimum value of λ which can be obtained by replacing the parameter A with its minimum value, giving $\lambda_{min} = \sqrt{\dfrac{\varepsilon_{si}}{6\varepsilon_{gate}}} \cdot \left(T_{Si}\right)^{\frac{1}{m}} \cdot \left(T_{gate}\right)^{\frac{1}{n}}$. This λ_{min} is an

important parameter to calculate the maximum E_m and evaluate the worst-case impact ionization for a given SDG structure and biasing condition, which has been shown to take place even at a low $V_D < 1$V [16].

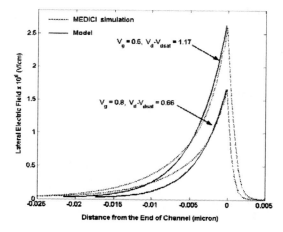

Fig. 6: Comparison of the results calculated from the model with the results obtained from the simulations for the SDG device with T_{ox} = 2nm, T_{si} = 10nm and E_{sat} = 3.5x10^4 V/cm.

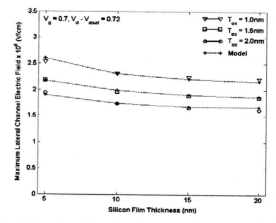

Fig. 7: The plot of the maximum lateral channel electric field (E_m) against the silicon film thickness (T_{si}) with different gate oxide thickness.

Fig. 8 shows the lateral electric field for two SDG devices with same I_{off} and V_T, but different T_{ox} and T_{Si} with value shown in the figure. The I_D-V_G characteristic of those SDG devices are shown as an insert in Fig. 9. The results show that the SDG devices have similar DC characteristic, but different E_m. Although scaling either the T_{Si} or the T_{ox} can give the same performance, peak channel E-field is more seriously affected by the scaling of T_{ox}.

The current formulation of channel E-field does not include quantum effects for simplicity. However, its effects in the VSR region is not serious because the transverse electric field is relative small. For the solution of the Poisson equation, it can be coupled to the exponential relationship between lateral electric field and position for the VSR region with continuous current flow.

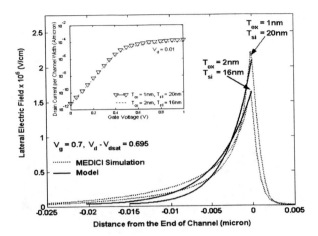

Fig. 8: Comparisons of the E_y for two SDG MOSFETs with same I_{off} and V_T with different T_{ox} 1nm, T_{Si} = 20nm and T_{ox} = 2nm, T_{Si} = 16nm.

In extremely short channel devices, velocity over-shoot (VOS) allows the carriers have a higher velocity than the classical saturation velocity (e.g. $7x10^6$ cm/s). In other words, the velocity of the carriers inside the VSR is not a constant, but increasing with the channel position. Since the current is continuous within the entire channel, the higher the carrier velocity, the lower the lateral electric field inside the VSR. Although the purposed model cannot predict the lateral electric field with VOS, the proposed model can still provide an upper bound on the trend on the maximum lateral electric field with vertical dimension scaling.

ACKNOWLEDGEMENT

The DG Modeling work is support by SRC under a customization program 2002-NJ-1001 and also Hong Kong Research Grant Council with an earmarked grant HKUST 6110/03E.

REFERENCES

[1] D. Frank, S. Laux, and M. Fischetti, "Monte Carlo Simulation of a 30-nm Dual-Gate MOSFET: How Far Can Silicon Go?" in IEDM Tech. Dig., 1992, pp. 553

[2] J. R. Brews, "A Charge Sheet Model of the MOSFET," *Solid-State Electron*, issue 21, 345 (1978).

[3] J. He, X. Xi, M. Chan, A. Niknejad, and C. Hu, "An Advanced Surface-Potential-Plus MOSFET Model", *Technical Proceedings of the 2003 Nanotechnology Conference*, pp. 262-665, February 23-27, 2003, San Francisco, California, USA.

[4] J. He, X. Xi, C.-H. Lin, M. Chan, A. Niknejad, and C. Hu, "A Non-Charge-Sheet Analytic Theory for Undoped Symmetric Double-Gate MOSFETs from the Exact Solution of Poisson's Equation using SPP Approach", *Technical Proceedings of the 2004 Nanotechnology Conference*

[5] M. Ieong, E. C. Jones, T. Kanarsky, Z. Ren, O. Dokumaci, R. A. Roy, L. Shi, T. Furukawa, Y. Taur, R. J. Miller, and H.-S.P. Wong, "Experimental evaluation of carrier transport and device design for planar symmetric/ asymmetric double-gate/ground-plane CMOSFETs," *IEDM Tech. Dig.*, pp. 19.6.1-19.6.4, 2001

[6] D. Esseni, A. Abramo, L. Selmi, and E. Sangiorgi, "Study of low field electron transport in ultra-thin single and double-gate SOI MOSFETs," *IEDM Tech. Dig.*, pp. 719-722, 2002.

[7] K. Uchida, H. Watanabe, A. Kinoshita, J. Koga, T. Numata, and S. Takagi, "Experimental study on carrier transport mechanism in ultrathin-body SOI n- and p-MOSFETs with SOI thickness less than 5 nm," *IEDM Tech. Dig.*, pp. 47-50, 2002.

[8] K. Uchida, J. Koga, R. Ohba, T. Numata, and S. Takagi, "Experimental evidences of quantum-mechanical effects on low-field mobility, gate-channel capacitance, and threshold voltage of ultrathin body SOI MOSFETs," *IEDM Tech. Dig.*, pp. 29.4.1 -29.4.4, 2001.

[9] D. Esseni, M. Mastrapasqua, G. K. Celler, C. Fiegna, L. Selmi, and E. Sangiorgi, "An experimental study of mobility enhancement in ultrathin SOI transistors operated in double-gate mode," *IEEE Trans. Electron Devices,* Vol. 50, No. 3, pp. 802-808, 2003.

[10] M. Shoji, and S. Horiguchi, "Electronic structures and phonon limited electron mobility of double-gate silicon-on-insulator Si inversion layers," *J. Appl. Phys.*, vol. 85, no. 5, pp. 2722–2731, 1999.

[11] M. S. Liang, J. Y. Choi, P. K. Ko, and C. Hu, "Inversion-Layer Capacitance and Mobility of Very Thin Gate-Oxide MOSFET's," *IEEE Trans. Electron Devices*, ED-33, pp. 409, 1986.

[12] Y. A. El Mansy and A. R. Boothroyd, "A simple two-dimensional model for IGFET operation in the saturation region," *IEEE Transactions on Electron Devices*, Vol. 24, No. 3, pp. 254-262, 1977.

[13] P. K. Ko, R. S. Muller and C. Hu, "A unified model for hot-electron currents in MOSFETs," *IEDM Tech. Dig.*, 1981, p. 600.

[14] Y. Taur, "Analytic solutions of charge and capacitance in symmetric and asymmetric double-gate MOSFETs," *IEEE Transactions on Electron Devices*, Vol. 48, No. 12, pp. 2861-2869, 2001

[15] K. W. Terrill, C. Hu, P. K. Ko, "An Analytical Model for the Channel Electric Field in MOSFET's with Graded-Drain Structures," *IEEE Electron Device Letter*, Vol. 5, No. 11, pp.440-442, 1984

[16] P. Su, K. Goto, T. Sugii and C. Hu, "A Thermal Activation view of Low Voltage Impact Ionization in MOSFETs", *IEEE Electron Device Letters*, Vol. 23, no. 9, pp. 550-552, September 2002

Compact, Physics-Based Modeling of Nanoscale Limits of Double-Gate MOSFETs

Qiang Chen[*], Lihui Wang, Raghunath Murali, and James D. Meindl

Microelectronics Research Center, Georgia Institute of Technology
791 Atlantic Dr. N.W., Atlanta, GA 30332-0269, U.S.A.
[*]Advanced Micro Devices, One AMD Place, MS 143, Sunnyvale CA 94088, USA, qiang.chen@amd.com

ABSTRACT

Compact, physics-based models of subthreshold swing and threshold voltage are presented for double-gate (DG) MOSFETs in symmetric, asymmetric, and ground-plane modes. Applying these device models, threshold voltage variations in DG MOSFETs are comprehensively and exhaustively investigated using a unique, scale-length based methodology. Quantum mechanical effects and fringe-induced barrier lowering effect on threshold voltage, caused by ultra-thin silicon film and potential use of high-permittivity gate dielectrics, respectively, have been analytically modeled giving close agreement to numerical simulations. Scaling limits projections indicate that individual DG MOSFETs with good turn-off behavior are feasible at 10 nm scale; however, practical exploitation of these devices toward gigascale integrated systems requires development of novel technologies for significant improvement in process control.

Keywords: double-gate, scaling, threshold voltage, subthreshold swing, MOSFET.

1 INTRODUCTION

The double-gate (DG) MOSFET, illustrated in Figure 1, has been considered as the most promising device structure to extend CMOS scaling into the nanometer regime [1]. The ultra-thin silicon channel is undoped (i.e., lightly doped with the background doping concentration less than 10^{16} cm^{-3}) to avoid random dopant placement effect and mobility degradation associated with high doping. Depending upon gate work functions and gate-bias conditions, a DG MOSFET can operate in symmetric (SDG), asymmetric (ADG), or ground-plane (GP) modes. Two key characteristics of a MOSFET, namely, subthreshold swing (S) and threshold voltage (V_{TH}), and their dependences on device parameters are usually exploited to gauge its immunity to short-channel effects (SCE), i.e., its scalability. In this paper, compact, physical models of subthreshold swing and threshold voltage for symmetric, asymmetric, and ground-plane DG MOSFETs are described, including quantum mechanical effects and fringe-induced barrier lowering (FIBL) effect. These new device models are applied to comprehensively analyze parameter variations, reveal device design insights, and project scaling limits and opportunities of DG MOSFETs.

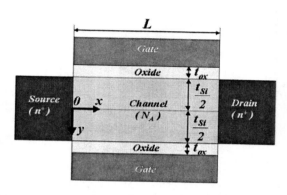

Figure 1: Cross-section schematic of a DG MOSFET.

2 SUBTHRESHOLD SWING MODELS

A two-dimensional (2-D) Poisson equation with the dopant term only is solved using the evanescent-mode analysis in the channel region to obtain the channel potential distribution [2]. To eliminate the uncertainty of choosing the most representative leakage path (channel surface versus channel center [3, 4]), it is assumed that the drain current is proportional to the *sheet* density of inversion carriers at the virtual cathode (i.e., the minimum potential point between the source and drain). Subthreshold swing, defined as the gate voltage (V_{GS}) change required for an order-of-magnitude change of the subthreshold current, is then obtained as [5],

$$S = \left[\int_{y=-t_{Si}/2}^{t_{Si}/2} n_m(y) \frac{\partial \varphi_{\min}(y)}{\partial V_{GS}} dy \middle/ \int_{y=-t_{Si}/2}^{t_{Si}/2} n_m(y) dy \right]^{-1} \frac{kT}{q} \ln 10, \quad (1)$$

where $n_m(y) = n_i \exp[\varphi_{\min}(y)q/kT]$, n_i is the intrinsic electron concentration, $\varphi_{\min}(y)$ is the potential profile at the virtual cathode.

2.1 Symmetric DG MOSFETs

Exploiting (1), a detailed study on subthreshold swing's dependence on the channel doping concentration [2] reveals that subthreshold swing in short-channel DG MOSFETs can be closely described by a concept of effective conducting path, i.e., its location with respect to the gate. In undoped symmetric DG devices, the effective conducting path is found in-between the channel surface and the channel center because of symmetry and the minuscule amount of ionized dopant atoms [2]. As a result, (1) can be greatly simplified without comprising much accuracy [2],

$$S = \left(1 - 2\Gamma_1 \cos \frac{t_{Si}}{4\lambda_1} e^{-\frac{L}{2\lambda_1}}\right)^{-1} \frac{kT}{q} \ln 10 \cdot \qquad (2)$$

The new S model is compared to previous models and Medici numerical simulations with improved agreement (Figure 2). Ideal subthreshold swing is achieved at large channel lengths, which is explained by ideal gate-to-gate coupling through the dielectric-like undoped channel [6]. The parameter λ_1 is a scale length and can be approximated as [2],

$$\lambda_1 = \frac{t_{Si} + \varepsilon_{Si} t_{ox} / \varepsilon_{ox}}{1 + \pi / 2} \quad or \quad \lambda_1 = \frac{t_{Si} + \sqrt{2}\varepsilon_{Si} t_{ox} / \varepsilon_{ox}}{\sqrt{2} + \pi / 2}, \qquad (3)$$

for $(\varepsilon_{Si} t_{ox} / \varepsilon_{ox} t_{Si}) \leq \pi/2$ and $> \pi/2$, respectively. It provides an efficient guideline in selecting appropriate t_{ox} and t_{Si} values for device designs (Figure 3).

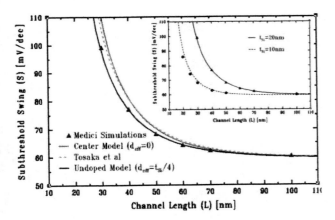

Figure 2: S roll-up in undoped SDG MOSFETs [2].

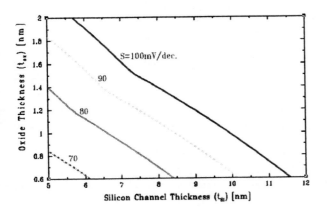

Figure 3: Design contours of a 15 nm undoped symmetric DG MOSFETs for different S requirements [2].

2.2 Asymmetric DG MOSFETs

Based on the concept of effective conducting path, (1) can be similarly simplified for asymmetric devices [5],

$$S = \left[1 - 2\Gamma_1 \frac{(V_1 + V_{DS}/2)}{\sqrt{V_1(V_1 + V_{DS})}} \cos \frac{d_{eff}}{\lambda_1} e^{-\frac{L}{2\lambda_1}}\right]^{-1} \frac{kT}{q} \ln 10, \qquad (4)$$

where d_{eff} is the location of effective conducting path, and is a more complex and sensitive function of gate work-functions and device geometry than in SDG devices [5]. In general, because of asymmetry of the channel potential profile, the effective conducting path is found to be closer to the gate in ADG MOSFETs than in symmetric ones of the same geometry, resulting in a stronger gate control over the channel, and consequently, a slightly smaller S (Figure 4). The slightly improved subthreshold swing of ADG devices may translate into a higher drive current than in SDG ones for a normalized off-current.

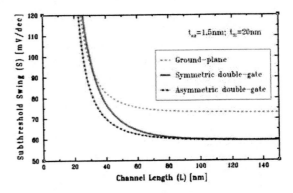

Figure 4: Comparison of subthreshold swing in SDG, ADG, and GP MOSFETs [5].

2.3 Ground-plane MOSFETs

The subthreshold swing of GP MOSFETs is obtained from (1) as,

$$S = \frac{kT}{q} \ln 10 \left[\frac{1}{2} - \frac{r}{r+2} \frac{d_{eff,linear}}{t_{Si}} - \Gamma_1 e^{-\frac{L}{2\lambda_1}} \frac{V_1 + V_{DS}/2}{\sqrt{V_1(V_1 + V_{DS})}} \cos \frac{d_{eff}}{\lambda_1}\right]^{-1}, \qquad (5)$$

and can be comprehended by a combination of capacitor divider model and effective conducting path [5]. For long-channel and moderately short-channel designs, GP MOSFETs demonstrate a significantly larger subthreshold swing than both SDG and ADG devices (Figure 4).

3 THRESHOLD VOLTAGE MODELS

It has been observed that the concentration of inversion carriers can exceed that of ionized dopant atoms under the threshold condition in undoped devices [7, 8, 9, 10]. Inversion carriers, thus, need to be included for threshold voltage calculations. Moreover, the conventional way of using the surface band bending equal to $2q\phi_B$, where $\phi_B = \ln(N_A/n_i)kT/q$, to define the threshold condition becomes irrelevant. An alternative is to define the threshold voltage as the gate voltage at which the sheet density of inversion carriers reaches a value of Q_{TH} adequate to identify the turn-on condition [11]. Such a definition is equivalent to the constant-current V_{TH} measurement widely used both in experiments and simulations.

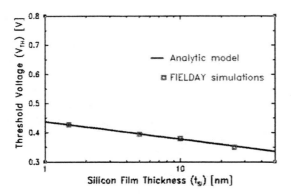

Figure 5: Long-channel V_{TH} vs. t_{Si} in SDG MOSFETs. Mid-gap gates are assumed [11].

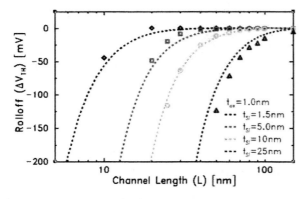

Figure 6: V_{TH} rolloff in mid-gap SDG MOSFETs [11].

3.1 Symmetric DG MOSFETs

A 2-D Poisson equation with only the mobile charge term included is analytically solved in the near-threshold region for SDG MOSFETs [11]. The potential profile at the virtual cathode is then determined, which, through the sheet density of inversion carriers, leads to a general short-channel threshold voltage model [11],

$$V_{TH} = \Phi_{MS,i} + \eta \frac{kT}{q} \frac{\cosh(\theta)}{\cosh(\theta/2)} \ln\left(\frac{Q_{TH}}{n_i t_{Si}}\right) - \left[\frac{\cosh(\theta)}{\cosh(\theta/2)} \eta - 1\right] \varphi_{0m}, \quad (6)$$

where $\Phi_{MS,i}$ is the gate work-function referenced to the intrinsic silicon. At large channel lengths, (6) readily reduces into a long-channel V_{TH} model,

$$V_{TH,long} = \Phi_{MS,i} + \frac{kT}{q} \ln\left(\frac{Q_{TH}}{n_i t_{Si}}\right). \quad (7)$$

Models (7) and (6) are compared with published FIELDAY numerical simulations [12] with close agreement (Figure 5 and Figure 6). The slight dependence of long-channel V_{TH} on t_{Si} is caused by the volume inversion effect [11].

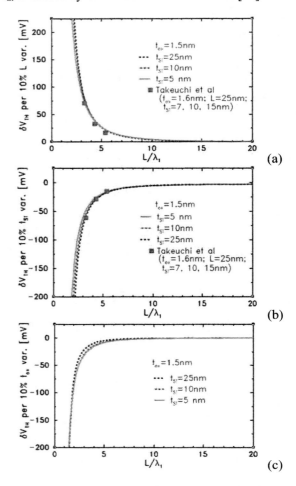

Figure 7: V_{TH} change versus the ratio of L/λ_1 caused by 10% increase of (a) L, (b) t_{Si}, and (c) t_{ox} [11].

An interesting application of the analytical *short-channel* V_{TH} model is to perform *quantitative* threshold voltage sensitivity analyses of DG MOSFETs in a more thorough and easier way than from numerical simulations, and so the effects of process variation can be assessed relatively easily. It was discovered [11] that the normalized V_{TH} sensitivities, $\delta V_{TH}/(\delta X/X)$, where X stands for L, t_{Si}, or t_{ox}, and $\delta X/X$ is its process tolerance expressed in percentage, can be represented, with reasonably good accuracy, by three unified, unique functional dependences on L/λ_1 for virtually all (L, t_{Si}, t_{ox}) designs (Figure 7). It enables a convenient and exhaustive study of the impact of process variations across technology nodes. For practical device designs (with L/λ_1 around ~4.5 to ~7) L causes 30%

to 50% more V_{TH} variation than does t_{Si} for the same process tolerance, while t_{ox} causes the least, relatively insignificant amount of V_{TH} variation.

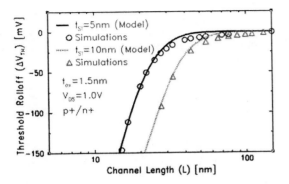

Figure 8: Threshold voltage rolloff in ADG devices [13].

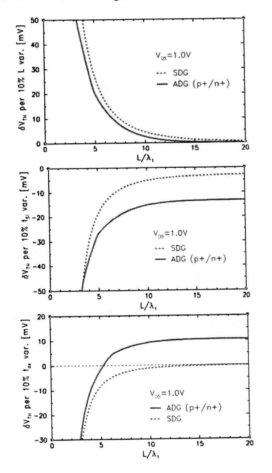

Figure 9: V_{TH} variations per 10% increase of L, t_{Si}, and t_{ox}: symmetric versus asymmetric devices [13].

3.2 Asymmetric DG MOSFETs

The effect of inversion carriers on threshold voltage calculations in asymmetric devices is included by a bi-sectional approach to computing the potential profile at the

virtual cathode [5], which leads to a simple, although explicit, short-channel V_{TH} model [13],

$$V_{TH} = \Phi_{MS,i,F} + \phi_{MAX} + \left(kT/q\right)n_i t_{Si}\exp\left(q\phi_{MAX}/kT\right)/Q_{TH}r \cdot \quad (8)$$

Model (8) is compared to Medici simulations with close agreement (Figure 8).

Applying the new V_{TH} variation analysis technique, V_{TH} sensitivities in ADG MOSFETs are investigated and compared to those in SDG devices (Figure 9). For practical designs of p+/n+ ADG devices, t_{Si} causes 35% to 100% more V_{TH} variation than L does for the same process tolerance. While ADG devices show a slightly smaller sensitivity to L than SDG devices, they may be more prone to t_{Si} and t_{ox} variations, particularly, in relatively long-channel designs.

3.3 Ground-plane MOSFETs

In ground-plane MOSFETs, the threshold voltage essentially represents a pair of signal-gate voltage and constant-bias voltage [5]. Depending upon the gate voltage combinations, the potential profile under the threshold condition can be strongly asymmetric, moderately asymmetric, or symmetric. Therefore, the threshold voltage model for GP MOSFETs is developed as a hybrid of V_{TH} models for symmetric and asymmetric devices (Figure 10). Region I seen in Figure 10 is undesirable because of very weak control of the signal gate over the channel, which is explained by the fact that strong inversion is found along the channel surface near the constant-bias gate. The moderate dependence of V_{TH} on constant-bias voltage found in Region II may be exploited to compensate for process induced V_{TH} variations.

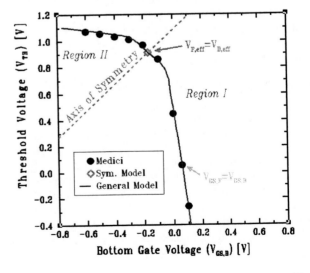

Figure 10: V_{TH} versus constant-bias gate voltage. t_{Si}=20 nm, t_{ox}=1.5 nm, and L=30 nm with p$^+$/n$^+$ polysilicon as front and bottom gates, respectively [5].

4 SCALING LIMITS PROJECTIONS

Scaling limits of DG MOSFETs are projected on the example of symmetric devices based on three scaling criteria: 1) an excellent turn-off behavior of S=70 mV/dec, 2) a moderate turn-off behavior of S=100 mV/dec, and 3) 70 mV maximum V_{TH} change caused by 30% L-equivalent process tolerance (to which all process variations are converted) [11]. As seen in Figure 11, individual DG MOSFETs with satisfactory turn-off characteristics are feasible with L as short as ~10 nm (~12 nm for S=70 mV/dec and ~7 nm for S=100 mV/dec). However, V_{TH} control, which is needed for gigascale integration of these devices, presents the biggest challenge for scaling, allowing L to be reduced only to ~16 nm.

5 QUANTUM MECHANICAL EFFECTS

It becomes clear in Figure 11 that 10 nm DG MOSFETs require ultra-thin silicon channel around 3 nm and ultra-thin gate oxide around 1 nm. Carrier confinement in such a thin silicon film becomes significant, leading to energy quantization and carrier re-distribution. Taking into account both the band structure of silicon and the quantization effect, a quantum mechanical threshold voltage model has been developed for symmetric DG MOSFETs [14],

$$\Delta V_{THLong} = \frac{E_g}{2q} + \frac{kT}{q}\ln\left(n_i t_{Si}\right) - \frac{kT}{q}\ln\left[\sum_{i=1}^{2} g_i \frac{m_{D,i}^*}{\pi\hbar^2}kT\sum_j \exp\left(-\frac{E_{j,i}}{kT}\right)\right], \quad (9)$$

which is compared to DESSIS numerical simulations with close agreement (Figure 12). In general, multiple subbands are needed for model calculations. As t_{Si} decreases below 3 nm, quantization becomes so strong that the lowest subband alone seems to suffice. Quantum mechanical effects dramatically increase V_{TH}'s sensitivity to t_{Si} as it decreases, and t_{Si} outgrows L as the largest source of parameter variations at t_{Si}=2 nm (Figure 13).

6 IMPACT OF HIGH-K DIELECTRICS

High-permittivity (high-κ) dielectrics have been proposed to replace SiO$_2$ as the gate oxide to alleviate the increasingly large gate tunneling current. A much thicker gate dielectric layer that results, however, leads to fringe fields in the non-ideal parallel-plate gate-insulator-channel structure, which weakens the gate control over the channel and consequently exacerbates SCEs. Incorporating the fringe-induced barrier lowering effect (FIBL), a short-channel V_{TH} model with high-κ dielectrics has been derived for symmetric DG MOSFETs [15],

$$V_{TH} = \Phi_{MS,i} + \eta\frac{kT}{q}\frac{\cosh(\theta)}{\cosh(\theta/2)}\ln\left(\frac{Q_{TH}}{n_i t_{Si}}\right) - \left[\frac{\cosh(\theta)}{\cosh(\theta/2)}\eta - 1\right]\left(\varphi_{0m} + 2.39\frac{kT}{q}\frac{t_I}{t_{Si}}\ln\frac{\varepsilon_I}{\varepsilon_{SiO_2}}\right), \quad (10)$$

where t_I and ε_I are the thickness and permittivity of gate dielectric. Applying a concerted analysis of FIBL-enhanced SCEs and gate direct tunneling current, candidate high-k gate dielectrics are assessed on their impact on DG MOSFETs' scaling limits (Figure 14). High-k gate dielectrics may extend DG MOSFET scaling beyond that with SiO$_2$, but the amount of channel length reduction is probably less than 20%.

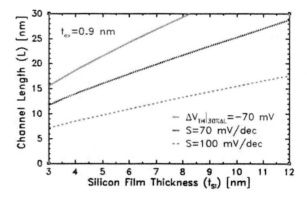

Figure 11: Scaling limits: L versus t_{Si}. [11].

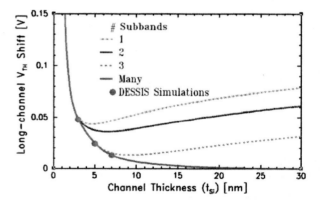

Figure 12: Quantization-based long-channel threshold voltage shift versus silicon channel thickness [14].

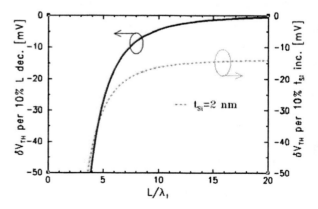

Figure 13: Overall V_{TH} change per 10% decrease of L (left) versus that per 10% increase of t_{Si} (right) as unified functions of the ratio of L/λ_1 [14].

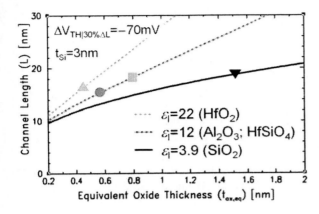

Figure 14: *L* versus EOT. Symbols mark minimum *L*'s determined by minimum EOTs (allowed for gate tunneling current): ▼ - SiO$_2$; ■ – HfSiO$_4$; ● - Al$_2$O$_3$; ▲ - HfO$_2$ [15].

7 CONCLUSIONS

Compact, physics-based models of subthreshold swing and threshold voltage are presented for undoped double-gate MOSFETs in symmetric, asymmetric, and ground-plane modes. While both symmetric and asymmetric DG MOSFETs have nearly ideal subthreshold swing at large channel lengths, ground-plane MOSFETs have significantly larger subthreshold swing. Subthreshold swing of ADG MOSFETs is slightly smaller than that of SDG devices. Based on a unique, scale-length based methodology, threshold voltage variations are analyzed comprehensively and exhaustively. In SDG MOSFETs, *L* causes 30% to 50% more V_{TH} variation than does t_{Si} for the same process tolerance. Contrarily, t_{Si} causes 35% to 100% more V_{TH} variation than *L* does in p+/n+ ADG devices. Quantum mechanical effects on threshold voltage have been analytically modeled. Multiple subbands are in general needed for model calculations, while the lowest subband alone seems to suffice for t_{Si} less than 3 nm. Fringe-induced barrier lowering effect has been modeled and included in the threshold voltage model for SDG MOSFETs. A concerted analysis of FIBL-enhanced SCEs and gate direct tunneling current shows that high-permittivity dielectrics may be helpful to reduce the channel length, but probably by less than 20% compared with SiO$_2$. Finally, scaling limits projections indicate that individual DG MOSFETs with good turn-off behavior are feasible at 10 nm scale; however, practical exploitation of these devices toward gigascale integrated systems requires development of novel technologies for significant improvement in process control.

REFERENCES

[1]   Int. Technol. Roadmap for Semiconductors: 2002, Semiconductor Industry Assoc., San Jose, CA, 2002, http://public.itrs.net/.

[2]   Q. Chen, B. Agrawal, and J. D. Meindl, "A comprehensive analytical subthreshold swing (*S*) model for double-gate MOSFETs," IEEE Trans. Electron Devices, 49, 1086-1090, 2002.

[3]   B. Agrawal, "Comparative scaling opportunities of MOSFET structures for Gigascale Integration (GSI)," Doctoral thesis, Rensselaer Polytechnic Institute, 1994.

[4]   Y. Tosaka, K. Suzuki, and T. Sugii, "Scaling-parameter-dependent model for subthreshold swing *S* in double-gate SOI MOSFET's," IEEE Electron Device Lett., 15, 466-468, 1994.

[5]   Q. Chen, "Scaling limits and opportunities for double-gate MOSFETs," Doctoral thesis, Georgia Institute of Technology, 2003.

[6]   Q. Chen, K. A. Bowman, E. M. Harrell, and J. D. Meindl, "Double jeopardy in the nanoscale court? – Modeling the scaling limits of double-gate MOSFETs with physics-based compact short-channel models of threshold voltage and subthreshold swing," IEEE Circuits and Devices Mag., 19, 28- 34, 2003.

[7]   C. T. Lee and K. K. Young, "Submicrometer near-intrinsic thin-film SOI complementary MOSFET's," IEEE Trans. Electron Devices, 36, 2537-2547, 1989.

[8]   K. Suzuki, T. Tanaka, Y. Tosaka, H. Horie, Y. Amimoto, and T. Itoh, "Analytical surface potential expression for thin-film double-gate SOI MOSFETs," Solid-State Electron., 37, 327-332, 1994.

[9]   P. Francis, A. Terao, D. Flandre, and F. Van de Wiele, "Modeling of ultrathin double-gate nMOS/SOI Transistors," IEEE Trans. Electron Devices, 41, 715-720, 1994.

[10] Y. Taur, "Analytic solutions of charge and capacitance in symmetric and asymmetric double-gate MOSFETs," IEEE Trans. Electron Devices, 48, 2861-2869, 2001.

[11] Q. Chen, E. M. Harrell, and J. D. Meindl, "A physical short-channel threshold voltage model for undoped symmetric double-gate MOSFETs," IEEE Trans. Electron Devices, 50, 1631 -1637, 2003.

[12] H.-S. Wong, D. J. Frank, and P. M. Solomon, "Device design consideration for double-gate, ground-plane, and single-gated ultra-thin SOI MOSFET's at the 25 nm channel length generation," in IEDM Tech. Dig., 1998, 407-410.

[13] Q. Chen and J. D. Meindl, "A comparative study of threshold variations in symmetric and asymmetric undoped double-gate MOSFETs," in Proc. IEEE Int. SOI Conf., 2002, 30-31.

[14] Q. Chen, L. Wang, and J. D. Meindl, "Quantum mechanical effects on double-gate MOSFET scaling," in Proc. IEEE Int. SOI Conf., 2003, 183-184.

[15] Q. Chen, L. Wang, and J. D. Meindl, "Impact of high-κ dielectrics on undoped double-gate MOSFET scaling," in Proc. IEEE Int. SOI Conf., 2002, 115-116.

Floating Gate Devices: Operation and Compact Modeling

Paolo Pavan[*], Luca Larcher[**] and Andrea Marmiroli[***]

[*]Dipartimento di Ingegneria dell'Informazione, Università di Modena e Reggio Emilia,
Via Vignolese 905, 41100 Modena, Italy, paolo.pavan@unimore.it
[**]Dipartimento di Scienze e Metodi dell'Ingegneria, Università di Modena e Reggio Emilia,
Via Fogliani 1, 42100 Reggio Emilia, Italy, larcher.luca@unimore.it
[***]STMicroelectronics, Central R&D, Via C. Olivetti 2, 20041 Agrate Brianza (MI), Italy,
andrea.marmiroli@st.com

ABSTRACT

This paper describes a possible approach to Compact Modeling of Floating Gate devices. Floating Gate devices are the basic building blocks of Semiconductor Nonvolatile Memories (EPROM, EEPROM, Flash). Among these, Flash are the most innovative and complex devices. The strategy followed developing this new model allows to cover a wide range of simulation conditions, making it very appealing for device physicists and circuit designers.

Keywords: compact model, nonvolatile memory, floating gate, reliability, circuit design.

1 INTRODUCTION

Flash Memories are one of the most innovative and complex types of high-tech, nonvolatile memories in use today, see for example [1]. Since their introduction in the early 1990s, these products have experienced a continuous evolution from the simple first ones to emulate EPROM memories, to the extreme flexibility of design application in today products. In the memory arena, Flash memory is the demonstration of the pervasive use of new electronic applications in our lives, exploiting this flexible and powerful memory technology either as a stand-alone component or embedded in a product. Flash are not just memories, they are "complex systems on silicon": they are challenging to design, because a wide range of knowledge in electronics is required (both digital and analog), and they are difficult to manufacture. Physics, chemistry, and other fields must be integrated; and conditions must be carefully monitored and controlled in the manufacturing process.

Compact Models (CMs) of Floating Gate (FG) devices are therefore needed and they have the same purpose of all compact models: *to be used within a program for circuit simulation*. The Floating Gate transistor is the building block of a full array of memory cells and a memory chip. In a first approximation, the reading operation of a FG device can be considered a single-cell operation. Nevertheless, CMs are fundamental to simulate the effects of the cells not directly involved in the operation under investigation and the effects of the parasitic elements. Furthermore, they allow the simulation of the interaction with the rest of the device, and hence they can be used to check the design of the circuitry around the memory array: algorithms for cell addressing, charge pump sizing taking into account current consumption and voltage drops, etc...

In this scenario, despite of the wide diffusion of FG-based Non-Volatile Memories, no complete CMs of FG devices were proposed and used in the industry until few years ago. Usually, MOSFET transistors whose threshold voltage was *manually* changed to model programmed and erased state of the FG memory cell were used in circuit simulations to reproduce (with poor accuracy) the FG memory behavior.

2 FLOATING GATE DEVICE MODEL

The FG device is the building block of a nonvolatile memory cell. The device is a MOS transistor with a conductive layer "floating" between gate and channel, Fig 1 [2]. The basic concepts and the functionality of this kind of device are easily understood if it is possible to determine the FG potential. The schematic cross section of a generic FG device is shown in Fig. 1. The FG acts as a potential well. If a charge is forced into the well, it cannot move from there without applying an external force: the FG stores charge. The presence of charge in this floating layer alters the threshold voltage of the MOS transistor (low threshold and high threshold, corresponding to "1" and "0").

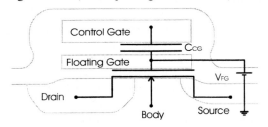

Figure 1. Cross section of a FG device and basic schematic of the CM subcircuit.

The FG device CM is the basic building block to model a single memory cell, full array, a memory chip. The simple idea underneath is to model the FG device as a circuit with a MOS transistor and a capacitor between the control gate and the FG node (which is the gate of the MOS transistor)

Fig 1. This CM exploits the MOS transistor model. Many MOS models have been developed (Philips MM11 [3], BSIM4 [4], EKV [5], SP [6], HiSIM [7]).

The approach for the FG device modeling followed here is independent of the specific MOS model adopted, thus exploiting all the improvements carried out for the basic MOS transistor models and the definition of their parameter extraction algorithms. The capacitance value is used, together with the charge injected in the floating gate, to calculate the FG node potential that is applied through a voltage controlled voltage source (V_{FG} in Fig. 1). This voltage generator is indispensable in DC conditions, as in circuit simulators there is no general solution to the calculation of the potential of a floating gate node in a DC conditions.

This model has also the advantage to allow the modeling of programming and erasing operations by simply adding a set of suitable current generators between the various electrodes. This modular approach enables the modeling of read/program disturbs, retention, leakage currents, in a rather simple way.

Two main limitations of this model can be foreseen. First, usually MOS compact models target thin gate oxide transistors with Lightly– or Medium– Doped Drain (LDD / MDD) diffusions. The oxide thickness of all FG devices is above the 7 nm, while the source and drain junctions are usually abrupt. It might become necessary to adapt the existing transistors models to this kind of devices. Second, there are a few coupling capacitances which are neglected: the coupling between the control gate node and the source, drain, and body nodes. Furthermore, as memory cells are getting smaller and closer one to the other, the coupling capacitance between the electrodes of two neighbor cells (which are not included in the model) may become more important.

2.1   DC Operation: Read

There are very few works in the literature to address the task of simulating the DC behavior of the FG memory cells, see for example [8]. With this specific CM, the FG node is biased to its correct value by an external source: the voltage-controlled voltage source, V_{FG}, which constitutes the core of the model in DC conditions.

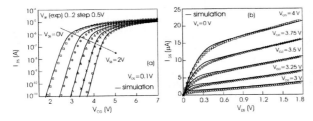

Figure 2. DC characteristics: experimental (symbols) and model (solid lines) for a 0.25μm Flash memory cell (W=0.25μm, L=0.375μm, C_{CG}=0.8fF).

The new approach gives many advantages compared to standard models [1]:

1. *scalability*: scaling rules are already included in the compact MOS model adopted and they do not affect directly the V_{FG} calculation routine;
2. *implementation*: it uses standard circuit elements whose parameters can be determined by applying the MOS parameter extraction procedure to the dummy cell, and the few additional parameters can be easily estimated from cell layout and cross section;
3. *accuracy*: it depends mainly on the compact MOS model adopted, taking advantage of the many efforts to improve and scale MOS CMs;
4. *computation time*: comparable to a MOS transistor;
5. *modularity*: it can be easily extended to simulate transient behaviors of FG memories by adding a suitable set of voltage controlled current sources to its basic structure.

2.2   Parameter extraction procedure

The procedure to extract the parameters of device CMs is not a "push-button" task. For FG devices, this task is even more complex than for standard MOS transistors. Reasonable results are obtained paying attention to the slightly different physics of the *dummy cell* (which is the cell where FG and CG are short-circuited) compared to a standard MOS transistor: narrow and short geometry, the lack of LDD and Pocket Implant determine a less ideal behavior such as larger DIBL effect and higher multiplication current. Care has to be devoted to extract the overlap capacitance values (overlap capacitances are very small: their evaluation is particularly critical).

Except the CG-FG capacitance (C_{CG}), additional parameters depend on the kind of FG memory considered. In EEPROM memory cell they are: 1) the area of the tunneling region; 2) the tunnel oxide thickness; 3) the doping levels of the drain well and the FG. In Flash memory cells they are: 1) the areas of S-FG, D-FG, and channel-FG overlap regions; 2) the doping levels of S and D wells, channel and FG. Generally, these parameters are either directly evaluated from the layout of the cell (C_{CG}, tunnel and overlap region areas), or straightly derived from the process recipe (doping). Sometimes, dedicated measurements performed on MOS capacitor test structures.

2.3   Transient Operations: Program and Erase

To simulate the program/erase operations of FG devices we have extended the model by adding a suitable set of voltage controlled current sources to the basic framework to implement compact formulae of program/erase currents. The number and position of current generators depend on the FG memory considered (Flash or EEPROM cells) and the writing mechanisms used to transfer charge to and from the FG. For example, to extend the DC model of Flash memory devices to account for writing operations, three voltage controlled current sources have to be added to reproduce program/erase currents [9]:

1. a voltage controlled current source between FG and S, I_{W1}, which models the Fowler Nordheim (FN) current flowing at the source side (needed when modeling Flash memories erased by FN tunnel at the source side);

2. a voltage controlled current source connected between FG and B, I_{W2}, which models the FN tunnel current flowing toward the substrate;

3. a voltage controlled current source connected between FG and D, I_{W3}, which models Channel Hot Electron (CHE) and Channel Initiated Secondary Electron (CHISEL) injection currents, via suitable compact formulae.

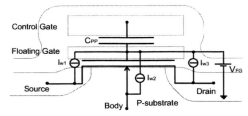

Figure 3. The complete CM of a Flash memory cell: basic framework plus three voltage controlled current sources, I_{W1}, I_{W2} and I_{W3}, that model P/E currents.

When using this model, the simulation accuracy of FG device program/erase operations depends strictly on the precision of CMs developed to describe FN, CHE and CHISEL currents. Therefore, great attention has to be devoted to develop effective CMs of these currents mechanisms.

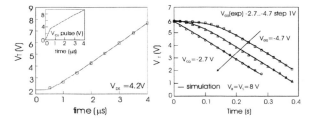

Figure 4. Threshold voltage (V_T) shifts measured (symbols) and simulated (solid lines) during program and erase of a Flash memory cell.

3 CIRCUIT SIMULATIONS

The schematic of the sense amplifier circuit simulated is shown in Fig. 5 [10]. It is a classic scheme where active load p-channel transistors are biased to provide the wanted constant current, thus allowing a controlled trip point voltage and temperature compensation. The structure is fully differential to have good noise immunity. Mn1 and Mn3, Mn2 and Mn4 provide the current/voltage conversion to bias the reference cell and the cell to be read in the matrix. V_{CELL} and V_{REF} are voltages deriving from the I-V

conversion of currents driven by the cell in the memory array and the reference cell, that are compared to generate the V_{SENSE_OUT} digital level.

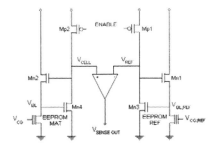

Figure 5. Schematic of the sense amplifier and the direct I-V conversion circuits of an EEPROM memory.

Simulation results in Figures 6 (a)-(b) [13] show that the output signal of the sense amplifier switches correctly according to the programmed/erased state of the EEPROM memory cell. This model is effective to simulate FG-based memory cells also in complex circuits, and therefore it can be used to simulate any circuit including a FG memory cell: read paths, non-volatile latches, X and Y decoders, voltage pumps.

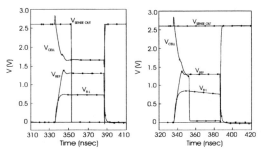

Figure 6. Control signals, and sense amplifier output obtained from read-path circuit simulations in the two cases of a programmed (a) and erased (b) EEPROM memory cell.

4 RELIABILITY SIMULATIONS

Usually, the reliability of FG memory devices is investigated through experimental techniques and the use of suitable ad-hoc models to describe leakage currents through their oxide layers. In fact, leakage currents through gate and interpoly oxides are the most serious concern for the reliability of FG memory devices, since they can strongly degrade data retention properties and increase program and read disturbs. In this scenario, we will show that the CM of FG devices (extended to include leakage current effects) can be a versatile and powerful tool for reliability predictions. CMs allow also to bridge the gap between the oxide quality characterization activity performed traditionally on MOS transistors and capacitors, and the actual impact of Stress Induced Leakage Current (SILC) on

FG memory reliability. This CM is an effective tool to predict FG memory reliability degradation, the influence on data retention of P/E cycles, P/E bias conditions, thickness and quality of tunnel oxide, and storage field.

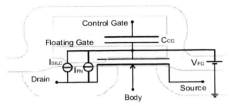

Figure 7. The CM of an EEPROM memory cell extended to simulate SILC-induced EEPROM reliability degradation by including the current generator, I_{SILC}, implementing an analytical SILC formula.

To this purposes, the CM of EEPROM memory cell can be extended by including a voltage-controlled current source implementing the empirical SILC expression proposed in [11]. As shown in Fig. 7, the SILC current generator is connected between the drain and the floating gate since in this region SILC is much larger due to the thinner thickness (~7nm) of tunnel oxide compared to the gate oxide one (~20nm).

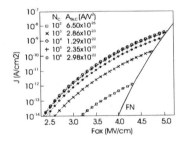

Figure 8. SILC curves (symbols) simulated through (1) in a 7 nm thick oxide on increasing the number of P/E cycles (N_C). The A_{SILC} corresponding to N_C is also indicated. The "classic" FN current is shown by a solid line.

To correlate the tunnel oxide degradation induced by high-field stress to the P/E cycles, N_C, and the P/E bias conditions, the charge exchanged during a P/E cycle and the P/E current density flowing through the tunnel oxide have been evaluated using the model. In Fig. 8, SILC curves simulated by considering typical P/E conditions (*Program*: D is ramped from 0 to V_R=12 V with a ramp rise time T_R=0.5 ms, CG and B are grounded, and S is left floating; *Erase*: CG is ramped from 0 to V_R=12 V with a ramp rise time T_R=0.5 ms, S, D, and B are grounded) are depicted for different P/E cycles.

Read disturb simulations have demonstrated that SILC is not a concern for EEPROM cells considered, since the oxide field and the time involved in read operations are too low to induce significant FG charge variations, i.e. V_T modification regardless the SILC magnitude. On the

contrary, data retention losses are strongly affected by Stress Induced Leakage Current, as predicted by reliability simulations and also confirmed by experimental data.

Fig. 9 shows V_T shifts simulated in an erased EEPROM cell left unbiased for ten years at room temperature. There are two aspects worth stressing. First, the threshold voltage reduction occurring after ten years increases with the number of P/E cycles (see dashed lines). This V_T trend is due to the SILC rise on increasing N_C (see Fig. 8), whereas the overlap of V_T-time curves for $N_C \leq 10$ is determined by the fact that in these storage field conditions the tunnel current is dominated by the FN component. Second, the ten year threshold voltage after 10^5 cycles does not depend on the initial V_T, i.e. on the initial storage oxide field, $F_{OX,S}$: as shown in Fig. 9, V_T-time curves simulated for EEPROM cells after 10^5 P/E cycles assuming different initial V_T (solid lines) converge to a similar value after 2-3 years, which depends on SILC magnitude, but it is independent on the initial V_T, i.e. $F_{OX,s}$.

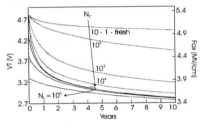

Figure 9. V_T decay for a single EEPROM cell left unbiased at room temperature on increasing the cycle number N_C (dashed lines). The oxide field is also indicated. All dashed lines start from the same initial storage field. For N_C=10^5, different initial storage fields have been assumed, and the decay curves are shown by solid lines.

REFERENCES

[1] VV.AA., Proceedings of the IEEE, Special Issue on: *Flash Memory Technology*, April 2003
[2] .L. Larcher, et al., IEEE Trans. on Electron Devices, Vol.49, N.2, pp. 301-307, 2002.
[3] http://www.semiconductors.philips.com/Philips_Models/
[4] W. Liu, X. Jin, K. M. Cao, an C. Hu, "BSIM4.0.0 MOSFET Model User's Manual," 2000.
[5] http://legwww.epfl.ch/ekv/
[6] G. Gildenblat, et al, Proc. 1997 International Semiconductor Device Research Symposium, p. 33, 1997.
[7] .http://home.hiroshima-u.ac.jp/usdl/HiSIM.html
[8] A. Kolodny, et al. IEEE Trans. Electron Dev., Vol. ED-33(6), pp.835-844, 1986.
[9] L. Larcher, P. Pavan, Proc. MSM2002, Puerto Rico, April 21-25, 2002, pp. 738-741
[10] P. Pavan, et al., IEEE Trans. on CAD, Vol. 22, N. 8, August 2003, pp. 1072 –1079
[11] L. Larcher, et al., IEEE Transactions on Device and Materials Reliability, Vol. 2, N. 1, March 2002, pp. 13 - 18

A Non-Charge-Sheet Analytic Theory for Undoped Symmetric Double-Gate MOSFETs from the Exact Solution of Poisson's Equation using SPP Approach

Jin He, Xuemei Xi, Chung-Hsun Lin, Mansun Chan*, Ali Niknejad, and Chenming Hu

Department of EECS, University of California at Berkeley, CA, 94720   jinhe@eecs.berkeley.edu
*Department of EEE, HKUST, Clear Water Bay, Kowloon, Hong Kong

ABSTRACT

A non-charge-sheet based analytic theory for undoped symmetric double-gate MOSFETs is presented in this paper. The formulation is based on the exact solution of the Poisson's equation to solve for electron concentration directly rather than relying on the surface potential alone. Therefore, carrier distribution in the channel away from the surface is also taken care, giving a non-charge-sheet model compatible with the classical Pao-Sah model. The formulated model has an analytic form that does not need to solve for the transcendent equation as in the conventional surface potentials or Pao-Sah formulation. The validity of the model has also been demonstrated by extensive comparison with AMD double-gate MOSFET's data

Keywords: Double-gate, MOSFETs, Compact modeling.

1   INTRODUCTION

As CMOS scaling continue to beyond the 50nm node, undoped (or lightly-doped) double-gate MOSFETs have become the most promise candidate device structure due to a number of unique features such as ideal 60 mV/decade subthreshold slope, volume inversion, free-dopant-associated fluctuation effects and so on . Compact modeling of this structure has been extensively studied. Most of these models, however, have to rely on the charge-sheet approximation. For thick film devices, carrier distribution with near a zero-thickness sheet can capture the essential part of double-gate MOSFET's characteristics. For an ultra-thin double gate MOSFET, the charge-sheet approximation cannot be used because the inversion layer thickness is comparable with silicon film thickness and a number of results from the charge-sheet approximation become invalid.

In a recent work, a one-dimensional (1-D) Poisson-Boltzamnn equation was solved for the double-gate MOSFET to derive analytical solution for surface potential and inversion charge density by Prof.Taur's group. This result can serve as a core foundation for developing non-charge-sheet-based model. Due to the numerical complexity and the need to solve a number of transcendental equations, the model is difficult to be implemented in circuit simulator. A simpler and yet equally accurate model is thus desired, not only for efficiency of simulation, but also to provide insight into the operation of the double-gate MOSFETs.

In this work, we have derived a non-charge-sheet based analytic theory for undoped symmetric double-gate MOSFETs by solving the exact carrier concentration based on both the surface potential and the quasi-Fermi potential in the channel to account for the vertical potential profile and distribution of the carrier. We referred this approach as surface potential plus (SPP), which is compatible with the Pao-Sah formulation. The analytical model has also been verified by AMD double-gate MOSFET's data.

2   ANALYTICAL SOLUTION

Fig. 1 shows the coordinate and energy level distribution diagrams of a symmetric double-gate MOSFET. We assume that the same voltage is applied to the two gates having the same work function. The formulation starts with the 1-D Poisson's equation along to the vertical direction of the silicon channel considering only the mobile charge (electron) density for the undoped body.

$$\frac{d^2\phi}{dx^2} = \frac{qn}{\varepsilon_{si}} \tag{1}$$

where q is the electronic charge, ε_{Si} is the permittivity of silicon, and n is the intrinsic carrier density. According to Boltzmann statistics, the mobile electron concentration can be expressed in term of potential

$$n = n_i \exp\left(\frac{q(\phi - v_{ch})}{kT}\right) \quad n_0 = n_i \exp\left(\frac{q(\phi_0 - v_{ch})}{kT}\right) \tag{2}$$

The spatial derivatives of the electron from (2) are

$$\frac{d\phi}{dx} = \frac{kT}{qn}\frac{dn}{dx} \tag{3}$$

$$\frac{d^2\phi}{dx^2} = \frac{kT}{qn}\frac{d^2n}{dx^2} - \frac{kT}{qn^2}\left(\frac{dn}{dx}\right)^2 \tag{4}$$

Substitution (4) into (1) gives an equation for electron concentration

$$\frac{d^2n}{dx^2} = \frac{1}{n}\left(\frac{dn}{dx}\right)^2 + \frac{q^2n^2}{\varepsilon kT} \tag{5}$$

This normal differential equation has two mathematical solutions, one is a trigonometric function and another is a hyperbolic function.

$$n(x) = \frac{c_0}{\cos^2\left[\left(\frac{q^2c_0}{2\varepsilon kT}\right)^{1/2}x\right]} \quad n(x) = \frac{c_0}{\cosh^2\left[\left(\frac{q^2c_0}{2\varepsilon kT}\right)^{1/2}x\right]} \tag{6}$$

Based on a physically reasonable consideration, the former

solution is taken. We choose the reference coordinate point as $x = 0$ and $n(x) = n_o$, (6) is further simplified into

$$n(x) = \frac{n_0}{\cos^2\left[\left(\frac{q^2 n_0}{2\varepsilon kT}\right)^{1/2} x\right]} \tag{7}$$

Substitution of (7) into (1) gives the corresponding electrical field and potential distributions in silicon film as

$$\phi(x) - \phi_0 = \frac{kT}{q} \ln \cos^{-2}\left[\left(\frac{q^2 n_0}{2\varepsilon kT}\right)^{1/2} x\right] \tag{8}$$

$$E(x) - E(x_0) = \left[\frac{2 n_0 kT}{\varepsilon}\right]^{1/2} \tan\left[\left(\frac{q^2 n_0}{2\varepsilon kT}\right)^{1/2} x\right] \tag{9}$$

The symmetry boundary condition of a symmetric double gate MOSFET set the electric field in the center of the silicon film to be zero. Thus, the reference coordinate takes the center of the film as the point of $x = 0$. The surface potential and the surface electric field are then given by

$$\phi_s = v_{ch} + \frac{kT}{q} \ln\left[\frac{n_0}{n_i} \cos^{-2}\left[\left(\frac{q^2 n_0}{2\varepsilon kT}\right)^{1/2} \frac{T_{si}}{2}\right]\right] \tag{10}$$

$$E_s = \left[\frac{2 n_0 kT}{\varepsilon}\right]^{1/2} \tan\left[\left(\frac{q^2 n_0}{2\varepsilon kT}\right)^{1/2} \frac{T_{si}}{2}\right] \tag{11}$$

If we define $Q_{in} = q \int_0^{T_{si}/2} n(x) dx$ as half of the total inversion charge, it can be expressed

$$Q_{in} = \left[2\varepsilon n_0 kT\right]^{1/2} \tan\left[\left(\frac{q^2 n_0}{2\varepsilon kT}\right)^{1/2} \frac{T_{si}}{2}\right] \tag{12}$$

The surface potential, field and carrier concentration are controlled by the applied gate voltage. According to Gauss's law, the gate voltage is

$$V_G - \Delta\psi_i = \phi_S + E_{ox} t_{ox} = \phi_s + \frac{Q_{in}}{\varepsilon_{ox}} t_{ox} \tag{13}$$

Substituting the surface potential and inversion charge into (13) results gives

$$V_G - \Delta\psi_i - v_{ch} = \frac{kT}{q} \ln\left[\frac{n_0}{n_i} \cos^2\left[\left(\frac{q^2 n_0}{2\varepsilon kT}\right)^{1/2} \frac{T_{si}}{2}\right]\right] + \frac{\varepsilon_{si}}{\varepsilon_{ox}} t_{ox} \left[\frac{2 n_0 kT}{\varepsilon}\right]^{1/2} \tan\left[\left(\frac{q^2 n_0}{2\varepsilon kT}\right)^{1/2} \frac{T_{si}}{2}\right] \tag{14}$$

Eq.(14) is the exact closed form expression for electron concentration at the center of the channel silicon (n_0) as a function of gate voltage, channel voltage, and silicon film. This expression is useful because of its exactness, but too complex to be implemented in compact model due to the transcendental equation. We will next simplify Eq. (14) to a more manageable form. The inversion charge expression given in Eq. (12) can be also be expressed as

$$Q_{in}^2 = 2\varepsilon_{si} n_0 kT \tan^2\left[\left(\frac{q^2 n_0}{2\varepsilon kT}\right)^{1/2} \frac{T_{si}}{2}\right] = 2\varepsilon_{si} n_0 kT \frac{\sin^2\left[\left(\frac{q^2 n_0}{2\varepsilon_{si} kT}\right)^{1/2} \frac{T_{si}}{2}\right]}{\cos^2\left[\left(\frac{q^2 n_0}{2\varepsilon_{si} kT}\right)^{1/2} \frac{T_{si}}{2}\right]} \tag{15}$$

We will use the inversion charge expression to eliminate the surface potential term in Eq.(13). This surface potential term comes into effect only in the sub-threshold

region where the potential profile in the vertical direction is almost flat and resulting in volume inversion. In the sub-threshold region case, the entire channel region contributes to conduction, and the inversion charge is almost proportional to the silicon film thickness. Therefore, we can express the inversion charge as

$$Q_{in} = q n_0 T_{si} / 2 \tag{16}$$

On the other hand, since a tangent function of the inversion charge expression is mainly determined by cosine function, an popular approximation is used to substitute sine function that can be combined with Eq. (16).

$$\sin^2\left[\left(\frac{q^2 n_0}{2\varepsilon_{si} kT}\right)^{1/2} \frac{T_{si}}{2}\right] \approx \left(\frac{q^2 n_0}{2\varepsilon_{si} kT}\right)\frac{T_{si}^2}{4} \approx \frac{q Q_{in}}{2\varepsilon_{si} kT} \frac{T_{si}}{2} \exp(f) \tag{17a}$$

This approximation works well for ultra-thin silicon case, leading to a negligible error. In fact, we will find in the following expression that this sine function term comes into effect in the sub-threshold region. In order to compensate the error leaded by this approximation, we use a dimensionless correction factor f. This factor should be a step function of the gate voltage, e.g. equal to unit below the threshold point and zero above the threshold point. An exact approximate solution of this factor is written as

$$f = \left[\frac{\tanh(v_G - \Delta\phi_i)}{v_G - \Delta\phi_i}\right]^2 \tag{17b}$$

via comparison with the numerical result and 2-D simulation. In this case, we have

$$\sin^2\left[\left(\frac{q^2 n_0}{2\varepsilon_{si} kT}\right)^{1/2} \frac{T_{si}}{2}\right] = \frac{q Q_{in}}{2\varepsilon_{si} kT} \frac{T_{si}}{2} \exp(f) \tag{17c}$$

Making use of this approximation, (16) is written as

$$Q_{in} = \frac{q n_0 T_{si/2} \exp(f)}{\cos^2\left[\left(\frac{q^2 n_0}{2\varepsilon kT}\right)^{1/2} \frac{T_{si}}{2}\right]} \tag{18}$$

Substitution of (18) into (13) and replacing the surface potential gives

$$V_G - \Delta\psi_i - v_{ch} + \frac{fkT}{q} \ln(q n_i T_{si} / 2) = \frac{kT}{q} \ln Q_{in} + \frac{Q_{in}}{\varepsilon_{ox} / t_{ox}} \tag{19}$$

This inversion charge equation is similar to that given by the bulk SPP model, except the sub-threshold slope n is replaced with by 60mV at room temperature. Thus, a similar approach as in the SPP can be used to obtain a consistent I-V and C-V characteristics

To simplify subsequent derivation, all voltages are normalized by kT/q and inversion charge is normalized by $C_{ox} kT/q$. (19) becomes after simplification

$$v_G - \Delta\phi_i - v_{ch} + f \ln\left(\frac{q^2 n_i T_{si} t_{ox}}{2 kT \varepsilon_{ox}}\right) = \ln q_{in} + q_{in} \tag{20}$$

The inversion charge in (20) is solved explicitly in terms of the Lambert W function to yield the following exact expression.

$$q_{in} = W_0\left[\exp\left(v_G - \Delta\phi_i - v_{ch} + f \ln\left(\frac{q^2 n_i T_{si} t_{ox}}{2 kT \varepsilon_{ox}}\right)\right)\right] \tag{21}$$

where W_0 stands for the usual short-hand notation used for the principal branch of the "**Lambert-W**" function.

The exact solution that we propose makes use of the principal branch of the **Lambert W** function which is defined as the solution to the equation $We^w = x$. The Lambert W function is a popular function in numerous physics applications. It has also been used to provide solutions to previously unsolved basic diode and bipolar transistor circuit analysis. Once the inversion charge is obtained, the I-V and C-V characteristics of undoped double gate MOSFET can be directly formulated.

We next proceed to describe the modeling of the IV characteristics. According to Pao-Sah model, the current including the diffusion and drift component is written as

$$I_{ds} = \mu w C_{ox} \left(\frac{kT}{q}\right)^2 q_{in} \frac{dv_{ch}}{dy} \qquad (22)$$

For constant gate voltage, the differential of (20) can be written as

$$dv_{ch} = dq_{in} / q_{in} + dq_{in} \qquad (23)$$

here we neglect the derivative contribution of f factor since it is a step function. Substitution of (23) into (22) gives

$$I_{ds} = \mu w C_{ox} \left(\frac{kT}{q}\right)^2 (q_{in} dq_{in} + dq_{in}) \qquad (24)$$

As a result, an analytical current expression is obtained.

$$I_{ds} = \frac{\mu w}{L} C_{ox} \left(\frac{kT}{q}\right)^2 \left[\frac{q_s^2 - q_d^2}{2} + (q_s - q_d)\right] \qquad (25)$$

As the charge formulation only accounts for half of the channel, the final current of a double-gate MOSFET should be doubled.

Note that the final current expression is similar to that of the SPP derived by the bulk model. Thus, a unified framework can be used for bulk and the undoped symmetric double-gate MOSFETs. This results in simple methodology to obtain both the I-V and C-V characteristics with already formulated such as gate tunneling current, short channel effect, QME and non-uniform lateral body doping [19]. In addition, the unified approach to treat the different device structures significantly simplify the study of the relative advantages of different device structure to be used to extend the CMOS scaling roadmap to the nanometer regime. .

3   RESULTS AND DISCUSSION

3.1 Electron and potential in the silicon film

Fig.2 illustrates the electron concentration and electrical potential distribution in the silicon film predicted by the analytical model for three different effective gate voltages. In low gate voltage or the subthreshold region, the electron distribution is almost constant along the vertical direction of the film as predicted by $Q_{in} = q n_0 T_{si} / 2$. With the increase of the effective gate voltage, the device goes into the strong inversion region where the surface electron concentration shows several order higher than that of the silicon film center. In this case, device behaves like a

surface channel bulk MOSFET and the surface electron concentration dominates the inversion charge density.

3.2 Surface potential and inversion charge dependence on the bias

Fig.3 shows the carrier charge per unit area in the channel versus the applied gate voltage, as calculated with (20), for three values of channel voltage. There are two distinct regions of operation in the undoped symmetric double-gate MOSFET, just like in a conventional MOSFET as shown in Fig.3(b). One significant result is the relation between the inversion charge density and the channel voltage. In contrast to the gate voltage, Fig.3 indicates that the channel voltage increases the inversion charge correspondingly decreases. One interesting observation is that the equivalent dielectric thickness has a slight effect on the surface potential in strong inversion region and has almost no effect on the sub-threshold region in the undoped bulk MOSFETs, as shown in Fig.4. This physical picture is in different from that of a conventional bulk MOSFETs where the sub-threshold region slope has a strong dependence on the oxide thickness.

3.3 Volume inversion

Fig.5 demonstrates the effect of the silicon film thickness on the inversion charge versus gate voltage characteristics. The silicon film thickness affects the amount of sub-threshold inversion charge but has no effect on the strong inversion charge density. This verified the existence of the volume inversion and implies that decrease of the silicon thickness can effectively control the sub-threshold region leakage current. It illustrates a potential weakness in the used of undoped symmetric double MOSFETs for nano-CMOS application. To optimize the device performance, the silicon film thickness should be reduced to suppress the off current.

3.4 I-V   characteristics

In order to verify the validity of the analytical model, we use this model to simulate a well-temped undoped double-gate MOSFET with the channel length of 10μm and gate oxide thickness of 3nm. A long channel device is used to avoid the complicated short channel effect to hinder the detail picture of 1-D analysis in the vertical direction of the channel. Short channel effects will be considered in future work. Fig.6 and Fig.7 show the DG MOSFET current voltage characteristics predicted by the derived analytical model with the constant mobility of $400 \, cm^2 / V.s$. The model is continuous from the sub to the strong inversion region at all drain bias conditions, as shown in Fig.6. One observation is that the drain voltage has no effect on the sub-threshold current but increases the strong inversion current. The saturation characteristics given by the model is also observed from Fig.7. Fig. 8 shows this model fitting for AMD double-gate MOSFET's data by using SPP modules on SCE, DIBL, QME, ect. In DG model. A good agreement is found between both.

4 CONCLUSION

In this paper, we have developed a non-charge-sheet based analytical theory for undoped symmetric double-gate MOSFET. We have derived the exact analytical solution of the electron concentration and density as an explicit function of the gate voltage and silicon film, thickness valid for all device operation regions. The result agrees well with the Pao-Sah current formulation and physically accurate. The presented model works with a more generic SPP framework that can be used to model a wide range of devices to be used in nano-CMOS technology and been verified by AMD double-gate data.

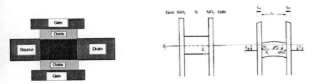

Fig.1. Diagrams of the structure and energy levels of undoped double-gate MOSFETs.

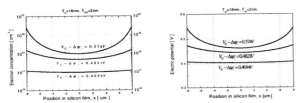

Fig.2. The mobile electron and potential distribution as a function of the silicon film thickness for undoped MOSFETs with symmetric double-gate structure.

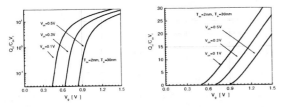

Fig.3. The mobile carrier density as a function of gate voltage with the different channel voltage for DG undoped MOSFETs.

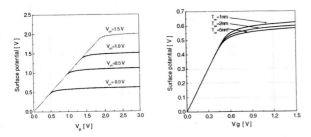

Fig.4 Surface potential versus gate voltage with the different oxide thickness and channel voltage for DG undoped MOSFETs.

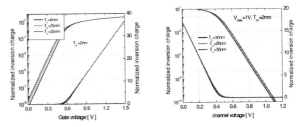

Fig.5 The mobile carrier density as a function of gate voltage and channel voltage with the different silicon thickness for undoped DG MOSFETs.

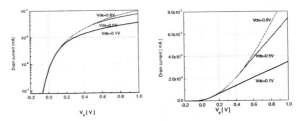

Fig.6. The drain current versus the gate voltage with the different drain voltage for a 10um undoped MOSFETs with symmetric double-gate structure.

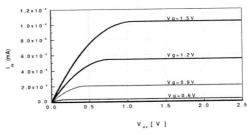

Fig.7. The drain current versus the drain voltage with the different gate voltage for a 10um DG undoped MOSFETs.

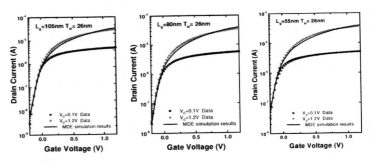

Fig.8 Comparison between the analytical model and AMD double-gate data.

Acknowledgement: This work was supported by SRC under the contract number 2002-NJ-1001 and partially supported by a grant from RGC of Hong Kong under the number HKUST6111/03E. We also thank AMD Corporation for providing experimental data and Prof. Taur from UC of San Diego for useful discussion.

An Exact Analytic Model of Undoped Body MOSFETs using the SPP Approach

Jin He, Xuemei Xi, Mansun Chan*, Ali Niknejad, and Chenming Hu

Department of EECS, University of California at Berkeley, CA, 94720   **jinhe@eecs.berkeley.edu**

*Department of EEE, HKUST, Clear Water Bay, Kowloon, Hong Kong

ABSTRACT

An exact analytic model for undoped (or lightly doped) body MOSFETs has been derived in this paper by incorporating the mobile charge density in the 1-D Poisson's equation. The formulation starts with deriving a close form solution that relates the channel band bending and inversion charge to gate and channel voltage using Gauss's law and the gradient equation of the field effect devices. A continuous current-voltage model is then developed based on the exaction solution to the channel charge. The preliminary model has been verified by comparing with long channel results generated by 2-D simulator. Very good agreements in all operation regions from sub-threshold to strong inversion, and from linear to saturation are obtained. By using simple geometry parameter alone, the model is capable to capture all features obtained from 2-D device simulator.

Keywords: *Undoped MOSFET*, charge carrier density, compact modeling, surface-potential-plus.

1 INTRODUCTION

The scaling law has been successfully applied to many generations CMOS technologies, resulting in a continuous improvement in device performance. However, new challenges are encountered in the scaling of conventional MOSFET structures to below 100 nm. For example, the high channel doping concentration required to suppress the short-channel effect (SCE) results in degraded mobility, serious threshold voltage fluctuation and enhanced junction leakage. To alleviate these problems, MOSFETs with undoped (or lightly doped) have been proposed. The idea is to avoid the variation of turn-on voltage (or threshold voltage in conventional language) due to the random fluctuation in the number of channel dopant atoms of nanoscale MOSFET's. In addition, it can also relieve the effect of mobility degradation due to impurity scattering. Due to the potential advantages of undoped body MOSFETs, there are a lot of research activities to realize it in the mainstream nano-CMOS technology in the near future. The implementation of undoped channel MOSFETs (sometimes referred to as ''intrinsic channel'' MOSFET) has been demonstrated in vertical structure [2], UTB[3] and several kinds of epitaxial channel MOSFETs [4,5]. A physically correct compact modeling of this kind structure, however, is still missing. A major reason for it is due to its physical incompatible with conventional

modeling approach that rely on the ionized dopant to establish the channel band structures and the classical definition of some channel doping related parameter such as threshold voltage (V_T).

In this paper, an exact analytic model of undoped MOSFETs is derived by solving electron density equation on both surface potential and quasi-Fermi potential. The approach is non-charge-sheet based that takes into account of the surface potential and channel potential below the surface [6-8], hence we called it a Surface-Potential-Plus (SPP) approach. The key step in the SPP formulation is the derivation of the surface potential in the term of the inversion electron density. We can obtain an exact and continuous analytical solution of inversion charge that can be used to replace the solution of the transcendent equation used in surface potential based models [9]. The analytical expression for inversion charge can then be used to develop a continuous current-voltage and capacitance-voltage model for undoped body MOSFETs. The results have been compared with the 2-D numerical simulator MEDICI and will be presented in this paper.

2 MODEL FORMULATION

Fig. 1 shows the energy band diagrams of an undoped body MOSFET. Here we only consider an nMOSFET where the holes are negligible (due to the intrinsic body used). However, the equation and derivation can easily be extended to pMOSFETs through changing the sign of the potential, field and polarity of carrier charge.

In the formulation, we first derive the gradient equation of the field effect device from the current continuity principle. This gradient equation was first applied by J.R.Bews and become one of the 2 major assumptions in deriving charge-sheet based model (the other assumption is the infinitely thin charge sheet approximation) [10]. As stated by Brews, the justification of this assumption relies on "its success in producing 'correct' I-V curves". Its prediction agrees well with that of Pao-Sah model though this equation appears to be somewhat drastic and the exact mathematical link with Pao-Sah model is not obvious.

Our derivation starts from the current density and drain current equations in the term of the quasi-Fermi-potential-level. The general 2-D current density can be written as

$$J_n = q\mu n \frac{dV}{dx} + q\mu n \frac{dV}{dy} \quad (1)$$

Strictly speaking, 2-D current density equation must be used to give exact I-V characteristics. Since the drain voltage mainly drives the channel current, we can assume that the quasi-Fermi-potential is constant along the vertical direction as in Pao-Sah model. In this case, Eq. (1) becomes 1-D and can be re-written as

$$J_n(x) = q\mu n(x)\frac{dV_{ch}}{dy} \qquad (2)$$

The total drain current is by integrating Eq.(2) over the active cross section of the channel region

$$I_{ds} \equiv W\int_0^\infty q\mu n \frac{dV_{ch}}{dy}dx = \mu W Q_{in}\frac{dV_{ch}}{dy} \qquad (3)$$

Rewriting Eq.(2) and (3), we can obtain

$$\frac{dJ_n}{dyJ_n} - \frac{dn}{n} = \frac{d}{dy}\left[\ln(q\mu\frac{dV_{ch}}{dy})\right] \qquad (4a)$$

and

$$\frac{dI_{ds}}{dyI_{ds}} - \frac{dQ_{in}}{Q_{in}} = \frac{d}{dy}\left[\ln(q\mu\frac{dV_{ch}}{dy})\right] \qquad (4b)$$

According to the current continuity, both the current density and total drain current are constant along the channel. Therefore, we have

$$\frac{dJ_n}{dy} = 0 \quad (5) \quad \text{and} \quad \frac{dI_{ds}}{dy} = \frac{d}{dy}\left[\mu W Q_{in}\frac{dV_{ch}}{dy}\right] = 0 \quad (6)$$

After some algebra, we can obtain

$$\frac{dQ_{in}}{dyQ_{in}} = \frac{dn}{dyn} = -\frac{d}{dy}\left[\ln(q\mu\frac{dV_{ch}}{dy})\right] \qquad (7)$$

Then, there holds

$$\frac{dQ_{in}}{Q_{in}} = \frac{dn_s}{n_s} \qquad (8)$$

To relate the electron concentration with channel potential, we use the Maxwell-Boltzmann relation, which is typically the case in the direction perpendicular to the channel. In such case, the Maxwell-Boltzmann electron distribution can be written as

$$n_s = n_i \exp\left(\frac{q\psi_s - qV_{ch} - q\Delta\psi_i}{kT}\right) \qquad (9)$$

Combining Eqs.(9) and (8) gives the gradient equation of the field effect device

$$\frac{dQ_{in}}{Q_{in}} = \frac{q}{kT}\left[d\psi_s - dV_{ch}\right] \qquad (10)$$

Next, we need to incorporate the applied voltage to Eq. (10). According to Gauss's law and the known boundary condition, we have

$$\varepsilon_{ox}\frac{(V_g - \Delta\psi_i - \psi_s)}{t_{ox}} = Q_{in} \qquad (11)$$

Differentiating Eq. (11), the following expression is obtained.

$$\varepsilon_{ox}\frac{(dV_g - d\psi_s)}{t_{ox}} = dQ_{in} \qquad (12)$$

Substituting (12) into (10) gives

$$dV_G - dV_{ch} = \frac{kT}{q}\frac{dQ_{in}}{Q_{in}} + \frac{dQ_{in}}{\varepsilon_{ox}/t_{ox}} \qquad (13)$$

Finally, the equation relating inversion charge to the apply voltage can be obtained by integrating Eq. (13)

$$V_G - \Delta\psi_i - V_{ch} + \frac{kT}{q}\ln(qn_iL_D) = \frac{kT}{q}\ln Q_{in} + \frac{Q_{in}}{\varepsilon_{ox}/t_{ox}} \qquad (14)$$

where L_D is the Debye length.

Eq. (14) can be used to derive both the I-V and C-V equations. To simplify the derivation, all voltages are normalized by kT/q and inversion charge is normalized by $C_{ox}kT/q$. After normalization, (14) is simplified to

$$v_G - \Delta\phi_i - v_{ch} + f\ln\left[\frac{q^2n_it_{ox}L_D}{\varepsilon_{ox}kT}\right] = \ln q_{in} + q_{in} \qquad (15a)$$

where a dimensionless correction factor f with the step function characteristics is introduced to exactly consider the sub-threshold current, e.g. equal to unit below the threshold point and zero above the threshold point. An exact approximate solution of this factor is written as

$$f = \left[\frac{\tanh(v_G - \Delta\phi_i)}{v_G - \Delta\phi_i}\right]^2 \qquad (15b)$$

The inversion charge in (15) can be solved explicitly in terms of the Lambert W function to yield the following exact expression

$$q_{in} = W_0\left[\left[\frac{q^2n_it_{ox}L_D}{\varepsilon_{ox}kT}\right]^f \exp(v_G - \Delta\phi_i - v_{ch})\right] \qquad (16)$$

where W_0 stands for the usual short-hand notation used for the principal branch of the ''Lambert-W'' function, which is the solution to the equation $We^w = x$. The Lambert W function is a popular function in numerous physics applications [11-12]. It has also been used to provide solutions to previously unsolved basic diode [13] and bipolar transistor circuit analysis [14].

Applying Gauss's law again, the analytical equation for surface potential can be obtained for the calculated inversion charge

$$\phi_s = V_G - \Delta\phi_i - q_{in}*C_{ox}kT/q \qquad (17)$$

Once the inversion charge is obtained, a continuous non-charge-sheet-based undoped body MOSFET model can be developed. In addition, following the quite same procedures, the accumulation charge equation for MOSFET's operation in the accumulation region can also be derived as

$$-(v_G - \Delta\phi_i) + f \ln\left[\frac{q^2 n_i t_{ox} L_D}{\varepsilon_{ox} kT}\right] = \ln q_a + q_a \quad (18)$$

Thus, the exact expression for the accumulation charge is

$$q_a = W_0\left[\left(\frac{q^2 n_i t_{ox} L_D}{\varepsilon_{ox} kT}\right)^f \exp\left(-v_G + \Delta\phi_i\right)\right] \quad (19)$$

Combining both inversion and accumulation charges, and taking into account of no depletion charge in the undoped body MOSFETs, the total gate charge in all operation regions can be expressed as

$$q_G = q_a + q_{in} \quad (20)$$

Assuming there is no channel voltage for simplification, the total inversion charge and accumulation charge can be related to the normalized gate capacitance with

$$C_{GG} = \frac{dq_a}{dv_G} + \frac{dq_{in}}{dv_G} \quad (21)$$

The derivative of the inversion charge and accumulation charge with respect to the gate voltage can be obtained from Eq. (15) and (18) respectively. The final result of the analytical gate capacitance expression is

$$C_{GG} = \frac{q_a}{1 + q_a} + \frac{q_{in}}{1 + q_{in}} \quad (22)$$

Substituting the expression for inversion and accumulation charge, the normalized gate capacitance characteristics can be obtained.

According to Pao-Sah model, the current including the diffusion and drift component is written as

$$I_{ds} = \mu w C_{ox}\left(\frac{kT}{q}\right)^2 q_{in} \frac{dv_{ch}}{dy} \quad (23)$$

For constant gate voltage, differentiating Eq. (13) gives

$$dv_{ch} = dq_{in} / q_{in} + dq_{in} \quad (24)$$

Substitution of (24) into (23), we obtain

$$I_{ds} = \mu w C_{ox}\left(\frac{kT}{q}\right)^2 (q_{in} dq_{in} + dq_{in}) \quad (25)$$

As a result, we get the analytical current expression

$$I_{ds} = \frac{\mu w}{L} C_{ox}\left(\frac{kT}{q}\right)^2\left[\frac{q_s^2 - q_d^2}{2} + (q_s - q_d)\right] \quad (26)$$

It should be noted that the final current expression is similar to the SPP model to be used with advanced CMOS technologies we reported before [6-8]. Thus, this similar of its framework to handle both doped and undoped body MOSFETs brings significant convenience and simplification for I-V and C-V modeling with different device structures. For example, small dimensional effects of undoped body MOSFET can be included in a unified current expression based on the SPP approach.

3   RESULTS AND DISCUSSION

Fig.2 shows the carrier charge per unit area in the channel versus the applied gate voltage calculated using Eq. (16), for three different of channel voltages. There are two distinct regions of operation in the undoped body MOSFET, just like in a conventional MOSFET. Below the turn-on voltage, the mobile charge density is low and increases exponentially with the gate voltage as shown Fig.2 (a). With the gate voltage above the turn on voltage, a linear increase of the carrier charge-sheet density with gate voltage is observed in Fig.2(b) as expected. One important effect to observe here is the relation between the inversion charge density and the channel voltage. In contrast to the gate voltage, Fig.2 indicates that the channel voltage causes the inversion charge to decrease exponentially in the sub-threshold region and linearly in the strong inversion region. This result comes directly from Eq. (14).

Fig.3 demonstrates the surface potential versus gate voltage characteristics obtained using (17) for an undoped-body MOSFETs. Again there are two distinct regions of operation. Below the turn-on voltage, the surface potential increases close to linearly with the gate voltage. In this region, the effect of the inversion charge is almost negligible. However, significant inversion charge is accumulated after turn-on, the surface potential only increases logarithmically with the gate voltage. One interesting observation is that the gate dielectric only has a slight effect on the surface potential in strong inversion region and has almost no effect on the sub-threshold characteristics in an undoped body MOSFETs. This phenomenon cannot be captured by the physics derived from conventional MOSFETs where the sub-threshold slope has a strong dependence on the oxide thickness.

To verify the validity of the derived analytical model, we use the numerical simulator MEDICI to simulate a well-controlled undoped MOSFET with the channel length of 5μm and gate oxide thickness of 3nm. In the simulation, the silicon film thickness is chosen to be larger than the room temperature intrinsic Debye length, e.g., 50um to give a general picture and avoid the effect of the finite silicon film. Fig. 4 and Fig. 5 show the comparison of the undoped MOSFET current voltage characteristics predicted by the presented analytical model and 2-D device simulation MEDICI with the constant mobility of $400 cm^2/V s$. The good agreement between the model prediction and numerical simulation can be found from the sub-threshold region to the strong inversion region for the different substrate bias conditions, as shown in Fig.4. The result shown in Fig.5 indicates that this new model is accurate enough not only in the current versus the drain voltage prediction but also in the output resistance calculation, verifying the validity of the analytical model presented in this paper.

Finally, Fig.6 shows the normalized gate capacitance curve in all operation regions predicted by

Eq.(22). This result is simple and shows the correct asymptotic characteristic.

4. CONCLUSION

The development of an analytical model for the family of undoped body MOSFETs can provide meaningful insight to the device design and optimization. Compared with numerical simulation, analytically derived model provides physical insight to the device operation as well as computation efficiency in simulation. In this paper, we have developed a general and yet exact analytical and continuous model of the undoped body MOSFETs, valid for all device operation regions. The model can form a basis for the analysis of a general class of undoped body MOSFETs including UTB devices as well as double-gate MOSFETs. The model has been extensively verified by 2-D device simulation and able to capture most of the physical behaviors predicted by the 2-D simulator.

Acknowledgement

This work was supported by SRC under the number #2003-NJ-1134 and partially supported by a grant from RGC of Hong Kong under the number HKUST6111/03E.

References

[1] Takeuchi K, Koh R, Mogami T. ED-48: pp. 1995–2001.

[2] Appenzeller J, Martel R, Avouris Ph, Knoch J, Scholvin J, del Alamo JA, et al. IEEE Electron Dev Lett 2002; EDL-23: pp.100–102.

[3] Taur Y. IEEE Electron Dev Lett 2000; EDL- 21: pp.245–247.

[4] Wong HS, Chan KK, Lee Y, Roper P, Taur Y. IEEE Trans Electron Dev 1997; EDL-44: pp.1131–1135.

[5] Ohguro T, Naruse H, Sugaya H, Morifuji E, Nakamura S, Yoshitomi T, et al. IEEE Trans Electron Dev 1999; ED-46: pp.1378–1383.

[6] Jin He, Xuemei Xi, Mansun Chan, Ali Niknejad, and Chenming. *Sixth International Conference on Modeling and Simulation of Microsystems,* San Francisco, CA, February, 2003: pp.23-27.

[7] Jin He, Xuemei Xi, Mansun Chan, Ali Niknejad, and Chenming Hu. "Non-charge-sheet based analytical model of undoped symmetrical double-gate MOSFETs using SPP approach". MSM'2004.

[8] Mansun Chan, Y.Taur, C.H.Lin, Jin He, Ali Niknejad, and Chenming Hu. Nanotech 2003, vol.2, MSM, 2003: pp.270-273.

Fig.1 Energy level diagram of an undoped bulk MOSFET.

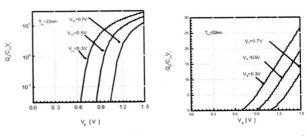

Fig.2 Normalized inversion charge versus gate voltage with the different oxide thickness and channel voltage.

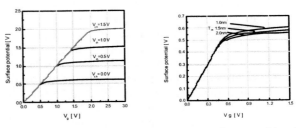

Fig.3 Surface potential versus gate voltage with the different oxide thickness and channel voltage.

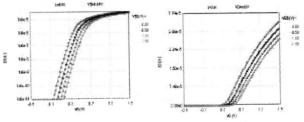

Fig.4 Drain current versus the gate voltage comparison between the analytical prediction(line) and MEDICI simulation(point).

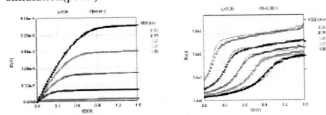

Fig.5 Drain current and resistance comparison between the analytical prediction (line) and MEDICI simulation(point).

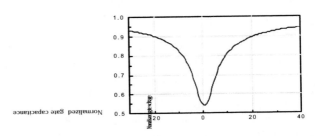

Fig.6 The analytical normalized gate capacitance with respect to the normalized gate voltage.

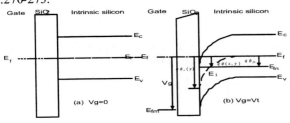

Linear Cofactor Difference Extrema of MOSFET's Drain Current and Their Application in Parameter Extraction

Jin He, Xuemei Xi, Mansun Chan*, Ali Niknejad, and Chenming Hu

Department of EECS, University of California at Berkeley, CA, 94720   jinhe@eecs.berkeley.edu
*Department of EEE, HKUST, Clear Water Bay, Kowloon, Hong Kong

ABSTRACT

The linear cofactor difference extrema of metal-oxide-semiconductor field effect transistor (MOSFET) drain current are presented in this paper and their application to extract MOSFET parameters is demonstrated. The extrema of the characteristic drain current are obtained by the applying the linear cofactor difference operator to the drain current versus gate voltage curve in the linear region. These extrema can be directly used to find the threshold voltage and mobility of a MOSFET. The method has been tested with experimentally fabricated MOSFETs and simulation data obtained by the device simulator DESSIS-ISE. The results agree well with those obtained with the standard second-derivative method, which demonstrates the validity of the method presented.

Keywords: Drain current linear cofactor difference, MOSFET, threshold voltage, parameter extraction.

1 INTRODUCTION

Threshold voltage and mobility are among the most important physical parameters in the analysis and modeling of metal-oxide-semiconductor field effect transistor (MOSFET) characteristics. The precise extraction of these parameters plays an important role in the accuracy of MOSFET models to predict the terminal currents in circuit simulations [1-3]. As a result, a number of methods have been developed in the attempt to provide a simple and accurate procedure to determine the values of these parameters [4-8]. However, simplicity and accuracy are difficult to achieve together, and most techniques are subjected to tradeoffs between the two extremes. Methods like linear extrapolation are simple, but do not provide enough accuracy. On the other hand, high precision methods like the regressive algorithm are very complex and time-consuming.

In this paper, a method to extract the threshold voltage and effective mobility of a MOSFET from the linear cofactor difference drain current of MOSFETs is proposed. By applying a newly developed linear cofactor difference operator (LCDO) on the transfer characteristics of a MOSFET measured in the linear region, the characteristic linear cofactor difference drain current extrema are obtained. Based on these extrema, the threshold voltage and the effective mobility can be directly determined. This

method has been tested on fabricated and simulated n-channel MOSFETs. Very good agreement from the extraction result using the LCDO method and the standard second-derivative method is observed. Compared with other reported methods, the LCDO approach is straightforward and can easily be extended to other complex MOSFET models.

2 FORMULATION OF LCDO METHOD

In this section, the application of the linear cofactor difference operator method to obtain the corresponding extrama of drain current linear cofactor difference is described. If a function $f(x)$ is strictly monotonic, non-linear, continuous over (x_0, x_1) and differentiable on $[x_0, x_1]$, then there exists a point x_p belongs to (x_0, x_1) such that

$$G'(x_p) = \frac{\partial G}{\partial x}\bigg|_{x=x_p} = 0 \tag{1}$$

where

$$G(x) = \Delta f(x) \equiv b + Kx - f(x) \tag{2}$$

is the linear cofactor difference of $f(x)$ and $\Delta f(x)$ is the linear cofactor difference operator (LCDO). b and K are the intercept and linear factor respectively.

The constants b and K can be determined by the end points of the interval (x_0, x_1) where the LCDO is applied, via the equations

$$G(x_1) = b + Kx_1 - f(x_1) = 0 \tag{3}$$

$$G(x_0) = b + Kx_0 - f(x_0) = 0 \tag{4}$$

The value of K are governed by $K > 0$ if $f(x)$ is a monotonically increasing function and $K \le 0$ otherwise. The above statement can be easily proven by using the Rolle's theorem [9]. We would like to emphasize that it is the choice of the LCDO region (x_0, x_1) that determines the value of K and b.

The drain current of an n-channel MOSFET in the linear region is usually modeled as [4]

$$\frac{V_{ds}}{I_{ds}} = R_s + R_d + \frac{1}{\beta \mu_{eff}(V_{gs} - V_{th})} \tag{5}$$

with $\beta = \dfrac{WC_{ox}\mu_0}{L}$, $\mu_{eff} = \dfrac{\mu_0}{1+\theta_0\left(V_{gs}-V_{th}\right)}$.

In equation (5), R_s and R_d are the source and drain parasitic resistances and θ_0 is the intrinsic mobility attenuation coefficient.

If we define $R_t = R_s + R_d$ and $\theta = \theta_0 + R_t\beta$, R_t carries the meaning of total parasitic source/drain series resistance, and θ represents the effective attenuation coefficient that includes the effects of the parasitic source/drain series resistance. Eq.(5) can then be simplified into a common expression

$$I_{ds} = \frac{\beta\left(V_{gs}-V_{th}\right)V_{ds}}{1+\theta\left(V_{gs}-V_{th}\right)} \qquad (6)$$

When we apply the LCDO to Eq. (6), the following equation is obtained

$$\Delta I_{ds}\left(V_{gs}\right) = \frac{\beta\left(V_{gs}-V_{th}\right)V_{ds}}{1+\theta\left(V_{gs}-V_{th}\right)} - \left(KV_{gs}+b\right) \qquad (7)$$

The region to apply the LCDO has to be carefully chosen to include the extrema of Eq (7). Afterward, K and b can be simultaneously determined from Eqs. (3) and (4).

To proceed, we defines

$$\left.\frac{\partial \Delta I_{ds}\left(V_{gs}\right)}{\partial V_{gs}}\right|_{V_{gs}=V_{GP}} = 0 \qquad (8)$$

where V_{GP} is the critical gate voltage that correspond to the extrema of the drain current linear cofactor difference $\Delta I_{ds}\left(v_{GP}\right)$ for a particular value of K. The extrema of the linear cofactor difference drain current versus the gate voltage can then be solved for different value of K and the result is shown in Fig.1. Substituting Eq.(8) into Eq.(7) at the critical gate voltage V_{GP}, we obtain

$$\frac{1}{1+\theta\left(V_{GP}-V_{th}\right)} = \sqrt{\frac{K}{\beta\,V_{ds}}} \qquad (9)$$

Substituting Eq.(9) into Eq.(6) gives

$$I_{ds}\left(V_{GP}\right) = \sqrt{Kv_{ds}\beta}\left(V_{GP}-V_{th}\right) \qquad (10)$$

By using two different K_1, K_2 corresponding to 2 different LCDO, Eq. (10) can be directly used to calculate the transistor gain β, and the result is a close form expression given by:

$$\beta = \frac{1}{V_{ds}}\left[\frac{\dfrac{I_{ds}\left(V_{GP\,1}\right)}{\sqrt{K_1}}-\dfrac{I_{ds}\left(V_{GP\,2}\right)}{\sqrt{K_2}}}{V_{GP\,1}-V_{GP\,2}}\right]^2 \qquad (11)$$

After evaluating β, the low field mobility μ_0 can be found with given gate oxide thickness, effective channel L and width W. V_{th} can then be determined using Eq. (10).

It is worth emphasizing that a significant feature of the method presented can be observed by studying Eqs. (11)

and (12). The derived V_{th} and β are independent of θ, which eliminated the effects of source and drain series resistance. It is in general not the case in other previously reported methods. The knowledge of V_{th} and μ_0 can in turn be used to calculate θ. Form Eq.(9), which we get

$$\theta = \frac{\sqrt{\beta V_{ds}}-\sqrt{K}}{\sqrt{K}\left[V_{GP}-V_{th}\right]} \qquad (12)$$

The intrinsic attenuation coefficient θ_0 can thus be deduced if the series resistance R_{sd} is known from a separate measurement, for instance, using the method described in [5].

3 RESULTS AND DISCUSSION

We first apply the LCDO method to output data from a fabricated n-channel MOSFET experimentally measured with the HP-4156B parameter analyzer. The experimental data are shown in the inset in Fig.1. The measured device was fabricated in the CMOS laboratory belongs to the Institute of Microelectronics, Peking University. The substrate doping concentration of the device is $1\times10^{16}\,\mathrm{cm^{-3}}$ and the gate oxide thickness is 16nm. The threshold voltage of the MOSFET was adjusted by ion-implant. The measured effective channel length and width are 5 μm and 4 μm respectively. Since we are interested in applying the LCDO to the linear region of the MOSFETs, the gate voltage has to be limited to $[x_0 \geq 1.2V, x_1 = 6V]$ as shown in Fig. 1. As a result, we get $K \geq 4\cdot10^{-6}$ and $b = 2\cdot10^{-6}$ for the I_{ds} versus V_{gs} curve with V_{ds}=0.1V.

Fig.1 presents a plot of the drain current linear cofactor difference versus V_{gs} with K assumed four different values of 4×10^{-6}, 4.2×10^{-6}, 4.4×10^{-6} and 4.6×10^{-6}. The same value of $b = 2\cdot10^{-6}$ is used in all 4 cases. As shown in the figure, the extrema of the drain current linear cofactor difference can be located for any of four the 4 different K values. The corresponding critical gate voltages are listed in Table I. From the result of Fig.1 and Table I, we obtain $\beta = 114.17\mu A/V^2$ and the threshold voltage V_{th} =1.5469V using Eqs. (11) and (10) with K_1 =4.0×10^{-6} and K_2=4.2×10^{-6}. From the value of β, we calculate the low field mobility and obtain μ_0=413.28 $cm^2/V\cdot s$. This result is consistent with previously measured inverse layer electron mobility [10]. The effective mobility attenuation coefficient is then give by $\theta = 0.2985V^{-1}$ after extraction using any one of the two different K, together with β. As the series resistance R_{sd} of our devices is measured to be about $R_s + R_d = 2\cdot810\Omega$, θ_0 is found to be around $0.1153V^{-1}$.

In order to investigate the sensitivity of the extracted parameters on the choice of the LCDO factor pairs of K_1 and K_2, six groups of calculated data using different combinations of K_1 and K_2 are used to examine the accuracy of the extracted values of V_{th}, β, μ_0 and θ. All the results are shown in Fig.2, which indicates the consistency of the parameter values despite the use of different combination of K_1 and K_2. Thus, the extracted parameters are relatively invariant on the choice of K_1 and K_2, indicating viability of the LCDO method. For instance, the extracted threshold voltage is 1.5469V by using the pair of $K_1 = 4.0 \times 10^{-6}$ and $K_2 = 4.2 \times 10^{-6}$, and V_{th}= 1.5479V for the combination of $K_1 = 4.0 \times 10^{-6}$ and $K_3 = 4.4 \times 10^{-6}$. In fact, decreasing the incremental step of gate voltage in the measurement can further reduce the difference in extracted threshold voltage between the choices of K pairs.

The LCDO method has been further verified MOSFETs characteristics data generated by the semiconductor device simulator DESSES-ISE [11]. The simulated device has a substrate doping concentration of 5×10^{16} cm^{-3}, a source/drain doping concentration of 10^{20} cm^{-3}, a junction depth 0.5 μm, and the gate oxide thickness is 20nm. In the simulations, the effective channel length is chosen to be 10 μm.

In order to obtain the extrema in both the subthreshold and the linear regions, the LCDO is applied to a range of operation regions covering the cut-off region, through the subthreshold region, and all the way to the linear region. The gate voltage is varied form zero to 6V to provide sufficient coverage in the simulation. As the absolute magnitude of the drain current at zero gate voltage is extremely small in this case, the value of the LCDO intercept, b, can be set to zero. The value of K is chosen to make the minimum and the maximum values of the gate voltage V_{gs} to approach the zero and 6V, respectively.

Fig.3 shows the plot the results, obtained by applying LCDO to the drain current, versus V_{gs} using three different values of K (1×10^{-6}, 1.05×10^{-6} and 1.1×10^{-6}). The original transfer characteristic of the MOSFET is also included in this figure. As shown in Fig.3, two extrema are observed, one locating in the subthreshold region and the other in the linear region, with the K values used. From the values of the extrema in the linear region, we obtain $\beta = 8.1908 \times 10^{-2} \mu A/V^2$ from Eq. (11). The V_{th} is then extracted from any one of the determined K together with the value of β, V_{GP} and $I_{ds}(V_{GP})$. We once again observe the consistency of the extracted threshold voltage using the LCDO method even with significantly different LCDO extrema values. The extracted threshold voltage of 0.8449V matches well with the value of V_{th} = 0.8459V extracted separately using a conventional second-derivative method. It should be noted that the extrema in the subthreshold region can also be used to study the subthreshold conduction of a MOSFET [12].

Using the same technique, linear cofactor difference operator with a different set of K values is also applied to the simulated n-channel MOSFETs with the same structure but different effective channel lengths. Fig.4 shows the threshold voltage obtained by the LCDO method for four devices with the effective channel length of 0.5, 1, 5 and 10 μm. As a control experiment, the second-derivative method has also been used to extract the threshold voltages of the devices, and the results are also included in the figure. Very good matching is observed between the results obtained using the 2 different methods, indicating the validity of the LCDO method to capture the threshold voltage roll-off with channel length scaling. Similarly, the room temperature low field electron mobility and its effective attenuation coefficient are also extracted from the transistor with different channel lengths. The results are shown in Fig.5. As observed from the figure, all mobility curves converges to a particular value of μ_0=478 $cm^2/V \cdot s$ at low V_{gs}. It is reasonable from the general understanding of the device physics. The data obtained from LCDO method correctly indicates the trend that the shorter the channel length of a MOSFET, the larger is the mobility degradation factor.

4 CONCLUSION

The use of the extreme in the drain current linear cofactor difference of a MOSFET to extract the threshold voltage and effective mobility of MOSFEFs has been presented in this paper. The principle of this extraction method is based on the application of linear cofactor difference operator to produce extrema in the current versus the gate voltage function after the transformation. Based on the extrema, different parameters from n-MOSFETs with the different channel length from simulation and experiment data have been extracted. The results are in good agreement with that obtained from the standard second-derivative method, showing the correctness of the method presented.

Acknowledgement:
This work was supported by SRC under the contract number 2002-NJ-1001 and partially supported by a grant from RGC of Hong Kong under the number HKUST6111/03E.

References
1. **Jin He**, Xing Zhang, Yangyuan Wang. IEEE Electron Device Letters, EDL, 2001, vol.18,Dec:597-9.
2. **Jin He**, Xuemei Xi, Mansun Chan, Chenming Hu, Zhang Xing, Wang Yangyuan. IEEE Electron Device Letters, EDL, 2002, vol.19, No.7.
3. Colin C. McAndrew and Paul A Layman. IEEE Trans. Electron Devices, 1992, ED-39: pp.2298-2311.

4. K.O.Jeppson. Microelectronic Engineering, 1998, vol.40: pp.181-186.

5. K.Terada, K.Nishiyama, K.I.Hatanaka. Solid State Electronics. 2001, vol.45: pp.35-40.

6. Z.X.Yan and M.J.Deen . in IEE Proc. Circuits, Devices and Systems, 1991, vol.138: pp.351-354.

7. A.Ortiz-Conde, E.D.Gouveia Fernandes. IEEE Trans. Electron Devices, 1997, ED-44: pp. 1523-1528.

8. F.J.Garcia Sanchez, A.Ortiz-Conde, G.D.Mercato, J.A.Salcedo, J.J.Liou, Y.Yue. Solid State Electronics, 2000, vol.44: pp. 673-675.

9. Nuogen Hua, Introduction to advanced mathematics, Science press. Beijing, 1979, pp. 165-166.

10. H. Shin, A. F.Tasch Jr., C.M.Maziar. IEEE Trans. Electron Devices, 1989, ED-36(6): pp. 1117-1124.

11. DESSIS-ISE.Ver.6.0, 2001, Integrated System Engineering Corporation, Switzerland.

12. **Jin He**, Xing Zhang, Ru Huang, Y.Y.Wang. IEEE Trans. Electron Devices, 2002, ED-49(1): pp.331-334.

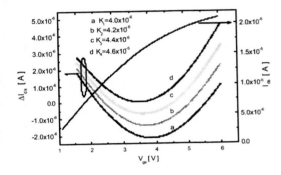

Fig.1 The drain current linear cofactor difference versus gate voltage for different value of K. The LCDO method is applied to the linear region I-V characteristics of a fabricated n-channel MOSFET.

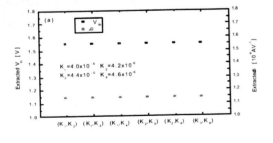

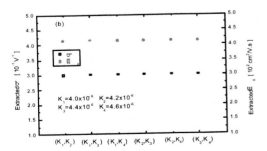

Fig.2 The dependence of the extracted parameters on the choice of the values of K_i and K_j: (a) Threshold voltage and transistor gain factor extracted for the different value of K and (b) Low field mobility and its degradation factor extracted for the different value of K.

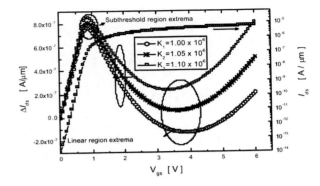

Fig.3 The drain current linear cofactor difference versus gate voltage for different values of K in the subthreshold and linear region of a simulated n-channel MOSFET.

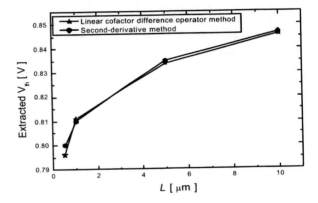

Fig.4 Comparison of threshold voltage variation obtained by LCDO the standard second-derivative methods.

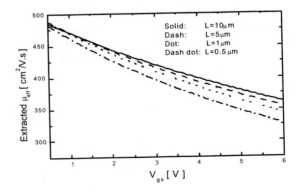

Fig.5 Mobility degradation characteristics of the simulated MOSFETs with the different channel length obtained by the linear cofactor difference operator method.

Extraction of Extrinsic Series Resistance in RF CMOS

M.S. Alam[*] and G.A. Armstrong

School of Electrical and Electronic Engineering
The Queen's University of Belfast, Ashby Building
Stranmillis Road, Belfast BT9 5AH N. Ireland (U.K.), g.armstrong@ee.qub.ac.uk
[*]Department of Electronics Engineering, Z. H. College of Engineering and Technology
A.M.U. Aligarh-202002 (U.P.) India, m_s_alam@lycos.com

ABSTRACT

An analytical approach for parameter extraction for CMOS incorporating substrate effect has been presented. The method is based on small-signal equivalent circuit valid in all region of operation, which uniquely extract extrinsic resistances used to extend the industry standard BSIM3v3 MOSFET model for RF applications. The verification of the model was carried out through the measurements of S-parameters and direct time-domain output voltage for a single device at 2.4 GHz in non-linear mode of operation. A circuit level evaluation of the model was carried out using 0.18μm CMOS amplifier and good results for power gain has been achieved.

Keywords: RF CMOS, parameter extraction, non-linear modeling

1 INTRODUCTION

Parasitic series resistance play an important role for RF circuit design and need to be accurately determined. There are many strategies for parameter extraction reported in the literature for CMOS technologies [1-10]. They utilize either full optimisation [1-3], partial parameter optimisation [4], or extraction assuming device symmetry [5]. Lee [4] provides the basic methodology for bulk CMOS, but ignores the effect of the substrate. Raskin [6] presents a comprehensive method for parameter extraction for SOI technology, whereas, Pengg [7] strategy is restricted to a particular region of transistor operation. The present work gives a full derivation of the relevant small signal parameters including the effect of the substrate. The analysis is valid in all regions of operation and does not pre-suppose any symmetry between source and drain [5].

2 EQUIVALENT CIRCUIT AND ANALYSIS

RF MOSFET equivalent circuit incorporating the extrinsic resistance, R_g, R_d, R_s, and R_{sub} is shown in Fig.1. R_g represents the effective distributed gate electrode resistance of multifingers devices connected in parallel and significantly affects the input admittance, whereas the effect of losses in the substrate is modeled by single resistance R_{sub} [8]. Extrinsic source and drain series resistances R_d and R_s included in BSIM3v3 [11] are treated only as virtual components in the current-voltage expression to account for the DC voltage drop across these resistances and therefore, invisible by the signal in the ac simulation. Therefore, external components for these series resistance and need to be included outside the intrinsic model to accurately describe the device AC impedance [2].

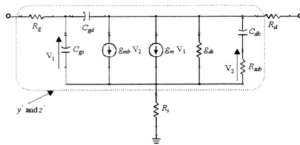

Figure 1: Small-signal equivalent circuit of a MOS transistor valid in all region of operation.

The small-signal circuit shown by dashed part in Fig. 1 can be analysed in terms of y-parameters [12] and the admittance matrix y' is given as

$$y' = \begin{bmatrix} y'_{11} & y'_{12} \\ y'_{21} & y'_{22} \end{bmatrix} \tag{1}$$

$$y'_{11} = \omega^2 R_g (C_{gs} + C_{gd})^2 + j\omega(C_{gs} + C_{gd}),$$
$$y'_{12} = -j\omega C_{gd}, \quad y'_{21} = g_m - j\omega C_{gd}$$
$$y'_{22} = g_{ds} + \omega^2 C_{db}^2 R_{sub}(1 + g_{mb} R_{sub}) + j\omega(C_{gd} + C_{db} + C_{db} g_{mb} R_{sub}).$$

Equation (1) has been derived subject to the assumptions: a) $\omega R_g (C_{gs} + C_{gd}) \ll 1$ for admittance parameters y'_{11}, y'_{12} and y'_{21}, and b) $\omega R_{sub} C_{db} \ll 1$ for y'_{22}.

By inverting the y'-matrix in (1), expanding terms and making appropriate simplifications, we obtain the z'-matrix

$$z' = \begin{bmatrix} z'_{11} & z'_{12} \\ z'_{21} & z'_{22} \end{bmatrix} \qquad (2)$$

$$z'_{11} = \frac{(g_{ds} + \omega^2 C_{db}^2 R_{sub} + j\omega(C_{gd} + C_{db})}{(-\omega^2 A + j\omega B)}$$

$$z'_{12} = \frac{j\omega C_{gd}}{(-\omega^2 A + j\omega B)}, \quad z'_{21} = \frac{-(g_m - j\omega C_{gd})}{(-\omega^2 A + j\omega B)}$$

$$z'_{22} = \frac{\omega^2 R_g (C_{gs} + C_{gd})^2 + j\omega(C_{gs} + C_{gd})}{(-\omega^2 A + j\omega B)}.$$

$A = C_{gd} C_{gs} + C_{db}(C_{gs} + C_{gd})$; $B = g_{ds}(C_{gd} + C_{gs}) + g_m C_{gd}$

Equation (2) has been simplified by applying the following conditions:

c) $\quad \omega << \dfrac{1}{\sqrt{R_g(C_{gs} + C_{gd})C_{db}R_{sub}}}$; d) $\quad \omega << \sqrt{\dfrac{g_{ds}}{C_{db}^2 R_{sub}}}$; and

e) $g_{mb}R_{sub} << 1$.

Assumptions (a) to (d) above are valid up to 10 GHz if the device has small number of fingers and condition (e) is true when the transistor is biased closed to threshold region.

Now, using (2) and incorporating the effect of R_s and R_d, the overall z-parameter of the equivalent circuit shown in Fig. 1 can be expressed as

$$z_{12} = \underbrace{R_s + \frac{BC_{gd}}{\omega^2 A^2 + B^2}}_{\mathrm{Re}(z_{12})} + \underbrace{\frac{-j\omega A C_{gd}}{\omega^2 A^2 + B^2}}_{\mathrm{Im}(z_{12})} \qquad (3)$$

and $\quad \mathrm{Re}(z_{22}) \cong R_s + R_d + \dfrac{(C_{gs} + C_{gd})B}{\omega^2 A^2 + B^2} \qquad (4)$

3 PARAMETER EXTRACTION

To test the validity of the model, S-parameter measurements were performed on a 0.18μm nMOS transistor, comprising 16 parallel gate fingers of 5μm width. After successful de-embedding [13] capacitances of input/output pads to the ground and associated resistors of the lossy substrate, intrinsic z and y-parameters were obtained from the measurements.

3.1 R_s and R_d

By plotting $\mathrm{Re}(z_{12})$ versus $-\mathrm{Im}(z_{12})/\omega$ and taking y-axis intercept of the best fit line gives R_s because at high frequency $-\mathrm{Im}(z_{12})/\omega$ tends to zero, as shown in Fig. 2(a). In (3) and (4) it is evident that $\mathrm{Re}(z_{12})$ and $\mathrm{Re}(z_{22})$ exhibit the same dependence on frequency [6], hence the slope m of the regression line can be derived as

$$m = \frac{d\,\mathrm{Re}(z_{22})}{d\,\mathrm{Re}(z_{12})} = \frac{(C_{gs} + C_{gd})}{C_{gd}} \qquad (5)$$

Similarly, by using (3)-(5) an expression for R_d can be derived from the regression line y-axis intercept c, slope m as shown in Fig. 2(b) and extracted value of R_s as

$$R_d = c + (m-1)R_s \qquad (6)$$

It has been suggested that the source/drain resistances become less bias dependent as gate length is reduced to deep sub-micron region [14]. This has been verified through extraction at a number of bias points as shown in Fig. 3, where at low-drain bias, the dependence of all four extracted resistance parameters on V_{gs} is minimal. Therefore, extracted resistance parameters were assumed bias independent and used to extend BSIM3 model [11] for RF applications. Values of $R_s = 2.9\,\Omega$ and $R_d = 6.4\,\Omega$ as obtained from Fig. 2(a) and Fig. 2(b) respectively, have been used for the extraction of rest of the small-signal equivalent circuit parameters.

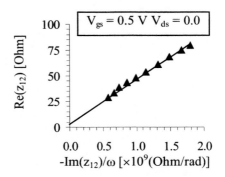

Fig. 2(a)

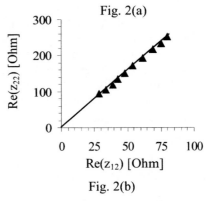

Fig. 2(b)

Figure 2: Extraction of (a) R_s (b) R_d. Measurement frequency 1-10 GHz

3.2 R_g and R_{sub}

Now de-embedding the effect of R_d and R_s in z-parameters domain, each element of the equivalent circuit shown by dashed part in Fig. 1 can be determined in

y-parameters domain [14]. After calculating g_{ds} from $\text{Re}(y'_{22})$ as $\omega \rightarrow 0$ and g_{mb} from the non-ideality factor of sub-thresold slope, R_g and R_{sub} can be extracted as

$$R_g = \frac{\text{Re}(y'_{11})}{(\text{Im}(y'_{11}))^2}$$

$$R_{sub} = \frac{\text{Re}(y'_{22}) - g_{ds}}{[\text{Im}(y'_{22}) + \text{Im}(y'_{12})]^2 - g_{mb}[\text{Re}(y'_{22}) - g_{ds}]} \quad (7)$$

Once again, the experimental results in Fig. 3 confirm that R_g and R_{sub} are not strongly bias dependent.

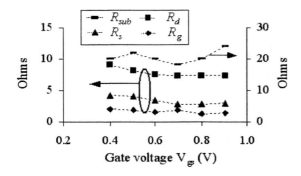

Figure 3: R_g, R_d, R_s and R_{sub} extracted data for $V_d = 0.2V$

4 SIMULATION RESULT

4.1 S-parameters

A comparison of simulation with measurement in Fig. 4 up to a maximum frequency of 10 GHz, shows that S-parameters based on the equivalent circuit fit the measurement even better than the extended BSIM3 model, indicating that all equivalent circuit model parameters are accurately extracted. The addition of R_g R_d improve the prediction S_{11} and S_{21} whereas R_{sub} significantly affect S_{22} at high frequency.

4.2 Time-domain waveform

In Fig. 5 the drain voltage at a frequency of 2.4 GHz is shown at 3 different power levels. It is evident that a better fit in the time domain is obtained at high power when the BSIM3 model is augmented by the four addition series resistences

4.3 Amplifier design

A CMOS amplifier designed using the device sizes shown in Fig. 6 was measured and the BSIM3 based RF model was shown to give good prediction of output power

and gain as shown in Fig. 7. Experimental results confirm that the extrinsic resistance parasitics are vital to predict power gain when applied to the CMOS power amplifier.

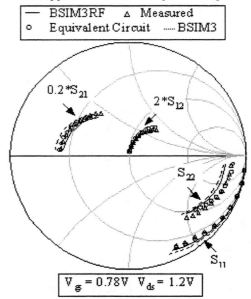

Figure 4: Comparison between the measured, BSIM3RF and equivalent circuit S-parameters

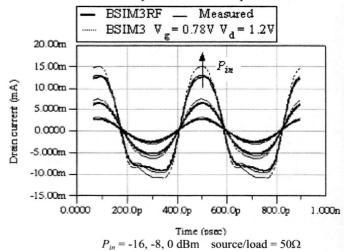

$P_{in} = -16, -8, 0$ dBm   source/load = 50Ω

Figure 5: Time-domain waveform for output drain voltage

5 CONCLUSION

A systematic approach for parameter extraction of extrinsic series resistance for RF CMOS transistor based on a small signal equivalent circuit has been presented. It was found that extraction works well over a wide range of bias conditions and excellent agreement has been achieved between all four simulated and measured S-parameters, indicating that all parameters have been accurately extracted. Adding the extracted resistance to the BSIM3 model gives the opportunity for more accurate non-linear time domain simulation at the device level and harmonic balance (HB) simulation of a CMOS power amplifier at 2.4 GHz

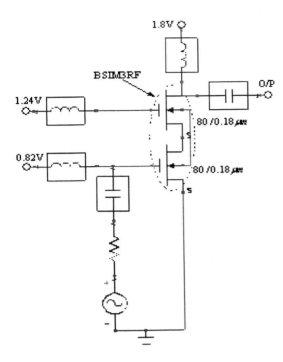

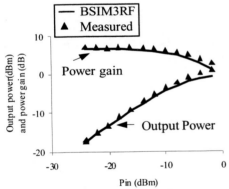

Figure 6: Schematic of CMOS power amplifier

Figure 7: Fundamental power sweep source/load = 50Ω

ACKNOWLEDGEMENTS

The authors would like to thank Mark Norton and Rawinder Dharmalinggam of Parthus Technologies (U.K.) for their assistance in the design of the cascode CMOS test MMIC.

REFERENCES

[1] R. Sung, R. P. Bendix, and M.B Das, "Extraction of High-Frequency Equivalent Circuit Parameters of Sub-micron Gate Length MOSFETs", IEEE Trans. on Electron Devices, vol. 45, no.8, p. 1769-1775, 1998

[2] C. C. Enz , and Yuhua Cheng "MOS Transistor Modeling for RF IC Design", IEEE Trans. on Solid-State Circuit, vol. 35, no. 2, p. 186-201, 2000.

[3] S.F. Tin, A. Asharaf, K. Mayaram, and Chenming Hu, "A Simple Sub-circuit Extension of the BSIM3v3 Model for CMOS RF Design", IEEE on Solid-State Circuits, vol. 35, no.4, p. 612-624, 2000.

[4] S. Lee, H.K. Yu, C.S. Kim,. J.G Koo, and K.S. Nam, "A Novel Approach to Extracting Small-Signal Model Parameters of Silicon MOSFETs", IEEE Microwave and Guided Wave Letters, vol. 7, no.3, p. 75-77, 1997.

[5] S.H.M. Jen, C.C. Enz, D.R. Pehlke, M. Schroter, and J. Sheu, "Accurate Modeling and Parameter Extraction for MOS Transistors Valid up to 10 GHz", IEEE Trans. on Electron Devices, vol. 46, no. 11, p. 2217-2226, 1999.

[6] J.P. Raskin, R. Gillon, J. Chen, D.V. Janvier, and J.P. Colinge, "Accurate SOI MOSFET Characterization at Microwave Frequencies for Device Performance Optimization and Analog Modeling", IEEE Trans. on Electron Devices, vol. 45, no.5, p. 1017-1025, 1998.

[7] F.X. Pengg, "Direct Parameter Extraction on RF CMOS", IEEE MTT-S Digest, p. 271-274, 2002

[8] Y. Cheng and M. Matloubin, "On the High Frequency Characteristics of Substrate Resistance in RF MOSFETs", IEEE Electron Device Letters vol. 21, no. 12, p.604-606, 2000

[9] S. Lee, C.S. Kim, and H.Y. Yu, "A Small-signal RF Model and Its Parameter Extraction for Substrate Effects in RF MOSFETs", IEEE Electron Devices Letters, vol. 48, no. 7, p. 1374-1379, 2001.

[10] C. Enz, "An MOS Transistor Model for RF IC Design Valid in All Regions of Operations", IEEE Trans. on MTT, vol. 50, no. 1, p. 342-359, 2002.

[11] Y. Cheng, M. Chan, K. Hui, M. Jeng, Z. Liu, J. Huang, K. Chen, P. Ko, and C. Hu, BSIM3v3 User's Manual, University of California, Berkeley 1997.

[12] M.S. Alam, B. Toner, G.A. Armstrong, and V.F. Fusco, "A Simple Small-signal Circuit Model for RF CMOS IC Design with Full Parameter Extraction", 9th International Symposium on Integrated Circuits, Devices & System (ISIC), Singapore, p. 149-152, 2001.

[13] T.E. Kolding, "On Wafer Caliberation Techniques for Giga-Hertz CMOS Measurements", Proc. IEEE Int. Conf. on Microelectronic Test Structures, vol. 12, p. 105-110, 1999.

[14] Y. Cheng, C.H Chen, M. Matloubian, M. J. Deen, "High-Frequency Small Signal AC and Noise Modelling of MOSFETs for RF IC Design", IEEE Electron Devices Letters, vol. 49, no.3, p. 400-408, 2002[1]

*Department of Electronics Engineering , Z. H. College of Engineering and Technology, A.M.U. Aligarh-202002 (U.P.) INDIA, Ph. +91-571-2721148, Fax. +91-571-2721148

Analytic formulae for the impact ionization rate for use in compact models of ultra-short semiconductor devices

H. C. Morris[a], M. M De Pass[b] and Henok Abebe[c]

a. Department of Mathematics, San Jose State University, San Jose, California CA 95192.
b. Department of Mathematics, Claremont Graduate University, Claremont, CA 91711
c. MOSIS Service, USC Information Sciences Institute, Marina del Rey, CA 90292

ABSTRACT

This paper presents a new approximate compact formulae for the impact ionization (I.I.) generation rate (G_{II}) in ultra-short-channel devices.

1 INTRODUCTION

For ultra short length devices it is vital to accurately account for hot electron effects such as impact ionization. By means of a Fermi Golden Rule calculation with a screened Coulomb interaction, Quade, Schöll and Rudan [1] have shown that, in the limit of large screening length, the impact ionization rate per unit time can be expressed in the form

$$G_{II} = \frac{n}{(2\pi)^3} \int_{k_{th}}^{\infty} \frac{f(k)}{\tau_{II}(k)} d^3k \qquad (1)$$

where n is the electron density

$$n = \frac{2}{(2\pi)^3} \int f(k) d^3k \qquad (2)$$

and $\tau_{II}(k)$ is the isotropic ionization scattering time of the kth moment given by

$$\frac{1}{\tau_{II}(k)} = \frac{1}{2\tau_0}(\frac{k}{k_{th}} + \frac{k_{th}}{k} - 2) \qquad (3)$$

where the parameter τ_0 is called the characteristic inverse time constant [1].

If the density function $f(k)$ is modelled by a heated Maxwellian of the form

$$f(k) = \frac{h^3}{2(2\pi m^* k_B T_e)^{3/2}} \exp(-\frac{\hbar^2 k^2}{2m^* k_B T_e}) \qquad (4)$$

in which T_e is the electron temperature, determined by the electric field within the device [2]. For a direct semiconductor, (1) yields the Schöll-Quade impact ionization rate formula

$$G_{II}(n,T_e) = \frac{n}{\tau_0}[\sqrt{\frac{u}{\pi}} \exp(-\frac{1}{u}) - \mathrm{erfc}(\frac{1}{\sqrt{u}})] = \frac{n}{\tau_0} G_{SQ}(u) \qquad (5)$$

involving the variable $u = {k_B T_e}/{E_{th}}$ where E_{th} is the threshold energy for impact ionization. This parameter is frequently approximated by the band gap energy $E_g \approx 1.12$ ev. The parameter $\tau_0 \approx 1.26 \times 10^{-14}$ sec. In general we expect τ_0 to be a function of the lattice temperature T_L. Selberherr et. al. [3], [4] have proposed new models for $f(k)$ that improve upon the heated Maxwellian (4) and provide a more accurate parameterization of the hot electron population. The analog of the integral (1) can be evaluated in terms of special functions and but these can be difficult to evaluate. We propose the use of the Gauss-Laguerre integration technique introduced in [5] as an effective way to generate simple analytic formulae that involve only elementary functions and provide a practical accurate formula for the impact ionization rate.

2 THE HEATED MAXWELLIAN

For the heated Maxwell case the impact ionization rate (1) is given by

$$G_{II}(n,T_e) = \frac{n}{\tau_0} G_{SQ}(u) \qquad (6)$$

where $G_{SQ}(u)$ is defined by the integral

$$\frac{2}{\sqrt{\pi}} \frac{\hbar^3}{(2m^* k_B T_e)^{\frac{3}{2}}} \int_{k_{th}}^{\infty} k^2 (\frac{k}{k_{th}} + \frac{k_{th}}{k} - 2) \exp(-\frac{\hbar^2 k^2}{2m^* k_B T_e}) dk \qquad (7)$$

By introducing the change of variable $z = \sqrt{\frac{\hbar^2}{2m^* k_B T_e}} k$ this integral can be expressed as the sum

$$\sqrt{u} I_3(u) + \frac{1}{\sqrt{u}} I_1(u) - 2 I_2(u) \qquad (8)$$

with the moment functions $I_n(u)$ defined by

$$I_n(u) = \frac{2}{\sqrt{\pi}} \int_{\frac{1}{\sqrt{u}}}^{\infty} z^n \exp(-z^2) dz \qquad (9)$$

The three moments required can be exactly evaluated

$$I_1(u) = \frac{2}{\sqrt{\pi}} \int_{\frac{1}{\sqrt{u}}}^{\infty} z \exp(-z^2) dz = \frac{1}{\sqrt{\pi}} e^{-\frac{1}{u}} \qquad (10)$$

$$I_2(u) = \frac{1}{2}(\mathrm{erfc}\left(\frac{1}{\sqrt{u}}\right) + \frac{2}{\sqrt{\pi}} \frac{1}{\sqrt{u}} e^{-\frac{1}{u}}) \qquad (11)$$

$$I_3(u) = \frac{1}{\sqrt{\pi}} e^{-\frac{1}{u}} \left(\frac{1}{u} + 1 \right) \qquad (12)$$

and substitution of these into (8) yields the Schöll-Quade function G_{SQ}.

3 AN INTEGRATION TECHNIQUE

In order to capture the parameter dependence of the integral (1) we cannot use regular quadrature technique such as an adaptive Simpson rule. Instead we use the Gauss-Laguerre formulae [6]. The general form of this integration technique is

$$\int_0^\infty e^{-x} f(x)dx = \sum_{k=1}^{n} \omega_k \cdot f(x_k) + \frac{(n!)^2}{(2n)!} f^{(2n)}(\zeta) \quad (13)$$

where $0 < \zeta < \infty$ and x_k's are the zeros (the nodes) of the Laguerre polynomials.

Using this integral approximation technique we can construct practical, closed-form, solutions that preserve the analytical dependence on parameters in the integral.

For the Schöll-Quade formula we need to evaluate three members of the family of integrals (9). Introducing the new variable $\zeta + \frac{1}{u} = z^2$ the integral $I_n(u)$ can be written in the form

$$I_n(u) = \frac{1}{\sqrt{\pi}} e^{-\frac{1}{u}} \int_0^\infty \left(\zeta + \frac{1}{u} \right)^{\frac{n-1}{2}} e^{-\zeta} d\zeta \qquad (14)$$

A two node approximation is given by

$$\widehat{I}_n^{(2)} = \frac{1}{\sqrt{\pi}} e^{-\frac{1}{u}} \left(\omega_1 \left(x_1 + \frac{1}{u} \right)^{\frac{n-1}{2}} + \omega_2 \left(x_2 + \frac{1}{u} \right)^{\frac{n-1}{2}} \right) \quad (15)$$

The parameters ω_1, ω_2, x_1 and x_2 are fixed numerical constants given by

$$\begin{aligned}
x_1 &= 0.585786437626905 \\
x_2 &= 3.414213562373095 \\
\omega_1 &= 8.535533905932735e-01 \\
\omega_2 &= 1.464466094067262e-01
\end{aligned}$$

The factor $e^{-\frac{1}{u}}$ enhances the convergence greatly so even the two node approximation (15) produces the fit shown in figure 1.

4 THE TAIL DISTRIBUTION

For short channel devices, the Maxwellian form (4) has to be modified. A possible generalization is given by

$$f = \frac{b\sqrt{\pi}}{4\Gamma(\frac{3}{2b})} \frac{h^3}{(2\pi m^* k_B T)^{\frac{3}{2}}} \exp\left(-\left(\frac{\hbar^2 k^2}{2m^* k_B T}\right)^b\right) \quad (16)$$

This form can accurately reproduce the Monte Carlo simulations of turned-on MOSFET's [3]. When the parameter $b = 1$ the expression (16) reduces to the heated Maxwellian form. However, when $b \neq 1$, T can no longer

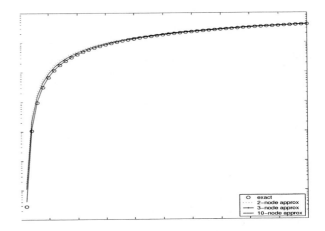

Figure 1: A fit of the exact Schöll-Quade function G_{SQ} with $u \in [0,5]$ by Gauss-Laguerre approximation with node numbers $N = 2, 3$ and 10.

be interpreted as the electron temperature. In this paper we will regard b and T as fitting parameters. The generalization G_{GSQ} of the Schöll-Quade function G_{SQ} to arbitrary b is given by

$$\begin{aligned}
G_{GSQ}(u,b) &= \frac{1}{2\Gamma(\frac{3}{2b})} [u^{\frac{1}{2}} \Gamma(\frac{2}{b})(1 - \Gamma_{\text{inc}}(u^{-b}, \frac{2}{b})) \\
&\quad + u^{-\frac{1}{2}} \Gamma(\frac{1}{b})(1 - \Gamma_{\text{inc}}(u^{-b}, \frac{1}{b})) \quad (17) \\
&\quad - 2\Gamma(\frac{3}{2b})(1 - \Gamma_{\text{inc}}(u^{-b}, \frac{3}{2b}))]
\end{aligned}$$

involving the incomplete gamma function $\Gamma_{\text{inc}}(x,a)$ defined by

$$\Gamma_{\text{inc}}(x,a) = \frac{1}{\Gamma(a)} \int_0^x \exp(-z) z^{a-1} dz \qquad (18)$$

The parameter $u = \frac{k_B T}{E_{th}}$ is determined by the tail distribution parameter T. The impact ionization rate is now given by

$$G_{II}^{\text{gen}}(n, T_e, b) = \frac{n}{\tau_0} G_{GSQ}(u,b) \qquad (19)$$

Figure 2. shows the a semilogarithmic plot of the function G_{II}^{gen}/n with $u \in [0,5]$ for three values of b close to the regular value $b = 1$. The plots are produced using the exact form of the incomplete gamma function. The incomplete gamma function is not trivial to evaluate and so we utilize the integration technique introduced in section three to obtain a simple analytic approximation to G_{GSQ}. Using our integration method we can approximate the incomplete gamma function (18) by the formula

$$1 - \exp(-\xi) - \frac{\exp(-\xi)}{\Gamma(a)} \sum_{k=1}^{N} w_k [(\xi + x_k)^{a-1} - x_k^{a-1}] \quad (20)$$

Formula (20) provides a source of simple formulae $G_{GSQ}^N(u,b)$ that approximate $G_{GSQ}(u,b)$. Figure 3 shows the the

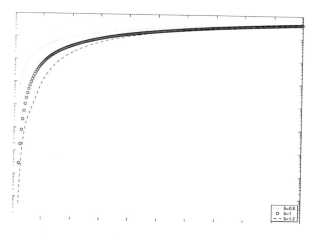

Figure 2: A semilogarithmic plot of the function G_{II}^{gen}/n with $u \in [0,5]$ for $b = 0.8, 1$ and 1.2.

Gauss-Laguerre approximations to $G_{GSQ}^{N}(u,b)$ with $u \in [0,5]$ for $N = 2, 3$ and 10 nodes when $b = 1.2$. From this

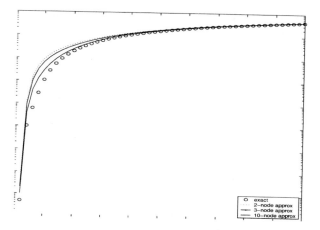

Figure 3: A fit of the function G_{GSQ}^{N} with $u \in [0,5]$ with $b = 1.2$ for Gauss-Laguerre approximation with node numbers $N = 2, 3$ and 10.

figure it is clear that we require a large numberof nodes if we require accuracy for small u but the qualitative shape is well captured by even the 2-node approximation.

5   A GENERALIZED MODEL

In [4] the following analytical expression for the distribution function

$$
\begin{aligned}
f(k) \;=\; & \frac{h^3}{2(2\pi m^* k_B T_e)^{\frac{3}{2}}} \Big[(1-c)\exp\Big(-\frac{\hbar^2 k^2}{2m^* k_B T_e}\Big) \\
& + c\frac{b\sqrt{\pi}}{2\Gamma(\frac{3}{2b})}\Big(\frac{T_e}{T}\Big)^{\frac{3}{2}}\exp\Big(-\Big(\frac{\hbar^2 k^2}{2m^* k_B T}\Big)^b\Big) \quad (21)
\end{aligned}
$$

has been proposed. When $c = 0$, $f(k)$ reduces to the heated Maxwellian (4) and when $c = 1$ it becomes the

tail distribution function (16). For $0 \le c \le 1$ it allows an interpolation between the two forms. The impact ionization rate is a linear function of f and so we obtain the new combined formula G_{II}^{new} given by

$$
G_{II}^{\text{new}}(u_1, u_2) = \frac{n}{\tau_0} G_{GSQ}^{\text{new}}(u_1, u_2) \quad (22)
$$

where

$$
G_{GSQ}^{\text{new}}(u_1, u_2) = (1-c)G_{SQ}(u_1) + cG_{GSQ}(u_2) \quad (23)
$$

and the variables u_1 and u_2 are defined by

$$
u_1 = \frac{k_B}{E_{th}}T_e, \; u_2 = \frac{k_B}{E_{th}}T \quad (24)
$$

This new model has four parameters c, b, T_e and T. In the next section we propose a special case of this general model that has fewer parameters and is easier to implement.

6   AN IMPROVED MODEL

To reduce the number of parameters we will make some reasonable assumptions about the values of the model parameters c, b, T_e and T. From the reasoning behind the ansatz (21), it is reasonable to suppose that T_e is the lattice temperature T_L and that the 'tail temperature' $T = \alpha T_L$ with $1 \le \alpha \le 2$. We set $c = \frac{1}{2}$ so that the distributions are combined in a balanced way. This means that we only have a two parameters b and α in our new model. The resulting expression for the impact ionization rate is given by

$$
G_{II}^{N}(u, \alpha) = \frac{n}{\tau_0} G_{GSQ}^{N}(u, \alpha) \quad (25)
$$

with

$$
G_{GSQ}^{N}(u, \alpha) = \frac{1}{2}(G_{GSQ}^{N}(u, b) + G_{GSQ}^{N}(\alpha u, b)) \quad (26)
$$

The variable u is the standard Schöll-Quade variable

$$
u = \frac{k_B}{E_{th}}T_L \quad (27)
$$

The approximation (26) with $b = 1$ was first introduced by Pop in [7]. Pop implemented the model in the context of the FIELDAY device simulation program and determined $\alpha = 1.8$ by data fitting. Figure 4 shows the improved model curves with $u \in [0,5]$ for the regular case $b = 1$ for $\alpha = 1$, the standard Quade formula, and $\alpha = 1.5$ and $\alpha = 2$.

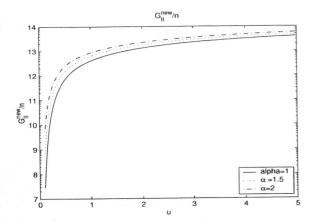

Figure 4: The generalized Schöll-Quade function $G_{II}^{new}(u, \alpha)$ with $u \in [0, 5]$ for $\alpha = 1$ (Quade) and $\alpha = 1.5$ and $\alpha = 2$.

7 CONCLUSION

We have presented a method for obtaining new accurate formulae for the impact ionization rate for use in compact modelling. We have combined an approach based on fundamental physics with the classical technique of Gauss-Laguerre integration. We have proposed both a general four parameter model and a simpler version we called the improved model. In our improved model, there are only two parameters. This latter model is a generalization of the model introduced by Pop [7]. In each case we have provided a family simple approximate formule for the impact ionization rate obtained by retaining a reduced number of Gauss-Laguerre nodes. This yields a series of models ranging from the lowest level 2-node approximation to the exact answer. We conjecture that the 2-node approximation may be adequate for most purposea as it captures the qualitative shape of the exact solution.

ACKNOWLEDGEMENTS

We wish to acknowledge Cesar Piña and Vance Tyree of the University of Southern California Information Sciences Institute for technical discussions.

REFERENCES

[1] W. Quade, E. Schöll, and M. Rudan. Impact Ionization within the Hydrodynamic Approach to Semiconductor Tranport. *Solid-State Electronics*, 36(10):1493–1505, 1993.

[2] M. Lundstrom, *"Fundamentals of Carier Transport"*, Addison-Wesley, 1990.

[3] T. Grasser, H. Kosina and S. Selberherr, Influence of the distribution function shape and the band structure on impact ionization modeling, *Journal of Applied Physics*, Vol. 90, Number 12, (2001) 6165-6171.

[4] T. Grasser, H. Kosina, Heitzinger and S. Selberherr, Characterization of the hot electron using six moments, *Journal of Applied Physics*, Vol. 91, Number 6, (2002) 3869-3879.

[5] H. C. Morris and M. M. de Pass, "Semi-analytical Boltzmann equation model for the substrate current of short channel MOSFET's with lightly doped drains", *Technical Proceedings of the Nanotech 2003 conference*, Vol. 2, 24-27. (2003).

[6] Philip J. Davis and Philip Rabinowitz, *"Numerical Integration"*, Blaisdell Publishing Company, Waltham, Massachusetts, Toronto, 1967.

[7] E. Pop, "CMOS Inverse Doping Profile Extraction and Substrate Current Modeling", download: http://www.stanford.edu/~epop/msthesis.pdf, June 1999.

On the correlations between model process parameters in statistical modeling

Jirí Slezák, Aleš Litschmann, Stanislav Banáš, Radim Mlcoušek, Martin Kejhar

SCG Czech Design Center, ON Semiconductor Czech Republic, B. Nemcové 1720, 756 61 Rožnov pod Radhoštem, Czech Republic, jiri.slezak@onsemi.com

1 ABSTRACT

Statistical modeling in the design of today's high performance integrated circuits (IC's) is a necessity to produce competitive products with short development time. The use of backward propagation of variance (BPV) [1,2] has proven its worth among other approaches proposed for the statistical modeling. This methodology introduces physically based process and geometry-dependent parameters (PGPs) for each device that is available to the design community in a model library. The goal of this brief contribution is to propose a simple method to establish mathematical relationships among correlated PGPs. The method conforms to syntax-based restrictions that have to be fulfilled in complex model libraries.

Keywords: statistical correlations, statistical modeling, process variations, Monte Carlo simulation

2 INTRODUCTION

Statistical modeling usually denotes a set of tools that enable designers to carry out sensitivity analysis in a particular design, generate case libraries and run Monte Carlo simulations. In the past two decades several methods have been developed to meet these demands. For instance, Principal Component Analysis [3] or Response Surface Modeling [4] can be mentioned. However, large number of measured lots and devices is required. This may be unacceptably time consuming. The methodology exploiting backward propagation of variance [1,2], which is sometimes called Linear Variance Model [5], profits from a sound mathematical background. It takes into account existing correlations between different measured device electrical characteristics *in a single device* (for instance beta, collector and base currents in a bipolar transistor). It guarantees accurate modeling of the measured device characteristics with respect to their mean values and standard deviations. The extra cost for the production of corner lots is avoided. Results can be obtained within one hour on workstations.

The modeling flow is as follows: Let's assume that the following parameters of a particular device have normal distributions described by mean values μ and standard deviations s. The measured $\mu(e_i)$ and $s(e_i)$ of the electrical parameters e_i (often correlated) form an input e-space for *each device* (MOSFETs, BJTs …). They are evaluated from process control (PC) tests. The e-space has a unique relationship to the space of uncorrelated model process parameters (p-space). This relationship between the independent p- and e-spaces is defined by mapping equations [6]. Therefore, it is possible to evaluate means $\mu(p_j)$ and standard deviations $s(p_j)$ of the PGPs. First, note that the resulting sets of $\mu(p_j)$ and $s(p_j)$ are obtained for *each device separately*. Second, for a single device the obtained PGPs are uncorrelated.

3 RESULTS AND DISCUSSION

It is evident that one cannot treat all of the PGPs of all devices as independent. Some of them are related to the same layer, mask or process option. For instance, the epitaxial layer forms the base of substrate PNPs, lateral PNPs and the collector of vertical NPNs as well. Variations in gate oxide thickness affect the n- and p-type MOSFETs and also some capacitors. A mask defining a BJT emitter has an impact on several different types of BJTs. Obviously, it's necessary to introduce a correlation among corresponding process parameters, i.e. among devices that have something in common. A simple transform is proposed. The transform makes sure that correlated PGPs "move in the same direction" and simultaneously it preserves their calculated distributions.

If random variables x_i have normal distributions $x_i = N(\mu_i, s_i^2)$ then random variable y

$$y = \left(\frac{x_i - \mu_i}{s_i} \right) \tag{1}$$

has a normalized Gaussian distribution $y = N(0,1)$. This statement is also true visa versa: When a "master" random variable y having $y = N(0,1)$ is introduced, y-dependent variables x_i

$$x_i = y * x_i + \mu_i \tag{2}$$

with *specific* distributions $x_i = N(\mu_i, s_i^2)$ can be generated. The effect of the transform is graphically demonstrated in Fig1. Note that non-Gaussian distributions can be also generated applying a suitable transform.

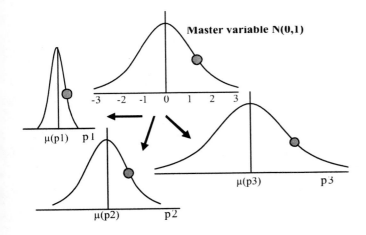

Fig1: Graphical demonstration of the master variable driving a set of correlated p_i (PGPs). Density distributions of p_i are preserved – see the transformed positions of the red point.

As an example we selected a substrate PNPS and a lateral PNPL. First, their base is the epitaxial layer. Therefore, a variation in its concentration has to be reflected in both the SPICE model of PNPS and PNPL. Next, the emitters of both devices and the collector of PNPL are formed by the same mask and implant. It is reasonable to assume that the geometry-dependent PGPs of PNPL and PNPS are also correlated. Again, the correlation is introduced by means of the proposed transform.

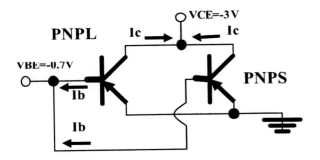

Fig2: Circuit schematics used for Monte Carlo simulations.

An operating point was defined by Vbe and Vce. Monte Carlo simulations were performed (see Fig2 for the circuit schematics). Fig3 shows the results with all the PGPs randomly generated, i.e. uncorrelated. Fig4 presents the outcome with the PGPs appropriately correlated via master variables and the transform described above.

Correlated parameters	Correlation coefficient
Ic_PNPS & Ic_PNPL	0.861
Ib_PNPS & Ib_PNPL	0.940

Table1: Correlation coefficients evaluated from PC data.

The latter graphs in Fig4 are in good agreement with the correlations evaluated from measured PC data in Table1. The message is clear: Fig3 presents combinations that are unlikely to appear. Simulation results with correlated PGPs are much closer to reality.

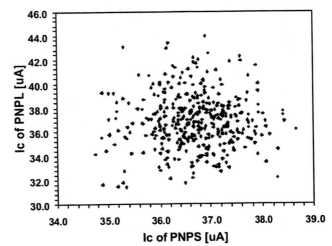

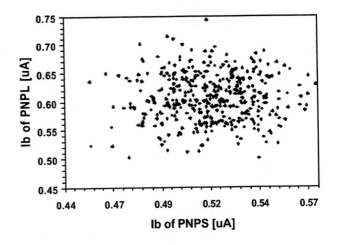

Fig3: Simulated collector and base currents of PNPS and PNPL with **uncorrelated PGPs**, i.e., all PGPs randomly generated according to their calculated distributions (Monte Carlo simulations).

4 CONCLUSIONS

The goal of this contribution was to find a solution to set up correlations among model process parameters. A straightforward transform was suggested. The neglect of correlations in the whole set of PGPs yields excessively pessimistic simulation results. Many of them are unlikely to occur. A new design, that has to accommodate all these situations, would be less competitive due to increased costs and prolonged development time.

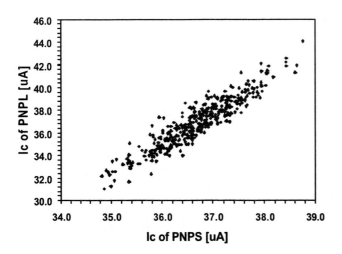

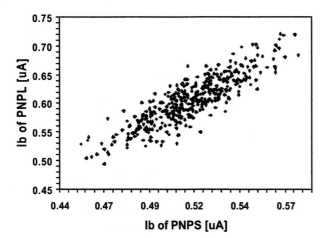

Fig4: Simulated collector and base currents of PNPS and PNPL with **correlated PGPs** via master variables (Monte Carlo simulations).

experimental designs, Proc. IEEE BCTM (1989), p. 262-265

SCG Czech Design Center, ON Semiconductor Czech Republic, B. Nemcové 1720, 756 61 Rožnov pod Radhoštem, Czech Republic, Fax: +420-571652099, Ph: +420-571667183, jiri.slezak@onsemi.com

REFERENCES

[1] C.C.McAndrew, J.Bates, R.T.Ida and P. Drennan, Efficient Statistical BJT Modeling, Why ß is More Than Ic/Ib, IEEE BCTM 1.2 (1997), p.28-31

[2] C.C.McAndrew, Statistical Modeling for Circuit Simulation, Proceedings ISQED'03 (2003)

[3] J.A.Power, B.Donellan, B.Mathewson and W.A.Lane., Relating Statistical MOSFET model parameter variabilities to IC manufacturing process fluctuations enabling realistic worst-case design, IEEE Trans. Semicond. Manufact. 7 (1994), p.306-318

[4] J.A.Power, S.Kelly, E. Griffith, D. Doyle, M. O'Neill, Statistical modeling for 0.6 μm BiCMOS Technology, Proc. IEEE BCTM (1997)

[5] G. Rappitsch, European IC-CAP User Meeting, Prague, 2003

[6] W.F.Davis, R.T.Ida, Statistical IC simulation based on independent wafer extracted process parameters and

A Trial Report: HiSIM-1.2 Parameter Extraction for 90 nm Technology

Y. Iino

Silvaco Japan
549-2 Shinano-machi Totsuka-ku, Yokohama, Japan, yoshihisa.iino@silvaco.com

ABSTRACT

This paper describes a parameter extraction procedure for HiSIM (Hiroshima-university STARC IGFET Model) [1] version 1.2 with the application result to practical devices. The procedure followed a traditional approach of the geometry selection for the parameter extraction. It starts with the long and wide channel (Large) device, then proceeds to the various channel length devices under the fixed wide channel width (L-array), finally to the various channel width devices under the fixed long channel length (W-array). No short and narrow channel devices (Small) are used. Semiconductor Technology Academic Research Center (STARC), Yokohama, Japan, permitted the paper author to use the practical device data which was said to be ready for 90 nm node. The result verified the scalability of HiSIM-1.2 down to 100 nm channel length with certain improvement required. The number of optimized parameters were 19. The details will be reported at Nanotech 2004 workshop on compact modeling.

Keywords: HiSIM, surface potential, parameter extraction, compact model

1 INTRODUCTION

As one of surface potential based Spice models, HiSIM version 1.0 was released and placed on a public domain on January 2002 [2]. Although the model was updated to the version 1.2 in April 2003, few parameter extraction result with the extraction procedure was published internationally [3], [4], [5]. This paper presents the extraction strategy with the application to practical devices. The strategy was developed with the consideration of HiSIM model derivation [6]. The device data which was said to be ready for 90 nm node was provided by STARC with the publication permission. Silvaco's UTMOST-III [7] with HiSIM-1.2 was used. In spite of using the specific extraction software, the procedure should be generic enough to be applicable to others.

2 EXTRACTION PREPARATION

2.1 Geometry Selection

The geometry selection should follow the traditional methodology for the scalable model parameter extraction

[8] with one exception: HiSIM model has no small size effect parameters.
Large device: the channel width and length were 10 um.
L-array devices: the fixed channel width was 10 um and the channel lengths were 10, 5, 1, 0.5, 0.3, 0.2, and 0.1 um.
W-array devices: the fixed channel length was 10 um and the channel widths were 5, 1, 0.3, 0.15, and 0.11 um.
Small devices: the fixed channel length was 0.1 um and the channel widths were 5, 2, 1, 0.56, 0.3, 0.15, and 0.11 um.

The small devices were used for the model validation.

2.2 Measurement Conditions

The measured DC characteristics follow.
Ids vs. Vgs: Vgs = 0 –> 1.0 (V), Vbs = 0 –> 0.5 (V),
    @ Vds = 0.05 and 1.0 (V)
Ids vs. Vds: Vds = 0 –>1.0 (V), Vgs = "near Vth" –>1.0 (V),
    @ Vbs = 0 and –0.5 (V)
Ids vs. Vgs: Vgs = 0 –> 1.0 (V), Vds = 0.1 –> 0.5 (V)

The third characteristic is optional and prepared in case for the extraction of high field mobility parameter of HiSIM [6]. In addition to the DC characteristics, Cgg capacitance measurement is required for the oxide thickness determination. The value for this study was provided by STARC. The Cgg characteristic is also required for such effects as poly depletion and quantum-mechanical effects which parameters were untouched for this extraction study.

3 EXTRACTION STRATEGY

Initial parameter values defined were the defaults described in HiSIM1.2.0 User's Manual [9]. In a following description, HiSIM model parameters are capitalized with the bold-faced letters.

3.1 NSUBC and VFBC: Idvg_large_HiSIM

Id/Vgs curves at the low drain voltage (50 mV) under the stepped body bias were used for **NSUBC** and **VFBC** parameters. The sub-threshold to onset of the strong inversion region for the large device (W/L = 10/10 um/um) was the optimizer target. The body bias effect region must be included at this step. Because there is no body bias effect parameter in HiSIM.

Prior to the execution, the **TOX** and **XWD** values were specified as 2.287 (nm) and 22 (nm), respectively. The **XWD** value was determined through the tentative study for the W-array devices. Also for this extraction study, the **LP** was changed [10] from the default (15 nm) to 300 nm which is the upper limit value. The modification was required to express the reverse short channel effect. And the reasonable **LP** determination needs the further study.

The **NSUBP** value was linked to the **NSUBC** during this optimization step in UTMOST to avoid the solver convergence problem. This is assumed that the effective substrate concentration in HiSIM coupled to the **NSUBC**, **NSUBP**, and **LP** might have violated certain criteria.

3.2 MUEPH1, MUECB0, MUECB1, MUESR1: idvg_lowMue_HiSIM

The required characteristic and geometry are exactly the same as in the previous one. These parameters are for HiSIM low field mobility which adopts the mobility universality [11]. Therefore, the target characteristic regions for the parameter optimization should be chosen according to the universality. The **MUEPH1** was optimized from the sub-threshold to onset of the strong inversion region of Ids/Vgs. The **MUECB0** and **MUECB1** were to the strong inversion. And the **MUESR1** was for the maximum Vgs region. The low field mobility parameters should be optimized at this step, especially for P-channel devices. Because the default set represents N-channel devices.

3.3 NSUBP: idvg_middle_HiSIM

The required DC characteristic is the same as in the previous strategies. The geometry should be selected from the L-array devices through the observation of Vth dependency on the channel length: the reverse short channel (RSC) effect devices. Such channel length devices as 5, 1, 0.5, and 0.3 (um) were used for this extraction. The **NSUBP** was capable of expressing the RSC effect for this extraction. And such HiSIM short-channel coefficients as the **SCP1** and **SCP3** had little influence.

3.4 PARL2, SC1, and SC3: idvg_short_HiSIM

The same Id/Vgs curves for the standard short channel (SC) effect devices are used for the **PARL2**, **SC1** and **SC3**. The **PARL2** and **SC1** were optimized first to express the Vth roll-off of the devices, and followed by the **SC2** which is related to the body bias (Vbs) influence on the short channel devices.

Four strategies described so far were applied to Ids vs. Vgs at the low Vds, and the fit result should be fairly acceptable. However, 0.2 um channel length device in this study showed that the simulated Ids/Vgs had rather large Vth compared to the measurement.

3.5 SCP2 and SC2: idvg_highVT_HiSIM

Both the **SCP2** and **SC2** are related to Vds. So Ids/Vgs at the high Vds (1.0 V) was used in the strategy. And the

sub-threshold to onset of strong inversion region was defined. The devices for the **SCP2** and **SC2** were 5, 1, 0.5, 0.3 um (RSC devices) and 0.2, 0.1 um (SC devices) channel length devices, respectively.

3.6 VOVER, VOVERP and VMAX: idvg_highVD_HiSIM

HiSIM high field mobility parameters such as the **VOVER**, **VOVERP** and **VMAX** were optimized to the Ids/Vgs at the large Vds (1.0 V). Also, Ids/Vgs curves at several Vds were tried to get the better dependency on the Vds.

3.7 WFC: idvg_narrow_HiSIM

The **WFC** was optimized to such W-array devices as 5, 1, 0.3, 0.15, and 0.11 um channel widths for Ids/Vgs at the low Vds. The Vth roll-off of the devices was expressed well.

4 ACKNOWLEDGEMENT

The author would like to thank Semiconductor Technology Academic Research Center for permitting the practical device data to be used with the publication.

REFERENCES
[1] M. Miura-Mattausch, et al., "HiSIM: A MOSFET Model for Circuit Simulation Connecting Circuit Performance with Technology," Tech. Digest IEDM, pp. 109-112, 2002.
[2] Semiconductor Technology Academic Research Center http://www.starc.or.jp/kaihatu/pdgr/hisim/index.html
[3] K. Tokumasu, et al., "Parameter Extraction of MOSFET Model HiSIM," IPSJ DA Symposium (in Japanese), A10-2, 2002.
[4] Y.Iino, "HiSIM Model Parameter Generation Flow using the Existing Model," IPSJ DA Symposium, pp.241-246, 2003.
[5] S. Ito, et al., "Parameter Extraction of HiSIM1.1 /HiSIM1.2," IPSJ DA Symposium, pp.247-252, 2003.
[6] Y.Iino, "Significant Aspects of HiSIM Parameter Extraction Derived from the Model Foundation with Several Tips," submission to the first International Workshop on Compact Modeling, ASP-DAC 2004 associated conference, Yokohama, Japan.
[7] Silvaco International, http://www.silvaco.com
[8] Y. Cheng and C. Hu, "MOSFET MODELING & BSIM3 USER'S GUIDE," p. 355, Kluwer Academic Publishers, 1999.
[9] HiSIM1.2.0 User's manual: http://www.starc.or.jp/kaihatu/pdgr/hisim/index.html
[10] Private communication with HiSIM developer.
[11] S. Takagi, et al., "On the Universality of Inversion -layer Mobility in N- and P-channel MOSFET's," Tech. Digest of IEDM, pp. 398-401, 1988.

(LEVEL = 111
VERSION = 1.2 **CORSRD** = 0 **COOVLP** = 0
COSMBI = 0 **TOX** = 2.287E-9* **XLD** = 0
XWD = 2.2E-8* **XPOLYD** = 0 **XDIFFD** = 0 **TPOLY** = 0
NSUBC = 1.223637E18* **VFBC** = -1.0788913*
LP = 3E-7* **NSUBP** = 1.866911E18* **XQY** = 0
KAPPA = 3.9 **SCP1** = 0 **SCP2** = 0.2063246*
SCP3 = 0 **PARL1** = 1 **PARL2** = 5E-8* **SC1** = 199.64881*
SC2 = 98.54520* **SC3** = 3.027419E-9* **PTHROU** = 0
WFC = 3.744748E-14* **W0** = 0 **COISTI** = 0
WVTHSC = 0 **NSTI** = 1E17 **WSTI** = 0 **QME1** = 4E-11
QME2 = 3E-10 **QME3** = 0 **PGD1** = 0.01 **PGD2** = 1
PGD3 = 0.8 **RS** = 1E-4 **RD** = 1E-4 **RPOCK1** = 1E-4
RPOCK2 = 0.1 **RPOCP1** = 1 **RPOCP2** = 0.5
BGTMP1 = 9.025E-5 **BGTMP2** = 1E-7

MUECB0 = 297.8752771* **MUECB1** = 35.8615598*
MUEPH0 = 0.3 **MUEPH1** = 2.327467E4* **MUEPH2** = 0
MUETMP = 1.5 **MUESR0** = 2 **MUESR1** = 2.180765E15*
NDEP = 1 **NINV** = 0.5 **NINVD** = 1E-9 **BB** = 2
VMAX = 2.034986E7* **VOVER** = 5E-4*
VOVERP = 6.095994E-4* **CLM1** = 0.7 **CLM2** = 2
CLM3 = 1 **COISUB** = 0 **SUB1** = 10 **SUB2** = 20
SUB3 = 0.8 **COGIDL** = 0 **COGISL** = 0 **GIDL1** = 5E-6
GIDL2 = 1E6 **GIDL3** = 0.3 **COIIGS** = 0 **GLEAK1** = 1E4
GLEAK2 = 2E7 **GLEAK3** = 0.3 **GLPART1** = 0
GLPART2 = 0 **VZADD0** = 0.01 **PZADD0** = 5E-3
CONOIS = 0 **NFALP** = 1E-16 **NFTRP** = 1E10
CIT = 0 **EF** = 0)

Table 1: Extracted HiSIM-1.2 model parameters.
The values marked with (*) were optimized.

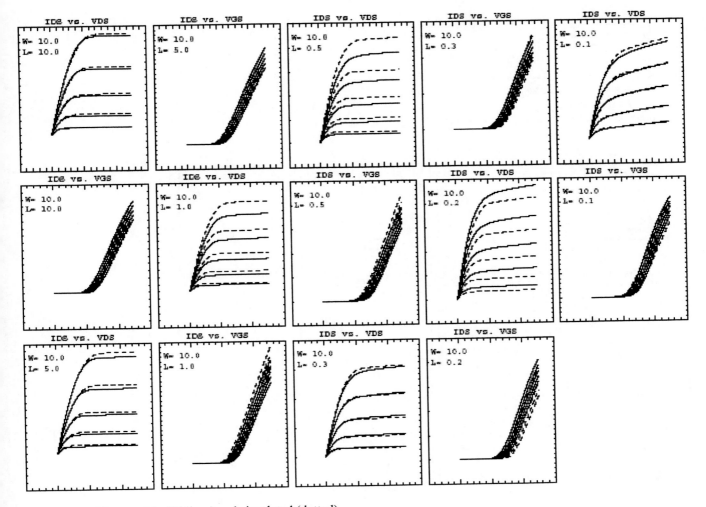

Figure 1: Measured (solid lines) and simulated (dotted)
L-array devices
Ids/Vgs @Vds = 50 (mV), Ids/Vds @Vbs = 0 (V)

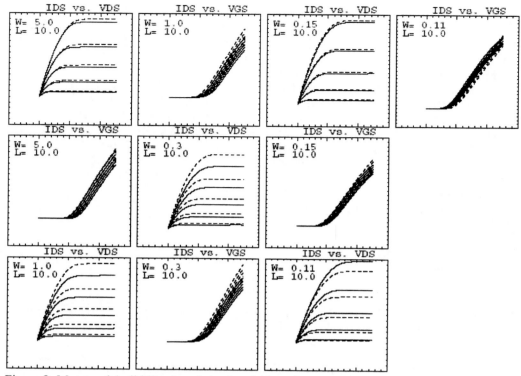

Figure 2: Measured (solid lines) and simulated (dotted)
W-array devices
Ids/Vgs @Vds = 50 (mV), Ids/Vds @Vbs = 0 (V)

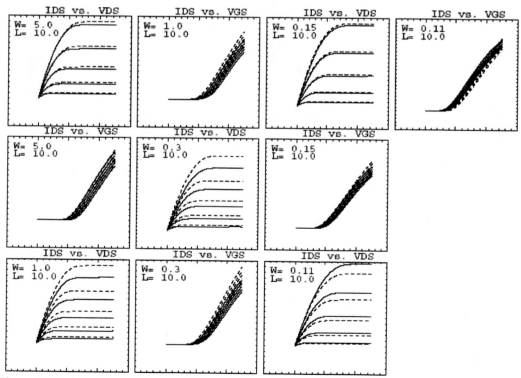

Figure 3: Measured (solid lines) and simulated (dotted)
Small devices
Ids/Vgs @Vds = 50 (mV), Ids/Vds @Vbs = 0 (V)

A Practical Method to Extract Extrinsic Parameters for the Silicon MOSFET Small-Signal Model

S. C. Wang*, G. W. Huang*, K. M. Chen*, A. S. Peng*, H. C. Tseng**, and T. L. Hsu**

*National Nano Device Laboratories
1001-1, Ta Hsueh RD, Hsinchu, 300, Taiwan, R. O. C., gwhuang@ndl.gov.tw
**United Microelectronics Corporation
No.3, Li-Hsin RD. 6, Science-Based Industrial Park, Hsinchu 300, Taiwan, R. O. C.

ABSTRACT

In this paper, the substrate parasitic has been added into the conventional MOSFET small-signal model for RFIC applications, and an extraction approach based on the curve-fitting method proposed by S. Lee [1] also has been developed to accurately determine the whole model parameters. The agreement between the measured and simulated S-parameters up to 40GHz proves the feasibility of this modified extraction method.

Keywords: MOSFET, substrate, small-signal modeling, parameter extraction

1 INTRODUCTION

An accurate small-signal model of Si MOSFET's is an urgent need for RF circuit designs, and some methods have been presented to extract these model parameters [1], [2], [3]. The curve-fitting method proposed by S. Lee [1], which is based on the analytical Z-parameter expressions derived from the conventional small-signal equivalent circuit shown in Fig. 1, is widely used to determine the series resistances and inductances [4], [5]. However, when the drain-side substrate parasitic is taken into account, careful attention must be taken into consideration, especially for RFIC applications.

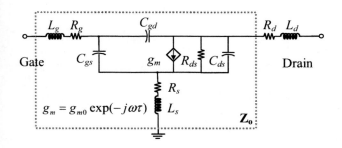

Figure 1: Conventional small-signal model for silicon MOSFET's

Consider the equivalent circuit shown in Fig. 2, where the drain-side substrate parasitic is coupled through the drain-to-substrate junction capacitance C_{jdb}. The substrate

parasitic coupled through the source-to-substrate junction is not included here because for most RFIC applications, intrinsic source and substrate nodes are tied together, and hence this associated substrate effect can be mitigated.

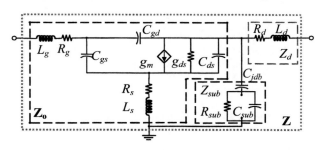

Figure 2: Small-signal model taking the drain-side substrate parasitic into account

If $\mathbf{Z}$ is the Z-parameter matrix for the dotted box shown in Fig. 2, and $\mathbf{Z_0}$ for that shown in Fig. 1, their relation can be derived directly from the matrix definition and are expressed in (1)-(4).

$$Z_{11} = Z_{o,11} - \frac{\left(Z_{o,12}/Z_{sub}\right)\cdot Z_{o,21}}{1 + Z_{o,22}/Z_{sub}} \tag{1}$$

$$Z_{12} = \frac{Z_{o,12}}{1 + Z_{o,22}/Z_{sub}} \tag{2}$$

$$Z_{21} = \frac{Z_{o,21}}{1 + Z_{o,22}/Z_{sub}} \tag{3}$$

$$Z_{22} = \frac{Z_{o,22}}{1 + Z_{o,22}/Z_{sub}} + Z_d \tag{4}$$

According to (1)-(4), it's obvious that, the original expressions derived for this conventional curve-fitting method [1] are valid only when $Z_{o,22}/Z_{sub}$ approaches zero. Besides, $Mag(Z_{o,12})$ is usually smaller than $Mag(Z_{o,22})$, so the assumption that $Z_{o,12}/Z_{sub}$ approaches zero when $Z_{o,22}/Z_{sub}$ approaches zero will automatically become true without any emphasis. Based on this modified equivalent circuit, because $Mag(Z_{sub})$ will roughly approach a constant (R_{sub}) and $Mag(Z_{o,22})$ will decrease with increasing frequency, we can predict that the conventional curve-fitting method can still be valid at higher frequencies. This assumption will be verified experimentally using the modified curve-fitting method proposed in this paper.

2 EXTRACTION PROCEDURE

The MOSFET device used in this experiment was fabricated in 0.18um RF CMOS process with 0.18um gate length and 8×5um gate width, and its S-parameter measurements were performed from 0.1GHz to 40GH at $V_{gs} = V_{ds} = 1.8V$. The measured S-parameters were then de-embedded by subtracting the OPEN dummy.

The original equations used to fit out the series resistances and inductances are listed in (5)-(10) [1].

$$\text{Re}(Z_{12}) = R_s + \frac{A_s}{\omega^2 + B} \tag{5}$$

$$\text{Re}(Z_{11} - Z_{12}) = R_g + \frac{A_g}{\omega^2 + B} \tag{6}$$

$$\text{Re}(Z_{22} - Z_{12}) = R_d + \frac{A_d}{\omega^2 + B} \tag{7}$$

$$\text{Im}(Z_{12})/\omega = R_s - \frac{E_s}{\omega^2 + B} \tag{8}$$

$$\text{Im}(Z_{11} - Z_{12})/\omega = R_g - \frac{E_g}{\omega^2 + B} - \frac{F_g}{\omega^2(\omega^2 + B)} \tag{9}$$

$$\text{Im}(Z_{22} - Z_{12})/\omega = R_d - \frac{E_d}{\omega^2 + B} \tag{10}$$

where B, A_s, A_g, A_d, E_s, E_g, F_g, and E_d are expressed as functions of other model parameters and can be considered as constants.

As mentioned above, these expressions are derived just based on the conventional model (Fig. 1). To take the substrate parasitic into account and still use Lee's method, the starting frequency point used for curve-fitting must be set higher at some critical frequency point (in this case, 15GHz). The fitting results for series resistances and inductances are shown in Fig. 3 and 4, respectively. It is obvious that the inconsistent fitting results at low frequency region are caused by the significant drain-side substrate parasitic. The selected critical frequency point seems arbitrary, but its rationality will be checked at the end of the whole modeling process.

To extract the parameters associated with the substrate parasitic, another curve-fitting method [5] can be utilized. After de-embedding R_g, L_g, R_d, and L_d from the modified small-signal model, the resulting network would become that shown in Fig. 5(a). In addition, if we neglect the voltage drop across R_s and L_s, this network can be simplified as that (with its Y-parameter, Y_{out}) shown in Fig. 5(b), and it will produce following equations:

$$\text{Re}(Y_{out,12} + Y_{out,22}) = \frac{1}{R_{ds}} + \frac{k_1\omega^2}{1 + k_2\omega^2} \tag{11}$$

$$\frac{1}{\omega}\text{Im}(Y_{out,12} + Y_{out,22}) = C_{ds} + C_{jdb}\left(\frac{1 + m_1\omega^2}{1 + m_2\omega^2}\right) \tag{12}$$

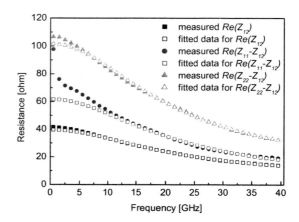

Figure 3: Curve-fitting results for series resistances ($V_{gs} = V_{ds} = 1.8V$)

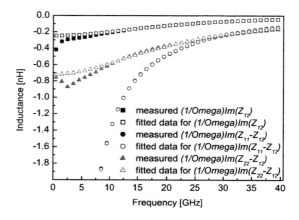

Figure 4: Curve-fitting results for series inductances ($V_{gs} = V_{ds} = 1.8V$)

$$R_{sub} = \frac{k_2}{k_1}\left(1 - \frac{m_1}{m_2}\right)^2 \tag{13}$$

$$C_{sub} = \frac{m_1 C_{jdb}}{m_2 - m_1} \tag{14}$$

where k_1, k_2, m_1, and m_2 are also expressed as functions of other model parameters and can be considered as constants.

The curve-fitting results based on (11) and (12) are shown in Fig. 6, and 7, respectively, and each curve-fitting was performed below 5GHz to suppress the fitting error likely caused by R_s and L_s at high frequencies.

Finally, when all the extracted substrate-related parameters and R_s, L_s are subtracted from the network shown in Fig. 5(a), the intrinsic network (with its Y-parameter, Y_i) shown in Fig. 5(c) will appear, and the intrinsic parameters can be directly extracted by the following equations [2]:

$$C_{gd} = -\frac{1}{\omega}\text{Im}(Y_{i,12}) \tag{15}$$

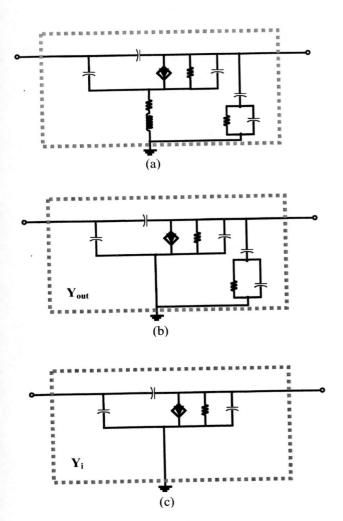

Figure 5: Equivalent circuits (a)after de-embedding R_g, L_g, R_d, and L_d, and then (b)neglecting the voltage drop across R_s and L_s .(c) is the intrinsic part of MOSFET devices

$$C_{gs} = \frac{1}{\omega} \text{Im}(Y_{i,11} + Y_{i,12}) \qquad (16)$$

$$C_{ds} = \frac{1}{\omega} \text{Im}(Y_{i,22} + Y_{i,12}) \qquad (17)$$

$$R_{ds} = \frac{1}{\text{Re}(Y_{i,22})} \qquad (18)$$

$$g_{m0} = Mag(Y_{i,21} - Y_{i,12}) \qquad (19)$$

$$\tau = -\frac{1}{\omega} Phase(Y_{i,21} - Y_{i,12}) \cdot \qquad (20)$$

All the extracted extrinsic and intrinsic model parameters are summarized in Table 1 for reference.

3   RESULTS AND DISSCUSSIONS

When we substitute these extracted parameters into the modified equivalent circuit, we can directly calculate the magnitude values for $Z_{o,12}$, $Z_{o,22}$, and Z_{sub}, which are shown

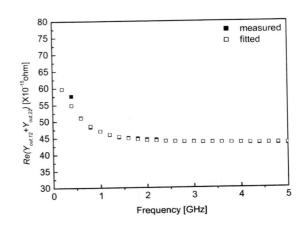

Figure 6: Substrate-related curve fitting results based on (11) ($V_{gs} = V_{ds} = 1.8V$)

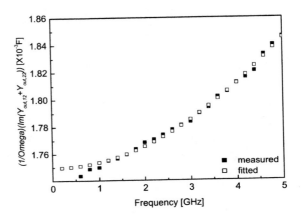

Figure 7: Substrate-related curve fitting results based on (12) ($V_{gs} = V_{ds} = 1.8V$)

in Fig 8. It indicates that with increasing frequency, $Mag(Z_{o,12})$ and $Mag(Z_{o,22})$ will decrease and $Mag(Z_{sub})$ will roughly approach a constant (R_{sub}). The values for $Mag(Z_{o,22}/Z_{sub})$ and $Mag(Z_{o,12}/Z_{sub})$ versus frequency are shown in Fig. 9, where we find that, above 15 GHz, these values will decrease as frequency increase and hence alleviate the influence of the substrate parasitic on the fitting accuracy of Lee's method, and it in turn proves the rationality of this selected critical frequency point. In addition, if more accurate modeling result is desired, the critical frequency may be selected higher to achieve this goal, at the cost of less data point can be used to perform extrinsic curve-fitting procedure for limited measurement points.

Figure 10 depicts the final simulated S-parameters, and they show good agreement with the measured data up to 40 GHz. It proves that, for MOSFET's with significant drain-side substrate effect, the proposed method in this paper indeed can accurately extract all the parasitic parameters.

Table I Extracted parameters for extrinsic and intrinsic parts

	$R_s(\Omega)$	L_s(H)	$R_g(\Omega)$	L_g(H)	$R_d(\Omega)$	L_d(H)	C_{jdb}(F)	C_{sub}(F)	$R_{sub}(\Omega)$	C_{gs}(F)	C_{ds}(F)	C_{gd}(F)	g_m (1/Ω)	Tau (sec.)	$R_{ds}(\Omega)$
value	6.9	11.9p	6.2	27.8p	8.5	0(<0.1f)	26f	0.67f	1271	47.8f	43.9f	18.2f	21.2m	50f	568

4   CONCLUSSIONS

The substrate effect was taken into account in this paper to reasonably model the high frequency behavior of silicon MOSFET's, and then a modified method based on the conventional curve-fitting method was proposed. The good modeling results up to 40GHz have verified the feasibility of this modified extraction method.

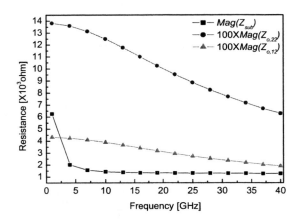

Figure 8 $Mag(Z_{sub})$, $Mag(Z_{o,22})$, $Mag(Z_{o,12})$ versus frequency

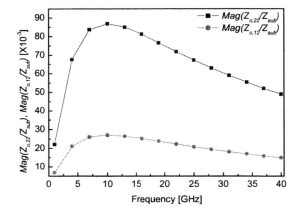

Figure 9 $Mag(Z_{o,22}/Z_{sub})$ and $Mag(Z_{o,12}/Z_{sub})$ versus frequency

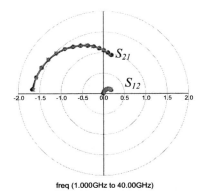

Figure 10 Measured (symbol 'x') and modeled (symbol 'o') S parameters from 1 to 40GHz

REFERENCES

[1] S. Lee et al., "A novel approach to extracting small-signal model parameters of silicon MOSFET's," IEEE Microwave Guided Wave Lett., Vol. 7, pp. 75-77, 1997.

[2] D. Lovelace et al., "Extracting small-signal model parameters of silicon MOSFET transistors," in IEEE MTT-S Tech. Dig., pp. 865-868, 1994.

[3] C. H. Kim et al., "Unique extraction of substrate parameters of common-source MOSFET's," IEEE Microwave Guided Wave Lett., vol. 9, pp. 108–110, Mar. 1999.

[4] J. –P. Raskin et al., "Accurate SOI MOSFET characterization at microwave frequencies for device performance optimization and analog modeling," IEEE Trans. Electron Devices, vol. 45, pp. 1017-1025, 1998.

[5] S. Lee et al., "A small-signal RF model and its parameter extraction for substrate effects in RF MOSFET's," IEEE Trans. Electron Devices, vol. 48, pp. 1374–1379, 2001.

Characterization and Modeling of Silicon Tapered Inductors

A.S. Peng[*], K.M. Chen[*], G.W. Huang[*], S.C. Wang[*], H.Y. Chen[**], and C.Y. Chang[**]

[*]National Nano Device Laboratories
1001-1 Ta Hsueh RD., Hsinchu 300, Taiwan, kmchen@ndl.gov.tw
[**]Department of Electronics Engineering, National Chiao Tung University
1001 Ta Hsueh RD., Hsinchu 300, Taiwan

ABSTRACT

The characteristics of tapered inductors and standard inductors have been compared, and an improved compact model for tapered inductors is presented. The measured data of spiral inductors shows that the tapered inductor has higher quality factor (Q) than standard ones above 2.3GHz and higher frequency in which the maximum Q occurs. Because the parasitic capacitances, such as parallel capacitance and oxide capacitance, have lower values for tapered inductors, the real part of the input impedance is smaller, leading to higher Q value. For accurately modeling the behavior of tapered inductors, we propose an improved compact model, which adds a branch to model the frequency-dependent resistance. This model has the advantage of easily acquiring a relative equation because of the simple parallel structure of the skin effect model. The modeled and measured results have excellent agreement.

Keywords: tapered inductor, quality factor, skin effect, frequency-dependent resistance,

1 INTRODUCTION

Spiral inductor technologies on Si substrate have been widely studied for integrating RF circuits into Si IC technologies. However, the lossy Si substrate contributes to the low quality factor (Q) of on-chip inductor. To enhance the quality factor, spiral inductors with width-tapered structure (as shown in Fig. 1), which metal width of the inductor is increasing from the center to the outer, have been proposed recently [1], [2]. It was presented that the width-tapered inductor has higher quality factor and higher frequency in which the maximum Q occurs [1], [2].

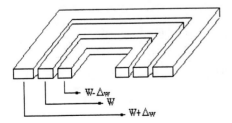

Figure 1: Geometrical structure of a tapered inductor

In this paper, we compared the measured data of the width-tapered inductors with various geometries in Section 2. By analyzing these measurement results, some guidelines for designing the layout geometries of tapered inductors have been obtained. Besides, we also present an improved inductor equivalent model for tapered inductors by adding additional branches to model the skin effect and substrate loss. The skin effect causes the series conductor resistance has frequency-dependent characteristic [3]. The equivalent single-π inductor model and extraction procedure will be described in Section 3. Experimental results show that the improved model can predict the frequency-dependent resistance, quality factor and inductance behaviors.

2 MEASURED RESULTS

To verify the advantage of tapered inductors, we compared the measured data of spiral inductors with metal space (S) = 3μm, inner diameter (ID) = 60μm, number of turns (N) = 4.5, but varied metal width. The results are listed in Table 1, including the geometric structures of a standard inductor and tapered inductors. We found that the tapered_1 inductor has the best performance of Q and inductance than the other tapered inductors. However the characteristics of tapered_2 and tapered_3 inductors do not have the improvement as expected. Because the size of the contact via is 4x4um^2 in our layout design, the tapered_2 and tapered_3 inductors have only one contact via connecting the center metal with underpass line due to narrower center metal. It would increase the contact resistance and thus degrade the performance of tapered_2 and tapered_3 inductors.

Inductor List	Metal Width Center to Outer (um)	Q max	L (nH)	Frequency @max Q (GHz)	Via NO.
standard	15/15/15/15/15	4.427	3.816	2.3	2x2
tapered_1	13/14/15/16/17	4.516	3.961	2.7	2x2
tapered_2	11/13/15/17/19	4.483	3.747	3	1x1
tapered_3	9/12/15/18/21	4.349	3.589	3	1x1

Table 1: Comparison of inductors with same metal space but different metal widths.

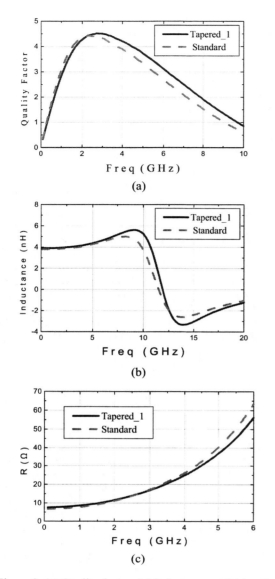

(a)

(b)

(c)

Figure 2: (a) Quality factor, (b) inductance, and (c) series resistance versus frequency for standard and tapered_1 inductors

Figure 2 compares the characteristics of standard and tapered_1 inductors. As shown in Fig. 2(a), the tapered_1 inductor has higher Q than the standard one above 2.3GHz. Also, the frequency at maximum Q shifts to a higher value. To understand the reason of Q-enhancement, the inductance and resistance are analyzed respectively.

Figure 2(b) illustrates the tapered_1 inductor has higher inductance and self-resonant frequency (f_{SR}). The thinner center metal width can naturally increase the inductance, but decrease the f_{SR}. The result of f_{SR} is different with Fig. 2(b), which means the tapered_1 inductor must have other reasons to increasing the f_{SR}.

Since the width-tapered inductors have narrower center metal, their underpass lines are thinner than standard inductors, which could effectively reduce the parasitic capacitance. It leads to higher f_{SR} for tapered-inductor in spite of higher inductance as compared with standard inductor.

The parasitic capacitances not only affect f_{SR}, but also raise the slope of series resistance. Figure 2(c) illustrates the influence of parasitic capacitances to the resistance. Standard inductor has lower resistance at low frequency, but higher resistance above 2.5G due to higher rising slope of resistance than tapered one.

With analyzing these measurement results of inductors, we propose some guidelines to enhance Q in the tapered structure for the RFIC designers: (1) Reducing parasitic capacitance, such as parallel capacitance (C_P) and oxide capacitance (C_{OX}), will cause slower rising of $Re(Z_{in})$ than standard device before resonance and enhance the Q value. (2) Contact via area may be reduced with narrowing center metal width; this is why the performance of tapered_2 and tapered_3 devices is not as good as expectation.

3 IMPROVED MODELING

3.1 Model Description

A conventional 9-element model for planar spiral inductors is extensively used in circuit simulation due to the advantage of easy parameters extraction due to the nature of equivalent single-π circuit structure. Additional elements are still needed to model some physical effects and to enhance the accuracy prediction of inductor behavior.

For accurately modeling the behavior of tapered inductors, we propose an improved compact model, as shown in Fig 3. Since the tapered inductors have apparent skin effect, we added a branch (L_{S2} and R_{S2}) to model the skin effect. Otherwise, a resistance R_P is series connected with the capacitance C_p. This takes conductor loss into account [4].

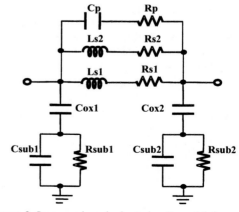

Figure 3: Improved equivalent circuit model for a spiral inductor.

The other element parameters: R_{S1} and L_{S1} denote the main resistance and inductance of the spiral inductor, respectively; C_P represents the direct capacitive coupling among the parallel turns and the coupling of the superimposition between the spiral and the underpass lines; C_{sub} and R_{sub} signify the capacitance and the resistance of the lossy substrate; and C_{OX} depicts the oxide capacitance between the metal line and silicon substrate.

3.2 Extraction Procedure

Because this model based on the equivalent single-π circuit structure is shown in Fig. 4, we can use this simple structure to extract the element parameters rather than the other model methods with complex skin effect components.

At low frequency, the branches with series capacitance can be neglected and the series inductance can be regarded as a short circuit, which leads to easy acquisition of the value of $R_{S1}//R_{S2}$

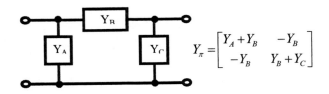

$$Y_\pi = \begin{bmatrix} Y_A + Y_B & -Y_B \\ -Y_B & Y_B + Y_C \end{bmatrix}$$

Figure 4: The single-π circuit structure and its relative Y parameters.

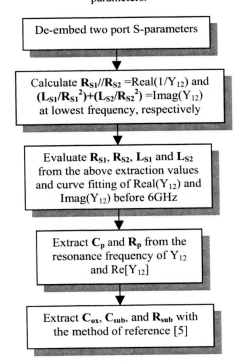

Figure 5: Extraction procedure for our improved model.

by extracting the $Re[1/Y_{12}]$ at low frequency. Besides, the Y_B of the improve model can be deduced by:

$$Y_B = \frac{1}{R_{S1} + j\omega L_{S1}} + \frac{1}{R_{S2} + j\omega L_{S2}} \quad (1)$$

After rationalizing (1), we obtain

$$Y_B = \frac{R_{S1} - j\omega L_{S1}}{R_{S1}^2 + \omega^2 L_{S1}^2} + \frac{R_{S2} - j\omega L_{S2}}{R_{S2} + j\omega L_{S2}} \quad (2)$$

Because $\omega L_{s1} << R_{s1}$, we can deduce the image part of Y_B:

$$Im[-Y_B] = Im[Y_{12}] = \frac{L_{S1}}{R_{S1}^2} + \frac{L_{S2}}{R_{S2}^2} \quad (3)$$

The extraction procedure is listed in Fig. 5, and is summarized as follow:
1) Measure the S-parameter of DUT and the "open" dummy and convert them to Y parameters. The parameters of devices can be obtained by calculating the admittance matrix $[Y_{DUT}]-[Y_{Open}]$.
2) Calculate $R_{S1} // R_{S2}$ and $L_{S1}/R_{S1}^2 + L_{S1}/R_{S1}^2$ from $Re[1/Y_{12}]$ and $Im[Y_{12}]$, respectively, at lowest frequency.
3) Extract R_{S1}, R_{S2}, L_{S1} and L_{S2} from the calculated values of step 2) and the curve fitting of $Re[Y_{12}]$ and $Im[Y_{12}]$ before 6GHz.
4) Extract C_p and R_p from the resonance frequency of Y_{12} and relative $Re[Y_{12}]$.
5) Extract the C_{ox}, C_{sub}, and R_{sub} using the method of [5].

3.3 Simulation Results

The improved model is mainly for modeling the frequency-dependence resistance. We make comparisons in terms of the real part of Z_{12} up to 6GHz between the simulation of improved and conventional model with measured data in Fig. 6. It can be shown that the improved model has an appreciable improvement in the frequency-dependence resistance. The improved model also has good prediction of Q and inductance, which is shown in Fig. 7

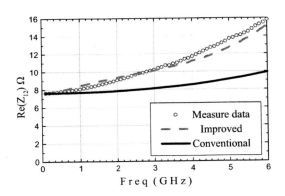

Figure 6: Simulated results of $Re(Z_{12})$ for a spiral inductor with conventional and our improved model. The measured data is also shown for comparison.

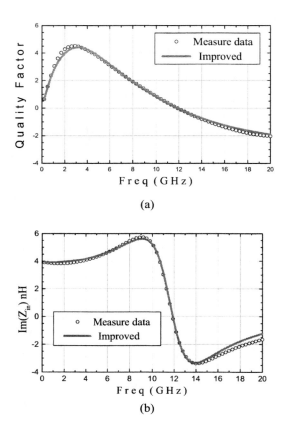

(a)

(b)

Figure 7: Measured and simulated (a) quality factor ,and (b) inductance for a spiral inductor using our improved model.

REFERENCES

[1] Chen, Y. -J.E. et al,"Investigation of Q enhancement for inductors processed in BiCMOS technology" IEEE Radio and Wireless Conference (RAWCON 99), pp. 263, Aug. 1-4 1999.

[2] Heng-Ming Hsu et al,"Silicon Integrated High Performance Inductors in a 0.18um CMOS Technology for MMIC" Symposium on VLSI Circuits Digest of Technical Papers, pp. 199, 2001.

[3] C. Patrick Yue, "Physical Modeling of Spiral Inductors on Silicon" IEEE Transactions On Electron Devices, Vol. 47, No. 3, March 2000.

[4] José R. Sendra et al, "Integrated Inductors Modeling and Tools for Automatic Selection and Layout Generation" International Symposium on Quality Electronic Design, pp. 400 –404, 2002.

[5] C. Y. Su et al., "A macro model of silicon spiral inductor," Solid-State Electronics, vol. 46, pp. 759-767, May 2002.

4 CONCLUSION

In this paper, we prove that spiral inductors with width-tapered structure can enhance the quality factor and find the quality factor of tapered inductors is dominated by via contact area and parasitic capacitances. The parasitic capacitance, such as parallel capacitance and oxide capacitance, will cause slower rising of $Re(Z_{in})$ than standard device before resonance and enhance the Q value.

An improved inductor model is proposed in this paper. These model parameters can be extracted based on a single-π model. This model can predict the frequency-dependent behavior of series resistance accurately. Further, we also can accurately model the quality factor and inductance of spiral inductors over a wide-band frequency.

Improved Compact Model for Four-Terminal DG MOSFETs

T. Nakagawa, T. Sekigawa, T. Tsutsumi[*], M. Hioki, E. Suzuki, and H. Koike

Electroinformatics Group,
Nanoelectronics Research Institute
National Institute of Advanced Industrial Science and Technology
1-1-1 Umezono, Tsukuba, Ibaraki, 305-8568 Japan
Phone: +81-298-61-5469 Fax: +81-298-61-5584
E-mail: nakagawa.tadashi@aist.go.jp

*Department of Computer Science,
School of Science and Technology, Meiji University
1-1-1 Higashi-mita, Tama, Kawasaki, Kanagawa, 214-8571, Japan

ABSTRACT

This paper reports improvements of the compact model for double-gate field-effect transistors (DG MOSFETs). The improvements make the model more accurate for wider range of device dimensions and device operation conditions.

Carrier density calculation at the source-end is modified to the model applicable for wide range of gate voltages. By this modification, iteration counts for Newton's method are also drastically reduced. Two approaches to construct the transport equations, unified and separate current methods, in the channel are compared. Conventional charge-sheet model leads to too thin carrier density at the drain-end with unphysical high carrier velocity. Transport equation with velocity saturation and explicit drain electric field is proposed. Simulation result based on this equation is demonstrated.

Keywords: MOSFET, double-gate, compact modeling

1   INTRODUCTION

Double-gate field-effect transistors (DG MOSFETs), which have two insulating gates sandwiching a Si-channel, as shown in Fig. 1, have gained much attention as a promising device structure with excellent scalability because of its minimum short channel effect [1]. Beside the use of DG MOSFETs as three-terminal transistors with one electrically-common gate, we have proposed and demonstrated that four-terminal operation of DG MOSFETs is promising [2], [3]. This operation mode enables fine-grain dynamic threshold-voltage (V_t) control and sophisticated use for analog signal processing. To evaluate these merits, we developed a compact four-terminal DG MOSFET model by adopting the double charge-sheet model [4]. This model deals with four-terminal operation of DG MOSFETs of both symmetrical and asymmetrical double-gate structure. We tested this model with the device simulator (ATLAS) results for long-channel DG MOSFETs, and found that it describes the characteristics of the transistor in good accuracy. In this presentation, we report the improvement of the model to make it more accurate for wider range of device dimensions and device operation conditions.

2   SOURCE CARRIER DENSITY

For the charge-sheet model, the first step is to calculate the potential ψ at the source-end. It is achieved by solving one dimensional Poisson equation along the axis x perpendicular to the silicon/oxide interface.

$$\frac{\partial^2 \psi}{\partial x^2} = \frac{q}{\varepsilon_s} n_i e^{\beta \psi} \tag{1}$$

where n_i is the intrinsic carrier density, q is the unit charge, ε_s is the dielectric constant of silicon, β is q/kT (k is Boltzmann constant, T is the temperature). By integrating this equation, we obtain an expression about the channel thickness d.

$$
\begin{aligned}
d = s2 - s1 &= (s2 - x_M) + (x_M - s1) \\
&= \int_{\psi_M}^{\psi_{s2}} \{F_2(\psi)\}^{-1/2} \partial \psi + \int_{\psi_M}^{\psi_{s1}} \{F_1(\psi)\}^{-1/2} \partial \psi
\end{aligned} \tag{2}
$$

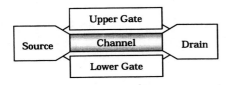

Fig. 1 Cross-sectional view of the DG MOSFET

where

$$F_1(\psi) = \left(\frac{C_{OX1}}{\varepsilon_s}(V_{G1} - \psi_{s1})\right)^2 + \frac{2qn_i}{\varepsilon_s\beta}\left(e^{\beta\psi} - e^{\beta\psi_{s1}}\right),$$

$$F_2(\psi) = \left(\frac{C_{OX2}}{\varepsilon_s}(V_{G2} - \psi_{s2})\right)^2 + \frac{2qn_i}{\varepsilon_s\beta}\left(e^{\beta\psi} - e^{\beta\psi_{s2}}\right).$$

In these equations, $s1$ and $s2$ stand for the upper and the lower silicon/oxide interface, x_M stands for the potential minimum. C_{OX1} (C_{OX2}) and V_{G1} (V_{G2}) are the gate capacitance and gate voltage of the upper (lower) gate. F_1 and F_2 become identical when the exact pair of the surface potentials ψ_{s1} and ψ_{s2} is given. These equations are used to determine two surface potentials.

The integrations in (2) give the analytical expressions as

$$\int_{\psi_M}^{\psi_{s2}}\{F_2(\psi)\}^{-1/2}\partial\psi$$
$$= \frac{2}{\beta\sqrt{-F_2(-\infty)}}\tan^{-1}\left(\frac{C_{OX2}(V_{G2} - \psi_{s2})}{\varepsilon_s\sqrt{-F_2(-\infty)}}\right), \tag{3}$$

which represents the distance between $s2$ and x_M, and (2) implies that the total channel thickness is divided to two charge sheets. But when there is only one charge-sheet, x_M is not in the channel, and (2) becomes subtraction of two positive value. Also if $F(-\infty)>0$, which occurs when there is only one lightly populated charge-sheet, the integration of (1) is

$$d = \frac{1}{\beta\sqrt{F(-\infty)}}\left|\log\left(\frac{(C_{OX1}/\varepsilon_s)|V_{G1} - \psi_{s1}| - \sqrt{F(-\infty)}}{(C_{OX1}/\varepsilon_s)|V_{G1} - \psi_{s1}| + \sqrt{F(-\infty)}}\right)\right.$$
$$\left. - \log\left(\frac{(C_{OX2}/\varepsilon_s)|V_{G2} - \psi_{s2}| - \sqrt{F(-\infty)}}{(C_{OX2}/\varepsilon_s)|V_{G2} - \psi_{s2}| + \sqrt{F(-\infty)}}\right)\right| \tag{4}$$

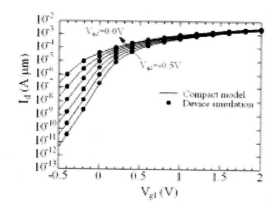

Figure 2 I-V characteristics of the DG MOSFET.

where F stands for either F_1 or F_2. Both equations (3) and (4) confront computational difficulty if $F(-\infty)\approx0$. In this case, another form

$$d = -\frac{2}{\beta}\left(\frac{\varepsilon_s}{C_{OX1}(V_{G1} - \psi_{s1})} + \frac{\varepsilon_s}{C_{OX2}(V_{G2} - \psi_{s2})}\right) \tag{5}$$

can be used. With this set of equations, calculation of (1) becomes robust accepting highly unbalanced pair of gate voltages. At the same time average count of iteration for the Newton's method drastically decreased to about 8.

The calculation results are then used as the boundary condition for the carrier transport equation for the channel. Since the relation between the surface potential and the charge-density obtained by this calculation is not hold under the charge-sheet approximation, some adjustments are needed. We found that we have to abandon the surface potential accuracy at the source-end to keep the charge density accuracy. Fig. 2 is the simulation result obtained after these improvements.

3 TRANSPORT EQUATION

Electron transport in the channel is described as

$$-I_D = I = I_1 + I_2$$
$$= q\left(-\mu_1\frac{\partial\psi_{s1}(y)}{\partial y}n_1(y) + D_1\frac{\partial n_1(y)}{\partial y}\right) \tag{6}$$
$$+ q\left(-\mu_2\frac{\partial\psi_{s2}(y)}{\partial y}n_2(y) + D_2\frac{\partial n_2(y)}{\partial y}\right)$$

where y is the axis along the channel with the origin $y=0$ at the source end, I is the total current, I_1 (I_2) and μ_1 (μ_2) are the current and the electron mobility of charge-sheet 1 (2), and D_i ($=\mu_i kT/q$) is the diffusion coefficient. Relationship between the surface potential and the carrier density in the channel is, under the gradual channel approximation:

$$\Delta\psi_{s1} = -qn_1/C_{11} - qn_2/C_{12}$$
$$\Delta\psi_{s2} = -qn_1/C_{21} - qn_2/C_{22}. \tag{7}$$

where C_{11}, C_{12}, C_{21} ($=C_{12}$), and C_{22} are the constants with the dimension of the capacitance. If $\mu=\mu_1=\mu_2$ holds, (6) can be rewritten as follows:

$$I = q\mu\left(-\frac{1}{C_{11}}\frac{\partial n_1}{\partial y}n_1 - \frac{1}{C_{22}}\frac{\partial n_2}{\partial y}n_2 - \frac{1}{C_{12}}\frac{\partial(n_1 n_2)}{\partial y} + \frac{kT}{q}\frac{\partial(n_1 + n_2)}{\partial y}\right). \tag{8}$$

This can be integrated easily.

Although this approach gives exact result, it is not practical for many situations. It is because the surface mobility is the strong function of the gate voltage. The mobility is also affected by the interface characteristics, and the interface characteristics may not be identical for device processing reason. To calculate two channels separately will be, therefore, more practical even if it is less accurate. In that case, relationship between n_1 and n_2 is necessary to simplify (7). Linear relationship as

$$n_1(y)/n_1(0) = n_2(y)/n_2(0) \tag{9}$$

was found to give accurate-enough approximation. Using this approximation, the transport equation becomes the sum of two transport equations.

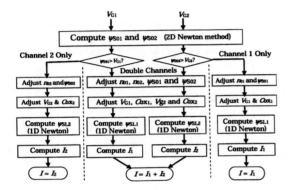

Figure 3 Computational algorithm of the double charge-sheet model of separate current method.

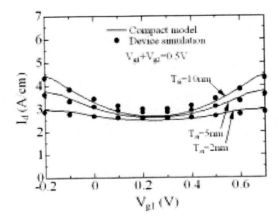

Fig. 4 Drain current change by the asymmetric gate voltages for 2, 5 and 10 nm thick Si channels. Drain voltage is 1V. Solid lines are for compact model results, and symbols are for the ATLAS results.

$$-I_D = I = I_1 + I_2$$
$$I_i = q\left(-\mu_i \frac{\partial \psi_{si}(y)}{\partial y} n_i(y) + D_i \frac{\partial n_i(y)}{\partial y}\right) \tag{10}$$

Fig. 3 shows the computational algorithm of the double charge-sheet model of separate current method. In the figure, adjustment stage of the gate voltages and capacitances are shown. It is necessary because the relation between the surface potential and the carrier density is, after adopting (9), described by an effective gate voltage and an effective gate capacitance.

Fig. 4 shows change in saturation current when the two gate voltages are varied while maintaining $V_{G1} + V_{G2}$ constant. Since the sum of two gate voltages is constant, the total channel charge does not change significantly, and the total current should remain relatively constant if the conduction mode is volume conduction. As shown in the figure, the total current has a distinct minimum value at $V_{G1} + V_{G2}$ for a thick Si channel is thick, whereas the total current remains relatively constant for thin Si channels. This shows that the proposed compact model accurately reproduces the transistor characteristics in both surface conduction and volume conduction regions.

Two kinks can be seen in the figure for all curves, corresponding to the single/double charge-sheet boundary. These kinks are caused by the approximation that the charge-sheet resides strictly at the interface, which is not rigorously true. Although these kinks do not have serious effects in the digital circuit simulation, care should be taken when the model is used for simulating analog circuits that use two gates for signal mixing.

4 SHORT CANNEL EFFECT

To make the model applicable to the short-channel device causes another problem. Conventional charge-sheet model assumes carrier density equilibrium between the drain and the channel at the drain-end as:

$$V_D = \psi_s(L) - \psi_s(0) + \frac{1}{\beta} \ln \frac{n(0)}{n(L)} \tag{11}$$

Although the expression describes small V_D subthreshold characteristics, it leads to the unphysical high carrier velocity when V_D becomes high. In other words, the conventional charge-sheet model is incompatible with the velocity saturation. To circumvent this problem we have to introduce the velocity saturation and the explicit drain electric field at the same time.

One simple form of the relation between carrier velocity v and the lateral electric field is

$$v = -\mu E / 1 + |E / E_c| \tag{12}$$

where E_c is the saturation electric field, and μ in this form is the low field mobility. This relation is reported to be accurate for PMOS and bulk n-type silicon. For NMOS channel electron, a more accurate form is known. In this paper, we adopt this form because of analytical simplicity. The transport equation including this effect is

$$-\mu \frac{\partial(\psi_s + \phi_s)}{\partial y} n + \mu \frac{kT}{q} \frac{\partial n}{\partial y} = \frac{I}{q} + \frac{I}{qE_c} \frac{\partial(\psi_s + \phi_s)}{\partial y}, \quad (13)$$

where ϕ_s is the drain electric field approximately have the form:

$$\phi_s(y) = \Delta\psi \exp(a\pi(y-L)/(d + t_{OX1} + t_{OX2}))$$
$$\Delta\psi = V_D - (\psi_s(L) - \psi_s(0)) \quad (14)$$

where L is the channel length, t_{OX1} and t_{OX2} are the gate insulator thickness, and a is a constant around 1. Equation (13) can be integrated much the same way as conventional transport equation, and the result is the same except by replacing the current term as

$$IL \rightarrow I\left(L + \frac{V_D}{E_c}\right) + \mu q \int_0^L \frac{\partial\phi_s}{\partial y} n \, dy. \quad (15)$$

The second term can be integrated using following approximation obtained by supposing that the diffusion current is negligible where the drain electric field is significant:

$$n(y) = n(L)\left(\frac{E(L)}{1 + |E(L)/E_c|}\right) \bigg/ \left(\frac{E(y)}{1 + |E(y)/E_c|}\right). \quad (16)$$

The result is as follows.

$$IL \rightarrow I\left(L_{eff} + \frac{V_D - \Delta\psi}{E_c}\right) \quad (17)$$

In the formula, L_{eff} is the point where $\psi_s/dy = \phi_s/dy$. Short channel effect is expressed that the channel length L is replaced to shorter L_{eff} causing larger drain current. On the other hand, $(V_D - \Delta\psi)/E_c$ describes the velocity saturation effect.

Fig. 5 shows transistor characteristics using this equation together with the separate current method. In the figure, the solid line is the result with no velocity saturation, whereas the dashed line is the result with velocity saturation and with explicit drain electric field. The result gives the smaller current mainly because of relatively low velocity near the velocity-saturation region inherent in (12). Still, excellent suppression of short channel effect characteristic of DG MOSFET is demonstrated distinctively.

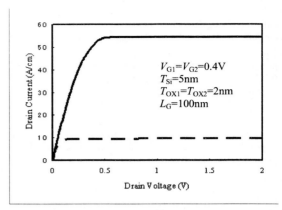

Fig. 5 Transistor characteristics without velocity saturation (solid line), and with velocity saturation and with explicit drain voltage (dashed line).

5 SUMMARY

Improvement of the compact modeling for the double-gate field-effect transistors was presented. Calculation of the carrier density at the source-end for a wider range of gate voltage was discussed. Reduction of the iteration count was achieved. Transport equations both for unified and separated current methods are compared. Velocity saturation effect was introduced in the transport equation, together with the short channel effect at the drain end. It was found that the analytical and simple form can be obtained. Compact model based on the equation was implemented.

REFERENCES

[1] Y. Liu, K. Ishii, T. Tsutsumi, M. Masahara, and E. Suzuki, "Ideal Rectangular Cross-Section Si-Fin Channel Double-Gate MOSFETs Fabricated Using Orientation-Dependent Wet Etching," IEEE Electron Devices Lett., vol. 24, pp. 484-486, 2003.

[2] T. Sekigawa and Y. Hayashi, "Calculated threshold-voltage characteristics of an XMOS transistor having an additional bottom gate," Solid State Electron., vol. 27, pp. 827-828, 1984.

[3] E. Suzuki, K. Ishii, S. Kanemaru, T. Maeda, T. Tsutsumi, T. Sekigawa, K. Nagai, and H. Hiroshima, "Highly Suppressed Short-Channel Effects in Ultrathin SOI n-MOSFET's," IEEE Trans. ED, vol. 47, 354-359, 2000.

[4] T. Nakagawa, T. Sekigawa, T. Tsutsumi, E. Suzuki, and H. Koike, "Primary Consideration on Compact Modeling of DG MOSFETs with Four-terminal Operation Mode," Tech. Digest Nanotech 2003, vol. 2, pp. 330-333, 2003.

Quantum-Mechanical Analytical Modeling of Threshold Voltage in Long-Channel Double-Gate MOSFET with Symmetric and Asymmetric Gates

J.L. Autran, D. Munteanu, O. Tintori, S. Harrison[*], E. Decarre, T. Skotnicki[*]

Laboratory for Materials and Microelectronics of Provence (L2MP, UMR CNRS 6137)
Bâtiment IRPHE, 49 rue Joliot-Curie F-13384 Marseille France (autran@newsup.univ-mrs.fr)
[*] STMicroelectronics, 850 rue J. Monnet, 38926 Crolles France

ABSTRACT

A quantum-mechanical (QM) full analytical model of the threshold voltage (V_T) for long-channel double-gate (DG) MOSFETs has been developed. This approach is based on analytical solutions for the decoupled Schrödinger and Poisson equations solved in the silicon region. Using this original model, a detailed quantitative comparison between symmetric (SDG) and asymmetric (ASG) architectures has been performed in terms of V_T dependence with film thickness and doping level.

Keywords: analytical modeling, quantum effects, threshold voltage modeling, double-gate MOSFET

1   INTRODUCTION

Double-Gate MOSFET is widely recognized as one of the most promising structures for meeting the roadmap requirements for ultimate deca-nanometer scale [1]. This is mainly due to the superior control of short channel effects, which allows low-doped channels, resulting in enhanced mobility and elimination of doping number fluctuation effects [2]. The operation of DG MOSFETs has been largely analyzed through numerical simulation and analytical models have been proposed for the classical (i.e. without carrier confinement) V_T [2-4]. However, these models only apply in particular cases (e.g. for intrinsic symmetric or asymmetric device). The impact of QM effects has been also largely addressed and quasi-analytical or numerical models have been proposed [5, 6]. However and to the best of our knowledge, no explicit analytical expression for V_T including QM in DG devices has been currently proposed in literature. The aim of this work is precisely to fill this lack and to develop an elementary QM V_T model dedicated to n-channel fully-depleted DG structures. The present approach is based on the decoupled Poisson and Schrödinger equations in the Si film. Firstly, a unified analytical model for the classical V_T, applying to both SDG and ADG with low and high doped films, is briefly described. Secondly, this model is enhanced for taking into account QM and a fully analytical expression of the "quantum" V_T is proposed. This model applies without any restriction to both ADG and SDG devices. We used the model to investigate V_T behavior in such structures and to quantify V_T-shifts with respect to the classical case.

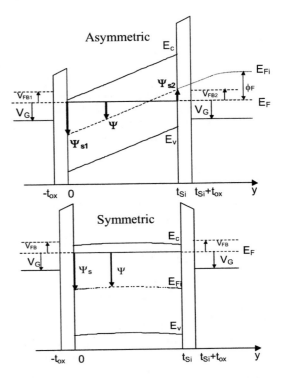

Figure 1: Schematic energy-band diagram for ADG (up) and SDG (down) structures at $V_G = V_T$.

2   ANALYTICAL CLASSICAL V_T MODEL

The first step of our analysis consists in modeling the V_T of a long channel fully-depleted DG MOSFET, with indifferently symmetric or asymmetric gates and doped or undoped channel. Typical band diagrams for SDG and ADG are schematically represented in Fig. 1. A parabolic dependence of the potential with the position in the silicon film at threshold is assumed [7]:

$$\Psi = \Psi_{s1} + \alpha y + \beta y^2 \qquad (1)$$

where Ψ is defined as the band bending with respect to the intrinsic Fermi level in the film (Fig. 1) and the y-direction is perpendicular to the Si/SiO$_2$ interfaces. The potential reference is defined as the electron Fermi level of the source region. Applying the Gauss law, the boundary conditions at the two oxide/silicon interfaces are:

$$V_G - V_{FB1} = \gamma t_{ox} E_1 + \Psi'_{s1} + \phi_F \qquad (2)$$

$$V_G - V_{FB2} = -\gamma t_{ox} E_2 + \Psi_{s2} + \phi_F \qquad (3)$$

where index 1 and 2 refer to the top (y=0) and bottom (y=t_{Si}) interfaces respectively, Ψ_s is the surface potential, V_{FB} is the flat-band voltage, t_{ox} is the oxide thickness, E is the electric field, ϕ_F is the bulk potential and $\gamma = \varepsilon_{Si}/\varepsilon_{ox}$. Potential and charge in the Si film are linked via the Poisson equation:

$$\frac{d^2\Psi}{dy^2} = \frac{qN_A}{\varepsilon_{Si}} + \frac{q}{\varepsilon_{Si}} n_i \exp(\frac{q\Psi}{kT}) \qquad (4)$$

where N_A is the film doping level. In the right side of Eq. (4), the first term represents the depletion charge and the second term is the mobile charge in the silicon film. Integrating Eq. (4) from y=0 to y=t_{Si} gives:

$$E_1 - E_2 = (qN_A t_{Si} + Q_{i,cl})/\varepsilon_{Si} \qquad (5)$$

where $Q_{i,cl}$ is the "classical" inversion charge density in the silicon film assuming a Maxwell-Boltzmann statistic:

$$Q_{i,cl} = \int_0^{t_{Si}} q n_i \exp(q\Psi(y)/kT)dy \qquad (6)$$

As discussed in Ref. [7], the usual definition of V_T as the gate voltage where $\Psi_s = 2\phi_F$ does no more apply to DG MOSFET, where two channels coexist. Therefore, for correctly defining the threshold voltage, we choose here the well-known definition where V_T is the gate voltage for which d^2I_D/dV_G^2 reaches a maximum. Then, the following relation holds at low V_D:

$$\frac{d^3I_D}{dV_G^3} = \frac{d^3Q_{i,cl}}{dV_G^3} = 0 \qquad (7)$$

This relation allows us to obtain the exact expression of the classical inversion charge, $Q^T_{i,cl}$, and the surface potential, $\Psi^T_{s1,cl}$, at threshold. Finally the $V_{T,cl}$ expression is given by:

$$V_{T,cl} = V_{FB1} \frac{\gamma t_{ox} + t_{Si}}{2\gamma t_{ox} + t_{Si}} + V_{FB2} \frac{\gamma t_{ox}}{2\gamma t_{ox} + t_{Si}} \qquad (8)$$
$$+ \Psi^T_{s1,cl} + \gamma t_{ox} t_{Si} \beta(Q^T_{i,cl}) + \phi_F$$

where $Q^T_{i,cl}$, $\Psi^T_{s1,cl}$, and $\beta(Q^T_{i,cl})$ are obtained as in [7].

This classical V_T model was extensively validated by numerical simulation for a wide number of long-channel SDG and ADG structures, with various values of t_{Si} and N_A (Fig. 2). A very good match is obtained between analytical and numerical results for all structures and different t_{Si}, with an error less than 10%, or even 5% for SDG.

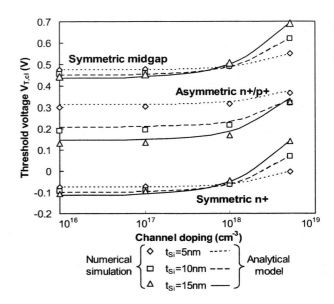

Figure 2: Comparison between V_T obtained with analytical model and numerical simulations in the classical case.

3 ANALYTICAL V_T MODEL - QUANTUM MECHANICAL MODELING

To take into account carrier confinement effects in the silicon film, the expression of the inversion charge must be quantum-mechanically (QM) reevaluated (we note $Q_{i,q}$ this new value). Considering the energy-band profile across the silicon channel, three different cases can be envisaged, as illustrated in Fig. 3 and detailed in the following:

• *Asymmetric structure*: for all N_A and t_{Si}, the conduction band profile is always triangular and the calculation of $Q_{i,q}$ reduces to the evaluation of the charge in a single triangular well using Eqs. (11) and (12).

• *Doped symmetric structure*:
(a) $t_{Si} \rightarrow 0$: the potential $\Psi(y) \approx \Psi_{S1} \approx \Psi_{S2}$ is almost constant in the film and the band diagram is close to the one of an infinite rectangular well for which energy levels and inversion charge can be calculated using Eqs. (9) and (12).
(b) $t_{Si} \rightarrow \infty$: The potential profile at each interface (y=0 and y=t_{Si}) is close to the one of a triangular well. Also in this case, analytical expressions (11) and (12) are used to evaluate $Q_{i,q}$ in each well. The total inversion charge present in the film must be then multiplied by a factor ×2.

• *Low doped symmetric structure*: in this limit case, the potential in the film is almost constant and the band diagram is close to an infinite rectangular well for any t_{Si}.

In a second step, a more carefully evaluation of these energy levels can be performed using a standard method for first-order perturbation [5]. Considering the parabolic nature of the electrostatic potential profile in the silicon film, as described by Eq. (1), the first-order correction to apply to the energy levels in the well is given by:

$$\Delta E^i = \langle \varphi^i | H | \varphi^i \rangle \qquad (16)$$

Asymmetric Structure			![Single triangular well] Single triangular well ($Q_{i,q}$)
Symmetric Structure	High doped and t_{Si} → / Low doped, any t_{Si} → Rectangular well ($Q_{i,q}$)	Rectangular well with parabolic profile at bottom ($Q_{i,q}$)	High doped and t_{Si} → Double triangular well ($2 \times Q_{i,q}$)
Energy levels	$E^i_{l,t} = \dfrac{\hbar^2 \pi^2 i^2}{2 q m^*_{l,t} t^2_{Si}}$ (9)	$E^i_{l,t} = \dfrac{\hbar^2 \pi^2 i^2}{2 q m^*_{l,t} t^2_{Si}} + \Delta E^i$ (10)	$E^i_{l,t} = \dfrac{1}{q}\left(\dfrac{\hbar^2}{2 m^*_{l,t}}\right)^{1/3}\left(\dfrac{3\pi q |\xi_S|}{2}\left(i + \dfrac{3}{4}\right)\right)^{2/3}$ (11)
Total inversion charge	$Q_{i,q} = \dfrac{qkT}{\pi\hbar^2}\sum_{l,t}\sum_i m^{t,l}_{2D} g_{t,l} \ln\left[1 + \exp\left(-\beta_T\left(E^i_{l,t} + \dfrac{E_g}{2} - \psi_S\right)\right)\right]$ (12)		$Q^* = \dfrac{qkT}{\pi\hbar^2}\sum_{l,t}\sum_i m^{t,l}_{2D} g_{t,l} \exp\left(-\beta_T\left(E^i_{l,t} + \dfrac{E_g}{2}\right)\right)$ (13)
$\Psi^T_{sl,q}$ and threshold voltage	$\Psi^T_{sl,q} = \dfrac{kT}{q}\ln\left(\dfrac{Q^T_{i,cl}}{Q^*}\right)$ (14)	$V_{T,q} = V_{FB1}\dfrac{\gamma t_{ox} + t_{Si}}{2\gamma t_{ox} + t_{Si}} + V_{FB2}\dfrac{\gamma t_{ox}}{2\gamma t_{ox} + t_{Si}} + \Psi^T_{sl,q} + \gamma t_{ox} t_{Si}\beta(Q^T_{i,cl}) + \phi_F$ (15)	

Figure 3: Band diagrams for SDG and ADG and expressions used to determine energy levels, $Q^T_{i,q}$ (integrated over the whole silicon film), $\Psi^T_{sl,q}$ and $V_{T,q}$. $g_l=2$, $g_t=4$, $\beta_T=q/kT$, $m_t^*=0.19\times m_0$, $m_l^*=0.98\times m0$, ξ_S is the interface electric field.

where $H = -q(\alpha y + \beta y^2)$ is the Hamiltonian of the perturbation and φ^i are the electron wave functions associated to the energy levels $E^i_{l,t}$. Finally, the first-order corrected energy levels are given by Eq. (10). Due to the analytical character of both φ^i and H, an elementary calculation gives ΔE^i in the case of an infinite rectangular well subjected to the above considered perturbation:

$$\Delta E^i = \frac{\beta t^2_{Si}}{6}\left[1 + \frac{3}{\pi^2 i^2}\right] \quad (17)$$

Once $Q_{i,q}$ has been evaluated, the QM V_T is calculated as the new V_G to be applied for verifying, this time, the QM inversion condition which is similar, at low V_D, to Eq. (7):

$$d^3 Q_{i,q} / dV_G^3 = 0 \quad (18)$$

From Eq. (12) and in the particular regime of $V_G \leq V_T$, $Q_{i,q}$ can be rewritten as $Q_{i,q} \approx Q^* \exp(\beta\Psi_S)$, where Q^* (given by Eq. (13)) is a constant which does not depend on V_G. Indeed, energy levels in both rectangular and triangular well do not depend on V_G. This is evident for Eq. (9) and (10). For Eq. (11), one can remark that, in the approximation of Eq. (1), and for $V_G \leq V_T$, ξ_S is does not depend on the gate voltage: $\xi_S = -(d\Psi/dy)|_{y=0} = -(\alpha + 2\beta y)$.

Thus, the surface potential at threshold ($\Psi^T_{sl,q}$) taking into account QM effects and verifying Eq. (18) can be expressed by Eq. (14), where the inversion charge at threshold is considered the same in both classical and QM cases. Eqs. (8) and (14) lead to the final expression of the "quantum" V_T, given by Eq. (15). Note that this expression is fully analytical and valid for both ADG and SDG architectures.

4 RESULTS AND MODEL VALIDATION

We have validated our model by an extensive comparison with quantum numerical simulation using a full 2-D quantum mechanical numerical simulation code (BALMOS, [8]). Fig. 4 shows that a very good match is obtained between analytical and numerical results in ADG device for all silicon thickness, with an error less than 5%. For SDG devices, V_{Tq} is evaluated in three different cases: infinite rectangular potential well (V_{Tq}^R), rectangular well with parabolic profile of the bottom (V_{Tq}^{RP}) and double triangular well (V_{Tq}^T), as shown in Fig. 5. For highly doped channels, as t_{Si} increases, the potential profile progressively evaluates from a single infinite rectangular well towards a double triangular well and therefore V_{Tq}^R decreases, while V_{Tq}^T increases.

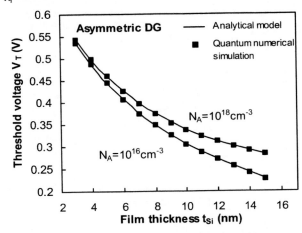

Figure 4: V_T calculated with analytical model and extracted from full quantum numerical simulation in ADG structures.

The film thickness for which the potential well behavior switches from rectangular to triangular is about 8nm for $N_A=1\times10^{24}$ m^{-3}. It is interesting to note that this switch between the two asymptotic behaviors is very sharp in highly doped channels and significantly slower in low doped channels, because in this last case, the potential well becomes triangular only in very thick films. Moreover, in very low doped channels, the triangular behavior is never reached (Fig. 5b). Figure 5 shows also that the quantum threshold voltage calculated with the analytical model using the perturbation method fits very well the numerical datas, for both low and high doped channels.

Figure 6 shows the V_T shift (defined as $\Delta V_T = V_{T,q} - V_{T,cl}$) for ADG and SDG devices with different t_{Si} and N_A. ΔV_T is shown to strongly increase when t_{Si} decreases and/or N_A increases because the potential well is deeper. At very thin films (<5nm), we note that V_T shift becomes almost independent of N_A, since the well profile is completely controlled by the channel thickness. On the contrary, as the film becomes thicker, the potential well is essentially controlled by the channel doping. For this reason, ΔV_T goes

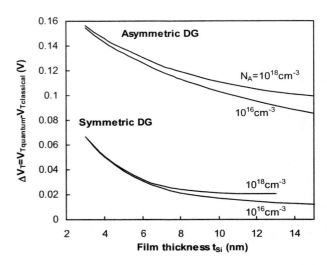

Figure 6: V_T shift due to carrier confinement in ADG and SDG structures as a function of t_{Si} and for different N_A.

to zero for very thick low-doped films whereas ΔV_T remains high in thick films. It is interesting to note that the impact of QM confinement is significantly more important in ADG than in SDG structures, due to a higher electric field at the interface. In SDG, ΔV_T is almost 3 times lower than in ADG and goes very rapidly to zero as the low doped film thickness increases above 5nm. In ADG, ΔV_T remains important even in thick undoped channels. However, both ADG and SDG structures operate at low electric fields compared with conventional bulk MOSFETs, which induces less energy quantization for thick films (>5nm) [3].

5 CONCLUSION

In summary, an original QM model of the threshold voltage applying for long-channel ADG and SDG devices has been developed. This work provides a full analytical and useful expression of V_T with a unified formalism employed in both classical and QM approaches. Our results highlight the fundamental differences that exist between ADG and SDG architectures in terms of QM effects and their impact on the threshold voltage.

REFERENCES

[1] D.J. Frank et al., IEDM Tech. Dig., 553 (1992).
[2] Y. Taur, IEEE Trans. Electron Dev. 48, 2861 (2001).
[3] P. Francis et al., IEEE Trans. Electron Dev. 41, 715 (1994).
[4] K. Suzuki et al., IEEE Trans. Electron Dev. 42, 1940 (1995).
[5] L. Ge, J. Fossum, IEEE Trans. Electron Dev. 49, 287 (2002).
[6] G. Baccarani et al., IEEE Trans. Electron Dev. 46, 1656 (1999).
[7] D. Munteanu et al., Proc. ULIS, 35 (2003).
[8] D. Munteanu, J.L. Autran, Solid State Electron. 47, 1219 (2003).

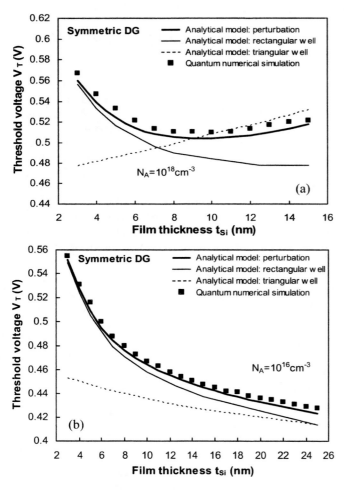

Figure 5: Quantum V_T predicted by the analytical model for SDG in (a) high and (b) low doped channels. Comparison with full 2-D QM numerical simulation.

Automatic BSIM3/4 Model Parameter Extraction with Penalty Function

Y. Mahotin and E. Lyumkis

Integrated System Engineering, Inc.
111 N. Market Street, Suite 710, San Jose, CA, USA, Yuri.Mahotin@ise.com

ABSTRACT

The BSIM3/4 compact models have been widely used in e microelectronic industry over the past decade. As the chnology scale down achieved the 0.09μ limit, the mplexity of the compact models and their parameter traction has increased incredibly. It has been clearly cognized that the automated parameter extraction ethodology can be beneficial for both semiconductor undries and IC design houses. In this work we report the st successful automatic extraction of BSIM3/4 model rameters based on the numerical optimization of a function, hich includes penalty functions. The reported algorithm is an tension of ISE simulator ISExtract [1].

eywords: BSIM3, BSIM4, parameter extraction, penalty nction.

1 INTRODUCTION

Numerical optimization algorithms, and, in particular, the evenberg-Marquardt algorithm have been commonly used in mpact model parameter extraction. This algorithm, invented most 60 years ago [2] and improved 20 years later [3], has a w known deficiencies, such as poor convergence without a od starting point for minimization, and significant merical difficulties to solve large dimensional optimization oblems. The limited numerical capability of optimization ocedures was one of the reasons why commonly used traction procedures consist of a sequence of many extraction eps, where each of them is a local optimization, or a direct traction step. The BSIM3/4 compact models have more than 0 parameters (if to account for binning parameters as well), d describe the complicated physics of micro- and nano-scale vices. Both the complexity of the models, and the deficiency the optimization procedure, lead to extremely complicated traction strategies, and even extraction experts have fficulties sometimes to achieve acceptable results.

To overcome this stalemate, new interest has recently isen around the global optimization methodology for model rameter extraction. Watts et al. [4] have successfully applied e Genetic Algorithm for BSIM3 model parameter extraction. ney concluded, in particular, that this optimization technique fficiently reduced the engineering effort to develop a model rd, and at the same time improved the model quality. Our rrent work also uses a global optimization approach. Our merical algorithm is based on the quasi-newton approach, d demonstrates good convergence with a large number of otimization parameters. The main details of the approach

have been reported in [5], where the method has been applied for BSIM3 model parameter extraction.

Besides numerical issues, the global optimization approach has another well-known problem: as a rule, the solution of the problem is not unique, which may lead to nonphysical values of extracted parameters. In traditional extraction methodologies this problem is often solved by direct extraction (not optimization!) of such parameters from physically appropriate sets of data. In our global optimization approach this problem is solved by the introduction of penalty functions, which keep values of extracted parameters within feasible range. Note, that because of the power of the optimization algorithm, the penalty functions did not really worsen the overall convergence of the optimization procedure.

2 OBJECTIVE FUNCTION

A well chosen objective function has a big impact on the quality of the obtained model parameters as well as on the success of the extraction process itself. The agreement between the measured and simulated data is measured by the objective function. Normally, the root mean square function (RMS) is used as the objective function for model parameter extraction:

$$F_{RMS} = \sum \left(\frac{Id_{meas} - Id_{sim}}{Id_{meas}} \right)^2 \qquad (1)$$

where the sum is over all measured data, Id_{meas} and Id_{sim} are measured and simulated (using appropriate compact model) drain currents, respectively. It is clear, that such optimization problems often might have a few, and sometimes, even an infinite number of solutions. Let us consider a simple example: the threshold voltage parameters dvt0, dvt1, and dvt2, are responsible for the short channel effect and enter into the expression for Vth as dvt0*f(dvt1, dvt2). If, for some reason, during the optimization process, the value of dvt0 becomes equal to zero, then dvt1 and dvt2 may have arbitrary values, and will never affect the value of the objective function. Therefore, the optimization task has an infinite number of solutions in the two dimensional hyperplane, dvt1*dvt2. Due to the complexity of the BSIM3/4 models, and, hence the complexity of the objective function, the N dimensional optimization problem, in general, may have an infinite number of solutions in the K dimensional hyperspace, where K < N. In other words, the F_{RMS} function gives too much freedom to the parameters which have to be optimized.

In order to overcome the above difficulties, and to keep the values of the parameters inside the desired intervals, we use the penalty functions approach. The contribution of the penalty functions to the objective function has been used in the following form:

$$F_{PEN} = \sum \sum \begin{cases} nf \times (x_0 - x)^2 & x < x_0 \\ 0 & x \geq x_0 \end{cases} \qquad (2)$$

where the first sum is over all penalties, the second sum over all data, nf is a normalization factor, x is a model parameter or an internal model variable, and x_0 is the boundary of the feasible interval of x.

The total objective function is defined as the sum of (1) and (2):

$$F = F_{RMS} + F_{PEN} \qquad (3)$$

3   EXAMPLES OF PENALTIES

Based on the physical and mathematical properties of the models, we introduce 49 penalties for the BSIM3 model and 66 penalties for the BSIM4 model. Because of limited space we can not describe each of these penalties, and we will narrow our discussion by considering a few examples.

Both BSIM3 and BSIM4 models impose a hard restriction on the value of the dvt1 parameter: if dvt1 < 0, the models terminate with a fatal error. This problem can be solved easily by imposing a linear constraint dvt1>0. However, it becomes more difficult to impose just such a rigid constraint if binning parameters for dvt1 have to be found as well: then dvt1 is already a function, and some restriction has to be imposed on this function. So, first of all, a smoothing function is used for dvt1 to avoid negative value, and, in addition, the following penalty function is introduced:

$$F_{PEN}(dvt1) = nf_{dvt1}\left(2 \times 10^{-3} - dvt1\right)^2 \qquad (4)$$

if $dvt1 < 2 \times 10^{-3}$. Here nf_{dvt1} is the normalization factor for the dvt1 parameter.

The above example shows, how a penalty function is applied to the model parameters. Similar expressions are used for imposing penalties on internal model variables. Let us consider, for example, an effective mobility model μ_{eff}, which can be written as:

$$\mu_{eff} = \frac{\mu_0}{Denomi} \qquad (5)$$

where μ_0 is the low field mobility, and $Denomi$ is a function of terminal voltages and more then 100 model parameters, including binning parameters, and describes the mobility degradation due to high electric field. If we want to impose a $Denomi > A$ restriction on the value of $Denomi$, then the appropriate penalty function will have the following form:

$$F_{PEN}(Denomi) = nf_{Denomi}\left(A - Denomi\right)^2 \qquad (6)$$

if $Denomi < A$, where nf_{Denomi} is the normalization factor for $Denomi$. The value of A can be chosen from numerical considerations, as it is done in the original BSIM3 model (A=0.2). We believe, that from the physical point of view, A=1 is a more appropriate value for silicon technology, and has been used in our simulations.

4   AUTOSELECTION OF BINNING PARAMETERS

It is well known that without binning the BSIM3/4 models can not provide the required accuracy to model modern technologies for various device sizes. There are no general rules or recommendations on how to choose an optimal set of binning parameters. Often the choice is based on previous experience, and is a very time consuming process as well. To automate the selection process, a procedure called "Autoselection of binning parameters" has been developed. The idea of autoselection is schematically illustrated in Fig.1. The procedure starts from the sub-optimization tasks (one-bin parameter optimization). For each parameter selected for extraction, it solves three two dimensional optimization tasks: 1) parameter plus length dependence binning parameter, 2) parameter plus width dependence binning parameter, and 3) parameter plus cross-term dependence parameter. Therefore three values of the objective function will be calculated for one parameter. After execution of all sub-optimization tasks, the 3N values of objective function are computed, where N is the number of non-binning model parameters. By selecting the minimum component of this vector, we assume, that the appropriate binning parameter is chosen. At the next step the full optimization over all non-binning parameters and the selected binning parameters is performed. The convergence criteria are checked after the full optimization procedure; if they are not satisfied, the procedure is repeated until convergence is reached.

5   PARAMETER EXTRACTION

Typically, the parameter extraction process begins by setting each parameter value to their "best guess" or "desired" value. In order to fit a set of simulated curves to measured data, a series of local optimization steps is performed. Each optimization step attempts to improve the

ting quality of some subset of measured data by adjusting a
small number of model parameters dominated on this subset of
data. This is the ideal case, but in reality, due to the
complicated physics of sub-micron devices, an interaction of
different physical effects is present for any subset of measured
data and therefore, the fitting quality depends not only on the
model parameters selected for local optimization, but on other
model parameters, which have the initial values or values
extracted before. These difficulties lead to the well-known
problem of sequential extraction strategies. In contrast, good
optimization algorithms can optimize many or even all model
parameters, but still it is crucial to choose a good initial guess.
It is well known in practical optimization, that the model
parameters can be divided into two groups: the first group
contains "work horse" parameters, and the second one – the
"ambitious" parameters. The "work horse" parameters are
very critical for the successful optimization of the objective
function and are easier to estimate. The "ambitious
parameters" are less important and/or are more nonlinear and
hence harder to estimate. In such cases, it is very useful first to
set the "ambitious parameters" to their default values in order
to obtain good initial values for the "work horse" parameters,
and then rerun the optimization with all model parameters
using these starting values.

Based on the above arguments, our extraction strategy can
be summarized as following. First, we execute a sequential
extraction strategy for "work horse" parameters. Second, we
optimize the "work horse" parameters over all measured data.
The next step is an optimization of all model parameters,
"work horse" and "ambitious parameters", over all measured
data. And if the fitting quality is not good enough, the
autoselection of binning parameters procedure can be
executed.

6   RESULTS

The new objective function (3) has been used to extract the
BSIM3/4 model parameters for three different technologies.
The BSIM3 model parameters were extracted for High-
Voltage 0.13μ technology, using available measured data for
3 devices with different geometry size. The "model" card
was extracted at the foundry using commercial extractors.
From the same measured data the "new model" card was
obtained using the proposed penalty function approach. A few
problems have been discovered in the "model" card, one of
them is illustrated in Fig. 2: the simulated Gmb becomes
negative when the channel length scales down. Due to a
penalty for negative Gmb, the same curves obtained using the
"new model" card (Fig. 3) do not show this wrong behavior.

The penalty function approach was also successfully
applied to solve similar problems in BSIM4 model parameter
extraction. For a new 0.09μ technology, the measurements of
2 devices were available. Again, the "model" card and "new
model" card were either provided by the foundry, or extracted
using our new approach. A comparison of Gmb for the
"model" and "new model" card is shown in Fig. 4: the
"model" card Gmb becomes negative when Vb is below -

1.5V, which may lead to bad convergence of circuit
simulators. To overcome this problem, the idea from [6]
was applied. In addition to measured data, we append a set
of dummy points with terminal voltages up to two times
higher than the normal operating voltage. For these dummy
points we do not optimize the drain current, and just keep
under control the penalty functions. As a result, in the "new
model" card Gmb is positive within twice the range of the
operating voltages. The effect of the penalty function
approach on internal BSIM4 variables is shown in Fig. 5,
where an effective mobility derivative over bulk voltage is
plotted. By introducing a penalty function which allows
only positive values of this derivative, we guarantee
appropriate (physically correct) results during extraction.

To illustrate the global fitting quality of the proposed
extraction techniques, Table 1 shows the RMS error
calculated using formula (1) and the maximum errors,
obtained for a standard 0.18μ technology with a BSIM4
model for the current range from 10^{-10} A to the maximum
current. We believe that achieving such a good quality is
impossible without using a method of numerical
optimization with penalty functions.

7   CONCLUSION

A newly developed penalty function approach was
applied to BSIM3/4 model parameter extraction. The
penalty functions always keep the values of the model
parameters within their physical range without manual
tuning and human influence on the extraction procedure. It
offers the possibility to get a high quality model card within
a short time.

REFERENCES

[1] ISE TCAD Release 9.0, Volume 4b, Integrated
System Engineering, 2003.
[2] Levenberg, K., "A Method for the Solution of
Certain Problems in Least Squares," Quart. Appl.
Math. Vol. 2, pp 164-168, 1944.
[3] Marquardt, D., "An Algorithm for Least-Squares
Estimation of Nonlinear Parameters," SIAM J.
Appl. Math. Vol. 11, pp 431-441, 1963.
[4] J. Watts et al., "Extraction of Compact Model
Parameters for ULSI MOSFETs Using A Genetic
Algorithm," Tech. Proc. of the Second Int'l Conf.
on Modeling and Simulation of Microsystems,
1999, 176-179.
[5] Y. Mahotin et al., "Parameter extraction of VLSI
MOSFET mathematical models by optimization
technique," All-Russia scientific and technical
conference "Micro- and Nano- electronics
engineering", 1998, V.2, P 3-47
[6] J. Watts, "How to Build an SOI MOSFET Compact
Model without Violating the Laws of Physics,"
Tech. Proc. of the 2002 Int'l Conf. on Modeling
and Simulation of Microsystems, 2002, 726-729.

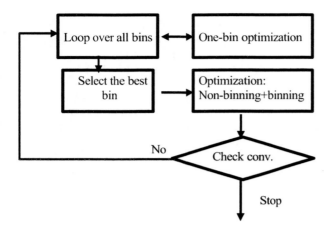

Fig. 1: Autoselection of binning parameters.

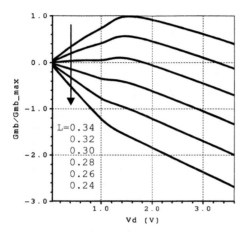

Fig. 2: The "model" card Gmb vs. Vd for different channel length.

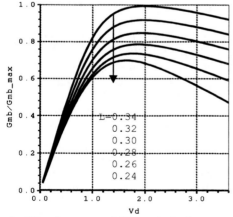

Fig. 3: The "new model" card Gmb vs. Vd for different channel length.

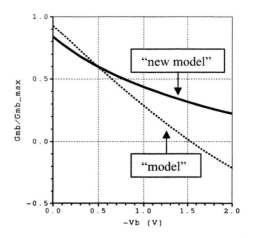

Fig. 4: Gmb vs. Vb at Vg-1.0V, Vd=1.1V

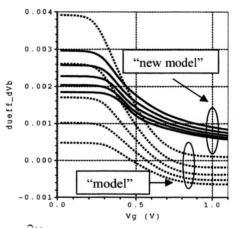

Fig. 5: $\dfrac{\partial \mu_{eff}}{\partial Vb}$ vs. Vg at Vd=0.05V and different Vb

Device	RMS (%)	MAX (%)
Large	1.6257	3.9397
Short1	1.9632	3.4686
Short2	0.9823	2.5600
Short3	2.4188	4.3845
Short4	2.8921	4.4679
Short5	3.2378	10.0014
Short6	1.6462	3.8046
Small	0.6841	1.3305
Narrow1	1.2388	3.5137
Narrow2	2.4042	4.9862
Narrow3	1.1989	3.2683
Narrow4	1.9566	2.8352
Narrow5	1.2715	3.5191

Table 1: RMS and maximum errors of IdVg curves at low Vd and Vb=0 for different devices.

An Analytical Subthreshold Current Model for Ballistic Double-Gate MOSFETs

J.L. Autran[*], D. Munteanu, O. Tintori, M. Aubert, E. Decarre

Laboratory for Materials and Microelectronics of Provence (L2MP, UMR CNRS 6137)
Bâtiment IRPHE, 49 rue Joliot-Curie, BP 146, F-13384 Marseille Cedex 13, France
([*]also with Institut Universitaire de France – autran@newsup.univ-mrs.fr)

ABSTRACT

The subthreshold characteristic of ultra-thin, ultra-short Double-Gate transistors (symmetric structures) working in the ballistic regime has been analytically modeled. This model takes into account short-channel effects, quantization effects and source-to-drain tunneling (WKB approximation) in the expression of the subthreshold drain current. Important device parameters, such as I_{off}-current or subthreshold swing, can be easily evaluated through this full analytical approach which also provides a complete set of equations for developing equivalent-circuit model used in ICs simulation.

Keywords: ballistic transport, Double-Gate devices, analytical modeling, subthreshold current model

1 INTRODUCTION

Double-Gate (DG) MOSFETs are extensively investigated because of their promising performances with respect to the ITRS specifications for decananometer channel lengths. One of the identified challenges remains the development of compact models [1-3] taking into account the main physical phenomena governing the devices at this scale of integration. In this work, an analytical subthreshold model of ultra-thin double-gate MOSFETs working in the ballistic regime is presented. As we show in the following, the present approach captures the essential physics of such ultimate DG devices (quantum confinement, thermionic current) and introduces two main novelties usually neglected in compact modeling: the 2D short-channel effects and the tunneling of carriers through the source-to-drain barrier. Results given by this analytical approach are finally compared with data obtained with a full-2D numerical code coupling the Schrödinger-Poisson system with the ballistic transport equation.

2 MODELING OF THE BALLISTIC SUBTHRESHOLD CURRENT

Figure 1 shows the schematic n-channel DG structure with symmetric gates considered in this work. Carrier transport in the ultra-thin silicon film (thickness t_{Si}, doping level N_A) is clearly 1D in the x-direction and the resulting

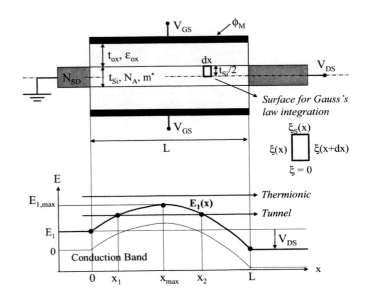

Figure 1: Schematic ultra-thin DG structure and its technological and electrical parameters considered in this work. The first energy subband profile $E_1(x)$ obtained from Eq. (11) is also represented. Turning points (x_1, x_2) and x_{max} have literal values due to the analytical character of $E_1(x)$.

current is controlled by the gate-to-source (V_{GS}) and drain-to-source (V_{DS}) voltages which impact the shape as well as the amplitude of the source-to-drain energy barrier. In the subthreshold regime, minority carriers can be neglected and Poisson's equation is analytically solved in the x-direction with explicit boundary conditions at the two oxide/silicon interfaces taking into account the electrostatic influence of V_{GS}. Considering the particular closed surface shown in Fig. 1 for Gauss's law integration, we can write:

$$-\xi(x)\frac{t_{Si}}{2} + \xi(x+dx)\frac{t_{Si}}{2} - \xi_S(x) = -\frac{qN_A t_{Si} dx}{2\varepsilon_{Si}} \quad (1)$$

where the electric field $\xi(x) \approx -\dfrac{d\psi(x)}{dx}$ (due to the 1D character of the electrostatic potential ψ in the Silicon film) and $\xi_S(x) = \dfrac{\varepsilon_{ox}}{\varepsilon_{Si}}\xi_{ox}(x)$.

Defining the electrostatic potential ψ as the band bending with respect to the intrinsic Fermi level in the silicon film and choosing the Fermi level in the source reservoir as the potential reference, the oxide electric field $\xi_{ox}(x)$ along the structure is then given by:

$$\xi_{ox}(x) = \frac{V_{GS} - V_{FB} - \psi(x) - \phi_F}{t_{ox}} \qquad (2)$$

where V_{FB} is the flat-band voltage and ϕ_F is the bulk potential of the Silicon film.

After some algebraic manipulations, we obtained the following differential equation for the electrostatic potential in the Silicon film:

$$\frac{d^2\psi}{dx^2} - \frac{2C_{ox}}{\varepsilon_{Si}t_{Si}}\psi = \frac{1}{\varepsilon_{Si}t_{Si}}\left[qN_A t_{Si} - 2C_{ox}(V_{GS} - V_{FB} - \phi_F)\right] \qquad (3)$$

where $C_{ox} = \varepsilon_{ox}/t_{ox}$ is the gate capacitance per unit area.

The analytical solution of Eq. (3) can be then expressed under the form:

$$\psi(x) = C_1 \exp(m_1 x) + C_2 \exp(-m_1 x) - \frac{R}{m_1^2} \qquad (4)$$

where the coefficients C_1 and C_2 are given by:

$$C_1 = \frac{\phi_S[1 - \exp(-m_1 L)] + V_{DS} + \dfrac{R[1 - \exp(-m_1 L)]}{m_1^2}}{2\sinh(m_1 L)} \qquad (5)$$

$$C_2 = -\frac{\phi_S[1 - \exp(+m_1 L)] + V_{DS} + \dfrac{R[1 - \exp(+m_1 L)]}{m_1^2}}{2\sinh(m_1 L)} \qquad (6)$$

with:

$$R = \frac{1}{\varepsilon_{Si}t_{Si}}\left[qN_A t_{Si} - 2C_{ox}(V_{GS} - V_{FB} - \phi_F)\right] \qquad (7)$$

$$m_1 = \sqrt{\frac{2C_{ox}}{\varepsilon_{Si}t_{Si}}} \qquad (8)$$

$$\phi_S = \frac{kT}{q}\ln\left(\frac{N_A N_{SD}}{n_i^2}\right) \qquad (9)$$

$$\phi_F = \frac{kT}{q}\ln\left(\frac{N_A}{n_i}\right) \qquad (10)$$

Considering the limit case of an ultra-thin Silicon film, the vertical confinement of carriers in the structure can be treated using the well-known approximation of an infinitely deep square well. The first energy subband profile $E_1(x)$ can be easily derived as:

$$E_1(x) = q(\phi_S - \psi(x)) + \frac{\hbar^2\pi^2}{2m_\ell^* t_{Si}^2} \qquad (11)$$

where $m_\ell^* \approx 0.98 \times m_0$ is the electron longitudinal mass.

Once $E_1(x)$ is known as a function of V_{GS} and V_{DS}, the ballistic current (per device width unit) is evaluated from the following equation for a two-dimensional gas of electrons [4]:

$$I_{DS} = \frac{2q}{\pi^2\hbar}\int_{-\infty}^{+\infty}dk_z \times \int_0^{+\infty}\left[f(E, E_{FS}) - f(E, E_{FD})\right]T(E_x)dE_x \qquad (12)$$

where $f(E_F, E_{FS})$ is the Fermi–Dirac distribution function, E_{FS} and E_{FD} are the Fermi-level in the source and drain reservoirs respectively, k_z is the electron wave vector component in the z direction, the factor 2 accounts for the two Silicon valleys characterized by m_ℓ^* in the y-direction (vertical confinement), $T(E_x)$ is the barrier transparency for electrons and E is the total energy of carriers in source and drain reservoirs given by:

$$E = E_1 + E_x + \frac{\hbar^2 k_z^2}{2m_t^*} \qquad (13)$$

where E_1 is the energy level of the first subband (given by Eq.(11)), E_x is the carrier energy in the direction of the current, m_t^* is the electron transverse effective mass.

Eq. (12) can be evaluated using a simple rectangular double integration method (transforming the integrals into discrete sums) while $T(E_x)$ can be calculated using the WKB approximation:

$$T_{WKB}(E_x \leq E_{1,max}) = \exp\left(-2\int_{x_1}^{x_2}\sqrt{\frac{2m_t^*(E_1(x) - E_x)}{\hbar^2}}dx\right)$$

$$T_{WKB}(E_x > E_{1,max}) = 1 \qquad (14)$$

where the turning point coordinates x_1 and x_2 have literal expressions due to the analytical character of the barrier:

$$x_{1,2}(E_x) = \frac{1}{m_1} \ln \left[\frac{A \pm \sqrt{\Delta}}{2C_1} \right] \qquad (15)$$

where the quantities A and Δ are defined as follows:

$$A = \phi_S + \frac{R}{m_1^2} + \frac{\hbar^2 \pi^2}{2qm_\ell^* t_{Si}^2} - \frac{E_x}{q} \qquad (16)$$

$$\Delta = A^2 - 4C_1C_2 \qquad (17)$$

We can alternatively consider the transfer matrix (TM) approach [5] to evaluate the barrier transparency (as we will see in Fig. 5). This method, applied to the analytical profile $E_1(x)$, requires a numerical subroutine to evaluate the transparency: the proposed model thus becomes numerical. Finally, if one remarks that the top (i.e the maximum) of the source-to-drain barrier is now analytically defined using Eqs. (11)-(15) ($E_x = E_{1,max}$), the thermionic component of the ballistic current can be alternatively derived from Eq. (12) using the well-known Natori's formula [1].

3 RESULTS AND MODEL VALIDATION

Figure 2 shows the first energy subband profile $E_1(x)$ in the Silicon film calculated with the analytical model for different gate voltages in a L=10nm device. In order to test the validity of the model, we compare these profiles with those obtained with a full quantum-mechanical 2D simulation code (BALMOS [6]). A good agreement is obtained between the two series of subband profiles, especially for low V_{GS} values, i.e. in the subthreshold regime. In particular, we note an excellent agreement for the positions of the maximum as well as the amplitude of the barrier between the analytical and numerical curves. The slight difference in the barrier width is due to the electric field penetration in the source and drain regions, only taken into account in the numerical approach. Concerning this point, the analytical model should perfectly reproduce the barrier profile for devices with metallic source and drain, as in Schottky-barrier transistors recently proposed [7]. When approaching the threshold voltage (e.g. $V_{GS} = 0.1$ V), the numerical and analytical profiles diverge, due to the presence of minority carriers in the channel, which is not taken into account by the analytical model. This limitation sets the validity domain of this later approach, as illustrated in Figures 3 and 4 (hachured area).

The subthreshold $I_{DS}(V_{GS})$ characteristics calculated with the analytical model for devices ranging from L=5 to 20nm are shown in Figure 3. In the subthreshold regime (typically $V_{GS} < 0.1V$), the curves very well fit numerical data obtained with BALMOS for devices in the deca-nanometer range. Below L=10nm, the effect of electric field penetration in source and drain regions on the barrier width becomes significant, leading to a slight overestimation of the drain current in the analytical case.

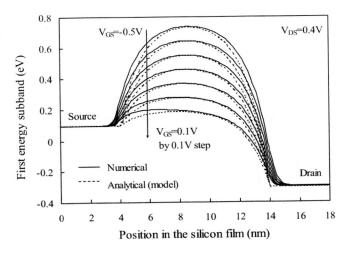

Figure 2: First energy subband profiles $E_1(x)$ in the silicon film calculated with the analytical model (Eq. (11)) and with the 2D numerical code BALMOS [6]. Devices parameters are: L = 10nm, t_{Si} = 2nm, t_{ox} = 0.8nm, ϕ_M = 4.3V, N_{SD} = 3×10²⁰ cm⁻³, N_A = 10¹⁵ cm⁻³.

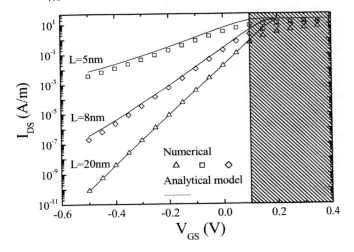

Figure 3: $I_{DS}(V_{GS})$ characteristics calculated with the analytical model. Values obtained with the 2D numerical code BALMOS [6] are also reported for comparison. The hachured area represents the near and above threshold regions where the analytical model losses its validity. Device parameters are the same as in Fig. 2.

Figure 4 shows subthreshold $I_{DS}(V_{GS})$ characteristics calculated with the analytical model. Two series of curves have been plotted, considering or not the WKB tunneling component of the ballistic current. These results highlight the dramatic impact of the source-to-drain tunneling on the subthreshold slope and also on the I_{off} current. In this subthreshold regime, the carrier transmission by thermionic emission is reduced or even suppressed due to the high channel barrier; consequently, when the channel length decreases the tunneling becomes dominant and constitutes the main physical phenomenon limiting the devices scaling, typically below channel lengths of ~10nm.

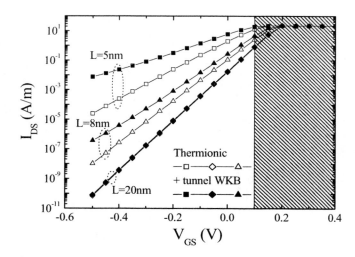

Figure 4: Subthreshold $I_{DS}(V_{GS})$ characteristics calculated with the analytical model when considering or not the WKB tunneling component in the ballistic current. Device parameters are the same as in Fig. 2.

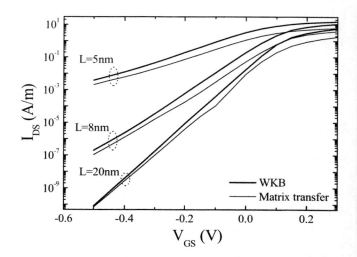

Figure 5: $I_{DS}(V_{GS})$ characteristics calculated with the compact model using two different evaluations of the source-to-drain barrier transparency. Device parameters are the same as in Fig. 2.

Finally, we report here how the present approach can be further transformed from an analytical to a compact model and then extended to all the device operation regimes. In this later approach, Eq. (3) is numerically solved using a 1D finite difference scheme with a uniform mesh Δx. The distribution of electrons in the channel n(x) can be also added in the second hand of Eq. (3) (assuming a realistic electron quasi-Fermi level profile in the channel) which can be rewritten under the form:

$$\frac{\psi_{i+1} + \psi_{i-1} - 2\psi_i}{2\Delta x^2} - \frac{2C_{ox}}{\varepsilon_{Si} t_{Si}} \psi_i =$$
$$\frac{1}{\varepsilon_{Si} t_{Si}} \left[q N_A t_{Si} - 2C_{ox} (V_{GS} - V_{FB} - \phi_F) + q t_{Si} n(x) \right] \qquad (18)$$

A similar equation can be considered to extend the simulation domain to source and drain regions, allowing us to take into account the electric field penetration in the reservoirs. After solving Eq.(18), the resulting potential profile $\psi_i(x_i)$ can be used to perform a more accurate calculation of the barrier transparency (TM method [5]) in order to correct T_{WKB} from the effect of carrier reflections on the barrier. On a computational point-of-view, the calculation of the drain current per bias point with this approach takes less than one second on a desktop computer. Figure 5 compares $I_{DS}(V_{GS})$ characteristics obtained with the compact model and with the two transparency calculation methods. As expected, I_{DS} evaluated using the TM method is lower than the WKB current, due to carrier reflections on the barrier. Note that this phenomenon not only impacts I_{DS} in the subthreshold regime but also above threshold, due to the presence of a residual barrier in the channel near the source. The difference between the two currents becomes however negligible in the subthreshold region for "long" channels (L=20nm).

4 CONCLUSION

An analytical model for the subthreshold drain current in ultra-thin symmetric double-gate MOSFETs working in the ballistic regime is presented. The model is particularly well-adapted for ultra-short DG transistors in the decananometer scale since it accounts for the main physical phenomena related to these ultimate devices: 2D short channel effects, quantum vertical confinement as well as carrier transmission by both thermionic emission and quantum tunneling through the source-to-drain barrier. The model is used to predict essential subthreshold parameters such as off-state drain current and subthreshold slope and can successfully be included in circuit models for the simulation of double-gate MOSFET-based ICs.

ACKNOWLEDGMENTS

The authors would like to thank T. Skotnicki and S. Harrison (ST-Crolles) for fruitful discussions.

REFERENCES

[1] K. Natori, J. Appl. Phys. **76** , 4879 (1994).
[2] A. Rahman, M.S. Lundstrom, IEEE Trans. Electron Dev. **49**, 481 (2002).
[3] Y. Naveh, K.K. Likharev, IEEE Electron Dev. Lett. **21**, 242 (2000).
[4] D.K. Ferry, S.M. Goodnick, Transport in nanostructures. Cambridge University Press; 1997.
[5] C. Cohen-Tannoudji, B. Diu, F. Laloë, Quantum mechanic, Hermann, Paris (1992).
[6] D. Munteanu, J.L. Autran, Solid-State Electron. **47**, 1219 (2003).
[7] J. Guo, M.S. Lundstrom, IEEE Trans. Electron Dev. **49**, 1897 (2002).

Threshold-Voltage-Based Regional Modeling of MOSFETs with Symmetry and Continuity

Siau Ben Chiah[*], Xing Zhou[*], Karthik Chandrasekaran[*],
Khee Yong Lim[**], Lap Chan[**], and Sanford Chu[**]

[*]School of Electrical & Electronic Engineering, Nanyang Technological University
Nanyang Avenue, Singapore 639798, exzhou@ntu.edu.sg
[**]Chartered Semiconductor Manufacturing Ltd, 60 Woodlands Industrial Park D, St. 2, Singapore 738406

ABSTRACT

This paper presents a unified threshold-voltage-based (V_t-based) MOSFET model, which maintains source–drain symmetry and allows accurate prediction of transconductance (g_m) and drain conductance (g_{ds}) and their derivatives (g_m' and g_{ds}') with smooth transitions across regions of operation. This has been achieved based on our previous unified source-extrapolated V_t-based model but re-derived with bulk reference for the drain current (I_{ds}). The unified model combines V_t-based model in strong inversion with surface-potential-based (Ψ_s-based) model in subthreshold with smooth transitions (in function as well as higher-order derivatives) across linear/saturation and weak/strong-inversion regions. It has been verified with the experimental data from a 0.18-μm CMOS shallow trench isolation (STI) technology wafer.

Keywords: Deep-submicron MOSFET, symmetry, threshold voltage, drain current, surface potential.

1 INTRODUCTION

Starting from the Meyer's model [1], V_t-based models have been the standard for decades in MOSFET compact models (CMs) for circuit simulation. However, in recent CM developments, more attention is focused on mixed-signal and low-power applications in which requirements for smooth transitions across different regions of operation become increasingly important. This becomes the challenge for V_t-based models to continue to survive as a standard for MOSFETs. Furthermore, a major problem associated with V_t-based models is symmetry, as seen in the BSIM model [2], which has been attributed to being source referenced. However, as pointed out in [3], source-reference and bulk-reference are essentially equivalent. A V_t-based model, in which V_t is the source-extrapolated threshold voltage, does not necessarily mean asymmetry as long as the drain-current equation is derived with bulk as the reference; likewise, a bulk-referenced model does not necessarily guarantee symmetry if velocity-saturation and lateral-field mobility are not handled properly.

In this paper, we focus on further development of the unified V_t-based model based on our previous source-referenced model, Xsim [4], but re-derive with bulk reference for the drain current (I_{ds}). In Section 2, formulation of the bulk-referenced I_{ds} and V_t models as well as parameter extraction is presented. Benchmark tests are carried out to verify the model continuity and symmetry, which are discussed in Section 3, together with experimental verification of the model.

2 MODEL FORMULATION AND PARAMETER EXTRACTION

The drain current equation based on charge-sheet approximation can be formulated as the drift current
$$I_{drift}(y) = \mu_{eff}W_{eff}Q_i(y)d\Psi_s(y)/dy \quad (1a)$$
and the diffusion current
$$I_{diff}(y) = \mu_{eff}W_{eff}v_{th}\,dQ_i(y)/dy \quad (1b)$$
where $\Psi_s(y)$ is the position-dependent surface potential,
$$Q_i(y) = C_{ox}[V_{gb} - V_{FB} - \Psi_s(y)] - Q_b(y) \quad (2a)$$
is the inversion charge along the channel, and
$$Q_b(y) = C_{ox}\gamma\sqrt{\Psi_s(y)} \quad (2b)$$
is the bulk charge. Rearranging (1a) and integrating across the channel from source ($y = 0$) to drain ($y = L_{eff}$), with the boundary conditions $\Psi_{s0} = 2\phi_F + V_{sb}$ and $\Psi_{sL} = 2\phi_F + V_{db}$:
$$\int_0^{L_{eff}} I_d(y)dy = \mu_{eff}W_{eff}\int_{\Psi_{s0}}^{\Psi_{sL}} Q_i d\Psi_s,$$
it can be shown [3] that the drift current is given by
$$I_{drift} = \beta_n\left[(V_{gs} - V_t)V_{ds} - \tfrac{1}{2}A_bV_{ds}^2\right], \quad (3)$$
in which
$$V_t = V_{FB} + \phi_{s0} + \gamma\sqrt{\phi_{s0}} - (A_b - 1)V_{bs} \quad (4)$$
$$A_b = 1 + \gamma/(2\sqrt{\phi_{s0}}) \quad (5)$$
for the bulk-referenced model, which are different in form from the source-referenced model. In the above equations, V_{FB} is the flat-band voltage, $v_{th} = kT/q$ is the thermal voltage, $\phi_{s0} = 2\phi_F$ is the ("pinned") surface potential at strong inversion, $\gamma = (2q\varepsilon_{Si}N_{ch})^{1/2}/C_{ox}$ is the body factor, and $\beta_n = \mu_{eff}C_{ox}(W_{eff}/L_{eff})$ is the gain factor.

Similarly, for the diffusion current, by rearranging (1b) and integrating from source to drain, and taking the source and drain end inversion charge as

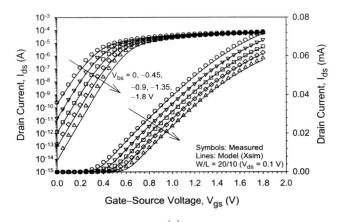

(a)

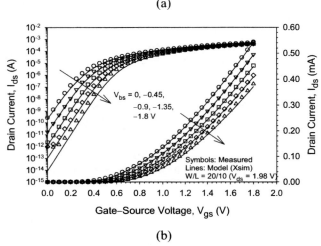

(b)

Figure 1: $I_{ds} - V_{gs}$ characteristics ($V_{bs} = 0/-1.8$ V at $V_{ds} = 0.1$ V fitted; others predicted) in (a) linear and (b) saturation regions for the $L = 10$-μm device.

$$Q_{i,s} \approx \frac{\gamma C_{ox}}{2\sqrt{\Psi_s}} v_{th} e^{(\Psi_s - 2\phi_F - V_{sb})/v_{th}} \qquad (6a)$$

$$Q_{i,d} \approx \frac{\gamma C_{ox}}{2\sqrt{\Psi_s}} v_{th} e^{(\Psi_s - 2\phi_F - V_{db})/v_{th}} \qquad (6b)$$

respectively, the compact diffusion current can be derived:

$$I_{diff} = \beta_n v_{th}^2 \left(C_d / C_{ox}\right) e^{(V_{gs} - V_t)/(nv_{th})} \left(1 - e^{-V_{ds}/v_{th}}\right) \qquad (7)$$

where

$$C_d = \gamma C_{ox} / 2\sqrt{\phi_{sub}} \qquad (8)$$

$$n = 1 + C_d / C_{ox} \qquad (9)$$

are the depletion capacitance and subthreshold slope, respectively. The bulk-referenced gate-bias-dependent subthreshold surface potential is given by

$$\phi_{sub} = \left[-\frac{\gamma}{2} + \sqrt{\frac{\gamma^2}{4} + V_{gb} - V_{FB}} \right]^2 \qquad (10)$$

Eqs. (3) and (7) are regarded as the piece-wise regional models, in which subthreshold is Ψ_s-based whereas strong inversion is V_t-based.

Since the bulk-referenced drain current (3) has the same form as the source-referenced model, our previously-developed analytical "drift + diffusion" model should still apply [4]:

$$I_{ds} = \beta_n V_{ge} = \beta_n \left(V_{gg} + V_{gd}\right), \qquad (11)$$

that can be decomposed into the sum of the "drift" ($I_{drift} = \beta_n V_{gg}$) and "diffusion" ($I_{diff} = \beta_n V_{gd}$) currents, which approach the correct asymptotes in the strong-inversion and subthreshold regions, respectively. The key "smoothing" function is the "effective gate/drain voltage product" [5]:

$$V_{gg} = \frac{2nv_{th} \ln\left(1 + e^{(V_{gs} - V_t)/(2nv_{th})}\right) V_{de}}{1 + V_{de} \dfrac{2n(C_{ox}/C_d) e^{-(V_{gs} - V_t - 2V_{off})/(2nv_{th})}}{v_{th}\left(1 - e^{-V_{ds}/v_{th}}\right)}} \qquad (12)$$

which includes the effect of bulk charge

$$V_{de} = \left(1 - \frac{A_b}{2} \frac{V_{deff}}{V_{geff}}\right) V_{deff} . \qquad (13)$$

V_{gd} is given by V_{gg}/W_{ge} to model the correct diffusion current in subthreshold without affecting drift current, where W_{ge} is derived to be [6]

$$W_{ge} = \frac{n}{A_b} \frac{e^{(V_{gs} - V_t)/(2nv_{th})}}{1 - e^{-V_{ds}/v_{th}}} . \qquad (14)$$

In the above equations, V_{geff} is the effective gate overdrive (BSIM interpolation function [2]), given by

$$V_{geff} = \frac{2nv_{th} \ln\left(1 + e^{(V_{gs} - V_t)/(2nv_{th})}\right)}{1 + 2n(C_{ox}/C_d) e^{-(V_{gs} - V_t - 2V_{off})/(2nv_{th})}} \qquad (15)$$

which approaches $V_{gs} - V_t$ in strong inversion. V_{deff} is the BSIM smoothing function [2]

$$V_{deff} = V_{dsat} - \frac{1}{2}\left[V_{dsat} - V_{ds} - \delta_s \right. $$
$$\left. + \sqrt{(V_{dsat} - V_{ds} - \delta_s)^2 + 4\delta_s V_{dsat}}\right] \qquad (16)$$

which approaches V_{ds} in linear region and

$$V_{dsat} = \frac{E_{sat} L_{eff} \left(V_{gs} - V_t\right)}{V_{gs} - V_t + A_b E_{sat} L_{eff}} \qquad (17)$$

in saturation region, where $E_{sat} = 2v_{sat}/\mu_0$ is the saturation field, and δ_s is a fitting (smoothing) parameter.

To extract the model parameters, a 0.18-μm CMOS STI wafer is measured. Long-channel measured threshold voltages for different body biases have been extracted from the $I_{ds} - V_{gs}$ curves at low drain bias based on the constant-current definition. The effective channel doping and the flat-band voltage (N_{ch}, V_{FB}) are calibrated to the $V_t - V_{bs}$ data, and the semi-empirical mobility model [7] is then calibrated to the $I_{ds} - V_{gs}$ data at low drain and body biases. The smoothing parameter δ_s in (16), together with the channel-length modulation parameter (negligible for long channel), can be tuned to the $I_{ds} - V_{ds}$ data at high gate and low body biases [4].

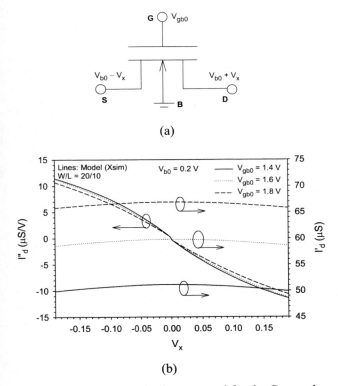

(a)

(b)

Figure 2: (a) The circuit diagram used for the Gummel symmetry test and (b) the first- and second-order derivatives of I_d with respect to V_x, exhibiting smooth transition at $V_x = 0$.

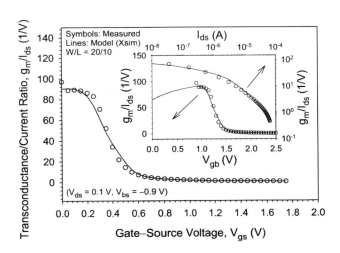

Figure 3: Predicted transconductance/current ratio at the terminal bias conditions indicated. The inset shows the same g_m/I_{ds} data versus V_{gb} (bottom–left) showing Gummel tree-top test; and versus I_{ds} (top–right) showing model smoothness.

3 RESULTS AND DISCUSSION

The unified I_{ds} model has been calibrated to a 0.18-μm CMOS technology wafer, and the results are shown in Figs. 1–4 for the long-channel (10-μm) device (short-channel results are available but not shown in this paper). Figs. 1(a) and 1(b) show the fitted (and predicted for intermediate V_{bs}) $I_{ds} - V_{gs}$ characteristics in linear and saturation regions, respectively. Fig. 2 shows the circuit diagram [8] used for the Gummel symmetry test [9], and the first and second derivatives of drain current with respect to V_x. The smooth I_d'' at $V_x = 0$ confirms model symmetry. The negligible "glitch" in I_d'' at $V_x = 0$ may come from the non-unity slope of V_{deff} as V_{ds} approaches zero. Although this is an intrinsic problem in symmetry, it is negligible for practical values of δ_s. Fig. 3 shows the prediction of transconductance-to-current ratio versus V_{gs} for $V_{bs} = -0.9$ V and $V_{ds} = 0.1$ V. The figure shows no discontinuity or kink effect between the weak and strong-inversion regions, which demonstrates model smoothness. The inset of Fig. 3 shows the same data versus V_{gb} (left–bottom axes) and versus I_{ds} (right–top axes), respectively. The V_{gb}-dependent g_m/I_{ds} in the subthreshold region is a result of the surface-potential-based modeling (ϕ_{sub}) in this region, which is consistent with the Gummel tree-top test [9].

Model prediction on higher-order derivatives of the drain current is shown in Fig. 4. Figs. 4(a) and 4(b) show the accurate prediction of the measured transconductance and output conductance versus gate and drain biases, respectively. Smooth transitions are observed between different regions [from weak to strong inversion in 4(a) and from linear to saturation region in 4(b)], even the second-order derivatives (g_m' and g_{ds}') are accurately predicted, as shown in the insets. The largest error occurs in g_m near $V_{gs} \approx V_t$, which is attributed to the interpolation function (V_{ge}) as well as inaccuracies in V_t. The intrinsic disadvantage of the regional model (as compared to the Ψ_s-based model) will be traded off with other advantages such as scalability and simplicity. Model continuity across regions of operation is an important criterion for analog circuit design, and our unified regional model has demonstrated model smoothness with reasonable accuracy as well as symmetry.

4 CONCLUSION

In conclusion, our previous source-referenced V_t-based (long-channel) drain current model is revised with bulk reference to preserve model symmetry, which uses only 9 parameters to characterize the entire range of operation and is capable of accurately predicting drain current and its higher-order derivatives with smooth transitions across different regions. Contrary to the general belief, our regional V_t-based model has demonstrated symmetry and continuity. This development maintains simple MOSFET equations that are familiar to circuit designers while removing major problems associated with the regional

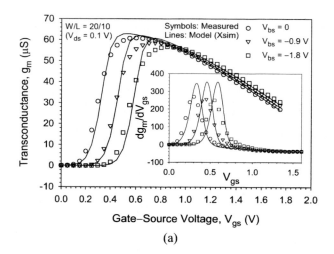

(a)

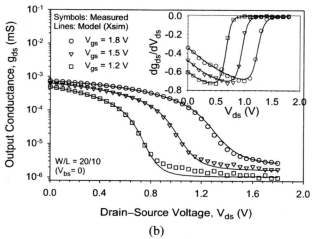

(b)

Figure 4: Prediction on the higher-order derivatives of drain current: (a) transconductance (and g_m' in the inset) and (b) output conductance (and g_{ds}' in the inset) at various terminal bias conditions indicated.

models. It serves as the starting point for the re-construction of our previously-developed unified scalable drain current model, Xsim [4]–[7], including short-channel effects, consistent charge-based intrinsic capacitances, and poly-depletion effect, to be presented elsewhere [10].

REFERENCES

[1] J. E. Meyer, "MOS models and circuit simulation," *RCA Rev.*, vol. 32, pp. 42–63, 1971.

[2] Y. Cheng, *et al.*, "BSIM3v3 Manual," Univ. of California, Berkeley, CA, 1997–1998.

[3] Y. Tsividis, "Operation and Modeling of the MOS Transistor," McGraw-Hill, 2nd ed., 1999.

[4] X. Zhou, S. B. Chiah, and K. Y. Lim, "A technology-based compact model for predictive deep-submicron MOSFET modeling and characterization," *Proc. Nanotech 2003*, Feb. 2003, vol. 2, pp. 266–269.

[5] X. Zhou and K. Y. Lim, "Unified MOSFET compact I–V model formulation through physics-based effective transformation," *IEEE Trans. Electron Devices*, vol. 48, no. 5, pp. 887–896, 2001.

[6] K. Y. Lim and X. Zhou, "MOSFET subthreshold compact modeling with effective gate overdrive," *IEEE Trans. Electron Devices*, vol. 49, no. 1, pp. 196–199, 2002.

[7] K. Y. Lim and X. Zhou, "A physically-based semi-empirical effective mobility model for MOSFET compact I–V modeling," *Solid-State Electron.*, vol. 45, no. 1, pp. 193–197, 2001.

[8] K. Joardar, K. K. Gullapalli, C. C. McAndrew, M. E. Burnham, and A. Wild, "An improved MOSFET model for circuit simulation," *IEEE Trans. Electron Devices*, vol. 45, no. 1, pp. 134–148, 1998.

[9] "Benchmarks for compact MOSFET models," Electronic Industries Alliance, http://www.eig.org/eig/cmc

[10] X. Zhou, S. B. Chiah, K. Chandrasekaran, K. Y. Lim, L. Chan, and S. Chu, "Unified regional approach to consistent and symmetric DC/AC modeling of deep-submicron MOSFETs," *Proc. Nanotech 2004*, Mar. 2004.

Physics-Based Scalable Threshold-Voltage Model for Strained-Silicon MOSFETs

Karthik Chandrasekaran, Xing Zhou, and Siau Ben Chiah

School of Electrical & Electronic Engineering, Nanyang Technological University
Nanyang Avenue, Singapore 639798, exzhou@ntu.edu.sg

ABSTRACT

In this paper, an analytical threshold-voltage (V_t) model derived from Poisson equation for NMOS devices with a strained-silicon channel is described in terms of band, material, doping, and structure parameters and validated with Medici simulations. The model equations are derived based on bulk reference to preserve the symmetry of the model, and extended to short-channel devices based on previously-developed bulk-Si V_t model.

Keywords: Threshold voltage, strained silicon, MOSFET, Poisson equation, bulk reference.

1 INTRODUCTION

Silicon-based MOSFETs have reached remarkable levels of performance through device scaling. However, with each technology generation, it is becoming increasingly harder to improve device performance through traditional scaling methods [1]. Consequently, innovative device structures and materials are actively being investigated to boost performance. Strained-silicon devices have been receiving considerable attention owing to their potential for achieving higher performance and compatibility with conventional silicon processing [2].

This recent strong interest calls for the extension of conventional MOSFET modeling to this promising device. Fig. 1 shows the cross-section of a strained-silicon NMOS device under study. A strained-silicon channel device has been simulated using TMA Medici. TCAD simulations can be used to study the physical insight into novel strained-silicon devices with sufficient tuning of material and mobility parameters. With change in Ge mole fraction, different material parameters such as energy bandgap, conduction- and valence-band offsets, and intrinsic concentration change accordingly [3].

In this paper, a threshold-voltage model for strained-silicon MOSFETs is developed and presented. The strained-silicon MOSFET exhibits a shifted threshold voltage from conventional silicon devices due to the band offset at the heterojunction between the strained silicon and the SiGe layers.

2 MODEL DERIVATION

The cross-section of the strained-silicon MOSFET is shown in Fig. 1. It consists of a strained-silicon channel on top of a relaxed silicon germanium (SiGe) substrate. In

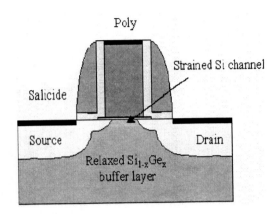

Figure 1: Surface-channel strained-silicon MOSFET.

strained-silicon MOSFETs, the silicon channel is very thin and usually small compared to the depletion width. For very small drain voltages and long-channel MOSFETs, an analysis can be carried out in one dimension, perpendicular to the surface. Poisson equation is solved for one-dimensional case, which is described below [4]

Depletion layer potential in the SiGe substrate can be written as

$$\phi_1(x) = \frac{q N_a x^2}{2 \varepsilon_{SiGe}} \quad (0 < x < x_d). \tag{1}$$

The potential across the strained-silicon layer can be written as

$$\phi_2(x) = \frac{q N_a x_d^2}{2 \varepsilon_{SiGe}} + \frac{q N_a t_{Si}^2}{2 \varepsilon_{SiGe}} + \frac{q N_a t_{Si} x_d}{\varepsilon_{Si}} \quad (0 < x < t_{Si}). \tag{2}$$

The potential across the oxide can be written as

$$\phi_3(x) = \frac{q N_a x_d^2}{2 \varepsilon_{SiGe}} + \frac{q N_a t_{Si}^2}{2 \varepsilon_{SiGe}} + \frac{q N_a t_{Si} x_d}{\varepsilon_{Si}} + \frac{q N_a t_{ox}(t_{Si} + x_d)}{\varepsilon_{ox}}$$
$$(0 < x < t_{ox}). \tag{3}$$

Since heavy inversion occurs at the Si/SiO₂ interface, the threshold potential can be solved from

$$\phi_{th} = \phi_2(t_{Si}) \tag{4}$$

where ϕ_{th} is given by the expression

$$\phi_{th} = \frac{2kT}{q} \ln\left(\frac{N_a}{n_i}\right) - \Delta E_c. \tag{5}$$

The intrinsic concentration changes with the effective density of states and the energy bandgap. It is given by the expression

$$n_i = \sqrt{N_C N_V} \exp\left(-E_g / 2kT\right). \tag{6}$$

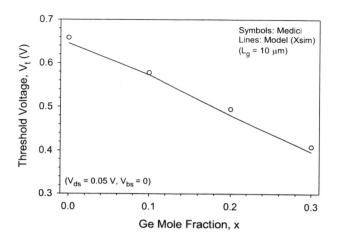

Figure 2: Prediction of the linear threshold voltage for different Ge mole fractions.

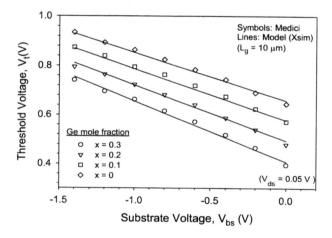

Figure 3: Prediction of the linear threshold voltage vs. substrate bias for different Ge mole fractions.

The width of the depletion region in the SiGe substrate can be obtained as

$$x_d = \sqrt{t_{Si}^2 - \frac{\varepsilon_{SiGe}}{\varepsilon_{Si}} t_{Si}^2 + \frac{2\varepsilon_{SiGe}}{qN_a}(\phi_{s0} - \Delta E_c)} - t_{Si} . \quad (7)$$

where $\phi_{s0} = 2\phi_F$ is the surface potential at strong inversion. In the above equations, N_a is the uniform substrate doping concentration, t_{si} is the thickness of the strained-silicon layer, x_d is the depth of the depletion layer from the strained Si/SiGe interface, ΔE_c is the conduction band offset, N_C and N_V are the effective density of states in the conduction and valence bands, respectively, n_i is the intrinsic concentration, E_g is the energy bandgap, ε_{SiGe} and ε_{Si} are the permittivities of strained silicon and relaxed SiGe, respectively.

The threshold voltage is obtained by substituting the boundary condition

$$V_t - V_{FB} = \phi_3(t_{ox}) . \quad (8)$$

Since the strained-silicon device is a heterostructure layer, there does not exist a flat-band voltage (V_{FB}). A reference voltage is defined such that no band bending occurs in the SiGe layer. The flat-band voltage in strained silicon is dependent on the Ge mole fraction of the SiGe substrate and the thickness of the strained-silicon layer [5]. The energy bandgap, electron affinity, and effective density of states in the material change with the strain. When the SiGe layer has zero charge or no band bending, a charge density equivalent to qN_a exists in the strained-silicon layer. The charge causes a voltage drop across the strained-silicon layer; the charge in the strained-silicon layer is imaged at the gate and is negative. These charges also affect the flat band in the strained silicon. The flat-band voltage in strained-silicon MOSFETs is given by the expression

$$V_{FB} = \phi_m - \left(\chi + \frac{E_g}{2} + \phi_F\right) + \frac{qN_a t_{Si}}{C_{ox}} - \frac{qN_a t_{Si}^2}{2\varepsilon_{Si}} . \quad (9)$$

The long-channel linear threshold voltage (zero body bias) is given by

$$V_{t0} = V_{FB} + (\phi_{s0} - \Delta E_c) + \frac{qN_a(x_d + t_{Si})}{C_{ox}} . \quad (10)$$

As bulk-referenced models are symmetric, continuous, and easy to handle in weak inversion and current saturation regions [6], the derivation is done based on bulk reference for the drain current. The model equation is given by

$$V_t = V_{t0} - (A_b - 1)V_{bs} \quad (11)$$

where V_{t0} is given by (10) and A_b is derived to be

$$A_b = 1 + \frac{\zeta \varepsilon_{SiGe}}{C_{ox}\sqrt{t_{Si}^2 - \frac{\varepsilon_{SiGe}}{\varepsilon_{Si}} t_{Si}^2 + \frac{2\varepsilon_{SiGe}}{qN_a}(\phi_{s0} - \Delta E_c)}} \quad (12)$$

where ζ is a fitting parameter introduced due to the Taylor series approximation. It is extracted from one set of V_t vs. V_{bs} data and can be used to predict strained-silicon devices with different Ge mole fractions.

A strained-silicon channel device has been simulated using Medici [7]. Linear threshold voltage has been extracted at different gate lengths using the maximum-g_m method and the threshold voltages for different substrate and drain biases have been extracted based on the constant-current definition [8].

The material and transport properties of different layers have to be specified in the Medici structure. The different properties are calculated as a function of Ge mole fraction and are fed in the materials statement. The energy bandgap of a $Si_{1-x}Ge_x$ alloy varies as a nonlinear function of Ge content [9], dropping off rapidly at Ge content above 85%.

$$E_g(x) = \begin{cases} 1.55 - 0.43x + 0.0206x^2 & (0 < x < 0.85) \\ 2.01 - 1.27x & [eV] \quad (0.85 < x < 1) \end{cases} \quad (13)$$

In the strained-silicon layer, there is a band offset in both the conduction band and the valence band. The bandgap in strained channel is calculated from the bandgap of relaxed SiGe and the band offsets. Comparing the results of various studies, the band offsets are given by [10], [11]

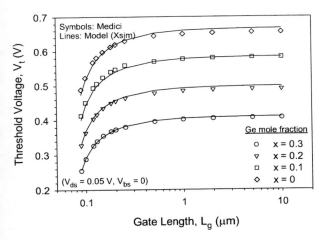

Figure 4: Prediction of the linear threshold voltage vs. gate length for different Ge mole fractions.

$$\Delta E_C(x) = 0.63x \quad (0 < x < 0.4) \quad [eV] \qquad (14a)$$

$$\Delta E_V(x) = 0.74x - 0.53x^2 \quad (0 < x < 0.4) \quad [eV] \qquad (14b)$$

The intrinsic concentrations for the different layers also vary and should be modeled properly. The intrinsic concentration is calculated from [12] based on the effective density of states and the energy bandgap. The density of states for $Si_{1-x}Ge_x$ can be expressed as

$$N_C(x) = [2.8 + x(1.04 - 2.8)] \times 10^{19} \quad [cm^{-3}] \qquad (15a)$$

$$N_V(x) = [1.04 + x(0.6 - 1.04)] \times 10^{19} \quad [cm^{-3}] \qquad (15b)$$

Further, the solution is extended to the two-dimensional case to account for the short-channel effects (SCEs). The SCEs on the threshold voltage of strained silicon are modeled by porting the above strained-silicon long-channel V_t model to our previously-developed short-channel V_t model for bulk silicon [13]. The general idea is to retain the simple one-dimensional form of the long-channel V_t equations (10) and (11) while building the 2-D short-channel effects in the effective channel doping and surface potential. This is done by changing N_a to N_{eff} and changing ϕ_{s0} to $\phi_s = \phi_{s0} - \Delta\phi_s$ in all of the above equations, where N_{eff} and $\Delta\phi_s$ include charge sharing, barrier lowering, and boron pile-up effects. Process-dependent fitting parameters (λ, α, β, κ) are used to account for the different SCEs [13].

To extract the fitting parameters, our compact model uses a very simple, four-step, one-iteration extraction procedure, which requires the measured or simulated V_t vs. L_g data from the same process with only four sets of measurements: V_t (long L_g) for all V_{bs}, V_t (low V_{ds}, low V_{bs}), V_t (low V_{ds}, high V_{bs}), and V_t (high V_{ds}, low V_{bs}). The $V_t - L_g$ data from bulk ("control") silicon was used to calibrate the model and the calibrated model was used to predict the strained-silicon short-channel threshold voltages [13] for different Ge mole fractions using the same physical parameters for Medici devices.

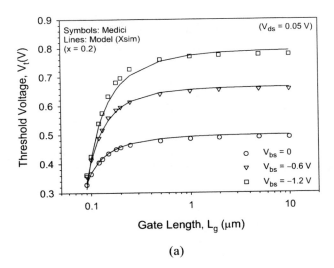

(a)

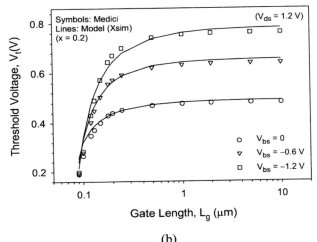

(b)

Figure 5: Prediction of the V_t vs. L_g curve for a Ge mole fraction of 20% for different substrate biases at (a) $V_{ds} = 0.05$ V and (b) $V_{ds} = 1.2$ V.

3 RESULTS AND DISCUSSION

Figure 2 shows the long-channel V_t model prediction for different Ge mole fractions. With change in mole fraction, the different material parameters such as bandgap, conduction band offset, intrinsic concentration, and permittivity changes, the model is able to predict the behavior quite well.

Figure 3 shows the threshold voltage model behavior for different substrate biases. The V_t vs. V_{bs} curve of bulk silicon was used to calibrate the model and the model was used to predict the V_t vs. V_{bs} behavior for different Ge mole fractions.

Figure 4 shows the predicted linear threshold-voltage model for different Ge mole fractions at different gate lengths. The strained-silicon behavior is predicted with just the control silicon data. The calibrated model can also be used to predict the V_t vs. L_g curve at different substrate and drain biases for different mole fractions. Figures 5(a) and 5(b) show the model prediction of the V_t vs. $\log(L_g)$ curve at different drain and substrate biases for a Ge mole fraction of 20%.

4   CONCLUSION

In conclusion, strained-silicon technology is a fast emerging technology that offers significant performance enhancement over bulk silicon as well as enjoys compatibility with robust and mature silicon technology. A threshold-voltage model has been developed for surface-channel strained-silicon MOSFETs. The model predicts the threshold-voltage shift with different Ge mole fractions. The model equations have been derived based on bulk reference to preserve the symmetry of the model. A set of threshold voltages for different gate lengths and bias conditions are needed to calibrate the short-channel model. The calibrated model can be used to predict the threshold voltages at different gate lengths and terminal bias conditions for all Ge mole fractions. The threshold-voltage shift is one of the most important changes in strained silicon from bulk silicon. This model will be a starting point in modeling other prominent parameters in strained-silicon MOSFETs, such as intrinsic charge/capacitance, inversion-layer mobility, and off-state leakage current.

REFERENCES

[1] A. G. O'Neil and D. A. Antoniadis, *IEEE Trans. Electron Devices*, vol. 43, p. 911, 1996.

[2] J. Welser, J. L. Hoyt, and J. F. Gibbons, *IEEE Electron Device Lett.*, vol. 15, p. 100, 1994.

[3] M. A. Armstrong, Ph.D. Thesis, MIT, 1999.

[4] K. Iniewski, S. Voinigescu, J. Atcha, and C. A. Salama, *Solid-State Electron.*, vol. 36, p. 775, 1993.

[5] Hasan M. Nayfeh, Ph.D. Thesis, MIT, 2003.

[6] Y. Tsividis, "Operation and Modeling of the MOS Transistor," McGraw-Hill, 2nd edition, 1999.

[7] MEDICI Users Manual, Synopsys.

[8] X. Zhou, K. Y. Lim, and W. Qian, *Solid-State Electron.*, vol. 45, n. 3, pp. 507–510, 2001.

[9] C. K. Maiti, N. B. Chakrabarthi, and S. K. Ray, "Strained silicon heterostructures: materials and devices," The Institution of Electrical Engineers, 2001.

[10] R. People and J. C. Bean, *Appl. Phys. Lett.*, vol. 48, p. 538, 1986.

[11] M. M. Rieger and P. Vogl, *Phys. Rev. B.*, vol. 48, p. 14276, 1993.

[12] ATLAS Users Manual, Silvaco.

[13] X. Zhou and K. Y. Lim, *IEEE Trans. on Electron Devices*, vol. 48, no. 5, pp. 887–896, 2001.

Predicting the SOI history effect using compact models

M.H. Na, J.S. Watts, E.J. Nowak, R.Q. Williams, W.F. Clark

IBM Corporation, 1000 River Road, MS/972F, Essex Junction, Vermont USA
phone: (802)769-5383, fax: (802)769-9659, email: myunghee@us.ibm.com

ABSTRACT

A simple and effective compact model methodology that predicts the history effect in silicon-on-insulator (SOI) is discussed. In this study we employ three physical parameters to modify the body-potentials of SOI FETs in an inverter during switching. These parameters are very challenging to measure accurately for sub-100nm devices, yet are very tightly correlated with the history effect. This methodology provides an effective means of adjusting history effects without significant alteration of the DC model. Furthermore this methodology enables construction of evaluation-level models in which DC device parametric, circuit performance and history goals exist but hardware meeting those goals is not yet available for model extraction.

Keywords: silicon-on-insulator, compact model, history, dynamic body effect, partially-depleted

INTRODUCTION

Partially-depleted SOI (PD-SOI) technology offers a performance advantage over bulk technologies [1]. However hysteresis effects due to the floating body in PD-SOI must be factored into SOI circuit design, since it causes a change in gate delay as pulses propagate through a logic path. History effect can range from 5 to 30 %, depending on circuit types [2-3]. Therefore, it is important that SOI circuit designers are able to simulate this effect in the early stages of design in order to ensure proper functionality of an SOI-based part.

To enable accurate timing of SOI circuits, the compact model should correctly predict the hysteresis caused by the varying potential of the floating transistor body [4]. This has previously been achieved by accurately modeling the body-to-source/body-to-drain diodes, impact ionization and oxide tunneling body currents from measured hardware [5].

As silicon body thickness continues to scale, such measurements become increasingly challenging due to the high body resistance and the parasitic currents associated with the body contact. As a result, accurate DC models may not necessarily guarantee accurate history prediction.

In this paper we present a methodology that enables model adjustments to match measured or expected history while preserving the DC model fit. We also discuss various physical effects that influence the time-dependant behavior of the body potential and the response of the MOSFET to the body potential.

PHYSICS OF SOI HISTORY EFFECT IN AN INVERTER

In equilibrium the body potential is determined by the balance of leakage currents from the body across the two pn junctions to source/drain, and to the gate electrode via tunneling through the gate dielectric [6-7]. During a transient condition, displacement currents through the capacitances associated with these interfaces are much larger than the leakage currents, and drive the body potential far from equilibrium. Once the body potential is known, the drive current, and therefore the switching performance, is determined by the sensitivities of the device to the body potential. In this paper, we describe how each of these effects can affect the body potential of a device in an inverter circuit and correlate them with the history effect of the circuit. This study is done using the BSIMPD SOI MOSFET model [8].

Figure 1 shows transient voltage waveforms for the nodes of three successive inverter stages in a chain. The applied voltage at *t=1.0* has existed long enough for the bodies of the transistors to be at equilibrium. In this study first switch is defined as a switching event with the body initially at its equilibrium potential and second switch as a switching event with the body initially at the potential it has immediately after a first switch event of the opposite sign. For simplicity, we call an output fall delay a pull-down delay, and an output rise delay a pull-up delay. First and second switch pull-down delays are indicated in the figure. First and second switch pull-up delays also appear in the figure but are not labeled.

The definition of the history effect is

$$History\ Effect = \frac{\tau_{pd,1} - \tau_{pd,2}}{\tau_{pd,1}} \times 100\ (\%) \qquad (1)$$

where $\tau_{pd,1}$ is the first-switch delay and $\tau_{pd,2}$ is the second-switch delay.

HISTORY CENTERING METHODOLOGY

There are three physical effects (and thus, parameters) that are difficult to measure directly in advanced SOI devices, but strongly affect history and weakly affect DC currents. Two parameters directly impact the body potential of a device, namely the gate-to-body tunnel current (*Igb*), and the body-to channel capacitance (*Cb*). The third affects the FET response to the body potential, namely, body sensitivity of threshold voltages. In this study, we adopt a strategy of adjusting these parameters to achieve required history effects. Junction diode currents are not selected since they can be measured accurately.

Figure 2 shows the body potential (*Vbs*) as a function of time for NFETs in two consecutive stages of an inverter chain.

At *t=0* both are in equilibrium. The solid curve for an NFET is in the off-condition. This NFET provides the drive current for a first-switch pull-down. The dashed curve is for an initial NFET in the on state. This NFET is passive during the first-switch pull-up condition, but provides the drive current for a second-switch pull-down. During a second-switch condition the drive strength is determined by the *Vbs* which results from a combination of the equilibrium *Vbs* and the *Vbs* shift due to capacitive coupling during the transition from first to second switch conditions. The equilibrium *Vbs* is determined by the balance between gate tunneling current into the body and diode current out of the body to the source and drain.

Figures 3 show how Cb and *Igb* modify the body potential that, in turn, modifies the switching delay. Fig. 3a shows the effect of varying the body-to-channel capacitive coupling. Differences of *Vbs* between first- and second-switch conditions result in SOI history effect in an inverter. There is no effect on the *Vbs* prior to first switch because this is set by the balance of diode currents. However, the body potential prior to second switch is modulated by the change of Cb due to the AC capacitive couplings. Fig.3b shows the effect of varying the gate tunneling current. The equilibrium potential in the off-state is not affected because with the gate off the tunneling current is small. Therefore, the change of *Vbs* at second-switch condition is responsible for the change of history effect. Adjusting body sensitivity of Vt does not change the body potential but does change how much history (as a percent of delay) results from given first- and second-switch body voltages.

Similarly, this methodology can also be applied to the PFET. Since the NFET is passive when the output signal rises, the *Vbs* at the PFET will modulate the pull-up history predominantly. Therefore the *Vbs* can be independently adjusted for NFET and PFET to achieve pull-down and pull-up history.

The methodology is illustrated in Fig. 4. The technique can be used to obtain a model when circuit performance and history do not match measurement or expectations, although the model matches DC characteristics accurately. The starting point for the methodology is an initial model that is extracted from hardware or centered according to DC device targets.

Using the initial model, the sensitivity of these parameters was carefully evaluated with a specific circuit (or circuits). Boundary conditions for the parameters should be physical and reasonable and take into consideration the uncertainty of the hardware measurement. Each parameter has a different effect on first- and second-switch delay so together they can be used to match overall performance. When these parameters were determined, the adjusted model should be checked to confirm whether the model still meets the DC and performance targets.

This method was evaluated using the compact model for an advanced SOI technology. The impact on the history effect due to the change of each parameter is shown in Table 1. For example, the inverter history increased from 2.3 to 13.8%. Although absolute history values vary by circuit types, the increase is shown to be consistent across the circuits.

CONCLUSION

The primary factors influencing the history effect in SOI circuits have been reviewed. A methodology for adjusting model parameters has been described, which allows the history effect to be adjusted within a finite range without significantly changing the DC model. This method can be used for tuning a model based on measured data, or a predictive model for a future technology that was created using extrapolated performance targets. This method will also be very useful to circuit designers, since it can provide a quick assessment of the circuit-level impact due to history effects.

REFERENCES

[1] J.P. Colinge, *SOI Technology*, Kluwer Press, 1991.
[2] R. Shiebel, et al., Proc. IEEE Int. SOI Conf., p.154, 1997.
[3] J. Jacobs, et al., Proc. IEEE Int. SOI Conf., p. 111, 1998.
[4] J. Gautier, et al., IEDM Tech. Dig., pp.407, 1997.
[5] S.K.H. Fung, et al, IEEE Int. SOI Conf. Proc., p.122-123, 2000.
[6] P. Su, et al., ISQED, pp. 487-491, 2002.
[7] D. Suh, et al., IEDM Tech. Dig., pp.661-664, 1994.
[8] BSIMPD2.2 User's manual, University of California, Berkeley, 1999.

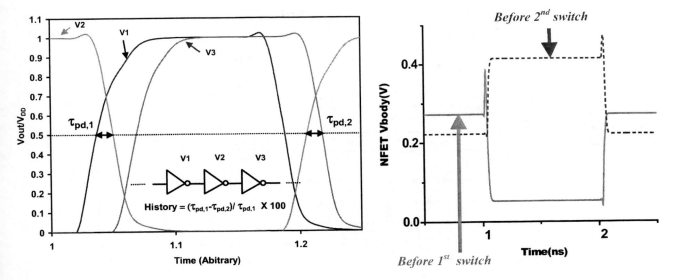

Fig.1. Typical Vin-Vout curves using three inverters are shown. The definition of the history effect is also shown in the plot.

Fig.2. Body potential trajectories of two NFETs in an inverter chain during switching are shown.

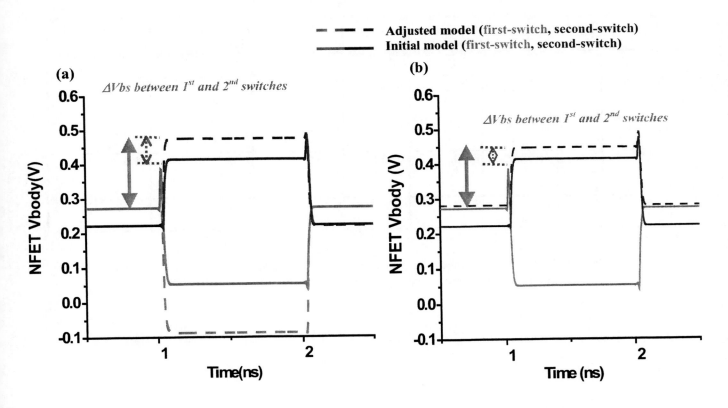

Fig. 3. Body potential trajectories of two NFETs in an inverter chain during switching are shown in various cases. The solid lines are for the cases with the initial model, and the dashed lines are for the cases with (a) the adjusted model with body-to-channel capacitance change only, and (b) the adjusted model with *Igb* change only. The dashed arrow lines show the increase of *ΔVbs* between the initial model and the adjusted models.

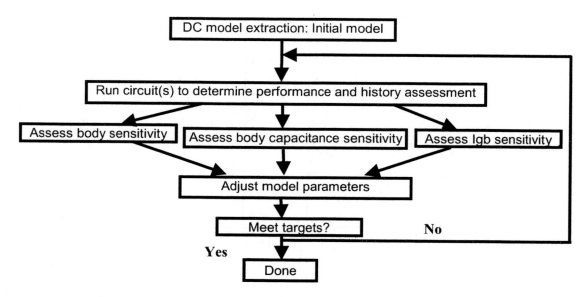

Fig. 4. The proposed modeling methodology for history adjustment is described.

Average History (%)						
Circuit	Description	Initial Model	Igb change only	Cbody change only	Body effect change only	Adjusted model with all changes
inverter	Fan out =3	2.29	6.88	7.95	3.01	13.80
nand3b	3-way input NAND, bottom NFET switching	3.16	8.12	8.42	3.63	14.57
nand3t	3-way input NAND , top NFET switching	2.64	7.54	8.26	3.33	14.06
nor2b	2-way input NOR, bottom PFET switching	6.08	10.54	10.74	7.12	16.84
nor3t	3-way input NOR , top PFET switching	2.02	5.26	6.74	2.45	13.92

Table 1. History effect is shown for the cases of the initial model, the models in which each parameter is adjusted separately, and the adjusted models with all changes

New Capabilities for Verilog-A Implementations of Compact Device Models

M. Mierzwinski,[*] P. O'Halloran,[*] B. Troyanovsky,[*] K. Mayaram,[**] R. Dutton[***]
[*]Tiburon Design Automation, Inc., Santa Rosa, CA, [**]Oregon State University,
[***]Stanford University

ABSTRACT

Historically, analog models and the simulators in which they are embedded form a single analog simulation kernel. This was true of SPICE and its predecessors and is true of most commercial and proprietary analog simulators in use today. As a consequence, while model interfaces generally serve the same function (allowing the model to define the differential equations of a system), the mechanics of each interface is specific to its simulator and the interface is complex and tightly interwoven with the analysis engine. Adding a new model to any analog simulator is a task that is measured in engineer-months requiring intimate knowledge of the simulator's architecture and, in some cases, thousands of lines of C code (BSIM4 is ~15k lines in SPICE3). Verilog-A, a language originally intended for behavioral modeling, has been shown to be a viable alternative to C-code [1-3] with comparable implementations typically providing an order of a magnitude reduction in the number of lines of code necessary.

Acceptance of a model requires its availability in main-stream simulators. Simulation vendors generally do not add unproven, immature models because the return on investment is never guaranteed. For these reasons the cycle of analog model development and maturation has never flourished in analog CAD – an area of CAD where perhaps having accurate predictive models is most crucial. The capability of Verilog-A to describe compact model behavior in a concise and portable fashion can only be realized if commercial simulators incorporate the interface in a consistent way.

In this paper we present simulation results using a new modular architecture implemented in both commercial and research simulators. We use industry standard models, including BSIMSOI and BSIM3, as well as MEMS models coupled into complex harmonic balance and device level simulators. This is the first demonstration of multiple commercial simulators sharing the same model binaries.

1 INTRODUCTION

Previously [1], we described a new model-simulator architecture that provided both Verilog-A OVI 2.0 language compliance as well as simulation speed of complex models comparable to C-coded models (figure 1). The architecture

is comprised of two key elements: a stand-alone Verilog-A compiler and a simulator-specific interface. The compiler generates an object file (shared library) from the Verilog-A source code. The simulator-specific interface, or run-time environment (RTE), provides the application program interface to each unique interface. The technology provides a way to effectively decouple the model specific description from the analysis requirements. The compiler and the RTE work together to provide the support necessary for a particular simulator's capabilities.

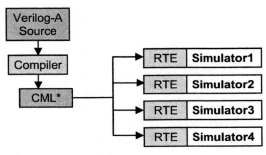

* CML = compiled model library

Figure 1: Tiburon Design Automation compiled Verilog-A uses simulator-specific interfaces (Run Time Environments, or RTEs) to couple the same compile model library into each simulator.

The architecture is engineered to be easily embedded into existing analog simulation engines and is currently being deployed in such environments as UC Berkeley SPICE, Agilent Technologies RF/MW design environments (ADS and RFDE) [5], Eagleware's GENESYS simulator product [6], in addition to other industry-proprietary and academic simulation engines such as Oregon State University's CODECS. The Verilog-A language has been shown to be fully capable of simulating complex models. Popular compact models such as BSIM3, BSIM4, BSIMSOI, HICUM, MEXTAM, MOS9, MOS11, and Angelov have been coded in Verilog-A and validated against their C-coded counterparts. Verilog-A has been demonstrated to be a fully-capable language for compact modeling and an active committee exists to address issues to improve the language for compact models. Figure 2 shows the output of a ring oscillator circuit comparing both the simulator's C-coded version of the UC Berkeley industry-standard BSIM3 model with a Verilog-A coded version. The results are identical, indicating the correct representation of the core model and parasitics. The simulation time is on the same order as the built-in version.

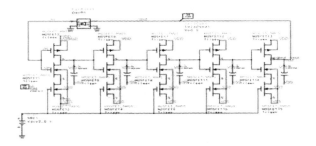

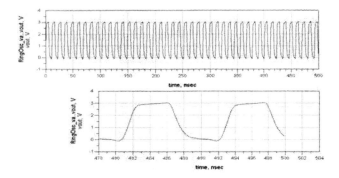

Figure 2: BSIM3 ring oscillator circuit and results showing identical behavior for the simulator "built-in" c-coded model as compared to the Verilog-A coded version.

2 VERILOG-A MODELING OF MEMS IN DEVICE SIMULATOR

CODECS is a *co*upled *de*vice and *c*ircuit *s*imulator that allows accurate and detailed simulation of semiconductor circuits [4]. The simulation environment of CODECS enables modeling of critical devices within a circuit by physical (numerical) models based upon the solution of Poisson's equation and the current-continuity equations. CODECS incorporates SPICE3 for the circuit-simulation capability and for analytical models of semiconductor devices. Analytical models can be used for the non-critical devices while numerical models are provided by a one- and two-dimensional device simulator. The numerical models in CODECS include physical effects such as bandgap narrowing, Shockley-Hall-Read and Auger recombinations, concentration- and field-dependent mobilities, concentration-dependent lifetimes, and avalanche generation.

Adding Verilog-A simulation capability to CODECS gives the user the ability to abstract certain models while simulating other key models at the physical level. In this example, development of a MEMS varactor diode (figure 3)

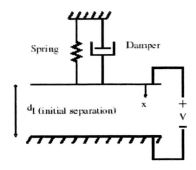

$$m \, d^2x/dt^2 + r \, dx/dt + k \, x = - \epsilon_0 A V^2 / 2(d_1 + x)^2$$

Figure 3: MEMS implementation of a varactor diode. The MEMS structure is modeled as a parallel plate where one plate can move and its motion is controlled by the balance of the static electric field and the equations of motion for a mass on a damped spring.

is aided by the use of differing levels of abstraction. It also allows the device to be characterized easily in more than one simulator. Figure 4 shows the setup and output for an S-parameter simulator sweep used to characterize the capacitance of the device over bias.

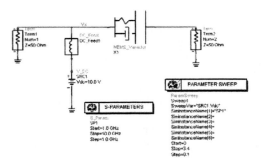

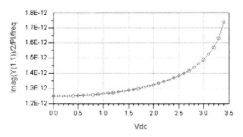

Figure 4: MEMS characterization using ADS for parameter swept S-parameter analysis.

The characterization is then used to develop a voltage-controlled oscillator circuit employing the varactors. The critical MOSFET devices are modeled at the numerical level while the MEMS device and other non-critical MOSFETs are modeled using Verilog-A. With CODECS the designer can simulate the MOS device at the numerical level, using the SPICE c-code implementation, or using a Verilog-A implementation. The Verilog-A implementation will give the same results as the C-coded version, but will give the developer the ability to quickly and easily edit the model to add or subtract model behavior. Figure 5 shows the results of the simulations.

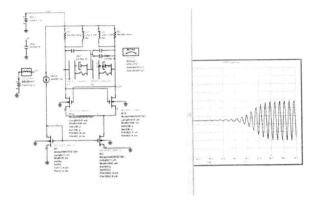

Figure 5: VCO circuit incorporating MEMS varactor diode and transient simulation of the VCO using a Verilog-A abstraction of a MEMS varactor device. These results show both MOSFETs simulated using the physical (numerical) equations as well as their Verilog-A compact model representation.

3 DEVICE SIMULATION IN MULTIPLE SIMULATORS

Since different simulators offer different analysis capabilities, modern circuit designs often require the use of multiple simulator platforms for development and verification. Agilent's Advanced Design System and Eagleware's GENESYS product are good examples of simulators that provide key simulation capabilities to RF IC and microwave designers. Sharing models between commercial simulators, even "standard" models, typically requires a great deal of verification.

Using the Tiburon-DA compiled interface, two separate commercial simulators can share the same compiled model library file (figure 6) as was used in CODECS. This dramatically reduces the chance of model implementation errors and differences as the model equations are for the most part independent of the simulator implementation. It also provides almost immediate access to new models, such as those developed using CODECS, to be distributed to end-users without loss of analysis capability or simulation performance. Compiling gives the added benefit that the shared libraries provide a good level of intellectual property protection by means of the compile process.

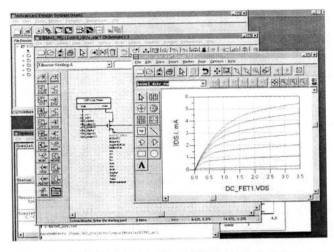

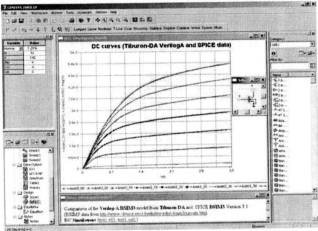

Figure 6: Agilent's ADS and Eagleware's GENESYS design suites sharing the same Verilog-A version of the BSIM3 MOSFET model that was used to simulate the CODECS MEMS VCO.

CONCLUSIONS

The ability to model at high levels of abstraction in Verilog-A, as well as to efficiently simulate complex transistor models, allows developers to easily shift between physical, compact, and abstract model domains. Verilog-A's acceptance is growing and most CAD vendors support or plan to support the language. The implementation described here provides a Verilog-A OVI 2.0 compliant solution in commercial products with simulation speeds close to that of traditional C-based models. For the first time, model developers now have a convenient development and release process for a wide range of analog models.

REFERENCES

[1] "Changing the paradigm for compact model integration in circuit simulators using Verilog-A," M. Mierzwinski, P. O'Halloran, B. Troyanovsky, and R. Dutton, Workshop on Compact Modeling at the 6th International Conference on Modeling and Simulation of Microsystems, 2003.

[2] "ADMS - Automatic Device Model Synthesizer," L. Lemaitre, C. McAndrew, S. Hamm, CICC 2002.

[3] "Portable high-performance analog models using Verilog-A," Troyanovsky B., et al., presented at the workshop "Fundamentals of Nonlinear Behavioral Modeling: Foundations and Applications" at MTT 2003 in Philadelphia, June 8 2003.

[4] "CODECS: a fixed mixed-level device and circuit simulator," Mayaram, K., Pederson, D.O., Computer-Aided Design, 1988, ICCAD-88 Digest of Technical Papers, IEEE International Conference on 7-10 Nov. 1988, pg. 112 -115.

[5] Agilent Technologies: http://eesof.tm.agilent.com.

[6] Eagleware Corporation: http://www.eagleware.com.

Simplified Half-Flash CMOS Analog-to-Digital Converter

P. B. Y. Tan[1, 2 *], A. V. Kordesch[1] and O. Sidek[2]

[1]Silterra Malaysia Sdn. Bhd. Kulim Hi-Tech Park,09000 Kulim, Kedah, Malaysia
[2]Universiti Sains Malaysia, 14300 Nibong Tebal, Pulau Pinang, Malaysia
*philip_tan@silterra.com

ABSTRACT

In this paper, we present a simplified method to construct a half-flash analog-to-digital converter (ADC) to achieve less die area consumption compared to the conventional half-flash ADC. Although the die area consumption is reduced but the conversion speed can still be maintained about the same as the conventional half-flash ADC. The simplified half-flash ADC performs two full-flash cycles as the conventional type but with the help of a voltage estimator (VE), the number of comparators used can be reduced up to 80 percent for 8-bit resolution ADC and more than 80 percent for higher resolution. This is because the VE can predict roughly the range where the input voltage, V_{in} resides on the resistor ladder and comparators are only placed at the predicted range. The major reduction in comparators count is the main factor that enables us to achieve less die area consumption.

Keywords: analog-to-digital converter, half-flash, CMOS, simplified method, voltage estimator.

1 INTRODUCTION

In very high-speed applications, flash type of analog-to-digital converter (ADC) is still the ideal option, but the large die area consumption makes it too expensive for many systems. In order to address this issue, half-flash ADC has been introduced, which has two conversion cycles instead of a single conversion cycle as the conventional full-flash ADC, but with much less die area consumption. The conversion speed is reduced by at least a factor of two compared to the full-flash ADC. The half-flash architecture sacrifices speed for die area.

As the resolution increases beyond 10 bits, even the half-flash ADC runs into limitations of die area consumption and thus making this architecture less attractive. A more efficient architecture known as multistep-flash ADC has been developed to address this limitation [1]. The multistep-flash architecture can save die area, while maintaining conversion speed similar to half-flash ADC, but its operation is more complicated compare to the conventional half-flash ADC.

We use the concept of multistep-flash that enable us to eliminate redundant comparators and apply it to the conventional half-flash ADC. By combining the advantages of both multistep-flash and half-flash architectures, a simpler, similar conversion speed as half-flash and die size efficient half-flash ADC is successfully developed.

2 COMPARISON

The basic operations of the four types of flash ADCs, full-flash, half-flash, multistep-flash and simplified half-flash is discussed in this section. Then, we will make a comparison in comparator count needed for each type of flash ADCs at various resolutions.

A full-flash ADC converts analog voltage to digital bits in one cycle. An N-bit full-flash ADC is made up of a resistor ladder of 2^N equal-value resistors between reference voltage, V_{ref} and ground. Comparators are placed along the resistor ladder at each tap point and compare the input voltage, V_{in} to the voltage that is tapped from the resistor ladder, V_{tap}. The output from the comparators is then encoded to digital bits. The comparator count, N_c for full-flash ADC is given as:

$$N_c(N) = 2^N - 1 \tag{1}$$

where N_c is comparator count and N is ADC resolution. For an 8-bit full-flash ADC, the comparator count is 255

The concept of operation for half-flash ADC is similar to full-flash except it performs the conversion in two cycles. An N-bit half-flash ADC consists of $(2^{N/2} - 1)$ MSB resistors of value R in series with $2^{N/2}$ LSB resistors of value $R/2^{N/2}$. The MSB resistors divide the reference into $(2^{N/2} - 1)$ equal pieces. The $(2^{N/2} - 1)$ comparators perform the first flash conversion along these MSB taps. The output is decoded as the first N/2 bits of data and a digital-to-analog converter (DAC) generates the analog equivalent of the N/2 bits of data. This analog equivalent is subtracted from V_{in}. The residual is compared with the lower $(2^{N/2} - 1)$ LSB tap points and decoded to yield the final N/2 bits. The comparator count, N_c of half-flash ADC is given as:

$$N_c(N) = 2^{(N/2)+1} - 2 \tag{2}$$

For an 8-bit half-flash ADC, 30 comparators are required.

Multistep-flash ADC consists of a 2-bit sense amplifier voltage estimator (VE_{SA}), an (N/2 – 1) bit flash converter, a DAC, and a digital correction circuit. The operation of a 10-bit multistep ADC [1] begins with the VE_{SA} determines the range of V_{in}. The outputs of the VE_{SA} are digitally corrected by the result of the first flash conversion to a 2-bit data. This 2-bit data is combined with the result of the first flash conversion to yield the first six MSBs of the final output data. These bits are converted back to analog voltage and subtracted from V_{in}. The residue is circulated back to the 4-bit flash, where the final four bits are determined. The comparator count, N_c for multistep-flash is given as [2]:

$$N_c(N) = 2^{(N/2)-1} \qquad (3)$$

A 10-bit ADC using this architecture needs 16 comparators, while for 8-bit resolution, requires eight comparators.

The simplified half-flash ADC is actually a modified version of the conventional half-flash ADC. The major modification that we have done is removing 80 percent of the comparators by introducing a voltage estimator (VE) which has a similar function as VE_{SA} in the multistep-flash ADC but with a different architecture. The detail architecture and operation of the simplified half-flash will be discussed in Section 3. From the simplified half-flash architecture, we have derived the comparator count, N_c as a function of ADC resolution as shown below:

$$N_c(N) = 2^{(N/2)-2} + 2 \qquad (4)$$

An 8-bit simplified half-flash only needs six comparators. This simplified half-flash ADC architecture is suitable for 8-bit and higher resolution.

It is interesting to compare the comparators count needed for each of the flash-type ADCs, when the resolution increases. Figure 1 shows the plot of comparator count versus ADC resolution for full-flash, half-flash, multistep-flash and simplified half-flash ADCs. Simplified half-flash requires the least number of comparators compared to the other three types of flash ADCs.

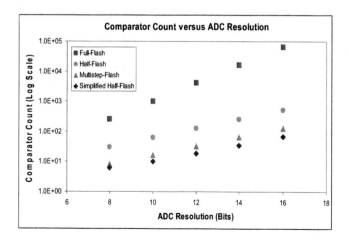

Figure 1: Plot of Comparator Count versus ADC Resolution.

3 ARCHITECTURE

An N-bit simplified half-flash ADC consists of six main modules. They are voltage estimator (VE), (N/2)-bit modified full-flash ADC (MFFADC), digital-to-analog converter (DAC), subtractor, latches and digital switch control (DSC). In this paper, we demonstrate the simplified half-flash ADC for an 8-bit resolution. In actual practice, resolution of the simplified half-flash ADC can be expanded beyond eight bits with similar reduction in comparator count as described in Equation (4). An 8-bit simplified half-flash architecture is shown in Figure 2

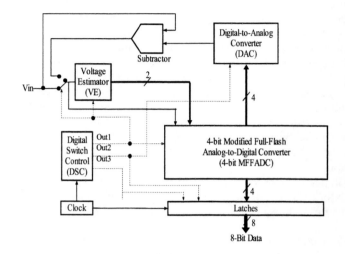

Figure 2: Simplified Half-flash ADC

The 8-bit simplified half-flash ADC performs a full conversion in two cycles. In the first cycle, the output signal, Out1 from DSC, selects V_{in} as the inputs to VE and the 4-bit MFFADC, and turn on the MSB reference voltage, $V_{R(MSB)}$. VE determines the range of V_{in} and the output of VE is supplied to a 4-bit MFFADC to choose the correct range of V_{in}. Then, the 4-bit MFFADC converts the analog signal, V_{in}. to the first four digital MSB bits. The 4-bit data is stored in latches as the first four MSB bits and also input to a DAC. The DAC then converts the 4-bit digital data back into analog voltage, which is controlled by Out2 signal from DSC. The analog voltage is then subtracted from the V_{in} and the residue is circulated back to the simplified half-flash ADC.

The second conversion cycle begins when Out1 signal switches the reference voltage to LSB reference voltage $V_{R(LSB)}$ and selects the residue of the subtraction as the inputs to VE and the 4-bit MFFADC where the final four LSB bits are determined. Out2 and Out3 signals from the DSC, control the storage of digital bits in the latches from the first and second conversion cycles. Finally, the 8-bit data is available at the output of the latches.

The two main modules that make the simplified half-flash ADC more efficient than the conventional half-flash

ADC are VE and (N/2)-bit MFFADC. In the following sub-sections, we will discuss these two modules in detail.

3.1 Voltage Estimator

The function of VE is to determine the range of the input voltage, V_{in}. The most important requirement for the VE is speed. In order to achieve this without significantly increasing the die size, a 2-bit full-flash converter is used as the VE, making it non-critical and fast. The 2-bit resolution is chosen for the VE because higher resolution will increase settling time [3].

The architecture of the VE is shown in Figure 3. The VE consists of a thermometer encoder, three comparators, and a resistor ladder with tap points at

$$V_{tap}(r) = (r/4) \cdot V_{ref} \qquad (5)$$

where (r = 1, 2, 3). Each of these tap points is tied to a comparator. The comparator compares V_{in} to the voltage that is tapped from the resistor ladder, $V_{tap}(r)$ and at the same time V_{in} is sampled at the input of the 4-bit MFFADC. Since both the operations are done in parallel to the acquisition of V_{in}, no extra time is required for the VE step, making this architecture as fast as the half-flash ADC.

VE determines one of the four ranges where V_{in} resides. These four ranges are given by:

$$[(p-1)/4] \cdot V_{ref} < V_{in} < (p/4) \cdot V_{ref} \qquad (6)$$

where (p = 1, 2, 3, 4). For example, when p = 1, the range of V_{in} is $[0 < V_{in} < (1/4 \cdot V_{ref})]$ and when p = 4, the range of V_{in} is $(3/4 \cdot V_{ref}) < V_{in} < V_{ref}$.

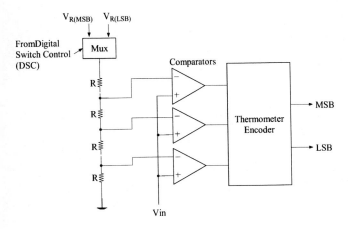

Figure 3: Voltage Estimator

3.2 (N/2)-bit Modified Full-Flash ADC

For 8-bit resolution, a MFFADC with 4-bit resolution is used. The 4-bit MFFADC has the same function as the conventional full-flash ADC, but the number of

comparators needed is reduced by a factor of four. This is because the comparators are only placed at one of the four predicted range for V_{in}.

The architecture of the 4-bit MFFADC for (N = 8), is shown in Figure 4. The 4-bit MFFADC consists of an address decoder, a resistor ladder, analog switches, three comparators and a thermometer decoder. The two input bits from VE are used as the first two MSB data bits and also decoded by the address decoder to predict the range of V_{in}. This range of V_{in} is tied to three comparators by analog switches. Comparators outside this range is not required because for flash converter, if a comparator output is "0", then all the comparators above it on the resistor ladder are also "0". If a comparator output is "1" then all the comparators below it will necessary have a "1" output due to the monotonic nature of a resistor ladder. The outputs of the comparators are then encoded to yield the other two MSB bits. Finally, all the four MSB data bits are obtained.

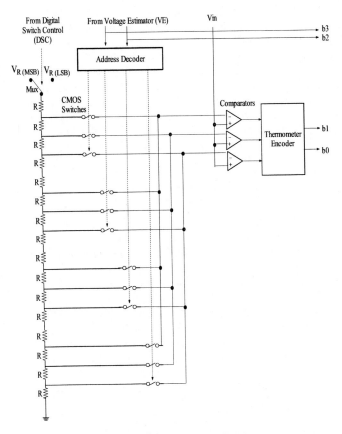

Figure 4: (N/2)-bit Modified Full-Flash ADC for (N = 8)

4 RESULTS AND DISCUSSION

The 8-bit simplified half-flash ADC was implemented using Tanner Tools and successfully simulated using Silterra 0.18μm Advance Mixed Signal CMOS technology. High performance (HP) NMOS and PMOS transistors with 0.18μm minimum channel length and 1.8V operating voltage, are used in the design. BSIM3 version 3.2 model (Level = 49) is used for the circuit simulation.

The simulation results show that the conversion time of the 8-bit simplified half-flash ADC is 1.8µs and the conversion rate is 555 kHz, which is approximately the same speed as the conventional half-flash ADC and about half of the speed of a full-flash ADC. The number of comparators used can be reduced up to 80 percent for 8-bit resolution ADC and more than 80 percent for higher resolution in comparison with the conventional half-flash ADC. For 8-bit resolution, the design of the simplified half-flash ADC requires approximately 1000 transistors whereas for the conventional half-flash ADC, it requires approximately 1500 transistors and the full-flash architecture requires approximately 3000 transistors. For the same process and resolution, the simplified half-flash architecture is estimated to reduce die area consumption by 33 percent compare to conventional half-flash architecture and 67 percent compare to full-flash architecture [4].

The reference voltage for the 8-bit simplified ADC in this project is 1.6V, which means each digital bit represents 0.00625V. Usually the performance of an ADC is determined by using different non-linearity (DNL), which represents the difference between the actual code step width and the ideal code step (1 LSB), and integral non-linearity (INL) known as linearity error that measures the deviation of the code transition point from their ideal position which represents the accuracy of the system. Both the DNL and INL of the 8-bit simplified half-flash ADC in this project are within ±0.5 LSB.

5 CONCLUSION

In conclusion, we have demonstrated that the 8-bit simplified half-flash CMOS ADC architecture can achieve less comparator count and thus less die area consumption compared to the conventional half-flash ADC. This is accomplished with the help of a voltage estimator, VE. By using the VE to predict the range of analog input voltage, V_{in}, the comparators no longer need to be placed on every resistor tap point. For conversion speed, the extra step for predicting V_{in} range runs in parallel with the conversion cycle, hence the conversion speed of the 8-bit simplified half-flash ADC remains approximately the same as the conventional half-flash ADC with the same resolution.

REFERENCES

[1] M. K. Mayes and S. W. Chin, "A Multistep A/D Converter Family with Efficient Architecture," IEEE J. Solid-State Circuits, 24, 6, 1989.

[2] M. K. Mayes and S. W. Chin, "A Multistep A/D Converter with Limited Range Spanning," Twenty-Third Asilomar Conference on Signals, Systems and Computers, 2, 975-979, 1989.

[3] K. Bacrania, "A 12-bit successive-approximation-type ADC with digital error correction," IEEE J. Solid-State Circuits, SC-21, 1016-1025, 1986.

[4] P. B. Y. Tan, B. S. Suparjo, R. Wagiran and R. Sidek, "An Efficient Architecture of 8-bit CMOS Analog-to-Digital Converter," ICSE Conference, 178-186, 2000.

Pareto Optimal Modeling for Efficient PLL Optimization

Saurabh Kumar Tiwary, Senthil Velu, Rob A Rutenbar and Tamal Mukherjee

Electrical and Computer Engineering Department
Carnegie Mellon University, Pittsburgh PA USA
{stiwary, snv, rutenbar, tamal}@ece.cmu.edu

ABSTRACT

Simulation-based synthesis tools for analog circuits [1,2] face a problem extending their sizing/biasing methodology to larger block-level designs such as phase lock loops or converters: the time to fully evaluate (i.e., to fully *simulate*) each complete circuit solution candidate is prohibitive inside a numerical optimization loop. In this paper, we show how to circumvent this problem with a careful mix of behavioral models for less-critical parts of the block, and pareto-optimal *trade-off models* for the critical components. In particular, we show how to adapt current circuit synthesis techniques to build the required tradeoff models. As a concrete example of the methodology, we show detailed simulation results from the synthesis of critical portions of a 500MHz digital frequency synthesizer PLL.

Keywords: behavioral modeling, pareto-optimal design, analog circuits, phase locked-loop, circuit optimization.

1 INTRODUCTION

Recent advances in design automation and increased computational capability of computers has led to a gradual transition from 'hand-calculation' based analog circuit design to a simulation-based sizing methodology. Simulation based synthesis uses efficient global optimization to visit many circuit candidates, and fully evaluates each candidate via detailed simulation. This methodology works very well for circuits having in the range of a few hundred devices. However, for larger circuits, the simulation time required for a single simulation is too large to do a practical simulator-in-the-loop circuit sizing. Also, due to the curse of dimensionality, the design space in which to search for optimal design points becomes too large for these circuits to be handled by the tools available today. Of late, much work has been done in the field of analog circuit *macromodeling* which aims at simulating these circuits faster [3], [4], [6]. This work proposes to adapt these behavioral modeling approaches to model the non-critical blocks of a large system level design and then use *pareto trade-off curves* to search for optimal designs of critical blocks to meet the overall system level performance specifications. A phase-locked-loop has been used to as a vehicle to show the results for our suggested methodology.

In section 2, an overview of the problem and the proposed methodology is presented. Section 3 gives a brief summary of the behavioral modeling methodology.

Generation of pareto trade-off curves are also discussed in this section. In section 4, the proposed method is applied to a 0.18um digital frequency synthesizer and results are presented. Finally, conclusions are presented in section 5.

2 OVERVIEW
2.1 Existing Methodologies

The simulation based circuit sizing methodology is a well researched field. The circuit designer chooses a circuit topology which is presented to the sizing tool with an approximate range in which the various transistor sizes should be. The sizing tool [7] uses global optimization techniques to search for an optimal design point which satisfies the circuit specifications as given by the designer. This approach uses a simulator "in-the-loop" to characterize each visited design point; hence, the time taken for convergence often is proportional to the time taken to complete the simulations for a single design point and the size of the design space. This approach thus, is not directly suitable for circuits with a very large number of design parameters. Also, circuits whose performance specifications require long simulation runs, would take too long a time for synthesis. The phase noise and settling time for a phase locked-loop are examples of such specifications which require a long time for evaluation for a single design point.

Researchers have looked at the problem of simulating large analog circuits efficiently. One of the methods suggested to tackle this problem is the use of behavioral models [3], [4]. Behavioral models capture the overall functionality of the circuit in terms of equations or simple circuit elements that are faster to simulate compared to the complete transistor level circuit. Designers often use the basic circuit sizing infrastructure to size the individual blocks and then do a system level simulation using the behavioral models. Designs have to go through this loop a number of times before the final tape-out. This methodology is obviously *not* the best way to design large, complex analog systems.

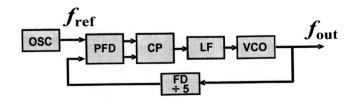

Figure 1   Block level schematic of the PLL

2.2 Proposed methodology

We suggest in this paper a methodology for automatic sizing of circuit elements while keeping system level design specifications under consideration. This is done by (a) using behavioral models for simulating circuits faster, (b) creating pareto trade-off models to capture the capabilities of the circuit across the whole design space and (c) optimizing the system level specs using performance behavioral models for individual blocks. We use a phase–locked-loop as an example to explain our methodology.

Simulating a phase locked loop takes a long time because of the large number of devices in the circuit and the closed loop feedback behavior of the circuit. Hence, calculating the lock-in time requires a large number of CPU cycles. Also, the phase noise specification for the PLL requires a transient noise simulation of the circuit with a small and very well controlled time step and extremely tight tolerances for the simulator to capture the effect of small noise sources. This takes considerable time to simulate-- days or weeks. [3] proposes a method of behavioral modeling for PLLs which is efficient while simulating the complete system. The noise models for the individual blocks e.g., VCO, PFD-CP, etc., are captured by simulating the particular circuit block individually. Behavioral models for the individual circuit blocks are created which capture circuit functionality as well as noise behavior. These models are now used for simulating the complete PLL. Since the models are designed in such a way that they insert the noise only at specific instances (zero crossings of the waveforms), it is not very expensive to simulate the whole circuit *including* the noise. Thus, the behavioral models make it possible to evaluate a design point for the PLL in a reasonable amount of time.

Since we have a method of creating the behavioral models which simulates much faster than the transistor level circuit, we can use the simulator-in-the-loop approach to synthesize the PLL. But, the new problem is that the behavioral models represent the circuit characteristics only at *one* specific design point where they were generated (modeled). At a *different* design point, the same behavioral models are *not* valid and we would need to construct a new set of models. Since the modeling procedure is expensive, we cannot afford to create models at each design point. To circumvent this problem, we create a *pareto* surface for the critical circuit block across its design space and use this pareto surface to optimize for the system level specifications.

2.3 Behavioral modeling

Figure 1 shows the block level diagram of the PLL. The circuit is a 0.18um CMOS 500MHz digital frequency synthesizer using a 100MHz reference signal. All the individual blocks--the phase frequency detector, charge pump, frequency divider and VCO--were behaviorally modeled based on [3]. The loop filter (Figure 2) is simply a

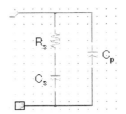

Figure 2   Schematic of the loop filter

2nd order RC filter and hence was not abstracted since modeling it will not give us much in terms of simulation speed-up.

The voltage controlled oscillator (Figure 3) was one of the more challenging blocks to model. Not only did we have to capture the output frequency behavior of the circuit as a function of the control voltage, but also the dynamics observed at the output with abrupt changes at the input. To capture the output frequency behavior as a function of control voltage, we used an equation of the form

$$f_{out} = F (V_{ctrl}) \tag{1}$$

where F is a polynomial equation of degree three. The coefficients for the non-linear equation are found using regression on the simulation data. Basically, the output frequency of the VCO is computed for a series of voltage values at the control voltage. Using this data, we find the coefficients of the polynomial equation via fitting. The dynamic behavior of the VCO is captured by adding a pole at the input of the VCO model. The parameters of the pole (R and C in parallel) are obtained by training the model against the circuit response for a square wave input control voltage. Figure 4 shows the model of the VCO. A similar procedure was adopted for modeling the functionality of all the other blocks as well.

Once we created the behavioral models for the individual blocks as suggested in [3], we simulated the modeled PLL and the transistor level circuit to check the efficacy of the model. We observed a close agreement between the responses of the two with significant simulation speed-ups for the model. The time required for simulating the PLL for 3µs was about 2 hours for the transistor level circuit. It was just 30 seconds for the behavioral models. Also, the behavioral models simulated the noise as well and predicted the phase noise behavior of the PLL which could not be done for the transistor level circuit without waiting for about a week for doing the transient noise simulation.

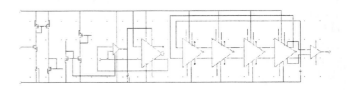

Figure 3 Current starved ring oscillator VCO schematic

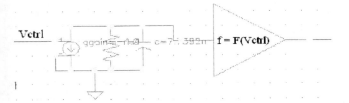

Vctrl

f = F(Vctrl)

Figure 4 Behavioral model for VCO

2.4 Pareto surface modeling

The VCO is a very important block in a PLL. It is a major contributor to the overall phase noise of the PLL. We want to find the optimum sizing for the transistors in the VCO so that it best meets the overall PLL specifications. As discussed earlier, we cannot directly synthesize the VCO circuit "outside" of the PLL. Since the amount of jitter produced by a VCO often trades-off with the power it consumes, we created a pareto curve that captures the trade-off between jitter and power for the VCO across its complete design space. Following are the steps involved in the creation of the pareto curve [5]:

- The pareto trade-off curve was generated with bias current (power) and jitter as the two variables.

- The other performance specifications for the VCO, for example, the minimum and maximum attainable frequency, linearity, area etc., were treated as constraints which need to be satisfied for any feasible point on the pareto.

- Two feasible design points were found through synthesis, one that minimized power and the other that minimized jitter.

- Using these two points, we used the synthesis tool to look for a design point which minimized both jitter and power such that for some small value ε,

$$| \text{Normalized jitter} - \text{Normalized power}| < \varepsilon \quad (2)$$

- Additional points were found using synthesis that lay *between* these prior points using the same approach as described in the previous step.

Using these points (that lie on the pareto) we fit a non-linear equation which captures the trade-off between jitter

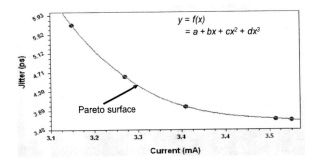

Figure 5 Pareto bias(power)/jitter tradeoff for VCO

and power for VCO across the design space (see resulting curve in Figure 5).

We note that in equation (2) the 'ε' term, which has the form of a "slack" specification, gives some flexibility to the synthesis tool to look at a large number of solution points and minimize the performance spec, instead of keeping on searching for an exact numerical equality for tradeoffs.

We can use the pareto curve and the simplified trade-off equations in the behavioral model for the VCO which to represent the oscillator's response across the *whole* design space (Figure 6). Since the phase noise and settling time behavior for the PLL are strongly dependent on the response of the loop filter and the VCO, we can use the circuit elements of the loop filter and performance variables for the VCO as design variables to optimize the overall PLL specifications.

```
module vcoModel (out, currentOut, in);
...
analog begin
  @(initial_step) begin
    // Pareto equation for jitter trade-off with power
    jitter = (1.03-(.87*cur)+(.24*cur^2)-(.02*cur^3))*1p;
  end

  //Output freq. of the VCO as function of control volt.
  freq = (-1.27 + 9.99*cos(1.10*V(in) + 4.29))*1e8;
...
  // Adding phase noise to the output
  dT = 1.414*jitter*$dist_normal(seed,0, 1);
  freq = freq/(1 + dT*freq);
...
  V(out) <+ transition(Vout, 0, tt);
I(currentOut) <+ cur;
```

Figure 6: Verilog-A code for the VCO model

3 SYSTEM LEVEL SIMULATION

All the behavioral models are combined together including the pareto trade-off model for the VCO. Since the behavioral models are much faster to simulate than their circuit level counterparts, it is now easy to look at system level performance trade-offs and optimize the entire PLL. We set up the synthesis run for the PLL using the behaviorally modeled circuit blocks (see Figure 7).

The variables that were used for optimization were R_s, C_s and C_p for the loop filter (Figure 2) and jitter and current for the VCO performance model. The elements of the loop filter were used as design variables because the loop filter plays a crucial role in determine the lock-in time of the PLL. Also, the loop-filter's frequency response shapes the noise from the individual PLL components at the PLL output. All the other circuit blocks were held as fixed. Table 1 shows the specs achieved by the circuit after optimization using commercial analog sizing tools [2,7] from Neolinear.

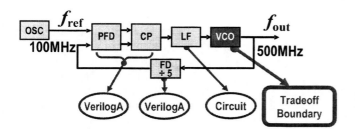

Figure 7 PLL model(s) used for system synthesis

Performance Specification	Achieved Value
PLL Settling Time	0.822 us
VCO bias current	3.37 mA
PLL jitter	0.45% (9.1ps)

Table 1: PLL performance specs after optimization

The PLL was simulated for 4000 cycles at each synthesis point and it took about 4 hours to synthesize the optimal circuit solution on a single CPU. The optimized variable values for the PLL include jitter and current for the VCO. We thus need to search for a design point for the VCO which would have the given jitter and current spec as obtained through PLL synthesis. Since the performance model for the VCO was built using the feasible pareto boundary for these two variables, it will not be very difficult to find a VCO design point that has the given jitter and current specs. This was verified by another synthesis run which obtained the widths and lengths of the transistors used in the VCO that satisfy *these* jitter and current requirements. In order to verify that our final design does indeed meet the specs for the PLL, we simulated the final transistor level circuit and its equivalent macromodel. Figure 8 shows the waveforms at the input control voltage of the voltage controlled oscillator for both the circuits. As can be seen from figure, the responses of the two circuits

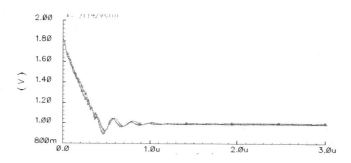

Figure 8 Comparison plot of the control voltage of the VCO for complete PLL simulation showing both transistor level circuit and efficient behavioral model.

match each other quite closely which confirms the accuracy of our models.

4 CONCLUSION

Simulation-based synthesis uses efficient global optimization to visit many circuit candidates, and fully evaluates each candidate solution via detailed simulation. The time to simulate difficult specifications of a PLL is the serious bottleneck: specifications such as jitter can easily require hours or days. The obvious solution is to replace each circuit in the PLL—the frequency divider, VCO, phase-frequency detector, charge pump, loop filter—with an appropriate macromodel. The problem is that we require more than just a model of one circuit instance. We need a model of the *entire* set of performance tradeoffs feasible for the circuit, since our goal is to synthesize each sub-circuit to optimize the performance of the *entire* PLL.

In this work, we showed how to adapt existing synthesis methods to build *pareto-optimal* tradeoff curves that represent the space of achievable specifications for each circuit in a block we wish to optimize. PLL synthesis selects *where* on the curve each circuit needs to be, to optimize the overall PLL behavior, which we evaluate by detailed circuit simulation. Optimization results for one 0.18um CMOS PLL achieved a jitter of 9.13ps (0.45%), a settling time of 0.82µs, and a VCO bias current of and 3.37mA, evaluated by simulation. In future work, we would like to replace *all* the behavioral models with their pareto optimal trade-off curves and optimize the *complete* system. This would give a globally optimal system level design for the circuit under consideration.

ACKNOWLEDGEMENTS

This work was funded in part by the DARPA NeoCAD program.

REFERENCES

[1] Phelps, M.J. Krasnicki, R.A. Rutenbar, L.R. Carley, J.R. Hellums, "A case study of synthesis for industrial-scale analog IP: Redesign of the equalizer/filter frontend for an ADSL CODEC," Proc. DAC'00, June 2000.

[2] A. H. Shah, S. Dugalleix, and F. Lemery. "Technology migration of a high performance CMOS amplifier using an automated front-to-back analog design flow," Proc. DATE'02, Mar. 2002.

[3] K. Kundert, "Predicting the phase noise and jitter of PLL based frequency synthesizers," http://www.designers-guide.com

[4] T., K. Ogawa, K. Kundert, "VCO jitter simulation and its comparison with measurement," Proceedings of the ASP-DAC '99. Asia and South Pacific, Jan. 1999.

[5] I.Das, J.E. Dennis, "Normal boundary intersection: A new method for generating the pareto surface in nonlinear multicriteria optimization problems," SIAM J. of Optimization, Vol 8,No 3, August 1998.

[6] D. Vasilyev, M. Rewinski, J. White, "A TBR-based trajectory piecewise-linear algorithm for generating accurate low order models for non-linear analog circuits and MEMS," DAC, June 2003.

[7] http://www.neolinear.com

Scalability and high frequency extensions of the vector potential equivalent circuit (VPEC)*

Bhaskar Mukherjee, Peiyan Wang, Lei Wang, and Andrea Pacelli

Department of Electrical and Computer Engineering, Stony Brook University
Stony Brook, NY 11794-2350, pacelli@ece.sunysb.edu

Abstract

We present a complete modeling technique for inductive parasitics, based on the vector potential equivalent circuit (VPEC) topology. Novel algorithms for layout extraction and sparsification are introduced. Examples are discussed in terms of CPU time, accuracy, and model complexity. Finally, extensions for high frequency applications are presented, including models for skin effect and full wave simulation.

1 Introduction

Performance estimation of high frequency analog and digital circuits requires the accurate modeling of large numbers of interconnection wires, including their inductive behavior. The widely accepted partial element equivalent circuit (PEEC) technique is based on a dense matrix of inductive couplings, which becomes unmanageable as circuit size grows. Sparsification techniques have been proposed to accelerate PEEC simulations, but they usually involve severe approximations, complex numerical calculations, or even modifications to the circuit simulator [1], [2].

We recently introduced an alternative approach, the vector potential equivalent circuit (VPEC) [3]. The VPEC model mimics inductive couplings by a network of resistive elements and current sources. Unlike the PEEC model, the VPEC topology is intrinsically sparse. VPEC circuit elements have a strong physical interpretation, and can either be derived from first principles, or as the reduction of a PEEC model [4], [5].

In this work we address two issues which are crucial to practical CAD applications, namely, scalability and high frequency modeling. After reviewing the general principles of the PEEC and VPEC models in Section 2, in Section 3 we describe a new layout extraction algorithm which generates a wireframe model. The PEEC model is converted into a VPEC model using an efficient iterative algorithm, described in Section 4. A new model for the skin effect is presented in Section 5, based on a novel direct zero-pole fitting algorithm. Finally, in Section 6, a possible extension of the VPEC model toward electromagnetic (full wave) models are discussed. Conclusions are drawn in Section 7.

*This work was supported by Defense Advanced Research Projects Agency (DARPA) and managed by the Sensors Directorate of the Air Force Research Laboratory, USAF, Wright-Patterson AFB, OH 45433-6543.

2 The PEEC and VPEC inductance models

The PEEC and VPEC circuit topologies represent two views of the same physical effect, modeled through integral and differential equations, respectively. We start from the equations for the vector potential in the Coulomb gauge:

$$\nabla^2 A_i + \mu J_i = 0, \qquad i = x, y, z. \tag{1}$$

This equation can be integrated to obtain an expression for the vector potential at every point as a function of the conduction current at every other point:

$$\mathbf{A}(\mathbf{r}) = \frac{\mu}{4\pi} \int d\mathbf{r}' \frac{\mathbf{J}}{|\mathbf{r}' - \mathbf{r}|}, \tag{2}$$

Equations (1) and (2) are the starting point for the VPEC and PEEC models, respectively. Integrating Eq. (2) over the wire length, one can obtain the familiar partial inductance equations:

$$V_{em}^k = \sum L_{kl} \frac{dI_l}{dt} \tag{3}$$

where L_{kl} is the partial inductance matrix (PIM) element, which requires the calculation of a double volume integral. Even when acceleration algorithms are used to efficiently compute the PIM entries [6], still the number of matrix entries is proportional to the square of the number of wires, and circuit simulation time explodes for systems with more than a few hundreds of wires. The VPEC model obviates this limitation, since it is based on Eq. (1) rather than Eq. (2). By integrating Eq. (1) over a finite volume Ω^k, one obtains the equation [3]

$$\sum_l \int_{S^{kl}} d\mathbf{S} \cdot \nabla A_i + \mu \int_{\Omega^k} d\Omega J_i = 0, \tag{4}$$

where the sum over l includes all nearest neighbors of Ω^k. This equation can be recast as a Kirchoff current law (KCL) for the circuit of Fig. 1:

$$\sum_l \frac{A_i^l - A_i^k}{R_i^{kl}} + I_i^k = 0. \tag{5}$$

In Eq. (5), the node voltages A_i^k and A_i^l represent the (suitably averaged) i-component of the vector potential in volumes Ω^k and Ω^l. For simplicity, we describe the proportionality between 'magnetic current' and conduction current by a linear

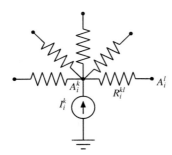

Figure 1: *Equivalent circuit for the i-component of the vector potential* **A**. *Voltages* A_i^k *and* A_i^l *correspond to the vector potentials in two neighboring control volumes* Ω^k *and* Ω^l. *Current source* I_i^k *models the effect of conduction current on the vector potential. Resistor* R_i^{kl} *represents the propagation of vector potential between the two regions.*

coefficient F_i^k:

$$I_i^k = F_i^k I_k, \qquad i = x, y, z \qquad (6)$$

By the same token, the electromotive force V_k is modeled by a single linear coefficient E_i^k:

$$V_k = \sum_{i=x,y,z} E_i^k \frac{dA_i^k}{dt}. \qquad (7)$$

The quantities E_i^k and F_i^k are parameters depending on the wire geometry only and can be easily evaluated [3].

In order to complete the model, one must determine the resistors R_i^{kl} to fit the surface integrals in Eq. (4). Finite-difference approximations, such as those proposed in [5], are much too crude to supply acceptable values for the VPEC resistors. In [3], more refined semi-analytical formulas were proposed based on the extraction methodology of [7]. In Section 4, we will describe a numerical technique that allows the direct extraction of VPEC resistance values from a pre-computed partial inductance matrix.

3 Layout extraction and inductance calculation

Resistance and inductance (*RL*) extraction requires the decomposition of a layout into a set of straight wires with constant cross-section, usually rectangular. Obviously, for general wire shapes, such decomposition is only an approximation. An algorithm for *RL* extraction was presented in [8], based on the identification of elementary rectangular regions which are connected at linear interfaces. We developed a novel algorithm which offers better generality when dealing with complex wire shapes. The algorithm is based on an 'explorer,' i.e., a rectangular shape which moves about the layout, and is completely contained within it. Every time the explorer changes size or direction of motion, a node in three-dimensional space is generated. Wires are generated

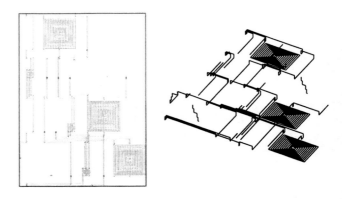

Figure 2: *Wireframe extraction from layout. Left: 2D layout. Right: Extracted three-dimensional wireframe representation, which can be fed into FastHenry or other magnetostatic solver.*

as straight lines connecting nodes, consistently with the geometric representation adopted by FastHenry [6]. The cross-section of a wire connecting two nodes is given by the size of the explorer, which is by definition constant along the length of the wire This algorithm has the key advantage of being independent of the particular layout representation (rectangular tiles, polygons, etc). Figure 2 shows an example of layout extraction for an RF circuit, including both interconnect wires and spiral inductors.

The wireframe model extracted can be directly fed into a magnetostatic solver such as FastHenry. Since our methodology only needs a subset of the complete partial inductance matrix, we developed a quick extractor using simplified analytical formulas for the partial inductance coefficients. The problem of computing partial inductance matrix for conductors with rectangular cross section was long ago solved exactly in the general case [9]. However the numerical stability of the exact solution is not satisfactory, especially for the case of very long wires [10]. On the other hand, stable simplified expressions are accurate only when the wire length l is much larger than the lateral distance d. We developed a hybrid expression that reduces to the asymptotic form

$$L_{kl}(l, d) = \frac{\mu l^2}{4\pi d}$$

for $d \gg l$, while following the stable simplified expression for $d \ll l$ [9]. The hybrid expression offers an almost exact match to the exact formula, without requiring complex calculations like the model of Ref. [10].

4 VPEC generation

In order to find a way to extract VPEC model parameters from PEEC partial inductances, we note that Eqs. (6) and (7) replace Eq. (3), using the vector potential as an intermediate step between conduction currents and induced voltages. Starting from this observation, it can be shown [3] that the inductance matrix can be written as

$$L^{lk} = \sum_{i=x,y,z} E_i^l \, \mathcal{Z}_i^{lk} \, F_i^k, \qquad (8)$$

where the quantity $\mathcal{Z}_i^{lk}$ is an entry in the impedance matrix* of the resistive network for the i-component of the vector potential, i.e., the vector potential at node l when a unit current is injected at node k. Under the assumption of equal propagation of vector potential in all three directions $\mathcal{Z}_i^{kl} = \mathcal{Z}^{kl}$, Eq. (8) becomes

$$L^{lk} = \mathcal{Z}^{lk} \sum_{i=x,y,z} E_i^l \, F_i^k \qquad \text{or} \qquad \mathcal{Z}^{lk} = \frac{L^{lk}}{\sum_{i=x,y,z} E_i^l \, F_i^k}.$$

From any given wire geometry and PIM, we can immediately determine the VPEC impedance matrix. The VPEC generation problem is then reformulated as *finding the resistors R^{kl} to reproduce a given impedance matrix Z.*

The unknowns in the sparsification problem are the n resistors to ground, plus all resistors connecting neighboring nodes. The calculation of all resistors requires the solution of a very large, nonlinear problem [4]. The size of the problem is due to the fact that the calculation of the impedance matrix requires a full circuit solution (i.e., n equations) for every node in the VPEC. Moreover, the problem is nonlinear, because the resistor values appear in the equations along with the node voltages. The problem can be greatly reduced in size, and linearized, by adopting a block iteration where at each step only the resistors connected to a single node are considered. At each iteration, a very sparse, $n \times n$ linear system results, which can be efficiently solved using standard sparse-matrix methods.

The sparsification algorithm has a complexity around $O(n^4)$, due to the need for solving $O(n)$ systems, each with $O(n^3)$ complexity of matrix inversion. We can reduce this complexity to $O(n)$, by adopting a windowing algorithm that splits a layout into regions which are separately extracted. VPEC models extracted for each of the windows are merged by discarding resistor values at the edges, which will display the largest errors [4]. Figures 3 and 4 report the accuracy and CPU time of the extracted VPEC model. An alternative method for VPEC extraction has been proposed in [5], based on matrix inversion. This approach relies on the fact that for a dense VPEC, where all elements R^{kl} have finite values, the problem of determining resistor values is linear rather than nonlinear. Therefore, the VPEC model is obtained by the inversion of a single $n \times n$ matrix. After VPEC generation, some resistors are discarded based on their magnitude, thus obtaining a sparsified model. This scheme scales with complexity $O(n^3)$, and is therefore potentially faster than the iterative algorithm described above. On the other hand, it requires the knowledge of the entire $n \times n$ PIM, which may

---

*The symbol $\mathcal{Z}$, and the term *impedance matrix*, will be employed although the network in question is purely resistive, to avoid confusion with resistor values in the following.

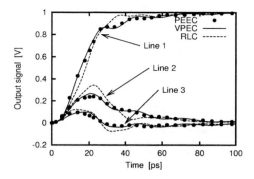

Figure 3: Transient simulation for 32-line bus, when a 1 V, 20 ps step input is applied to the first line. The transient response is shown at the far end of the first three lines.

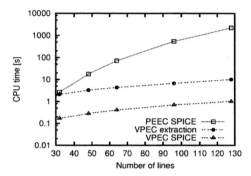

Figure 4: Total CPU time needed for extraction and simulation of a parallel-line bus, with two segments per line, as a function of the number of lines. All simulations were performed using HSPICE on a 2 GHz Pentium-4.

pose storage problems of its own. Moreover, even a complexity $O(n^3)$ is too high for large scale circuit extraction, so that some form of windowing will be required anyway.

5 Skin effect

In the multi-gigahertz frequency range, skin effect causes current to flow only at the surface of conductors, leading to a decrease of wire inductance and an increase of resistance. The latter effect causes a signal attenuation which increases with frequency, thus limiting the available interconnect bandwidth. Existing skin effect models are based either on a ladder circuit [11] or a parallel-*RL* topology [12]. Both models were originally developed for wires with circular cross section and are inadequate for on-chip applications. We developed a skin effect model based on the direct synthesis of a transfer function for the wire impedance [13]. The transfer function can be directly implemented in the VPEC model, or synthesized as an *RL* equivalent circuit for inclusion in a PEEC framework. Figure 5 shows a comparison between the new skin effect model and FastHenry results. With respect to previous works, the new model accounts for a wide range of the wire aspect ratios, and can be easily tailored to fit numer-

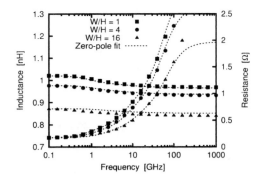

Figure 5: Comparison between FastHenry results (symbols) and the new model based on the direct zero-pole synthesis (dashed lines). Three zero-poles pairs were employed.

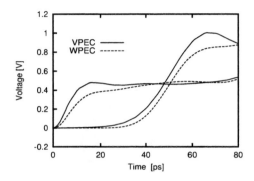

Figure 6: Transient simulation of a 5-mm transmission line, composed of two parallel wires with open-circuit termination. The wave propagation equivalent circuit (WPEC) is compared to VPEC, showing the elimination of the slow rising edge before the time of flight (30 ps).

ical results over any given frequency range.

6 Wave propagation equivalent circuit

Since the VPEC topology is derived from the equations of magnetostatics, wave propagation effects can be accounted for by including the displacement current term in the equations. The resulting wave propagation equivalent circuit (WPEC) promises to unify for the first time transmission line, magnetostatic, and electromagnetic models. Figure 6 shows the application of the new model to the simulation of a simple differential transmission line, comparing WPEC to VPEC models. It is apparent that the improved model, including time-of-flight of the traveling waves, removes the unphysical rising edge before the arrival of the signal at the end of the line.

7 Conclusions

We have reviewed the foundations of the VPEC inductance modeling methodology, discussing several key issues

for large-scale and high-frequency CAD applications. Future work will address the full integration of the VPEC model with electromagnetic simulation, which is key to system performance prediction in the multi-gigahertz range.

REFERENCES

[1] M. W. Beattie and L. T. Pileggi, "Inductance 101: Modeling and extraction," in *Proc. Design Automation Conference*, 2001, pp. 323–328.

[2] A. Devgan, H. Ji, and W. Dai, "How to efficiently capture on-chip inductance effects: Introducing a new circuit element K," in *Proc. IEEE/ACM International Conference on Computer Aided Design*, 2000, pp. 150–155.

[3] A. Pacelli, "A local circuit topology for inductive parasitics," in *Proc. International Conference on Computer Aided Design*, San Jose, CA, Nov. 2002, pp. 208–214.

[4] A. Pacelli, P. Wang, and B. Mukherjee, "Vector potential equivalent circuit (VPEC) for PEEC model reduction," submitted to *IEEE Trans. Computer Aided Design*.

[5] H. Yu and L. He, "Vector potential equivalent circuit based on PEEC inversion," in *Proc. Design Automation Conference*, 2003, pp. 718–723.

[6] M. Kamon, M. Tsuk, and J. White, "FASTHENRY: A multipole accelerated 3-D inductance extraction program," *IEEE Trans. Microwave Theory and Techniques*, vol. 42, no. 9, pp. 1750–1758, Sept. 1994.

[7] A. Pacelli, M. Mastrapasqua, and S. Luryi, "Generation of equivalent circuits from physics-based device simulation," *IEEE Trans. Computer Aided Design of Integrated Circuits*, vol. 19, no. 11, pp. 1241–1250, Nov. 2000.

[8] P. M. Xiao, E. Charbon, A. Sangiovanni-Vincentelli, T. Van Duzer, and S. R. Whiteley, "INDEX: an inductance extractor for superconducting circuits," *IEEE Trans. Applied Superconductivity*, vol. 3, no. 1 part 4, pp. 2629–2632, Mar. 1993.

[9] E. B. Rosa, "The self and mutual inductances of linear conductors," *Bulletin of the Bureau of Standards*, vol. 4, no. 2, pp. 301–344, Jan. 1908.

[10] G. Zhong and C.-K. Koh, "Exact closed form formula for partial mutual inductances of on-chip interconnects," in *Proc. IEEE Inl. Conf. on Computer Design: VLSI in Computers and Processors*, 2002, pp. 428–433.

[11] S. Kim and D. P. Neikirk, "Compact equivalent circuit model for the skin effect," in *IEEE Intl. Microwave Symposium Digest*, 1996, pp. 1815–1818.

[12] B. Sen and R. L. Wheeler, "Skin effects models for transmission line structures using generic SPICE circuit simulators," in *IEEE Topical Meeting on Electrical Performance of Electronic Packaging*, 1998, pp. 128–131.

[13] B. Mukherjee, L. Wang, and A. Pacelli, "A practical approach to modeling skin effect in on-chip interconnects," submitted to the 2004 Great Lakes Symposium on VLSI.

A Full-System Dynamic Model for Complex MEMS Structures

L.A. Rocha*, E. Cretu** and R.F. Wolffenbuttel*

*Delft University of Technology, Faculty EEMCS
Dept. for Micro-Electronics, Delft, The Netherlands, l.rocha@ewi.tudelft.nl
**Melexis, Transportstr. 1, Tessenderloo, Belgium

ABSTRACT

For full characterization of a surface micromachined MEMS device, where the thickness of the moving layer is just a few times the gap size, the modeling has to take large-signal behavior and end effects into account. In this work, a numerical method using finite differences is implemented in Simulink to solve the Reynolds equation. The spatial derivatives are solved using the finite differences. The use of the Simulink capabilities for time integration allows solving of the time derivatives at any mesh point. To increase efficiency, a low-level language is used inside Simulink to solve the parameterized finite differences and the time derivatives. As the Reynolds equation is being solved inside a high-level language description, other system parameters (such as: mass, spring constant, non-trivial geometry) can be easily incorporated. Measurements on a complex 2DOF MEMS device are compared with simulation results, and the agreement validates the full system approach proposed.

Keywords: MEMS modeling, squeeze film damping, large-signal analysis, macro model

1 INTRODUCTION

The penetration of microelectromechanical system MEMS technology in an increasing number of applications calls for advanced modeling tools to deal with the complexity of the system on the microscale and the commercial drive towards first-time-right and fast turnaround design [1].

The dynamics of a MEMS microstructure is governed by inertia and the squeeze-film damping. The non-linear behavior of the Reynolds equation and the frequency dependence of the gas film present a modeling challenge.

For simple geometries and simple movements, the Reynolds equation can be solved and efficient full-system models based on analytical solutions can be implemented [2,4] accounting for large signal behavior [3,4] and including end effects [5,6]. For more complex geometries, the Reynolds equation can not be solved analytically and numeric solutions have to be used. Usually these solutions are based on Finite Element Modeling [6], or Finite Differences [7]. However, for simulation at the full-system level, while accounting for large signal and end effects, the calculation effort required to accurately describe the motion of the overall system becomes excessive. In this paper we

report a flexible way to model complete MEMS systems. Reynolds equation is solved using the finite differences method, implemented in such a way that other relevant system parameters can be introduced. Simulations of a model of a 2-degree-of-freedom (2DOF) structure are compared with measurements.

2 MOMENT ACTUATED ACCELEROMETER

The device used in this work is a 2DOF moment-actuated accelerometer (Fig. 1). The movement of the structure is fully characterized by two state variables: displacement w_1, and angle φ_1 (Fig. 2). Such a moment actuated device may compare favorably to a normal 1DOF accelerometer in terms of damping coefficient (higher quality factor). The modeling difficulties of the structure arise from the fact that it presents two distinct movements (translational and rotational) with cross-couple terms in between.

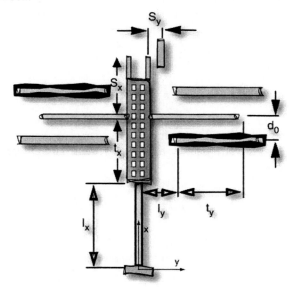

Figure 1: Schematic of the moment actuated accelerometer.

2.1 Fabricated Device

The epi-poly process was used for the fabrication of the test structures [8]. This process is very suitable for the fabrication of relatively thick and high aspect ratio free-standing beams on top of a silicon wafer. This device is basically a free-standing lateral beam (200 μm long, 3 μm wide and depth of 10.6 μm) anchored at one end (the base) only (Fig. 3).

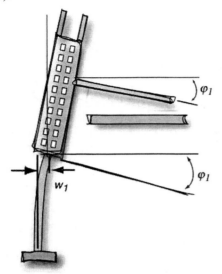

Figure 2: Identification of the state variables used in the 2DOF model.

The beam can be deflected by electrostatic actuation in the plane of the wafer using a voltage applied across parallel plate capacitors (2μm gap). These are composed of two sets of electrodes located alongside the free-standing tip, with counter electrodes anchored to the substrate. The deflection can be measured using a set of differential sense capacitors located alongside the free-standing tip. Finally, there are electrically isolated stoppers to limit the lateral motion.

3 SQUEEZE FILM MODELING

For structures in which only the width of the small gap between two plates changes in time, the pressure changes p relative to the wall velocity are described by the modified Reynolds equation [3]:

$$\frac{\partial}{\partial x}\left[\rho h^3 Q_{pr} \frac{\partial p}{\partial x}\right] + \frac{\partial}{\partial y}\left[\rho h^3 Q_{pr} \frac{\partial p}{\partial y}\right] = 12\eta \frac{\partial(\rho h)}{\partial t}, \quad (1)$$

where ρp^{-n} is constant, and the pressure p, gas density ρ and gap size h are functions of space and time. The gas has a viscosity η, and Q_{pr} is the relative flow rate coefficient. When an isothermal process is assumed (n=1), density ρ can be replaced with pressure p.

The modified Reynolds equation is used, because rarefaction effects have to be included. For transitional and molecular damping regimes, Q_{pr} is a function of the Knudsen number K_n, the ratio between the mean free path of the gas molecules and the gap separation. In this work, the flow rate coefficient is given by [2]:

$$Q_{pr} = 1 + 9.638(K_n), \quad K_n = \frac{\lambda_0 P_0}{ph}, \quad (2)$$

where λ_0 denotes the mean-free-path at pressure P_0.

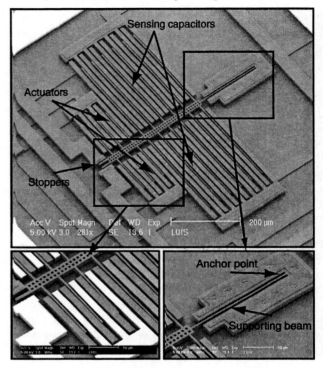

Figure 3: Photograph of the fabricated device.

3.1 Finite Differences Model

For the finite differences model, the surface is first divided into a rectangular grid of M×N elements (x=m.Δx, y=n.Δy, m=0…M-1, n=0…N-1). At each mesh point, equation (1) is implemented. The spatial derivatives are solved using the finite difference method [9]. Using this method, we end up with a set of M×N time differential equations.

Very important in any squeeze-film model, is the inclusion of large signal effects (already accounted for in the modified Reynolds equation) and end effects [5,6]. Incorporation of the end effects in this model implies that the pressure on the plate edges is not simply assumed at ambient pressure, but rather that the system dynamics are also considered at the device edges [6].

3.2 Model Implementation

In order to solve the time differential equations, Simulink was used. A parameterized model was built in a low level language (C language) and introduced in Simulink.

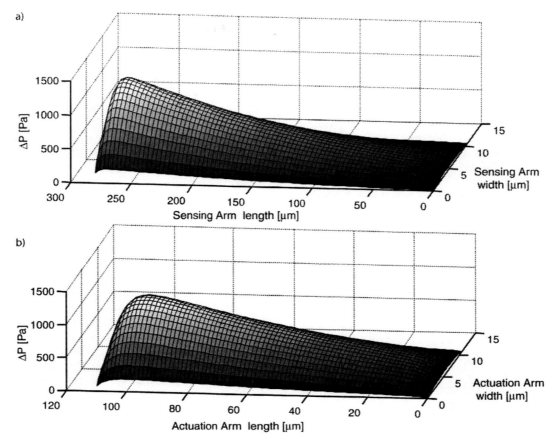

a)

b)

Figure 4: Pressure distribution in a a) sensing and b) actuation arm using a 80×20 finite difference mesh.

For each time interval, the spatial derivatives are solved using the finite difference method, and the time derivatives are solved by the methods already implemented in Simulink. The use of a high-level language description enables the introduction of other system properties (full system functionality), and a very good parameter flexibility. Once the model is implemented is very easy to study the influences of the various parameters (mesh size, structure dimensions, pressure changes, gap sizes, etc.).

The implemented squeeze film model was tested for the different arms of the structure (actuation and sensing arms). As these have different sizes, different mesh grids have to be used for each of the arms. The pressure distribution is presented in Fig. 4 for a single actuation and sensing arm, when the structure oscillates with a maximum angle (φ_1=0.0051 radians) and displacement (w_1=0.685μm) at 400 kHz.

3.3 Reduced-Order Model Generation

Another significant advantage of the use of a high level language description is the increase in flexibility. Reduced-order modeling techniques [10] have been introduced to solve dynamic problems. Based on some FEM or FD simulations, a reduced model can be build having the same response of the original gas film full model (even for complex geometries).

For testing the finite difference modeling approach and evaluation of the advantages of reduced-order models, a reduced-order model was build with just a spring-damper network [10]. For a large-signal behavior of the reduced-order model, simulations have to be performed for several gap sizes, since the values of the spring and damper are gap dependent. The huge advantage of the reduced model is a large decrease in computer time per simulation.

Moreover, the flexibility of the finite difference model enables the automatic implementation of the squeeze-film reduce model: the simulations are performed in a programmed sequence, and all the fitting that is needed is automatically generated.

After the generation of the reduced-order model, some results of the finite difference model are compared with the reduced model. Fig. 5 shows the damping moment of both models, when the structure oscillates with a maximum angle (φ_1=0.0051 radians) and displacement (w_1=0.685μm) at 500 kHz.

4 FULL SYSTEM MODEL

The movement of the 2DOF structure, in the absence of an external acceleration, can be described by the non-linear differential equations (a voltage V is applied to the actuation arms):

$$\begin{cases} F_m + F_b(w_1, \varphi_1) + F_k(w_1, \varphi_1) = F_{elect}(w_1, \varphi_1, V) \\ M_m + M_b(w_1, \varphi_1) + M_k(w_1, \varphi_1) = M_{elect}(w_1, \varphi_1, V) \end{cases} \quad (3)$$

A translation and a rotational equation describe the full system. Most of the forces and moments depend on both state variables (w_1 and φ_1) – cross-couple terms are present. All these dependencies can be easily implemented within Simulink. The full system (with the finite difference method used to compute the damping force and moment) is thus implemented. Simulations were performed for various input voltages.

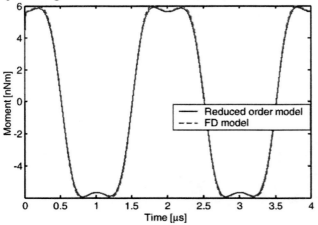

Figure 5: Comparison of the damping moment between full and reduced order models.

5 EXPERIMENTAL RESULTS

Measurements on fabricated devices have been compared with simulations (Fig. 6). The results are in good agreement, thus demonstrating the validity of the model.

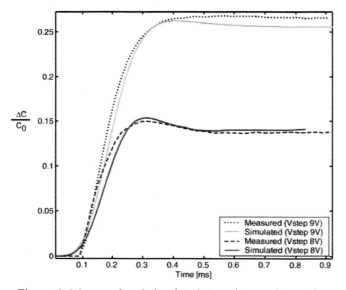

Figure 6: Measured and simulated capacitance change for an input step of 8 and 9 volts.

6 CONCLUSIONS

As demonstrated, MEMS devices can generally be modeled with Finite Differences. Even the motion of very complex geometries is adequately described by one translation – perpendicular to the gap – and two rotational movements.

The full-system can be simulated with this approach. The capability of automatic generation of reduced-order models is another advantage of the proposed model. As shown in Fig. 5, a simple spring-damper network presents the same behavior of the modified Reynolds equation. This may lead to a fast development time in the design phase, and also to a much better understanding of the full dynamics of complex MEMS structures.

REFERENCES

[1] S. D. Senturia, "Perspectives on MEMS, past and future: The tortuous pathway from bright ideas to real products", in *Proc. Transducers'03*, Boston, USA, 2003, pp. 10-15.
[2] T. Veijola, H. Kuisma, J. Lahdenperä and T. Ryhänen, "Equivalent-circuit model of the squeezed gas film in a silicon accelerometer", *Sensors and Actuators*, vol A48, pp. 239-248, 1995
[3] T. Veijola, H. Kuisma, and J. Lahdenperä, "Compact large-displacement model for capacitive accelerometers", in *Proc. MSM'99*, (San Juan), Apr. 1999, pp. 218-221.
[4] L.A. Rocha, E. Cretu and R.F.Wolffenbuttel "Displacement model for dynamic pull-in analysis and application in large-stroke electrostatic actuators" in *Proc. Eurosensors XVII*, Guimarães, Portugal, 2003, pp. 448-451.
[5] T. Veijola, "End effects of rare gas flow in short channels and in squeezed-film dampers", in *Proc. MSM'02*, San Juan, Puerto Rico, USA, 2002, pp. 104-107.
[6] S. Vemuri, G.K. Fedder and T. Mukherjee, "Low-order squeeze film model for simulation of MEMS devices", in *Proc. MSM'00*, San Diego, USA, 2000, pp. 205-208.
[7] T. Veijola, "Finite-difference large-displacement gas-film model", in *Proc. Transducers'99*, Sendai, Japan, 1999, pp. 1152-1155.
[8] http://www.europractice.bosch.com/en/start/index.htm.
[9] H. Levy and F. Lessman, "Finite difference equations" New York: Dover, 1992
[10] J. E. Mehner, W. Doetzel, B. Schauwecker and D. Ostergaard, "Reduced-order modeling of fluid structural interactions in MEMS based on model projection techniques", in *Proc. Transducers'03*, Boston, USA, 2003, pp. 1840-1843.

Mixed-Domain Simulation of Step-Functional Voltammetry with an Insoluble Species for Optimization of Chemical Microsystems

Steven M. Martin, Timothy D. Strong, Fadi H. Gebara, Richard B. Brown

Department of Electrical Engineering and Computer Science
University of Michigan, Ann Arbor, MI 48109
{stevenmm, strongtd, fadi, brown}@umich.edu

ABSTRACT

Microinstruments with integrated sensors and electronics are well adapted for remote environmental monitoring applications. Optimization of these microinstruments, however, has been difficult due to a previous lack of mixed-domain simulation environments. This work presents a complete electrode model for step-functional voltammetry with an insoluble species. The model includes the electrochemical response, sensor parasitics, sensor noise, and electrical instrumentation, and can be simulated in a standard electronic CAD environment. The simulation results are verified against experimental data taken using microfabricated sensors. Finally, the design environment is used to analyze system-wide design trade-offs.

Keywords: mixed-domain simulation, voltammetry, microsystem design, toxic metals

1 INTRODUCTION

An important application of electrochemical microsensors is the analysis of trace levels of toxic metals [1]. Electrochemical analyses using anodic stripping voltammetry (ASV) can provide detection limits down to 10 picomolar for some heavy metals [2]. For remote monitoring applications, it may be necessary to miniaturize or even integrate the electronic instrumentation for ASV directly with the sensors [1]. Furthermore, the long-term deployment of these microinstruments requires that both sensor and electronics be optimized to operate on small power budgets, occupy little area, and provide adequate detection limits. Unfortunately, the trade-offs among these goals are not always obvious and simulation tools for system optimization have not existed, making chemical microinstrument optimization, or even designing to specification, a difficult task at best.

While physics-based models for sensor parasitics [3], sensor noise sources [4,5], and electrochemical faradaic currents [6] have previously been developed, there has been limited work in combining the components into a cohesive model and analyzing the trade-offs for detecting heavy metals on bare electrodes. Morgan [4] performed analyses for voltammetric sensors and their instrumentation when used to detect analytes with soluble products–a problem which has a closed-form solution. Electrochemical detection of heavy metals, however, involves voltammetry with a species that has an insoluble product. Voltammetry at bare electrodes with an insoluble species has no closed-form solution and requires numerical methods to solve.

This work presents a complete model and method for the system-level simulation of a mixed-domain, bare-electrode, heavy-metal microdetector. First, a sensor model is presented which combines linear electrical components and Verilog-AMS code to model the electrode process, parasitics, and noise. The model is then combined with schematics of the instrumentation and simulated using the circuit simulator, *Spectre®*. The model is verified against experimental results, and design optimization analyses employing the simulation environment are discussed.

2 SENSOR FUNDAMENTALS

2.1 Electrode Processes with Insoluble Species

Voltammetry is an electrochemical technique in which a series of potentials, E, are applied between an electrode and solution, and the resulting faradaic current, I_f, is measured [2]. This electrode is called the working electrode (WE), and the faradaic current is given by

$$I_f(t) = nFAD_o \frac{\partial C_o(x, t)}{\partial x}\bigg|_{x=0} \quad (1)$$

where x is the distance away from the electrode surface, n is the number of electrons transferred in the redox reaction, F is Faraday's constant, A is the area of the electrode, and D_o and C_o are the diffusion coefficient and concentration of the oxidized form of the analyte, respectively. The concentration change is defined by Fick's 2nd Law of Diffusion which is the partial differential equation,

$$\frac{\partial C_o(x, t)}{\partial t} = D_o\left(\frac{\partial^2 C_o(x, t)}{\partial x^2}\right). \quad (2)$$

The boundary conditions for voltammetry with an insoluble species on a bare electrode were first reported in [6] for a cyclic voltammetry experiment. Altering the boundary conditions of [6] to apply to any arbitrary step-functional voltammetry yields the three boundary conditions:

$$C_o(x, t)\big|_{\substack{t=0 \\ x \geq 0}} = C_o^* \quad (3)$$

$$\lim_{x \to \infty} C_o(x, t) = C_o^* \quad \text{for } t \geq 0 \quad (4)$$

$$\frac{-D}{K_s}\frac{\partial C_o}{\partial x}\bigg|_{x=0} = a\exp\left[\frac{\beta nF}{RT}(E-E^0)\right]$$
$$-C_o(0,t)\exp\left[\frac{-\alpha nF}{RT}(E-E^0)\right] \quad (5)$$

where $a = a_\infty\left\{1-\exp\left[-\gamma\left(Q-\int_0^t i\,dt\right)\right]\right\}$. (6)

C_o^* is the initial concentration, K_s is the heterogeneous rate constant for the redox reaction, a is the activity of the insoluble species on the electrode, α and β are transfer coefficients, E^0 is the reduction potential of the analyte, Q is the total charge of the moles plated onto the electrode, R is the gas constant, γ is the activity coefficient, and T is temperature. This partial differential equation has no closed-form solution but can be solved numerically using a difference equation technique. The solution method used here follows the derivation presented in [6] except that a step-functional input waveform is assumed instead of a cyclic waveform. This slightly alters the change of variables in the derivation.

2.2 Parasitics

Application of the desired potential to the test solution requires the use of additional electrodes. Typical microamperometric sensors consist of three electrodes—the auxiliary (AE), reference (RE), and working electrodes (WE) [3]. Unfortunately, each of these electrodes introduces parasitics into the electrochemical cell. These parasitics can be adequately modeled using linear electrical components as shown in Figure 1 [3]. C_a and C_w represent the double-layer capacitances of both the AE and WE respectively, and are given by [7]

$$C_j \approx 20\left(\frac{\mu F}{cm^2}\right)\cdot A_j \quad (7)$$

for each electrode j. R_s and R_t represent the solution resistance between the various electrodes and are given by [3]

$$R_j = \frac{d_j}{\kappa A_j} \quad (8)$$

where d_j is the distance between the two electrodes and κ is the solution conductivity. The charge transfer impedances at the AE and WE, R_a and R_w, are given by [2]

$$R_j = \frac{1}{i_{o_j}nFA_j} \quad (9)$$

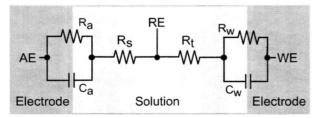

Figure 1: Parasitics and noise model of a three-electrode voltammetric sensor.

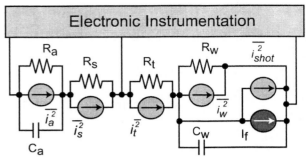

Figure 2: Microsystem model including faradaic current source, sensor parasitics, noise sources, and electronic instrumentation.

where i_o is the equilibrium current density. Of the parasitics, C_w is of utmost concern since any current needed to charge/discharge the double-layer capacitance is summed with the concentration dependent faradaic current. Large charging currents mask the desired redox current and lead to poor detection limits. Equations (7)-(9) demonstrate that the parasitics have dependencies on the areas of the electrodes and their separations. Furthermore, the relationships between the various parasitics and their areas create trade-offs in electrode sizing.

2.3 Noise

Noise in the electrochemical cell is generated from both the faradaic current and from the parasitics. Similar to any resistor, the solution impedances and charge transfer resistances produce a thermal current noise given by

$$\overline{i_{thermal}^2} = \frac{4kTB}{R} \quad (10)$$

where k is the Boltzmann constant, T is temperature, B is the bandwidth, and R is the impedance [4]. Additional noise in the form of shot noise arises from the discrete electron transfer from ion in solution to electrode and is given by

$$\overline{i_{shot}^2} = 2qB(I_f+I_c) \quad (11)$$

where q is the charge of an electron and I_c is the charging current [4]. In the electrical model, these noise sources can be added in parallel with their respective sources. Since the magnitudes of R, I_f and I_c are dependent on area, electrode dimensions directly affect the total noise generated in the system.

3 SYSTEM SIMULATION

3.1 System Model

With the various components of the sensor defined, the modeling and simulation of the entire chemical microsystem can be achieved. Combining the faradaic current, parasitics, noise sources, and necessary electronics yields the electrically equivalent model shown in Figure 2. The electrical simulator *Spectre®* was chosen as the simulation environment since it can operate on both electrical netlists and Verilog-AMS code. The electrical parts of the system model

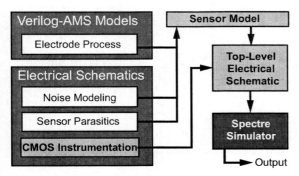

Figure 3: Design environment for the simulation of the heavy-metal microdetector.

Variable	Value	Description
A_w	0.00375	Area of working electrode (cm^2)
A_a	1	Area of auxiliary electrode (cm^2)
κ	0.01	Conductivity of solution (S/cm)
d_w	1	Distance between RE and WE (cm)
d_a	2	Distance between RE and AE (cm)
B	100	Circuit bandwidth (kHz)
Q	0	Initial charge deposited (C)
D_o	9.9e-6	Pb^{2+} diffusion coefficient (cm2/s)
n	2	Electrons transferred
K_s	5e-3	Rate constant (cm/s)
E^0	-0.5	Standard potential (V)
α, β	0.5	Transfer coefficients
γ	0.5	Activity coefficient
l	4000	Max diffusion distance (unitless)
N_x	1000	Number of Δx bins
$Max_{\Delta t}$	15	Maximum timestep (ms)

Table 1: Simulation parameters used.

were implemented in schematic form with variables representing key design parameters like electrode areas and separation. Verilog-AMS was used to find the numerical solution to step-functional voltammetry with insoluble species present. Noise analysis was accomplished using the standard *Spectre®* noise solver with inputs derived from the transient simulations of sensor currents. Figure 3 details the implementation of the system model for mixed-domain simulation.

3.2 Simulation Time

Since the solution of the partial differential equation defining the faradaic current involves solving a difference equation in two variables (x and t), the simulation time is on the order of N^2 as shown in Figure 4. To speed the simulation, Δx and Δt can be scaled, but scaling these values ultimately degrades the accuracy of the solution. Figure 4 shows the percent error incurred by scaling Δx and Δt.

4 VERIFICATION

To verify the accuracy of the simulation environment, results from the simulations were compared with experimental results. The experimental setup consisted of a microfabricated, 0.375mm^2 Pt working electrode [8], a saturated calomel reference electrode, and a Pt wire auxiliary elec-

trode submerged in a solution of 10mM HNO_3, 10mM NaCl with various concentrations of Pb^{2+}. A commercial potentiostat (FAS2^{TM} Femtostat, *Gamry Instruments*) was used to generate and record the voltammetric sweeps. The experiments were run in a faraday cage to limit the effects of noise on the system. Table 1 lists the simulator's input parameters.

Figure 5 shows the results of total electrolyzed charge versus initial concentration of Pb^{2+} during a cyclic voltammogram. Both the experimental and simulated data show a linear trend versus concentration and compare reasonably well. Figure 5 also shows the dynamic response of the system during a cyclic voltammetric sweep. The simulated and experimental waveforms have similar shapes, but the experimental peak current is larger than the simulation predicts. A simple fitting parameter will be added to the simulation in future experiments to correct this inaccuracy.

Figure 6 compares the individually normalized peak currents versus voltammetric scan rate during a linear sweep voltammogram. These curves were generated in 1mM Pb^{2+}.

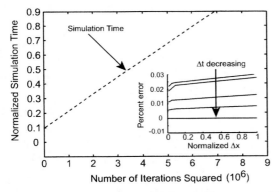

Figure 4: Normalized simulation time vs. the square of the number of discrete Δx and Δt bins chosen for solving the differential equation. (Inset) Percent error in the voltammetric peak current for various values of Δx and Δt.

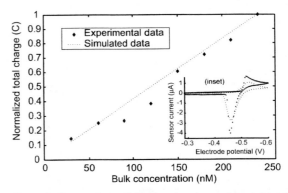

Figure 5: Comparison of simulated and experimental total electrolyzed charge during a cyclic voltammogram. (Inset) Dynamic response of simulation and experiment during a cyclic voltammogram.

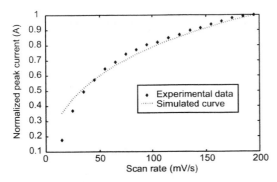

Figure 6: Comparison of simulated and experimental voltammetric peak current versus scan rate of the applied potential.

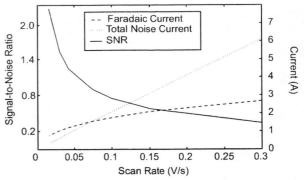

Figure 8: Simulation results for the analysis of signal-to-noise ratio versus scan rate for a heavy-metal microdetector.

Again, the simulated and experimental results agree favorably. Finally, Figure 7 compares the individually normalized peak current versus deposition time [2] during ASV with linear scan stripping. The simulated curve has a natural logarithm shape due to the anticipated absence of stirring during deposition, while the experimental results show a more linear tendency. This discrepancy can be attributed to unwanted convection in the experimental setup during deposition.

5 MICROSYSTEM ANALYSIS

The design environment detailed here can be used to alter various system parameters and analyze their effects on overall system performance. One such analysis is the effect of the applied potential's scan rate on the signal-to-noise ratio of the system. The results of this analysis are given in Figure 8. The results show an increase in the total noise current as scan rate increases. This is due to the increasing charging currents required to charge/discharge the working electrode's double-layer capacitance. As verified in section 4, the faradaic current also increases with scan rate. The faradaic current, however, increases at a much slower rate than the total noise creating the signal-to-noise ratio curve shown. For the optimization of SNR, the simulations show that slow scan rates are better. The design environment presented is useful for all types of chemical microsystem optimization including power, area, bandwidth, and detection limit.

6 CONCLUSION

A complete model for electron transfer between electrodes and analytes with insoluble products has been developed through the synthesis of several existing models. The model was implemented in Verilog-AMS and was simulated within the framework of a standard CAD tool allowing for complete mixed-domain simulation. Results were experimentally verified by analyzing Pb^{2+} on microfabricated Pt electrodes. The simulation environment was then used to investigate system-wide trade-offs. This work aids in the development of near optimal chemical microsystems. This optimization is required if microsystem technology is to become a cost-effective, environmental monitoring alternative.

ACKNOWLEDGEMENTS

This work was supported in part by the Engineering Research Centers Program of the US National Science Foundation under Award Number EEC-9986866 and under a National Science Foundation Graduate Research Fellowship.

REFERENCES

[1] S. Martin, T. Strong, F. Gebara, et. al., "Integrated Microtransducers and Microelectronics for Environmental Monitoring," *Proceedings of the 15th Biennial University/Government/Industry Microelectronics Symposium (UGIM 2003)*, Boise, ID, pp. 300-303, June 2003.

[2] P. Kissinger, and W. Heineman Eds. *Laboratory Techniques in Electroanalytical Chemistry Second Edition, Revised and Expanded*, Marcel Dekker, Inc., New York, 1996.

[3] A. Bard, and L. Faulkner, *Electrochemical Methods*, New York: John Wiley & Sons, 1980.

[4] D. Morgan, and S. Weber, "Noise and Signal-to-Noise Ratio in Electrochemical Detectors," *Analytical Chemistry*, vol. 56, pp. 2560-2567, 1984.

[5] R. VandenBerg, A. DeVos, and J. DeGoede, "Resistivity fluctuations in ionic solutions," in *8th Intl. Conf. on Noise in Physical Systems*, A. D'Amico, and P. Mazzetti, Eds., North Holland, New York, 1985.

[6] K. Brainina, *Stripping Voltammetry in Chemical Analysis*, John Wiley & Sons, New York, NY, 1974.

[7] M. Madou, and S. Morrison, *Chemical Sensing with Solid State Devices*, Academic Press, Inc., New York, 1989.

[8] M. Poplawski, H. Cantor, et.al., "Microfabricated amperometric biosensors," *Intl. Conf. on Solid-State Sensors and Actuators (Transducers '91)*, San Francisco, Ca, June24-28, 1991, pp. 51-53.

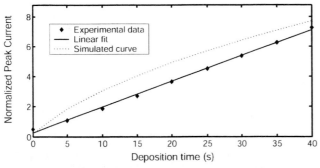

Figure 7: Comparison of simulated and experimental normalized peak current versus deposition time.

System-Level Optical Interface Modeling for Microsystems

Timothy P. Kurzweg[*], Ankur S. Sharma[*], Shubham K. Bhat[*],
Steven P. Levitan[**], and Donald M. Chiarulli[**]

[*]Drexel University, Department of Electrical and Computer Engineering,
Philadelphia, PA 19104, kurzweg@ece.drexel.edu
[**]University of Pittsburgh, Departments of Electrical Engineering and Computer Science,
Pittsburgh, PA 15261

ABSTRACT

In this paper, we present an accurate and computationally efficient system-level optical propagation technique suitable for the modeling of optical interfaces. Our technique is based on extensions to the angular spectrum technique used to solve the Rayleigh-Sommerfeld formulation. By using a FFT, the angular spectrum technique is efficient and suitable for system-level modeling of the complete system. To support the reflection and transmission at optical interfaces, we implement a semi-vector technique, taking into account the polarization of the optical wavefront. The polarization is used to determine the reflection and transmission coefficients through the use of Berreman's 4x4 matrix. Solutions are provided for typical TE and TM waves, however, wavefronts with arbitrary linear polarization are also supported. In this paper, we present a system-level simulation of a Silicon on Sapphire (SOS) interface.

Keywords: angular spectrum, semi-vector, Fresnel coefficients, FDTD, system-level simulation.

1 INTRODUCTION

Current optical systems are becoming smaller with rapid developments in the field of device fabrication. [1]. In addition, optical coatings and thin films are being used advantageously to increase the performance of optical microsystems. As optical devices approach the size of optical wavelengths, full vector solutions, such as the Finite Difference Time Domain (FDTD) technique, are required for accurate solutions. However, the distance of propagation computationally limits the FDTD approach, as distances on the order of tens to hundreds of wavelengths require excess solutions for the electromagnetic boundary conditions [1]. To overcome this costly computation, a scalar approach is typically used to solve for propagation of distances larger than the wavelength of light. In previous work, we have shown the use of a scalar approach for accurate and efficient modeling of systems using the angular spectrum technique, used to solve the Rayleigh-Sommerfeld formulation [3]. The limiting condition on scalar techniques is the distance of propagation, which must be greater than the wavelength of operation. Therefore, scalar method cannot be used for optical modeling at optical interfaces due to its inherent drawback to account for the reflected and transmitted rays. Typically, a full-vector approach is used to solve for the reflected and transmitted wavefronts, however, as discussed earlier, this technique is computationally inefficient for full system simulation. In this paper, we present a semi-vector scalar solution used to model optical interfaces accurately and efficiently. Our system-level modeling requires a two-step approach: the angular spectrum technique is used to solve for propagation, and a semi-vector approach is used to solve for the optical interfaces.

2 TECHNIQUE

In this section, we present our semi-vector modeling of an optical interface. We first show the angular spectrum technique, followed by our extension for the reflection and transmission waves.

2.1 Angular Spectrum Technique

The angular spectrum technique, used to solve the Rayleigh-Sommerfeld formulation, decomposes a complex wavefront into a set of plane waves through a Fourier transform. Each plane wave is multiplied by the medium transfer function and recombined into a propagated complex wave by taking the inverse Fourier transform. We next present brief details on the technique.

Consider the propagation of a monochromatic optical wave of wavelength λ and complex amplitude $U(x,y,z)$ in free space between the planes $z = 0$ and $z = d$, called the aperture and observation planes respectively. d denotes the distance between the two planes in the z direction. Let us regard $f(x,y)=U(x,y,0)$ as the complex amplitude at the aperture plane and $g(x,y)=U(x,y,d)$ as the complex amplitude of the observation plane. These planes are shown in Figure 1.

The Fourier transform of the input wave, $A(v_x,v_y,0)$, decomposes the complex optical wavefront into plane waves defined by the spatial frequencies v_x and v_y. The Fourier Transform is show in Equation (1).

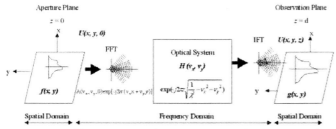

Figure 1: Angular spectrum method

$$A(v_x, v_y, 0) = \iint U(x,y,0) \exp[-j2\pi (v_x x + v_y y)] \delta x \delta y \quad ...(1)$$

The exp $\{-j2\pi (v_x x + v_y y)\}$ term represents the plane waves propagating with different angles related to the spatial frequencies: $\theta_x = \sin^{-1}(\lambda v_x)$ and $\theta_y = \sin^{-1}(\lambda v_y)$. This is also graphically shown in Figure 1.

The product of the free space/propagation transfer function and the wave function in the frequency domain, describes the position of each of the plane waves after propagation. The plane waves are transformed back into the spatial domain using the inverse Fourier transform seen below:

$$U(x,y,z) = \iint A(v_x, v_y, 0) \exp[-j2\pi (v_x x + v_y y)] \exp(-jk_z z) \delta v_x \delta v_y$$
$$... (2)$$

In this equation $k_z = 2\pi (1/\lambda^2 - v_x^2 - v_y^2)^{1/2}$. Further details can be found in [3].

2.2 Angular Spectrum Extension for Reflected and Transmitted waves

To support reflection and transmission at the optical interface, we must find coefficients for the reflected and transmitted waves. This is achieved in three steps, which are applied to each plane wave in the frequency domain.
1. Calculate the spatial frequencies and transmission angles (v_{ty}, v_{tx}, θ_{tx}, θ_{ty}) using Snell's law.
2. Determine the reflection and transmission coefficients $R(v_x, v_y)$ and $T(v_x, v_y)$.
3. Multiply $R(v_x, v_y)$ and $T(v_x, v_y)$ to obtain the reflected and transmitted waves respectively.

Details are provided in the following sections.

2.2.1 Calculation of $v_{ty}, v_{tx}, \theta_{tx}, \theta_{ty}$

At the boundary between two planar isotropic media, the incident plane wave in the frequency domain is decomposed into a reflected and a transmitted ray, as shown in Figure 2.

In order to calculate the reflection and transmission coefficients we first need to determine the reflected, θ_{rx} and θ_{ry}, and transmitted angles, θ_{tx} and θ_{ty}, which are related to the spatial frequencies for each plane wave in both the x and y directions. By applying Snell's law we calculate the transmitted angles with the following relations:

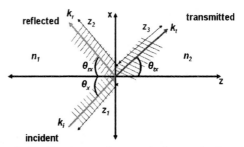

Figure 2: Reflection and transmission at the boundary between two planar isotropic media.

$$\theta_{tx} = \arcsin(n_i / n_t \sin \theta_x) \quad ...(3)$$
$$\theta_{ty} = \arcsin(n_i / n_t \sin \sin \theta_y) \quad ...(4)$$
$$v_{tx} = \sin(\theta_{tx})/\lambda \quad ...(5)$$
$$v_{ty} = \sin(\theta_{ty})/\lambda \quad ...(6)$$

For reflected angles, we have, $\theta_{rx} = \theta_x$, $\theta_{ry} = \theta_y$, $v_{rx} = \sin(\theta_{rx})/\lambda$, $v_{ry} = \sin(\theta_{rx})/\lambda$.

2.2.2 Calculation of Reflection and Transmission Coefficients

To calculate the coefficients of the reflection and transmission waves, the incident polarization must be known. Scalar techniques do not inherently support polarization. Therefore, we need to apply a semi-vector theory to solve for the boundary conditions, which are required by electromagnetic theory to obtain the reflection and transmission coefficients $R(v_x, v_y)$ and $T(v_x, v_y)$. Each plane wave in the frequency domain will be multiplied by these coefficients to find the reflected and transmitted ray respectively.

The reflection coefficients are found using the 4X4 matrix developed by Berreman for multilayered structure [2]. We assume reflection from a single planar interface and since the mediums are isotropic, the reflection and transmission coefficients are simply the Fresnel's coefficients [5].

Let us regard θ_x and θ_y to be the incident angles for TE (x-polarized) and TM (y-polarized) waves. n_i and n_t to be the refractive indices of the incident and transmitted mediums separated by the planar interface. Using Snell's law, we obtain θ_{tx} and θ_{ty} as the transmitted wave vector angles in y-z and x-z planes respectively, as seen in the previous section.

The reflection and transmission coefficients for transverse electric (TE) and transverse magnetic (TM) polarized waves are given by r_x, r_y, t_x and t_y respectively.

$$r_x = \frac{n_i \cos(\theta_x) - n_t \cos(\theta_{tx})}{n_i \cos(\theta_x) + n_t \cos(\theta_{tx})}, \text{ TE Reflection} \quad ...(7)$$

$$r_y = \frac{n_i \cos(\theta_{ty}) - n_t \cos(\theta_y)}{n_i \cos(\theta_{ty}) + n_t \cos(\theta_y)}, \text{ TM Reflection} \quad ...(8)$$

$$t_x = \frac{2n_i \cos(\theta_x)}{n_i \cos(\theta_x) + n_t \cos(\theta_{tx})} = 1 + r_x, \text{ TE Transmission} \quad \dots (9)$$

$$t_y = \frac{2n_i \cos(\theta_y)}{n_i \cos(\theta_{ty}) + n_t \cos(\theta_y)} = n_i / n_t (1 + r_y), \text{ TM Trans} \dots (10)$$

The reflection and transmission coefficients for TE and TM polarized waves are:

$$R_{TE} = r_x^2, R_{TM} = r_y^2 \qquad \dots (11)$$

$$T_{TE} = \left(\frac{n_t \cos(\theta_t)}{n_i \cos(\theta_i)} \right) t_x^2, T_{TM} = \left(\frac{n_t \cos(\theta_t)}{n_i \cos(\theta_i)} \right) t_y^2 \qquad \dots (12)$$

For general linear polarization cases, we support waves with any polarization orientation angle. Let θ_0 be the orientation angle, specifying the direction of the electric field vector, $\mathbf{E}$, relative to the positive y-axis. Since both regions are isotropic, there is only one reflected and one transmitted wave. If we observe that the two normal modes for this system are linearly polarized waves with polarization along the x and y directions, the solutions for the transmitted and reflected wave are completely separable [4]. It can be determined that the incident, the reflected, and the transmitted wave with their electric field vectors pointing in the x-direction are self–consistent with the boundary conditions [7]. Hence, the x-polarized or the TE portion of the wave and the y-polarized or the TM portion of the wave can be considered independently and then recombined through linear superposition. As seen in the equations below, the terms are completely separable and there is no coupling of waves at the interface:

$$r = \{ [r_x \cos(\theta_0)]^2 + [r_y \sin(\theta_0)]^2 \}^{1/2} \qquad \dots (13)$$

$$t = \{ [t_x \cos(\theta_0)]^2 + [t_y \sin(\theta_0)]^2 \}^{1/2} \qquad \dots (14)$$

$$R = r^2 \qquad \dots (15)$$

$$T = \left(\frac{n_t \cos(\theta_t)}{n_i \cos(\theta_i)} \right) t^2 \qquad \dots (16)$$

From these equations we verify that the results obtained for TE ($\theta_o = 0^0$) and TM ($\theta_o = 90^0$) waves are the same as those obtained in Equations (11) and (12).

2.2.3 Implementation

As seen previously in our discussion of the angular spectrum technique, the inverse Fourier transform is used to recombine the complex wavefront in the spatial domain. However, to solve for the reflected, U_r, and transmitted, U_t, complex wavefronts, the reflection and transmission coefficients must be multiplied to each of the angular frequencies as seen in the equations below:

$$U_r(x, y) = \iint A(v_x, v_y) \exp[-j2\pi (v_x x + v_y y) R(v_x, v_y)] \delta_{vx} \delta_{vy} \qquad \dots (17)$$

$$U_t(x, y) = \iint A(v_x, v_y) \exp[-j2\pi (v_x x + v_y y) T(v_x, v_y)] \delta_{vx} \delta_{vy} \qquad \dots (18)$$

We have proved that these equations support:

$$U_i(x, y) = U_r(x, y) + U_t(x, y) \qquad \dots (19)$$

$$R + T = 1. \qquad \dots (20)$$

In the case of a transmitted wave, the spatial frequencies are changed due to Snell's Law, as seen previously. In the spatial domain, it is necessary to compensate for this frequency change by remapping the corresponding spatial points.

3 EXAMPLE

In our first example, we demonstrate a Gaussian beam, with a spot size of 10 μm and wavelength of 1550 nm, at the interface between two optical planes having refractive indices of 1 and 1.5, respectively. In Figure 3, we can see the incident beam, the reflected beam and the transmitted beam at the interface. Figure 3, verifies Equation (19), with peak intensity values of: U_i=1, U_r= 0.04 and U_t=0.96.

In a second, more complex example, we model and simulate the air-sapphire interface found in the silicon-on-sapphire (SOS) technology [6]. The SOS process for the electrical portion of an opto-electronic multi-chip module (MCM) is growing in use. This SOS chip is bump-bonded to a bottom-emitting GaAs VCSEL/Detector chip, creating a complete MCM module, which can be used for source and detector components in short-haul optical interconnects [1]. The advantage of using a SOS process is that the sapphire substrate is transparent, allowing optical signals to pass through the substrate. An open area is left in the Si layer, creating a hole for the optical signal to pass into the sapphire substrate. A diagram of the MCM is shown in Figure 4(a), and a photograph of a detector chip taken through the sapphire is seen in Figure 4(b).

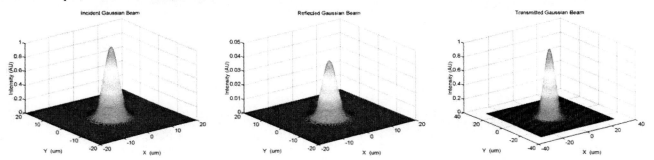

Figure 3: Incident, Reflected, and Transmitted complex optical wavefront at interface (n1=1, n2=1.5)

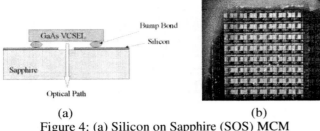

(a)   (b)

Figure 4: (a) Silicon on Sapphire (SOS) MCM
(b) Photodetector seen through the sapphire substrate

We simulate a TE Gaussian optical wavefront, emitting from a VCSEL. The emitted Gaussian beam intensity is normalized and has a wavelength of 850 nm and a spotsize of 10 μm. This beam is propagated 20 μm through free-space, past the sapphire (n=1.77) optical interface, and through 250 μm of sapphire. Details of this system can be seen in Figure 5.

In the first stage of simulation, this beam is propagated through a free space distance of 20 μm, which represents the gap between the GaAs VCSEL and the sapphire block. As seen in the intensity diagrams in Figure 5, the intensity drops below the normalized value of 1 with a spread in the original waist.

In the next stage, we simulate the beam propagation in the sapphire block. The reflected and transmitted Fresnel coefficients are calculated using the technique discussed in this paper. Using the reflected and transmitted Fresnel coefficients, we determine the reflected and transmitted rays at the air-sapphire interface. The transmitted ray at the interface is then propagated through a distance of 250μm in sapphire. Intensity plots at the different stages are shown in Figure 5.

4   CONCLUSION

In this paper we have presented a semi-vector technique to perform system-level simulation of optical interfaces.

This work builds on the combined advantages of using scalar methods for efficient optical propagation and vector techniques to solve for the reflection and transmission coefficients. Future work includes polarized beam propagation through complete systems and birefringent materials. Also, experimental validation of the technique will be performed for the simulated silicon-on-sapphire system.

REFERENCES

[1] Bakos, et. al, "Optoelectronic Multi-Chip Module Demonstrator System," OC2003, Wash., D.C.

[2] Berreman, "Optics in stratified and anisotropic media: 4x4-Matrix formulation," J. Opt. Soc. Am., Vol 62, No.4., pp. 502-510

[3] Kurzweg, et. al, "A Fast Optical Propagation Technique for Modeling Micro-Optical Systems," DAC, New Orleans, LA, June 10-14, 2002.

[4] Landry, Maldonado, "Gaussian beam transmission and reflection from a general anisotropic multiplayer structure," Applied Optics, Vol. 35, No. 30, October 1996, pp. 5870-5879.

[5] Landry, Maldonado, "Complete method to determine transmission and reflection characteristics at a planar interface between arbitrarily oriented biaxial media," J. Opt. Soc. Am. A, Vol. 12, No.9, September 1995, pp. 2048-2063.

[6] Liu, et. al, "High bandwidth ultra thin Silicon-on-Sapphire multi-channel optical interconnects," 23 Army Conference, Session B, BP-02.

[7] Prather, et. al "Electromagnetic analysis of finite-thickness diffractive elements," Opt. Eng. 41(8), (August 2002), pp.1792-1796.

[8] Scarmozzino, et. al, "Comparison of finite-difference and Fourier-transform solutions of the parabolic wave equation with emphasis on integrated-optics applications," JOSA A, Vol. 8, No. 5, May 1991, pp. 725-731.

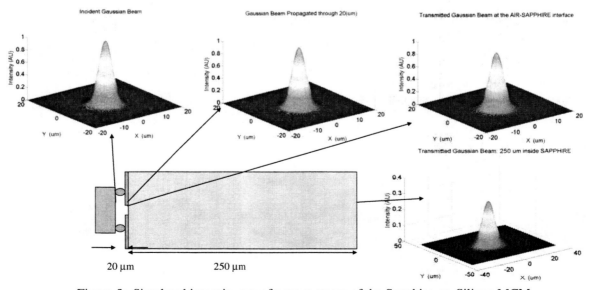

Figure 5: Simulated intensity wavefronts at stages of the Sapphire-on-Silicon MCM

Free surface flow and acousto-elastic interaction in piezo inkjet

Herman Wijshoff

Océ Technologies B.V., St.Urbanusweg 43, P.O.Box 101, 5900 MA Venlo, the Netherlands,
tel +31(0)77 3593425, fax +31(0)77 3595472, hwij@oce.nl

ABSTRACT

Modeling plays an essential role in our research on new inkjet technologies. Structural modeling with Ansys includes piezo-electricity. Acoustic modeling in Ansys and Matlab involves fluid-structure interaction. CFD modeling with Flow3D includes wall-flexibility and free surface flow with surface tension. Added to our measurements this reveals the phenomena involved in our main goal: firing droplets of ink at a very high rate with any desired shape, velocity, dimension and a reliability as high as possible.

Keywords: inkjet, acoustics, drop formation, two-phase flow, fluid-structure interaction

1 INTRODUCTION

Inkjet is an important technology in color document printing. Océ applies inkjet in its wide format color printing systems, Figure 1.

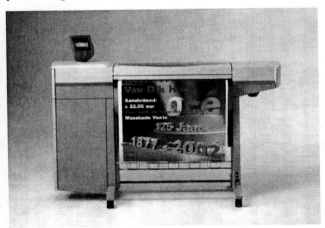

Figure 1: A recently introduced wide format inkjet printer, the Océ TCS400.

New piezo technologies provide opportunities towards higher productivity, quality and reliability.

2 PRINTHEAD BASICS

A long ink channel with a nozzle is the basic geometry of the inkjet device, Figure 2. A piezo actuator element drives each channel. To fire a droplet, an electric voltage is applied deforming the structure resulting in pressure waves inside the channel. The pressure waves propagate in both directions and will be reflected at changes in cross-section and compliancy of the channel structure. The channel structure is designed to get an large incoming positive pressure wave at the nozzle. In the small cross-section of the nozzle acceleration of the ink movement results in drop formation.

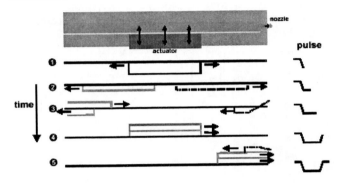

Figure 2: Actuation taking into account channel acoustics.

3 MEASURING

Droplets are measured by means of various optical methods like stroboscopic illumination at drop formation rate and high-speed camera recordings with a Phantom V7 camera. With laser-Doppler measurements meniscus movements without drop formation are recorded. These measurements give details on ink flow outside the printhead. The phenomena inside the channels are difficult to measure. The only suitable method uses the actuator also as a sensor. Switching the piezo elements from the electronic driving circuit to a measuring circuit gives a recording of the average pressure inside the ink channel.

4 MODELING

We need more information on the phenomena preceeding the drop formation for a better understanding of the physical processes. Details on ink flow and acoustic pressure waves are only available through modeling.

Two main aspects in modeling piezo inkjet are the acousto-elastic interaction and free surface flow. To model printhead dynamics and the interaction with ink acoustics the FEM code Ansys is used. Matlab is used for acoustic calculations based on the "narrow gap" theory. For free surface flow with surface tension (drop formation) and its impact on channel acoustics we use the VOF code Flow3D.

4.1 Structural modeling

When a voltage is applied to one piezo element also the neighbouring channels will be deformed. This results in a direct cross-talk effect. Secondly the generated pressure waves will deform the structure too and this results in an induced cross-talk effect. Bridge structures with balanced elongational and bending deformation components are designed using Ansys simulations and eliminate these cross-talk effects. To take into account structural details like glue layers, a static full structural analysis in a 2D plain strain model is used with standard structural and piezo-electric elements. Figure 3 shows the deformation of the piezo finger structure below a channel block in a semi cross channel configuration.

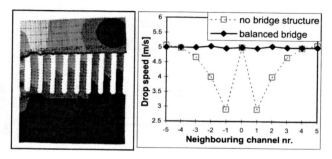

Figure 3: Calculated deformation of the printhead structure and measured resulting cross-talk effect on drop speed.

On actuating more piezo elements a range of vibrational modes of the printhead structure is excited. Modal analysis with 3D Ansys models of the complete printhead structure helps us to identify the individual modes of these printhead dynamics.

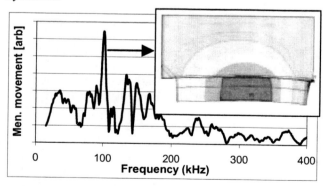

Figure 4: Measured meniscus movement and calculated shape and strain of the dominating mode.

In Figure 4 the frequency characteristic of the meniscus movement from a sinusoïdal small signal actuation of 128 piezo elements on the other side of the printhead is shown as measured with our laser-Doppler equipment. For the dominating frequency the shape of the corresponding mode

from modal analysis is shown in a cross-section along the channels. The piezo actuator is located at the bottom and the nozzles are located at the left hand side.

A special application is the simulation of impedance spectroscopy spectra of the piezo elements [1]. Ansys simulations identify the relevant modes. Mathematical modeling give the effective macroscopic material parameters derived from the true micoscopic material properties and for example a multilayer structure. We developed these mathematical models in Matlab based on coupling the Maxwell equations for the electro-quasi-static case and linear elasticity for the specific modes.

4.2 Acoustic modeling

We implemented a mathematical model in Matlab based on the "narrow gap" theory [2]. A very important effect of the flexible channel structure is the decrease of the effective speed of sound. Only a quasi-static compliance of the channel cross-section is taken into account and this results in a typical effective speed of sound inside the channels of 800 m/s instead of 1200 m/s.

For efficiency, the actuation pulse has to be tuned to the main frequency characteristics of the meniscus movement. The amplitude of the meniscus movement resulting from a small sinusoïdal actuation as a function of the frequency is shown in Figure 5.

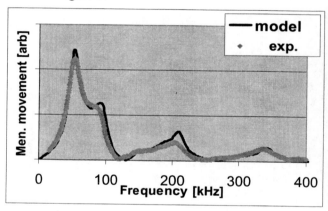

Figure 5: Frequency characteristics of meniscus movement as calculated and measured with laser-Doppler.

Complex interactions of printhead structure dynamics with channel acoustics are identified by acoustic modeling with Ansys. The 3D structural models are extended with FLUID80 elements for the ink domain, connected by constraint equations. The mesh structure of the ink domain is adapted to the velocity profile as predicted by the shear wave number [2] to reduce the number of elements. An example is given in Figure 6. Maximum acoustic coupling is reached on matching frequencies and deformation shapes. Identifying the most disturbing modes is a first step in suppressing unwanted cross-talk effects. The most complete version of this model also includes the entire ink reservoir. From this complete model design specifications for the reservoir can be derived to avoid acoustic cross-talk effects through the ink domain in the reservoir.

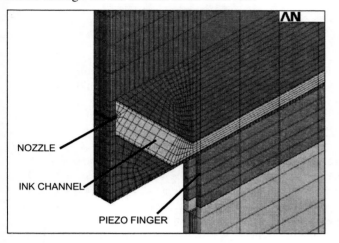

Figure 6: Element distribution near the nozzle and the first part of the channel structure in a half channel module.

4.3 Modeling Drop Formation

A first order approximation of the drop properties is calculated with a mathematical model in Matlab based on a balance of energies [3]. To model the details of the drop formation process we use a 2D rotational symmetric Flow3D model. Both models use the nozzle pressure or flow from acoustic modeling as input.

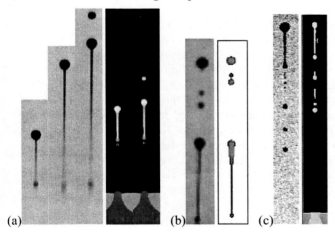

Figure 7: Measured and calculated drop formation at high and low drop speed and repetition rates.

The first part of the drop formation is highly affected by the actuation force. Tail properties and breakup are mainly determined by ink properties as surface tension and viscosity. At higher driving voltage the velocity of the head increases but the tail is not affected at all, Figure 7a, resulting in longer tails at higher drop speed. At very high drop speed, a satellite drop will be formed from the head of the drop. Simulations revealed the mechanism: during the first microseconds of the actuation the acceleration of the ink has to stay below a critical level. Above this level surface tension forces are no longer capable of holding the amount of ink together. In figure 7b the satellite drop formation at high drop repetition rates is shown. The satellite droplets are now formed by breakup of the tail. In Figure 7c the breakup of a very long tail in many small droplets is shown, the Rayleigh instability.

At tail breakup from the meniscus surface a cascade of structures can be seen as described in [4], Figure 8a. Since noise is the source of this structure it is not seen in Flow3D simulations unless the grid is too course giving numerical noise. Exciting a higher order drop formation results in smaller droplets, Figure 8b. Flow3D simulations revealed the mechanism.

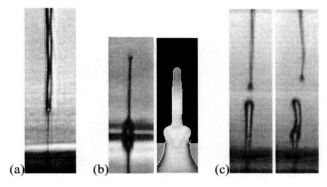

Figure 8: Cascade of structure at tail breakup, higher order drop formation and a distortion of the drop formation

Meniscus movement affects the channel acoustics by changing the reflection conditions in the nozzle region. To model the impact of the drop formation process on channel acoustics and refill an extension of the Flow3D code is developed to include wall flexibility. The obstacle velocity in this version is a function of the local instantaneous pressure. Printhead dynamics are incorporated by means of a relaxation time and acoustic waves are calculated with an adiabatic motion in an ideal liquid. Free surface flow in the nozzle and the acoustics in the ink channels are designed to give a good refill of the nozzle making high drop repetition rates possible.

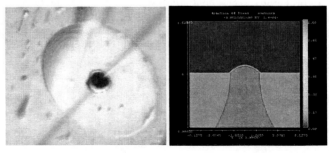

Figure 9: Overfill of nozzle at 10 kHz as measured with a standard camera and simulated with Flow3D.

A strong refill proces however can give overfill at low repetition rates as shown in Figure 9. The overfill situation can result in wetting of the nozzle plate surface. Modeling the wetting phenomena is difficult due to very small contact angles between ink and nozzle plate surface. For a better understanding of the refill mechanism we use mathematical modeling. At short time scales the asymmetric acceleration of the ink due to the variable filling of the nozzle is the main mechanism. This is opposed by resistance terms from viscous, inertial and acoustic coupling effects. At longer time scales capillary forces take over.

4.4 Modeling bubble dynamics

During drop formation, sometimes air is entrapped at the surface resulting in nozzle failure. Air entrapment is always preceeded by a distortion of the drop formation, like by an accidental particle as recorded with the Phantom V7 camera, Figure 8c. Because of the statistical nature of this process no direct Flow3D simulations are done.

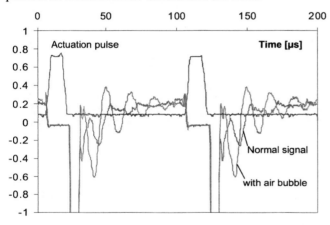

Figure 10: The impact on channel acoustics as measured.

The effects of air bubbles on drop formation can be measured directly. Much more sensitive is the measurement of the change in the acoustics since air bubbles change the reflection conditions at the nozzle. It turns out that starting bubbles are very small, showing only resonances at their Minnaert frequency as can be seen from FFT-analyses of the piezo signal. Also cyclic movement patterns can be seen from principle component analysis. In a large acoustic pressure field air bubbles can grow by rectified diffusion to larger dimensions and will then disturb the drop formation and the basic channel resonance as shown in Figure 10.

Modeling small bubbles and their possible recovery by dissolving or jetting out gives problems with Flow3D because of the very fine grid size needed. Large bubbles can be modeled with adiabatic bubble motion of an ideal gas. An example is shown in Figure 11. Understanding the behavior, growth and displacement of air bubbles in an acoustic field is of major importance for improving the reliability of the drop formation process.

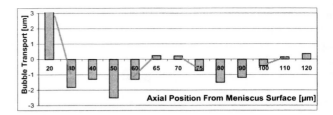

Figure 11: Net bubble displacement as function of axial position of a 10 μm bubble as calculated with Flow3D.

5   FURTHER RESEARCH

We are working now on an analytical model describing the behavior of (small) air bubbles in the nozzle region based on a balance of forces (Bjerknes, drag etc.). On system level we are developing an ILC model [5] based on a chain of two-ports. For understanding the wetting phenomena we need a better model to describe contact line movements with very low contact angles. Because of the very high interation level between all phenomena we continue to look for one complete model, combining detailed CFD calculations with structural simulations.

6   CONCLUSIONS

Modeling in addition to our measurements plays an essential role in understanding the inkjet proces. A major part of the phenomena preceeding drop formation, drop formation itself and the disturbing air bubbles can be well described now, but a lot of work still need to be done.

ACKNOWLEDGEMENTS

Thanks to my Océ colleagues Hans Reinten, Marc van den Bergh, Wim de Zeeuw and Jos de Jong. Thanks also the group of Henk Tijdeman of Twente University for their work on the "narrow-gap" theory, the group of Detlef Lohse of Twente University for their work on bubble behavior, Mustafa Megahed of Logica-PDV in Bochum for his work on the Flow3D models, Ken Williams of Flow Sim. Services in Albuquerque for his work on flex-walls in Flow3D and Wybo Wagenaar of Infinite Simulation Systems in Breda for his work on the Ansys models.

REFERENCES
[1] L.H. Saes and F.R. Blom, Proceedings Materials Week 2001.
[2] Marco Beltman, "Viscothermal wave propagation including acousto-elastic interaction", Thesis Twente University, 1998.
[3] J.F. Dijksman, J.Fluid.Mech, 139, 173, 1984.
[4] X.D. Shi, M.P. Brenner and S.R. Nagel, Science, 265, 219, 1994.
[5] S.H. Koekebakker, S.P. van de Geijn and M.B. Groot Wassink, Poc. Symposium on Mechatronics and Microsystems, Delft, April 2003, 45.

Contact Force Models, including Electric Contact Deformation, for Electrostatically Actuated, Cantilever-Style, RF MEMS Switches

R. A. Coutu, Jr.* and P. E. Kladitis*

* Air Force Institute of Technology
AFIT/ENG 2950 Hobson Way, Wright Patterson AFB, OH, USA
ronald.coutu@afit.edu, ronald.coutu@wpafb.af.mil, paul.kladitis@afit.edu

ABSTRACT

Electrostatically actuated, cantilever-style, metal contact, radio frequency (RF), microelectromechanical systems (MEMS) switches depend on having adequate contact force to achieve desired, low contact resistance. In this study, contact force equations that account for beam tip deflection and electric contact material deformation are derived. Tip deflection is modeled analytically using beam bending theory and contact material deformation is modeled as elastic, plastic, or elastic-plastic. Contact resistance predictions, based on Maxwellian theory and newly derived contact force equations, are compared to experimental results. Contact force predictions not considering tip deflection or material deformation overestimate contact force and result in underestimated contact resistance. Predictions based on the new contact force models agree with measurements.

Keywords: microelectromechanical systems, microswitch, contact force, contact resistance

1 INTRODUCTION

This paper reports on contact force models for electrostatically actuated, cantilever-style, radio frequency (RF), microelectromechanical systems (MEMS) metal contact switches like that shown in Figure 1.

This new contact force modeling approach considers beam tip deflection and electric contact material deformation that occurs after switch closure. Prior studies focused on mechanical designs to achieve the high contact force needed for low contact resistance connections [1]. The effects of tip deflection and material deformation on contact force in electrostatically actuated, microswitches has not been addressed. Previous work by this author, showed that contact force, bounded by pull-in and collapse voltages, could be analytically modeled using the beam illustrated in Figure 2 [2].

Equation 1 is the resulting contact force equation [3]:

$$F_c = \frac{F_a}{2l^3}a^2(3l - a) \qquad (1)$$

**The views expressed in this article are those of the authors and do not reflect the official policy or position of the United States Air Force, Department of Defense, or the U.S. Government.

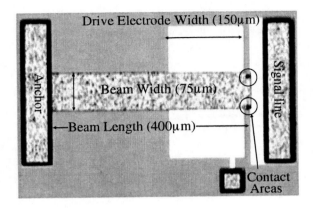

Figure 1: A captured video image of a 75 μm-wide by 400 μm-long micro-switch.

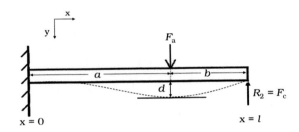

Figure 2: Cantilever beam model with a fixed end at $x = 0$, a simply supported end at $x = l$, and an intermediately placed external load (F_a) at $x = a$.

where F_c is contact force (μN), F_a is applied electrostatic force (μN), a is the location of the applied electrostatic force (μm), and l is beam length (μm). This simple model does not consider either beam tip deflection or contact material deformation after switch closure or pull-in.

In this study, contact force equations are derived by assuming elastic, plastic, and elastic-plastic electric contact material deformation [4] and modeling tip deflection [3]. Beam bending equations, material deformation models, and the principle of superposition are used to derive the new contact force equations.

2 IMPROVED SWITCH MODEL

In metal contact micro-switches, initial switch closure is defined by the pull-in voltage. At pull-in, physical contact between the switch's upper and lower electric contacts is first established, with minimal contact force. Increased actuation voltage causes higher contact force and increased pressure at the surface asperities resulting in contact material deformation.

After pull-in, the switch is best modeled as a deflected beam with a fixed end, a simply supported end, an intermediate external load (F_a), and contact material deformation, as illustrated in Figure 3.

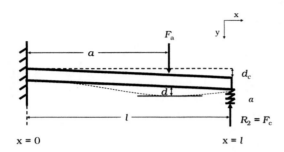

Figure 3: Deflected cantilever beam model with a fixed end at $x = 0$, a simply supported end at $x = l$, an intermediately placed external load (F_a) at $x = a$ and contact material deformation (α) at $x = l$.

The external load (F_a), modeled as electrostatic force, was found by neglecting fringing fields and using a first-order parallel plate capacitor description:

$$F_a = \frac{\epsilon_o A_{sa} V^2}{2g^2} \qquad (2)$$

where ϵ_o is dielectric permittivity ($\frac{F}{\mu m}$), A_{sa} is surface area (μm^2) of one plate, V is applied voltage (V), and g is the gap (μm) under the beam.

Like Figure 2, the beam in Figure 3 is statically indeterminate and another equation is needed to supplement the static equilibrium equations (i.e. $\sum F_x = 0$, $\sum F_y = 0$, and $\sum M = 0$). Using superposition, tip deflection is represented by the sum of two statically determinate systems. The first is a beam with a fixed end, a free end, and an intermediate load (F_a) and the second is a beam with a fixed end, a free end, and an end load ($-R_2$).

Using the method of moments, Equation 3 results for the maximum tip deflection for the first statically determinate system (beam with intermediate load) [3]:

$$d = \frac{F_a a^2}{6EI_z}(3l - a) \qquad (3)$$

where d is maximum cantilever beam tip deflection (μm), E is elastic modulus (GPa), and I_z is the area

moment of inertia about the z-axis ($\frac{m^4}{12}$). The method of moments is also used to determine Equation 4 or the maximum tip deflection for the second statically determinate system (beam with end load) [3]:

$$d' = \frac{-R_2 l^3}{3EI_z} \qquad (4)$$

where d' is maximum cantilever beam tip deflection (μm).

The final equation, needed to solve the indeterminate system in Figure 3, is found by summing Equations 3 and 4, setting that sum equal to overall tip deflection (i.e. the sum of the distance between the electric contacts and the amount of material deformation), and solving for the reaction force, R_2, results in:

$$R_2 = \left[\frac{F_a a^2}{2l^3}(3l - a)\right] - \left[\frac{3EI_z}{l^3}(d_c + \alpha)\right] \qquad (5)$$

where d_c is the gap between the electric contacts (μm) and α is contact material vertical deformation (μm). In micro-switches, the beam reaction force (R_2) is equal to contact force and the external load is equal to electrostatic force. Substituting F_c for R_2, Equation 2 for F_a, and $g = g_o - d$, in Equation 5 results in:

$$F_c = \left[\frac{\epsilon_o A_{sa} V^2}{4l^3(g_o - d)^2}a^2(3l - a)\right] - \left[\frac{3EI_z}{l^3}(d_c + \alpha)\right]. \qquad (6)$$

3 MATERIAL DEFORMATION

Contact force is a compressive force that causes material deformation similar to that predicted by conventional material tensile testing. The difference being tensile loads cause material deformation by necking out and compressive loads cause deformation by bulging. Elastic, plastic, and elastic-plastic material deformation models are discussed next.

3.1 Elastic

When electric contact surfaces initially come together, they experience elastic or reversible material deformation. Equations 7 and 8 define the contact area and force as a function of vertical deformation [4]:

$$A = \pi R \alpha \qquad (7)$$

where A is contact area (μm^2), R is radius of curvature (μm), and α is vertical deformation (μm) and

$$F_c = \frac{4}{3}E'\alpha\sqrt{R\alpha} \qquad (8)$$

where F_c is contact force (μN) and E' is Hertzian modulus (GPa) defined by:

$$\frac{1}{E'} = \frac{1 - \nu_1^2}{E_1} + \frac{1 - \nu_2^2}{E_2} \qquad (9)$$

where E_1 is elastic modulus (GPa) and ν_1 is Poisson's ratio for contact one and E_2 and ν_2 are for contact two. For circular areas (i.e. $A = \pi r^2$), Equations 7 and 8 are related to the effective contact area radius (r_{eff}) (μm) through Hertz's contact radius model [5]:

$$r_{\text{eff}} = \sqrt[3]{\frac{3F_c R}{4E'}}. \qquad (10)$$

Vertical deformation, in terms of contact force, is derived when $A = \pi r_{\text{eff}}^2$ is substituted into Equation 7, solved for R, and then substituted into Equation 10, resulting in:

$$\alpha = (\frac{3F_c}{4E'r_{\text{eff}}}). \qquad (11)$$

When deformation is no longer reversible and the material has gone passed the yield point (Y), plastic deformation begins. Fully plastic deformation, discussed next, occurs when the contact load exceeds $\sim 3Y$ [5].

3.2 Plastic

Plastic material deformation is modeled using the well known fully plastic contact model [6]. Equations 12 and 13 define the contact area and force as a function of vertical deformation [4]:

$$A = 2\pi R\alpha \qquad (12)$$

and

$$F_c = HA \qquad (13)$$

where H (GPa) is the Meyer hardness [5]. Vertical deformation, in terms of contact force, results when Equation 12 is substituted into Equation 13 and solved for α, resulting in:

$$\alpha = \frac{F_c}{2\pi HR}. \qquad (14)$$

When using the elastic model from section 3.1 and this plastic model, an area discontinuity exists during the transition from elastic to fully plastic behavior [5]. Chang addressed this by assuming volume conservation of deformed asperities and developing an elastic-plastic material deformation model suitable from initial plasticity onset to fully plastic behavior [4].

3.3 Elastic-Plastic

Elastic-plastic material deformation refers to the situation when portions of the contact area are plastically deformed but completely surrounded by elastic material. This phase of material deformation occurs beyond the elastic limit but prior to fully plastic behavior. Equation 15 is used to define the critical vertical asperity deformation limit where initial elastic-plastic material deformation begins [4]:

$$\alpha_{\text{c}} = R(\frac{H\pi}{2E'})^2 \qquad (15)$$

where α_{c} is critical vertical deformation (μm). Equations 16 and 17 are the contact area and force equations from Chang's elastic-plastic model [4]:

$$A = \pi R\alpha(2 - \frac{\alpha_{\text{c}}}{\alpha}) \qquad (16)$$

and

$$F_c = [3 + (\frac{2}{3}K_{\text{Y}} - 3)\frac{\alpha_{\text{c}}}{\alpha}]YA \qquad (17)$$

where K_{Y} is the yield coefficient given by Equation 18:

$$K_{\text{Y}} = 1.282 + 1.158\nu. \qquad (18)$$

Since most metals exhibit a yield point equal to one third the material's hardness $(\frac{H}{3})$ and a yield coefficient (K_{Y}) approximately equal to three [4], Equation 17 can be rewritten as:

$$F_c = [1 - (\frac{\alpha_{\text{c}}}{3\alpha})]HA. \qquad (19)$$

Vertical deformation, in terms of contact force, is derived when $A = \pi r_{\text{eff}}^2$ is substituted into Equation 19 and then solved for α, resulting in Equation 20:

$$\alpha = \frac{1}{3}(\frac{H\pi r_{\text{eff}}^2\alpha_c}{H\pi r_{\text{eff}}^2 - F_c}). \qquad (20)$$

New contact force models are derived next using the above material deformation models.

4   NEW CONTACT FORCE MODELS

When contact areas deform elastically, a contact force model is derived by substituting Equation 11 into Equation 6 resulting in:

$$F_{\text{cE}} = [\frac{\epsilon_{\text{o}} A_{\text{sa}} V^2}{4l^3(g_{\text{o}} - d)^2}a^2(3l - a)] - [\frac{3EI_z}{l^3}(d_{\text{c}} + \frac{3F_{cE}}{4E'r_{\text{eff}}})]. \qquad (21)$$

Solving Equation 21 for F_{cE} results in:

$$\boxed{F_{\text{cE}} = 2E'r_{\text{eff}}[\frac{(\frac{\epsilon_{\text{o}} A_{\text{sa}} V^2}{2(g_{\text{o}}-d)^2})a^2(3l-a)-(6E'I_z d_{\text{c}})}{(9EI_z+4E'r_{\text{eff}}l^3)}].} \qquad (22)$$

For plastically deforming contact areas a contact force model is derived by substituting Equation 14 into Equation 6 resulting in:

$$F_{\text{cP}} = [\frac{\epsilon_{\text{o}} A_{\text{sa}} V^2}{4l^3(g_{\text{o}} - d)^2}a^2(3l - a)] - [\frac{3EI_z}{l^3}(d_{\text{c}} + \frac{F_{cP}}{2\pi HR})]. \qquad (23)$$

Solving Equation 23 for F_{cP} results in:

$$F_{cP} = \pi H R \left[\frac{(\frac{\epsilon_o A_{sa} V^2}{2(g_o - d)^2}) a^2 (3l - a) - (6 E I_z d_c)}{(3 E I_z + 2 \pi H R l^3)} \right]. \quad (24)$$

For elastic-plastic deformation, a contact force model is derived by substituting Equation 20 into Equation 6 resulting in:

$$F_{cEP} = \left[\frac{\epsilon_o A_{sa} V^2}{4 l^3 (g_o - d)^2} a^2 (3l - a) \right] - \left[\frac{3 E I_z}{l^3} (d_c + \frac{1}{3} (\frac{H \pi r_{eff}^2 \alpha_c}{H \pi r_{eff}^2 - F_{cEP}})) \right]. \quad (25)$$

Solving Equation 25 for F_{cEP} results in:

$$F_{cEP}^2 + C_1 F_{cEP} = C_2 \quad (26)$$

where

$$C_1 = \frac{1}{2 l^3} [(6 E I_z d_c) - (2 \pi H l^3 r_{eff}^2) - (\frac{\epsilon_o A_{sa} V^2}{2(g_o - d)^2}) a^2 (3l - a)]$$

and

$$C_2 = \frac{H \pi r_{eff}^2}{2 l^3} [2 E I_z (3 d_c + \alpha_c) - (\frac{\epsilon_o A_{sa} V^2}{2(g_o - d)^2}) a^2 (3l - a)].$$

5 RESULTS

The micro-switches shown in Figure 1 were designed, fabricated, and tested. The electroplated gold structural layer was $\sim 5\ \mu m$-thick and the gold-on-gold electric contacts consisted of two hemispherical upper contacts $\sim 8\ \mu m$ in diameter ($\sim 3\ \mu m$ radius of curvature) and a flat lower contact. The elastic modulus, hardness, and resistivity used in the analytic predictions were $80\ GPa$, $2\ GPa$, and $2\ \mu\Omega - cm$, respectively [5]. An Alessi Rel-4100A probe station, HP 6181B DC voltage source and HP 3455A digital voltmeter were used to actuate the switches with $\sim 60\ V_{DC}$ and measure the contact resistance using a four-point probe configuration.

Contact resistance measurements and predictions, based on Maxwellian spreading resistance theory [5], elastic, plastic, and elastic-plastic contact material deformation [4], and the new, improved contact force models (Equations 22, 24, and 26) are summarized in Table 1 for devices with three different drive electrode widths.

Note from Table 1, the elastic-plastic based predictions agree best with the measurements. This indicates that the contact surface asperities were deformed beyond the elastic limit and material deformation was best characterized using an elastic-plastic model [4].

Contact resistance, in terms of the new contact force model (Equation 26) and the original contact force model

Table 1: Contact resistance measurements and analytic predictions for switches with varying drive electrode widths, contact materials with elastic, plastic, and elastic-plastic material deformations, and the new, improved contact force models (Equations 22, 24, and 26).

Width	Measured R_c	R_{cE}	R_{cP}	R_{cEP}
50 μm	0.353 Ω	0.089 Ω	0.179 Ω	0.189 Ω
100 μm	0.253 Ω	0.067 Ω	0.111 Ω	0.123 Ω
150 μm	0.263 Ω	0.056 Ω	0.085 Ω	0.096 Ω

(Equation 1), are compared in Table 2. Note that contact resistance predictions, as a function of the new contact force model agree with measurements better than predictions based on the original contact force model. Also from Table 2, note that contact force (per contact) found using Equation 1 (F_c) is substantially higher than that found using Equation 26 (F_{cEP}).

Table 2: Contact resistance predictions for switches with varying drive electrode widths. The contact force values were found using a new contact force model (Equation 26) and the original contact force model (Equation 1).

Width	$R_{cEP}(F_{cEP})$	F_{cEP}	$R_{cEP}(F_c)$	F_c
50 μm	0.189 Ω	2.95 μN	0.121 Ω	7.13 μN
100 μm	0.123 Ω	6.89 μN	0.084 Ω	15.00 μN
150 μm	0.096 Ω	11.32 μN	0.066 Ω	23.86 μN

6 CONCLUSIONS

Contact force models not accounting for beam tip deflection and material deformation (Equation 1) overestimate micro-switch contact force. Measurement and prediction discrepancies, normally attributed to contaminant film resistance, result when contact force is overestimated. The new micro-switch contact force models contain both mechanical beam parameters and contact material properties that are important for evaluating switch designs and selecting contact materials.

REFERENCES

[1] D. Peroulis, et al., IEEE MTT, 51, 259–70, 2003.
[2] R. Coutu, Jr., et al., Proceedings: Eurosensors XVII, Guimarães, Portugal, 64–67, 2003.
[3] J. Shigley and C. Mischke, "Mech. Engineering Design, 5th Ed.," McGraw-Hill, New York, 1990.
[4] W. Chang, J. of Wear, 212, 229–27, 1997.
[5] R. Holm, "Electric Contacts: Theory and Applications," Berlin: Springer, 1969.
[6] E. Abbot and F. Firestone, ASME, 55, 569, 1933.

Electromechanical buckling of a pre-stressed layer bonded to an elastic foundation

Samy Abu-Salih and David Elata

Technion - Israel Institute of Technology, Haifa 32000, Israel
samyas@tx.technion.ac.il, elata@tx.technion.ac.il

ABSTRACT

The ElectroMechanical Buckling (*EMB*) of a pre-stressed layer bonded to an elastic foundation is analyzed. A new analytic solution of the mechanical post-buckling is presented. In addition, it is shown that electrostatic forces can precipitately instigate buckling even when the pre-stress is lower than the critical value that allows mechanical buckling.

Keywords: buckling, electrostatic instability, electromechanical buckling.

1 INTRODUCTION

Mechanical buckling is a well-known phenomenon that occurs in thin elastic structures which are subjected to compressive loads. Mechanical buckling develops only if the compressive loads are larger than a critical value. In most structures, reduction of a post-critical compressive load to a sub-critical level will eliminate the buckling deformation. In thin sheet-like elastic solids that are bonded to an elastic foundation, a compressive stress can cause a dense occurrence of buckling flexures.

A different kind of instability that may develop in deformable solids is due to the application of electrostatic forces. The inherent nonlinearity of electrostatic attraction forces can cause the well known pull-in electromechanical instability.

This work investigates the electromechanical response of an electrically conducting, pre-stressed elastic layer, bonded to an elastic foundation, which is subjected to both electrostatic forces and compressive loads. The analysis suggests that for a sub-critical compressive stress, buckling can be precipitately instigated by application of electrostatic forces that destabilize the structure. Elimination of these forces eliminates the buckling deformations.

The presented analysis considers a one dimensional model of the problem. Figure 1 describes an electrically conductive, pre-stressed thin layer bonded to a dielectric elastic foundation that is fixed to a conductive solid substrate.

2 MECHANICAL BUCKLING

The governing equation of the problem is given in the following normalized form

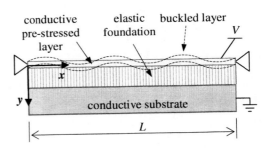

Figure 1: A pre-stressed conductive layer bonded over a dielectric elastic foundation.

$$\frac{1}{(2\pi)^4}\frac{d^4\tilde{y}}{d\tilde{x}^4} + 2\beta\frac{1}{(2\pi)^2}\frac{d^2\tilde{y}}{d\tilde{x}^2}$$
$$-\frac{1}{\alpha}\left[\int_0^\alpha \frac{1}{2}\left(\frac{d\tilde{y}}{d\tilde{x}}\right)^2 d\tilde{x}\right]\frac{1}{(2\pi)^2}\frac{d^2\tilde{y}}{d\tilde{x}^2} + \tilde{y} = \tilde{q} \quad (1)$$

where

$$\tilde{x} = \frac{x}{\Lambda_{cr}}, \qquad \tilde{y} = \frac{\sqrt{S}\,y}{\Lambda_{cr}}, \qquad \beta = \frac{\sigma}{\sigma_{cr}}, \qquad \tilde{q} = \frac{\sqrt{S}\,q}{\Lambda_{cr}k_f},$$

$$S = \frac{EA}{\sqrt{k_f}\sqrt{E^*I}}, \quad \sigma_{cr} = -\frac{2\sqrt{k_f E^* I}}{A}, \quad \Lambda_{cr} = 2\pi\left(\frac{E^*I}{k_f}\right)^{\frac{1}{4}}$$

Here β is the load parameter, S is a non-dimensional stiffness ratio that governs the postbuckling configuration, A, and I are the cross-section area and second moment, per unit width, E and $E^* = E/(1-v)$ are Young's modulus and the effective modulus in bending, v is Poisson's ratio, k_f is the elastic foundation stiffness, and q is a distributed load. The physical meaning of the parameters $\alpha, \sigma_{cr},$ and Λ_{cr} is revealed in the following. Throughout section 2 it is assumed that $\tilde{q} = 0$.

2.1 Critical State

To extract the critical state (i.e., the verge of buckling) the linear form of the equation is obtained by omitting the third term (with the square brackets), that accounts for axial strains due to axial variations in vertical deflection. It is postulated that the buckling deformation is of the form

$$\tilde{y} = B\sin\left(2\pi\, \tilde{x}\,/\,\alpha\right) \qquad (2)$$

Here B is the amplitude of the deformation, and the variable α is the normalized wavelength. Later (in 2.2-3), the buckling problem will have to be solved numerically. Since an infinite layer can not be modeled numerically, periodic solutions of layers with finite length will be considered. The specific layer length that is associated with a true period of the solution of an infinite layer, will be identified using energy considerations. In this respect, α in (2) ensures a periodic solution for a layer of length $L = \alpha\Lambda_{cr}$.

Substituting this postulated function in to the linear form of (1) yields,

$$B\frac{\sin(2\pi\, \tilde{x}\,/\,\alpha)}{\alpha^4}\left(\alpha^4 - 2\beta\alpha^2 + 1\right) = 0 \qquad (3)$$

The nontrivial solution of this equation is

$$\alpha = \sqrt{\beta \pm \sqrt{\beta^2 - 1}} \qquad (4)$$

α is real only if $\beta \geq 1$ and therefore the critical state of the system is $\beta_{cr} = 1$, $\alpha_{cr} = 1$. Accordingly, it is deduced that Λ_{cr} and σ_{cr} are the critical wavelength and critical pre-stress, respectively, at which buckling occurs in an infinite layer. A form of this result was previously derived by Hetényi [1].

Figure 2 illustrates the critical states for layers with various lengths $L = \alpha\Lambda_{cr}$. However, for a sufficiently long layer ($\alpha > 1$) two deformation waves with wavelength $\alpha/2$ may develop. For a layer that is longer still, a third mode with wavelength $\alpha/3$ may also develop (and so on).

2.2 Post-Buckling State

The post-buckling state is governed by the nonlinear equation (1). Due to this nonlinearity, the governing equation is often solved using approximation methods such as approximate analytical solutions (e.g., Rayleigh-Ritz [2]) or numerical solutions (e.g., finite differences).

In this work we present an exact analytical solution of the nonlinear equation. To this end we postulate a solution of the form (2). Substituting this postulated deformation in (1) yields

$$B\sin(2\pi\, \tilde{x}\,/\,\alpha)\frac{1}{\alpha^4}\left[\alpha^4 - 2\alpha^2\beta + B^2\pi^2 + 1\right] = 0 \qquad (5)$$

The nontrivial solution of the above equation is given by $B = (2\beta\alpha^2 - \alpha^4 - 1)^{1/2}/\pi$, and therefore, the postbuckling solution is

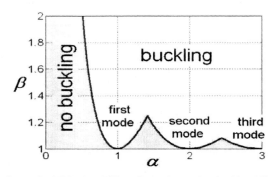

Figure 2: Critical stability of purely mechanical buckling.

$$\tilde{y} = \frac{\sqrt{2\beta\alpha^2 - \alpha^4 - 1}}{\pi}\sin(2\pi\, \tilde{x}\,/\,\alpha) \qquad (6)$$

The actual normalized wavelength associated with a given value of β, minimizes the total strain energy (per unit length) $U = U_B + U_R + U_{EF}$ of the system. The total strain energy is the addition of three components associated with bending deformation of the layer (U_B), axial deformation of the layer (U_R), and deformation of the elastic foundation (U_{EF}). Normalizing the strain energy components by the (axial) strain energy at the verge of buckling ($U_{cr} = \sigma_{cr}^2 A/2E$), yields

$$\tilde{U}_B = \frac{1}{16\pi^2}\frac{1}{\alpha}\int_0^\alpha\left(\frac{d^2\tilde{y}}{d\tilde{x}^2}\right)^2 d\tilde{x} \qquad (7a)$$

$$\tilde{U}_A = \frac{1}{\alpha}\int_0^\alpha\left(\beta - \frac{1}{2\alpha}\int_0^\alpha\frac{1}{2}\left(\frac{d\tilde{y}}{d\tilde{x}}\right)^2 d\tilde{x}\right)^2 d\tilde{x} \qquad (7b)$$

$$\tilde{U}_{EF} = \frac{\pi^2}{\alpha}\int_0^\alpha \tilde{y}^2 d\tilde{x} \qquad (7c)$$

Substituting (6) into (7) yields

$$\tilde{U} = \tilde{U}_B + \tilde{U}_A + \tilde{U}_{EF} = \frac{\alpha^4 + 1}{4\alpha^4}\left(4\alpha^2\beta - \alpha^4 - 1\right) \qquad (8)$$

For arbitrary loads $\beta \geq 1$, the normalized total strain energy has a minimum at $\alpha = 1$ (the other roots of $d\tilde{U}/d\alpha = 0$ are non physical).

Accordingly, it is deduced that the postbuckling wavelength is identical to the critical wavelength, and that the postbuckling deformation is given by

$$\tilde{y} = \frac{\sqrt{2\beta - 2}}{\pi}\sin(2\pi\, \tilde{x}) \qquad (9)$$

Figure 3 illustrates the total strain energy of finite layers with various normalized wavelength α. For a given pre-stress β, the minimal energy is associated with $\alpha = 1$ (and the integer multiplications $\alpha = 2,3,...$).

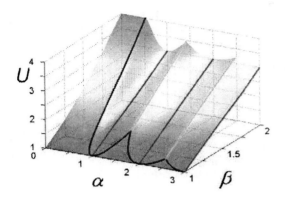

Figure 3: Total strain energy U of the pre-stressed layer. For a given pre-stress β, the wavelength α related to the minimal strain energy, is presented by the straight red lines.

In previous studies *approximate* analytic solutions were obtained by minimizing the elastic energy of the system (e.g., [3]). Although these approximate solutions resemble (9) they do not satisfy the equilibrium equation as was shown here.

2.3 Nonlinear foundation

When the displacements in a buckled layer become sufficiently large relative to the thickness of the elastic foundation, the foundation can no longer be modeled as linear (due to the physical inconsistency that the thickness of a linear foundation can be reduced to zero by applying a finite force). A more realistic model of elastic foundation should include of stiffening in compression and possible softening in extension (e.g., [3]).

The effect of stiffening and softening elastic foundations on the *postbuckling* behavior of the system is investigated. To this end, the governing equation (1) is rewritten, for stiffening or softening foundation, and is numerically solved by means of the finite differences implemented in a MATLAB code. The non-linear softening and stiffening elastic foundations are modeled here by replacing the last term ($\tilde{y}$) in (1) by one of the following terms:

$$\tilde{b}\sinh^{-1}\left(\tilde{y}/\tilde{b}\right) \qquad \text{(softening foundation)} \qquad (10a)$$

$$\tilde{b}\sinh\left(\tilde{y}/\tilde{b}\right) \qquad \text{(stiffening foundation)} \qquad (10b)$$

Here $\tilde{b}$ is a dimensionless parameter that determines the intensity of the stiffening and softening elastic foundations.

In case of small deformation, the above non-linear foundations asymptotically approach the linear foundation.

Figure 4 presents the normalized wavelength α as function of the normalized pre-stress β for a stiffening, linear, and softening foundations. α decreases with increasing β for a stiffening foundation, is constant for a linear foundation, and increases with increasing β for a softening foundation.

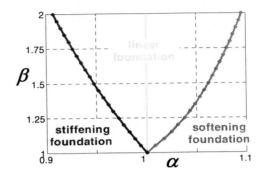

Figure 4: Post-buckling pre-stress β vs. wavelength α for a linear, stiffening and softening elastic foundations.

3 ELECROMECHANICAL BUCKLING

Mechanical buckling due to internal compressive stress is one kind of instability. Another type of instability that is prevalent in MEMS is the well known electromechanical pull-in phenomenon.

In this work we investigate the instability that arises when a thin layer is subjected to *both* a compressive pre-stress and electrostatic forces. The electrostatic forces are induced by applying a voltage difference between the conductive pre-stressed layer and the conductive substrate (Fig. 1). This force takes the form

$$\tilde{q} = \frac{1}{2}\frac{\varepsilon_0 V^2}{(g-y)^2}\frac{\sqrt{S}}{\Lambda_{cr}k_f} \qquad (10)$$

where g is the foundation thickness.

In this section we are interested in the affect of the electrostatic forces on the critical state. Accordingly, the term in square brackets in (1) is omitted, and the equation is re-normalized in the form

$$\frac{1}{(2\pi)^4}\frac{d^4\bar{y}}{d\bar{x}^4} + 2\beta\frac{1}{(2\pi)^2}\frac{d^2\bar{y}}{d\bar{x}^2} + \bar{y} = \frac{\bar{V}^2}{(1-\bar{y})^2} \qquad (11)$$

where

$$\bar{y} = \frac{\tilde{y}}{\tilde{g}}, \quad \tilde{g} = \frac{\sqrt{S}}{\Lambda_{cr}}g, \quad \bar{V}^2 = \frac{\varepsilon b V^2 \Lambda_{cr} S^{3/2}}{2E^* I \tilde{g}^3 (2\pi)^4}$$

Here we postulate a deflection in the form

$$\bar{y} = \bar{y}_0 + B\sin(2\pi\tilde{x}/\alpha) \tag{12}$$

where $\bar{y}_0$ is the average electromechanical displacement, and B is the amplitude of the structural waves that develop due to buckling. If no buckling occurs, the problem reduces to the case of a parallel-plates electrostatic actuator. At the verge of buckling, B is small, and the electrostatic force is approximated by a Taylor expansion. Substituting the postulated solution (12) and the first two terms of the Taylor expansion into (11) yields

$$B\frac{\sin(2\pi\tilde{x}/\alpha)}{\alpha^4}\left[\alpha^4(1-\delta) - 2\alpha^2\beta + 1\right] = \frac{\bar{V}^2}{(1-\bar{y}_0)^2} - \bar{y}_0 \tag{13}$$

where δ is the normalized electrostatic stiffness given by

$$\delta = \frac{2\bar{V}^2}{(1-\bar{y}_0)^3} \tag{14}$$

On the verge of buckling (i.e., $B = 0$), the trivial solution is

$$\bar{V}^2 = \bar{y}_0(1-\bar{y}_0)^2 \tag{15}$$

The deflection of the layer in this case is uniform, $\bar{y} = \bar{y}_0$. For incipient buckling, (15) holds and can be subtracted from (13) to yield the electromechanical buckling equation

$$\alpha^4(1-\delta) - 2\alpha^2\beta + 1 = 0 \tag{16}$$

The solution of the last equation is

$$\alpha = \sqrt{\frac{\beta \pm \sqrt{\beta^2 - (1-\delta)}}{1-\delta}} \tag{17}$$

At the critical state $\beta_{cr} = \sqrt{1-\delta}$, and consequently the critical value of the wavelength parameter is found to be

$$\alpha_{cr} = 1/\sqrt{\beta_{cr}} \tag{18}$$

In the previous discussion it was shown that in the absence of electrostatic forces buckling cannot occur for $\beta < 1$. In contrast, from (17) it is clear that due to the destabilizing affect of the electrostatic forces, buckling can occur for $\beta < 1$ provided that $\delta > 0$.

For a given normalized pre-stress β, the critical voltage $\bar{V}_{cr}$ that switches the flat layer into its buckled state is extracted by solving (14) with (15) to yield

$$\bar{V}_{cr}^2 = 4\frac{1-\beta_{cr}^2}{(3-\beta_{cr}^2)^3} \tag{19}$$

Figures 5 and 6 present the normalized voltage $\bar{V}_{cr}^2$ and normalized wavelength α_{cr}, at the critical states, for various values of normalized pre-stress β. At the limit of zero pre-stress, the critical wavelength is infinite and the voltage approaches an asymptotic value. This state is in essence the pull-in state of an infinite parallel-plates actuator.

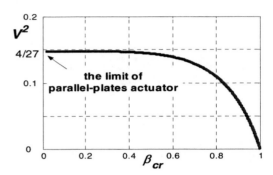

Figure 5: The normalized critical voltage $\bar{V}^2$ and critical pre-stress β_{cr} of electromechanical buckling. At zero pre-stress the normalized critical voltage is the same as for a parallel-plates actuator.

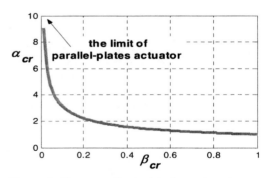

Figure 6: The normalized critical wavelength α_{cr} and critical pre-stress β_{cr} of electromechanical buckling. At zero pre-stress the normalized wavelength is infinite.

REFERENCES

[1] M. Hetényi, *Beams on Elastic Foundation*, U. of Mich. Press, Ann Arbor, 1946.

[2] D.O. Brush and B.O. Almroth, *Buckling of Bars, Plates and Shells*, Mc-Graw Hill, 1975.

[3] HUI D., "Postbuckling behavior of infinite beams on elastic foundations using Koiter's improved theory", *Int. J. non-linear Mech.*, 23(2), 113-123, 1988.

Three-Dimensional CFD-Simulation of a Thermal Bubble Jet Printhead

T. Lindemann[*], D. Sassano[**], A. Bellone[**], R. Zengerle[*] and P. Koltay[*]

[*]IMTEK, Institute of Microsystem Technology, Lab for MEMS Applications, University of Freiburg
Georges-Koehler-Allee 103, 79110 Freiburg, Germany, lindeman@imtek.de
[**]Olvetti I-Jet, Arnad, Italy, D.Sassano@olivettitecnost.com

ABSTRACT

This paper reports on a three-dimensional simulation of a commercial, thermally actuated bubble jet printhead using an appropriate pressure boundary condition for the bubble nucleation and expansion. The ink jet system has been used as a realistic show case for the application of the volume of fluid (VOF) method that is implemented in the simulation package ACE+ of CFDRC [1]. Comparing the results obtained by computational fluid dynamics (CFD) simulations with stroboscopic and gravimetric measurements, excellent agreement with experimental findings has been obtained. The differences in ejected droplet volume and droplet speed between simulation and experiment have been lower than 5 %. The simulation tool applied has been used to derive quantitative predictions not only in simplified two-dimensional test geometries, but also for the real three-dimensional device. It proves to be a valuable tool for cost- and time-efficient optimisation of printhead performance.

Keywords: bubble jet printhead, computational fluid dynamics (CFD), volume of fluid (VOF)

1 INTRODUCTION

High quality colour image, low machine cost and low printing noise are basically the main advantages of ink jet printers. This has led to a rapid expansion of this technology in the recent years. The two competing actuation principles of these, so-called drop-on-demand printers are the piezoelectric driven printhead invented in 1972 [2] and the thermally actuated bubble jet printhead developed in 1984 [3]. Currently the bubble jet printer outstands with low manufacturing costs at comparable print quality. The aim of manufacturers and many research establishments is the further optimisation concerning maximum print frequency, active area of the printhead, resolution and quality of ejected droplets. This requires a further miniaturisation and optimisation of the printhead geometry including fluid channels, heaters and nozzles to achieve an ideal printhead performance. An important tool for the optimisation is the simulation of the complete device which is very cost- and time-effective compared to experimental hardware optimisation according to the trial and error principle. With CFD a variety of relevant parameters like geometry effects and ink properties can be investigated.

2 SIMULATION PROCEDURE

Due to the complexity of the ejection process of a thermally actuated bubble jet printer including heating of a micro-heater, bubble nucleation, collapse of the vapour bubble and the actual droplet ejection no complete physical simulation could be accomplished so far. The simulation task is typically divided into smaller sub problems like actuation and ejection. To model the actuation it is necessary to find an appropriate pressure boundary condition to substitute the complicated bubble nucleation, expansion and collapse. Using this as input for the simulation package ACE+ of CFDRC a computational fluid dynamics (CFD) simulation has been set up and the droplet ejection process has been studied. Using a complete three-dimensional model, instead of a simplified two-dimensional one, provides the opportunity to examine the influences occurring only in the three-dimensional case. For instance this can be the effect of asymmetric ink inlet channels or a non-spherical bubble shape. But also the effects appearing at problematic parts of the printhead geometry, for example problems of the capillary refilling at convex edges or transitions between the angular nozzle chamber and the round nozzle, can be examined in 3D simulations. A further advantage of the three-dimensional case is, that the realistic flight path of the droplet can be studied in dependence of the whole geometry.

2.1 Pressure Boundary Condition

Asai et al [4,5] presented a model for the conversion of a given current pulse through a micro-heater of a conventional bubble jet printer into an equivalent pressure pulse using the Clausius-Clapeyron equation.

$$P_v[T_v] = P_{atm} \exp\left[\frac{w \cdot Q_{vap}}{R}\left(\frac{1}{T_b} - \frac{1}{T_v}\right)\right] \quad (1)$$

Where P_v and T_v are the approximately uniform pressure and temperature in the vapour bubble, P_{atm} is the atmospheric pressure (= 100 kPa), R is the gas constant (= 8.3148 J mol^{-1} K^{-1}), T_b is the boiling point of the ink, and w and Q_{vap} are molecular weight and heat of vaporization, respectively. Adding a time-dependent heating pulse leads to the following exponentially decreasing pressure function $P_v[t]$ which is displayed in figure 1.

$$P_v[t] = P_t(T_t) \exp\left[-\left(\frac{t}{t_0}\right)^\lambda\right] + P_s(T_{amb}) \qquad (2)$$

Where $P_t(T_t)$ is the initial bubble pressure depending on the maximum heating temperature T_t. In this case the initial bubble pressure is 9 MPa. P_s is the bubble pressure in the later stage depending on ambient temperature T_{amb}. The parameter t_0 is a time constant, which has been estimated from the bubble dynamics to be 0.17 µs [4,6]. λ is a coefficient, expected to be between 0.5 and 1.5, but should be determined by experimental bubble growth data [4-6]. It was taken to be 0.5.

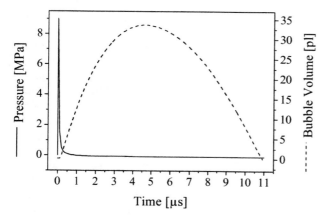

Figure 1: Pressure pulse used as pressure boundary condition calculated by applying the Clausius-Clapeyron equation and the resulting bubble volume.

2.2 Resulting bubble volume

The presented pressure boundary has been applied to the model of a commercial printhead design by substituting the heater area of the real design with an inlet pressure boundary in the simulation using air as incoming medium. The pumped air form the bubble displaces the ink and creates a flow through the nozzle and back into the reservoir. The exponential decreasing pressure function results in a parabolic time-dependency of the bubble volume as displayed also in figure 1. Whereas maximum bubble volume and time range depend on the printhead geometry and ink properties.

This approach and different variants of it have often been used in literature to simulate simple 2D devices [7-9]. Results for such 2D geometries could easily be reproduced using the VOF module including surface tension as implemented in ACE+. Following these initial tests a more complicate 3D model, depicted in figure 2, has been set up to simulate a complete dosage cycle of a printhead developed by the company Olivetti I-Jet [10] including first priming, printing and refilling. Thus it was possible to simulate the print frequency for what otherwise a second

tool like a network simulation would have been necessary. The simulation of such a complete ejection process takes about two days of computing time on a state of the art PC.

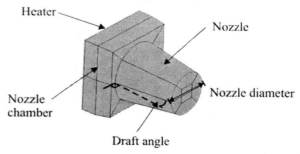

Figure 2: Simplified picture of the 3D model of one nozzle of the bubble jet printhead showing only the fluidic part (inlet channels not shown).

3 SIMULATION VS. EXPERIMENTS

Comparing the simulations with experimental results like stroboscopic pictures as displayed in figure 3, good qualitative agreement has been obtained. The shape of the droplet and the tail look very similar. A detailed look at the simulation output also reveals the non-spherical shape of the bubble, which affects the droplet trajectory. The gravimetrically measured droplet volume of 26 pl agrees very well with the simulated volume of 24.8 pl as well as the measured droplet velocity of 11 m/s, which is extracted from the stroboscopic measurements, agrees with the simulated one of 10.5 m/s. The duration of an ejection cycle of 70 µs in the simulation leads to a theoretical achievable print frequency of about 15 kHz whereas the experiment shows a maximum frequency of about 12 kHz due to the fact that the electronic circuit limits a faster execution. Thus the used pressure boundary condition, presented for the first time by Asai for a 2D case, can also be applied successfully for the 3D case. The applicability of the simulation model is validated with tolerable deviations.

4 SIMULATION RESULTS

After having validated the simulation model further simulations have been performed to optimise the printhead with fixed guidelines by varying parameters like geometry dimensions, heating pulse or ink properties.

4.1 Assembly and packaging tolerances

The determination of the influences of geometrical tolerances on the droplet volume and velocity, are depicted in figure 4. Knowledge of the allowable tolerances is very important for the assembly and packaging of the printhead. Especially the effect of lateral adjustment deviations or differences of chamber height are interesting for the current research project: A new one inch large printhead shall be

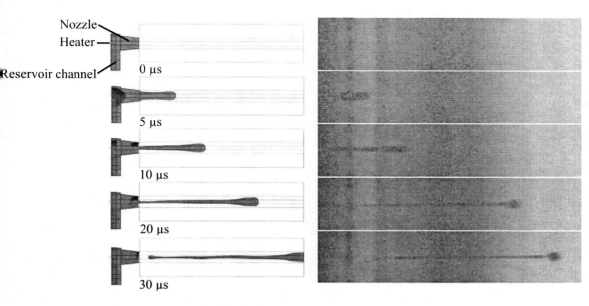

Figure 3: Comparison of the 3D simulations of the droplet ejection (only the fluidic part is shown) and the corresponding stroboscopic pictures.

developed which is a technical challenge for the assembling and packaging. Results of this examinations are, for instance, that a small variation of the nozzle chamber height or a minimal deviation of the adjustment leads to a negligible change of the resulting droplet volume and velocity whereas a deviation of the nozzle diameter or its draft angle induces a more significant change of the ejected droplet volume and velocity as depicted in figure 4.

4.2 Ink properties

It is well known that the ink properties play a major role in determining the print volume. Tuning of the ink is the usual method ink jet manufacturer apply to adjust the volume. Varying the ink parameters like density, surface tension or dynamic viscosity leads to a significant change in the droplet volume as displayed in figure 5. The droplet volume can be halved by increasing the density six fold. But reducing the droplet volume by varying ink parameters induces further changes in the ejection behaviour. The droplet velocity is also very sensitive to fluid properties especially to the ink density (cf. also figure 5). High density ink leads to very low droplet velocities, which is not desirable. Furthermore also altering surface tension or viscosity causes a significant change of the velocity. Another result of the VOF simulation can be educed by surveying the shape of the ejected droplet and its tail. Tail length and potential satellites are an important criterion for

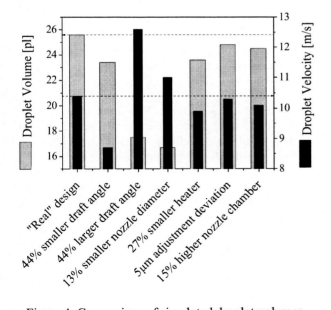

Figure 4: Comparison of simulated droplet volumes considering different geometrical tolerances.

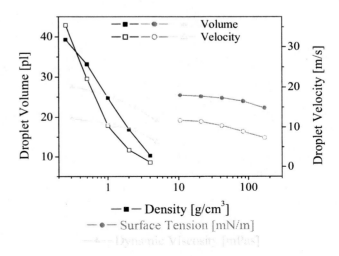

Figure 5: Simulated droplet volume and velocity depending on the density, the surface tension and the dynamic viscosity of the ink, respectively.

the quality of later printouts. The dependence of the tail length on the different ink properties is illustrated in figure 6. The bandwidth of the tail length ranges between very long tails (1000 µm) and very short tails. In a very long tail, the tail breaks in some satellites whereas in very short ones, the tail contracts with the droplet so that no satellites occur. Beyond increasing the print resolution [11] and realizing a monolithic printhead [12,13] preventing satellite droplets is an important part of current research and development activities [8,9].

Admittedly regarding this results one has to keep in mind that the ink properties cannot be tuned independently. In reality density, surface tension and dynamic viscosity are intimately connected. Combining all this simulation results the optimum interaction between print head geometry and ink parameters can be found.

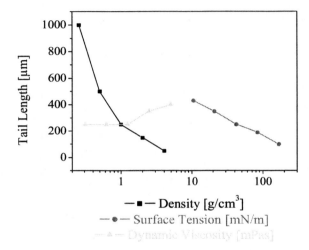

Figure 6: Simulated tail length of the droplet depending on the density, the surface tension and the dynamic viscosity of the ink, respectively.

5 CONCLUSIONS

The presented three-dimensional simulation model of a thermal ink jet printhead provides a valuable approach to optimise thermal bubble jet printheads regarding droplet volume, droplet velocity, droplet quality and print frequency also including the consideration of 3D sensitive problems. The correctness of the used pressure boundary condition and the simulation model in the three-dimensional case was verified by comparing simulation, gravimetrical and stroboscopic results. For the optimisation or the designing of a new printhead a variety of specific models may be investigated by adjusting all relevant parameters like geometry effects and ink properties. Nevertheless it is still inevitable to build prototypes to validate the simulations and to test the functionality. After having established a model CFD is however very helpful to shorten further design cycles significantly.

ACKNOWLEDGEMENT

This work was supported by Federal Ministry of Education and Research (BMBF), Germany (grant no. 16SV1607) within the EURIMUS program (IDEAL EM 42).

REFERENCES

[1] CFD Research Corporation, http://www.cfdrc.com/, 2003.
[2] S. I. Zoltan et al., "Pulsed Droplet Ejecting System", US Patent 3,683,212, 1972.
[3] J. L. Vaught et al., "Thermal Ink Jet Printer", US Patent 4,490,728, 1984.
[4] A. Asai et al., "One-Dimensional Model of Bubble Growth and Liquid Flow in Bubble Jet Printers", Japanese Journal of Applied Physics, vol. 26, no. 10, pp. 1794 – 1801, 1987.
[5] A. Asai et al., "Three-Dimensional Calculation of Bubble Growth and Drop Ejection in a Bubble Jet Printer", ASME Journal of Fluids Engineering, vol. 114, pp. 638 – 641, 1992.
[6] A. Asai et al., "Bubble Dynamics in Boiling Under High Heat Flux Pulse Heating", ASME Journal of Heat Transfer, vol. 113, pp. 973 – 979, 1991.
[7] P.-H. Chen et al., "Bubble Growth and Ink Ejection Process of a Thermal Ink Jet Printhead", International Journal of Mechanical Science, vol. 39, no. 6, pp. 683 – 695, 1997.
[8] F.-G. Tseng et al., "A High Resolution High Frequency Monolithic Top-Shooting Microinjector Free of Satellite Drops - Part I: Concept, Design, and Model", Journal of Electromechanical Systems, vol. 11, no. 5, pp. 427 – 436, 2002.
[9] F.-G. Tseng et al., "A High Resolution High Frequency Monolithic Top-Shooting Microinjector Free of Satellite Drops - Part II: Fabrication, Implementation, and Charakterization", Journal of Electromechanical Systems, vol. 11, no. 5, pp. 437 – 447, 2002.
[10] Olivetti I-Jet, http://www.olivettii-jet.it, 2003.
[11] R. Nayve et al., "High Resolution Long Array Thermal Ink Jet Printhead Fabricated by Anisotropic Wet Etching and Deep Si RIE", Proc. of IEEE-MEMS 2003, Kyoto, Japan, pp. 456 - 461, 2003.
[12] S. S. Baek et al., "T-Jet: A Novel Thermal Inkjet Printhead with Monolithically Fabricated Nozzle Plate on SOI Wafer", Proc. of IEEE-Transducers 2003, Boston, USA, pp. 472 - 475, 2003.
[13] Y.-J. Chuang et al., "A Thermal Droplet Generator with Monolithic Photopolymer Nozzle Plate", Proc. of IEEE-Transducers 2003, Boston, USA, pp. 472 - 475, 2003.

Circuit Modeling and Simulation of Integrated Microfluidic Systems

Aveek N. Chatterjee[*] and N. R. Aluru[**]

Beckman Institute for Advanced science and Technology - University of Illinois, Urbana-Champaign
[*]Doctoral student, University of Illinois at Urbana-Champaign
[**] Corresponding Author, Associate Professor, University of Illinois at Urbana-Champaign, e-mail:
aluru@uiuc.edu, http://www.staff.uiuc.edu/~aluru.

ABSTRACT

A combined circuit-device model for the analysis of integrated microfluidic system is presented. The complete model of an integrated microfluidic device incorporates modeling of the fluidic transport, chemical reaction, reagent mixing and separation. The microfluidic flow can be caused by an applied electrical potential gradient and/or a pressure gradient. In the proposed compact model, the fluidic network has been modeled by a circuit based representation and the other modules of the µ-TAS have been represented by a device model. As an example, we present the modeling and simulation of a lab-on-a-chip.

Keywords: microfluidics, lab-on-a-chip, circuit-modeling, simulation

1 INTRODUCTION

In this paper, we report on the advancement of a compact model for micro/nanofluidic transport driven by a combined electric field and pressure gradient, which had been introduced earlier by Qiao et. al.[1]. The new model has incorporated a number of additional elements, which were not included in the simple model that was proposed earlier (e.g. inclusion of capacitive elements, complete circuit representation of fluidic transport, circuit model for non-slip flow etc.). In addition, the new model is able to simulate mixing, reaction and separation, which are some of the most important functions of an integrated µ-fluidic system. Finally, an example is considered, where the combined circuit-device model has been demonstrated successfully.

2 MODEL DEVELOPMENT

The derivation of the circuit model for a flow driven by combined pressure gradient and electrical potential gradient is described in the beginning of this section. The compact model for the fluidic transport is composed of two parts, namely, the electrical part and the fluidic part.

2.1 Electrical Model

For microfluidic devices that rely on electrokinetic force as the driving force, the electric field must be solved first. In the case of electroosmotic flow, the potential field due to an applied potential can be computed by solving the Laplace equation [2]. Therefore, the axial potential variation can be represented by linear electrical resistances [1]. The EDL can be decomposed into the stern layer and the diffuse layer [3]. As the stern layer and the diffuse layer store charge, the capacitance associated with these layers is important. Figure 1 and 2 illustrate a typical cross shaped channel segment in a microfluidic system and its circuit representation, respectively. The expression for computing the electrical resistance of a solution filled channel is given in [1]. The expression for the effective capacitance is given by:

$$\left(C_{eff,i}\right)^{-1} = \left(C_{st,i}\right)^{-1} + \left(C_{dl,i}\right)^{-1} \quad (1)$$

where, $C_{st,i}$ is the capacitance of the stern layer of the ith channel (shown in figure 2), $C_{dl,i}$ is the capacitance of the diffuse layer of the ith channel and $C_{eff,i}$ is the effective capacitance of the ith channel. The expression for C_{st} and C_{dl} are given in [3] and [4], respectively. Using the relation between capacitance of a layer and the charge stored in a layer we get the following expression:

$$C_{eff,i}\psi_{0,i} = q_{st,i} = \sigma_{T,i}A_{s,i} \quad (2)$$

or

$$\psi_{0,i} = \frac{\sigma_{T,i}A_{s,i}}{C_{eff,i}} \quad (3)$$

where $\psi_{0,i}$ is the surface potential on the ith channel and $q_{st,i}$ is the total charge stored in the EDL of the channel. From the principle of conservation of charge it can be shown that the net charge stored in the EDL neutralizes the total surface charge.

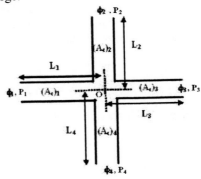

Figure 1: A typical cross-shape channel segment of a microfluidic system. The electrical potentials, ϕ_{1-4}, and pressures, P_{1-4}, are given.

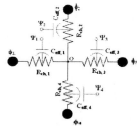

Figure 2: The electrical network representation for the electrical part. $R_{ch,1-4}$ are the electrical resistances, Ψ_{1-4} are the surface potentials of the channel walls and $C_{eff,1-4}$ are the capacitances of the EDLs.

2.2 Fluidic Model

For the fluidic transport driven by an electrical field and/or a pressure gradient, the "through quantities" are the flow rates through the channels and the "across quantities" are the electrical potential differences and the pressure differences imposed on the fluidic channels. In this section we present a derivation of the constitutive equation relating the "through quantities" to the "across quantities".

In those cases, where the thickness of the EDL is insignificant compared to the depth or diameter of the channel the effect of the electrokinetic force can be represented by a slip velocity at the wall given by the Helmholtz-Smoluchowski equation [1]. The slip velocity is a function of the ζ potential of the channel wall [2] and the ζ potential can be related to the surface potential of the channel wall [5] through the Debye-Huckel theory. The capacitance model described in section 2.1 can be used for computing the surface potential of the channel wall. The velocity profile across a capillary slit is a function of only the slip velocity and the pressure gradient [1], i.e.,

$$u = -\frac{1}{2\mu}\frac{dp}{dx}\left(y^2 - \frac{h^2}{4}\right) + u_{slip}$$
(4)

where x denotes the stream direction of the channel, y denotes the transverse direction across the channel and h is the channel depth. In the region where the flow is fully developed, it can be shown [1] that the pressure calculation is reduced to Laplace equation. Integrating the velocity profile given in equation (4) across the cross-section of the capillary slit, we get the following expression for the flow rate per unit width:

$$Q = \left(\frac{h^3}{12\mu L}\right)\Delta p + \left(\frac{\varepsilon\zeta}{\mu L}h\right)\Delta\phi$$
(5)

For the ith channel in an array of channels, equation (5) can be represented as:

$$Q_i = H_i\Delta p_i + E_i\Delta\phi_i$$
(6)

where H_i is the hydraulic conductance of the ith channel and E_i is the electro-hydraulic conductance of the ith channel. The expressions for H_i and E_i for the capillary slit are given in equation (5).

The slip velocity model discussed above can be employed when the Debye length is thin compared to the

channel width. However, when the Debye length is comparable to the channel width, (typically, for channels smaller than 100 nm) the slip velocity model may not be accurate. For a rectangular channel or a capillary slit, the velocity profile is given by the following expression [5]:

$$u_{(y)} = -\frac{1}{2\mu}\frac{dp}{dx}\left(y^2 - \frac{h^2}{4}\right) - \frac{\varepsilon}{\mu}\nabla\phi\left(\psi_0 - \psi_{(y)}\right)$$
(7)

where

$$\psi_{(y)} = \frac{\psi_0\,cosh\left(y/\lambda_D\right)}{cosh(h/\lambda_D)}$$
(8)

Integrating the velocity profile given in equation (7) across the cross section and using equation (8), we get the following expressions for the hydraulic conductance and the electro-hydraulic conductance of the ith channel:

$$H_i = \frac{h_i^3}{12\mu_i L_i}$$
(9)

$$E_i = \frac{\varepsilon_i}{\mu_i L_i}\psi_0\left(2h_i - 2\lambda_D\frac{sinh\left(h/\lambda_D\right)}{cosh\left(h/\lambda_D\right)}\right)$$
(10)

where λ_D is the Debye length [5].

Figure 3 shows the circuit representations of the fluidic domain for the cross-shaped channel segment shown in figure 1. It is to be noted that the total flow is the sum of the electrokinetically driven flow and the pressure driven flow. In the case of channels with integrated elastic parts (e.g. a flexible membrane), a capacitive element needs to be included in the circuit model of the fluidic domain as shown in figure 3. The fluidic capacitor can be modeled as:

$$C_{fl} = \frac{\iint_\Gamma w(x,y)d\Gamma}{p}$$
(11)

where C_{fl} is the fluidic capacitance, w is the deflection, Γ is the total surface area of the flexible membrane and p is the pressure.

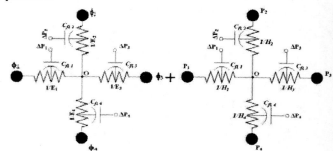

Figure 3: Circuit representation of the electrokinetically driven flow is on the left and circuit representation of the pressure driven flow is given on the right. The "plus" sign between the two figures indicates that the total flow is the superimposition of the electrokinetically driven flow and the pressure driven flow.

2.3 Device Model: Reactions

Consider a scheme in which the chemical species A and B are transported to the reaction chamber, where they undergo a second order reversible reaction process to produce specie C. The governing equations for this reaction process are given by [6]:

$$\frac{\partial m_A}{\partial t} = Q_A C_A - k_1 (m_A)(m_B) + k_2 (m_C) \tag{12}$$

$$\frac{\partial m_B}{\partial t} = Q_B C_B - k_1 (m_A)(m_B) + k_2 (m_C) \tag{13}$$

$$\frac{\partial m_C}{\partial t} = k_1 (m_A)(m_B) - k_2 (m_C) \tag{14}$$

where Q_i is the flow rate of the ith specie, which is computed from the fluidic transport model (or known from the design specifications), C_i is the concentration of the ith specie, m_i is the number of moles of the ith specie present in the reaction chamber, k_1 is the forward reaction rate and k_2 is the backward reaction rate. Trapezoidal method is used to discretize the ODEs given in equations (12) to (14). The non-linear discretized equations are then solved by employing the Newton-Raphson scheme.

2.4 Device Model: Separation

Consider an example, where two species A and B are to be separated using the separation channel. Assume that, specie A is unit-positively charged and specie B is unit negatively charged, while the surface of the channel has a negative fixed charge. In this case, the electro-osmotic flow through the channel would be from the anode side to the cathode side. The electrophoretic flow for A would be from anode to cathode but that for B it would be in the opposite direction. This is due to the difference in the electrophoretic velocity of these two species. Thus, the ratio of the rate of molar increment at the outlet of the separation channel for the two species is given by the following expression:

$$Separation\ Ratio = \frac{\left(Q_{os} + sign(z_A) \times |Q_{ph}|_A\right) c_A^{in}}{\left(Q_{os} + sign(z_B) \times |Q_{ph}|_B\right) c_B^{in}} \tag{15}$$

where, c_A^{in} is the concentration of specie A at the inlet, c_B^{in} is the concentration of specie B at the inlet and Q_{os} is the bulk flow, which is computed from equation (6). The electrophoretic flux can be represented by a circuit model [6]:

$$|Q_{ph}|_i = \left(\frac{FD_i |z|_i}{RT}\frac{A_c}{L}\right)\Delta\phi = \Im_i \Delta\phi \tag{16}$$

Where, $\Im_i$ is the electrophoretic flux conductance of the ith specie in the separation channel.

3 RESULTS

As an example, we consider a lab-on-a-chip system (figure 4), which is designed based on the "Nanochip" reported by Becker et. al.[4]. The various chemical species are transported to the different modules on the chip from their sources by using electrokinetic transport. One third of the channels (marked as set A1 in figure 4) perform the dual role of fluid transport and passive mixing. These channels have been designed in such a way that the characteristic dimension at a given level is half of that at the previous level. Thus, the homogeneity of the sample being transported increases. Figure 5 shows the dependence of the homogeneity of the mixture [7], e_{mix}, on the number of split levels [6]. Electrophoretic separation and electrokinetic transport is the governing mechanism through the set of channels marked as A2 (in figure 4), while electrokinetic transport is the governing mechanism through the set of channels marked as A3. The specie in set A1 (say A) is then transported to the detection module (D), where it reacts with specie B (already present in the detection chamber) to produce specie C, which can be used for off-chip detection. Figure 6 shows the variation in the rate of formation of specie C with time for different applied potentials. The chemical species (G and H) transported through the channels A2 and A3 are transported to the reactor module (R in figure 4), where they undergo a second order reversible chemical reaction to produce another chemical specie, F. Figure 7 shows the effect of the number of input ports on the variation of the concentration of F with time.

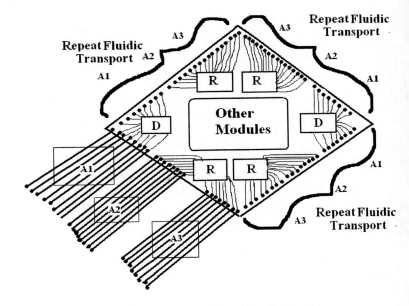

Figure 4: The schematics of the microfluidic chip considered as an example. The fluidic transport system represented on the south-west side of the chip is duplicated on the other sides.

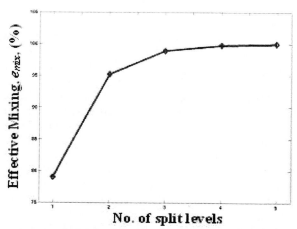

Figure 5: The dependence of the effectiveness of the mixing (i.e. homogeneity of the mixture) on the number of split levels used.

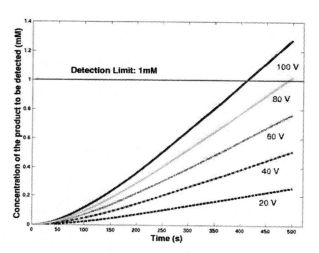

Figure 6: Concentration (C) versus time for various applied potentials.

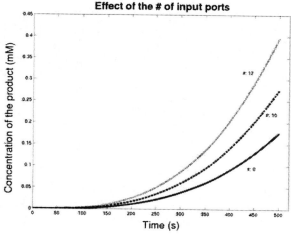

Figure 7: Concentration (F) versus time for different numbers of input ports per side of the microfluidic chip.

4 CONCLUSION

An advanced compact model for rapid analysis of microfluidic systems is presented. The fluidic transport system is represented by an electrical circuit network. Therefore, the fluidic variables (e.g. flow rate, pressure etc.) can be computed using the same schemes, which are used for computing the electrical variables. The new circuit model is significantly self sufficient and considers a number of additional elements compared to the compact model presented by Qiao et. al.[1]. As a result the circuit model presented in this paper can capture the physics of the fluidic transport process in much greater detail. Device models of the significant modules of a μ-TAS have also been presented in this paper. The integration of the circuit model for the fluidic transport system and the device models for the modules of the lab-on-a-chip example has been demonstrated. In conclusion, the circuit model based microfluidic-CAD tool presented here can be used to simulate and design next generation very large scale integrated (VLSI) microfluidic chips.

REFERENCES

1. R. Qiao and N. R. Aluru, "A compact model for electroosmotic flows in microfluidic devices", *J. Micromech. Microeng.*, Vol. **12**, pp. 625-635, 2002.
2. M. Mitchell, R. Qiao and N. R. Aluru, "Meshless analysis of steady state electroosmotic transport", *J. Microelectromech. Syst.*, Vol. **9**, pp. 435-449, 2000.
3. K. B. Oldham and J. C. Myland, *Fundamentals of Electrochemical Science*, Academic Press Inc: San Diego, 1994.
4. H. Becker and L. E. Locascio, "Polymer microfluidic devices", *Talanta*, Vol. **56**, pp. 267-287, 2002.
5. N. A. Patankar and H. H. Hu, "Numerical simulation of electroosmotic flow", *Anal. Chem.*, Vol. **70**, pp. 1870-1881, 1998.
6. A. N. Chatterjee and N. R. Aluru, "Combined circuit/device modeling and simulation of integrated microfluidic systems", *to be submitted for Journal Publication*.
7. P. Huang and K. S. Breuer, "Performance and scaling of an electro-osmotic mixer", *The 12th International Conference on Solid-State Sensors, Actuators and Microsystems (Transducers 2003, Boston)*, pp. 663-666, 2003.
8. S. C. Jacobson, T. E. McKnight and J. M. Ramsey, "Microfluidic devices for electrokinetically driven parallel and serial mixing", *Anal. Chem.*, Vol. **71**, pp. 455-4459, 1999.

Extending the Validity of Existing Squeezed-Film Damper Models with Elongations of Surface Dimensions

T. Veijola*, A. Pursula** and P. Råback**

*Helsinki University of Technology, P.O. Box 3000, FIN-02015 HUT, Finland, timo.veijola@hut.fi
**CSC – Scientific Computing Ltd., P.O. Box 405, FIN-02101 Espoo, Finland.

ABSTRACT

Border flow effects in squeezed-film dampers having a gap separation comparable with the surface dimensions are studied with 2D and 3D FEM simulations and with analytic models derived from the linearized Reynolds equation for small squeeze-numbers. Surface elongations are extracted with 2D FEM simulations for 1D squeezed-film dampers for variable surface topologies for both linear and torsional modes of motion. To model 2D squeezed-film dampers, these elongations are used directly in the compact models, and the results are verified with 3D FEM simulations. FEM simulations show that a simple surface-elongation model gives excellent results, and extend the validity range of existing compact models. To improve the model, drag forces acting on the upper surface and the sidewalls are approximated with simple equations based on FEM simulations.

Keywords: compact model, gas damping, Reynolds equation, squeezed-film damper

1 INTRODUCTION

Compact squeezed-film damper models have been presented for cases where the gap height is very small compared to the surface dimensions [1]–[4]. These idealized models underestimate the damping force, since the open border effects at the surface borders are ignored. Vemuri et al. [5] showed with 3D flow simulations and measurements that the open border effects considerably increase the damping force (35 %), even at surface width / gap height ratios of 20.

In [6] an end effect model was derived based on an assumption for acoustic boundary conditions at the damper borders. FEM simulations presented in this paper will reveal that a simple model with trivial boundary conditions and augmented surface dimensions gives much better results than the model presented in [6].

2 SQUEEZED FILM DAMPER

In the squeezed film problem illustrated in Fig. 1, the force caused by a flat gas film between moving surfaces is modeled with the modified Reynolds equation [2], [7]. Assuming a small pressure change p compared to the

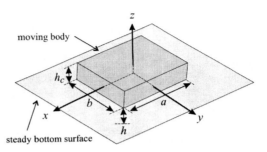

Figure 1: Dimensions of a squeezed film damper. The rectangular body moves causing fluid flow in the air gap below it.

ambient pressure P_A, and a small displacement z compared to the static height h of the gap, a linearized form of the modified Reynolds equation results. The linearized equation for isothermal conditions is

$$\frac{P_A h^2}{12\eta_{\text{eff}}} \nabla^2 \left(\frac{p}{P_A}\right) - \frac{\partial}{\partial t}\left(\frac{p}{P_A}\right) = \frac{v}{h}, \qquad (1)$$

where pressure change p, and the velocity $v = \partial z/\partial t$ are functions of time t and position (x, y), and η_{eff} is the effective viscosity coefficient [2].

An important measure for the squeezed-film effect is the squeeze number $\sigma = 12\eta a^2 \omega/(P_A h^2)$ that specifies the ratio between the spring force due to the gas compressibility, and the force due to the viscous flow. In this paper, only the case $\sigma \ll 1$ is considered.

Traditionally, the squeezed-film problem has been solved applying the trivial boundary condition where the pressure p vanishes at the borders [1], [4], [8]. This condition is justified if the surface dimensions are large compared to the gap height.

However, in practical squeeze-film dampers, the flow escaping from the damper borders might have a significant effect on the damping coefficient. This is true especially when the length and width of the damper are comparable with the air gap height.

2.1 Acoustic Boundary Conditions

The elongation model for a rectangular channel is suggested for calculating the end effect at the borders of a squeezed-film damper in [6]. The border flow can be modeled in a simple way by assuming that the flow channel continues outside the damper borders. The boundary conditions are derived simply from the fact that the pressure changes linearly in this fictitious channel of length $\Delta a/2$:

$$\left.\frac{\partial p}{\partial n}\right|_{\text{border}} = \mp \frac{p}{\Delta a/2}, \qquad (2)$$

where n is a coordinate normal to the border.

2.2 Simple Model with Elongations

An alternate model uses solutions calculated with trivial boundary conditions, but applies effective surface dimensions: the length and width of the surface, a and b, are replaced with effective dimensions $a + \Delta a$ and $b + \Delta b$, respectively.

3 1D SQUEEZED-FILM DAMPERS

3.1 Compact Models

The border effects change the behaviour of the viscous flow close to the borders, but their contribution to the gas compressibility is insignificant. This is the reason why the elongation model is derived first ignoring the gas compressibility ($\sigma \ll 1$). The time dependency in the Reynolds equation, Eq. (1), will be ignored, and it is assumed that the width b is infinite and the length a is finite.

For linear motion the velocity is independent of position and we can write $v(x) = v_0$. The trivial boundary conditions are $p(a/2) = p(-a/2) = 0$. This results in a parabolic pressure distribution $p(x)$ and the mechanical resistance $R_{\text{ml},0} = -F/v_0$ is

$$R_{\text{ml},0} = \frac{1}{v_0} \int\limits_{-a/2}^{a/2} \int\limits_{-b/2}^{b/2} p(x)\mathrm{d}x\mathrm{d}y = \frac{ba^3\eta_{\text{eff}}}{h^3}. \qquad (3)$$

When applying the acoustic boundary conditions specified in Eq. (2), the mechanical resistance is

$$R_{\text{ml}} = \frac{ba^3\eta_{\text{eff}}}{h^3}\left(1 + 3\frac{\Delta a}{a}\right). \qquad (4)$$

In the torsion motion about the y-axis the velocity of the surface can be written as $v(x) = 2v_0 x/a$, where v_0 is the surface velocity at $x = a/2$. Applying the trivial boundary conditions, the pressure distribution $p(x)$ is solved and the mechanical resistance (force F reduced to $x = a/2$) becomes

$$R_{\text{mt},0} = \frac{2}{av_0} \int\limits_{-a/2}^{a/2} \int\limits_{-b/2}^{b/2} xp(x)\mathrm{d}x\mathrm{d}y = \frac{ba^3\eta_{\text{eff}}}{15h^3}. \qquad (5)$$

When applying the acoustic boundary conditions given in Eq. (2) the mechanical resistance becomes

$$R_{\text{mt}} = \frac{ba^3\eta_{\text{eff}}(a + 6\Delta a)}{15h^3(a + \Delta a)}. \qquad (6)$$

3.2 Extraction of Elongations

2D FEM simulations have been carried out to test the accuracy of the end-effect models. The multiphysical simulation software Elmer [9] has been used to solve the incompressible Navier-Stokes equations using stabilized finite element formulation. Isothermal conditions and no-slip boundary conditions are assumed (continuum flow regime). The dimensions of the simulated 1D squeezed-film dampers are shown in Fig. 2 for linear and torsional motion. Generation of the geometry, meshing and FEM simulations were automated with shell scripts. The simulated 2D volume consists of 35000 mesh points.

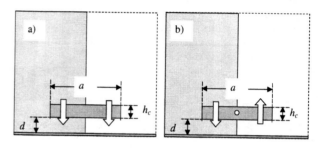

Figure 2: Simulated 2D volumes for a) linear motion and b) torsional motion. The symmetry was exploited to simulate only half of the structure.

A set of simulations was performed, where the dimensions of the structure were varied and the simulated forces (F_{top}, F_{bot}, and F_{side}) or twisting moments (τ_{top}, τ_{bot}, and τ_{side}) acting on all surfaces were recorded. The height of the damper h_c was varied from 4 to 32. A nominal surface width b of 1 m was assumed. Flow velocities at the damper surfaces were set to 1 m/s or in case of torsional motion, the boundary conditions (v_x, v_y) at the damper surfaces were set to imitate the rotation of the damper about the y-axis.

The actual squeezed-film force acts on the bottom of the surface. Forces acting on other surfaces are first excluded from the squeezed-film damper model since they do not obey the typical relation $\approx h^{-3}$. However, these drag forces, that depend also on h_c, can be approximated and added to the flow resistance of the damper (see section 3.3).

In the following, the effective elongation is extracted from the simulated force F_{bot} (τ_{bot}) for each ratio a/h. First, a simple elongation model based on the trivial boundary conditions is used: a is replaced with $a + \Delta a_S$ and Δa_S is solved from Eq. (3). Next, the elongation based on the acoustic boundary conditions is used. Solving for Δa from Eq. (4) results in the elongation Δa_A. Similarly, the elongations for torsion motion are solved from Eqs. (5) and (6). The extracted elongations are shown in Table 1 as a function of a/h.

Table 1: Relative elongations Δa_S and Δa_A extracted from 2D FEM simulations.

	Linear		Torsional	
a/h	$\Delta a_S/h$	$\Delta a_A/h$	$\Delta a_S/h$	$\Delta a_A/h$
2.0	1.307	2.346	1.654	21.202
4.0	1.307	1.781	1.662	1.009
8.0	1.299	1.522	1.662	0.103
16.0	1.294	1.401	1.650	-0.790
32.0	1.295	1.348	1.640	-2.312

Comparing the extracted elongations Δa_A and Δa_S shows clearly that the simple elongation model with the trivial boundary conditions is much better than the acoustic resistance boundary condition model. The data shows that the extracted elongations $\Delta a_S/h$ for torsional motion are systematically about 30 % larger than in the linearly moving case.

3.3 Design Equations with Corrections

Based on the presented FEM simulations, novel models are suggested here for 1D squeezed-film dampers where the end effects and drag forces are included. They are given in the form of mechanical resistances for both linear and torsional motion.

For linear movement the average elongation in Table 1 is 1.3. An improved model that accounts for the additional mechanical resistance due to the drag forces $R_{ml,drag}$ acting on the upper surface and on both of the sidewalls of the damper is $R_{ml} = R_{ml,elo} + R_{ml,drag}$, and

$$R_{ml} = \frac{b(a + 1.3h)^3 \eta_{eff}}{h^3} + 5.0\eta_{eff}b\sqrt{\frac{a + 2.8h_c}{h}}. \quad (7)$$

Equation (7) is derived heuristically and the coefficients were found by fitting R_{ml} to the simulated total mechanical resistance $R_{ml,FEM} = (F_{bot} + F_{top} + 2F_{side})/v_y$.

Table 2 shows the importance of the end effect model and the refinement of the model due do the drag force corrections as a function of a/h. If the end effects are ignored, the errors are quite large, from 11 % up to 82.6 %. The elongation model alone is good for $a/h > 8$, but the error of $R_{ml,elo}$ is as large as 20.4 % for small a/h ratios.

For torsional movement the average extracted elongation in Table 1 is 1.65. According to the simple model with modified dimensions in Eq. (5), the resistance $R_{mt} = R_{mt,elo} + R_{mt,drag}$ is

$$R_{mt} = \frac{b(a + 1.65h)^4 \eta_{eff}}{15ah^3} + 3.2\eta_{eff}b\sqrt{\frac{a + 2.7h_c}{h}}, \quad (8)$$

where $R_{mt,drag}$ is an additional term due to the drag forces. R_{mt} has been fitted to the total mechanical resistance $R_{mt,FEM} = (\tau_{bot} + \tau_{top} + 2\tau_{side})/v_y/(a/2)$ given by the FEM-simulations. Table 2 shows the relative errors of the models for the torsional squeezed-film damper.

Table 2: Maximum relative errors in 1D squeezed-film damper models compared to the FEM simulation results $R_{ml,FEM}$ and $R_{mt,FEM}$ ($4 < a/h_c < 32$).

	Linear			Torsional		
a/h	$R_{ml,0}$	$R_{ml,elo}$	R_{ml}	$R_{mt,0}$	$R_{mt,elo}$	R_{mt}
2.0	-82.6	-20.4	-1.5	-95.4	-49.3	-1.9
4.0	-60.7	-8.7	-0.6	-82.5	-30.3	2.1
8.0	-38.4	-2.5	0.8	-59.6	-14.5	1.5
16.0	-21.6	-0.5	0.7	-35.4	-4.7	-0.8
32.0	-11.4	0.2	0.4	-19.2	-1.2	-0.4

From the large errors in $R_{ml,0}$ and $R_{mt,0}$ in Table 2 it can be seen that the contribution of the border flow is considerable in squeezed-film dampers even with moderate a/h ratios. For torsional motion, the border effects are even stronger, for $a/h = 32$ the relative error of $R_{mt,0}$ is still about 20 %. It can also be seen that in the torsional mode the elongation model $R_{mt,elo}$ alone is not sufficient for moderate a/h ratios, the drag force correction is necessary for an accurate model.

4 2D SQUEEZED-FILM DAMPERS

4.1 Compact Models

Assuming trivial boundary conditions, the linearized Reynolds equation Eq. (1) can be solved analytically for rectangular surfaces. If the flow in the corners is ignored, the elongation model derived in the previous section can be applied.

For operation at low squeeze numbers the mechanical resistance can be used. For linear motion [2]

$$R_{ml,elo} = \sum_{m,n} \frac{1}{G_{m,n}(a_{eff}, b_{eff})}, \quad \left\{ \begin{array}{l} m = 1,3,5,\ldots \\ n = 1,3,5,\ldots \end{array} \right. \quad (9)$$

where $a_{eff} = a + 1.3h$, and $b_{eff} = b + 1.3h$. For torsional motion

$$R_{mt,elo} = \sum_{m,n} \frac{1}{G_{m,n}(a_{eff}, b_{eff})}, \quad \left\{ \begin{array}{l} m = 2,4,6,\ldots \\ n = 1,3,5,\ldots \end{array} \right. \quad (10)$$

where $a_{eff} = a + 1.65h$, and $b_{eff} = b + 1.3h$ and

$$G_{m,n}(a,b) = \left(\frac{m^2}{a^2} + \frac{n^2}{b^2} \right) \frac{m^2 n^2 \pi^6 h^3}{768 \eta_{eff} ab}. \quad (11)$$

The trivial models $R_{ml,0}$ and $R_{mt,0}$ result from $a_{eff} = a$ and $b_{eff} = b$.

4.2 FEM Simulations

A set of 3D FEM simulations was performed for a rectangular damper in linear and torsional motion to verify the novel elongation model for 2D dampers. The relative length of the surface a/h was varied from 2 to 32, the relative height of the damper a/h_c was varied from 4 to 32, and a/b was varied from 0.5 to 2. The symmetry of the structure was exploited in the FEM simulations; a quarter of the structure was simulated with 250000 elements for each set of the geometries. The simulations were performed with a 1 GHz Compaq AlphaServer.

4.3 Design Equations with Corrections

To estimate these drag forces, an approximation was used to include the drag force into the mechanical resistances R_{ml} and R_{mt} adding $R_{ml,elo}$ and $R_{mt,elo}$ to Eqs. (9) and (10), respectively:

$$R_{ml,drag} = 2.2\eta_{eff}a\sqrt{\frac{b+7.1h_c}{h}} + 2.2\eta_{eff}b\sqrt{\frac{a+7.1h_c}{h}}$$

$$R_{mt,drag} = 0.72\eta_{eff}a\frac{0.28a+b+h_c}{h} + 3.2\eta_{eff}b\sqrt{\frac{a+2.7h_c}{h}}.$$

the resulting relative error is below 6.5 %.

Table 3: Maximum relative errors in 2D squeezed-film damper models compared with the FEM simulation results $R_{ml,FEM}$ and $R_{mt,FEM}$.

	Linear			Torsional		
a/h	$R_{ml,0}$	$R_{ml,elo}$	R_{ml}	$R_{mt,0}$	$R_{mt,elo}$	R_{mt}
2.0	-90.6	-28.5	1.1	-96.7	-68.6	-4.5
4.0	-72.8	-15.1	-0.7	-86.7	-52.7	4.7
8.0	-47.3	-10.2	1.7	-64.8	-30.3	-2.9
16.0	-26.7	-5.5	6.2	-36.1	-11.0	-3.1
32.0	-11.3	6.0	6.3	-18.9	-2.3	3.0

Table 3 gives the maximum relative errors of various models compared with the simulated total force and twisting moment. The error of the trivial models $R_{ml,0}$ and $R_{mt,0}$, Eqs. (9) and (10) with $a_{eff} = a$ and $b_{eff} = b$, is quite large, from 20 % up to 90 %. The elongation model ($R_{ml,elo}$) with a_{eff} and b_{eff}, is reasonable good for $a/h > 4$. The results for $a/h = 2$ are strongly affected by the drag forces acting on the sidewalls and on the top surface.

The remaining errors in R_{ml} and R_{mt} are larger than in the case of the 1D damper. More accurate drag force approximations could be derived, but it is not justified here, since the accuracy of the 3D FEM simulation is not better than 5% for $a/h \geq 16$. This is due to the insufficient number of elements that could be simulated.

5 CONCLUSIONS

Design aids for squeezed-film dampers were presented in the form of compact models. With the aid of these models, the Reynolds equation has been extended to cases where the air gap size is comparable to the surface dimensions. It was shown that the simple surface-elongation model gives superior results compared to the acoustic elongation model. FEM simulations proved that the drag forces acting on the top surface and on the sidewalls become important, when small a/h-ratios are modeled, especially for torsional motion. The simulations also show clearly that the contribution of the end effects is considerable even at relatively high a/h-ratios.

The elongation model for 1D dampers was usable also in modeling rectangular 2D dampers. Could this approach be extended to general geometry? This would enable the replacement of the 3D Navier-Stokes equation with a 2D Reynolds equation with effective elongations, and the calculation times could be reduced considerably.

This paper considers only the case $\sigma \ll 1$. The extension of the model to higher squeeze numbers is basically straightforward, since the compressibility is not changed by the edge effects. However, in practical topologies where a/h is not small, the inertial [10], acoustic radiation, and rare gas effects become important. Additionally, if rare gas effects are modeled, the length a might be comparable to the mean free path of the gas. In this case the elongation becomes a function of a/h [11].

REFERENCES

[1] W. S. Griffin et al. Journal of Basic Engineering, Trans. ASME, vol. 88, pp. 451–456, June 1966.

[2] T. Veijola et al. Sensors and Actuators A, vol. 48, pp. 239–248, 1995.

[3] F. Pan et al. Journal of Micromechanics and Microengineering, vol. 8, pp. 200–208, 1998.

[4] R. B. Darling et al. Sensors and Actuators A, vol. 70, pp. 32–41, 1998.

[5] S. Vemuri et al. Proceedings of MSM'2000, (San Diego), pp. 205–208, April 2000.

[6] T. Veijola Proceedings of MSM 2002, (San Juan), April 2002.

[7] S. Fukui and R. Kaneko Journal of Tribology, Trans. ASME, vol. 110, pp. 253–262, April 1988.

[8] J. B. Starr Solid-State Sensor and Actuator Workshop, IEEE, (Hilton Head Island), pp. 44–47, June 1990.

[9] "Elmer – Finite Element Solver for Multiphysical Problems." http://www.csc.fi/elmer.

[10] T. Veijola Journal of Micromechanics and Microengineering, 2004. Submitted for publication.

[11] F. Sharipov and V. Seleznev J. Phys. Chem. Ref. Data, vol. 27, no. 3, pp. 657–706, 1998.

Feature Length-Scale Modeling of LPCVD and PECVD MEMS Fabrication Processes

Lawrence C. Musson[†], Pauline Ho[‡], Steven J. Plimpton[†] and Rodney C. Schmidt[†]

† Sandia National Laboratories, P.O. Box 5800, Albuquerque, NM 87185-0316
lcmusso@sandia.gov, sjplimp@sandia.gov, rcschmi@sandia.gov

‡ Reaction Design, 6440 Lusk Blvd., Suite D209, San Diego, CA 92121
pho@reactiondesign.com

ABSTRACT

The surface micromachining processes used to manufacture MEMS devices and integrated circuits transpire at such small length scales and are sufficiently complex that a theoretical analysis of them is particularly inviting. Under development at Sandia National Laboratories (SNL) is Chemically Induced Surface Evolution with Level Sets (ChISELS), a level-set based feature-scale modeler of such processes. The theoretical models used, a description of the software and some example results are presented here. The focus to date has been of low-pressure and plasma enhanced chemical vapor deposition (LPCVD & PECVD) processes. Both are employed in SNL's SUMMiT V technology though as of this writing the PECVD process model includes only unbiased wafers. Examples of step coverage of SiO_2 into a trench by each of the LPCVD and PECVD process are presented.

Keywords: MEMS, CVD, LPCVD, PECVD, level-set method, feature scale.

1 SOME CHALLENGES IN MEMS SMM TECHNOLOGY

In the surface micromachining (SMM) approach to the fabrication of MEMS devices three-dimensional (3D) structures are formed by deposition and etching of thin films. Careful construction of the lithographic masks that control these steps and the application of a final selective "release" etch permits the creation of a variety of free-standing movable parts. SMM fabrication can involve a wide variety of chemical processes in the deposition and etching steps. For example, deposition processes in the SUMMiT V [1] technology developed by Sandia National Labs include low-pressure chemical vapor deposition (LPCVD) of undoped polysilicon, P-doped polysilicon, silicon dioxide from TEOS [$Si(C_2H_5O)_4$], and Si-rich silicon nitride from $SiCl_2H_2$ and NH_3, as well as steam oxidation for the initial SiO_2 layer and the plasma deposition (PECVD) of SiO_2 from SiH_4 and O_2. Etching processes include a plasma etch of oxide and nitride using C_2F_6 and CHF_3, of polysilicon using Cl_2, He, and/or HBr, and a wet etch using aqueous HF.

Many of the above mentioned deposition and etching processes can result in non-ideal device geometries at the feature scale. For example, CVD processes give near-conformal films, which yield rounded corners and dimples. Step coverage can range from perfectly conformal to non-conformal, and lower step coverage can result in sloped sidewalls. Under some conditions non-uniformities and irregularities in surface coverage occur. The unexpected appearance of any of these types of geometric irregularities can be particularly costly in the design, analysis, and batch fabrication cycle associated with the development of a new MEMS device. Thus a thorough understanding of the detailed chemistry and physics which lead to these geometric variations is essential to the development of improved SMM fabrication equipment, higher yield and more reliable fabrication processes, and more useful MEMS designs.

2 THEORETICAL MODELING AT THE FEATURE SCALE WITH ChISELS

Theoretical modeling of the detailed surface chemistry and concomitant surface evolutions during microsystems fabrication processes is recognized as having a great potential for improving SMM process fabrication technologies. The viability of computational simulations for these types of problems has been clearly demonstrated by earlier researchers and advances have been made in developing transport models, chemical mechanisms, and surface evolution modeling (*e.g.*, see [2] [3] [4]). However, currently available computer codes have not been designed to use massively parallel architectures efficiently, nor fully exploit all of the modeling advances that different researchers have made. Thus speed and robustness factors have unduly limited the size and complexity of problems that can be treated with available tools.

To overcome these limitations, we are developing a computer code called ChISELS (Chemically Induced Surface Evolution with Level-Sets), a parallel, 2D and fully 3D feature-scale modeler to explore the time development of material deposition/etch on patterned wafers at low pressures. ChISELS can be viewed as a platform to build and improve on previous simulation tools while taking advantage of the most recent advances in dynamic mesh refinement, load balancing, and highly scalable solution algorithms.

There are three inter-related aspects to modeling the overall physics of the problem: (1) transport of chemical species, (2) gas phase and surface chemistry, and (3) the dynamic evolution of the solid surface.

Currently, all gas-phase transport is assumed to occur in the free-molecular flow regime (i.e., particle-particle collisions are negligible). In ChISELS, we adopt the ballistic transport and reaction model (BTRM) developed and described by Cale and coworkers [2] [5]. An important aspect of this method is the need to calculate view-factors for each surface element of the discretized feature surface. Chaparral [6], a radiation heat transfer modeler, is used for this purpose. The flux of species k to surface element i is computed from

$$F_{ik} = F_{ik}^0 + G_{ij}(F_{jk} - R_{jk}) \qquad (1)$$

where F_{ik}^0 is the direct flux from the bulk of the reactor, G_{ij} is the view factor between surface i and j, F_{jk} is the flux of species k to surface j, R_{jk} is the reaction rate of species k on surface j, and by Einstein's convention repeated indices in the product imply summation.

Deposition or etching occurs through the chemical reaction of gas phase species with bulk and surface phase species at the feature surface. The thermodynamics and heterogeneous surface chemistry of these reactions are modeled in ChISELS by coupling with Surface Chemkin [7]. This requires the specification of a chemical reaction mechanism for each surface reaction to be modeled in the simulation.

Feature scale microsystem fabrication modelers such as ChISELS are at heart topology modelers, *i.e.* they model the evolution of a free boundary according to the physics that cause it to move. ChISELS uses an implicit surface-tracking approach called the level set method [4]. In the level-set method, a domain-spanning function, ϕ is defined; the zero-value contour, or level set, of which conforms to the feature surface. The level-set function is evolved by solving the scalar partial-differential equation,

$$\frac{\partial \phi}{\partial t} + \mathbf{v} \cdot \nabla \phi = 0 \qquad (2)$$

over the volume and integrating through time. The velocity, $\mathbf{v}$ is called the extension velocity and is defined over the entire domain. The extension velocity must be chosen so that the level set of ϕ evolves in such a way that it mimics the evolution of the feature surface; *i.e.* it is chosen based on the velocity of the surface—the deposition or etch rate. The level set method avoids the debilitations of explicit conform-and-track methods because the mesh which is used to solve Equation 2 does not deform, so distortion effects are avoided. Likewise,

because a volume-defined function is evolved, merging surfaces do not create problems in the method. Equation 2 is solved by an augmented method of characteristics [8].

Additional details of the methods and models that ChISELS uses can be found in [9] or on the ChISELS web site, http://www.cs.sandia.gov/~wchisels.

3   EXAMPLES

Two two-dimensional examples are provided here of ChISELS results. One is of the deposition of SiO_2 into a 10x1 trench using silane by a PECVD process. The other is the deposition of SiO_2 into an identical trench using TEOS. In each case, the deposition rate as a function of depth in the trench is presented.

3.1   PECVD of SiO$_2$ from SiH$_4$

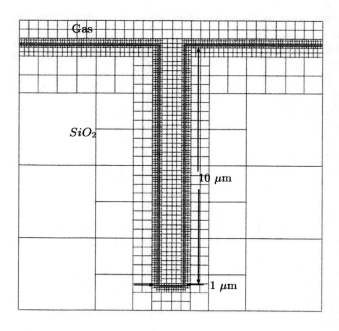

Figure 1: The trench used in the examples is 1 μm wide and 10 μm deep.

The deposition of silicon dioxide into a trench of aspect ratio 10:1 (*cf.* Figure 1) from a mixture of SiH_4, O_2, Ar and derivative compounds in a high-density plasma reactor is modeled by ChISELS. The grid used to model the process is also shown in Figure 1. In this case, there is no biasing of the substrate, so, as a first approximation, ions and electrons approach it in a diffuse way just as the neutrals do.

The chemistry mechanism used in this model is a subset of the one published by Meeks *et al.* [11]. All the reactions in their mechanism are used except for the ion sputtering reactions. This leads to a system with

46 gas-phase species, 13 surface-phase species and 185 surface reactions.

The gas composition in the reactor immediately above the wafer surface is a required input to ChISELS. Because that composition is unlikely to be identical to that introduced into the reactor, a reactor-scale model must be employed to provide the mole fractions of the introduced reactants and their products near the wafer. In this case, Aurora, a a zero-D reactor-scale modeler and Reaction Design product, part of the Chemkin suite, was used to determine those compositions.

The inlet conditions provided to Aurora are a temperature of 300K, and mole fractions of 0.0097, 0.0223 and 0.9680 respectively of SiH_4, O_2 and Ar at a total flow rate of 0.0028 g/s. The output of Aurora is the mole fractions of all reactants and products at the wafer as well as surface, ion and electron temperatures. All of these are inputs to ChISELS.

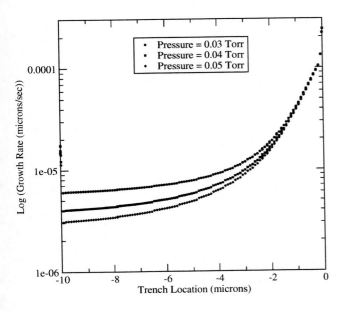

Figure 2: Growth rates as a function of depth of SiO_2 in a PECVD process using SiH_4 and O_2.

Figure 2 shows the deposition rate in the trench as a function of its depth for the indicated pressures of 0.03, 0.04 and 0.05 Torr. As shown in the figure and perhaps contrary to intuition, the lowest pressure of 0.03 Torr actually produced the highest growth rate of all pressures tried. This is due to the difference in the gas-phase composition at the reactor due to a change of reactor pressure. This underscores the importance of coupling ChISELS to a reactor-scale model.

As Figure 2 shows, the deposition rate, and thus step coverage, in this process is not uniform. The deposition rate in the bottom corner of the trench is about two orders of magnitude smaller than that at the top.

The source of this non-uniformity is most likely due to the plasma forming highly reactive radicals that get depleted almost entirely near the top of the trench and are hardly present near the bottom. Note also that there are some higher growth rates on the ordinate of Figure 2. These points correspond to surface elements near the center of the bottom plane of the trench. These higher growth rates are owed to greater visibility of the reactor and its constituents above.

3.2 LPCVD of SiO₂ from TEOS

The second example is the deposition of SiO_2 into a trench from TEOS. The trench geometry and grid are identical to the first example. There are two chemistry mechanisms used in this example. The first is a mechanism published by Coltrin et al. [12], and the second by IslamRaja et al. [13]. Once again, Aurora is used to compute gas-phase composition. The mechanism used in Aurora is Coltrin's. The IslamRaja mechanism is defined only for heterogeneous reactions. The input conditions are a reactor temperature of 1003K—the temperature at which the IslamRaja mechanism was tuned and the only temperature for which it is truly valid—pressures of 0.3, 0.4 and 0.5 Torr and an inlet flow rate of 0.046 g/s of TEOS and N_2 with mole fractions of 0.13 and 0.87 respectively.

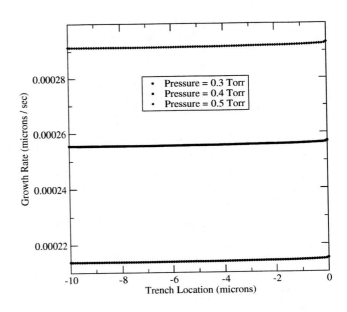

Figure 3: Growth rate as a function of depth of SiO_2 in a LPCVD process using TEOS—Coltrin's chemistry.

Figures 3 and 4 show the deposition rate of SiO_2 as a function of the depth in the trench. Unlike the previous example, the deposition rate as predicted by both mechanisms increases with the reactor pressure. The difference between the two TEOS mechanisms is the magnitude and uniformity of the deposition rate. Coltrin's mechanism predicts a lower overall rate and one which is exceptionally uniform. The IslamRaja mechanism produces a very non-uniform deposition rate varying two orders of magnitude from the bottom of the trench to the top.

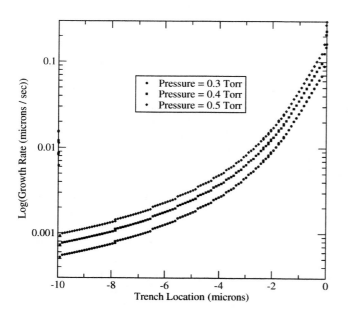

Figure 4: Growth rate as a function of depth of SiO_2 in a LPCVD process using TEOS—IslamRaja's chemistry.

Experiments have shown the deposition of SiO_2 indeed to be non-uniform. So Coltrin's mechanism clearly does not work at the feature scale though it has been shown to work on the reactor scale. The IslamRaja mechanism is not even defined on the reactor scale, so there is no way to couple ChISELS to a reactor-scale model with that mechanism in use. The next step is to refine Coltrin's mechanism so that it remains working on the reactor scale and also produces realistic growth rates at the feature scale.

SUMMARY

A cursory description of the ChISELS feature length-scale modeler has been given. Two examples were discussed of the deposition of SiO_2 by LPCVD through TEOS and by PECVD through silane and oxygen in plasma. Step coverages for each process and how they are affected by reactor operating conditions were shown. More details of the ChISELS software can be found in [9] and [10].

4 ACKNOWLEDGEMENTS

The authors would like to thank Sandia National Laboratories and the Department of Energy for supporting this work. Sandia is a multiprogram laboratory operated by Sandia Corporation, a Lockheed Martin Company, for the United States Department of Energy under Contract DE-AC04-94AL85000.

REFERENCES

[1] M. S. Rodgers, J. J. Sniegowski, "Designing Microelectromechanical Systems-on-a-Chip in a 5-Level Surface Micromachine Technology," 2nd Int. Conf. Eng. Design and Automation, 1998.

[2] T. S. Cale, T. P. Merchant, L. J. Borucki and A. H. Labun, Thin Solid Films, 356, 152-175, 2000.

[3] A. H. Labun, H. K. Moffat and T. S. Cale, J. Vac. Sci. Technol. B., 18, 267-278.

[4] J. A. Sethian, "Level Set Methods", Cambridge Univ. Press, Cambridge, 1996.

[5] T. S. Cale and V. Mahadev, in Thin Films: Modeling of Film Deposition for Microelectronic Applications, Vol. 22, Ed. S. Rossnagel, Academic Press, 176-277, 1996.

[6] M. W. Glass,"CHAPARRAL: A library for solving enclosure radiation heat transfer problems", Sandia National Laboratories, Albuquerque, NM, 2001.

[7] SURFACE CHEMKIN: A Software Package for the Analysis of Heterogeneous Chemical Kinetics at a Solid-Surface – Gas-Phase Interface Interface, Reaction Design Inc. 2001 (See also: CHEMKIN@ReactionDesign.com).

[8] J. Strain, J. Comput. Phys., 161, 512–528, 2000.

[9] Musson, Lawrence C., Plimpton, Steven J., and Schmidt, Rodney C. MEMS Fabrication Modeling with ChISELS: A Massively Parallel 3D Level-Set Based Feature Scale modeler, Technical Proceedings of the 2003 Nanotechnology Conference and Trade Show, Volume 3 Nanotech 2003 Grand Hyatt San Francisco, San Francisco, CA, U.S.A.February 23-27, 2003

[10] http://www.cs.sandia.gov/~wchisels

[11] Ellen Meeks, Richard Larson, Pauline Ho, Christopher Apblett, Sang M. Han, Erik Edelberg and Eray S. Aydil, J. Vac. Sci. Technol. A, 16(20), 544–563, 1998.

[12] Michael E. Coltrin, Pauline Ho, Harry K. Moffat and Richard J. Buss, Thin Solid Films, 365, 251–263, 2000.

[13] M. M. IslamRaja, C. Chang, J. P. McVittie, M. A. Cappelli and S. C. Saraswat, J. Vac. Technol. B 11(3), 720–726, 1993.

Compact Models for Squeeze–Film Damping in the Slip Flow Regime

Robert Sattler and Gerhard Wachutka

Institute for Physics of Electrotechnology,
Munich University of Technology
Arcisstr. 21, 80290 Munich, Germany; sattler@tep.ei.tum.de

ABSTRACT

We propose a mixed–level simulation scheme for squeeze film damping effects in microdevices, which makes it possible to include damping effects in system-level models of entire microsystems in a natural, physically–based and flexible way. Our approach allows also for complex geometries, large deflection and coupling to other energy domains. In this work, we focus on the extension of our model to the slip flow regime. To this end, the flow problem is separated into appropriate blocks. For each block the slip factor is derived analytically or extracted from FEM simulations based on the Navier–Stokes equation with slip boundary conditions. Our fully parametrized and, therefore, predictive mixed–level model could be verified by FEM and experimental analysis.

Keywords: squeeze–film damping, mixed–level modeling, compact model, slip flow, finite network.

1 INTRODUCTION

Movable structural parts in MEMS devices without expensive vacuum packaging are strongly affected by air damping. It is therefore that squeeze–film damping effects are an inherent part of the design and, hence, have to be taken into account to allow a predictive simulation of the transient behavior. However, it is prohibitive to use a continuous field solver for devices of real–world complexity. Instead, we follow a mixed–level approach, which makes it possible to reduce the model complexity by incorporating damping [1] and electrostatic [2] effects in system–level models in a physically correct and accurate manner. The damping model is based on the Reynolds equation, a simplified version of the Navier–Stokes equation (NSE), which is solved using the Finite Network method. The basic idea of our approach has been presented in [3]. In the terminology of Kirchhoffian Network variables, we consider the fluidic mass flow as generalized flux ("through"" variable), which is driven by the gradient of pressure acting as generalized force ("across" variable). Physically–based, scalable compact models are integrated in the network to correct for finite size effects and perforations in the structure (Fig.1).

All these compact models have recently been enhanced and validated. They are now thoroughly parametrized and can be rigorously derived from the Stokes equation [4].

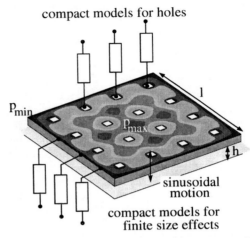

Fig 1: Elementary test structure: The colors indicate the pressure distribution below the actuating plate [3].

With this approach, we are able to perform predictive simulations of squeeze–film damping forces, which include physical parameters only and needs no longer the adjustment of fit parameters. Fig. 2 gives a more detailed view of the compact models and their connectivity with the Finite Network.

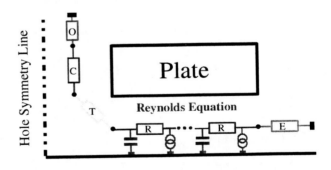

Fig 2: Schematic view of the mixed–level damping model. The Finite Network solves the nonlinear Reynolds equation for arbitrary geometry of the plate. The compact models account for corrections at the edge (E) and at perforations. The latter incorporates orifice flow (O), channel flow (C) and the transit flow (T) between the plate and the hole

2 RAREFACTION

With progressive miniaturization (e.g. the air gap underneath our demonstrator in Fig. 7 is only 1 μm), the effect of rarefaction has to be taken into account. Macroscopic fluids can be described by the NSE and the so–called "no–slip" boundary condition at the fluid–solid interface. As no discontinuities are allowed at the fluid–solid interface, the fluid velocity must be zero relative to the surface. But this boundary condition is only valid in thermodynamic equilibrium. In dilute gases the collision frequency between the fluid and the solid surface is not high enough to ensure equilibrium. Therefore a tangential velocity slip must be admitted:

$$u_{gas} - u_{wall} = \frac{2-\alpha}{\alpha} \cdot \lambda \frac{\partial u}{\partial n} \qquad (1)$$

This simple first–order slip model relates the slip velocity at the wall to the mean free path of the gas λ and the gradient of the gas velocity in the normal direction of the wall [5]. The tangential–momentum–accommodation coefficient α is defined as the fraction of molecules which are diffusively reflected. This coefficient has been introduced by Maxwell and depends on the gas, the solid and the surface finish and has to be determined experimentally. It lies typically between 0.2 and 0.8, where the upper limit is attained for most practical surfaces. For quantifying the degree of rarefaction, the Knudsen number $Kn = \lambda/\mathcal{L}$ is used, where $\mathcal{L}$ is a characteristic length of the flow (e.g. the diameter of a fluidic channel). If the Knudsen number exceeds the order of magnitude of 10^{-1}, the continuity assumption breaks down and the Boltzmann transport equation has to be used to describe the flow.

$Kn<0.01$	$0.01<Kn<0.1$	$0.1<Kn<10$	$Kn>10$
Continuum Flow	Slip Flow	Transit Flow	Free Molecular Flow

Tab. 1: Empirical classification of Knudsen number regimes

3 COMPACT MODELS

In the following we investigate the reduction of the fluidic resistance of each compact model due to slip flow.

3.1 Channel Flow

The compact models R and C in Fig. 2 are based on a Poiseuille channel flow. Under the constraints of a one dimensional channel flow the incompressible NSE reduces to: $\eta\, d^2 u_x / dn^2 = dp/dx$

(viscosity η, pressure p, flow direction x, velocity u).

Solving this equation with the slip boundary condition (eq. 1) results in eq. 2 for the fluidic resistance of a channel of length L and height h. If we transform this equation to cylindrical coordinates, we find the fluidic resistance for a circular tube with radius r (eq. 3).

$$R_{Channel} = \frac{\Delta p}{\rho \cdot v \cdot h} = 12\frac{\eta}{\rho}\frac{L}{h^3} \cdot (1+6\,Kn)^{-1} \qquad (2)$$

$$R_{Tube} = \frac{\Delta p}{\rho \cdot v \cdot \pi\, r^2} = \frac{8}{\pi}\frac{\eta}{\rho}\frac{L}{r^4} \cdot (1+8\,Kn)^{-1} \qquad (3)$$

The last term in brackets gives the correction of the compact models due to slip flow. This slip factor is equals one for continuous flow (Kn=0). The fluidic resistance in the Reynolds equation (R, Fig. 2) is represented by eq. 2, as firstly derived by Burgdorfer [6]. The fluidic resistance for a circular tube (C, Fig. 2) is represented by eq. 3. This slip factor has also be named "effective viscosity" η_{eff} [10].

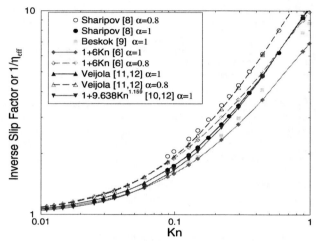

Fig. 3: Inverse slip factor for channel flow as calculated from various authors from atomistic simulations.

Fig. 3 illustrates that, with increasing Knudsen number, there is an increasing discrepancy between the simple slip model and the solution of the Boltzmann equation as reported by various authors. We used the simple model, on the one hand because of the uncertainties in the atomistic simulations, on the other hand because for intermediate Knudsen numbers the proper choice of the accommodation coefficient has much more impact than a higher order slip model. Hence, in the following diffusive reflection (α=1) is assumed.

3.2 Orifice Flow

The compact models E and O (Fig. 2) are based on viscous orifice flow, which can be derived from the Stokes equation. The velocity distribution in an elliptic orifice with the semi–axes a and b, the perimeter s and the area A is given by eq.4 [7]:

$$u(x,y) = \frac{A \cdot \Delta p}{\pi\, s\, \eta} \cdot \sqrt{1 - \frac{x^2}{a^2} - \frac{y^2}{b^2}} \qquad (4)$$

$$\dot{m} = \rho \iint_A u(x,y)\, dA \qquad (5)$$

Applying a slip velocity boundary condition and calculating the mass flow from eq. 5, we derived the fluidic resistance for two special cases of an elliptic orifice: A circular orifice of radius r (eq. 6) and a narrow slit of width h and length L with h << L (eq. 7).

$$R_{circular} = \frac{\Delta p}{\dot{m}} = \frac{3}{r^3}\frac{\eta}{\rho}\left[(1+2\,Kn)^{3/2}-(2Kn)^{3/2}\right]^{-1} \quad (6)$$

$$R_{slit} = \frac{16}{L\cdot h^2}\frac{\eta}{\rho}\left[\sqrt{2Kn}+(1+2Kn)\arcsin(1+2Kn)^{-1/2}\right]^{-1} \quad (7)$$

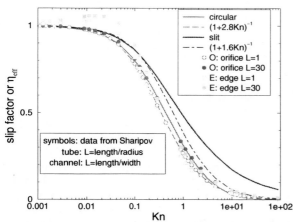

Fig. 4: Slip factor for the orifice calculated from eq. 5, 6 in comparison with data from Sharipov [8].

Sharipov solved the Boltzmann equation for channel flow and for tube flow of finite length. He describes the pressure drop at the orifice by an extra length of the channel. His tabulated results [8] have been converted into a slip factor for the orifice flow and are depicted together with our results in Fig. 4.

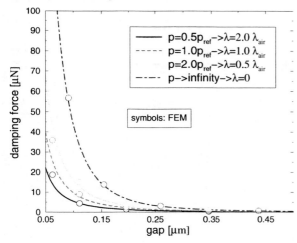

Fig. 5: Damping force on a disc (R=50 μm) moving with constant velocity towards ground in dependance of the mean free path.

In order to validate the slip factors for our compact models the slip flow boundary condition (eq. 1) was implemented in the NSE–solver of ANSYS. The computational expense of the FEM simulations was reduced by choosing a circular disc as our theoretical demonstrator. The Knudsen number depends on the mean free path which is inverse proportional to the ambient pressure. Fig. 5 demonstrates the influence of the ambient pressure on the damping force of the disc. Our mixed–level model agrees well with the solution of the NSE with slip flow boundary conditions. But for small gaps the Knudsen number increases and the slip factor $(1+6Kn)^{-1}$ underestimates the damping force on the disc. This is due to the moving boundary condition of the upper channel wall. All results in Fig. 3 are based on a channel with fixed walls, but this is usually not the case for a moving part of a MEMS device.

3.3 Transit Flow

So far only elementary structures such as tubes and channels have been investigated by numerical atomistic simulation. Therefore the slip factor for the transit resistances T in Fig. 2 cannot be found in literature. For this region it is difficult to define a Knudsen number, because the characteristic length scale $\mathscr{L}$ could be the diameter of the hole (2r in eq. 3) or the gap under the plate (h in eq. 2). In Fig. 6 the latter was chosen. The slip factor (eq. 8) in the compact model for the transit resistance (T) has been adjusted with reference to the FEM simulation of the NSE with slip boundary conditions (eq. 1). Figure 6 also illustrates that our mixed–level model can reproduce the damping force on a punched disc for various hole sizes.

$$\frac{R_{Transit}^{Slip}}{R_{Transit}^{no\,Slip}} = (1+1\,Kn)^{-1} \quad (8)$$

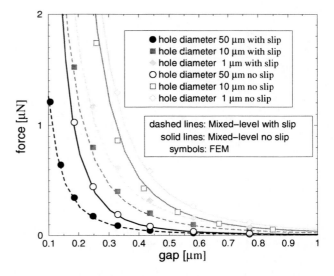

Fig. 6: Damping force acting on a punched disc which moves with constant velocity towards ground (R=50 μm).

4 EXPERIMENTAL DEMONSTRATOR

Finally, we compare our mixed–level model with experimental data extracted from measurements on industrial prototypes of the microrelay shown in Fig. 7:

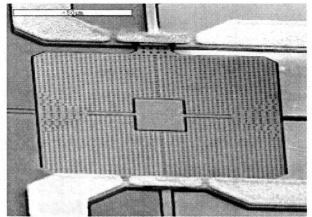

Fig. 7: SEM image of an electrostatic microswitch perforated with about 3000 etch holes.

The plate of the torsional switch is highly perforated with small etch holes. Three different types of perforations have been fabricated. They differ in the ratio of the hole area to the total cell area. Fig. 8 shows the influence of slip flow on the fluidic resistance of a single cell. With and without slip flow the mixed–level model agrees well with the FEM simulation of the NSE.

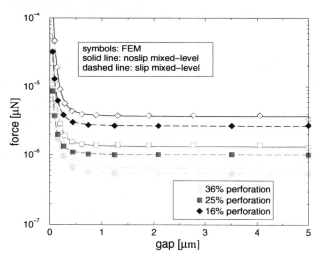

Fig. 8: Damping force on a single cell of the microswitch in Fig. 7 moving with constant velocity towards ground.

Fig. 9 shows the off–to–on and on–to–off switching transients of the same three device variants. The simulation transients are in excellent agreement with the measured characteristics, which demonstrates the high quality and predictive capability of our modeling approach.

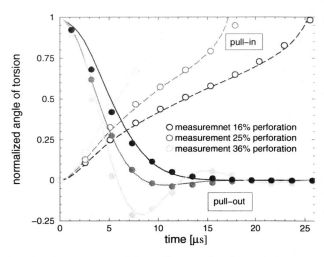

Fig. 9: Measurement of the pull–in and pull–out behavior of the microswitch in comparison with our mixed–level model.

CONCLUSION

We demonstrated that our mixed–level model based on a physical device description in terms of a standard hardware description language (VHDL–AMS) can successfully be extended to the slip flow regime. For every elementary structural part of the plate (Fig. 2) the proper model has to be chosen. Therefore it is misleading to use the term "effective viscosity", as this suggests using one single formula everywhere. All slip flow models in conjunction with Reynolds equation presented so far (Fig. 3) are based on the channel flow with fixed walls. But for real devices at least one wall is movable. With increasing degree of rarefaction the correct choice of the boundary conditions gain importance. Therefore atomistic simulations have to be done that take care of this effect.

REFERENCES

[1] R. Sattler, G. Schrag, G. Wachutka, MSM'02, pp. 124–127.
[2] R. Sattler, G. Schrag, G. Wachutka, MSM'03, pp. 284–287.
[3] G. Schrag, G. Wachutka, Sensors and Actuators A 97–98 (2002), pp. 193 200.
[4] R. Sattler, G. Wachutka, Improved Physically–Based Mixed–Level Damping Model, submitted to DTIP'04, Montreux, Switzerland.
[5] S.A. Schaaf, F.S. Sherman, "Skin Friction in Slip Flow", Journal of Aeronautical Sciences, vol. 21, no. 2, 1953,pp. 85–90.
[6] A. Burgdorfer, "ASME Journal of Basic Engineering, March 1959, pp. 94–100.
[7] H. Hasimoto, J. Phys.Soc.Japan, 13, no 5,1958, pp. 633–639.
[8] F. Sharipov, V. Seleznev, Data on Internal Rarefied Gas Flow, J.Phys.Chem.Ref.Data,Vol.27, No.3, pp.657–706, 1998.
[9] P. Bahukudumbim A. Beskok, A Phenomenological Lubrication Model for the Entire Knudsen Regime, J. Micromech. Microeng. 13, No. 6, November 2003, pp.873–884.
[10] Veijola,..,Sensors and Actuators A 48 (1995), pp. 239–248.
[11] Veijola,..,Sensors and Actuators A 66 (1998), pp. 83–92.
[12] Fukui,Kaneko, J.Tribol.Trans.ASME, 112(1990), pp.78–83.

Dynamic Simulation of an Electrostatically Actuated Impact Microactuator

X. Zhao, H. Dankowicz, C.K. Reddy, and A.H. Nayfeh

Department of Engineering Science and Mechanics, MC 0219
Virginia Polytechnic Institute and State University
Blacksburg, Virginia 24061, USA, anayfeh@vt.edu

ABSTRACT

We study the dynamics of an electrostatically driven impact microactuator reported by Mita and associates. The microactuator is modeled as a two-degree-of-freedom rigid multibody system. The impact phenomenon is described by a simple impact law based on Newton's coefficient of restitution. The system dynamics is intensively studied. We investigate the influence of various parameters including excitation frequency and voltage.

Keywords: microactuators, electrostatic, impact, dynamic simulation

1 INTRODUCTION

Precise displacement control and manipulation is required in microscopes, optical devices, nanoscale data storage and micro surgery. Microdevices are ideal for micropositioning due to their small sizes. Microactuators used to produce small displacements would need large actuation forces and a long driving distance. Actuators based on impulsive forces provide a solution to this problem, because impact actuation can generate relatively large motion from small displacement [1]. Recently, impact microactuators have attracted a lot of attention due to ease of fabrication, capability of batch processing, robustness to environment, high accuracy and high power [2]–[5].

Microactuators based on impact can generate linear or rotational motion without mechanical linkages such as ball screws or complicated clamping systems. However, impacts make the dynamical system nonsmooth [6]–[8]. The study of nonsmooth dynamical systems is challenging and one cannot just use traditional dynamical system tools to understand their behaviour. It is important to understand the underlying dynamics in order to design a better microactuator.

In the present work, we explore the dynamics of the impact microactuator reported in [5] through numerical simulation. The nonlinear dynamic model is derived, using a rigid multibody approach[8] and an impact law based on Newton's coefficient of restitution. We examine the influence of the system parameters such as the excitation frequency and the amplitude, on the dynamics of the actuator for specified values of the other system parameters. The objective of the present work is to illustrate the importance of understanding the underlying dynamics in an impact microactuator in order to design a more robust and consistent device.

2 DYNAMIC MODEL

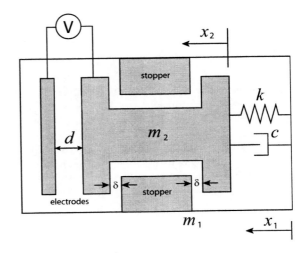

Figure 1: The schematic of the impact microactuator.

A schematic planform of the microactuator is shown in Fig. 1. The system consists of a movable block m_2, which is connected to m_1 by a linear spring with viscous damping. The stoppers and the electrode are rigidly fixed to the frame. The movable mass acts as one of the electrodes. The frame rests on the horizontal ground. Friction between the frame and the ground is assumed to be Coulomb. We denote the coefficient of static friction by μ_s, and that of dynamic friction by μ_d. When a driving voltage, $v(t)$, is applied between the electrodes, m_2 is accelerated towards the stopper till there is an impact. The impact impulse is large enough to overcome the static friction between the frame and the ground. This produces a small displacement of the frame. The minimum stepsize determines the resolution of the microactuator. When a periodically varying voltage is applied, there are repeated impacts and the frame moves by a certain amount during each cycle, thereby producing the needed displacement over a period of time. The equations of motion of the system can be written as

$$m_1 \ddot{x}_1 = F \qquad (1)$$

$$m_2(\ddot{x}_1 + \ddot{x}_2) + kx_2 + c\dot{x}_2 = \frac{\alpha v(t)^2}{(d - x_2)^2} \qquad (2)$$

where x_2 is the displacement of m_2 relative to m_1, and x_1 is the displacement of m_1 relative to the ground. We note that $\alpha = \frac{1}{2}\epsilon_0 A$, where ϵ_0 is the permittivity of free space, A is the overlap area, and d is the zero-voltage gap between the electrodes. As m_1 moves, the total force on m_1, F is

$$F = kx_2 + c\dot{x}_2 - \frac{\alpha v(t)^2}{(d - x_2)^2} - \mu_d N \operatorname{sign}(\dot{x}_1) \qquad (3)$$

where N is the normal reaction. We assume that gravity is the only external force, in which case $N = (m_1 + m_2)g$. When m_1 is stationary,

$$F = 0 \text{ and } |kx_2 + c\dot{x}_2 - \frac{\alpha v(t)^2}{(d - x_2)^2}| \leq \mu_s N \qquad (4)$$

As shown in Fig. 1, the separation between m_2 and the stoppers is δ. When $|x_2| = \delta$, m_2 impacts with the stopper. Let v_1 and v_2 be the velocities of m_1 and m_2 respectively, just before impact, and v_1' and v_2' be the velocities after impact. Assuming that the collision between m_1 and m_2 is inelastic with a restitution coefficient e, conservation of linear momentum yields [9]

$$v_1' = v_1 + \frac{(1 + e)m_2}{m_1 + m_2} v_2 \qquad (5)$$

$$v_2' = -ev_2 \qquad (6)$$

3 NUMERICAL SIMULATION

The equations of motion derived in the previous section are piecewise smooth due to impacts and friction. The dynamics can be divided into two different phases. In the first phase only m_2 is in motion, described by Eqs. (1), (2), and (4). In the second phase both m_1 and m_2 are in motion, described by Eqs. (1)-(3). The integration procedure is shown in Fig. 2, where impact map is described by Eqs. (5) and (6).

Direct numerical integration generates results qualitatively consistent with those reported in [5]. A sinusoidal voltage of 100V at 1HZ is applied in the experiments [10] and the average displacement is found to be 20nm/impact. Parameter values reported by Mita et al. are used for the numerical simulation. The parameter values not reported are fitted to obtain a displacement per impact of approximately 20 nm. Figure 3 shows the relationship between the displacement and voltage. We note that there are two impacts in one excitation cycle, consistent with the experimental observations.

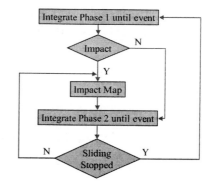

Figure 2: Flowchart of the algorithm.

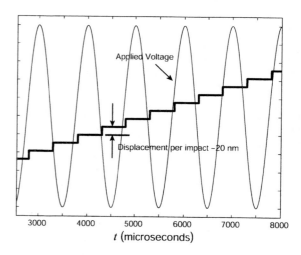

Figure 3: Displacement per impact.

4 SWITCH ON AND OFF MECHANISM

In this paper, we apply a sinusoidal voltage of the form $v(t) = f\cos(\omega t)$. We treat the amplitude f and frequency ω as our design variables and study the dynamics with respect to them. The system parameter values used for simulations are $m_1 = 5$, $m_2 = 1$, $k = 1$, $c = 0.04$, $d = 1$, $\delta = 0.5$, $e = 0.8$, $\mu_s = 0.4$, $\mu_d = 0.27$, and $\alpha = 1$.

The system possesses two types of steady orbits: those with impacts and those without impacts. In a previous paper [9], we have investigated the bifurcation of the impact orbits. One interesting question is how the system transfers back and forth between impact and non-impact orbits when changing parameters. In this paper, we call the impact orbits "on" status, and the non-impact orbits "off" status. Therefore the switch on and off mechanism is the interest of this study. Three different types of switch on and off mechanism are found for the microactuator.

To visualize the switch on and off scheme, a Poincaré

section [11] is chosen at $x_2 = 0$ in the state space. One example of type I switch on and off mechanism is shown in Fig. 4 for $\omega = 0.5$. Figure 4 shows the variation of v_2 on the Poincaré section with respect to the applied voltage f. When $f = 0.16$, the voltage is not strong enough to drive the microactuator. Here, the frame is held still by static friction, and the amplitude of the movable block is less than δ. As the voltage is slowly increased, the maximum displacement of m_2 approaches δ. At a critical voltage, a grazing periodic solution is established as zero-velocity contact occurs between the movable block and the stopper. As seen in the diagram, a further increase in f results in a transition of the asymptotic dynamics to an impacting solution with relatively large impact velocity. On the other hand, when $f = 0.24$ the system exhibits a periodic orbit with one impact per period, which is called period one impact orbit. Reducing the voltage decreases the impact velocity and therefore the step movement of the microactuator. When a critical voltage is reached, an eigenvalue of the Jacobian of the Poincaré map corresponding to the impacting solution equals 1. This corresponds to a cyclic fold bifurcation and no impacting solution exists if the voltage is reduced further. Instead, a further reduction in f results in a transition of the asymptotic dynamics to a non-impacting solution with relatively small amplitude. Figure 4 shows the presence of hysteresis when the system is switched on and off.

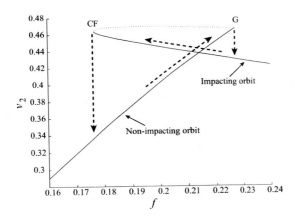

Figure 4: Type I switch on and off mechanism. G indicates a grazing bifurcation and CF a cyclic fold bifurcation.

Figure 5 shows the type II switch on and off mechanism for $\omega = 0.485$. When $f = 0.171$, m_1 is held still by static friction, and m_2 is oscillating with an amplitude smaller than δ. As the voltage is slowly increased, the amplitude of m_2 increases. When a critical voltage is reached, grazing impact occurs. A narrow impact chaotic band immediately follows the grazing when the voltage is increased slightly. On the other hand when

$f = 0.177$, the system exhibits a period one impact orbit. When the voltage is slowly decreased, the system exhibits a period doubling route to chaos. When the voltage is further decreased, the chaotic band touches the non-impact periodic orbit at grazing. Type II does not show hysteresis.

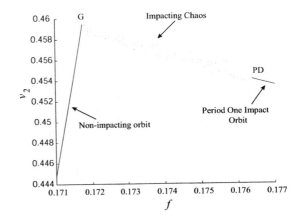

Figure 5: Type II switch on and off mechanism. G indicates a grazing bifurcation and PD a period-doubling bifurcation.

An example of type III switch on and off mechanism is shown in Fig. 6, for $\omega = 0.45$. At low voltages, the microactuator is "off". As the voltage is slowly increased, the oscillation amplitude of m_2 increases. Before the amplitude of m_2 is large enough to touch the stopper, there is a cyclic fold bifurcation at the critical voltage $f = 0.2356$. Slightly increasing the voltage beyond the critical value causes the system jump to a chaotic impact orbit. When the voltage is decreased, the chaotic impact orbit persists until it touches an unstable non-impact orbit at $f = 0.2355$. On the other hand, when $f = 0.29$, the system exhibits a period one impact orbit. When the voltage is slowly decreased, the system becomes chaotic through a period doubling sequence.

The different behavior regimes are shown in the $f - \omega$ parameter space in Fig. 7. The solid line denoted the switch on boundary and the dotted line denoted the switch off boundary. When $\omega > 0.489$, the switch on and off mechanism is type I. When $0.480 < \omega < 0.489$, the system undergoes type II switch on and off mechanism. Type III mechanism is experienced when $\omega < 0.480$. A chaotic band follows the switch on voltage in both type II and type III. Both type I and type III have hysteresis. To produce accurate and predictable displacement, it is necessary to operate the microactuator in a periodic manner with impact occuring once each period. Because the existence of chaos in type II and type III, excitation frequencies in the type I range is favorable.

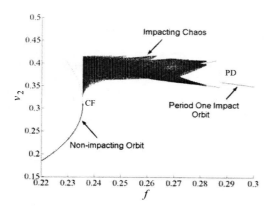

Figure 6: Type III switch on and off mechanism. CF indicates a cyclic fold bifurcation and PD a period-doubling bifurcation.

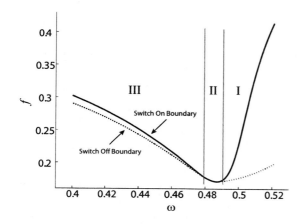

Figure 7: Variation of switch on and off voltages with respect to the excitation frequency.

5   CONCLUSIONS

Micro impact actuators are suitable for high resolution with large travel range. We have studied the dynamics of an impact microactuator modeled as a system of two coupled rigid bodies. A periodically varying excitation voltage is applied. Impacts make the vector field discontinuous and render the system nonsmooth. The numerical simulations presented above clearly show the complex dynamics exhibited by the simple model. The influence of the excitation frequency and amplitude on the dynamic behavior is examined. The numerically obtained displacement/impact values are consistent with the experiment. The numerical results bring out the importance of understanding the underlying dynamics so as to ensure a robust and consistent device operation.

REFERENCES

[1] Ragulsikis, K., Bansevicus, Barausleas, R., and Kulvietis, G., *Vibromotors for Precision Micro-robots*, New York: Hemisphere, pp. 289-295, 1988.

[2] Lee, A.P., Nikkel, D.J. Jr., and Pisano, A.P., "Polysilicon Linear Microvibromotors", in *Proc. 7th Int. Conf. Solid-State Sensors and Actuators*, Vol. 1(2), pp. 46-49, 1993.

[3] Daneman, M.J., Tien, N.C., Solgaard, O., Pisano, A.P., Lau, K.Y., and Muller, R.S., "Linear Microvibromotor for Positioning Optical Components", *IEEE Journal of Microelectromechanical Systems*, Vol. 5(3), pp. 159-165, 1996.

[4] Saitou, K., Wang, D.A., and Wou, S.J., "Externally Resonated Linear Microvibromotor for Microassembly", *IEEE Journal of Microelectromechanical Systems*, Vol. 9(3), pp.336-346, 2000.

[5] Mita, M., Arai, M., Tensaka, S., Kobayashi, D., and Fujita, H., "A micromachined impact microactuator driven by electrostatic force", *IEEE Journal of Microelectromechanical Systems*, Vol. 12(1), pp. 37-41, 2003.

[6] Shaw, S. W., and Holmes, P. J., A periodically forced piecewise linear oscillator, *Journal of Sound and Vibration*, Vol. 90(1), pp. 129-155, 1983.

[7] Nordmark, A.B., *Grazing Conditions and Chaso in Impacting Systems*, Ph.D. Thesis, Royal Institute of Technology Stockholm, Sweden, 1992

[8] Brogliato, B., *Nonsmooth Mechanics*, Springer-Verlag, New York, 1999.

[9] Zhao, X., Reddy, C.K., Nayfeh, A.H., "Bifurcations and Chaotic Dynamics in an electrostatically Actuated Impact Microactuator: a Numerical Exploration", *ASME International 19th Biennial Conference on Mechanical Vibration and Noise*, DETC2003/VIB-48519, Chicago, Illinois, 2003.

[10] Personal communication with Dr. Makoto Mita.

[11] Nayfeh, A. H., and Balachandran, B., *Applied Nonlinear Dynamics: Analytical, Computational, and Experimental Methods*, Wiley, New York, 1995.

Microplate Modeling under Coupled Structural-Fluidic-Electrostatic Forces

Mohammad I. Younis and Ali H. Nayfeh

Department of Engineering Science and Mechanics, MC 0219, Virginia Polytechnic
Institute and State University, Blacksburg, Virginia 24061, USA, anayfeh@vt.edu

ABSTRACT

We present a model for the dynamic behavior of microplates under the coupled effects of squeeze-film damping, electrostatic actuation, and mechanical forces. The model simulates the dynamics of microplates and predicts their quality factors under a wide range of gas pressures and applied electrostatic forces up to the pull-in instability. The model utilizes the nonlinear Euler-Bernoulli beam equation, the linearized dynamic von-Kármán plate equations, and the linearized compressible Reynolds equation. The static deflection of the microplate is calculated using the beam model. Perturbation techniques are used to derive analytical expressions for the pressure distribution in terms of the plate mode shapes around the deflected position. The static deflection and the analytical expressions are substituted into the plate equations, which are solved using a finite-element method.

Keywords: Microplates, squeeze-film damping, electrostatic forces, quality factors.

1 INTRODUCTION

MEMS devices widely employ a parallel-plate capacitor, in which one plate is actuated electrically and its motion is detected by capacitive changes. To increase the efficiency of actuation and the sensitivity of detection, the distance between the capacitor plates is minimized and the sizes of the plates are maximized. Under such circumstances, squeeze-film damping becomes increasingly pronounced.

We present a model for microplates under the effect of squeeze film damping, electrostatic actuation, and mechanical forces. The majority of the literature is dedicated to rigid structures. Few papers treat flexible structures. However, they use approximations in the structural problem while solving the fluidic problem [1] and vice versa, despite the fact that both problems are coupled. Other models require many iterations between different energy-domain CAD solvers, making the simulation process cumbersome. Further, no model has been presented to calculate the plate mode shapes, their natural frequencies, and quality factors, while sweeping the DC voltage up to the pull-in instability. Such problems

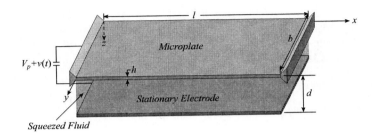

Figure 1: Electrically actuated microplate under the effect of squeeze-film damping.

are of significant importance for many MEMS applications, such as electrostatic projection displays.

In a previous work [2], we presented a beam model that simulates the behavior of electrically actuated microstructures up to the pull-in instability. However, the model does not account for squeeze-film damping. In [3], we presented a model utilizing a linearized plate equation coupled with the compressible Reynolds equation to simulate the dynamic behavior of microstructures under the effect of squeeze-film damping and small DC loading. However, the model does not apply for plates actuated by large electrostatic forces or exhibiting large deflections.

In this work, we utilize the beam model in [2] to simulate the static behavior of the microplate under the effect of electrostatic forces. We derive from the compressible Reynolds equation analytical expressions for the pressure distribution in terms of the plate mode shapes around the deflected position, and hence eliminate the pressure as a variable in the solution procedure. We substitute the static deflection and the analytical expression of the pressure in the dynamic von-Kármán plate equations, linearized around the deflected position. We then solve the resulting equations using a finite-element method.

2 PROBLEM FORMULATION

We consider a microplate, Figure 1, actuated by an electric load composed of a DC component V_p and an AC component $v_e(t)$ and subject to a net pressure force $\bar{P}(x,y,t)$ per unit area due to squeeze-film damping. The electrostatic force varies along the length of the plate and is constant across its width. The electrostatic force deflects the microplate to a new equilibrium position. We assume that the pressure around the deflected plate is redistributed such that $\bar{P} = 0$. Because of the symmetric nature of the electrostatic forces and the boundary conditions, we use the nonlinear Euler-Bernoulli beam model in [2] to calculate the static deflection w_s of the plate. We use the dynamic version of the von-Kármán plate equations to represent the microplate motion.

To derive the linear damped eigenvalue problem describing the free vibration of the microplate, we linearize the von-Kármán plate equations around w_s, drop the forcing term $v_e(t)$, and obtain

$$\nabla^4 W_n - N W_{n,xx} - \alpha_1 U_{n,x} w_{s,xx} - \alpha_2 W_{n,xx} w_{s,x}^2$$
$$- \alpha_3 V_{n,y} w_{s,xx} - \alpha_4 W_{n,yy} w_{s,x}^2 - \omega_n^2 W_n - \alpha_5 \frac{V_p^2 W_n}{(1-w_s)^3}$$
$$+ P_{non} P_n = 0 \tag{1}$$

$$U_{n,xx} + \alpha_6 W_{n,x} w_{s,xx} + \alpha_6 w_{s,x} W_{n,xx} + \nu V_{n,xy}$$
$$+ \frac{1}{2}(1-\nu)(U_{n,yy} + V_{n,xy} + \alpha_6 w_{s,x} W_{n,yy}) = 0 \tag{2}$$

$$V_{n,yy} + \nu U_{n,xy} + \nu \alpha_6 w_{s,x} W_{n,xy} + \frac{1}{2}(1-\nu)(U_{n,xy}$$
$$+ V_{n,xx} + \alpha_6 w_{s,x} W_{n,xy} + \alpha_6 W_{n,y} w_{s,xx}) = 0 \tag{3}$$

where $W_n(x,y)$, $V_n(x,y)$, and $U_n(x,y)$ are the nth transverse, lateral, and longitudinal complex mode shapes of the plate, respectively, at the nondimensional positions x and y, $P_n(x,y)$ is the corresponding nth pressure mode shape, and ω_n is the nth complex nondimensional eigenvalue. The nondimensional parameters appearing in equations (1-3) are

$$N = \frac{N_1 \ell^2}{D}, \quad P_{non} = \frac{P_a \ell^4}{dD}, \quad T^2 = \frac{\rho h \ell^4}{D}$$
$$\alpha_1 = \frac{\ell d d_3}{D}, \quad \alpha_2 = \frac{d_3 d^2}{2D}, \quad \alpha_3 = \frac{\nu \ell d}{D}$$
$$\alpha_4 = \frac{\nu d_3 d^2}{2D}, \quad \alpha_5 = \frac{\epsilon \ell^4}{D d^3}, \quad \alpha_6 = \frac{d}{\ell} \tag{4}$$

where d is the initial gap width, $D = \frac{E h^3}{12(1-\nu^2)}$ is the plate flexural rigidity, h is the microplate thickness, E is Young's modulus, ν is Poisson's ratio, $\hat{N}_1$ is the axial force per unit length in the x direction, ϵ is the dielectric constant of the gap medium of the airgap, ρ is mass density of the plate, and P_a is the static pressure in the airgap. The symbol ∇^4 denotes the nondimensional biharmonic operator ($\nabla^4 = W_{n,xxxx} + 2W_{n,xxyy} + W_{n,yyyy}$), where a comma denotes partial differentiation with respect to the corresponding coordinate. The parameter d_3 is defined as

$$d_3 = \frac{Eh}{1-\nu^2} \tag{5}$$

The parameter ω_n is related to the dimensional complex frequency ω by $\omega = \omega_n/T$. The structural boundary conditions for the case in Figure 1 are

At $y = 0$ and $y = b/\ell$

$$V_{n,y} + \nu \alpha_6 U_{n,x} + \nu w_{s,x} W_{n,x} = 0 \tag{6}$$
$$U_{n,y} + V_{n,x} + \alpha_6 w_{s,x} W_{n,y} = 0 \tag{7}$$
$$W_{n,yy} + \nu W_{n,xx} = 0 \tag{8}$$
$$W_{n,yyy} + (2-\nu) W_{n,xxy} = 0 \tag{9}$$

Clamped edges at $x = 0$ and $x = 1$

$$U_n = 0 \tag{10}$$
$$V_n = 0 \tag{11}$$
$$W_n = 0 \tag{12}$$
$$W_{n,x} = 0 \tag{13}$$

We assume small variations of pressure around P_a and write the following nondimensional Reynolds equation, linearized around w_s and P_a, governing the pressure distribution underneath the microplate:

$$\frac{\partial}{\partial x}\left[(1-w_s)^3 P_{n,x}\right] + \frac{\partial}{\partial y}\left[(1-w_s)^3 P_{n,y}\right]$$
$$= i\sigma \omega_n \left[(1-w_s)P_n - W_n\right] \tag{14}$$

where $\sigma = \frac{12\eta_{eff}\ell^2}{d^2 P_a T}$ is the squeeze number and η_{eff} is the effective viscosity [4] of the fluid in the gap. The pressure boundary conditions for the case in Figure 1 are zero-flux pressure at the clamped edges of the plate and trivial pressure at the open edges; that is,

$$P_{n,x}(0,y) = P_{n,x}(1,y) = 0 \tag{15}$$
$$P_n(x,0) = P_n(x,b/\ell) = 0 \tag{16}$$

Because typically $\omega_n \sigma \gg 1$, the boundary-value problem represented by equations (14-16) is a singular-perturbation problem [5]. In this problem, the pressure changes sharply near the free edges. We use the method of matched asymptotic expansions [5] to derive a uniform approximation to the solution of equations (14-16), which yields

$$P_n(x,y) = \frac{1}{1-w_s}\left[W_n(x,y) - W_n(x,b/\ell)e^{\frac{-1+i}{\sqrt{2}}\frac{b/\ell - y}{\sqrt{\epsilon(1-w_s)}}}\right.$$
$$\left. - W_n(x,0)e^{\frac{-1+i}{\sqrt{2}}\frac{y}{\sqrt{\epsilon(1-w_s)}}}\right] \tag{17}$$

where $\epsilon = 1/(\omega_n \sigma)$. Equation (17) gives an approximate analytical expression for the nth complex pressure mode shape in terms of the nth complex plate mode shape and eigenvalue.

3 RESULTS

We substitute equation (17) into equation (1) and obtain an equation, which along with equations (2,3,6-13), represent a linear distributed-parameter system for the dynamic behavior of the microplate under the coupled effect of squeeze-film damping, structural forces, and electrostatic forces. We solve this system for the nth complex mode shape and eigenvalue using a finite-element method. The real part of the complex eigenvalue yields frequency of the microplate, whereas the ratio between its real and twice the imaginary part yields the quality factor.

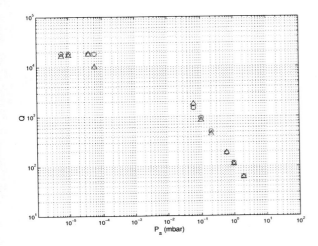

Figure 2: A comparison of the quality factors calculated using our model (circles) and the model in [3] (squares) with the experimental data (triangles) of Legtenberg and Tilmans [6].

In Figure 2, we show a comparison of the calculated quality factors for a microplate of lengths $310\mu m$ [6] using our model (circles) and a linear plate model [3] (squares), which neglects the in-plane displacements, with the experimental data (triangles) of Legtenberg and Tilmans [6]. The microbeam is actuated at $V_p = 2V$. We note that the model gives results very close to the model of Nayfeh and Younis [3] because the plate is actuated by a low DC voltage, in which the in-plane deflections are negligible. Therfore, we conclude that the simplified linear model in [3] produces accurate results at low DC loadings.

The operating conditions of many MEMS devices exceed the range of small DC loadings, such as in projection display arrays. In such cases, the in-plane displacements become significant, and hence a simple plate model that neglects the in-plane displacements as in [3] is suspect. In Figure 3, we show a comparison of the calculated quality factors and fundamental natural frequencies of a $210\mu m$ length microplate [6] using the

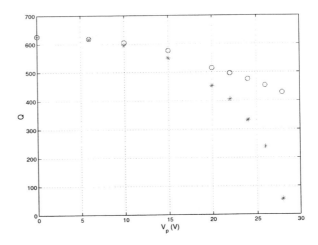

(a) Quality factors.

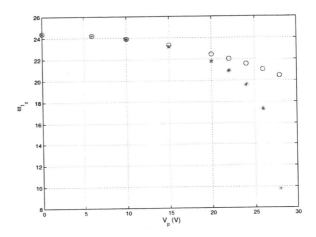

(b) Fundamental natural frequencies.

Figure 3: Variations of the quality factors and natural frequencies of a $210\mu m$ length microplate for various values of the DC voltage.

present model with those calculated using the simplified model in [3]. We note that the simplified model overestimates the quality factors and the natural frequencies at nearly 30% of the pull-in voltage. This is because, as the DC force increases, the static deflection increases, the gap width decreases, and hence the squeeze-film damping increases. Further, the increase in the electrostatic force decreases the natural frequencies. This effect is not represented correctly in the simplified model.

Figure 4 shows variation of the quality factor of the microplate of Figure 3 with V_p for various values of P_a. Clearly, the quality factor is a strong function of the gas pressure. Figure 5 shows the spatial variation of the pressure distribution P_{1_0}, corresponding to the first

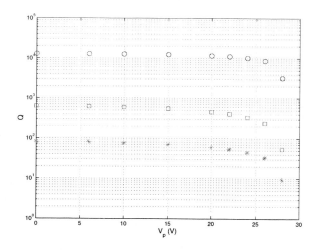

Figure 4: Variations of the quality factor with V_p for $P_a = .01mbar$ (circles), $P_a = 1mbar$ (squares), and $P_a = 10mbar$ (stars).

mode shape of the microplate of Figure 3, for various values of the electrostatic force and gap pressure. The function P_{1_0} is normalized such that its integral over the area of the plate is unity. We note that varying the pressure has a slight effect on P_{1_0} and that increasing the electrostatic loading has a much significant effect.

4 CONCLUSIONS

We presented a model for the dynamic behavior of microplates under the coupled effects of squeeze-film damping, electrostatic actuation, and mechanical forces. The model simulates the dynamics of microplates and predicts their quality factors under a wide range of gas pressures and applied electrostatic forces. The model presents a novel approach to the simulation of coupled-energy systems, which reduces the computational cost. The results show that the electrostatic force has more significant influence on the mechanical behavior of microplates than the encapsulation pressure. However, the pressure does affect the quality factor of microplates significantly.

REFERENCES

[1] Yang, Y. J., Gretillat, M. A., and Senturia, S. D., "Effect of air damping on the dynamics of nonuniform deformations of microstructures," Int. Conf. Solid State Sens. Actuat., TRANSDUCERS '97, Chicago, Illinois, Vol. 2, 1997, pp. 1093–1096.

[2] Younis, M. I., Abdel-Rahman, E. M., and Nayfeh, A. H., "A Reduced-order model for electrically actuated microbeam-based MEMS," J. Microelectromech. Sys., Vol. 12, 2003, pp. 672–680.

[3] Nayfeh, A. H. and Younis, M. I., "A new approach to the modeling and simulation of flexi-

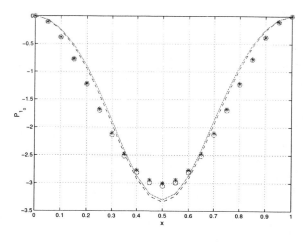

(a) P_{1_0} at $y = b/2\ell$.

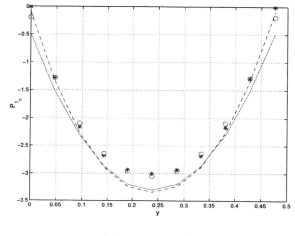

(b) P_{1_0} at $x = 1/2$.

Figure 5: The spatial variation of P_{1_0} when $V_p = 6V$ (discrete points) and $V_p = 28V$ (solid and dashes lines). The data shown in circles and dashed line corresponds to $P_a = 0.01mbar$ and the data in stars and solid line corresponds to $P_a = 2mbar$.

ble microstructures under the effect of squeeze-film damping," J. Micromech. Microeng., Vol. 14, 2004, pp. 170–181.

[4] Veijola, T., Kuisma, H., Lahdenperä, J., and Ryhänen, T., "Equivalent-circuit model of the squeezed gas film in a silicon accelerometer," Sens. Actuat. A, Vol. 48, 1995, pp. 239–248.

[5] Nayfeh, A. H., Introduction to Perturbation Techniques, Wiley, New York, 1981.

[6] Legtenberg, R. and Tilmans, H. A., "Electrostatically driven vacuum-encapsulated polysilicon resonators. Part I. Design and fabrication," Sens. Actuat. A, Vol. 45, 1994, pp. 57–66.

A Model for Thermoelastic Damping in Microplates

Ali H. Nayfeh and Mohammad I. Younis

Department of Engineering Science and Mechanics, MC 0219, Virginia Polytechnic Institute and State University, Blacksburg, Virginia 24061, USA, anayfeh@vt.edu

ABSTRACT

We present a model and analytical expressions for the quality factors of microplates due to thermoelastic damping. We solve the heat equation for the thermal current across the thickness of a microplate, and hence decouple the thermal equation from the plate equation. We utilize a perturbation method to derive an analytical expression for the quality factor of microplates, of general boundary conditions under electrostatic loading and residual stresses, in terms of their structural mode shapes. For the special case of no electrostatic and in-plane loadings, we derive a simple analytical expression for the quality factor, which is independent of the mode shapes. We verify the model by comparing the calculated quality factors of the special case of microbeams to theoretical results obtained using the microbeam models in the literature. We present several results for various modes of microplates with various boundary conditions.

Keywords: Microplates, thermoelastic damping, electrostatic forces, quality factors.

1 INTRODUCTION

Thermoelastic damping has been shown recently to be a dominant source of intrinsic damping in MEMS. Thermoelastic damping results from the irreversible heat flow generated by the compression and decompression of an oscillating structure. A structure in bending has one side under tension and the other side under compression of higher temperature. This variation in temperature causes a thermal gradient inside the material of the structure, which adjusts itself to allow for a thermal equilibrium. However, the energy used in this adjustment cannot be restored (irreversible process) even if the structure returns to its original state of zero-bending stress. Hence, thermoelastic damping is also referred to as internal friction.

The first to realize the importance of thermoelastic damping and analyze it rigorously is Zener [1], who gave an analytical approximation for the quality factor of metallic beams due to thermoelastic damping. Roszhart [2] showed experimentally that thermoelastic damping can limit the quality factor of devices in the micro scale. In a recent work, Lifshitz and Roukes [3] solved exactly

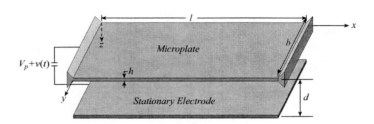

Figure 1: Electrically actuated microplate.

the problem of thermoelastic damping in beams and derived an analytical expression for their quality factors.

Microplates have been increasingly tested and used in various MEMS devices, such as micropumps and pressure sensors. While the models of Lifshitz and Roukes [3] and Zener [1] have been proven to yield quantitatively good results for microbeams, their ability to predict thermoelastic damping in microplates is questionable. This is particularly important in cases, such as fully-clamped plates, where a beam model does not apply. Also both models [1,3] cannot predict the quality factors of structural modes, which vary spatially across the plate width, and hence cannot be captured using a beam model.

In previous works [4], we presented a plate model to simulate the dynamic behavior of microstructures under the effect of squeeze-film damping. In this paper, we derive an analytical expression for the quality factor of a microplate, under the effect of small electrostatic forces and in-plane loading, due to thermoelastic damping.

2 PROBLEM FORMULATION

We consider a microplate (Figure 1) actuated by an electric load composed of a DC component V_p and an AC component $v_e(t)$. We assume that all forms of extrinsic damping, such as viscous damping, can be neglected. Assuming small strains and displacements, we obtain the following linear equation of motion governing the transverse deflection of the microplate including the effect of thermoelastic damping:

$$D\nabla^4 w - \hat{N}_1 \frac{\partial^2 w}{\partial x^2} + \rho h \frac{\partial^2 w}{\partial t^2} = \frac{\epsilon V_p v_e}{d^2} + \frac{\epsilon V_p^2}{d^3}w - \nabla^2 w N^T - \nabla^2 M^T \quad (1)$$

where $w(x, y, t)$ is the transverse deflection of the plate at the position x and y at time t, ρ is the plate material density, d is the initial gap width, $D = \frac{Eh^3}{12(1-\nu^2)}$ is the plate flexural rigidity, h is the microplate thickness, E is Young's modulus, ν is Poisson's ratio, $\hat{N}_1$ is the axial force per unit length in the x direction, ϵ is the dielectric constant of the gap medium, and N^T and M^T are the in-plane axial force and bending moment due to thermoelastic damping, respectively. The operators ∇^2 and ∇^4 are the Laplacian and biharmonic operators, respectively. The thermal axial force and bending moment are defined as

$$N^T = \frac{E\alpha_t}{1-\nu} \int_{-h/2}^{h/2} \Delta T dz \qquad (2)$$

$$M^T = \frac{E\alpha_t}{1-\nu} \int_{-h/2}^{h/2} z\nabla^2 \Delta T dz \qquad (3)$$

where $T(x, y, z, t)$ is the temperature distribution and α_t is the coefficient of thermal expansion.

The temperature distribution can be expressed by the linearized version of the heat conduction equation [5]:

$$k\nabla^2 T + Q = \left[\rho C_p + \frac{E\alpha_t^2 T_0(1+\nu)}{(1-2\nu)(1-\nu)} \right] \frac{\partial T}{\partial t} - \frac{E\alpha_t T_0}{1-\nu} \frac{\partial}{\partial t} (z\nabla^2 w) \qquad (4)$$

where C_p is the heat capacity coefficient at constant pressure, Q is the heat flux, and T_0 is the stress-free temperature.

For the linear damped eigenvalue problem, we let

$$w(x, y, t) = \phi_n(x, y)e^{i\omega_n t} \qquad (5)$$

$$\Delta T(x, y, z, t) = \theta_n(x, y, z)e^{i\omega_n t} \qquad (6)$$

where $\phi_n(x, y)$ and $\theta_n(x, y, z)$ are the nth complex mode shapes of the plate and the associated temperature variation, respectively, and ω_n is the nth complex eigenvalue. Substituting equations (5) and (6) into equations (1) and (4), dropping the Q and the forcing terms v_e, and noting that the temperature variation across the plate thickness is much larger than its variation across the plane of the plate, we obtain

$$\nabla^4 \phi_n \quad -\hat{N}_1 \frac{\partial^2 \phi_n}{\partial x^2} + \frac{E\alpha_t}{1-\nu}\nabla^2\phi_n \int_{-h/2}^{h/2} \theta dz$$
$$+ \frac{E\alpha_t}{1-\nu} \int_{-h/2}^{h/2} z\nabla^2\theta dz - \frac{\epsilon V_p^2}{d^3}\phi_n = \omega_n^2 \phi_n \qquad (7)$$

$$k\frac{\partial^2 \theta}{\partial z^2} = i\omega_n \left[\rho C_p + \frac{E\alpha_t^2(1+\nu)T_0}{(1-\nu)(1-2\nu)} \right] \theta$$
$$- i\omega_n \frac{E\alpha_t T_0}{1-\nu} z\nabla^2\phi_n \qquad (8)$$

The temperature conditions are zero-heat flux from the plate to the ambient [1]; that is,

$$\frac{\partial\theta}{\partial z} = 0 \quad \text{at} \quad z = \frac{1}{2}h \quad \text{and} \quad z = -\frac{1}{2}h \qquad (9)$$

The solution of equation (8) can be expressed as

$$\theta = \frac{E\alpha_t T_0}{(1-\nu)\chi} z\nabla^2\phi_n + c_1\sin(K_p z) + c_2\cos(K_p z) \qquad (10)$$

where

$$\chi = \rho C_p + \frac{E\alpha_t^2(1+\nu)T_0}{(1-\nu)(1-2\nu)}, \qquad K_p = (1-i)\sqrt{\frac{\omega_n\chi}{2k}}$$

Next, we impose the boundary conditions, equation (9), on equation (10) and obtain

$$\theta = \frac{E\alpha_t T_0}{(1-\nu)\chi}\nabla^2\phi_n \left(z - \frac{\sin(K_p z)}{K_p\cos(\frac{1}{2}K_p h)} \right) \qquad (11)$$

Substituting equation (11) into equation (7), carrying out the integrations, and retaining the linear terms, we obtain

$$D^T\nabla^4\phi_n - \hat{N}_1\frac{\partial^2\phi_n}{\partial x^2} - \frac{\epsilon V_p^2}{d^3}\phi_n = \omega_n^2\phi_n \qquad (12)$$

where $D^T = D + D_t$ and D_t is given by

$$D_t = \frac{E^2\alpha_t^2 T_0}{(1-\nu)^2\chi} \left(\frac{h^3}{12} + \frac{h}{K_p^2} - \frac{2\tan(0.5K_p h)}{K_p^3} \right) \qquad (13)$$

Equations (12) and (13) can be used, along with any appropriate set of boundary conditions, to simulate the behavior of an electrostatically actuated microplate accounting for thermoelastic damping.

2.1 Perturbation Analysis

Next, we derive an analytical expression for the quality factor of microplates due to thermoelastic damping. Because $D_t << D$, we apply the method of strained parameters [6] to equation (12). The analysis is general for any boundary condition. To this end, we seek a first-order solution to equation (12) and associated boundary conditions in the form

$$\phi_n \approx \phi_{n_0} + \epsilon\phi_{n_1} \qquad (14)$$

$$\omega_n \approx \omega_{n_0} + \epsilon\omega_{n_1} \qquad (15)$$

$$D^T \approx D + \epsilon D_t \qquad (16)$$

where ϵ is a scaling parameter. Substituting equations (14-16) into equations (12) and separating terms of different order yields

$O(\epsilon^0)$

$$D\nabla^4\phi_{n_0} - \hat{N}_1\frac{\partial^2\phi_{n_0}}{\partial x^2} - \frac{\epsilon V_p^2}{d^3}\phi_{n_0} - \omega_{n_0}^2\phi_{n_0} = 0 \qquad (17)$$

$O(\epsilon)$

$$D\nabla^4\phi_{n_1} \quad -\hat{N}_1\frac{\partial^2\phi_{n_1}}{\partial x^2} - \frac{\epsilon V_p^2}{d^3}\phi_{n_1} - \omega_{n_0}^2\phi_{n_1} =$$
$$2\omega_{n_0}\omega_{n_1}\phi_{n_0} - D_t\nabla^4\phi_{n_0} \qquad (18)$$

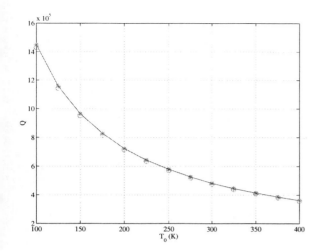

Figure 2: Comparison of the quality factors calculated using our model (−) to those of Lifshitz and Roukes [3] (∗) and Zener [1] (○) for various values of T_0.

Next, we apply the solvability condition [6], which demands that the right-hand side of equation (18) be orthogonal to ϕ_{n_0}, and obtain

$$\omega_{n_1} = \frac{\frac{D_t}{D}\left[\hat{N}_1\phi_{n_0}\frac{\partial^2\phi_{n_0}}{\partial x^2} + (\frac{\epsilon V_p^2}{d^3} + \omega_{n_0}^2)\phi_{n_0}^2\right]dxdy}{2\omega_{n_0}\int_0^1\int_0^{b/\ell}\phi_{n_0}^2 dxdy} \quad (19)$$

Equation (19) gives the imaginary part of the complex natural frequency, which yields the damping of the system. The ratio between ω_{n_0} and twice ω_{n_1} yields the quality factor. We note from equation (19) that the axial force and the electrostatic load amplify the effect of thermoelastic damping.

For the special case of $V_p = 0$ and $\hat{N}_1 = 0$, and assuming ϕ_{n_0} normalized such that $\int_0^1\int_0^{b/\ell}\phi_{n_0}^2 dxdy = 1$, equation (19) reduces to the following simple analytical expression for the quality factor:

$$Q = \frac{h^3\chi(1-\nu)}{12(1+\nu)E\alpha_t^2 T_0\left(\frac{h^3}{12} + \frac{h}{K_p^2} - \frac{2\tan(0.5K_ph)}{K_p^3}\right)} \quad (20)$$

3 RESULTS

In this section, we present numerical results for the quality factor of silicon plates due to thermoelastic damping for the special case of $V_p = 0$ and $\hat{N}_1 = 0$. We start with a plate clamped across its length and free across it width. In Figure 2, we set $\nu = 0$ in equation (20) (to approximate the case of a beam), calculate Q for various values of temperature T_0, and compare the results (−) to those of Lifshitz and Roukes [3] (∗) and Zener [1] (○). The results shown are for a microbeam with $\ell = 200\mu m$, $b = 20\mu m$, and $h = 2\mu m$. We note full agreement among the results.

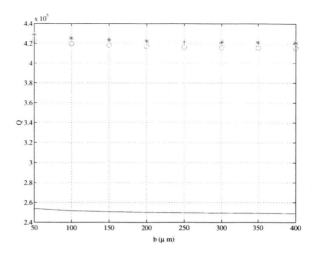

Figure 3: Comparison of the quality factors of the first mode calculated using our model (−) to those of Lifshitz and Roukes [3] (∗) and Zener [1] (○) for various values of b.

Figures 3 and 4 show the effect of varying the width of a plate on Q of the first and second modes, respectively. The plate specifications are $\ell = 200\mu m$, $h = 2\mu m$, $\nu = 0.3$, and $T_0 = 300K$. Figure 3 shows a comparison between the results of the plate model (solid line) to the models of Lifshitz and Roukes [3] (∗) and Zener [1] (○). Clearly, the plate model predicts less quality factors than the beam model. Figure 3 shows that Q of the first mode changes slightly with the plate width. On the other hand, Figure 4 shows that Q of the second mode depends directly on the width of the plate. This is expected because, unlike the first eigenvalue, the second eigenvalue varies with the plate aspect ratio.

Next, we consider thermoelastic damping in clamped plates. These plates cannot be simulated using a beam model, and hence the beam models [1,3] for thermoelastic damping might not yield accurate results. Figures 5 and 6 show variation of Q of the first mode of a plate with T_0 and h, respectively. The plate specifications are $\ell = 200\mu m$, $b = 100\mu m$, and $\nu = .25$. We use $h = 1.5\mu m$ in Figure 5 and $T_0 = 300K$ in Figure 6. Figure 7 shows variation of Q of the first (○) and second (∗) vibration modes with ℓ for the same microplate of Figure 6. Here we note that Q of both modes are strong functions of the plate length because the corresponding natural frequencies are dependent on the plate aspect ratio.

4 CONCLUSIONS

We derived an analytical expression for the quality factors of microplates due to thermoelastic damping. The analytical expression shows that the electrostatic loading and residual stresses amplify the effect of thermoelastic damping. For the special case of no electrosta-

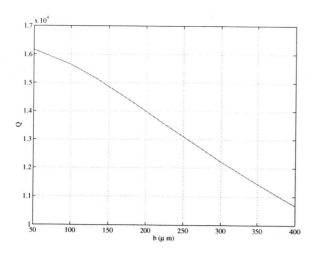

Figure 4: Variation of the quality factor of the second mode of a plate for various values of b.

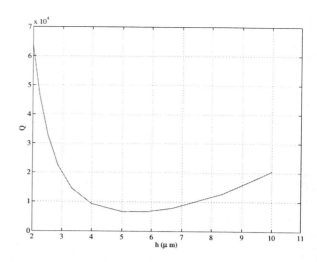

Figure 6: Variation of Q of the first mode of a fully clamped plate with h.

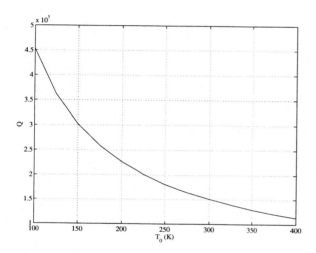

Figure 5: Variation of Q of the first mode of a fully clamped plate with T_0.

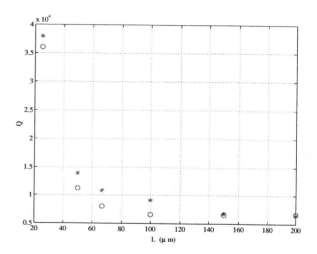

Figure 7: Variation of Q of the first ($\circ$) and second ($*$) modes against various values of ℓ.

tic and in-plane loadings, we derived a simple analytical expression for the quality factor, which is independent of the mode shapes. Comparing the calculated quality factors of plates using our model to the beam models in the literature shows that beam models overestimate the quality factors.

REFERENCES

[1] Zener, C., "Internal friction in solids: I. Theory of internal friction in reeds," Physical Review, Vol. 52, pp. 230–235, 1937.

[2] Roszhart, T. V., "The effect of thermoelastic internal friction on the Q of micromachined silicon resonators," in Proceedings of the IEEE technical Digest of Solid-State Sensors and Actuators Workshop, Hilton Head, South Carolina, pp. 13–16, 1990.

[3] Lifshitz, R. and Roukes, M. L., "Thermoelastic damping in micro- and nanomechanical systems," Physical Review B, Vol. 61, pp. 5600–5609, 2000.

[4] Nayfeh, A. H. and Younis, M. I., "A new approach to the modeling and simulation of flexible microstructures under the effect of squeeze-film damping," Journal of Micromechanics and Microengineering, Vol. 14, pp. 170–181, 2004.

[5] Boley, B. A. and Weiner, J. H., Theory of Thermal Stresses, Wiley, New York, 1960.

[6] Nayfeh, A. H., Introduction to Perturbation Techniques, Wiley, New York, 1981.

Design and modeling of a 3-D micromachined accelerometer

S.H.Ghafari [*], M.F.Golnaraghi [**] and R.Mansour [***]

[*]Graduate student, University of Waterloo, shasanza@engmail.uwaterloo.ca
[**]Professor, Canada Research Chair, University of Waterloo, mfgolnar@mecheng1.uwaterloo.ca
[***]Professor, NSERC/COM DEV Industrial Research Chair, rrmansou@ece.uwaterloo.ca

ABSTRACT

This paper presents the operation principles, modeling methods, design, and fabrication considerations of a 3-D micromachined accelerometer. MEMS technology in this work combines small size, low cost and low power consumption to create a sensor that is suitable for wide usage in different applications such as automotive industry and inertial navigation systems.

Keywords: accelerometer, inertial sensors, silicon sensors, micromachining, micro-fabrication technology.

1 INTRODUCTION

In several applications such as inertial navigation, vibration monitoring and robotics control, measurement of acceleration in 3-D space is required. In an integrated set of 1-D accelerometers, aligning the axes of sensors normal to each other is a difficult task. In these applications utilizing a single chip for measuring acceleration in three directions instead of three 1-D accelerometers, significantly increases the accuracy of measurement. In addition micromachining technology shrank the sensor size, reduce the fabrication cost, and allow its electronics to be integrated on the same silicon chip. Accelerometers generally consist of a proof mass which is suspended to a reference frame by anchored beams as spring elements. Acceleration causes a relative displacement of the proof mass with respect to the frame. The displacement of the mass, which can be measured by various methods, is proportional to driving acceleration.

2 PRINCIPLES OF ACCELERATION MEASUREMENT

Accelerometers are electro-mechanical transducers which change mechanical displacement to DC voltage. Figure 1 shows the lumped element model of an accelerometer. The accelerometer can be modeled by a second order mass-spring-damper system. External acceleration displaces the proof mass relative to the reference frame which changes the internal stress in suspension springs. Both the relative displacement of the proof mass and the beam stress can be used as a measure of external acceleration.

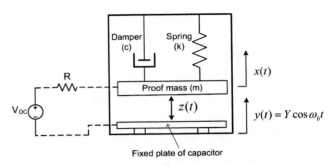

Figure 1: Lumped model of a capacitive accelerometer

The equation of the motion can be written in terms of relative displacement $z(t)$

$$\ddot{z}(t) + 2\xi\omega_n\dot{z}(t) + \omega_n^2 z(t) = -\ddot{y}(t) \qquad (1)$$

Where $\xi(= c/2m\omega_n)$ is the damping ratio, and $\omega_n(= \sqrt{k/m})$ is the natural frequency of the system. The solution of this expression can be obtained by applying a forced harmonic vibration analysis for this base excitation problem [1]. The steady state response of the system is shown to be

$$z(t) = \frac{-1}{\omega_n^2 \sqrt{\left(1-r^2\right)^2 + \left(2\xi r\right)^2}} \ddot{y}(t) \qquad (2)$$

Figure 2 shows the plot of coefficient of base acceleration versus the frequency ratio $(r = \omega_b/\omega_n)$. Within the region $0 < r < 0.1$, the coefficient of base acceleration is approximately equal to unity for a variety of values of ξ. As a result, this range defines the suitable bandwidth of the accelerometer within ten percent of its natural frequency.

Static response of an accelerometer regarding Newton's second law is equal to $z_{static} = (m\ddot{y}/k) = (\ddot{y}/\omega_n^2)$. These equations clearly illustrate the trade of between sensitivity and the sensor's bandwidth. For a high DC sensitivity, a low natural frequency is required which results in a short bandwidth.

Force feedback method is applied to eliminate this limitation in most accelerometers.

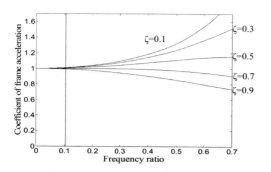

Figure 2: Proportionality of relative displacement of the proof mass and base acceleration.

2.1 Spring elements

Micromachined accelerometers mostly contain micro-springs in form of rectangular cross section beams. Spring elements are usually anchored to the ground layer in one end, and attached to a free moving mass at the other end. The configuration of the beams is extremely important to satisfy the off-axis sensitivity. This property, which defines the sensitivity of the sensor in different directions, plays an important role in the performance of the accelerometer. To minimize the off axis sensitivity the effect of residual strain in beams and their thermal expansion should be considered in designing the geometry of suspension structures.

Determining natural frequency of the beam is an essential requirement for designing accelerometers. One end or double end clamped beams are widely used as spring elements in inertial sensors. Solving forced response of a distributed-parameter beam results the natural frequency of the system. The deflection of the beam for a general load in the form of $Q(x,t) = f(x)q(t)$ by utilizing separation of variables technique $(w(x,t) = X(x)T(t))$ is equal to [2]

$$w(x,t) = \sum_{i=1}^{\infty} \frac{X_i}{\omega_n} \int_0^l f(x)X_i dx \int_0^t q(\tau)\sin\omega_n(t-\tau)d\tau \quad (3)$$

Table 1 summarizes the dynamic properties of one end and double end clamped beams.

Table 1

Parameters	Natural frequency	Beam stiffness	Effective mass
Cantilevered	$\dfrac{3.516}{2\pi}\sqrt{\dfrac{EI}{ml^3}}$	$\dfrac{3EI}{l^3}$	24%
Double clamped	$\dfrac{22.337}{2\pi}\sqrt{\dfrac{EI}{ml^3}}$	$\dfrac{192EI}{l^3}$	38%

A nonlinear characteristic of the electrostatic force between parallel plates of a capacitor creates a virtual spring. This effect causes a shift of natural frequency in the system. *Spring softening* would allow the possibility of controlling natural frequency of the sensor, which can actively control the sensitivity of the device to a specific range of excitation frequencies. This nonlinear phenomenon can be linearized in an equilibrium point and can be defined as a negative stiffness equal to [3]

$$k_e = -\frac{C_s}{z_0^2}V_{DC}^2 \quad (4)$$

Where C_s is equivalent capacity, and z_0 is initial gap between two plates of capacitor.

2.2 Viscous damping

Major sources of energy dissipation in an accelerometer are viscous damping of the moving proof mass in the surrounding fluid (usually air), and squeezed film damping between sensing fingers. Couette flow model can be used for estimating damping coefficient of steady viscous flow existing between moving proof mass parallel with ground layer [4]. The damping constant c can be obtained from $c = \eta A / h$ where, η is the viscosity of the surrounding fluid, A is the overlapped area and h is the air gap between moving mass and ground layer. The squeezed film damping happens when a gas fills the space between transverse moving parallel plates. This scenario usually happens for sensing fingers of a sensor. The following equation is an estimation of squeezed film damping factor [5].

$$c = n\frac{96\eta lW^3}{\pi^4 h^3} \quad (5)$$

Where n is the number of fingers. As a result of micro scale effects, theoretically estimated c is 35 to 40% less than measured values.

2.3 Noise

MEMS sensor noise typically arises from the damping of fluid surrounding the proof mass, which is called "*Brownian motion*" noise. According to the thermodynamics the spectral density function of the force noise is $4k_B Tc$, where k_B is the Boltzman constant. An equivalent acceleration spectral density, which is called the Total-Noise-Equivalent-Acceleration, is commonly used to express the noise level of the system.

$$TNEA = \sqrt{\frac{4k_B T\omega_n}{Qm}} \quad (6)$$

t is obvious that to achieve low noise levels, a large proof mass and a high quality factor are required.

3 MODELING CONCEPTS

Accelerometers can be modeled at two different levels: system level and physical level. Finite element methods generally are used to simulate the device behavior in a 3D coupled field for physical level simulation. Regardless the geometry of the sensor its dynamic characteristic can be simulated in the system level as an energy transducer device. The main methods for system simulation are *multi port* device simulation and *equivalent circuit*. The goal of *multi port* theory is to develop the links between dependant and independent variables of the transducer. A set of differential equations can be obtained by applying energy methods and thermodynamics for defining the characteristics of the system state variables. Then these equations can be analyzed for various inputs and initial conditions by mathematical software. Since the capacitance transduction method utilized for the designed accelerometer, for instance we consider the capacitive accelerometer model shown in figure 1. Applying thermodynamics equilibrium for the stored energy through the mechanical and electrical ports of the transducer results the following governing equations of the capacitive accelerometer.

$$\frac{\partial q}{\partial t} = \frac{1}{R}\left(V_{in} - V\right) = \frac{1}{R}\left(V_{in} - \frac{qz}{\varepsilon A}\right) \qquad (7)$$

$$\frac{q^2}{2\varepsilon A} + m\frac{\partial^2 z}{\partial t^2} + c\frac{\partial z}{\partial t} + k(z - z_0) = \frac{\partial^2 y}{\partial t^2} \qquad (8)$$

The above set of nonlinear differential equations can be solved by means of mathematical software numerically.

The other quick way of getting insight into the dynamic behaviour of accelerometers is the *equivalent circuit* approach. In this method both electrical and mechanical elements of transducer are represented by electrical equivalents. In this analogy mechanical force plays the same role as voltage, the velocity as the current and the displacement as the charge. The mass in the mechanical system corresponds to the inductance, viscous damping to the resistance and the flexibility (k^{-1}) to capacitance. Also this analogy defines possible equivalent circuits, which represent electromechanical transducers [6].

4 FABRICATION AND PACKAGING

Manufacturability of an accelerometer should be considered as an important issue in the design of sensors. The first micromachined capacitive accelerometer used bulk micromachining and wafer bonding to fabricate a thick large proof mass in order to increase the sensitivity of the device. Although this design has the advantage of large proof mass, the wafer bonding process is required to realize an air gap which causes high damping factor. The thin structures fabricated using surface micromachining techniques contains array of holes to reduce damping. These devices can also be easily combined with their electronic circuitry.

There are various commercialized surface machining processes which provide design rules and guidelines to achieve a desired micromachined structure. In addition to the process design rules, there are critical manufacturing issues which should be considered in all micromachining processes to achieve defect free devices. The intrinsic film stress generated during the deposition process plays important role in the structural stability of the multi layer polysilicon sensors. Annealing will be required to release the residual stress created during fabrication process.

All accelerometers require an etch step to release the moving mass from the supporting layer or substrate. This process is usually a wet etch of a sacrificial layer. Release stiction typically occurs during the drying step when the surface tension forces in the liquid draw the micromachined structures into close contact due to van der Waals attraction. Super CO_2 drying can avoid the stiction failure. Controlling of etch properties and profile is the on-going challenge of etching technology. The variation of structural dimensions, which extremely impact design parameters, should be avoided by considering etching properties. Accelerometers contain fragile components such as thin beams. Wafer level hermetic packages which provide electrical connections and mechanical support are suitable for protecting the MEMS structure of accelerometer.

5 MEMS 3-D ACCELEROMETER

The geometric design of the proposed open loop sensor is similar to the forced balanced accelerometer introduced by Lemkin [7]. Figure 3 shows the design layout of the 3-D accelerometer along with the FEM simulation of the out of plain resonance mode, and figure 4 shows the fabricated sensor using MUMPs®[1] foundry process.

Figure 3: FEM stress and modal analysis of the 3-D accelerometer for out of plain acceleration.

The deposition of the first polysilicon layer creates the fixed electrode for sensing z-axis acceleration. Proof mass, sensing fingers and folded springs are patterned on the second 2μm-thick doped polysilicon. The nominal gap between capacitor plates and fingers is 2μm and the maximum length of overlap region is 150 μm. It is possible to assume sensing fingers as parallel plate capacitors to get the correct order of the at-rest total capacitance. The mass of the main electrode with attached sensing fingers including the effective mass of the springs is equal to 5.8×10^{-10} kg. The stiffness of the lateral motion including the spring softening effect of the applied voltage is 4.5 N/m. In addition by adding squeezed film damping with Couette flow effect, a quality factor equal to 5 can be obtained.

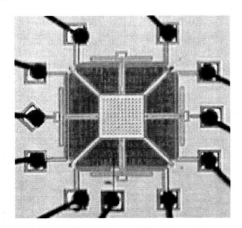

Figure 4: A 3-D micromachined accelerometer

Quad symmetry of the proof-mass about the z-axis minimizes sensitivity to off-axis accelerations. When a lateral acceleration is applied to the substrate, finger gaps change from their nominal 2μm value causing an imbalance in the capacitive sensing half bridge. The output voltage of the bridge is linearly proportional to the proof mass displacement. Under an applied z axis acceleration the proof mass moves out of plane causing a change in the parallel plate capacitance formed between the center of the proof mass and a bottom plate made from first polysilicon layer. Figure 5 shows the frequency response of the sensor in the range of 0 to 1200 Hz.

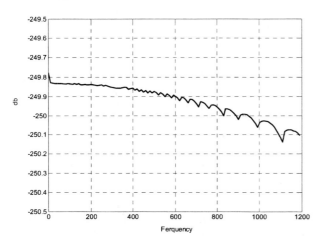

Figure 5: Frequency response of the 3-D accelerometer

Table 2 summarizes the other electrical and mechanical properties of the proposed accelerometer.

Parameter	x-axis	y-axis	z-axis
Natural frequency	15.2 kHz	15.2 kHz	4.6 kHz
Full scale range	±15g	±15g	±10g
Total sensing capacitance	116 fF	116 fF	242 fF
Noise level (TENA)	0.75 $mg/\sqrt{Hz}$	0.75 $mg/\sqrt{Hz}$	0.1 $mg/\sqrt{Hz}$

6   CONCLUSIONS

This article has outlined the modeling considerations of MEMS accelerometers and emphasized the important role of mechanical concepts in this area. A 3-D accelerometer has been designed and simulated using *multi port* theory. Simulation results show that the concept and the design are feasible and accurate. The process required to fabricate the proposed accelerometer, MUMPs®, is a commercialized micromachining process and a prototype has been fabricated to prove the concept and test the on-chip sensor.

REFERENCES

[1] D.J. Inman, *Engineering Vibration*. New Jersey: Prentice-Hall, 2001.
[2] S.Timoshenko, D.H.Young and W.Weaver, *Vibration Problems in Engineering*. 4th ed. New York: John Wiely, 1974.
[3] Y.He, J.Marchetti, C.Gallegos, F.Maseeh, "*Accurate fully-coupled natural frequency shift of MEMS actuators due to voltage bias and other external forces*," Proc. MEMS'99, 321-325,1999.
[4] F.K.Moore, *Theory of Laminar Flows*. New Jersey: Princeton University, 1964, Section D.
[5] Stephen D.Senturia, *Microsystem Design*. Massachusetts: Kluwer Academic Publishers, 2001.
[6] H.A.Tilmans, "*Equivalent circuit representation of electromechanical transducers: 1. Lumped-parameter systems*," Journal of Micromech. Microeng. vol.6, 1996.
[7] Mark A.Lemkin, *et. al.*, "*A 3-Axis Force Balanced Accelerometer Using a Single Proof Mass*," 9th Int. Conf. Solid-State Sensors and Actuators (TRANSDUCERS '97), Chicago, June 97.

[1] MUMPs is a registered trademark of Cronos Integrated Microsystems, a JDS Uniphase Company.

Effect of Thermophysical Property Variations on Surface Micromachined Polysilicon Beam Flexure Actuators

Amarendra Atre and Stephen Boedo

Department of Mechanical Engineering
Rochester Institute of Technology, Rochester, NY, 14623, sxbeme@rit.edu

ABSTRACT

Electrically heated, thermally driven, surface micromachined polysilicon beam flexure thermal actuators have been investigated using analytical methods that employ constant material properties either taken at room temperature or based on a set of averaged temperatures over the device operational range. In this paper, we present a comprehensive finite element analysis approach to examine the relative importance of temperature dependent material properties of heavily doped polysilicon on the static response of thermal actuator systems in air and vacuum environments. The results of the comprehensive analysis, which includes conduction, convection and radiation, are validated by comparing the predicted actuator deflection to that obtained experimentally.

Keywords: thermal microactuators, finite element analysis, thermal conductivity, ANSYS.

1 INTRODUCTION

The basic electro-thermal beam flexure actuator uses the principle of Joule heating for thermal expansion and movement. As shown in Figure 1, the beam flexure actuator design consists of a thin arm, wide arm, and flexure arm connected together at one end and constrained elastically at the anchors. The anchors are rigidly attached to the substrate. Application of a potential difference at the anchors generates a non-uniform electric field. The larger current density in the thin arm causes a greater thermal expansion than that in the wide arm, leading to motion of the actuator tip towards the wide arm.

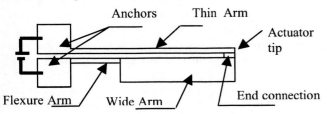

Figure 1. Basic beam flexure actuator

These actuators are typically fabricated by a MUMPs surface micromachining process that utilizes heavily doped polysilicon as the structural layer [1]. An array of such actuators has been widely employed in optical MEMS applications [2,3]. Efficient MEMS development not only requires reliable fabrication processes, but it also requires flexible design and analysis tools. Recently, these actuators have been analyzed using analytical [4,5] or finite element methods that use constant material properties based either on room temperature or averaged over a range of temperatures [6,7,8]. Typically, these actuators have been characterized for their force, deflection and current characteristics.

In this paper, the response of a thermal microactuator is analyzed with a comprehensive finite element model that includes full temperature dependencies of material properties and the means to impose all heat transfer modes. Actuator deflection characteristics are examined in air and vacuum environments, and the importance of thermophysical property variations that affect the analysis are investigated. The model is partially validated by comparing computed actuator tip deflection with experimental data [2,3,5,6,8]. This comparison will provide insight to the material properties that influence the behavior of the actuator and whether more research is warranted on temperature dependent parameters of heavily doped polysilicon.

2 ANALYSIS

The analysis of the actuator requires the solution of a generally nonlinearly coupled electro-thermal-elastic boundary value problem.

2.1 Boundary Conditions

Figure 2 shows a 3-D view of the actuator suspended above the substrate after release. The actuator arms are separated from the substrate by a 2 μm air gap, while the anchors remain attached to the substrate which acts as a heat sink at an assumed ambient temperature [1]. A thin nitride layer separates the substrate from the air gap. The substrate, nitride layer, and the surrounding air are not modeled directly but their effects are included indirectly through various boundary conditions. An electrical potential difference V is applied across the anchors, which causes non-uniform Joule heating. The deflection of the actuator due to this heating is parallel to the substrate.

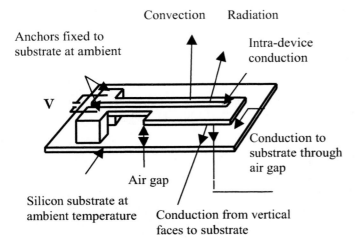

Convection   Radiation

Anchors fixed to
substrate at ambient

Intra-device
conduction

V

Conduction to
substrate through
air gap

Air gap

Silicon substrate at
ambient temperature

Conduction from vertical
faces to substrate

Figure 2. Boundary Conditions

Under normal modes of operation, the actuator will transfer heat to the surroundings and substrate by all three basic modes of heat transfer (conduction, convection and radiation), though some modes will likely dominate over the others under different conditions.

The conductive heat loss to the substrate through the air gap is modeled as an effective conductive heat transfer coefficient h defined as

$$h = 1/(t_a/k_a + t_n/k_n + t_{Si}/k_{Si}) \qquad (1)$$

Here t_a, t_n, t_{Si} represent the thickness of the air gap, nitride layer, and silicon substrate, respectively, and k_a, k_n, k_{Si} represent thermal conductivities of air, nitride layer, and silicon, respectively. A conductive shape factor S is used to account for the heat loss from the vertical faces of the actuator to the substrate, where S is expressed as [9]

$$S = (t/w)[2t_a/t + 1] + 1 \qquad (2)$$

where t is the thickness of the element vertical face, t_a is the height of the air gap between the element and the substrate, and w is the width of the element vertical face.

The convective heat loss from the top horizontal faces of the actuator to the (upward) ambient surroundings is represented by heat transfer coefficients taken from correlations developed for a heated horizontal face facing upwards [10].

The effects of radiation heat transfer are included as the device operates at high temperatures and in vacuum environments.

Table 1 lists available constant and temperature-dependent material property parameters taken from the literature and used in this analysis. Note that the constant value for each material property does not necessarily coincide with that computed at room temperature using the corresponding equation.

Table 1. Material property parameters

Parameter	Constant Value	Ref.
Elastic modulus of polysilicon	169 GPa	16
Poisson's ratio of polysilicon	0.22	16
Emmisivity	0.6	14
Stefan-Boltzmann constant	5.67×10^{-8} W-m^{-2}-°C^{-4}	14
Thermal cond. of air (k_a)	0.026 W-m^{-1}-°C^{-1}	9
Thermal cond. of nitride (k_n)	2.25 W-m^{-1}-°C^{-1}	9
Thermal cond. of silicon (k_{Si})	150 W-m^{-1}-°C^{-1}	11
Thermal conductivity of polysilicon (k_p)	32 W-m^{-1}-°C^{-1}	12
Coeff. of thermal expansion of polysilicon (α)	2.7×10^{-6} K^{-1}	14
Electrical resistivity of polysilicon (ρ_0)	2×10^{-3} Ω-cm	14
Resistivity coefficient of polysilicon (α_r)	1.25×10^{-3} °C^{-1}	14

$$k_p(T)=[(-2.2*10^{-11})T^3 + (9.0*10^{-8})T^2 \\ +(-1.0*10^{-5})T+0.014]^{-1} \text{ W-m}^{-1}\text{-}°C^{-1} \qquad 13$$

$$\rho(T) =\rho_0[1+\alpha_r(T-T_0)] \qquad 14$$

$$\alpha(T)=(3.725\{1-\exp(-5.88*10^{-3}(T-125))\} \\ +5.548*10^{-4}T)*10^{-6} \text{ K}^{-1} \qquad 14$$

$$k_a(T) = 3.9539 \times 10^{-4} + (9.886 \times 10^{-5}) T \\ - (4.367 \times 10^{-8}) T^2 + (1.301 \times 10^{-11}) T^3 \text{ W-m}^{-1}\text{-}°K^{-1} \qquad 14$$

The boundary conditions for the structure consist of constraining the displacement of the bottom faces of the anchors. For the results that follow, l_t, l_f, w_t, w_f, w_w, and g denote the length of thin arm, length of flexure arm, width of thin, flexure and wide arms, and gap between thin and wide arms, respectively, in microns.

2.2 Simulation

The coupled electro-thermal-elastic simulation is performed in ANSYS 5.6 using the direct method for analysis [17]. Thermal radiation is modeled using the radiosity solver method available in ANSYS [17]. As a baseline calculation, the finite element model employs constant material properties listed in Table 1 for the thermal conductivity of polysilicon and the thermal conductivity of the air gap. Temperature variations listed in Table 1 are then introduced. Simulation results are compared with measured data from six geometrically different thermal actuators operating in air and with measured data from a single thermal actuator operating in vacuum. Results from the air models also compared with results from an analytical model that assumes temperature averaged material property parameters. In vacuum, there is no conductive heat loss to the substrate and no free convection, so the conduction and convection parameters are eliminated from the model.

3 RESULTS

Figure 3 compares the steady state tip deflection for one actuator from the finite element simulations with that obtained from experimental testing and with that obtained from a simplified analytical approach [2]. It is observed that constant material assumptions of k_p and k_a in the finite element model ("Constant k_p, k_a") overestimates experimentally determined deflections at high voltages, while the analytical model predicts otherwise. At low voltages (and hence low temperatures), there is no significant difference among the model assumptions. Improved prediction of tip deflections at high voltage is observed by including either the temperature dependent thermal conductivity of polysilicon alone ("k_p (T)") or that of the air gap alone ("k_a (T)"), but not both ("k_p (T),k_a (T)"). In all the finite element models employing air, full temperature dependency of thermal expansion and electrical resistivity are included.

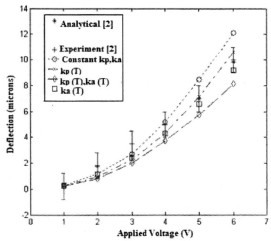

Figure 3. Deflection vs. voltage in air for actuator 1, l_t = 250, l_f = 34, w_t = w_f = 2, w_w = 15, g = 4

At higher voltages, and hence higher temperatures, the thermal conductivity of air increases and more heat is transferred from the thin arm to the substrate that, in turn, limits deflection. Figure 4 shows that the performance of the second actuator agrees well with experiment using the same model assumptions employed with the previous actuator.

For the third actuator, Figure 5 shows that the "Constant k_p, k_a" model again overestimates experimental deflection predictions. However, in this case, both "k_a(T)" and "k_p(T),k_a(T)" models give acceptable agreement with measured data, while the "k_p(T)" model also overpredicts (to a lesser extent) the measured data at higher voltages.

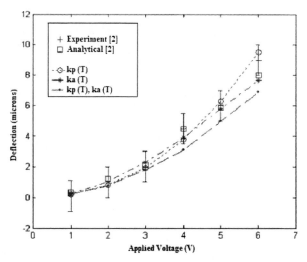

Figure 4. Deflection vs. voltage in air for actuator 2, l_t = 200, l_f = 27, w_t = w_f = 2, w_c = 15, g = 4

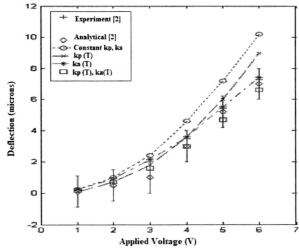

Figure 5. Deflection vs. voltage in air for actuator 3, l_t= 200, l_f = 34, w_t = w_f = 2, w_w = 15, g = 4

For the remaining three actuators, it is difficult to choose one model that can be employed for accurate predictions due to the observed variations in agreement with the experimental data (Table 2). These variations might be caused by the use of the shape factor equation (2), which might be overestimating the heat loss, leading to errors in the actuator temperature profiles.

Figure 6 shows comparisons of the finite element results with experiments obtained in vacuum [6]. Constant material property assumptions (including thermal expansion and electrical resistivity) are observed to be completely insufficient in predicting actuator deflections. The thermal profiles in this case (not shown) also predict melting temperatures that are inconsistent with experimental evidence. These results indicate that, in the absence of an air layer, the thermal conductivity of polysilicon plays an important role in predicting actuator deflection, and full temperature dependency should be included for accurate model predictions.

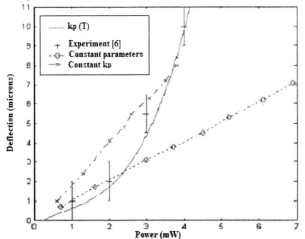

Figure 6. Deflection vs. power in vacuum, $l_t = 200$, $l_f = 30$, $w_t = w_f = 2$, $w_c = 14$, $g = 2$

4 CONCLUSIONS

A comprehensive finite element analysis of a beam flexure thermal microactuator has been performed in air and vacuum environments. Steady-state actuator tip deflection data has been compared with available experimental evidence, taking into account different actuator designs and temperature-dependent material property variations. In air at relatively large voltages, the finite element model predicts significant differences in deflection depending upon the thermophysical properties of polysilicon. A unique finite element model that consistently agrees with experiment could not be characterized. The variation among the different formulations might arise from the assumption of shape factors that approximate the heat loss from the vertical sidewalls of the thermal actuator. Another possibility for such variations might arise from the material property variations of thin film polysilicon. The thermal actuator behavior in vacuum was found to be highly dependent on material property variations in polysilicon, and full temperature dependencies must be employed to accurately characterize actuator behavior.

REFERENCES

[1] Koester, D., Mahadevan, R., Hardy, B., Markus, K., "Multi-user MEMs processes (MUMPs) Design Handbook," Cronos Integrated Microsystems, JDS Uniphase, NC, Revision 6.0, 2001.
[2] Comtois, J .H., Bright Victor M., Phipps, M.W., "Thermal microactuators for surface-micromachining processes," *Proc. SPIE*, Vol. 2642, 1995, pp. 10-21.
[3] Butler, J.T., Bright, V.M., Reid, R.J., "Scanning and rotating micromirrors using thermal actuators," *Proc.SPIE*, Vol. 3131, 1997, pp. 134-144.
[4] Huang, Q.A., Lee, N.K.S., "Analysis and design of polysilicon thermal flexure actuator," *Journal of Micromechanics and Micrengineering*, vol. 9, 1999, pp. 64-70.
[5] Hickey, R., "Analysis and Optimal Design of Micro-machined Thermal Actuators," M..Sc. Thesis, Dalhousie University, Nova Scotia, CA, 2001.
[6] Butler, J.T., Bright, V.M., Cowan, W.D., "SPICE modeling of polysilicon thermal actuators," *Proc. SPIE*, vol. 3224, 1997, pp. 284-293.
[7] Reid, J.R., Silversmith, D.J., "Joule heating simulation of polysilicon thermal micro-actuators," International Conference on Modeling and Simulation of Microsystems, vol. 2, 1999, pp. 613-616.
[8] Kolesar, E.S., Allen, P.B., Howard, J.T., Wilken, J.M., Boydston, N., "Thermally-actuated cantilever beam for achieving large in-plane mechanical deflections," *Thin Solid Films*, 1999, vol. 355-356, pp 295-302.
[9] Lin, L., Chiao, M., "Electrothermal responses of lineshape microstructures," *Sensors and Actuators A*, vol. 55, 1996, pp.35-41.
[10] Mills, A., *Basic Heat and Mass Transfer*, 1999, Prentice Hall.
[11] Mankame, N.D., "Modeling of electro-thermal-compliant mechanisms," M.S. Thesis, University of Pennsylvania, Philadelphia, PA, 2000.
[12] Tai, Y.C., Mastrangelo, C.H., Muller, R.S., "Thermal conductivity of heavily doped LPCVD polycrystalline silicon films," *J. App. Physics*, vol. 63, no.5, 1988, pp.1442-1447.
[13] Manginell, R.P., "Polycrystalline-silicon microbridge combustible gas sensor," Ph.D Thesis, University of New Mexico, Albuquerque, NM, 1997.
[14] Lott, C.D., "Electrothermomechanical modeling of surface-micromachined linear displacement microactuator," M.S. Thesis, Brigham Young University, Provo, Utah, 2001.
[15] Cronos Integrated Microsystems, www.memsrus.com, JDS Uniphase.
[16] Sharpe, W. N., Jr., Eby, M. A., and Coles, G., "Effect of temperature on mechanical properties of polysilicon," *Proc.Transducers'01*, pp 1366-1369, 2001.
[17] ANSYS Online Help Manual, Version 5.6.

Table 2.Summary of deflection comparisons
4: $l_t=200$, $l_f=40$, $w_t=w_f = 2$, $w_w=15$, $g=2$, $t=2$, $V=6.5V$
5: $l_t=200$, $l_f=35$, $w_t=w_f = 2$, $w_w=14$, $g=2$, $t=2$, $V=6.7V$
6: $l_t=230$, $l_f= 50$, $w_t =w_f = 2.5$, $w_w=16$, $g=2.5$, $t=2$, $V=4.3V$

Actuator	Expt.	Constant k_p, k_a	$k_p(T)$	$k_a(T)$	$k_p(T)$, $k_a(T)$
4	12.8[8]	15	13.7	10.2	9.1
5	11 [2]	16	14.5	11.1	10.2
6	8 [3]	6	5.3	4.6	3.4

A New 3D Model of the
Electro-Mechanical Response of Piezoelectric Structures

D. Elata, E. Elka and H. Abramovich

Technion - Israel Institute of Technology,
Faculty of mechanical Engineering, Haifa 32000, Israel
elata@tx.technion.ac.il

ABSTRACT

The constitutive equations of multi-layered piezoelectric structures are derived in a new form. In this form, the electromechanical coupling is presented as an additional stiffness matrix. This matrix is a true property of the piezoelectric structure and is independent of specific mechanical boundary conditions that may apply to the structure. A novel model of the electromechanical response of such structures is presented. This model accounts for the 3D kinematics of the structure deformation. Solution of example problems using the new model shows excellent agreement with full 3D finite element simulations. These solutions are also compared with the results of previous 2D model approximations presented in literature, and the inaccuracies associated with these previous models are discussed.

Keywords: piezoelectric, multimorphs actuators, electromechanical coupling.

1 INTRUDUCTION

In recent years new piezoelectric materials and advance manufacturing technologies have been developed, that enable the deposition of thin films and their integration into MEMS devices [1]. The electro-mechanical response of piezoelectric materials is complex as it involves a mechanical response, an electrical response, and a mutual coupling between the mechanical and electrical domains.

In most MEMS applications, Multi-Layered Piezo-Electric Structures (MLPES) can be considered as thin structures. Many models of the electro-mechanical response of MLPES have been proposed in recent years.

Krommer [2], derived the constitutive equations of MLPES, by assuming 2D kinematics of the structure. In his work the electro-mechanical coupling appears as a scalar term. By ignoring this coupling term, the Small Piezo-Electricity (SPE) approximation is obtained [2]. In MEMS applications, the substrate may be as thin as the piezoelectric layer. Under these conditions, the SPE approximation may lead to considerable errors, as is shown in the present study.

The constitutive equations of the multilayer structure may be simplified by using the SPE approximation, and may be further simplified by considering specific kinematics. Smits developed SPE constitutive equations for bimorphs and for unimorphs [3], assuming a 2D plane stress kinematics for each individual layer. A more general model for multimorphs was later presented by DeVoe and Pisano [4], and by Weinberg [5].

The assumption that plane stress is applicable to each individual layer in a MLPES, leads to inconsistency in the sense that the lateral displacements are not compatible between the layers of the structure. Accordingly, the assumption of plane strain was preferred in some models (e.g., [4], [5]).

In the present study the constitutive equation of MLPES are re-derived and presented in a new form. Furthermore, a novel 3D model for the electromechanical response of MLPES that are electrically actuated is presented. Accordingly the 3D model is used for a comparative study of previous approximated models presented in the literature [8].

2 FORMULATION

The constitutive equations of a linear piezoelectric material in the "S-E" form are given by [6]

$$T_i = C_{ij}^E S_j - e_{ij} E_j$$
$$D_i = e_{ji} S_j + \varepsilon_{ij}^S E_j \tag{1}$$

In these equations the vectors T, S, D, and E, are the stress, strain, electric displacement, and electric field, respectively. The matrices C^E, e, and ε^S, are the mechanical stiffness at constant electric field, the stress piezoelectric coefficient, and the permittivity at constant strain, respectively.

A schematic description of a multi-layer structure is shown in Figure 1. The x, y, and z axes are the axial, lateral and vertical (transverse) directions, respectively. In general the structure may be subjected to an axial force Fx, a vertical force Fz, and a bending moment M_B, at the free edge. In this structure, each layer may have a different thickness, different electrical and different mechanical properties, which are uniform in each layer. All layers in the structure have the same width and length and perfect bonding is assumed between layers.

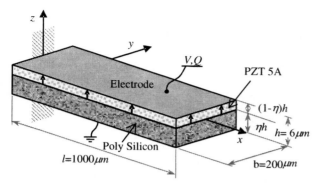

Figure 1: Schematic view of a piezoelectric unimorph.

Assuming small strains and small displacements, the resulting axial strain S_1, and the lateral strain S_2, are given by (Kirchhoff-Love)

$$S_1(x,z) = \frac{\partial u}{\partial x} = \alpha - \frac{d^2 w_{0x}}{d^2 x} z = \alpha - \beta(x)z$$

$$S_2(x,z) = \frac{\partial v}{\partial y} = \gamma - \frac{d^2 w_{0y}}{d^2 y} z = \gamma - \delta(x)z \quad (2)$$

Here $\beta(x)$ and $\delta(x)$ are the axial and lateral curvatures, respectively. The kinematics is 3-dimensional, and the *double-curvature* of the new model refers to β and δ. Through out this work it is assumed that S_1, S_2 and the transverse stress T_3, are independent of y.

In terms of the structure level mechanical variables, S_1, S_2, T_3, and for the described structure, the piezoelectric equations may be reformulated in the effective form [7]

$$\begin{pmatrix} T_1 \\ T_2 \\ S_3 \end{pmatrix}_k = \begin{pmatrix} C_{11}^* & C_{12}^* & C_{13}^* \\ C_{12}^* & C_{11}^* & C_{13}^* \\ -C_{13}^* & -C_{13}^* & C_{33}^* \end{pmatrix}_k \begin{pmatrix} S_1 \\ S_2 \\ T_3 \end{pmatrix} - \begin{pmatrix} e_{31}^* \\ e_{31}^* \\ e_{33}^* \end{pmatrix}_k E_{3k} \quad (3)$$

$$D_{3k} = \begin{pmatrix} e_{31}^* & e_{31}^* & -e_{33}^* \end{pmatrix}_k \begin{pmatrix} S_1 \\ S_2 \\ T_3 \end{pmatrix} + \begin{pmatrix} \varepsilon_{33}^{S*} \end{pmatrix}_k E_{3k} \quad (4)$$

where $c_{11}^* = C_{11} - \frac{C_{13}^2}{C_{33}}$; $C_{12}^* = C_{12} - \frac{C_{13}^2}{C_{33}}$; $C_{13}^* = \frac{C_{13}}{C_{33}}$; $C_{33}^* = \frac{1}{C_{33}}$;

$e_{31}^* = e_{31} - \frac{C_{13}e_{33}}{C_{33}}$; $e_{33}^* = -\frac{e_{33}}{C_{33}}$; $\varepsilon_{33}^{S*} = \varepsilon_{33}^S + \frac{e_{33}^2}{C_{33}}$, for each layer k.

The resultant mechanical forces per unit length in the structure are given by

$$N_x = \int_z T_1 dz \qquad N_y = \int_z T_2 dz = \sum_{k=1}^{n} \int_{z_{k-1}}^{z_k} T_{2k} dz \qquad M_x(x) = -\int_z T_2 z dz$$

$$M_y(x) = -\int_z T_1 z dz - F_x \left(w_{0x}(l) - w_{0x}(x) \right) \quad (5)$$

where $w_{0x}(x) = \int_0^x \left(\int_0^{x'} \beta(x'') dx'' \right) dx'$ is the vertical displacement along the structure centerline, k is the layer number, z_k, z_{k-1} are the upper and lower vertical coordinates of the k-th layer, T_1, T_2 are the axial and lateral stresses, N_x is the resultant axial force per unit width, N_y is the resultant lateral force per unit length, M_y is resultant moment per unit width and M_x is resultant moment per unit length.

Equilibrium of the entire structure requires that the following relations be satisfied

$$N_x = F_x; \quad N_y = 0; \quad M_y(x) = M_B; \quad M_x(x) = 0 \quad (6)$$

3 THE MECHANICAL RESPONSE OF MLPES

When electric boundary conditions are directly applied on each individual layer, the layers are only mechanically coupled. For this case, the electric field equation (4) is solved together with the electrostatic equations [7],[8]. By substituting the relevant electric field into (3) the mechanical response of the layer is found. For compactness, the following notations are introduced

$$\begin{pmatrix} \bar{S}_1 \\ \bar{S}_2 \\ \bar{T}_3 \end{pmatrix}_k = \frac{1}{h_k} \int_{z_{k-1}}^{z_k} \begin{pmatrix} S_1 \\ S_2 \\ T_3 \end{pmatrix} dz; \quad \begin{pmatrix} \tilde{S}_1 \\ \tilde{S}_2 \\ \tilde{T}_3 \end{pmatrix}_k = \begin{pmatrix} S_1 \\ S_2 \\ T_3 \end{pmatrix}_k - \begin{pmatrix} \bar{S}_1 \\ \bar{S}_2 \\ \bar{T}_3 \end{pmatrix}_k ;$$

$$\begin{pmatrix} \bar{\bar{S}}_1 \\ \bar{\bar{S}}_2 \\ \bar{\bar{T}}_3 \end{pmatrix}_k = \frac{1}{A_e^k} \int_{A_e^k} \left(\frac{1}{h_k} \int_{z_{k-1}}^{z_k} \begin{pmatrix} S_1 \\ S_2 \\ T_3 \end{pmatrix} dz \right) dA_e^k \quad (7)$$

The over-bar denotes the transverse average, the over-tilde denotes the transverse variance, and the over-double-bar denotes the spatial average.

For an applied voltage or a closed electrodes condition, the electro-mechanical response is given in (8), and for charge actuation or open electrodes condition, the electro-mechanical response is given in (9).

The last two equations relate the mechanical response of a single layer $(T_1, T_2, S_3)_k$, to the independent mechanical variables (S_1, S_2, T_3) and the geometrical parameters (h_k, A_e). The matrix in the last term on the RHS of (8) and (9) includes terms that have the dimensions of mechanical stiffness. This matrix is therefore referred to as the *Electro-Mechanical Coupling Stiffness* (EMCS).

In equation (8), the EMCS is multiplied by the transverse variance of the mechanical variables. Characteristically, this variance is nonzero for bending deformations of the structure. In equation (9), the EMCS is multiplied by the transverse variance and by the spatial average of the mechanical variables. In most cases of interest, the spatial average does not vanish, however, in some anti-symmetric vibration modes the spatial average may vanish [2],[7]. The EMCS affects the electromechanical response whenever the mechanical variables are non-uniform.

In some previous studies that consider a 2D kinematics, the electromechanical coupling stiffness is a *scalar*, known as the Electro-Mechanical Coupling Factor (EMCF) [5]. Furthermore, it is only considered after specific mechanical boundary conditions are enforced. The EMCS matrix that appears here in the constitutive equations of a piezoelectric layer within a structure (8),(9) is a generalization of the scalar EMCF. It represents the *structure response* for *any* of the applicable mechanical boundary conditions.

4 EXAMPLE PROBLEM

The capabilities and importance of the new 3D double-curvature model is demonstrated on a unimorph beam. The geometrical dimensions used for the presented calculations are given in Fig. 1. η is the relative ratio of substrate to total thickness, and various values of this parameter are considered. The material constants of the piezoelectric layer are (PZT-5A): C_{11}=120.32 *GPa*; C_{12}=75.16 *GPa*; e_{31}=-5.353 C/m^2; e_{33}=15.782 C/m^2; $\varepsilon^T_{33}/\varepsilon_0$=1700 *F/m*. The material properties for the unimorph substrate are (Poly-Silicon): *E*=160 *GPa*; ν=0.2.

The analytical solution of a quasi-static electrically actuated piezoelectric structure is examined here.

For comparison, the problem is solved by three solution methods: a) analytically, assuming small piezoelectricity (SPE); b) analytically with *Full Piezo-Electric Coupling* (FPEC); and c) numerically using ANSYS™ 6.1 finite element code.

In some studies, the response of MLPES is assumed to be plane strain (e.g., [4],[5]) or plane stress (e.g., [2]). These assumptions simplify the solution process but are inconsistent as sown in the following. For comparison with these modeling approaches, each of the solution methods listed above is performed for the following kinematics: 1) plane strain; 2) plane stress; 3) the double-curvature 3D kinematics described by α, β, γ and δ.

In a plane strain kinematics the lateral strain vanishes (S_2=0), and accordingly $\gamma = \delta$ =0. In a plane stress kinematics, the lateral stress in each layer vanishes (T_2=0). According to this last approach, the lateral displacements are incompatible between the layers.

The influence of η on the response of the unimorph is presented in Figs 2 and 3.

Figure 2a presents the curvature of the voltage actuated unimorph as a function of η, for the different modeling approaches. Similarly, Fig. 2b presents the unimorph curvatures associated with charge actuation. The errors relative to the analytic 3D-FPEC solutions are presented in Fig. 3a and 3b, for voltage and charge actuation, respectively. For the electrically actuated unimorph, it is emphasized that in the analytical 3D solution $\alpha=\gamma$ and $\beta=\delta$. This stresses the importance of the 3-dimensional response of MLPES.

The analytic 3D-FPEC solutions agree well with the 3D Finite element solutions of the problem. All other solutions show errors. The plane strain kinematics yields considerable relative errors ranging from 20% to more than 100%, and overestimates the deflection and curvature. For the plane strain kinematics, the SPE approximation results in higher errors than the FPEC. The plane stress kinematics yields better results, but the errors are still considerable. The SPE approximation may increase or decrease the errors associated with the plane stress kinematics.

$$
\begin{pmatrix} T_1 \\ T_2 \\ S_3 \end{pmatrix}_k = \begin{pmatrix} C^*_{11} & C^*_{12} & C^*_{13} \\ C^*_{12} & C^*_{11} & C^*_{13} \\ -C^*_{13} & -C^*_{13} & C^*_{33} \end{pmatrix}_k \begin{pmatrix} S_1 \\ S_2 \\ T_3 \end{pmatrix}_k + \begin{pmatrix} e^*_{31} \\ e^*_{31} \\ e^*_{33} \end{pmatrix}_k \frac{V_k}{h_k} + \begin{pmatrix} \frac{\left(e^*_{31}\right)^2}{\varepsilon^{s*}_{33}} & \frac{\left(e^*_{31}\right)^2}{\varepsilon^{s*}_{33}} & -\frac{e^*_{31}e^*_{33}}{\varepsilon^{s*}_{33}} \\ \frac{\left(e^*_{31}\right)^2}{\varepsilon^{s*}_{33}} & \frac{\left(e^*_{31}\right)^2}{\varepsilon^{s*}_{33}} & -\frac{e^*_{31}e^*_{33}}{\varepsilon^{s*}_{33}} \\ \frac{e^*_{31}e^*_{33}}{\varepsilon^{s*}_{33}} & \frac{e^*_{31}e^*_{33}}{\varepsilon^{s*}_{33}} & -\frac{\left(e^*_{33}\right)^2}{\varepsilon^{s*}_{33}} \end{pmatrix}_k \begin{pmatrix} \tilde{S}_1 \\ \tilde{S}_2 \\ \tilde{T}_3 \end{pmatrix}_k
\tag{8}
$$

$$
\begin{pmatrix} T_1 \\ T_2 \\ S_3 \end{pmatrix}_k = \begin{pmatrix} C^*_{11} & C^*_{12} & C^*_{13} \\ C^*_{12} & C^*_{11} & C^*_{13} \\ -C^*_{13} & -C^*_{13} & C^*_{33} \end{pmatrix}_k \begin{pmatrix} S_1 \\ S_2 \\ T_3 \end{pmatrix}_k + \begin{pmatrix} e^*_{31} \\ e^*_{31} \\ e^*_{33} \end{pmatrix}_k \frac{Q_k}{\left(\varepsilon^{s*}_{33}\right)_k A^k_e} + \begin{pmatrix} \frac{\left(e^*_{31}\right)^2}{\varepsilon^{s*}_{33}} & \frac{\left(e^*_{31}\right)^2}{\varepsilon^{s*}_{33}} & -\frac{e^*_{31}e^*_{33}}{\varepsilon^{s*}_{33}} \\ \frac{\left(e^*_{31}\right)^2}{\varepsilon^{s*}_{33}} & \frac{\left(e^*_{31}\right)^2}{\varepsilon^{s*}_{33}} & -\frac{e^*_{31}e^*_{33}}{\varepsilon^{s*}_{33}} \\ \frac{e^*_{31}e^*_{33}}{\varepsilon^{s*}_{33}} & \frac{e^*_{31}e^*_{33}}{\varepsilon^{s*}_{33}} & -\frac{\left(e^*_{33}\right)^2}{\varepsilon^{s*}_{33}} \end{pmatrix}_k \left[\begin{pmatrix} \tilde{S}_1 \\ \tilde{S}_2 \\ \tilde{T}_3 \end{pmatrix}_k + \begin{pmatrix} \bar{\bar{S}}_1 \\ \bar{\bar{S}}_2 \\ \bar{\bar{T}}_3 \end{pmatrix}_k \right]
\tag{9}
$$

For voltage actuated unimorphs with a large substrate thickness fraction η, the analytic 3D SPE solution almost coincides with the analytic 3D FPEC solution (Fig. 2a).

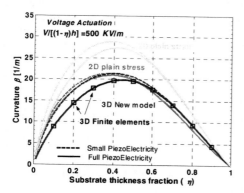

Figure 2a: The curvature of a voltage actuated unimorph as a function of η, for different modeling approaches.

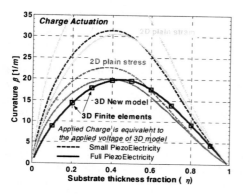

Figure 2b: The curvature of a charge actuated unimorph as a function of η, for different modeling approaches.

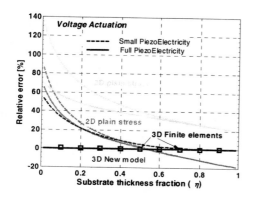

Figure 3a: The relative error in the curvature of a voltage actuated unimorph as a function of η.

In contrast, for charge actuated unimorphs with a large substrate thickness fraction η, the analytical 3D SPE

solution is not as close to the analytical 3D FPEC solution (Fig. 2b).

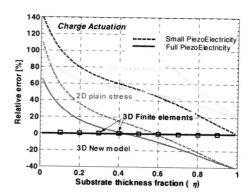

Figure 3b: The relative error in the curvature of a charge actuated unimorph as a function of η.

5   CONCLUSIONS

In this work the 3D nature of the deformation that develops in piezoelectric structures is analyzed. Specifically, it is shown that both the axial and the lateral curvatures, which characterize the deformed piezoelectric structure, must be considered. A new modeling approach that accounts for this double-curvature deformation is derived and is used to predict the electromechanical response. To this end, the electromechanical coupling is presented in new form as an additional stiffness matrix in the constitutive equations.

The predictions are in excellent agreement with 3D finite element simulations. The new model is used to explain why the prevalent 2D kinematic assumptions of plane stress and of plane strain, lead to wrong results. The model provides physical insight and predicts the behavior of such structures in actuation as well as in sensing mode operation.

REFERENCES

[1] P. Muralt, J. Micromech. Microeng., **10**, 136-146, 2000.
[2] M. Krommer, Smart Mat. Struct., **10**, 668-680, 2001.
[3] J.G. Smits and W-S. Choi, IEEE Trans. Ultrasonic, Ferroelectric, and Freq. Cont.., **38**(3), 256-270, 1991.
[4] D.L. DeVoe and A.P. Pisano, J. Micromech. Microeng, **6**(3), 266-270, 1997.
[5] M.S. Weinberg, J. Micromech. Microeng, Vol. 8(4), pp.529-533, 1999.
[6] IEEE Standard on piezoelectricity, ANSI/IEEE Standards, 176-1987.
[7] E. Elka, D. Elata and H. Abramovich, ETR No. 2003-2, http://meeng.technion.ac.il/Research/TReports/.
[8] E. Elka, D. Elata and H. Abramovich, J. Micromech. Microeng 2004 (in press).

Analytical Model for the Pull-In Time of Low-Q MEMS Devices

L.A. Rocha[*], E. Cretu[**] and R.F. Wolffenbuttel[*]

[*]Delft University of Technology, Faculty EEMCS
Dept. for Micro-Electronics, Delft, The Netherlands, l.rocha@ewi.tudelft.nl
[**]Melexis, Transportstr. 1, Tessenderloo, Belgium

ABSTRACT

A meta-stable transient region just beyond pull-in displacement that ultimately governs the pull-in time in critically damped systems is identified in this paper. Since the pull-in displacement time is basically governed by this second region (almost 90% of the pull-in time), the modeling of this region largely determines the reliability of the overall calculation of the pull-in dynamic transition. An analytical model for this region is derived and compared with measurements. The model accuracy, despite its simplicity, makes it a valuable tool for the design of MEMS switches outside vacuum and for sensors based on measuring pull-in time.

Keywords: Pull-in, MEMS dynamics, squeeze film damping, large-signal analysis

1 INTRODUCTION

Efficient and fast models for anticipating the dynamic and large-signal behavior of Microelectromechanical systems (MEMS) in the design phase are a topic of on-going research. The driver is the reduced design time and the increased probability of a first-time-right design. Complications are due to the various physical phenomena acting on MEMS.

A key characteristic of MEMS is the coupling between the different energy domains (mechanical, electrical, thermal) at the microscale level. Associated with the coupling between the mechanical and electrical domains, is the pull-in phenomenon. This instability in parallel plate electrostatic actuators has been subject of various studies [1,2]. If a quasi-static regime is assumed, mass and damping are neglected. The model describing the displacement of the MEMS structure in the direction normal to the electrode area, due to voltage applied, reduces to finding the equilibrium between mechanical and electrostatic forces. This results in a sudden pull-in at a well-defined pull-in voltage at 1/3 of the gap for 1 degree-of-freedom displacement structures [2].

A key issue in the design of MEMS-based switches and sensors based on pull-in time measurements is the dynamics of pull-in. In this case, the static regime does not apply, and for a meaningful study of the dynamic pull-in behavior, damping forces and mass inertia need to be included in the modeling. Some studies have considered the dynamic effects to enhance understanding of the pull-in time [3], or proposed a pressure sensor based on measuring the pull-in time [4]. More recently the pull-in time was shown to be sensitive to acceleration [5].

Studies on the dynamics of pull-in have generally assumed a smooth pull-in displacement. A more careful analysis reveals a meta-stable deflection for low-Q devices just beyond pull-in displacement. The pull-in time of realistic devices is determined by this particular region. It is therefore of crucial importance in the study of the dynamics of pull-in in general. Moreover, it is required for the design of the pressure sensor and accelerometer described, and for MEMS switches operated outside vacuum. This paper identifies the meta-stable regime. An analytical model is derived and compared with measurement results that confirm the modeling.

2 PULL-IN ANALYSIS

The structure used for analysis and experimental verification is basically a laterally movable beam with folded beam suspension at both ends and electrodes extending perpendicular to the axial direction. Three sets of stator electrodes in the same plane are used. One for electrostatic actuation in the direction normal to the electrode area and two sets of electrodes are used for capacitive displacement measurement (Fig. 1).

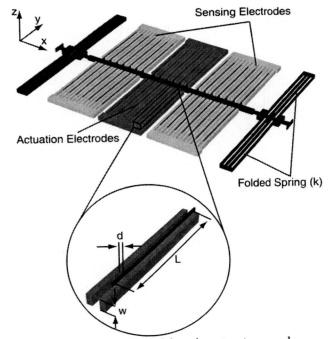

Figure 1: Drawing of the microstructure used.

The displacement of a MEMS device with 1DOF (Fig. 1) subject to an external acceleration and an input voltage is described by the differential equation:

$$m\frac{d^2x}{dt^2} + b(x,\frac{dx}{dt})\frac{dx}{dt} + kx = ma_{ext} + F_{elect} \qquad (1)$$

Here m represents the movable mass, b the (nonlinear) damping coefficient, k the spring constant, and $F_{elect} = \frac{C_0 d_0 V^2}{2(d_0-x)^2}$ is the electrostatic force due to the voltage V applied across a capacitor C_0 with initial gap d_0. At around critical damping ($Q\approx$ ½) the displacement x for an applied voltage V higher than the static pull-in voltage, $V_{pi} = \sqrt{\frac{8}{27}}d_0\sqrt{\frac{k}{C_0}}$ [2] ($V=\alpha V_{pi}$ with $\alpha>1$), proceeds as shown by curve in Fig. 2.

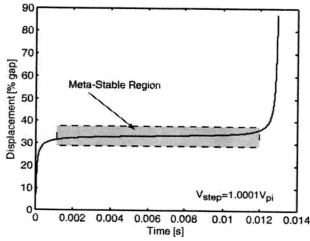

Figure 2: Pull-In displacement characteristic of low-Q (overdamped) microdevices.

2.1 Meta-stable Region

A qualitative analysis of the graph for the low-Q device, enables to distinguish three regions: a first region where the structure moves fast until close to the static pull-in displacement, a second meta-stable region where the movement is very slow, and finally a third region that takes the structure to the stoppers. Generally the first part of the curve is extrapolated to full deflection.

The fact that is often overlooked is that many realistic structures operate in the low-Q (Q< 1) mode and the curve presented in Fig. 2 applies. In such a system, the first region is where the electrostatic force imposed is compensated for by the mechanical and the damping forces. At the start of the step response the damping force dominates, but with deflection the mechanical force increasingly compensates for the electrostatic force. At the onset of the second region the structure moves very slowly and the mechanical force is almost the same as the electrostatic. This results in a kind of meta-stable equilibrium. Finally, due to the non-linear nature of the electrostatic force, the mechanical force cannot indefinitely compensate for the electrostatic force and the structure snaps.

The damping force is the crucial element in the first region, and largely determines the duration of the meta-stable region. Using a large-displacement dynamic model [6], the effect of the damping on the dynamic displacement was checked, and the different curves are presented in Fig. 3.

It is interesting to see that there is a threshold for which the meta-stability exists (at $Q_{max}\approx 1.4$). This is of crucial importance for understanding the mechanisms causing the meta-stable region, and can be of help when pull-in time is a crucial factor.

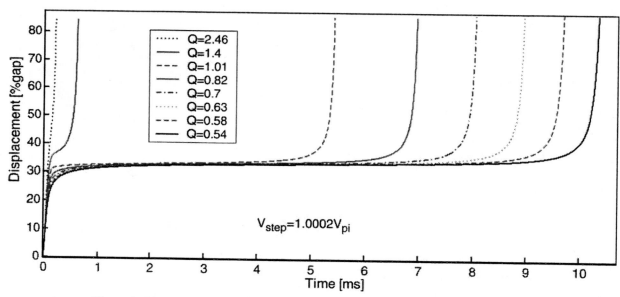

Figure 3: Simulations of the dynamic pull-in displacement for different Q values.

3 MODELING

The previous section shows that the pull-in displacement time for a low-Q device is basically governed by the second region (almost 90% of the pull-in time), thus the modeling of this region only can be assumed to reasonably predict the pull-in dynamic transition.

The meta-stable region occurs around a small region around the static pull-in displacement $x_{pi} = d_0/3$. If a local linearization of (1) around x_{pi} is realized, a linear second-order system results. The two non-linear terms being linearized are the electrostatic force, and the damping coefficient.

The electrostatic force is approximated in Taylor series around x_{pi}:

$$F_{elect} = \frac{9C_0V^2}{8d_0} + \frac{27C_0V^2(x - x_{pi})}{8d_0{}^2} \qquad (2)$$

3.1 Damping Coefficient

For structures in which only the size of the small gap between two plates changes in time, the pressure changes relative to the wall velocity are described by the Reynolds equation. An analytical solution for the forces acting on the surfaces can be found if some conditions are assumed [7]. The solution is frequency dependent and is not suitable for transient analysis. A very suitable approach is presented in [7] where the damping force can be represented by a network of frequency independent spring-damper elements, which have the same transfer function of the initial solution:

$$b_{m,n} = \frac{768lw\eta}{\pi^6 d^3 Q_{pr}} \frac{1}{(mn)^2 \left(\frac{m^2}{w^2} + \frac{n^2}{l^2} \right)} \qquad (3)$$

and:

$$k_{m,n} = \frac{64lwp_a}{\pi^4 d} \frac{1}{(mn)^2}, \qquad (4)$$

where m and n are odd integers, d is the gap, η is the viscosity of the medium, P_a the ambient pressure and w and l are the width and length of the surfaces (Fig. 1), respectively. Q_{pr}, the relative flow rate coefficient is a function of the Knudsen number K_n, the ratio between the mean free path of the gas molecules and the gap separation. In this work, the flow rate coefficient is given by [7]:

$$Q_{pr} = 1 + 9.638(K_n), \quad K_n = \frac{\lambda_0 P_0}{P_a d}, \qquad (5)$$

where λ_0 is the mean free path at pressure P_0.

If end effects are included [8], the frequency response of Fig. 4 is obtained for the gas film. For low frequencies the

damping force depends linearly on the velocity of the plate (frequency), so that the film acts as a pure damper. At higher frequencies, the relation becomes non-linear and the spring force increases, indicating that at these frequencies the film is acting like a spring.

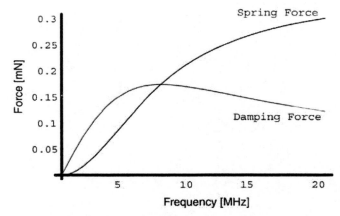

Figure 4: Damping and spring forces of the gas film.

For the analysis of the meta-stable region, the spring effect of the gas is neglected because the movement is very slow (low-frequency movement). The damping coefficient is linear in the frequency range of interest and can be computed for a gap size of $\frac{2d_0}{3}$.

3.2 Pull-In Time

Linearization of equation (1) around x_{pi}, while applying the variable transformation, $y = x - x_{pi}$ and assuming $V = \alpha V_{pi}$, gives the transfer function for the meta-stable regime:

$$H(s) = \frac{a_{ext} + \frac{kd_0}{3m}(\alpha^2 - 1)}{s^2 + \frac{b}{m}s + \frac{k}{m}(1 - \alpha^2)} \qquad (6)$$

For $\alpha > 1$ the linearized transfer function presents two poles (a positive and a negative). As the dominant one is the positive pole (the system is unstable), the effect of the negative pole can be neglected without introducing a big error. With this simplification, an expression for the transition time, during which the structure moves from an initial starting point x_1 (start of the meta-stable region) to a final point x_2 (end of the meta-stable region) is given by the following formula ($\Delta x = x_2 - x_1$):

$$t_{pi} = \frac{Log\left[\frac{\left(k(\alpha^2-1)(d_0+3\Delta x)+3ma_{ext}\right)^2}{\left(d_0 k(\alpha^2-1)+3ma_{ext}\right)^2} \right](b + Root)}{4k(\alpha-1)(\alpha+1)}, \qquad (7)$$

$$with\ Root = \sqrt{4km(\alpha^2 - 1) + b^2}$$

4 EXPERIMENTAL RESULTS

Measurements of the pull-in time for different voltages and external accelerations have been performed on a surface micromachined accelerometer fabricated using an epi-poly process [9] (Fig. 5).

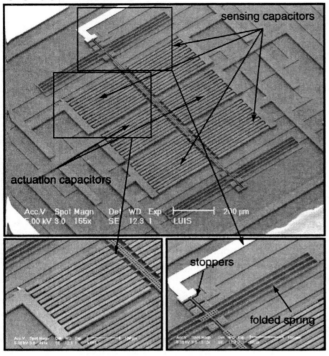

Figure 5: Photo of the fabricated device.

Fig. 6 compares the experimental results with the predictions based on (7). The important aspect is the high sensitivity of the transition time to an external acceleration. In fact, as the equilibrium of forces is the best description of the meta-stable region, it is intuitive that any small change acts as a perturbation of that meta-stable equilibrium, thus providing a means for achieving a very high sensitivity. This is valid only for small values of α.

5 CONCLUSIONS

In this paper a dominant region during dynamic pull-in displacement for low-Q devices has been identified and modeled. This meta-stable equilibrium region is very sensitive to external mechanical forces.

The meta-stable region has been studied using a simple linearized model. Since the dominant region occurs around a well-known displacement, linearization of the second-order equation of motion allows the derivation of a simple analytical expression for the pull-in time.

The model is confirmed by experiment. The model simplicity makes it a valuable tool for the design of MEMS switches operating outside vacuum, as well as accelerometers based on measuring the pull-in transition time.

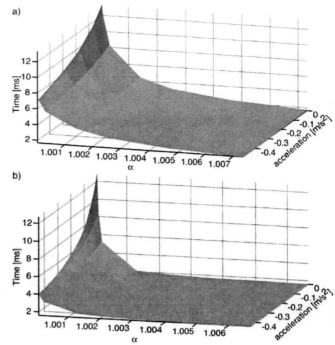

Figure 6: : Pull-in time changes with voltage step amplitude (α) and external acceleration. a) Experimental and b) Linear model.

REFERENCES

[1] P.M. Osterberg and S.D. Senturia, "M-TEST: A test chip for MEMS material property measurement using electrostatically actuated test structures", J. Microelectromech. Syst., vol. 6, pp. 107-118, 1997.

[2] H.A.C. Tilmans, and R. Legtenberg, "Electrostatically driven vacuum-encapsulated polysilicon resonators, Part 2, Theory and performance", Sensors and Actuators, vol A45, pp. 67-84, 1994.

[3] M.H.H. Nijhuis, T.G.H. Basten, Y.H. Wijnant, H.Tijdeman and H.A.C. Tilmans "Transient Non-Linear Response of 'Pull-in MEMS Devices' Including Squeeze film Effects", in Proc. Eurosensors XIII, The Hage, The Netherlands, 1999, pp. 729-732.

[4] R.K. Gupta and S.D. Senturia "Pull-in time dynamics as a measure of absolute pressure" in Proc. MEMS'97, Nagoya, Japan, 1997, pp. 290-294.

[5] H. Yang, L.S. Pakula and P.J. French, "A Novel Pull-in Accelerometer" in Proc. Eurosensors XVII, Guimarães, Portugal, 2003, pp. 204-207.

[6] L.A. Rocha, E. Cretu and R.F. Wolffenbuttel, "Displacement Model for Dynamic Pull-In Analysis and Application in Large-Stroke Electrostatic Actuators" in Proc. Eurosensors XVII, Guimarães, Portugal, 2003, pp. 448-451.

[7] T. Veijola, H. Kuisma, J. Lahdenperä and T. Ryhänen, "Equivalent-circuit model of the squeezed gas film in a silicon accelerometer", Sensors and Actuators, vol A48, pp. 239-248, 1995.

[8] T. Veijola, "End Effects of Rare Gas flow in Short Channels and in Squeezed-Film Dampers", in Proc. MSM'02, San Juan, Puerto Rico, USA, 2002, pp. 104-107.

[9] http://www.europractice.bosch.com/en/start/index.htm.

Finite element validation of an inverse approach to the design of an electrostatic actuator

J. Juillard, M. Cristescu, S. Guessab
Department of Measurement, Supélec,
91192, Gif-sur-Yvette CEDEX, FRANCE
jerome.juillard@supelec.fr

ABSTRACT

We present an approach to the design of electrostatically-actuated micro-structures and discuss its implementation in a software tool called IDEA. The main advantage of this approach is that it considerably reduces problems associated to coupling and to large-displacement non-linearities. The results obtained with IDEA are then compared with ANSYS simulations of an electrostatic micro-mirror.

Keywords: MEMS design, optimization, coupled-field problems

1 INTRODUCTION

From simple RF micro-switches to high-precision adaptive micro-mirrors [1], electrostatic actuation is one of the most common principles in the field of MEMS. However, the fact that the pressure-voltage and displacement-pressure relationships are non-linear and the strong coupling between these relationships, makes such MEMS very difficult to design accurately. For example, in one of the most elaborate of these models [4], the deformable membrane of the MEMS under consideration can be described by two purely linear models corresponding to whether bending stresses or tensile stresses are dominant. This paper, based on the results of [6], aims at validating an approach to the design of electrostatic MEMS and of deformable micro-mirrors in particular, where both kinds of stresses may be present.

After showing how the inverse mechanical problem, yielding a pressure distribution as a function of prescribed displacements, can be described by the equations of Von Karman and how these can be treated as a set of two equations, the more complex of which is a linear partial differential equation (PDE), the electrostatic problem is addressed: starting from the calculated ideal pressure distribution and from a given set of electrodes, we use a linearized model of the membrane to compute the voltages that give out the smallest residual deformation from the desired shape. The main advantage of this approach is that electromechanical coupling can be neglected inside the optimization loop, leading to efficient and simple calculations.

We then go on to discuss the implementation of this approach into a graphical tool using Matlab. Finally, in the last part of this paper, we validate the approach by comparing the results obtained with IDEA in the case of an electrostatic deformable micro-mirror and those obtained with fully non-linear ANSYS simulations of the device.

2 DIRECT AND INVERSE APPROACH TO THE DESIGN OF ELECTROSTATIC ACTUATORS

Let us consider the problem of imposing a displacement to a thin deformable structure with a given set of electrodes, such as can be found in micro-relays, micro-pumps or micro-mirrors. It is particularly difficult to solve this problem accurately because of the electromechanical coupling and also because of the non-linearity arising from large displacements (i.e. the prescribed displacement is not small compared to the thickness of the structure).

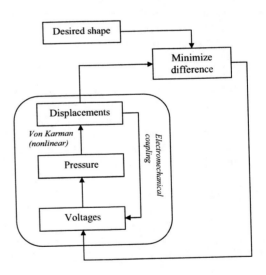

fig. 1: classical optimization approach. The optimization loop contains complex non-linear calculations which can only be solved iteratively.

A typical approach is illustrated in fig. 1. Starting from an initial guess of the voltages, the corresponding displacements are computed through a non-linear coupled-field simulation; they can then be compared to the desired shape and the initial guess can be modified accordingly. The main difficulty with this approach is that the optimization loop contains a non-linear coupled-field problem which must be solved at every iteration.

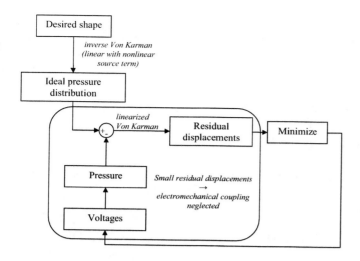

fig. 2 : inverse optimization approach. The only delicate calculation is placed outside the optimization loop. Since residual displacements are supposedly small, electro-mechanical coupling involves no iteration within the optimization loop.

We have proposed in [6] another approach to this problem, which consists in (fig. 2):

- solving the non-linear mechanical inverse problem to calculate the ideal pressure distribution corresponding to the desired displacements,
- for a given set of electrodes, solving the quadratic programming problem of finding the voltages that give the best approximation to this ideal pressure distribution.

This can be translated into a more mathematical language; let us consider a thin two-dimensional structure described by the equations of Von Karman:

$$\begin{cases} \dfrac{Eh^3}{12(1-v^2)}\Delta^2 w \\ -h\left(\dfrac{\partial^2 F}{\partial y^2}\dfrac{\partial^2 w}{\partial x^2}+\dfrac{\partial^2 F}{\partial x^2}\dfrac{\partial^2 w}{\partial y^2}-2\dfrac{\partial^2 F}{\partial x\partial y}\dfrac{\partial^2 w}{\partial x\partial y}\right)=P(x,y) \\ \Delta^2 F + E\left(\dfrac{\partial^2 w}{\partial x^2}\dfrac{\partial^2 w}{\partial y^2}-\left(\dfrac{\partial^2 w}{\partial x\partial y}\right)^2\right)=0 \end{cases}$$

(1)

As shown in [6], it is possible to solve these equations for the pressure $P(x,y)$ knowing the displacements $w(x,y)$. For simple membrane shapes, it is even possible to find analytical solutions. For example, for a circular membrane, the quasi-parabolic shape:

$$w(r)=W_0\left(1-\dfrac{r^2}{R_0^2}\right)^2 \quad (2)$$

can be obtained with the following pressure distribution:

$$P(r)=\dfrac{16}{3(1-v^2)}\dfrac{Eh^3 W_0}{R_0^4}$$
$$+\dfrac{4}{3}\dfrac{EhW_0^3}{R_0^4}\left(\dfrac{5-3v}{1-v}-\dfrac{22-18v}{1-v}\dfrac{r^2}{R_0^2}+30\dfrac{r^4}{R_0^4}-20\dfrac{r^6}{R_0^6}+5\dfrac{r^8}{R_0^8}\right)$$
(3)

Once the ideal pressure distribution corresponding to $w(x,y)$ is known, it can be injected into the optimization loop. Assuming small residual displacements $\delta w(x,y)$ and a plane capacitance approximation, one may neglect part of the electro-mechanical coupling by writing the electrostatic pressure created by the kth electrode with potential V_k as:

$$P_k=\dfrac{\varepsilon_0}{2}\dfrac{\tilde{V}_k^2}{(g-w-\delta w)^2}\approx\dfrac{\varepsilon_0}{2}\dfrac{\tilde{V}_k^2}{(g-w)^2} \quad (4)$$

Finding the voltages that yield the best approximation to the desired shape can then be re-formulated into a quadratic programming problem with non-negativity constraints, using (4) and the first equation of Von Karman linearized in the neighborhood of $w(x,y)$. This problem can then be solved using Matlab's *quadprog* function or any other quadratic programming algorithm.

There are two advantages to this approach :

- part of the mechanical problem is linearized (the inverse Von Karman equations are linear with non-linear source terms) and placed outside the loop. For some simple shapes, it also has analytical solutions.
- the optimization loop is initialized very close to the solution. It is thus possible to linearize the inner loop calculations and, assuming small residual displacements, to neglect electro-mechanical coupling.

This method has been implemented using Matlab into a graphical tool called IDEA (for Inverse Design of Electrostatic Actuators). Symbolic calculations are used to compute the ideal pressure distribution corresponding to the desired displacements. Then, for a given electrode set, the voltages can be computed using either a Galerkin method (with the membrane mode shapes as basis functions) or a collocation method: using a collocation method rather than a Galerkin method makes it possible to minimize the residual displacements locally rather than globally. It is also possible to solve the regularized problem and look for an approximate solution giving out small residual displacements together with low actuation voltages.

3 FINITE ELEMENT VALIDATION

To illustrate our approach, we consider, as in [6], a clamped micro-mirror with radius 1cm and thickness 3μm and a set of electrodes of increasing outer radii [1.5, 3, 4.5, 6, 7.5, 9]mm. The electrostatic gap is 50μm and the spacing between neighbouring electrodes is 100 μm.

An equivalent ANSYS model of the device is built using *membrane51* elements (axisymmetric shell) and *plane121* elements. A fine regular mesh is chosen for the air region with a typical element size of 10 μm. The principle of the validation is the following: starting from a given shape of the membrane, the corresponding optimal voltage distribution is taken from IDEA and used as an input of the ANSYS model. The resulting deflection is then computed with ANSYS under large deflection/stress-stiffening hypotheses and compared to the objective displacement. As long as the residual displacements computed with IDEA remain small, the two models compare very well (fig. 3 and fig. 4).

There is more discrepancy when IDEA is used to compute the voltage distribution corresponding to an 'impossible' shape, such as a parabola – which is 'impossible' considering the mirror is supposedly clamped on its edges (fig. 5).

This points out the necessity to take bending stresses into account even when large displacements are involved – in the case of fig. 5, the mirror's maximum deflection is 20 μm, which must be compared to the mirror's thickness of 3 μm. However, this problem may be solved by finding first an approximation to the 'impossible' shape which verifies the boundary conditions and then using IDEA to compute the voltage distribution corresponding to the approximation.

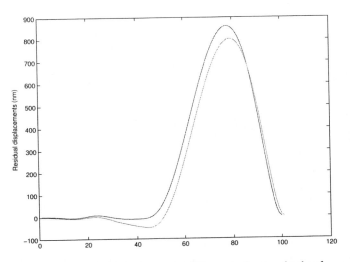

fig. 4 : residual displacements of the membrane obtained with IDEA (blue curve) and ANSYS (red curve) for the quasi-parabolic shape with maximum deflection W_0=20 μm.

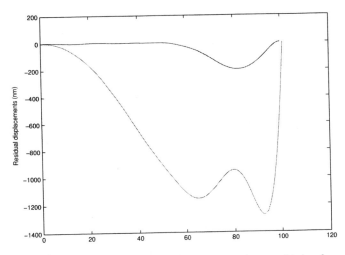

fig. 5 : residual displacements of the membrane obtained with IDEA (blue curve) and ANSYS (red curve) for the perfect parabolic shape with maximum deflection W_0=20 μm.

4 CONCLUSION

We have given the outline of a novel approach to the design of electrostatic actuators for thin structures undergoing large displacements and we have discussed its implementation in a software tool called IDEA. We have then given results concerning the actuation of a micro-mirror and compared them with ANSYS simulations of the device. Both models compare well as long as the desired mirror shape is a possible solution of the Von Karman equations with the good boundary conditions. If this is not the case, an approximation to the desired shape must first be found prior to using IDEA.

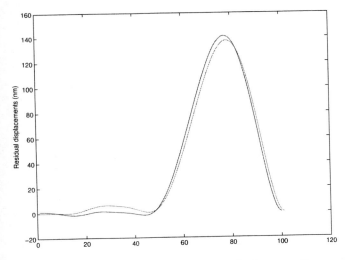

fig. 3 : residual displacements of the membrane obtained with IDEA (blue curve) and ANSYS (red curve) for the quasi-parabolic shape (eq. 2) with maximum deflection W_0=10 μm.

IDEA is currently being extended to two-dimensional problems and to dynamical problems – an advantage of the inverse approach being that, assuming small residual displacements, the strong coupling typical of non-linear damping phenomena, such as squeeze-film damping, can also be neglected.

REFERENCES

[1] G. Vdovin, P. M. Sarro, "Flexible mirror micromachined in silicon", Applied Optics, vol. 34, pp. 2968-2972, 1995

[2] I. Dufour, E. Sarraute, "Analytical modelling of beam behavior under different actuations: profile and stress expressions", Journal of Modeling and Simulation of Microsystems, vol. 1, issue 1, pp. 57-64, 1999

[3] J. Chen, S. Kang, "Techniques for coupled circuit and micromechanical simulation", Proceedings of MSM 2000 International Conference on Modeling and Simulation of Microsystems, San Diego, USA, 2000

[4] P. K. C. Wang, F. Y. Hadaegh, "Computation of static shapes and voltages for micromachined deformable mirrors with nonlinear electrostatic actuators", Journal of Microelectromechanical Systems, vol. 5, issue 3, pp. 205-219, 1996

[5] E. S. Claflin, N. Bareket, "Configuring an electrostatic membrane mirror by least square fitting with analytically derived influence functions", Jounal of the Optical Society of America A, vol. 3, issue 11, pp.1833-1839, 1986

[6] J. Juillard, M. Cristescu, "Inverse approach to the design of adaptive micro-mirrors", Journal of Micromechanics and Microengineering, vol. 14, pp. 347-355, 2003.

mor4ansys:
Generating Compact Models Directly from ANSYS Models

Evgenii B. Rudnyi, Jan Lienemann, Andreas Greiner, Jan G. Korvink

IMTEK – Institute for Microsystem Technology, Albert Ludwig University
Georges Köhler Allee 103, D-79110 Freiburg, Germany
Tel. +49 761 203 7383, Fax. +49 761 203 7382, Email: rudnyi@imtek.de

ABSTRACT

Model reduction of linear large-scale dynamic systems is already quite an established area [1]–[3]. In a number of papers (see references in [3]), the advantages of model reduction have been demonstrated. In the present paper, we describe a software tool to perform moment-matching model reduction via the Arnoldi algorithm directly to ANSYS finite element models. We discuss the application of the tool to a structural mechanical problem with a second order linear differential equation (ODE). Its successful application to the first order case of electro-thermal modeling is demonstrated elsewhere [4], [5].

Keywords: model order reduction, damped second order system, Rayleigh damping, Arnoldi process, Krylov subspace

1 INTRODUCTION

mor4ansys is a command-line tool built on top of the ANSYS-supplied library to read ANSYS binary files [7] and the TAUCS library for sparse linear algebra [8]. After a model is built and meshed in ANSYS, an ODE system is obtained. Consider for example the PDE for an elastic body

$$\boldsymbol{f}_I + \boldsymbol{f}_D + \boldsymbol{f}_S = \mathbf{b}u(t), \tag{1}$$

with $\boldsymbol{f}_I$ the force caused by inertia, $\boldsymbol{f}_D$ the damping force, $\boldsymbol{f}_S$ the elastic force and $\mathbf{b}u(t)$ external forces depending on user input and varying in time [6]. The *scatter matrix* $\mathbf{b}$ distributes the inputs $\boldsymbol{u}$ to the domain. By discretization with n spatial degrees of freedom x_i, $1 \le i \le n$, ANSYS transforms the PDE to n ordinary differential equations,

$$\boldsymbol{f}_S = \mathbf{K}\boldsymbol{x}, \qquad \boldsymbol{f}_D = \mathbf{C}\dot{\boldsymbol{x}}, \qquad \boldsymbol{f}_I = \mathbf{M}\ddot{\boldsymbol{x}} \tag{2}$$

$$\implies \quad \mathbf{M}\ddot{\boldsymbol{x}}(t) + \mathbf{C}\dot{\boldsymbol{x}}(t) + \mathbf{K}\boldsymbol{x}(t) = \mathbf{B}\boldsymbol{u}(t), \tag{3}$$

where $\mathbf{M}$, $\mathbf{C}$, and $\mathbf{K}$ are the system matrices, $\mathbf{B}\boldsymbol{u}(t)$ is the load vector, and $\boldsymbol{x}$ is a vector with unknown degree of freedoms, its dimension being routinely from 10 000 to 500 000. The outputs of the system can in principle be an arbitrary linear combination of states

$$\boldsymbol{y} = \mathbf{L}^T \boldsymbol{x}, \tag{4}$$

but usually $\mathbf{L}$ is only used to pick certain degrees of freedom.

The information from ANSYS is transferred as a file with element matrices (EMAT file) and lists for Dirichlet boundary conditions, nodal forces and output degrees of freedoms. The developed software uses these files as input and produces a reduced model by means of the Arnoldi algorithm. The user may choose a maximum order m for the reduced model. Because of the iterative nature of the Arnoldi algorithm, one obtains all possible reduced models with dimensions ranging from 1 to m as specified by the user. The postprocessing, that is, the solution of a reduced ODE system as well as computing its transfer function is currently performed in Mathematica. It is worth mentioning that it can be done in any other environment as the reduced model is stored as an ASCII file.

A conventional approach to model reduction is to find a low-dimensional subspace $\mathbf{V}$

$$\boldsymbol{x} = \mathbf{V}\boldsymbol{z} + \boldsymbol{\varepsilon} \tag{5}$$

that can well approximate the trajectory of the state vector and then project (3) on that subspace:

$$\mathbf{M}_r\ddot{\boldsymbol{z}} + \mathbf{C}_r\dot{\boldsymbol{z}} + \mathbf{K}_r\boldsymbol{z} = \boldsymbol{b}_r \tag{6}$$

where $\mathbf{M}_r = \mathbf{V}^T\mathbf{M}\mathbf{V}$, $\mathbf{C}_r = \mathbf{V}^T\mathbf{C}\mathbf{V}$, $\mathbf{K}_r = \mathbf{V}^T\mathbf{K}\mathbf{V}$, $\boldsymbol{b}_r = \mathbf{V}^T\boldsymbol{b}$. In mechanical engineering, the subspace $\mathbf{V}$ is usually chosen from the eigenstates of (3) or by the Guyan method [9].

Moment matching via Krylov subspaces is a new technique [1], [3], [10] that allows us to find a low-dimensional subspace with excellent approximating properites for relatively low computational effort. For example, the time for model reduction in mor4ansys is comparable with the time required for a stationary solution or for a single timestep during an ANSYS transient simulation process. Other advantages mentioned above are as follows:

1. User intervention is minimal: one has just to specify the maximum dimension for the reduced system. There is no selection of dominant eigenmodes, master degree of freedoms or the like.

2. Iterative nature of the algorithm: one can change the dimension of the reduced model without additional computations.

A straightforward application of Krylov subspace methods to second order ODEs produces a reduced system in the form of a first order system of ordinary differential equations [1], [3], [10], and this is undesirable for structural mechanics. Su and Craig have suggested a modified version of the Arnoldi algorithm that preserves the second order in the reduced model [11]. In both cases, the damping matrix $\mathbf{C}$ takes part in the process of generation of the matrix $\mathbf{V}$.

The main difference of our approach with those in Refs [10], [11] is that the damping matrix is not employed at all during the generation of a low-dimensional basis $\mathbf{V}$, that is, the latter is built as the orthogonolized Krylov subspace $\mathcal{K}\left(\mathbf{K}^{-1}\mathbf{M}, \mathbf{K}^{-1}\boldsymbol{b}\right)$. Nevertheless, the reduced damping matrix has been computed as a projection in (5). Such an approach is based on the engineering intuition that the damping matrix should not play a major role in finding a good subspace $\mathbf{V}$ as the most essential information is contained within the mass and stiffness matrices. Unfortunately, we cannot prove this mathematically. From a pragmatic viewpoint, such an approach allows us a great deal of advantage in the most often encountered case in structural mechanics when a damping matrix is built up as a linear combination of mass and stiffness matrices [6], i.e., the damping is chosen as mode preserving Rayleigh damping

$$\mathbf{C} = \alpha\mathbf{M} + \beta\mathbf{K}. \tag{7}$$

The unit of α is $1/\mathrm{s}$, the unit of β is $1\,\mathrm{s}$.

The motivation for this choice comes from the fact that by choosing $\mathbf{M}$ and $\mathbf{K}$ such that

$$\mathbf{C} = \mathbf{M}\sum_b a_b \left(\mathbf{M}^{-1}\mathbf{K}\right)^b, \tag{8}$$

we gain the following properties [6]:

- Damping orthogonality, thus the different modes of the system do not couple through the damping.

- The vibration mode shapes are the same for the damped and undamped system.

- The essential dynamic response is associated with the lowest few modal coordinates and thus suitable for reduction.

There are also intuitive interpretations of this form, basically saying that the damping contributions come from internal friction and the surrounding air, and it often happens that the resulting behavior is sufficiently accurate for many applications.

In this case, one can show that the reduced damping matrix $\mathbf{C}_r = \mathbf{V}^T\mathbf{C}\mathbf{V}$ can be computed directly from the reduced mass and stiffness matrices as

$$\mathbf{C}_r = \alpha\mathbf{M}_r + \beta\mathbf{K}_r, \tag{9}$$

that is, the parameters α and β remain as parameters during the model order reduction process.

1.1 Moment matching for second order systems

After Laplace-transformation of (3) and (4), the transfer function $\boldsymbol{H}(s) = \mathcal{L}(\boldsymbol{y}(s))/\mathcal{L}(U(s))$ can be written as

$$\boldsymbol{H}(s) = \mathbf{L}^T\left(s^2\mathbf{M} + s\mathbf{C} + \mathbf{K}\right)^{-1}\mathbf{B}. \tag{10}$$

For simplicity, let us assume that we only have one output and one input terminal, so that $\boldsymbol{H}(s)$ becomes a scalar, $\mathbf{L}^T$ a vector $\boldsymbol{l}^T$ and $\mathbf{B}$ a vector $\boldsymbol{b}$. For our method, we also drop the damping term. We expand $H(s)$ by a Taylor series for s^2 at $s_0 = 0$

$$H(s) = \boldsymbol{l}^T\left(s^2\mathbf{M} + \mathbf{K}\right)^{-1}\boldsymbol{b} \tag{11a}$$

$$= \boldsymbol{l}^T\left(s^2\mathbf{K}^{-1}\mathbf{M} + \mathbf{I}\right)^{-1}\mathbf{K}^{-1}\boldsymbol{b} \tag{11b}$$

$$= \sum_{i=0}^{\infty} s^{2i}\boldsymbol{l}^T\left(\mathbf{K}^{-1}\mathbf{M}\right)^i\mathbf{K}^{-1}\boldsymbol{b} = \sum_{i=0}^{\infty} m_i s^{2i}. \tag{11c}$$

The m_i are called the moments of the transfer function. We now seek a projection $\mathbf{V}$ that provides a Padé approximation, i.e., that yields the same first q moments for the transfer function of the reduced system.

The Arnoldi algorithm reduces the $n \times n$ matrix $\mathbf{K}^{-1}\mathbf{M}$ to a small $q \times q$ block upper Hessenberg matrix $\mathbf{H}_q$ and during this transformation creates a matrix $\mathbf{V}$ such that

$$\mathrm{colspan}(\mathbf{V}) = \mathcal{K}_m(\mathbf{K}^{-1}\mathbf{M}, \mathbf{K}^{-1}\boldsymbol{b}) \tag{12a}$$

$$\mathbf{V}^T\mathbf{K}^{-1}\mathbf{M}\mathbf{V} = \mathbf{H}_q \tag{12b}$$

$$\mathbf{V}^T\mathbf{V} = \mathbf{I}_q \tag{12c}$$

It can be shown that by using these matrices the corresponding moments in the full and reduced system match up to the mth moment [2].

2 WIRE BOND MODEL

As a benchmark for the algorithm we use a model of a gold wire bond needed for the packaging of micro devices (see fig. 1). The material properties are listed in table 1. The application area for this benchmark is the design of wire bonds and the configuration of bonding machines.

The model was created in ANSYS and meshed with tetrahedral 10-node elements (SOLID187). It features 32877 degrees of freedom. A step load is applied to the first bend of the wire, with direction parallel to the z-axis.

Young's modulus	$E = 78\,\mathrm{GPa}$
Poisson's ratio	$\nu = 0.44$
Density	$\varrho = 19300\,\mathrm{kg/m^2}$

Table 1: Material properties for the bond wire.

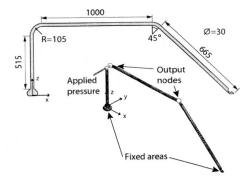

Figure 1: Sketch of a wire bond used for micro chip connection (all dimensions in μm). Inset: Mesh and applied loads for wire bond model.

3   RESULTS

The model was simulated in the time domain (transient simulation) and in the frequency domain (harmonic simulation). We investiged different settings for β, α was set to zero for all simulations.

ANSYS has a built-in model order reduction tool using the Guyan method [12]. This tool requires choosing "master nodes" from the complete set of degrees of freedom. These nodes can be chosen by hand, but since our aim is to provide a reduction method which should not need to rely on the experience of the designer, we only assigned the output nodes as master nodes and use the automatic master node selection from ANSYS. The model was order reduced using both the Guyan method and the Arnoldi method and the transfer functions were compared.

3.1   Transient simulation

Figure 2 shows the transient response at the output node marked with an arrow (fig. 1) to a step load for the ANSYS model (simulation performed in ANSYS) and two reduced models for a damping of $\alpha = 0$ and $\beta = 1\,\mu s$ (fig. 2a) or $\beta = 0.01\,\mu s$ (fig. 2c). It is remarkable that even a model with three (higher damping) or five (lower damping) degrees of freedom is able to catch the transient behavior almost perfectly. The curves for higher order reduced models are indistinguishable from the ANSYS curves for the chosen resolution of the graph.

Figures 2d, e show the same simulation with the Guyan method. To achieve similar results, the order of the reduced system needs to be considerably higher than for the Arnoldi method.

3.2   Harmonic simulation

The difference is also clearly visible in the frequency plot (fig. 3). The figures show the response of the output node for harmonic excitation of the beam at the arrow location. While the Guyan model of order 10 in fig. 3a leaves the curve of the full model near the second peak, the Arnoldi model of

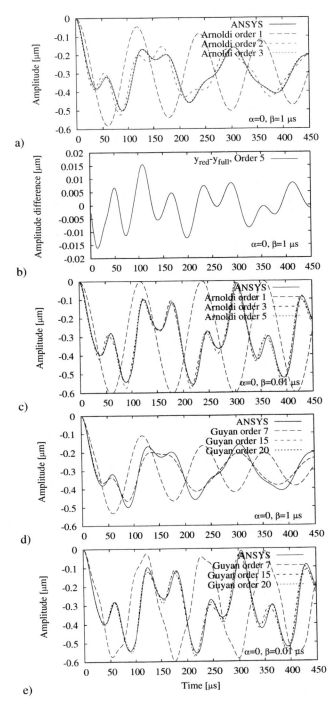

Figure 2: Transient simulation with full and reduced models of various order. a) Damping 1 μs, Arnoldi. b) Difference between reduced model of order 5 and full model. c) Damping 0.01 μs, Arnoldi. d) Damping 1 μs, Guyan. e) Damping 0.01 μs, Guyan

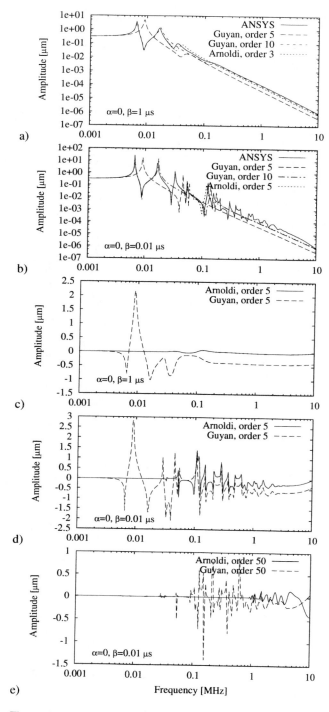

a)

b)

c)

d)

e)

Figure 3: Harmonic spectrum for full and reduced models.
a) Spectrum for Damping 1 µs. b) Spectrum for 0.01 µs. c)
Logarithmic difference $\log_{10}|H_r(s)| - \log_{10}|H(s)|$ between
reduced models and full model for damping 1 µs, d,e) for
damping 0.01 µs.

order 3 matches the curve for a wide range of frequencies.
The same behavior can be seen for lower damping. However, for lower damping, the deviations are usually larger
than for lower damping because of the slower decay of high
frequency modes.

4 CONCLUSIONS

We have presented a novel approach to compute reduced
order models of second order damped systems. We showed
that this approach works very well for systems where the
damping matrix is a linear combination of the stiffness and
mass matrix, and that it greatly outperforms the Guyan method.

5 ACKNOWLEDGMENTS

Partial funding by the DFG project MST-Compact (KO-
1883/6), by the German Ministry of Research BMBF, and by
an operating grant of the University of Freiburg is gratefully
acknowledged.

REFERENCES

[1] A. C. Antoulas, D. C. Sorensen, "Approximation
of large-scale dynamical systems: An overview",
Technical Report, Rice University, Houston, 2001,
http://www-ece.rice.edu/~aca/mtns00.pdf.

[2] R. W. Freund, "Krylov-subspace methods for reduced-
order modeling in circuit simulation", Journal of Computational and Applied Mathematics 123, 395–421,
2000.

[3] E. B. Rudnyi, J. G. Korvink, "Automatic Model Reduction for Transient Simulation of MEMS-based Devices", Sensors Update 11, 3-33, 2002.

[4] C. Bohm, T. Hauck, E. B. Rudnyi, J. G. Korvink,
"Compact Electro-thermal Models of Semiconductor
Devices with Multiple Heat Sources", submitted to EU-
ROSIME 2004.

[5] T. Bechtold, E. B. Rudnyi, J. G. Korvink, C. Rossi, "Efficient modelling and simulation of 3D electro-thermal
model for a pyrotechnical microthruster", submitted to
PowerMEMS 2003.

[6] R. W. Clough, J. Penzien, "Dynamics of Structures",
McGraw-Hill, 194–195, 1975.

[7] "Guide to interfacing with Ansys", ANSYS corp.

[8] S. Toledo, D. Chen, V. Rotkin, "TAUCS - A library of sparse linear solvers", http://www.tau.ac.il/
~stoledo/taucs/.

[9] J. Guyan, "Reduction of stiffness and mass matrices",
AIAA J. 3, 380, 1965.

[10] Z. J. Bai, "Krylov subspace techniques for reduced-
order modeling of large-scale dynamical systems", Applied Numerical Mathematics 43, 9-44, 2002.

[11] T. J. Su and R. R. Craig, "Model-Reduction and Control
of Flexible Structures Using Krylov Vectors", Journal
of Guidance Control and Dynamics 14 260–267, 1991.

[12] ANSYS theory manual

Piecewise perturbation method (PPM) simulation of electrostatically actuated beam with uncertain stiffness

J. Juillard, H. Baili, E. Colinet

Department of Measurement, Supélec,
91192, Gif-sur-Yvette CEDEX, France
jerome.juillard@supelec.fr

ABSTRACT

We present a new approach to the simulation of uncertainties in micro-electromechanical systems, based on the same principle as perturbation methods. This approach is valid for large variations of the uncertainties and requires much less simulations than a Monte-Carlo method. An implementation in the case of an electrostatically actuated beam with uncertain stiffness is presented and compared with obtained with Monte-Carlo.

Keywords: uncertainties, simulation, perturbation

1 INTRODUCTION

There are many difficulties to the simulation and modelling of micro-electromechanical systems : coexistence of many coupled physical phenomena, non-linearities and important uncertainties. They can also rise from simple lack of knowledge, for example, regarding the mechanical properties of a material. In fact, from an experimental point of view, such uncertainties are a major limiting factor to the credibility of MEMS models.

Taking them into account is most usually done via "stochastic" simulation methods, the most famous of which is the Monte-Carlo method [1]. This costly but robust approach requires an important number of realizations in order to obtain a correct statistical description of the system. It also relies on the ability to generate a random variable with a probability density function (pdf) that may be neither gaussian nor uniform. Other methods include the stochastic finite-element method [2], second-moment analysis [3] or resolution of the Fokker-Planck equation [4].

We propose in this paper a completely deterministic approach to the problem of simulating a possibly nonlinear system with one uncertain parameter, based on the principle of perturbation analysis and we illustrate this method with the case of a beam of uncertain stiffness undergoing electrostatic actuation. Results obtained with our method are then compared with Monte-Carlo simulations of the device.

2 CLASSICAL AND PIECEWISE PERTURBATION METHODS

2.1 Classical perturbation methods

Let us consider a system $(S)_\varepsilon$ with input x, output y and uncertain parameter ε, with zero mean and small variance.

Let us consider also that the input-output relationship can be put in the form:

$$y = f(x, \varepsilon) \quad (1)$$

Supposing small perturbations of the system, an nth-order Taylor expansion of (1) can be made:

$$y \approx f(x,0) + \varepsilon \frac{df}{d\varepsilon}\bigg|_{x,\varepsilon=0} + ... + \frac{\varepsilon^n}{n!}\frac{d^n f}{d\varepsilon^n}\bigg|_{x,\varepsilon=0} = F_x(\varepsilon) \quad (2)$$

In the classical approach [3], the order of the expansion is usually limited to $n=2$. Relation (2) can then be used to determine the first and second moments of y, knowing those of ε. This approach requires the simulation of the original, unperturbed system plus n simulations for the higher-order terms of the series expansion.

Although this method is quite straightforward, there are some cases in which the moments do not hold much information and for which it is preferable to express the pdf of the system's output. This can be done using the following method.

Starting from (2), the pdf of the system's output can be calculated with the following formula [5]:

$$\text{pdf}_y(y_0) = \sum_p \frac{\text{pdf}_\varepsilon(\varepsilon_p)}{\left|\dfrac{dF_x}{d\varepsilon}\right|_{\varepsilon=\varepsilon_p}} \quad (3)$$

with ε_0 the real roots of $y_0 = F_x(\varepsilon)$.

Provided the degree of F_x is small enough, finding its roots is a relatively simple task. The probability density of the system's output can then be known, since the pdf of the uncertain parameter is supposedly known too (fig. 1).

This scheme can be extended to the case of several uncertain parameters: the multidimensional equivalent of (3) involves calculating convolution-type integrals of the uncertain parameters pdfs on domains depending on the degree of F_x [5]. This is only practical for $n=1$, which restricts the usefulness of this method to simple low-dimensional cases. Also, both methods are restricted to the case of small variations of the uncertain parameter(s), because the nth-degree Taylor expansion is only valid within a limited range. This drawback can be surmounted by using the following scheme, which is valid for arbitrarily large variations of the uncertain parameters.

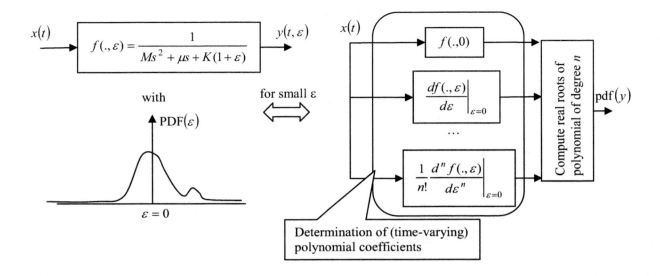

fig. 1: the implementation of the perturbation method requires the simulation of 1+n systems. Each of these systems corresponds to a derivative of the original system near the central value $\varepsilon=0$

2.2 Piecewise perturbation method

The piecewise perturbation method, whose algorithm is presented in fig. 2, consists in dividing the domain of the uncertain parameter in P sub-domains (which need not be the same size) with central values $\{\varepsilon_1,...,\varepsilon_P\}$ and applying to each of these sub-domains the pdf-based perturbation method described in the previous part.

For each sub-domain, the system and its derivatives are simulated using the corresponding central value and a truncated pdf that is zero outside the sub-domain: the P resulting partial pdfs can then be summed to yield the total pdf of the system's output.

It is clear that no matter how large the variation of the uncertain parameter, the sub-domain decomposition can be made fine enough to make the Taylor approximation valid. Another advantage of the method is that the size of the sub-domains can be adapted locally to account for discontinuities of the pdf of ε or for possible critical values of ε for which a bifurcation occurs in the system, as in the following case.

3 IMPLEMENTATION IN THE CASE OF A BEAM OF UNCERTAIN STIFFNESS

Let us consider a beam-mass system undergoing electrostatic actuation. Using modal analysis, it is a simple matter to deduce the position of the mass from the Bernouilli beam equation. Introducing mass M, damping coefficient D, stiffness K, voltage V, gap g_0, it is possible to approximate position y_ε as:

$$M\ddot{y}_\varepsilon + D\dot{y}_\varepsilon + K(1+\varepsilon)y_\varepsilon - \frac{\alpha V^2}{(g_0-y_\varepsilon)^2} = 0 \quad (4)$$

In (4), ε is a random variable with mean 0 which is not necessarily Gaussian. Applying a first-order PPM, we write

$$y_\varepsilon = y_{e_k} + (\varepsilon - \varepsilon_k)y_{e_k}^{(1)} \quad (5)$$

Using a Taylor expansion of the non-linear term in (4), it is possible to show that, in the neighbourhood of ε_k, (4) is equivalent to the following system:

$$\begin{cases} M\ddot{y}_{\varepsilon_k} + D\dot{y}_{\varepsilon_k} + K(1+\varepsilon_k)y_{\varepsilon_k} = \dfrac{\alpha V^2}{(g_0-y_{\varepsilon_k})^2} \\[2ex] M\ddot{y}_{\varepsilon_k}^{(1)} + D\dot{y}_{\varepsilon_k}^{(1)} + K(1+\varepsilon_k)y_{\varepsilon_k}^{(1)} = -Ky_{\varepsilon_k} + 2\dfrac{\alpha V^2}{(g_0-y_{\varepsilon_k})^2}\dfrac{y_{\varepsilon_k}^{(1)}}{g_0-y_{\varepsilon_k}} \end{cases}$$
(6)

Each of these equations must then be solved in turn for the different sub-domains of ε. The corresponding partial pdfs can then be calculated with (3) and (5) and then summed. It must be noted that there exists a certain value of ε below which the system is pulled-in.

4 RESULTS

To illustrate our method, we consider the case of a system with the following parameters: K=15 kg.s^{-2}, D=4.10^{-4} kg.s^{-1}, M=2.10^{-7} kg, V=2.5 V, g_0=2,5.10^{-6} m and α=5.10^{-18}.

For the sake of simplicity, we consider a Gaussian density for ε, with standard deviation 0,1. For the PPM, we split the interval [-0,4 0,4] into 40 sub-domains and obtain the results of fig. 3.

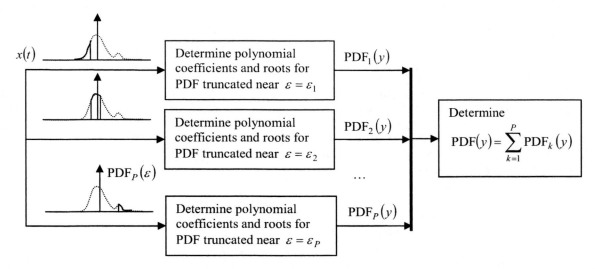

fig. 2 : schematic of the piecewise perturbation method. The total number of simulations is equal to the number of sub-domains (P) times the order of the Taylor expansion plus one ($N+1$).

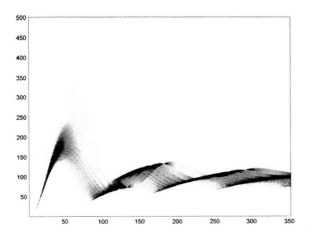

fig. 3: pdf(y) obtained with first-order PPM.

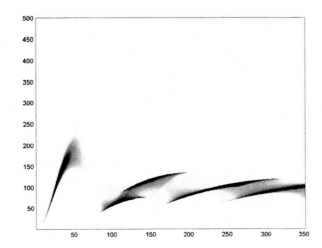

fig. 4: pdf(y) obtained with third-order PPM.

One can see clearly that the beam-mass system is pulled-in for certain values of ε, near which the sub-domain decomposition could be refined. An example of a higher-order PPM calculation is presented in fig. 4, with the same number of sub-domains as in fig. 3. We present in fig. 5 and 6 the comparison of these results with a Monte-Carlo simulation of the system with 10^5 realizations of ε.

These figures show that the PPM gives roughly the same results as the Monte-Carlo method for a much lower computational cost, even for low PPM orders.

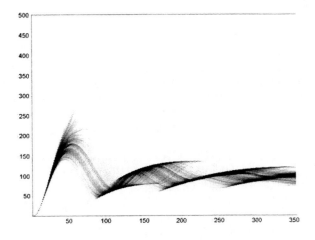

fig. 5: pdf(y) obtained with Monte-Carlo method.

Although the Monte-Carlo method is much more general and does not necessitate a Taylor expansion of the system with respect to the uncertain parameters, it is clear that the PPM also has advantages, such as the possibility to focus (i.e. to reduce the sub-domain size) locally in order to account for bifurcations in the solution and it has a good accuracy even for low expansion orders. It also makes it possible to account for large variations of the uncertain parameters, as opposed to classical perturbation methods. Finally, even though this approach is less immediate than the Monte-Carlo approach, it is much less costly: this should be even truer as the number of uncertain parameters increases.

5 CONCLUSION

We have introduced in this paper a novel approach to the simulation of uncertainties: this approach is based solely on deterministic simulations of the perturbed system and of its derivatives with respect to the uncertain parameter. As opposed to other perturbation methods, it is valid for arbitrarily large variations of the uncertain parameter. We have shown how this method applies in the case of an electrostatically actuated beam with uncertain stiffness. The results were compared with those obtained with a Monte-Carlo simulation of the system. This method is currently being extended to the case of many uncertain parameters.

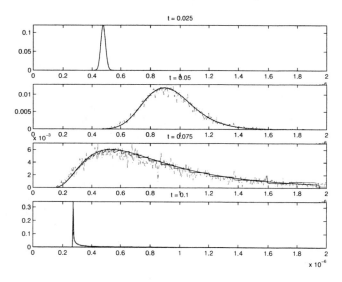

fig. 6: comparison of first-order PPM (40 sub-domains, blue line), third-order PPM (40 sub-domains, black line) and Monte-Carlo method (10^5 realizations, red line) at four different moments.

REFERENCES

[1] E. Vanmarcke, "Random fields, analysis and synthesis", MIT Press, Cambridge, 1984
[2] W. K. Liu, T. Belytschko, A. Mani, "Random field finite elements", International Journal of Numerical Methods in Engineering, vol. 23, pp. 1831-1845, 1986
[3] F. S. Wong, "First-order, second-moment methods", Computers and Structures, vol. 20, pp. 779-791, 1985
[4] H. Risken, "The Fokker-Planck equation: methods of solution and application", Springer-Verlag, Berlin, 1984
[5] P. Z. Peebles, "Probability, random variables and random signal principles", McGraw-Hill, New York, 1993

Dynamic Simulations of a Novel RF MEMS Switch

Mohammad I. Younis, Eihab M. Abdel-Rahman, and Ali H. Nayfeh

Department of Engineering Science and Mechanics, MC 0219, Virginia Polytechnic
Institute and State University, Blacksburg, Virginia 24061, USA, anayfeh@vt.edu

ABSTRACT

We present a dynamic analysis of a novel RF MEMS
switch utilizing the dynamic pull-in phenomenon. We
study this phenomenon and present guidelines about its
mechanism. We propose to utilize this phenomenon to
design a novel RF MEMS switch, which can be actuated by a voltage load as low as 40% of the traditionally
used static pull-in voltage. The switch is actuated using a combined DC and AC loading. The AC loading
can be tuned by altering its amplitude and/or frequency
to reach the pull-in instability with the lowest driving
voltage and fastest response speed. The new actuation
method can solve a major problem in the design of RF
MEMS switches, which is the high deriving voltage requirement.

Keywords: Dynamic pull-in, RF switches, microbeams,
electric actuation.

1 INTRODUCTION

The emerging technology of microelectromechanical
systems (MEMS) has enabled the design and fabrication
of an important class of devices, RF MEMS switches,
which promises breakthrough advances in telecomuni-
cations and radar systems. RF MEMS switches overcome the limitations of existing conventional switches
and present many attractive features, such as low-power
consumption and high isolation. However, a major drawback of these devices is the requirement of high driving
voltages [1]. It is highly desirable to bring the actuation voltage to a level compatible or close to that of the
device circuits. Unfortunately, the state of the art of
RF MEMS switches has not achieved this requirement,
which forms a barrier toward the development of this
technology.

In this work, we propose a novel RF MEMS switch
that can be actuated by a voltage load as low as 40% of
the static pull-in voltage. The new actuation method is
based on the dynamic pull-in phenomenon, in which the
switch is brought to pull-in by a voltage lower than the
static pull-in voltage. The dynamic pull-in phenomenon
has been previously reported and analyzed for switches
actuated by a step voltage [2,3] and various ramping
rates [2]. Both studies [2,3] indicate that the dynamic

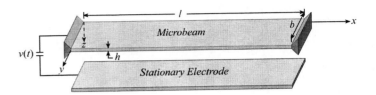

Figure 1: A schematic of an electrically actuated microbeam.

pull-in voltage can be as low as 91% of the static pull-
in voltage. In the presence of squeeze-film damping,
the dynamic pull-in voltage is shown to approach the
static pull-in voltage [3]. Here, we propose to actuate
the switch using a combined DC and AC loading. The
AC loading can be tuned by altering its magnitude or
frequency to reach the pull-in instability with the lowest
driving voltage and fastest response speed.

In [4-7], we presented a model, which predicts the
static pull-in phenomenon. In [8], we utilized perturbation methods to predict the dynamic behavior of resonators undergoing small motions near the equilibria.
In [6,7], we developed a reduced-order model to simulate the static and dynamic behaviors of resonators and
switches undergoing small or large motions. In this paper, we use the reduced-order model in [6,7] to simulate the dynamic behavior of the proposed RF MEMS
switch. We utilize a shooting technique [9] and long-
time integrations of the equations of motion to predict
periodic motions. This approach can be applied to a
wide range of loadings and initial conditions, and hence
it can be used to study the 'global' dynamics of switches
(unlike the model presented in [8], which is based on
perturbation methods and applicable for small motions
near the equilibria). We use the global approach to
demonstrate the new actuation method.

2 RESULTS

We consider a clamped-clamped microbeam, Figure
1, actuated by an electric load $v(t) = V_{DC} + V_{AC} \cos(\Omega t)$,
where V_{DC} is the DC polarization voltage, V_{AC} is the
amplitude of the applied AC voltage, and Ω is the excitation frequency. We utilize the reduced-order model
in [6,7] to calculate the equilibria of the microbeam

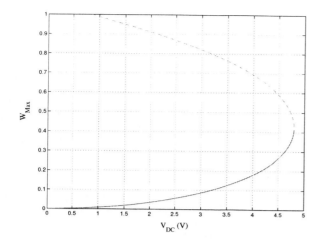

Figure 2: Equilibria of an electrostatically actuated microbeam.

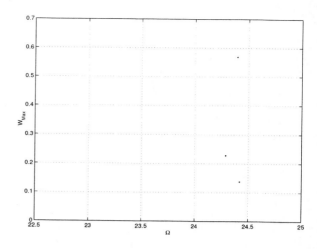

Figure 3: Frequency-response curves showing the onset of the dynamic pull-in.

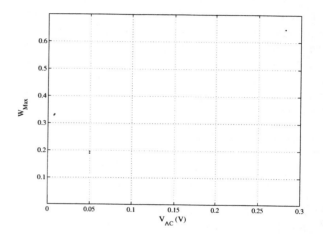

Figure 4: Force-response curves showing the onset of the dynamic pull-in.

under a constant DC loading. Details of the calculations are presented in [6,7]. We study a microbeam of $\ell = 510\mu m$, $h = 1.5\mu m$, $b = 100\mu m$, a capacitor gap width $d = 1.18\mu m$, and subject to a nondimensional axial load of $N = 8.7$. The parameter N relates to the dimensional axial force $\hat{N}$ by $N = \frac{\hat{N}\ell^2}{EI}$, where I is the moment of inertia of the cross section and E is Young's modulus. Figure 2 shows a lower stable branch (solid line) and an upper unstable branch (dashed line), both collide in a saddle-node bifurcation at the static pull-in voltage, which is $V_{DC} \approx 4.8V$. Hence, the static analysis shows that MEMS devices should be designed to operate below this value to ensure stability.

For all the following results, we assume the microbeam to be actuated by a DC loading $V_{DC} = 2V$ and subject to a viscous damping with a quality factor $Q = 1000$. Figures 3 and 4 show variation of the mid-point deflection W_{Max} of the microbeam with Ω (Figure 3) and V_{AC} (Figure 4): $V_{AC} = 0.1V$ in Figure 3 and $\Omega = 24.15$ in Figure 4. The dynamic pull-in instability observed in both figures is characterized by a slope approaching infinity. In both cases, a Floquet multiplier approaches unity near pull-in. However, the magnitude of the Floquet multiplier can be used as a pull-in criterion for force sweep cases, like that in Figure 4, only because Floquet multiplier always approaches unity around the maximum point on the curve of a frequency sweep regardless of whether the point corresponds to pull-in or a cyclic-fold bifurcation [8].

Figures 3 and 4 show that the dynamic pull-in instability occurs much earlier than the static pull-in limit. For this case, the static pull-in limit is at $V_{DC} \approx 4.8V$ while the total voltage loads where the dynamic pull-in occurs are less than $2.25V$ in both cases. This significant results suggest a novel actuation method, in which a switch is brought to pull-in at a voltage level as low as

40% of the static pull-in voltage. This is a very promising result in the search to reduce the actuation level of RF MEMS switches. The result also illustrates the importance of designing RF MEMS filters and resonant sensors based on a dynamic analysis to avoid possible failure of these devices.

Figure 5 shows phase portraits (plots of velocity versus displacement) for selected points in Figure 4. The solid lines correspond to stable periodic orbits and the dashed lines correspond to unstable periodic orbits. As V_{AC} increases, the size of the orbit increases until $V_{AC} = 0.2891V$ where it becomes large enough to trigger pull-in.

Figures 6 and 7 show a time history evolution and a phase portrait demonstrating the onset of the dynamic pull-in for the microbeam at $V_{AC} = 0.08V$ and $\Omega = 24.4$. The time is nondimensionalized with respect to $T =$

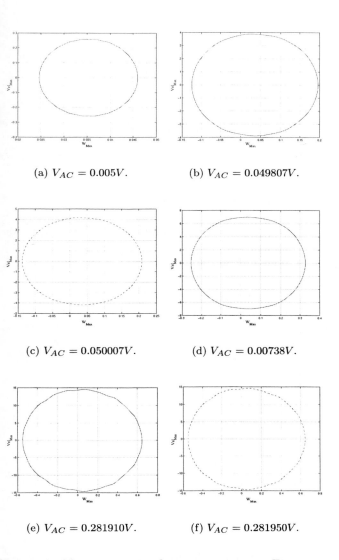

(a) $V_{AC} = 0.005V$.

(b) $V_{AC} = 0.049807V$.

(c) $V_{AC} = 0.050007V$.

(d) $V_{AC} = 0.00738V$.

(e) $V_{AC} = 0.281910V$.

(f) $V_{AC} = 0.281950V$.

Figure 5: Phase portraits for some points on Figure 4. The dashed lines are unstable periodic orbits.

$\sqrt{\frac{\rho b h \ell^4}{EI}}$, where ρ is the material density. The figures are generated from a long-time integration of the reduced-order model equations in [6,7]. We note that, for this case, the dynamic pull-in occurs when the orbit collides with the saddle (the unstable equilibrium solution) and its stable manifold, which is at $W_{Max} \approx 0.91$ (compare the location of the saddle in Figures 2, 6, and 7). Figures 8 and 9 are generated from the same data of Figures 6 and 7 except that the sign of one initial condition is changed from positive to negative. We note that the motion is stable, which indicates a fractal dynamical behavior that is sensitive to initial conditions [9].

In light of the above results, we make the following remarks about the dynamic pull-in phenomenon. In the absence of the AC forcing and damping, a homoclinic orbit [9], which starts and ends at the saddle, encircles the

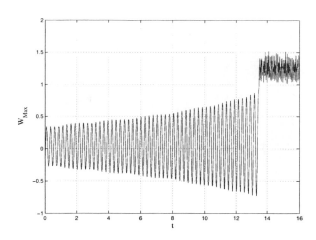

Figure 6: Time history showing the onset of dynamic pull-in.

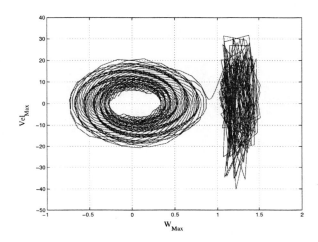

Figure 7: Phase portrait showing the onset of dynamic pull-in.

stable fixed point (center). This orbit is composed of a stable manifold and an unstable manifold that intersect transversely at the saddle. Any initial conditions inside the closed orbit result in a stable oscillatory motion and any initial conditions outside it result in an unstable motion. In the presence of damping, the center becomes a stable focus, the saddle remains a saddle, and the homoclinic orbit is destroyed. Here, the stable and unstable manifolds do not intersect. If the system is excited by an AC force with initial conditions near the stable focus, the motion will be a stable periodic motion. As the amplitude of excitation increases, the response amplitude increases and the stable manifold approaches the unstable manifold. Eventually, both manifolds intersect each other transversally infinitely many times, resulting in a complex dynamic behavior called homoclinic tangles [9]. One implication of this complex behavior is the sensitivity to initial conditions or the unpredictability

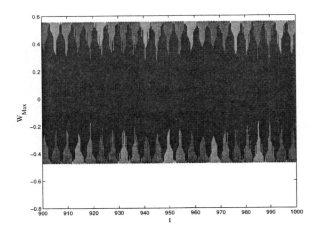

Figure 8: Time history for the same parameters of Figures 6 and 7 and different initial conditions.

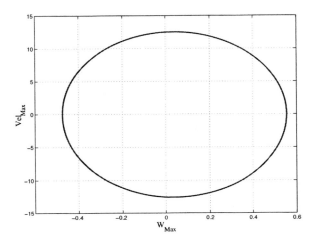

Figure 9: Phase portrait for the same parameters of Figures 6 and 7 and different initial conditions.

of motion. Increasing the amplitude of excitation further results in an erosion of the basin of attraction of bounded motions; and the possibility of finding a set of initial conditions that leads to a stable motion decreases. It is worth noting that the dynamic pull-in instability in this case is to a great extent similar to the phenomenon of capsizing of ships [9,10].

3 SUMMARY AND CONCLUSIONS

We proposed a novel actuation method for RF MEMS switches and presented a dynamic analysis of this method. We studied the dynamic pull-in instability and showed the danger inherent in device designs based on static analysis only. On the other hand, we showed that this phenomenon can be used to advantage to solve a challenging dilemma in the design of RF MEMS switches, namely the high deriving voltage requirement. Fur-

thermore, it holds the promise of realizing switches of faster response by tuning the AC excitation amplitude to higher magnitudes than the pull-in limit and the frequency of excitation to be as close as possible to the resonance frequency. Experimental work is planned to investigate the feasibility of the new actuation method.

REFERENCES

[1] Tilmans, H. A., Raedt, W. D., and Beyne, E., "MEMS for wireless communications: 'from RF-MEMS components to RF-MEMS-Sip', " J. Micromech. Microeng., Vol. 13, 2003, pp. 139–163.

[2] Ananthasuresh, G. K., Gupta, R. K., and Senturia, S. D., "An approach to macromodeling of MEMS for nonlinear dynamic simulation," in Proc. ASME International Conference of Mechanical Engineering Congress and Exposition (MEMS), Atlanta, GA, 1996, pp. 401–407.

[3] Krylov, S. and Maimon, R., "Pull-in dynamics of an elastic beam actuated by distributed electrostatic force," in Proc. 19th Biennial Conference in Mechanical Vibration and Noise (VIB), Chicago, IL, 2003, paper DETC2003/VIB-48518.

[4] Younis, M. I., Abdel-Rahman, E. M., and Nayfeh, A. H., "Static and dynamic behavior of an electrically excited resonant microbeam," in Proc. 43rd AIAA Structures, Structural Dynamics, and Materials Conference, Denver, Colorado, 2002, AIAA Paper 2002-1305.

[5] Abdel-Rahman, E. M., Younis, M. I., and Nayfeh, A. H., "Characterization of the mechanical behavior of an electrically actuated microbeam," J. Micromech. Microeng., Vol. 12, 2002, pp. 795–766.

[6] Abdel-Rahman, E. M., Younis, M. I., and Nayfeh, A. H., "A nonlinear reduced-order model for electrostatic MEMS," in Proc. 19th Biennial Conference in Mechanical Vibration and Noise (VIB), Chicago, IL, 2003 (a), paper DETC2003/VIB-48517.

[7] Younis, M. I., Abdel-Rahman, E. M., and Nayfeh, A. H., "A Reduced-order model for electrically actuated microbeam-based MEMS," J. Microelectromech. Sys., Vol. 12, 2003, pp. 672–680.

[8] Younis, M. I. and Nayfeh, A. H., "A study of the nonlinear response of a resonant microbeam to an electric actuation," Nonlinear Dynamics, Vol. 31, 2003, pp. 91–117.

[9] Nayfeh, A. H. and Balachandran, B., Applied Nonlinear Dynamics, Wiley, New York, 1995.

[10] Nayfeh, A. H. and Sanchez, N. E., "Chaos and dynamic instability in the rolling motion of ships," in Proc. 17th Symposium on Naval Hydrodynamics, Hague, Netherlands, 1988, pp. 617–631.

Using Topology Derived Masks to Facilitate 3D Design

R. Schiek* and R. Schmidt**

Sandia National Laboratory, Computational Sciences Department,
P.O. Box 5800, Albuquerque, New Mexico, 87122-0316, U.S.A.
*rlschie@sandia.gov, **rcschmi@sandia.gov

ABSTRACT

To accelerate MEMS design for surface microma-chining applications, an algorithm and associated design tool have been created which translates designers' 3D-models into 2D lithographic production masks. Typi-cally, designing a surface micromachined, MEMS device requires the creation of a two-dimensional mask set de-scribing how layers of material are used to construct the three-dimensional object. Mask sets are specific to a fixed production process and are effectively the tool-ing required to manufacture a device. This design tool was developed and implemented such that when given a three-dimensional object it can infer from the object's topology the two-dimensional masks needed to produce that object with surface micromachining. The masks produced by this design tool can be generic, process in-dependent masks or, if given process constraints, specific for a target process allowing 3D designs to be carried across multiple processes.

Keywords: topology, mask, mems, design, optimiza-tion

1 Design Issues

Designing a device for production by silicon micro-machining is very different from macro-scale mechanical design. In the macro-scale it is often sufficient for a de-signer to create a 3D model of their device, which a de-sign program then translates into the tool paths needed for production. For a silicon micromachined device how-ever, the designer must create a set of process specific, lithographic masks needed to fabricate the device. Cre-ating such masks is similar to requiring the macro-scale designer to design the tools needed to fabricate their product as well as the product itself. Because masks are dependent on the process in which they are used and can have complex dependency interactions within a production system, creation of the masks is a significant challenge to innovative device design and the manufac-ture of a device on multiple processes. Thus it is nec-essary and desirable to develop a tool for translating a designer's 3D model of a product directly into the masks needed to produce their product.

Earlier efforts on this problem have leveraged ex-isting technology in process simulators, i.e. programs which when supplied with a mask set for a given pro-cess can simulate fabrication from those masks. Typi-cally, this approach uses a trial mask set to produce a 3D object that is then compared to the desired object. Differences between the two objects are used to alter the trial mask set and then the process is repeated until a mask set is found which correctly produces the desired part. When coupled with a sophisticated optimization scheme, this approach works well for anisotropic etching processes. [1] Being computationally intensive however, optimization trial masks through a process simulator has yet to produce masks for complex, multi-layer sur-face micromachined devices. Another approach starts from a 3D model that is annotated with data which de-scribes when in the process each section of it will be made and from each annotated section a mask is de-rived. [2] More recently progress has been made on a geometric approach where a 3D model is interrogated for features that can be made via surface micromachining, and a mask set is derived for these features. [3] While promising, these techniques cannot produce masks for specific processes nor handle isotropic etching processes such as wet etches.

2 A Topology Based Approach

Surface micromachining builds a MEMS device with the successive deposition and controlled etching of ma-terials on a silicon substrate. Motivation for an alter-native approach can be found in consideration of the process steps typical of surface micromachining. For ex-ample, to produce the simple part shown in figure 1a, two layers of deposited material and two masks are used. First, a layer of silicon dioxide is deposited (silicon diox-ide is commonly used as a supporting material since it can easily be removed at the end of the process) and a mask is used to define the region of silicon dioxide to be retained as shown in figure 1b. Unmasked silicon dioxide is etched away resulting in the structure shown in figure 1c once the mask is removed. Next, a layer of polysilicon is deposited. As before, a mask is used to define the region of polysilicon to retain during the next etch; see figure 1d. Etching the extraneous polysil-

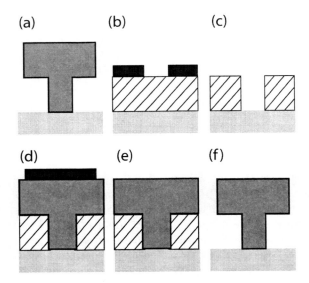

Figure 1: Surface Micromachining of a simple part. (a) Target polysilicon part on a silicon substrate. (b) A sacrificial oxide layer is deposited and a mask is placed. (c) The oxide is etched and the mask removed. (d) Polysilicon is deposited and a mask is placed. (e) Extraneous polysilicon is removed and the mask is removed. (f) The sacrificial silicon oxide is removed.

icon and removing the mask produces the part shown in figure 1e. Finally, removal of the sacrificial silicon via chemical dissolution reveals the final, desired part as depicted in figure 1f.

Considering the production of this simple device, one can identify two, horizontal cross-sections in the 3D object which directly correlate to the masks used to manufacture the device. First, the narrow cross section of the post relates to the mask used to etch the sacrificial oxide. Second, the cross section of the larger top directly correlates to the mask used to produce the top section. Therefore, if important cross sections can be identified in a 3D model, then these cross sections can be used to create masks to manufacture the device.

2.1 Cross Sections

Considering the example of figure 1, the horizontal cross sections of a device can be used to identify the masks. Given a body, let z represent the scaler distance from a reference ground plane and let $C(z)$ denote the cross section of a body at the height z. The function $C(z)$ is not necessarily a continuous function of height as a part with exactly vertical sides will create discontinuities in $C(z)$ when the cross section changes.

While a cross section itself is infinitely thin, one can identify a range of heights within which a given cross section is constant. Thus, if one defines $\mathcal{C}$ as a constant

cross section, one can then write:

$$P_i = C(z_i) : C(z) = \mathcal{C} \ \forall \ z \in [z_i, z_{i+1}) \qquad (1)$$

In defining P_i, one has implicitly subdivided the z domain into intervals within which a given cross section is constant. Since $C(z)$ may not be continuous, the range of heights within which $C(z)$ is constant cannot easily be defined as closed. Thus a range of acceptable z values can be written as either $[z_i, z_{i+1})$ or $(z_i, z_{i+1}]$ and the sequence z_i may be increasing or decreasing. In this analysis, it is assumed that the range of allowed heights is traversed from top to bottom implying $z_i > z_{i+1}$ and that any discontinuities in $C(z)$ are placed at the lower height yielding the closure defined in equation 1. Now, given P_i a set can defined as follows:

$$U = \{P_1, P_2, ...P_N\} \qquad (2)$$

This defines U as the sequential set of all unique cross sections for a given body. Note that unique here only implies that P_i is not equivalent to either P_{i-1} or P_{i+1}, i.e. unique relative to ones neighbors. With a notation on hand to describe a body's cross sections and where they arise, attention next will be directed to organizing the cross sections into a useful topology tree.

2.2 Topology Graph Analysis

Considering again the example of figure 1a, one can find two unique cross sections for such a part yielding $U = \{P_1, P_2\}$. Since neither cross section is composed of multiple subcomponents, one can see that P_1 is connected to P_2. This topological relationship can be represented by $P_1 \rightarrow P_2$. In general, a given cross section may contain multiple subcomponents, islands or lumps. To account for this one can expand the definition of P_i as:

$$P_i = L_{ij} : j = 1, J \qquad (3)$$

where J is the number of subcomponents or lumps of cross section P_i. Using the notation L_{ij} to denote a lump of a given cross section, a graph or tree can be constructed relating the connectivity of the lumps of the various cross sections. For example, the following tree could relate three cross sections where the middle cross section has two lumps:

$$
\begin{array}{ccc}
 & L_{11} & \\
\swarrow & & \searrow \\
L_{21} & & L_{22} \\
 & & \downarrow \\
 & & L_{31}
\end{array}
\qquad (4)
$$

Next, nodes within the tree are categorized. For each tree node, its surface area is calculated and then compared to child and parent nodes to determine if the current node is a local maxima or local minima in cross sectional area. Local minima in particular are important as they typically indicate where one deposition layer of material joins a material deposited at a later time. If an extrema in cross sectional area occurs at a head or terminal node then special process masks may be required. No masking decisions are made at this stage; rather these nodes are just marked so that they can receive attention during the mask reconciliation stage.

Once the nodes are categorized, the tree is traversed to find the cross sections required to build the device. It is assumed at this stage that the surface micromachining process proceeds by depositing a layer of material, using a mask to etch away unwanted material, removing the mask and then repeating this process. This is a simplification that real processes do not necessarily follow which can be accounted for at a later stage as covered in the next section. While traversing each branch of the tree, first the locations of local minima nodes are recorded and between any pair of local minima on a given branch, a local maxima is sought. Local maxima nodes are typed as *poly* masks as they typically represent how a structural layer like a polysilicon layer was masked before etching; *poly* is purely a name of convenience as this method would work for any material. Similarly, local minima nodes are typed as *sac-ox* masks as they typically correspond to masks used in etching sacrificial layers like the sacrificial oxide layers, SiO_2. Again, this nomenclature is for convenience. Terminating nodes that end in local extrema are typed as *dimples* if they are local minima or *undercuts* if they are local maxima. These two mask types are almost equivalent to *sac-ox* and *poly* masks respectively, however their use in a fabrication process is different from *sac-ox* and *poly* masks so they are singled out at this stage.

The masks thus far identified have an additional attribute associated with them. Each masks has a *thickness* which corresponds to the difference in height between the node where the mask was identified and either the next extrema on a child branch or the end of the current branch. When attempting to match or reconcile the masks found from the topology analysis with masks required for given process, this thickness is used to determine if a given process step is compatible with a given mask. Next, these masks will be converted to production masks.

2.3 Creating Production Masks

The candidate masks found in the previous section apply only to an idealized version of surface micromachining as was assumed earlier. If one were only given a model of a part, and the part's designer did not have a specific production process in mind for that part, then the candidate masks together with their thicknesses and material types would define a new, idealized production scheme for this device.

However, if the designer of this part had a specific production process in mind then the candidate masks must be reconciled with process mask specifications to yield valid masks as follows. First the process specification is searched for the materials and material thicknesses it uses, masks names and their locations in the process stream. Next, the target process is searched for places where the assumed deposition-mask-etch process order does not occur. With these parameters known, the candidate masks can be searched for masks that match the function of those used in a given process step. If a candidate mask corresponds to a layer which is thicker than layers in the target process, then that mask can be duplicated and used to produce two laminated layers in the actual process. If all of the candidate masks cannot be fit to the target process then the designer can be informed of what feature is blocking this fit.

3 Method

The analysis described in the previous section forms the basis of the following algorithm, which successfully infers 2D mask sets from complex 3D models. Aspects of the algorithm that have not yet been discussed concern largely logistical points. For example, a given 3D model will have many non-intersecting bodies. It is efficient to work on one body at a time, so initially the model is divided into its non-intersecting components. Compensation for this division occurs later when the mask sets are summed. This summation is straightforward as the non-intersecting bodies will have non-overlapping masks. Finally, a simplification of the topology tree is conducted where redundant nodes are joined, a process where by nodes that topologically connect the same nodes are combined to one node. Given a 3D model, the algorithm is:

1. Locate independent bodies.

   (a) find all non-intersecting bodies
   (b) separate bodies made of different materials
   (c) separate bodies only connected via. ground

2. For each body

   (a) Generate a topology tree.
   (b) Categorize the nodes of the tree.
   (c) Combine redundant nodes.
   (d) Locate deposition boundaries.

3. For each deposition domain

   (a) Locate masks
   (b) Save masks in candidate mask set

4. Sum all candidate masks

5. Reconcile masks with the target process.

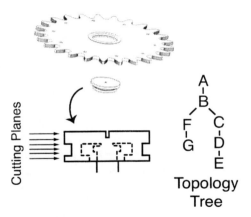

Figure 2: Locating the unique cross sections and building the topology tree for a hub which holds a gear in place.

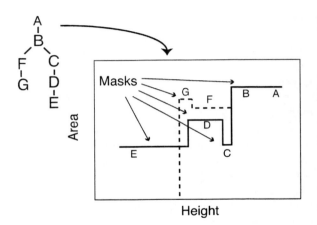

Figure 3: Analyzing the topology tree allows one to locate masks.

It is significant to note that specific process details do not enter the algorithm until the final step. Allowing most of the algorithm to operate independently of process details keeps the algorithm flexible to process changes.

3.1 Implementation

The algorithm was implemented in a C++ program called faethm using the ACIS geometric modeling library version 11 (http://www.spatial.com) for import and manipulation of the 3D models. Models were both manually generated and provided by Sandias SUMMiT V 3D Modeler [4].

3.2 Example

As an illustrative example of this method, figure 2 depicts a hub which is used to hold a gear in place. A single hub is an example of an independent, non-intersecting body found in step one of the method listed previously. The hub is cut into horizontal cross sections and the unique cross sections are assembled into a topology tree. Note that the hub's topology tree is branched and non-symmetric as the center post has a different topology than the outer ring of the hub. Figure 3 demonstrates the analysis of the topology tree. After the area of each topology node is calculated, an area versus height graph is created where the vertical lines connect the nodes to indicate topological relationships. Since the hub's topology tree is branched, the branch for the outer part of the hub is drawn with a dashed line. Using the area data and the topological connectivity of the nodes, candidate masks can be selected. Reconciliation of the masks with process constraints produces a set of production masks for the hub.

4 Conclusion

The algorithm presented here and coded in the *faethm* program is capable of generating accurate mask sets for complex 3D devices. By focusing on a models topology first, this work can identify masks for anisotropic and isotropic (dry and wet) etching processes.

5 Acknowledgments

Sandia is a multiprogram laboratory operated by Sandia Corporation, a Lockheed Martin Company, for the United States Department of Energy, National Nuclear Security Administration under Contract DE-AC04-94AL85000.

REFERENCES

[1] L. Ma & E. Antonsson, "Robust mask-layout and process synthesis" *J. of Microelectromechanical Systems* vol. 12, no. 5, pp. 728–729 (2003)

[2] V. Venkataraman, R. Sarma & S. Ananthasuresh, "Part to Art: Basis for a Systematic Geometric Design Tool for Surface Micromachined MEMS" *Proceedings of the ASME Design Engineering Technical Conferences 2000* pp. 1–14 (2000)

[3] S. Cho, K. Lee & T. Kim, "Development of a geometry-based process planning system for surface micromachining" *Int. J. of Production Research* vol. 40, no. 5, pp. 1275–1293 (2002)

[4] C. Jorgensen & V. Yarberry, "A 3D Geometry Modeler for the SUMMiT V MEMS Designer" *Modeling and Simulation of Microsystems 2001* pp. 594–597 (2001)

Modeling, Fabrication and Experiment of a Novel Lateral MEMS IF/RF Filter

Mehrnaz Motiee[*], Amir Khajepour[*] and Raafat R. Mansour[**]

University of Waterloo, Waterloo, Ontario, Canada
[*]Mechanical Eng. Dept., mehrnaz@mems.uwaterloo.ca, akhajepour@uwaterloo.ca
[**]Electrical and Computer Eng. Dept., rmansour@maxwell.uwaterloo.ca

ABSTRACT

MEMS based mechanical resonators and filters have shown promising characteristics in achieving high Q values and good stability. This paper introduces a novel V-shape coupling element that is used to mechanically couple two clamped-clamped MEMS resonators laterally. The stiffness of the proposed V-shape coupling element is adjustable via changing the length of the V sidelines and/or the V conjunction angle to flatten the filter passband. In previous literature, only a single resonator in lateral vibration is considered. No suggestions were given on the coupling of such type of resonators. In this work a V-shape coupling and two beam elements are used to construct a 2-pole bandpass filter operating in the intermediate frequency (IF) range. It is fabricated using the PolyMUMPs process. A lumped modeling approach is presented, which allows a fast and accurate modeling and optimizing of the structures. With the help of finite element analysis, the validity and accuracy of the lumped modeling is investigated. Filters have been fabricated and tested. Presented filters have center frequencies varying from 700 kHz to 1.7 MHz, quality factors of 300 to 1500 when tested in a non-vacuum chamber.

Keywords: RF MEMS, Bandpass Filters, MUMPs, Intermediate Frequency MEMS.

1 INTRODUCTION

Currently, the global interest in the wireless communication field demands the integration of RF transceivers on a single silicon chip [1,2]. Filters have a major role in the construction of high performance transceivers. Bandpass filters are one of the most important elements in frequency selective circuits. This paper introduces a novel V-shape coupling element that is used to mechanically couple two clamped-clamped MEMS resonators in lateral vibration, as shown in Figure 1. The V-shaped coupling element imparts the vibrational energy from the input resonator to the output resonator by compression and expansion. In [2], only a single resonator in lateral vibration is considered. No suggestions were given on the lateral vibrating coupling elements.

The stiffness of the proposed V-shape coupling element is adjustable via changing the length of the V sidelines and/or the V conjunction angle to flatten the filter passband.

Figure 2 shows a picture of the fabricated chip. It is fabricated using the PolyMUMPs process [3].

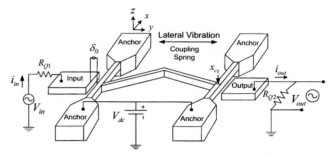

Figure 1: Schematic of a two-resonator filter with a typical measurement circuit.

Due to the multi-domain operation of these structures, a fast and accurate design tool, using general lumped modeling approach is needed. This model should include the effects of distributed and multi-domain nature of MEMS filters. In this paper, we present a lumped-element model for the modeling and design of such type of MEMS filters. The model is developed using direct electromechanical analogy [4]. It has been used for primary simulation and parametric optimization of the design. The accuracy of this model has been verified by comparing its result with finite element results. An excellence agreement between the lumped and the finite element analysis show the applicability of the lumped model to MEMS filter analysis and design. The comparison between simulation and test results under air damping is presented.

2 DESIGN CONCEPT AND OPERATION

Figure 1 presents a two-resonator filter, along with appropriate bias, excitation and input-output circuitries.

The conductive electrodes serve as capacitive transducer electrodes to induce resonator vibration in lateral, parallel with substrate, direction. A 2 μm resonator to electrode gap, δ_0, is used. To operate this filter, a *dc* voltage is applied to the suspended filter structure, which is called *dc*-bias voltage, V_{dc}, while an *ac* input voltage, V_{in}, is applied through resistor R_{Q1} to the input electrode, as shown in Fig. 1. The application of this input creates an electrostatic force in y direction and when the frequency of the input voltage reaches the first natural frequency of the resonator, the beam shows maximum oscillation amplitude.

As the frequency of the input voltage increases, the resonator beam intends to vibrate in its second mode of vibration. However, according to selected mode excitation theory [5], the straight-line electrode shape mainly excites the first mode and the first natural frequency becomes the dominant mode of vibration. This vibration energy is imparted to the output resonator via the coupling spring. Vibration of the output resonator creates a time varying capacitor between the conductive resonator and output electrode, which sources an output current that causes a variable output voltage, V_{out}.

3 MATHEMATICAL MODELING

Following is a lumped modeling approach, including the effects of the distributed and multi-domain nature of the problem. At each step, the results are compared with FEM simulation from CoventorWare [6] software to verify the model.

3.1 Resonator Design

Figure 3(a) shows the schematic of a single beam resonator in lateral vibration. The natural frequency of this clamped-clamped beam resonator is not dependent on its thickness, T. Assuming the first mode, frequency vs. dimensions can be plotted in a contour plot as shown in Fig. 4(a), which defines the center frequency of the resonator with respect to W and L [4].

We have selected two different beams for two different ranges of frequencies. The dimensions of the beams and their center frequencies are given in Table 1.

For the filter design, it is more convenient to define an equivalent lumped parameter mass-spring-damper mechanical circuit for this resonator. Knowing the frequency of the beam and its first mode shape, using the energy equations, its equivalent lumped mass, stiffness and damping parameters at any location x along the beam length are found [7].

3.2 Electro-mechanical operation

The electrostatic force between the resonator and electrode is nonlinear. Assuming small vibration amplitude, the force equation is linearized around its operation point with respect to voltage and displacement. As a result, an extra term with negative sign in the stiffness part arises, which is called electrical spring, denoted with K'. The total stiffness of the resonator, $K_e(x)$, is the summation of $K(x)$ and K', which decreases the total stiffness [7]. The center frequency is shifted duo to electrical stiffness. The shifted original and frequencies are listed in Table 2.

Also there is a transformer coefficient between force F and the voltage V, which illustrates the energy transformation between mechanical and electrical domain.

This is called electromechanical transformer ratio, denoted by η. The η strongly depends on gap spacing δ_0 [4].

With direct analogy, mechanical lumped parameters are transferred to electrical domain to form the electrical circuit shown in Fig. 5. Calculated lumped parameters for Beam1 and Beam2 are listed in Table 3.

3.3 Stability and Pull-in Voltage Analysis

The operation of the filter introduced above is based on applying a dc-bias voltage to the initial gap between the resonator and electrode to pull down the beam to its stable equilibrium position. The resonator is at stable equilibrium position if

$$K > K' \tag{1}$$

Eq. (1) can be solved for the proper dc-bias voltage. The effect of distributed electrical stiffness is considered by writing the electrical stiffness on a small differential element of the beam and integrating it over the electrode width [4]. Clearly, with increasing the voltage there will be a specific point at which the stability of the equilibrium is lost and structure collapses toward substrate. This is called *pull-in* voltage, denoted by V_{PI}. At the pull-in voltage:

$$K' = K \tag{2}$$

Figure 6 shows the pictures of fabricated Beam1 in neutral and pull-in positions. When the pull-in voltage is known, the deflection of the beam for Beam1 and Beam 2 can be calculated from, [7]:

$$y_{PI}(x) = \delta_0 - \int_{L_1}^{L_2} \frac{V_{PI}^2 \varepsilon_0 W}{2K(x)\delta(x)^2} dx \tag{3}$$

In Fig. 4(b), y_{PI} vs. V_{PI} for Beam1 and Beam 2 is plotted. The analytical solution of Eq. (2) is $V_{PI} = 315$ Volts for Beam1 and $V_{PI} = 115$ Volts for Beam2. These values can be compared with the pull-in graphs from FEM analysis in CoventorWare software [6] as superimposed in Fig. 4(b) and listed in Table 2.

3.4 Coupling Beam Design

If the resonators are identical, the passband of the resultant filter will be centered around these resonators's frequency. The coupling beam provides the needed stiffness to shift the resonator frequencies, creating two close resonance modes that form the passband. For a given filter, center frequency f and bandwidth B, the required coupling beam stiffness can be found from:

$$K_{cij} = K_e(c)k_{ij}\frac{B}{f} \tag{4}$$

where K_{cij} is the needed stiffness, c is the coupling location along the beam length, as can be seen in Fig. 1, k_{ij}

is the normalized coupling coefficient, and $K_e(c)$ is the resonator stiffness at location x_c [8]. Using the transmission line theory, the coupling beam is modeled as a T network of mechanical impedances [9]. In this case, half of the mass of the coupling beam is added to each resonator, modeled as series inductors, and its stiffness is modeled as a shunt capacitor as shown in Fig. 7.

$$Ls_a = Ls_b = \frac{1}{2}M_s = \rho T_s W_s L_s \qquad (5)$$

$$Cs_c = \frac{1}{Ks_c} = \frac{L_s^3}{6EI_s}$$

where L_s is the coupling beam length, I_s is the moment of inertia of the beam and other parameters are shown in Fig. 3(b). The thickness of the structure layer and the minimum width of the coupling beam are usually fixed by fabrication process limitations. From Fig. 3(b) we can see that

$$L_s = \frac{d_s}{2\cos\theta_s}, \quad R_s = L_s \sin\theta_s \qquad (6)$$

Applying the physical limits, equations (4), (5) and (6) are solved simultaneously to find the best values for coupling beam length, L_s, and angle θ_s [4].

The quality factor of the filter is proportional to the ratio of resonator stiffness at the coupling point and coupling beam stiffness. The quality factor of the filter increases when the stiffness of the resonator at coupling beam location increases [4]. By coupling the beams at locations closer to the anchor higher filter quality factor is achievable with same resonator and coupling dimensions.

We used the two beams in Table 1 to create two filters with different center frequencies and coupling locations. The resonator beams, lumped parameters and the correspondent coupling beam dimensions are listed in Tables 3 and 4.

4 TEST RESULTS

Adding the coupling beam network and low velocity coupling transformer to the lumped model of a single resonator in Fig. 5, the complete equivalent electrical circuit of filter is constructed as Fig. 7. The micromechanical filter should be terminated with the proper impedance values. Without a proper termination, the filter passband consists of distinct ripples. In Figs. 5 and 7, resistors R_{Q1} and R_{Q2} serve this function.

The V_{out}/V_{in} in frequency domain forms the filter response shape as presented in Figs. 8(a) and 9(a). These plots are based on operation in vacuum condition. The air damping effect is added to the system as a series resistor. Figs 8(b) and 9(b) show the plot of $dBm(V_{out})$ with respect to frequency in non-vacuum conditions, considering air damping. The fabricated Filters have been testes in air. Figs 8(c) and 9(c) show the $dBm(V_{out})$ from spectrum analyzer at output. Comparison between simulation and test result in air show a good agreement, which serves as proof of concept. The center frequencies are as predicted from analytical and FEM analysis.

5 CONCLUSIONS

In this paper, a novel V-shape coupling beam was presented which extends the possibility of coupling resonators in lateral direction. It has the advantage of being adjustable with changing V angle to flatten the passband.

A lumped model was derived for fast filter design. Experimental and simulation results were presented to corroborate the modeling and design concept.

REFERENCES

[1] C. T. -C. Nguyen, Dig. of Papers, Topical Meeting on Silicon Monolithic Integrated Circuits in RF Systems, Sept. 12-14, 2001, pp. 23-32.

[2] W.-T. Hsu, J. R. Clark, and C. T.-C. Nguyen, Tech. Dig., IEEE Int. Elect. Devices Meeting, San Francisco, 2000, pp. 493-496.

[3] David A. Koestar, et. al. "PolyMUMPs Design Handbook,", JDS Uniphase Company, Revision 10.

[4] M. Motiee," MEMS IF/RF Filters," M.A.Sc. thesis, University of Waterloo, 2003.

[5] Albert park, et. al, J. of Microelectromechanical Sys., Vol. 1, No. 4, Dec. 1992, pp. 179-186.

[6] www.coventor.com

[7] M. Motiee, et. al., ASME-IMECE, Washington DC, November 16-21, 2003, # 41568.

[8] Anatol I. Zverev, "Handbook of Filter Synthesis," John Wiley and Sons, Inc., New York, 1958

[9] Harrie A. C. Tilmans, J. Mircomech. Microeng., Vol. 6, 1996, pp 157-176, UK.

Beam Sample	L μm	W μm	T μm	f_0
Beam 1	100	2	3.5	1.67 MHz
Beam 2	150	2	3.5	723 KHz

Table 1: Dimension of resonators and center frequency.

Beam1				
	f_0	Shifted f	M (Kg)	V_{PI} (V)
Lumped	1. 71MHz	1.69 MHz	6.38E-13	315
FEM	1.70 MHz	1.67 MHz	6.26E-13	277
Beam2				
Lumped	724 KHz	723 KHz	9.58E-13	115
FEM	723 KHz	722 KHz	9.62E-13	95.7

Table 2. Comparison between analytical lumped model and FEM simulation result.

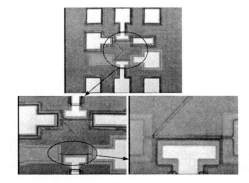

Figure 2: Photo of the fabricated filter sample, Filter1.

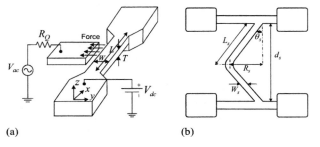

(a)   (b)

Figure 3: (a) Schematic of a single resonator clamped-clamped beam with proper electrode and bias, (b) coupling beam schematic and dimensions.

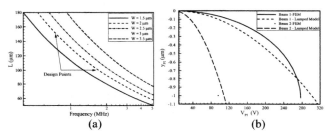

(a)   (b)

Figure 4: (a) Design curve for clamped-clamped beam in lateral vibration and selected design points, (b) Pull-in voltage graphs for Beam1 and Beam2.

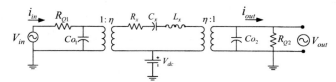

Figure 5: Equivalent lumped circuit for a single resonator.

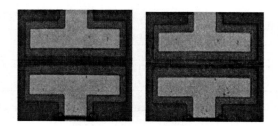

Figure 6: The clamped-clamped beam sample, deformed under pull-in voltage.

Parameter	Filter 1	Filter 2	Units
C_o	6.20E-16	8.85E-16	F
η	3.40E-08	5.07E-08	-
L_x	6.38E-13	9.58E-13	H
C_x	1.36E-02	4.86E-02	F
R_x	3.00E-09	2.50E-09	Ohms
η_c	8.4	8.4	-
$Ls_a=Ls_b$	1.85E-12	2.77E-12	F
Cs_c	6.50E-01	2.27E00	F
R_Q	8.5E+05	2.50E+05	KOhms

Table 3: Equivalent lumped parameters, refer to Fig. 7.

Filter	Beam	L_s	R_s	θ_s	W_s	x_c
Filter 1	Beam1	115	103	65.15	2	10
Filter 2	Beam1	172	155	64.15	2	15

Table 4: Coupling beam dimensions (μm)

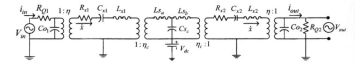

Figure 7: Equivalent lumped model.

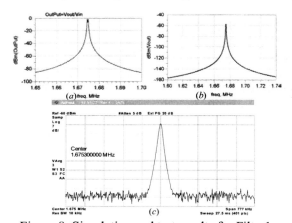

(a) freq. MHz   (b) freq. MHz

(c)

Figure 8: Simulation and test results for Filter1.

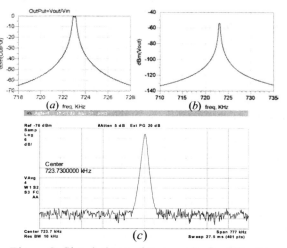

(a) freq. KHz   (b) freq. KHz

(c)

Figure 9: Simulation and test results for Filter2.

Characterization of an Electro-thermal Microactuator with Multi-lateral Motion in Plane

C. H. Pan[*], Y. K. Chen[**], C. L. Chang[**]

[*]National Chin-Yi Institute of Technology, Taiwan, ROC
Fax: 886-4-23930681, E-mail: pancs@chinyi.ncit.edu.tw
[**] National Yunlin University of Science & Technology, Taiwan, ROC

ABSTRACT

We present a new electro-thermal microactuator to have multi-lateral motion in plane by only varying voltage potentials at the contact pads. To extend the operating function, the larger operating range or multi-mode switch, relay and optic tweezer can be achieved. For focusing on the characterization of the microactuator, the finite element software ANSYS is used to perform the electro-thermo-mechanical behaviors of the microactuator to demonstrate the feasibility of the design principle. Design parameters (including structural dimensions, selective doping and thermal boundary conditions) significantly influencing the performance are studied. According to the analysis results, it is found that low voltages (0~7V) are required to achieve displacements in microns with the operating temperatures below 300℃. The optimal structure can be obtained by varying geometric dimensions and resistivity of the beams to meet the proper performance.

Keywords: microactuator, electro-thermal, multi-lateral motion

1 INTRODUCTION

Although there are different types of microactuators, most of them perform one-direction motion in plane (parallel to the substrate) or out-of-plane (vertical to the substrate). Only some of them may have bi-direction motion [1-5], which are achieved by mechanically coupling an array of microactuators, by assembling 3D microactuators or by utilizing bi-stable effect. Very few microactuators can perform multi-direction motion in plane or out-of-plane by itself. In 2001, Pan and Hsu [6] proposed an electro-thermally and laterally driven microactuator, which had bilateral motion in plane. It was symmetric combination of two basic microactuators presented by Pan and Hsu in 1997 [7]. In 1992, Guckel et al. [8] presented an electro-thermal microactuator which operating principle is based on unequal thermal expansion of the structure with different beam widths. This paper will present a new microactuator that combines the traits of the two basic microactators proposed by Pan and Guckel, respectively, to achieve multi-lateral motion in plane. The design principle is based on: (1) the asymmetrical thermal expansion of the beams with different lengths and cross sections and (2) the selective doping of varying resistivity within the structure.

2 DESIGN CONCEPT

A schematic diagram of the new microactuator is shown in Figure 1. Figure 2 displays four operating modes (mode (a), (b), (c) and (d)) of the microactuator with its current path and effective heating beams, and the deformed shapes of simulation are presented also. In mode (a), both inside hot beams (we call them hot beam II) are effective heating beams. The two tips of microactuator will move outward simultaneously. In mode (b), at another input voltage mode, both outside hot beams (we call them hot beam I) are effective heating beams. The two tips will move inward simultaneously. In mode (c), the hot beam (I) on the right side and the hot beam (II) on the left side are effective heating beams. Both two tips will move to the left. Finally, in mode (d), the hot beam (I) on the left side and the hot beam (II) on the right side are effective heating beams. The two tips will move to the right.

3 FINITE ELEMENT MODELING

The commercial finite element code ANSYS is used to perform coupled-field analysis of the electro-thermo-mechanical behaviors by modeling the microactuator as a 3-D shape structure. The solid45 mechanical element type and solid69, solid70 thermal element types are used. Besides, the coupled field element type solid5 is used also. Figure 3 displays the 3D solid model and meshing model of the microactuator for the entire structure (including the suspended beams, the anchor layer and the substrate). In simulation, the material properties that are temperature dependent, such as the thermal expansion coefficient, thermal conductivity, resistivity, Young's modulus and convective film coefficient, are treated as constant values here for the low operating temperatures (< 300℃).

4 CHARACTERIZATION

4.1 Optimal Structure Approach

In order to ensure that the microactuator can realize the four operating motion and to gain maximum lateral displacement at the same input voltage but with maximum temperature below 300℃, some design rules motivate the structure design focusing on the optimal lengths of cold beam (Lc), hot beam II (Lh2), flexure beam (Lf) and the optimal width of bridge beam (Wb). Other design dimensions are choose and fixed at Lh1=500 μm, Lb=50 μm, Wt=10 μm, Wc=24 μm, Wt=10 μm, W=4 μm, h=2 μm, g=3 μm, air gap=3 μm, contact pads=80 μm x 80 μm and the Silicon substrate referred as a thicker block.

Furthermore, the influence of selective doping (varying resistivity) of the beams is also considered. Figures 4, 5 and 6 show the variations of themaximum displacement of the microactuator with different dimensions of the cold beam (Lc), hot beam II (Lh2), flexure beam (Lf) and the bridge beam (Wb). Figure 7 shows the influence of varying resistivity in heavily doped area on the performance. According to the above results, it is found that the geometrical parameters and varying resistivity have a strongly influence on the performance of the microactuator, and thus the optimal structure can be obtained by varying the dimensions and resistivity of the beams to meet proper performance.

4.2 Thermal Boundary Condition Effect

After optimal structure approach, one optimal dimesions (Lh1=500 μ m, Lb=50 μ m, Lh2=150µm, Lc=250µm, Wb=30µm, Lf=70µm, Wt=10 μ m, Wc=24 μ m, Wt=10 μ m, W=4 μ m, h=2 μ m, g=3 μ m, air gap=3 μ m, pads=80 μ m x 80 μ m) is adopted for inquiring into the effect of thermal boundary conditions. The steady state temperature distribution and the tip displacements of the microactuator under various applied voltages, different operating modes and different thermal boundary conditions are simulated in this section. By feasible assumption, under low operating temperatures(< 300℃), the heat dissipated through conduction and radiation to the ambient can be neglected as compared to the heat loss by conduction to the substrate via the pads/anchors (Here the substrate is referred as a heat sink with a large thermal mass at the ambient temperature) [9-12]. Besides, the conductive heat loss through thin (2~3 μ m) air gap to the substrate seems to be not neglected [12,13]. Here, various trenches (air gaps) under the suspended beams are included in the numerical analysis. Furthermore, because of the ratio of surface area to volume of a solid increases with diminishing size as the structure size reduces, the convection heat loss from the surfaces, especially the large surface areas of cold beam, bridge beam and pads, are accounted for in the model. The results are summarized from figure 8(a)-(c). Gathering the results of the figures, it is indicated that the thermal boundary conditions really play an important role in the performance of the microactuator. Although the performance of the microactuator may be affected by various thermal boundary conditions, the multi-lateral motions are going quit well and can be controlled accurately at an invariant environment. However, for widely and accurately utilizing the microactuator, comprehensive heat transfer analysis (such as the temperature dependence of thermophysical and heat transfer properties, heat loss by radiation) under a high temperature range should be studied further.

5 CONCLUSIONS

This paper presents a new electro-thermal microactuator to have four operating motions in plane by only varying voltage potentials at the contact pads. According to the characterization, it is found that only low input voltages (0~7V) are required to achieve displacements in microns with the operating temperatures below 300℃. The parameters that influence the design limitation and performance of the microactuator have been studied. It is revealed that the optimal structure can be obtained by varying dimensions and resistivity of the beams to get proper performance of the microactuator. The performance of the microactuator may be affected by various thermal boundary conditions, but the multi-lateral motion behaviors are going quit well and can be controlled accurately at an invariant environment. To extend the operating function, we can manipulate the microactuator to generate a versatile path motion in plane by varying operating modes and with various input voltages, that will enlarge applications in micro devices with larger operating range or acting as a multi-mode switch, relay or optic tweezer.

REFERENCS

[1] Jaecklin V. P., C. Linder, N. F. de Rooij, and J. -M. Moret, Sensors and Actuators A 39, pp.83-89, 1993.
[2] Matoba H, Ishikawa T, Kim C J and Muller R S IEEE Transducers'94, pp 45-50, 1994.
[3] Yeh R., Kruglick E. J. J. and Pister K. S. J., J. Micromech. Syst., 5, 1, pp.10-17, 1996.
[4] Comtois, J. J. and Bright, V. M., Sensors and Actuators A **58**, pp.19-25, 1997.
[5] Que L, Park J S and Gianchandani Y B, J Microelectromech. Syst., 10, 2, pp.255-262, 2001.
[6] Pan C. S. and Hsu W., J. of the Chinese Society of Mechanical Engineers, 22, 1, pp.71-78, 2001.
[7] Pan C. S. and Hsu, W., J. Micromech. Microeng., 7, pp.7-13, 1997.
[8] Guckel, H., Klein, J., Christenson, T., Skrobis, K., Landon, M., and Lovell, E. G., IEEE Solid State Sensor and Actuator Workshop, pp.73-75, 1992.
[9] Fedder G. K. and Howe R. T., *Proc. IEEE Micro Electro Mechanical System Workshop* pp.63-68, 1991
[10] T ai Y. C., Mastrangelo C. H., Muller R. S., *J. Appl. Phys.* 63 (5) 1 pp.1441-1447, 1988.
[11] Lin L and Pisano A P, *ASME DSC-32* pp 147-163, 1991.
[12] Lin L., Chiao M, *Sensors and Actuators* A55 pp.31-41, 1996.
[13] Mankame N D and Ananthasuresh G K, *J. Micromech. Microeng.* 11 pp. 1-11, 2001.

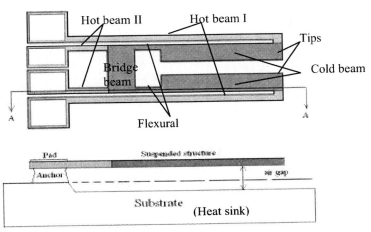

Figure 1: the schematic diagram of the new microactuator

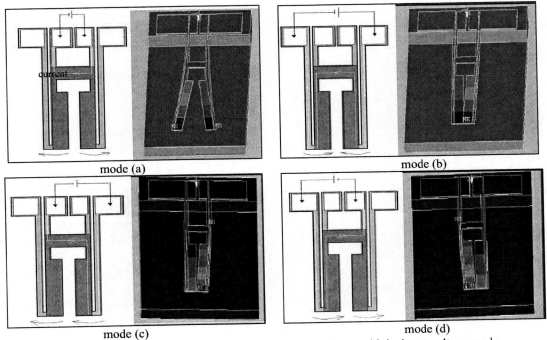

mode (a)

mode (b)

mode (c)

mode (d)

Figure 2(a)-(d): four operating motions of the microactuator with its input voltage mode

Figure 3: 3D solid and meshing models for finite element analysis

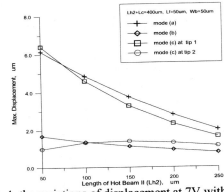

Figure 4: the variations of displacement at 7V with various dimensions of the hot beam II (Lh2) and cold beam (Lc)

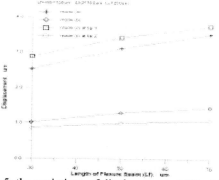

Figure 5: the variations of displacement at 7V with various dimensions of the flexure beam (Lf) and bridge beam (Wb)

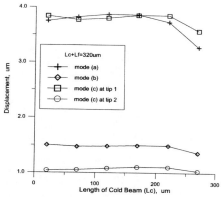

Figure 6: the variations of displacement at 7V with various dimensions of the flexure beam (Lf) and cold beam (Lc)

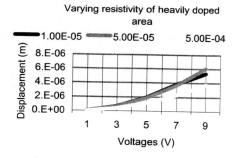

Figure 7: the influence of varying resistivity on performance (mode (c) as example)

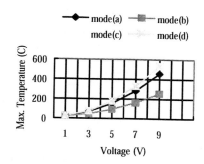

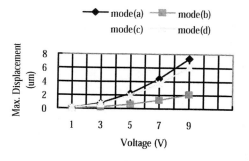

(a) only conduction

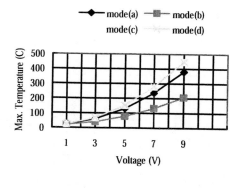

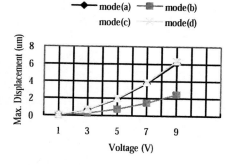

(b) conduction and convection (hf=50 [W · m^{-2} · $^{\circ}$C^{-1}])

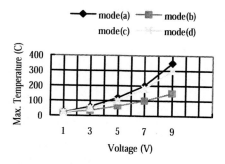

(c) conduction, convection and conduction to the substrate via air gap (3 μ m)

Figure 8(a)-(c): the influence of different thermal boundary conditions on the temperature and displacement of the microactuator

NSTI-Nanotech 2004, www.nsti.org, ISBN 0-9728422-8-4   Vol. 2, 2004

MEMS Compact Modeling Meets Model Order Reduction: Examples of the Application of Arnoldi Methods to Microsystem Devices

Jan Lienemann, Dag Billger*, Evgenii B. Rudnyi, Andreas Greiner, and Jan G. Korvink

IMTEK – Institute for Microsystem Technology, Albert Ludwig University
Georges Köhler Allee 103, D-79110 Freiburg, Germany
Tel. +49 761 203 7386, Fax. +49 761 203 7382, Email: lieneman@imtek.de
*The Imego Institute, Arvid Hedvalls Backe 4, SE-411 33 Göteborg, Sweden
Tel. +46 317 501 853, Fax. +46 317 501 801, Email: dag.billger@imego.com

ABSTRACT

Modeling and simulation of the behavior of a system consisting of many single devices is an essential requirement for the reduction of design cycles in the development of microsystem applications. Analytic solutions for the describing partial differential equations of each component are only available for simple geometries. For complex geometries, either approximations or numerical methods can be used. However, the numerical treatment of the PDEs of thousands of interconnected single devices with each exhibiting a complex behavior is almost impossible without reduction of the order of unknowns to a lower-dimensional system. We present a fully automatic method to generate a compact model of second-order linear systems based on the Arnoldi process, and provide an example of successfull model order reduction to a gyroscope.

Keywords: Arnoldi process, model order reduction, compact modeling, second order differential equations, butterfly gyroscope

1 INTRODUCTION

For the computational treatment of electronics and microsystems (MEMS[1]), different approaches can be employed. In this section, we review a conventional approach, that is, simplifying a system to an equivalent circuit by hand-made or semiautomatic compact models. We suggest a new way to automate the generation of low-dimensional systems of equations by means of mathematical techniques.

What separates MEMS from purely electronic devices (such as very large scale integration or VLSI transistors and other circuit elements) is that MEMS devices are transducers that convert signals between electronics and all other energy domains. For example, most microgyroscopes and accelerometers found in automobiles are currently produced using MEMS technology. Their coupling functionality results in special requirements for the modelling of MEMS. But also undesired coupling – parasitic effects – need a thorough consideration, since on this small scale mutual influence can become a severe problem.

---

[1] We will call all microsystems MEMS, although more functions than only micro-electromechanics are possible.

Thus, the engineers need to simulate the system as a whole. By experience, they are able to define coupling effects between devices, which can be probed at certain terminals. For electrical devices, there is often a natural choice of these terminals, e.g. the emitter, collector and base of a bipolar transistor. However, e.g., for the temperature transport from a computer microchip , this choice is not so obvious.

Once these terminals are identified, the microsystem can be partitioned into a number of devices and energy domains, each coupled by terminals.

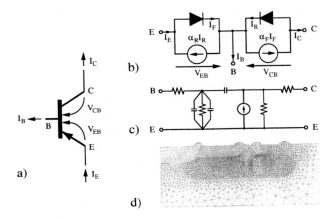

Figure 1: Different modeling approches for an p-n-p transistor. a) Transistor representation for circuit diagram. b) Ebers-Moll compact model of a transistor. c) Compact model for small signal dynamical behaviour analysis. d) Mesh for numerical discretization of PDEs. b) and c) adapted from [1], d) own model (unpublished).

1.1 Compact Modelling vs. Model Order Reduction

In electrical engineering, the common approach is to find a "compact model" of a single device in an analytical form. Whereas there is almost no problem to write down a relationship for simple circuit elements such as resistors and capacitors, the modeling of semiconductor devices was a challenge right from the start. In principle, to accurately describe the transistor operation one should solve the transport PDEs for electrical carriers coupled with a Poisson-Boltzmann equation.

This is possible in analytic form for some special cases. However, as technology develops, the old compact model cannot be applied any more to a newly developed device, and newer models must be employed (see fig. 1).

For MEMS, due to the large number of possible devices, working principles and design freedoms for the engineer, there is no "transistor" device, so that hand-made models are not a viable solution for the long term.

On the other hand, automatic model order reduction (MOR) aims at providing reduced models only with minimal intervention by the designer. The goal is to provide a software which - based on a spatial discretization of the PDE, e.g. by the finite element method - is capable to return ODEs with a far lower number of state variables than the previous discretized system without sacrificing too much acuracy. These ODEs can then be used in SPICE-like simulators, allowing for system simulations in acceptable time.

The designer does not need to worry about the details of the reduction process, and the software should be robust enough for use in industrial applications. Model order reduction thus provides "Compact Modeling on Demand".

1.2 State of the art and the future of automatic model reduction

At present, MOR of first and second order linear ordinary differential equation can be considered as solved. These equations occur in a large number of cases in microsystem engineering. In electronic circuits, during a small signal analysis, linearization and replacement by a simpler equivalent circuit is also often possible. Very often possible nonlinearities are mostly suppressed by a suitable feedback circuit, and so the assumption of a linear system is quite valid. MOR can be an important part for the design of those components.

For some cases like bilinear or quadratic nonlinearities (occuring e.g. in fluid dynamics), or for nonlinearities occuring near a given state trajectory, recently solutions were presented [2], [3]. However, to be able to compete with sophisticated nonlinear transistor compact models, more research is certainly needed. At the moment, the usefulness lies especially in coupling multiphysics simulations with highly depelopped compact circuitry models.

2 ARNOLDI PROCESS FOR SECOND ORDER SYSTEMS

The first step for finding a reduced order model is to formulate a discretized version of the system's PDE. For example, considering the force equilibrium for a linear time invariant elastic system, we obtain the linear system

$$\mathbf{M}\ddot{x}(t) + \mathbf{C}\dot{x}(t) + \mathbf{K}x(t) = \mathbf{B}u(t), \tag{1}$$

where $\mathbf{M}$, $\mathbf{C}$ and $\mathbf{K}$ are called the mass, damping, and stiffness matrix and $\mathbf{B}$ is the scattering matrix to distribute the inputs $u(t)$ on the domain.

Often the engineer is only interested in the solutions of a few degrees of freedom or linear combinations thereof. A selector matrix $\mathbf{L}^T$ yields the output vector $y = \mathbf{L}^T x$.

After Laplace-transformation of (1) the transfer function $H(s) = \mathcal{L}(Y(s))/\mathcal{L}(U(s))$ can be written as

$$H(s) = \mathbf{L}^T \left(s^2\mathbf{M} + s\mathbf{C} + \mathbf{K}\right)^{-1} \mathbf{B}. \tag{2}$$

The goal of MOR is to find a new system of equations

$$\mathbf{M}_r\ddot{z}(t) + \mathbf{C}_r\dot{z}(t) + \mathbf{K}_r z(t) = \mathbf{B}_r u(t), \quad y = \mathbf{L}_r^T z \tag{3}$$

with a lower number of equations n_r and a low dimensional state vector z such that the transfer function is near to the transfer function of the original system.

A number of mathematical procedures are available to achieve this, the most popular probably projection algorithms that replace x by a lower dimensional state vector z such that $x = \mathbf{V}z$. Our approach is based on the Arnoldi process. This algorithm returns a projected system whose first terms of the Taylor series of the transfer function match those of the full system. Details are presented elsewhere [5] (and references in there).

3 THE BUTTERFLY GYRO

Figure 2: Finite element mesh of the gyro with a background photograph of the gyro wafer pre-bonding.

The *Butterfly* gyro is developed at the Imego Institute in an ongoing project with Saab Bofors Dynamics AB. The *Butterfly* is a vibrating micro-mechanical gyro that has sufficient theoretical performance characteristics to make it a promising candidate for use in inertial navigation applications. The goal of the current project is to develop a micro unit for inertial navigation that can be commercialized in the high-end segment of the rate sensor market. This project has reached the final stage of a three-year phase where the development and research efforts have ranged from model based signal processing, via electronics packaging to design and prototype manufacturing of the sensor element. The project has also included the manufacturing of an ASIC, named μSIC, that has been especially designed for the sensor (fig. 4).

The gyro chip consists of a three-layer silicon wafer stack, in which the middle layer contains the sensor element. The sensor consists of two wing pairs that are connected to a common frame by a set of beam elements (figure 2); this is the reason the gyro is called the *Butterfly*. Since the structure is manufactured using an anisotropic wet-etch process, the connecting beams are slanted. This makes it possible to keep all electrodes, both for capacitive excitation and detection, confined to one layer beneath the two wing pairs. The excitation electrodes are the smaller dashed areas shown in fig. 3. The detection electrodes correspond to the four larger ones.

By applying DC-biased AC-voltages to the four pairs of small electrodes, the wings are forced to vibrate in anti-phase in the wafer plane. This is the excitation mode. As the structure rotates about the axis of sensitivity (fig. 3), each of the masses will be affected by a Coriolis acceleration. This acceleration can be represented as an inertial force that is applied at right angles with the external angular velocity and the direction of motion of the mass. The Coriolis force induces an anti-phase motion of the wings out of the wafer plane. This is the detection mode. The external angular velocity can be related to the amplitude of the detection mode, which is measured via the large electrodes.

When planning for and making decisions on future improvements of the *Butterfly*, it is of importance to improve the efficiency of the gyro simulations. Repeated analyses of the sensor structure have to be conducted with respect to a number of important issues. Examples of such are sensitivity to shock, linear and angular vibration sensitivity, reaction to large rates and/or acceleration, different types of excitation load cases and the effect of force-feedback.

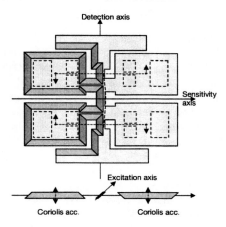

Figure 3: Schematic layout of the *Butterfly* design.

The use of model order reduction indeed decreases runtimes for repeated simulations. Moreover, the reduction technique enables a transformation of the FE representation of the gyro into a state space equivalent formulation. This will prove helpful in testing the model based Kalman signal processing algorithms that are being designed for the *Butterfly* gyro.

Figure 4: The *Butterfly* and μSIC mounted together.

4 RESULTS

We reduced an *ANSYS* model of the *Butterfly* gyroscope from 17361 degrees of freedom to models with different lower orders. Due to the properties of the Arnoldi process, all reductions with a lower order are contained in a higher order reduced model, model, so it is sufficient to perform the reduction for the largest order desired. In this case, we reduced the model to 40 degrees of freedom.

The time to create this model is about the same as the calculation of a single timestep in *ANSYS*.

All calculations for the full model were performed in *ANSYS*. The reduction process itself is performed by an external C++ program, which operates on the *ANSYS* .emat files and outputs the reduces matrices as well as projection matrices. The postprocessing is done im *Mathematica*.

4.1 Time domain

Figure 5 shows a comparison of the transient behavior for the full model and some examples of a reduced model. We see that while order 5 is not good enough (fig. 5a), the reduced model of order 10 is already very good (fig. 5b,c).

4.2 Frequency domain

Figure 6 shows a comparison of the transfer functions. While the reduced models of order 5 up to 15 show considerable deviations for the low frequency range, the model with order 20 shows a perfect match for a larger extend. The order 40 model is even closer for higher frequencies, though this is not so important for the gyroscope. For a step input with its large portion of low frequencies, and the timescale considered in fig. 5, order 10 yields already very satisfying results.

5 DISCUSSION

The exceptionally good results were also demonstrated for other energy domains. Every linear problem in *ANSYS* can be model order recuced this way. It is also possible to extend the tool to other simulation packages as long as the system matrices can be recovered. Various examples linear first and second order systems were succesfully reduced, thereby showing distinct advantage over commercially available methods:

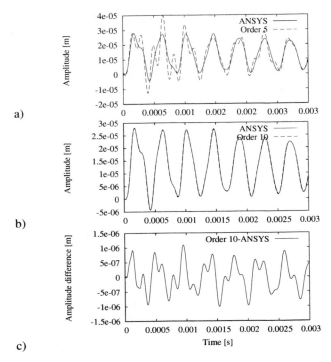

a)

b)

c)

Figure 5: Comparison of transient behavior for full and reduced models: a) model order 5 vs. *ANSYS*, b) model order 10, c) difference between model order 10 and *ANSYS*.

- First order thermal and electro-thermal systems [6]

- Second order mechanical systems

- Piezoelectric actuation of a surface acoustic waves device

- Acoustic simulations

- Electromagnetic systems.

Some results for these systems are published [5], [6], thereby showing distinct advantage over e.g. the Guyan method implemented in *ANSYS*, others are in preparation.

6 CONCLUSIONS

Model order reduction techniques provide a valuable tool for the designer of coupled multiphysics system. Especially for applications with a large number of similar devices, as often encountered in microsystem applications, the method facilitates a low time to market and a increase of design quality due to the possibility to simulate whole systems. As an example, we successfully reduced the order of an ANSYS model by four orders of magnitude. Even for a low number of degrees of freedom, low frequency transient curves showed an excellent match.

This method works perfectly for first or second order linear time invariant systems. But also for nonlinear or time variant systems, research results are coming and raise hope

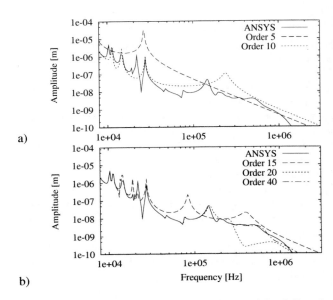

a)

b)

Figure 6: Comparison of transfer functions of the full and reduced models.

to be able to simulate nonlinear elements like transistors in the future.

7 ACKNOWLEDGMENTS

Partial funding by the DFG project MST-Compact (KO-1883/6), by the Italian research council CNR together with the Italian province of Trento PAT, and by an operating grant of the University of Freiburg is gratefully acknowledged.

REFERENCES

[1] S. M. Sze "Semiconductor Devices – Physics and Technology", John Wiley & Sons, New York, 1985

[2] M. Rewienski, J. White, "A trajectory piecewise-linear approach to model order reduction and fast simulation of nonlinear circuits and micromachined devices", IEEE Transactions on Computer-Aided Design of Integrated Circuits and Systems 22, 155–170, 2003.

[3] J. R. Phillips, "Projection-based approaches for model reduction of weakly nonlinear, time-varying systems", IEEE Transactions on Computer-Aided Design of Integrated Circuits and Systems 22, 171–187, 2003.

[4] R. W. Clough, J. Penzien, "Dynamics of Structures", McGraw-Hill, 1975.

[5] E. B. Rudnyi, J. Lienemann, A. Greiner, J. G. Korvink, "mor4ansys: Generating Compact Models Directly from ANSYS Models", accepted for publication in the *Technical Proceedings of the 2004 Nanotechnology Conference and Trade Show*, Boston, USA, March 7–11 2004.

[6] T. Bechtold, E. B. Rudnyi, J. G. Korvink, "Automatic Generation of Compact Electro-Thermal Models for Semiconductor Devices", IEICE Transactions

Guidelines of Creating Krylov-subspace Macromodels for Lateral Viscous Damping Effects

Po-Ching Yen and Yao-Joe Yang
Department of Mechanical Engineering
National Taiwan University, Taipei, Taiwan, ROC
TEL: +886-2-23646491   FAX: +886-2-23631755   E-mail: yjy@ccms.ntu.edu.tw

ABSTRACT

In this work, the guideline of generating *accurate* lateral viscous damping macromodels by the Krylov-subspace algorithm is described. A three-dimensional (3-D) finite-difference (FDM) Stokes flow solver for simulating lateral damping effects was developed. The system matrices generated by the solver was then reduced to low-order macromodels that can be easily inserted into a system-level modeling simulators, such as Saber, Simulink or SPICE for transient and frequency analysis. Based on physical and numerical constraints, the required orders of macromodels as well as the appropriate sizes of computational meshes are proposed. Finally, the experimental results for comb-drive devices show that the error of the results estimated by the macromodels are within 15%.

Keywords: lateral viscous damping, macromodel, model order reduction, comb-drive, system-level analysis

INTRODUCTION

Many MEMS devices such as accelerometers, gyroscopes, switches, micro-mirrors, and resonant sensors need fully understanding of lateral gas damping effects for accurate dynamic modeling. It is well known that modeling 3-D lateral viscous damping effects using FEM or FDM not only requires intensive solid-modeling work, but also require significant computational resources even for a steady incompressible-flow simulation. Therefore, earlier works on lateral viscous damping were based on the 1-D analytical Stokes and Couette flow solutions [1,2]. The 1-D analytical approaches can easily provide the first-order estimation of lateral damping effect, and require almost negligible computational resources so that they are intrinsically compatible with any system-level simulators. However, the 1-D approaches over-simplify the geometrical complexity of typical MEMS devices with lateral damping effects, and thus the error is very large in most cases.

Recently, Aluru and Wang [3,4] developed 3-D Stokes solvers using boundary-element method (BEM), and demonstrated that the BEM approaches not only require much less computational cost than typical FEM/FDM approaches [5], but also significantly reduce the works on creating solid models. However, the solutions of the BEM approaches are in frequency domain, and hence are not completely compatible with transient analysis. In this work, we develop model-order-reduction (MOR) methodology [6,7,8,9] for generating accurate *time-domain* macromodels from 3-D FDM/FEM Stokes solvers, and explore the characteristics of the macromodels under various conditions.

Figure 1 outlines the concept of the model-order-reduction procedure for lateral damping effects. The initial step of generating the macromodels is to generate 3D solid models. This step is very similar to the typical procedure of performing FEM//FDM fluidic simulation. However, we used the commercially-available MEMS modeling packages, such as Coventorware, IntelliSuite and MEMS-Pro, to generate the 3D solid model of the air film from 2-D mask layout, by considering the air-film surrounding the structures as the fictitious sacrificial layer. After creating the solid model of the air film surrounding a MEMS laterally-movable structure, the FEM/FDM techniques are used to discretize the solid model. Since the governing equation is in time domain, the discretization creates a system (set) of ordinary differential equations whose state variables are in fact the velocity distribution of the air film. Typically the system is so large that huge computational resources for time-domain integration are required. Without direct integration of the system, the Arnoldi algorithm is applied to reduce the system of differential equations into a low-order system, the so-called macromodel. The macromodel can be readily inserted into system-level simulators, such as Saber® or Simulink®, for transient and frequency-response analysis.

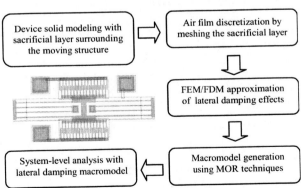

Figure 1: *Procedure of extracting lateral damping macro-model for MEMS devices*

In the following section, the theory of the fluidic damping effects and the application of the model-order-reduction technique are presented. Based on physical and numerical constraints, the required orders of macromodels as well as the appropriate sizes of computational meshes are proposed. Finally, the measured and simulated results are demonstrated.

THEORY

The governing equation of lateral viscous damping is the Stokes equation. The 3-D Stokes' equation is:

$$\rho \frac{D\vec{V}}{Dt} = -\nabla p + \mu \nabla^2 \vec{V} \qquad (1)$$

where p is pressure, ρ is the density of the gas, μ is the viscosity coefficient, and $\vec{V} = \begin{bmatrix} u & v & w \end{bmatrix}^T$ is the velocity vector. For our case, the imposed pressure gradient is assumed to be zero, so the first term on the right-hand side can be eliminated. Since the damping contributed by the surfaces, whose normal vectors are parallel to the direction of plate motion, is negligible under our assumptions of $u >> v \approx w$, the velocity components perpendicular to the direction of the in-plane motion are ignored. As a result, the continuity equation is not considered [1], and Equation (1) can be simplified to:

$$\frac{\partial u}{\partial t} = v \nabla^2 \mu = v \left(\frac{\partial^2 u}{\partial x^2} + \frac{\partial^2 u}{\partial y^2} + \frac{\partial^2 u}{\partial z^2} \right) \qquad (2)$$

where $v = \mu / \rho$ is the kinetic viscosity. Based on this equation, a finite-difference solver is developed.

Furthermore, the simplified governing equation, as shown in Equation (2), is a linear equation, so the system matrices generated by the FDM approximation process can be reduced by an Arnoldi-based model-order-reduction (MOR) technique. The dynamic system equation formulated by the FDM approximation of Equation (2) can be written as:

$$\dot{\vec{u}} = \mathbf{A} \cdot \vec{u} + \mathbf{B} \cdot v_{in}$$
$$\vec{y} = \mathbf{C}^T \cdot \vec{u} + \mathbf{D} \cdot v_{in} \qquad (3)$$

where $\mathbf{A}$ is an n by n matrix and n is the total number of nodes, $\vec{u}$ is the vector which contains the unknown velocity distribution on each node, and the input function v_{in} is the imposed velocity on the moving boundary of the computational domain. In this case, we carefully formulate $\mathbf{C}$ and $\mathbf{D}$, so that the output $\vec{y}$ will be the frictional shear (calculated by Newtonian law of viscosity) on the plate. In Laplace domain, the transfer function of the system is:

$$T(s) = \mathbf{C}^T (\mathbf{I}s - \mathbf{A})^{-1} \mathbf{B} + \mathbf{D} = \mathbf{C}^T (\mathbf{I} - s\mathbf{A}^{-1})^{-1} \vec{b} + \mathbf{D}$$
$$\vec{b} = -\mathbf{A}^{-1}\mathbf{B} \qquad (4)$$

After expanding the transfer function in Taylor series about $s=0$, we obtain:

$$T(s) = \mathbf{C}^T (\mathbf{I} + s\mathbf{A}^{-1} + s^2\mathbf{A}^{-2} + \ldots)\vec{b} + \mathbf{D} = \sum_{k=0}^{\infty} m_k s^k + \mathbf{D} \quad (5)$$

where m_k are the coefficients of the Taylor series, and are equal to $m_k = \mathbf{C}^T (\mathbf{A}^{-k}) \vec{b}$.

The Taylor expansion can be truncated to approximate the transfer function $T(s)$. Since $\mathbf{A}^{-k}\vec{b}$ quickly line up with a single eigenvector, this moment

matching procedure is usually numerically unstable. Therefore, we apply the Arnoldi-based algorithm to stably compute orthogonal bases v_i that spans the Krylov subspace:

$$K_q \left(\mathbf{A}^{-1}, \vec{b} \right) = span \left\{ \vec{b}, \mathbf{A}^{-1}\vec{b}, \mathbf{A}^{-2}\vec{b}, \cdots, \mathbf{A}^{-(q-1)}\vec{b} \right\} \qquad (6)$$

Given the matrix $\mathbf{V_q}$ whose columns are $\{v_i\}$, the Arnoldi algorithm reduces the system matrix $\mathbf{A}$ to a small upper Hessenberg matrix $\mathbf{H_q}$ whose entries are the Gramm-Schmidt orthogonalization coefficients:

$$\mathbf{V_q}^T \mathbf{A} \mathbf{V_q} = \mathbf{H_q} \qquad (7)$$

Finally, the reduced transfer function can be written as:

$$T_q(s) = \mathbf{C}^T \mathbf{V_q} \left(\mathbf{I_q} - s\mathbf{H_q} \right)^{-1} \mathbf{V_q}^T \vec{b} + \mathbf{D} \qquad (8)$$

Note that the reduced system transfer function, as shown in Equation (8), has the same input (v_{in}) and output ($\vec{y}$) as those in Equation (3). Since the typical sizes of the system matrices are very small, the computational efficiency for simulating transient responses and frequency responses of the reduced models are significantly increased [7].

COMB-DRIVE DEVICE SIMULATION

The picture of the simulated and measured comb-drive structure fabricated by MUMPs® process is shown in Figure 2. A DC bias of up to 70 V and a sinusoidal driving voltage with amplitude (peak to peak) up to 20 V are used. The lateral fluid damping effect surrounding the comb drive is simulated and then compared with the experimental results to verify the accuracy of the developed 3-D Stokes solver.

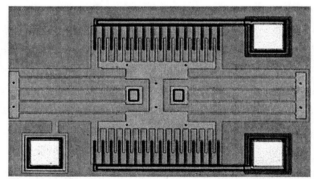

Figure 2: *The CCD picture of a comb-drive measured in this work*

I. Study on Discretization Convergence

Figure 3 shows the schematic of velocity profiles induced by a laterally oscillating plate for different frequencies. Figure 3(a) shows that when the structure oscillates at relatively low frequency, the air film above the structure is assumed to be of Stokes-type, and the air film underneath the structure is assumed to be of Couette-type. As the oscillating frequency increases (as shown in Figure 3(b)), the velocity profiles will extend only to a short distance from the structure surface (i.e. short penetration depth of the velocity profile). In this case, even for small-amplitude motion, the air film will introduce a considerable amount of damping [1].

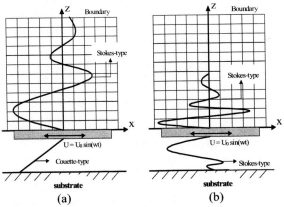

(a)   (b)

Figure 3: *Velocity profiles induced by a laterally oscillating structure with (a). low frequency, (b). high frequency.*

Therefore, under higher oscillating frequency, finer discretization of the FDM calculation in z-direction is required because the spatial variance of the Stokes-flow's velocity profiles is much higher than the counterpart of the Couette-type flow. Figure 4 shows that the required maximum discretization length in z-direction is about 0.2 μm for simulating converged shear force results at frequency below 1 MHz. In other words, if the operating frequency is higher than 1 MHz, the discretization length has to be less than 0.2 μm since the penetration depth of the velocity profile decreases.

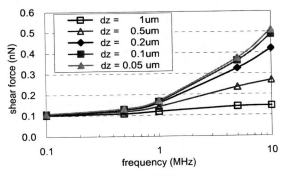

Figure 4: *The total shear force vs. oscillating frequency for a comb structure. The simulation accuracy for high frequency model (>1 MHz) is strongly relative to the distance between the meshing nodes.*

II. Required Extent of Fluidic Domain

Figure 5 shows the relationship between the upper air film thickness and the simulated shear force of the device. When oscillating at a frequency higher than 1MHz, the shear forces calculated by the FDM/FEM models with different upper air-film thickness are the same. However, as the frequency decreases, the model with thin computational domain overestimates the damping, since the assumption of zero-velocity boundary

on the top of the air film is no longer valid. This figure also indicates that the minimum air-film thickness is about 10 μm for a frequency as low as 10 kHz.

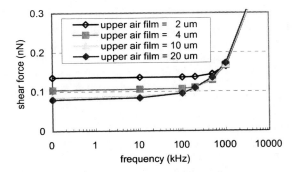

Figure 5: *Shear force versus oscillating frequency for different air film thickness above the moving structure.*

The shear forces contributed by different parts of the air film are shown in Figure 6. The damping contributed by the underneath film increases as the oscillating frequency increases, and finally becomes comparable to the damping contributed by the top film when the frequency is higher than 500 kHz.

Figure 7 is the frequency response of the system damping shear force for different order macromodels, and indicates that the macromodels with orders greater than 15 are required for the converged results under a wide range of operating frequencies (from 10 kHz to 10 GHz).

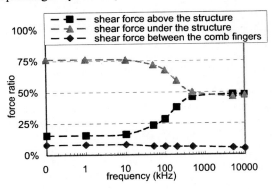

Figure 6: *The contribution of the shear force ratios by the air film above the structure, under the structure and between the comb fingers.*

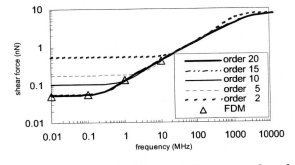

Figure 7: *Shear force versus frequency for the macromodels with different orders.*

SYSTEM-LEVEL SIMULATION COMPARED WITH MEASUREMENT

The macromodels generated by the Arnoldi-based MOR algorithm can be readily inserted into system-level simulators, such as Saber® or Simulink®, for transient and frequency-response analysis. Figure 8 presents the experimental results of the comb drive device with folded beam length of 228 μm as well as the system simulation results of the 20th order damping macromodel. The macromodel underestimates the damping by about 10% in this case, and we speculate the major source of this error comes from the fact that the macromodel neglects the pressure back force on the tip-ends of the comb figures. Figure 8 shows the simulated and measured comb-drives quality factors vs. different folded-beam lengths. The results by the 1-D Stokes/Couette analytical models are also presented in the Figure. The 1-D analytical model over-predicts the quality factors by 30~40%, while the discrepancy between the measured results and the macromodels is within 15%.

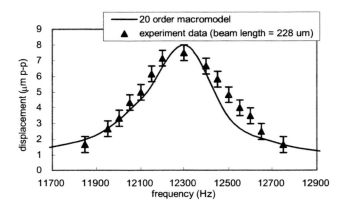

Figure 8: *Frequency response of the 20 order macromodel of the comb drive device with 228 μm folded beam compared with the experimental data.*

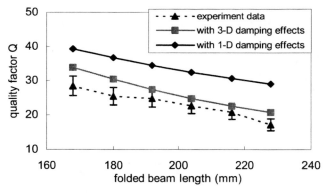

Figure 9: *Comb-drive quality factors vs. different folded-beam lengths are shown.*

CONCLUTION

This paper presented a 3-D FDM Stokes' solver, and a macro-model generation methodology for lateral damping effects based on the application of Arnoldi-based

model-reduction technique. The theory of the Arnoldi-based model-order-reduction is described. The studies on the FEM/FDM mesh convergence and the appropriate size of computational domain were also discussed. The macromodels generated by the technique were successfully inserted into the Simulink for system level analysis, and the results were also compared with the experimental data. The discrepancy of the simulated and measured quality factors are within 15%.

Acknowledgement

This research was supported by the National Science Council (Grant No. 91-2218-E-002-021), Taiwan, R.O.C.

REFERENCES

1. Y.-H. Cho, A. P. Pisano and R. T. Howe, "Viscous damping model for laterally oscillating microstructures," *J. Microelectromechanical Systems*, Vol. 3, No. 2, Jun. 1994, pp. 81-87.
2. T. Veijola and M. Turowski, "Compact damping models for laterally moving microstructures with gas rarefaction effects," *J. Microelectromechanical Systems*, Vol. 10, No. 2, Jun. 2001, pp. 263-273.
3. N. R. Aluru and J. White, "A fast integral equation technique for analysis of microflow sensors based on drag force calculations," in *Proc. of MSM*, Santa Clara, USA, Apr. 1998, pp. 283-286.
4. X. Wang, M. Judy, and J. White, "Validating fast simulation of air damping in micromachined devices," in *Proc. IEEE 15th International Conference on Micro Electro-mechanical Systems Workshop (MEMS 2001)*, Las Vegas, USA , Jan. 2002, pp. 210-213.
5. W. Ye, X. Wang, W. Hemmert, D. Freeman, and J. White, "Viscous drag on a lateral micro-resonator: fast 3-d fluid simulation and measured data," in *Tech. Dig. of 2000 Solid-State Sensor and Actuator Workshop*, Hilton Head Island, SC, Jun. 2000, pp 124-127.
6. P.-C. Yen and Y.-J. Yang, "Time-domain reduced-order models of lateral viscous damping effects for 3d geometries," in *Proc. of MSM*, 2002, San Juan, Puerto Rico, USA, Apr. 2002, pp. 190-193.
7. P.-C. Yen and Y.-J. Yang, "Macromodels of 3d lateral viscous damping effects for mems devices", *in* Proc. 12th International Conference on Solid-State Sensors and Actuators (Transducers '03), *Boston, USA, June, 2003, pp. 1848-1851.*
8. Y.-J. Yang, M. Kamon, V. L. Rabinovich, C. Ghaddar, M. Deshpande, K. Greiner and J. R. Gilbert, "Modeling gas damping and spring phenomena In mems with frequency dependent macromodels," in *Proc. IEEE 14th International Conference on Micro Electro-mechanical Systems Workshop (MEMS 2001), Interlaken, Switzerland, January 2001, pp. 365-368.*
9. A. Odabasioglu, M. Celik, and L. T. Pileggi, "PRIMA: passive reduced-order interconnect macro-modeling algorithm," in *Proc. IEEE Transaction on Computer-Aided Design of Integrated Circuits and Systems,* Vol. 17, No. 8, Aug. 1998, pp. 645-654.
10. J. A. Fay, Introduction to Fluid Mechanics, MIT Press, Cambridge, 1994.

Computationally Efficient Dynamic Modeling of MEMS

Dan O. Popa[†], James Critchley[†], Michael Sadowski[†], Kurt S. Anderson[†], George Skidmore[‡]

[†] Center for Automation Technologies, Rensselaer Polytechnic Institute,
Troy, New York 12180, USA, e-mail: popa@cat.rpi.edu
[‡] Zyvex Corporation, Richardson, Texas

ABSTRACT

Traditional modeling work in MEMS includes simplified PDE/ODE formulation, based on physical principles, and Finite Element Analysis. More recently, reduced order modeling techniques using Krylov subspace decomposition have been proposed in the context of nodal analysis. This modeling technique makes it possible to predict the dynamic behavior of more complex MEMS, but the computational engine is still a traditional cubic order solver.

In this paper we apply a new modeling approach for complex MEMS based on a linear $O(n+m)$ (n- number of bodies, m – number of constraints) solver for rigid multibody dynamics. As direct applications, we present simulation and experimental results of models for thermally driven MEMS actuators, compared against established simulation tools, namely FEA (Intellisuite), AUTOLEV, and SUGAR 3.0.

Keywords: MEMS modeling and simulation, compact modeling, O(N) simulation

1. Introduction

During the last 15 years modeling and simulation tools for Micro-Electro-Mechanical Systems (MEMS) have evolved from simplified PDE/ODE formulations based on physical principles, to MEMS specific layout, process and FEA tools (Intellisuite, CoventorWare). The algorithm complexity associated with these tools in dynamic simulation is cubic in the number of mesh elements, $O(N^3)$, with additional overhead and numerical stability problems associated with the treatment of intermittent constraints. More recently, reduced order modeling techniques using Krylov subspace decomposition have been proposed in the context of nodal analysis (SUGAR) [7,8]. This modeling technique makes it possible to predict the dynamic behavior of more complex MEMS, but the computational engine is still a traditional $O(n^3)$ solver (that is, the number of mathematical operations need to perform the simulation at each integration time step increases as a cubic function of the number of beam elements n). Compared to FEA, however, the MEMS system is described by n large beam elements, with n<<N.

In this paper we apply a different modeling approach for complex MEMS for use with a fully nonlinear $O(n+m)$ (where n is the number of generalized coordinates and m

the independent algebraic loop constraints) Recursive Coordinate Reduction (RCR) solution method for rigid body dynamics [1,5]. Such computationally efficient multibody simulation methods have been receiving increasing attention since the first functional $O(n)$ algorithm was developed by Armstrong in 1979 [2]. Recent advances in this area have extended linear order solutions to include kinematic loops, link flexibility, and intermittent constraints which are needed to characterize the operation of very complex MEMS devices. In this paper, we illustrate the benefits of an RCR formulation for MEMS actuated via thermal flexure actuator banks.

Our approach involves approximating the continua (the flexible MEMS parts) via a series of rigid bodies which are interconnected by stiff springs. Once the rigid body model is obtained, the RCR formulation is applied to the approximate system to realize a simulation. Nodal analysis uses an analogous approach, decomposing the flexible components into a series of interconnected beam elements, but the manner of solving the resulting set of differential equations is conventional (i.e. mass-matrix inversion by direct methods).

We pay particular attention to two issues that need addressing specifically in the context of MEMS, namely:

- How to best approximate the actual structural and dynamical characteristics of the flexible MEMS device using a rigid body/flexible link model.
- How to organize the computation of dynamical equations of motion so that they can be solved in linear time at each iteration.

The resulting simulation method represents a significant improvement in performance relative to what is possible for MEMS simulation using established methods. The performance is not only reflected in a linear execution time, but also in the numerical stability of the algorithm.

2. Motivation for an O(n+m) simulation tool for complex MEMS

Figure 1 shows an example of a MUMPS rotary stage actuated by orthogonal banks of bimorphs [6]. If we consider a single quarter stage arm of the rotor, namely an XY stage, we could approximate it as a series of rigid bodies and flexible joints, or as a set of interconnected beam elements. As an example, Figure 2 shows a quarter stage approximate model that includes n=123 generalized coordinates. This model is an interconnection of a simpler single thermal bimorph models, each containing 12 generalized coordinates. The number of additional constraint equations is also large, three being associated with each kinematic loop present in the system (m~½n). The manner in which the discretization is obtained is in itself a non-trivial task [3,4], and will be presented in the next section of this paper. To this multibody model, electro-thermal and damping effects may be added through applied forces or, in the case of thermal expansion, prescribed displacements.

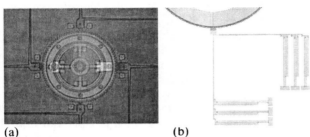

(a)   (b)

Fig 1: Actual MUMPS rotary stage (a), using 4 orthogonal thermal bimorph banks (b).

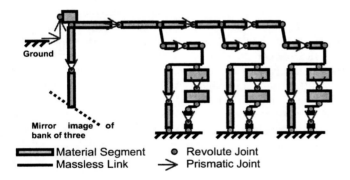

Ground

Mirror   image   of
bank of three

■▬▬ Material Segment   ● Revolute Joint
— Massless Link   ➔ Prismatic Joint

Fig 2: Rigid body/flexible link discretization model for XY stage

After approximating the flexible system by a set of interconnected rigid bodies, the new efficient linear order constrained multi-rigid-body solution scheme of Recursive Coordinate Reduction (RCR) may be applied to the model.

The dynamic behavior of the XY stage can be simulated using tools of different computational complexity such as Finite Elements, Nodal Analysis, and Rigid Body Dynamical solvers, as shown in Table 1. It is apparent that for situations where n and/or m are large, a cubic computational cost is required for each temporal integration step. A truly linear O(n+m) formulation becomes extremely attractive for complex systems.

Methods	Pro(s)	Con(s)
F.E.M. (Abaqus, Ansys, Intellisuite, CoventorWare)	Available now High fidelity Multi-role	Very computationally expensive Poor Contact Element performance
$O(n^3)$ (ADAMS, DADS, AUTOLEV, SUGAR)	Available now Moderate cost	Still computationally expensive for complex structures. Do not handle unilateral constraints
$O(n+nm^2+m^3)$ $O(m+m)$ (RCR)	Lowest cost. Most unilateral constraints Minimum Contact Model	Not generally used Not all unilateral constraints

Table 1: Advantages and disadvantages of various simulation tools available for dynamical simulation

3. Rigid Body – Flexible Link Model

The model discretization shown in Figure 2 assumes that a set of stiffnesses and masses can be found, and therefore, the bending moment at each of the interconnection point is directly proportional to the angular displacement between joints. In fact, this is only true for constant cross section beams of equal lengths, and therefore we make use of a more general joint spring formulation in which the bending moment at the i-th joint is expressed solely in terms of its displacement and the displacement of its immediate neighbors.

In order to approximate the MEMS XY stage with a rigid body/flexible link model it was necessary to address the following questions:

Q1: *If a flexible beam is split into N rigid bodies, what is the best position of the split points such that the first N fundamental modes match the FEA, SUGAR (nodal analysis), and/or the analytical model?*

If N=2, for a beam with length L and cross-sectional moment I. Theoretically,

$$K = E\frac{I}{L} \text{ and } I = fW^2T$$

where T is the beam thickness, W is the beam width, E is Young's Modulus, and f is a correction factor. The correction factor is theoretically equal to 0.5, but from SUGAR and AUTOLEV simulations it was best fit to 0.43.

Q2: *Given the layout of a MEMS device, how does one obtain the values of inertias, stiffnesses, and position of the links in the rigid body/flexible link model?*

Most MEMS device layouts are described in standard formats, such as .gds file formats. A particular format, the Caltech Interchangeable Format (.cif) has been used in the past to describe a MEMS device geometry composed of beams of constant width. It has also been used in conjunction with SUGAR by translating it into a SUGAR-specific net list. We use the same method to describe the geometry of a MEMS device.

Q3: *What kind of interconnects are necessary to describe the discretized approximate model for a MEMS device composed of beams?*

We have identified and modeled four types of beam interconnects necessary to describe the quarter stage shown in Figure 2, namely:
- Rotary flexible joint between beams with same cross-section (type 1).
- Rotary flexible joint between beams with different cross-section (type 2).
- Rotary flexible T-joint (type 3).
- Prismatic flexible joint (type 4).

Q4: *How are backbending, and other nonlinear thermomechanical or electrostatic effects included in the model?* Backbending effects (i.e. thermally induced plastic deformations), thermal expansion effects (i.e. the basis for thermal actuation, proportional with the electrical input power), or electrostatic deflections (i.e. comb drive forces or attraction between plates) can be modeled as externally imposed reduced order nonlinear "gap" functions. For example, a thin MEMS beam expanding by thermo-electrics behaves similarly to a gap function of a type 4 joint proportional to the square of the voltage applied through the link. Moreover, a backbent bimorph bank shown in Figure 3 is equivalent to a negative spring constant shortening the beam.

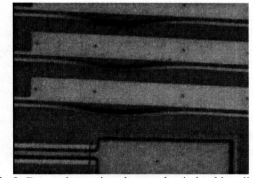

Fig 3: Beam shortening due to plastic backbending deformation of bimorph hot arms.

Q5: *Using the continua to rigid body rigid body/flexible link conversion rules outlined below, how close are the structural modes of the resulting structure to those computed using FEA?*

A three-bimorph bank structure was transformed to a rigid body/flexible link model. Table 2 shows modal analysis results comparing SUGAR, FEA, and AUTOLEV, showing a reasonable match for the first seven flexible modes.

Mode	Intellisuite	AUTOLEV	SUGAR3.0
i	Hz	Hz	Hz
1	83200.30	87004.09	86506.01
2	405413.00	316201.84	N/A
3	405926.00	316424.22	N/A
4	421218.00	330743.72	N/A
5	613638.00	653664.79	649154.89
6	626903.00	664621.89	662160.46
7	773568.00	846293.33	829454.79
8	1130000.00	2002156.90	1879437.60
9	1140000.00	2119784.30	2084951.20
10	1160000.00	2543336.50	3073013.30

Table 2: Comparison of modal analysis results using different simulation tools.

4. Implementation of RCR

The recursive coordinate reduction was applied to the rigid body/flexible link model of the quarter stage using code written in MATLAB, and compared to results obtained in AUTOLEV, SUGAR 3.0 as well as FEA. The RCR method is already the topic of several journal articles and it is to these articles [1,5] that the reader is referred for the equations and detailed derivations.

However during the verification of the numerical exactness of the RCR implementation within MATLAB it was observed that AUTOLEV introduces numerical instabilities that prohibit accurate computation of anything more complex than a single bimorph. This instability is attributed to the symbolic nature of AUTOLEV that implements a constraint solution via Gaussian Elimination without regard to the numerical values of the matrix elements (it does not pivot). The instability is observed as a divergence from the RCR solution and an inability to temporally integrate the equations of motion with an error controlled variable step integrator.

As a result, we compared RCR against dynamic simulations obtained using FEA analysis (namely Intellisuite) as well as Sugar 3.0. Dynamic simulations were performed with the quarter stage model released from rest after bending to a horizontally applied 100µN force.

The X time-dependent coordinate of the complete quarter stage is shown in Figure 4. Figure 5 further illustrates the accuracy of the approximation by showing the planar trajectories of the simulations plotted on 1:1 scaled axis.

Also shown with the trajectories are the results obtained from SUGAR 3.0. These SUGAR curves have been obtained by running a simulation with the constant mass and stiffness matrix that are generated for modal analysis. This linear simulation is not representative of the intended dynamic capabilities of SUGAR, however such functionality is not currently present in that package, possibly due to numerical integration instability for stiff systems. It is interesting to note that for our system the mass matrix inversion of the SUGAR matrices generates a condition number warning in MATLAB.

By observing that the character of the FEA result is maintained, the results provide solid evidence that both the RCR implementation is correct and that the rigid body modeling technique is a meaningful approximation.

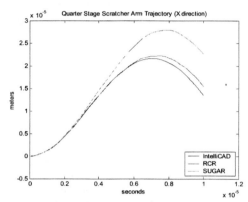

Fig 4: Comparison between FEA, RCR and Sugar 3.0, showing XY stage displacement vs. Time.

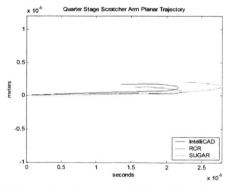

Fig 5: Comparison between FEA, RCR and Sugar 3.0, showing XY stage tip 2D trajectory.

5. Conclusion and Future Work

In this paper we presented a new approach for modeling and dynamic simulation of MEMS devices based on the linear complexity Recursive Coordinate Reduction (RCR) algorithm. The algorithm has been tentatively used in the past for simulating rigid multibody dynamics with a large number constraints and degrees of freedom.

Applying this algorithm to MEMS requires the conversion of a MEMS layout consisting of interconnected beams into a rigid body and stiff spring model. Currently, the conversion from a .cif layout is done through a SUGAR .net file using four types of joints. In the current implementation the RCR is used as an add-on MATLAB toolbox, and can be used in parallel with SUGAR. Further work is necessary in automating the MEMS layout conversion, model conversion (i.e. parameter calculations), and the general application of the RCR.

Acknowledgement
This work was performed under the support of the U.S. Department of Commerce, National Institute of Standards and Technology, Advanced Technology Program, Cooperative Agreement Number 70NANB1H3021, and the National Science Foundation Grants Number CMS-9733684 and CMS-0219734.

References
[1] K. S. Anderson and J. H. Critchley. Improved order-n performance algorithm for the simulation of constrained multi-rigid-body systems. Multibody System Dynamics, 9:185-212, 2003.
[2] W. W. Armstrong. Recursive solution to the equations of motion of an n-link manipulator. In Fifth World Congress on the Theory of Machines and Mechanisms, volume 2, pages 1342–1346, 1979.
[3] A. K. Banerjee and S. Nagarajan. Efficient simulation of large overall motion of beams undergoing large deflection. Multibody System Dynamics, 1, 1997.
[4] Arun K. Banerjee. Dynamics and control of the wisp shuttle-antennae system. The Journal of the Astronautical Sciences, 41(1):73–90, January-March 1993.
[5] J. H. Critchley and K. S. Anderson. A generalized recursive coordinate reduction method for multibody system dynamics. Journal of Multiscale Computational Engineering, 1(2&3):181-200, 2003.
[6] Dan O. Popa, Byoung Hun Kang , John T. Wen, Harry E. Stephanou, George Skidmore and Aaron Geisberger. Dynamic modeling and open-loop control of thermal bimorph MEMS actuators. IEEE International Conf. in Robotics and Automation, 2003.
[7] J. Clark J. Demmel K. Pister N. Zhou Z. Bai, D. Bindel. New numerical techniques and tools in SUGAR 3d MEMS simulation. Tech Proc 4rd Intl Conf On Modeling and Simulation of Microsystems, 2001.
[8] J.V. Clark, et.al,, "Addressing the Needs of Complex MEMS Design", Proc. IEEE International MEMS Conf., Las Vegas, NV, Jan. 20-24, 2002.

Dynamic Modeling and Input Shaping for MEMS

Dan O. Popa[†], John T. Wen[†], Harry Stephanou[†], George Skidmore[‡], Matt Ellis[‡]

[†] Center for Automation Technologies, Rensselaer Polytechnic Institute,
Troy, New York 12180, USA, e-mail: popa@cat.rpi.edu
[‡] Zyvex Corporation, Richardson, Texas

ABSTRACT

One of the common characteristics of Silicon MEMS actuators is that they operate in open-loop, due to difficulties and cost of integrating sensors at small scales. A proper dynamic response (trajectory following, vibration suppression, etc.) is difficult to obtain using open-loop control, as it is much better suited for quasi-static device operation. As a result, the dynamic performance of MEMS devices in many applications is not fully achieved. In this paper we propose to use input-shaping techniques, along with an identified reduced-order model to generate suitable input profiles for MEMS actuators. This systematic approach is a major improvement over the current Finite Element/Design of Experiments practices in the MEMS community.

Keywords: Reduced order modeling, input shaping for MEMS, open-loop control.

1. Introduction

During the last 15 years several actuation technologies at the micro-scale have gained wide acceptance, including piezo materials, shape memory alloys, electro-thermally driven actuators, and electrostatic comb drives. The last two are the most popular driving principles for Silicon MEMS, using common fabrication techniques such as MUMPS surface micromachining, Sandia SUMMIT, and DRIE etching.

When micromachined from silicon, thermal flexure actuators can be made alongside other passive micro-components such as flexure joints, beams, and gears, and, unlike piezo and electrostatic drives, they do not require large operating voltages.

Electrostatic comb-drives are also popular microactuator alternatives to thermal bimorphs. Some of the advantages of using MEMS comb-drives are their superior repeatability, low power consumption, higher bandwidth, and increased number of cycles, but disadvantages include high driving voltages and reduced displacement and force outputs.

If MEMS actuators are used in high-speed actuation using flexures or other compliant mechanical structures, suppressing residual vibrations is highly desirable. Other types of performance requirements are maximizing the actuation speed of lower bandwidth thermal MEMS devices, minimizing overshoot, following a certain motion profile, etc. In optical switching systems, for example, we have to control point-to-point motion of micro-mirrors used for redirecting optical signals. Typical settling times are a few milliseconds [3].

Increased precision and speed requirements placed on MEMS structures are traditionally addressed using a combination of quantitative analysis, modeling, and experimentation [1,2]. Modeling techniques include simplified PDE/ODE formulation from physical principles, Finite Element Analysis, and Nodal Analysis. More recent work includes reduced-order modeling using Krylov subspace techniques [7]. Some previous modeling work is simply based on extraneous vision and laser-based position measurement [4]. In addition, finite element analysis and direct experimentation have also been used to generate open-loop driving signals for these actuators [2].

While analytical tools such as MEMS motion analyzers are extremely useful for characterizing and measuring actuator response to different input stimuli, they do not address the more practical inverse control problem: *given a MEMS actuator output displacement profile, what should the input voltage (or current) be in order to achieve the desired output.* Note that if measurable outputs (displacement, force, stress) were directly available, the inverse problem can be best addressed by using closed-loop control. In this paper we aim to address the inverse control problem without using additional sensors for feedback.

Input shaping is a popular control method for vibration reduction, particularly well suited in applications where feedback signals are not available. In the context of vibration suppression, the zero-vibration-derivative (ZVD) method, introduced by Singer and Seering [5,6] (1989) is well-known. The ZVD method essentially consists of finding an FIR filter which applied to any input suppresses vibration, using the first few resonant modes of the system. Other input-shaping work [8,9] using a model matching technique makes it possible to enforce realistic input and state constraints.

In this paper, we use this time-domain model matching method for the synthesis of open-loop driving signals for MEMS actuators. The identification method is done through an ARX (Auto-Regressor-with eXtra-input) reduced order model fit from actual experimental data. Finally, we apply the optimal inputs back to the actuators in order to test the performance improvement.

2. Reduced-Order Models for MEMS

The geometry of a basic thermal bimorph and an electrostatic comb drive is shown in Figure 1.

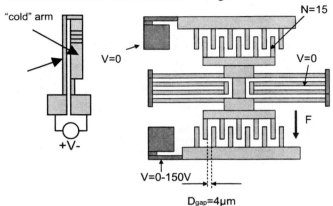

Fig 1: Discretized thermal bimorph (left) and electrostatic comb drive (right)

The MUMPS rotary stage shown in Figure 2 has a hot arm with a width of 8μm, about a third of the width of the cold arm. The angular velocity of the stage depends on the motion profile of each of the actuated arms and the size of the teeth. Electrical current applied on the square contact pads provides the input for the thermal bimorph actuator. Its deflection is governed by thermal expansion resulting from heat dissipation, according to the heat equation.

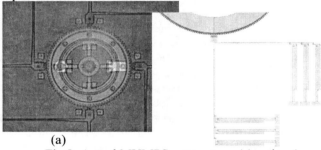

(a)

Fig 2: Actual MUMPS rotary stage (a), using 4 orthogonal thermal bimorph banks (b).

In [25], it was shown that a spatial discretization of the heat equation leads to an expected 3-rd order ARX model for a simple thermal bimorph:

$$x_k + a_1 x_{k-1} + a_2 x_{k-2} + a_3 x_{k-3} = b_1 V_{k-1}^2 \quad (1)$$

In equation (1), V is the voltage applied to the actuator, and x is the actuator deflection at its tip. As a result, by considering that $u=V^2$ is the plant input, we expect to fit a linear ARX model to experimental data.

For thinner (2-4 μm) MUMPS bimorphs, bandwidths between 4 and 27 KHz, and displacements of up to 14 μm

are reported, and depend on the geometry and number actuators, as well as on the environmental conditions. For thicker substrates, the force generated can increase from a few μN to a few mN, while the bandwidth will also decrease to hundreds of Hz [1,2].

Similarly, a good reduced order model for a comb drive actuator is a second order model dependent on the number and geometry of the comb fingers:

$$F = N\varepsilon_o \frac{h}{d_{gap}} V^2, K = 4Eh \frac{W^3}{L^3}, B = M \frac{\omega_o}{Q_{exp}}$$

$$M \frac{d^2 x}{dt^2} + B \frac{dx}{dt} + Kx = F$$

This is in fact a second order model between the square of the applied comb voltage and the shuttle displacement.

3. Input shaping using the time-domain output matching method

Using the reduced-order models, we pose the following constrained optimal control problem:

Given plant model G, and a desired output trajectory $y_d(t)$, find the optimal, constrained control input u(t) minimizing the 2-norm:

$$\min_{u \in L_2[0,T]} \| Gu - y_d \|_2$$
$$\max_{u \in L_2[0,T]} | u(t) | \le u_{max}, u(t) \ge 0 \quad (2)$$

Note that the input constraints in (2) are very important since in our case u(t) is the square of the voltage applied to the MEMS actuator. Too high a voltage will cause plastic deformation or device burnout.

We can pose the optimization problem (2) in terms of a set of M basis functions to express our input:

$$u(t) = \sum_{k=1}^{M} a_k \Phi_k(t), 0 \le t \le T,$$

where the basis functions could be any independent set, including the sinusoidal, Schroeder-phase, etc. bases, or the natural basis of R^M. For the sinusoidal basis functions

$$\Phi_k(t) = \begin{cases} 1, k=1 \\ \sin(\frac{(2(k-1)-1)}{2T} \pi t), k>1, t<T, \\ (-1)^k, k>1, t<T \end{cases}$$

because the plant G is linear, if

$y_k = G\Phi_k$, then the plant output y for input u will be given by

$$y(t) = \sum_{k=1}^{M} a_k y_k(t) = Y^T(t)X, X = (a_1 a_2 ... a_M)^T.$$

A discrete-time equivalent of equation (5) for N samples of the interval [0, T], is given by:

$$\min_{X \in R^M} \sum_{i=1}^{N} (Y^T(t_i)X - y_d(t_i))^2, t_i = (i-1)\frac{T}{N}$$

$$\max_{X \in R^M, 1 \le i \le N} \| \Phi^T(t_i)X \| \le u_{max}, \Phi^T(t_i)X \ge 0$$

The optimization problem reduces to the constrained least-square problem

$$\min_{X \in R^M} (\frac{1}{2} X^T H X - fX), AX \le b, \text{ where}$$

$$H = \sum_{i=1}^{N} Y(t_i)Y^T(t_i), f = -\sum_{i=1}^{N} y_d(t_i)Y^T(t_i),$$

$$A = \begin{bmatrix} [u_k(t_i)]_{1 \le k \le M, 1 \le i \le N} \\ -[u_k(t_i)]_{1 \le k \le M, 1 \le i \le N} \end{bmatrix}, b = \begin{bmatrix} u_{max,k}(t_i) \\ 0^{Mx1} \end{bmatrix}.$$

The solution of the numerical optimization problem depends on the choice of basis, number of basis elements, and the desired input shape.

3. Experiments

Using a UMECH networked probe station instrumented with a stroboscopic, high resolution camera, we performed transient displacement experiments on 3.5 µm thick, 250 µm long, MUMPS single bimorph actuators and bimorph banks, 50 µm thick DRIE rotation and translation drives and MUMPS electrostatic comb drives. The transient dynamic displacement data was used to fit appropriate reduce-order ARX models according to the theoretical predictions. Figure 3 shows the input response of a XY rotator stage to a square pulse train with 25% duty cycle, at 500 Hz, along with a first order ARX model fit. We notice that the system has an overdamped behavior, and, in fact, the thermal bandwidth of the actuator is 4.33 KHz, an order of magnitude lower than the first mechanical resonant mode. As a result, the system behaves essentially like a first order pole.

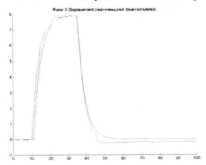

Fig 3: Experimental dynamic response of an XY stage, showing both the measured x displacement (noisy), and the model fit response (smooth).

The first order model fit corresponding to experimental I/O data is non-minimum phase, but it is still first order:

$$G_p(s) = \frac{0.01088s - 1088}{s + 12490}, Y = G_p V^2, T_r = 80\mu s$$

Another set of experiments was performed on a more complex MEMS device, consisting of a DRIE mirror assembled into an actuated socket, similar to the one shown in Figure 4.

200µm

Fig 4: Deep Reactive Ion Etching (DRIE) mirror-stage assembly used as a Variable Optical Attenuator (VOA). Picture courtesy of Zyvex Corp.

In the case of the actuated socket-mirror assembly, the system response to an input pulse (50% duty cycle at 200 Hz) is highly oscillatory and shown in Figure 5, and the ARX model fit is 9-th order.

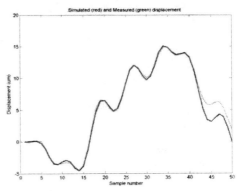

Fig 5: 50% duty cycle pulse response showing measured data (top), and the simulated 9-th order ARX fit (bottom).

Using the fitted models, we generated optimal input shapes to increase the actuator performance. For the rotary stage, by performing an actuation sequence shown in Figure 7 (top) using the input profiles in Figure 7 (bottom), we experimentally increased the rotor RPM from the initial 1390 RPM to 1666 RPM, a 20% increase, as shown in Figure 6.

For the scanning mirror, the response in Figure 5 can also be improved by shaping the input voltage according to Figure 8.

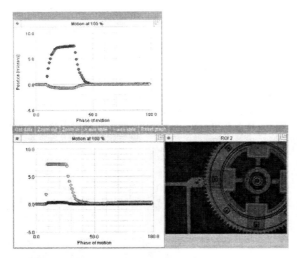

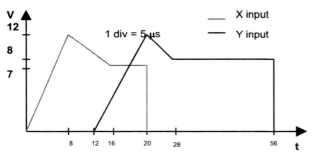

Fig 6: Experimental response of XY stage to a pulse (top), and to a shaped input (bottom).

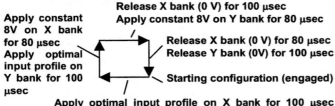

Release X bank (0 V) for 100 μsec
Apply constant 8V on Y bank for 80 μsec

Apply constant 8V on X bank for 80 μsec

Release X bank (0 V) for 80 μsec
Release Y bank (0V) for 100 μsec

Apply optimal input profile on Y bank for 100 μsec

Starting configuration (engaged)

Apply optimal input profile on X bank for 100 μsec
Apply 0V on Y bank for 80 μsec

Fig 7: Actuation sequence necessary to achieve a single tooth rotation (top), using shaped inputs (bottom).

5. Conclusion

In this paper we presented a model-based approach to generating optimal inputs for controlling displacement of thermal MEMS actuators. An open-loop approach is advantageous for these devices because of difficulties in integrating simple and reliable sensors within the actuators. Once ARX models were derived from experimental data, shaped inputs were generated using the time-domain input matching method. We are currently applying the input shaping method to improve the performance of many other MEMS devices.

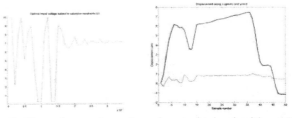

Fig 8: Shaped scanning mirror input obtained with a 10V maximum voltage constraint, and M=40 basis elements. (left) and Experimental scanning mirror displacement measured at the top after aplying the shaped input (right).

Acknowledgement

This work was performed under the support of the U.S. Department of Commerce, National Institute of Standards and Technology, Advanced Technology Program, Cooperative Agreement Number 70NANB1H3021, and the National Science Foundation Grant #CMS-0301827. We wish to thank Kimberly Tuck, Corina Nistori, and Aaron Geisberger of Zyvex Corporation for their help with the UMECH experiments.

References

[1] John H. Comtois, Victor M. Bright, Mark W. Phipps, "Thermal Microactuators for Surface Micromachining Processes", *Proc. Of SPIE*, vol2642, 1995, pp.10-21.
[2] David M. Burans, Victor M. Bright, "Design and performance of a double hot arm polysilicon thermal actuator", *Proc. Of SPIE*, vol3224, 1997, pp.296-306.
[3] J. C. Chiou, Yu-Chen, Yi-Cheng Chang, "Dynamic Characteristics Measurement System for Optical Scanning Micromirror", *Proc. Of SPIE, Micromachining and Microfabrication*, vol4230, 2000, pp.180-186.
[4] Y. Luo and B.J. Nelson, "Fusing Force and Vision Feedback for Manipulating Deformable Objects," *Journal of Robotic Systems*, 18(3), pp. 103-117, 2001.
[5] N.C. Singer, W.P. Seering, "Design and Comparison of Command Shaping Methods for Controlling Residue Vibrations", *IEEE Intl. Conf. on Robotics and Automation*, Scottsdale, AZ, 1989.
[6] W.E. Singhose, N.C. Singer, W.P. Seering, "Shaping Inputs to Reduce Vibration", *IEEE Intl. Conf. on Robotics and Automation*, Cincinatti, OH, 1990, pp. 922-927.
[7] J. Clark J. Demmel K. Pister N. Zhou Z. Bai, D. Bindel. New numerical techniques and tools in SUGAR 3d MEMS simulation. Tech Proc 4rd Intl Conf On Modeling and Simulation of Microsystems, 2001.
[8] J.T. Wen, B. Potsaid, "Input Shaping for Motion Control", CAT Report, Rensselaer Polytechnic Institute, May 2002.
[9] Dan Popa et. al., "Dynamic Modeling and Input Shaping of Thermal Bimorph Actuators", in *Proc. IEEE ICRA'03*, Taipei, Taiwan.

Function-Oriented Geometric Design Approach
To Surface Micromachined MEMS

Feng Gao   and   Y. Steve Hong

Department of MIME, University of Toledo, Toledo, OH, USA, 43606
{ fgao, yhong } @eng.utoledo.edu

ABSTRACT

Geometric modeling is an important aspect of MEMS design. It not only creates geometric model for visual valuation, but also supplies input for device performance analysis. This paper focuses on developing a feature-based geometric design methodology that enables designers to create fabrication-ready 3D models of MEMS devices without concerning the mask layout. Compared with present geometric design routine, which builds 3D device model through simulating the fabrication process from the photolithography masks, the function-oriented geometric design method allows designers to establish 3D model by using a set of pre-defined volumetric primitives associated with geometric constraints. The fabrication information is derived from corresponding function-oriented data specified by designers. Hence, designers are released from the downstream fabrication planning, and can focus on creative design. This research is the application of feature modeling and constraint-based design to the micro world.

Key Words: Surface Micromachined MEMS, 3D Geometric Design, Feature Modeling, Design by Feature

1   FUNCTION-ORIENTED PARADIGM

As we know, the microsystem is originally developed from the semiconductor industry. From the structural material, fabrication methods, to design paradigm, the microsystem inherits the characters of integrated circuits. Design a new device starts from figuring the masks. The design scheme is represented as a set of photolithograph masks. In this mode, the factors of the fabrication, rather than the function implementation, are the preferentially concerned topics. Therefore, the paradigm is fabrication-oriented. The 3D solid device model is generated after planning fabrication data. This paradigm, mainly concerning device fabricationbility, makes the modeling procedure un-intuitive and cumbersome. Present MEMS geometric modeling approaches actually are simulation-based design verification tools, not design aid tools. Along with the development of microfabrication, especially the surface micromachining, more structural layers are involved, which make it possible that complex spatial structure can be made. However, under this situation, it is more and more difficult to handle function implementation and fabrication planning at the same time and by the same person. This difficulty imposes designer a heavy burden which may diverge them from creative contribution.

In view of the drawbacks of the fabrication-oriented design paradigm, an alternative scenario is expected to facilitate fundamental improvement on microsystems design. The concept of "structured design method" for microsystems [2, 3, 4] was proposed to "develop and integrate methodologies, tools, environments and technologies needed to be able to automate the rapid, efficient and accurate design and construction processes, artifacts and systems of artifacts". A key long-term objective is to develop a methodology that can be applied generally to mechanical and electromechanical systems. After several years of effort, although the situation has been greatly improved, this kind of demand still remains. Aiming to comprehensively solve the problems existing in designing surface micro-machined microsystems, the function-oriented design paradigm is proposed. The paradigm flow chart is shown in Figure 1. At the beginning phase of product development cycle, the 3D model can be built directly through the feature-based geometric modeler according to the designer's intents (desired functions). Following it, the 3D device model is visual evaluated and fed to the domain analysis tools to predict device performance before it is physical fabricated. Model modification need not refer to changing masks any more; instead, any adjustment of design can be completed directly on the 3D geometric model. The photolithograph mask is generated on the refined device model. This phase is moved from the upstream stage to downstream stage.

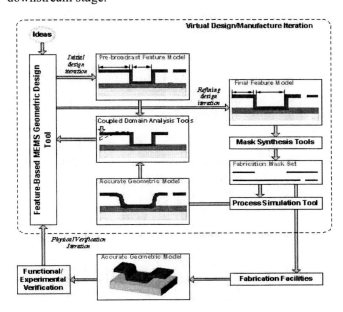

Figure 1: Function-Oriented Design Paradigm

In the function-oriented design paradigm, designers are allowed to directly manipulate and visualize the 3D model of the microdevice that they are working on, without detouring through the steps of mask layout and process simulation. Besides the precise geometric representation, the feature model of MEMS is compatible to carry comprehensive information that is valuable for most design activities in the whole product developing life circle. Finally, the mask synthesis tool generates masks from 3D device model. The two key points of implement this routine are *how to decouple the task of design and fabrication planning, and how to find the interrelation between them in order to automatically derive fabrication data from design information.*

2 DECOUPLING OF DESIGN AND FABRICATION PLANNING

Because the coupling of design and fabrication planning is the major reason leading current difficulties of modeling and maintaining the 3D MEMS device model [1], decoupling is the natural way to solve this problem. When designing a RF switch schematized in Figure 2, designer's concern is concentrated on some spatial structures which execute desired functions (such as the bending of the polysilicon beam, the dimple between two beams) and some geometric dimensions which may considerably influence the device performance (beam length $d1$ and $d2$, width $d4$, gap distance $d3$, and dimple width $d5$). It will be intuitive and efficient for designer if the core component, polysilicon-1, can be separately and directly constructed without considering the etching operations on all involved layers.

It is assumed that a predefined fabrication process should be selected before a surface micromachined MEMS device is designed. Among the fabrication steps of surface micro machining, some (all kinds of deposit) are un-controllable for designer because the process parameters of deposit (type, material, thickness) are fixed when a standard fabrication process is selected to make the device. Designers have no authority to change them. On the contrary, some steps are controllable for designers since designers can specify the acting area of selective etching or doping, although some other process parameters cannot be modified by designers. Currently, the major task of MEMS geometric design is to specify the acting area of etching, the task of fabrication planning.

From the cross section of RF switch in Figure 2, the device geometry can be categorized into two folds. The first one is the geometry carved by high-aspect-ratio etching. For instance, the etching on the polysilicon-0 layer makes the square electrode. Apparently, as having a strong relation with fabrication, it is called fabrication-oriented (FAO) geometry. The other fold is the geometry generated when material is deposited on the previous etched layers. For instance, the bending and dimple on polysilicon-1 layer are formed by depositing material on the etched underneath layers. Sometimes, these types of shapes and their geometric parameters directly express design intents somehow. Hence,

these shapes can be seen as function-oriented (FUO geometry. These two forms are adopted to classify the task c design and fabrication from the view of geometry.

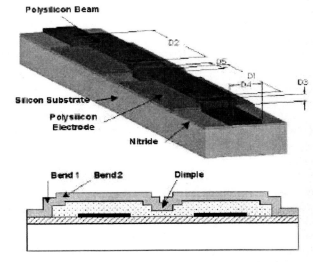

Figure 2: RF Switch (a), 3D model and constituent layers; (b), Cross-section and function-oriented geometry.

In previous research on feature modeling for MEMS [7], two sets of features, design feature and fabrication featu , are defined to construct MEMS device model. Five gener design features, including bend, protrusion, cut, anchor ar transferring, are used to directly construct 3D device mod A feature instance is initialized by unambiguously specifyi feature type, planar outline and depth value. Then, t volume of feature instance is merged into the original mod to change device shape. The feature planar shape and locatic can be defined by imposing geometric constraints. The shap of most surface micromachined microdevices can represented in the terms of these five design features and th combination. Two fabrication features are the two design controllable fabrication steps, etching and doping. The features are used to represent fabrication operations.

The feature structure is diagramed in Figure 3. Desi features are the interface between fabrication operations a designers. They are the engineering meaningful primitives which designers construct device model according to desir function. Fabrication features are connected with fabricati process. As not transparent to designers, they are n explicitly specified. Actually, they are automatically deriv from corresponding design features. Both the volumes c design feature and fabrication feature instances compose t MEMS feature model.

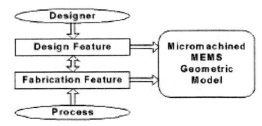

Figure 3: Bi-level Feature Structure

3 DESIGN FOR MANUFACTURING (DFM) FOR MEMS

The purpose of decoupling design and fabrication is to handle the fabrication planning automatically and systematically, rather than ignore the issue of fabrication in design phase. Manufacturability is one of the major factors evaluating design scheme. For surface micromachined MEMS, due to the restricted fabrication measures, arbitrary spatial shape cannot be made. This is the reason why present MEMS designers start a new device from figuring masks. Design for manufacturing (DFM), a key component of Current Engineering (CE), has been introduced into the MEMS system optimization [5]. In this research, DFM will be addressed to guarantee the manufacturability of a microdevice from the perspective of geometry.

Since MEMS devices are fabricated by the iterations of layer deposit and etching, tight contacts exist among the materials of each layer. Hence, the geometries carved by high-aspect-ratio etching in the lower layer must influence the shape of the upper layers. Based on this phenomenon, called geometric dependency, the fabrication feature can be derived from the inputted design feature. If a fabrication feature can be sought, the corresponding design feature is valid and can be attached on the object. Otherwise, it cannot be initialized and attached to the device model.

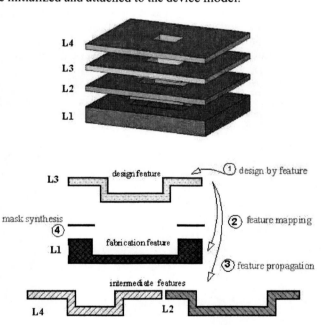

Figure 4: Feature Mapping and Feature Propagation

For the example schematized in Figure 4, the model consists of four structural layers which are labeled as L1, L2, L3, and L4, respectively. In the first step, a design feature "bend" is imposed on layer L3. The shape of layer L3 is updated immediately by merging the feature instance into the original layer. Secondly, a fabrication feature acted on layer L1 can be derived to achieve the "bend" shape because there is an etching acted on the layer L1 according to the process.

This deriving process is called feature mapping. Besides updating the layer L3 and L1, all other involved constituent layers should also be updated, otherwise there will be volumetric interference among layers. Thirdly, two similar bend shapes should be initialized and attached on layer L2 and L4 respectively. This process is called feature propagation. Finally, the mask for the etching can be derived by extracting the planar outline of the fabrication feature attached on L1. In summary, the information transforms as the flowchart diagramed in figure 5. First, designers transfer intents into design features. Then, fabrication feature is derived through feature mapping. Some new design features may be generated through feature propagation. Finally, the fabrication feature is transferred to a photolithograph mask layout.

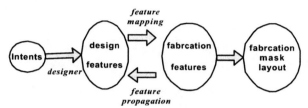

Figure 5: Information Transformation Flow

4 DEMO

Figure 6 shows the procedure applying the function-oriented geometric modeling method on RF switch design. Originally, there are four untouched constituent layers, *nitride, polysilicon-0 (PS0), sacrificial oxide(SO)* and *poly-silicon-1 (PS1)*. In the first step, a "bend" feature is imposed on *PS1*. Its corresponding fabrication feature is the etching on *SO*. The bend forms the approximate shape of beam. Then, a "cut" is imposed on *PS0* to form the electrode. This feature will be propagated up to layer *SO* and *PS1* forming bend shape. In the third step, a bend feature, which forms the dimple in the middle position of the beam bottom, is attached on *PS1*. Its corresponding fabrication feature is also the etching on *SO*. But this time it is to-depth etching, not previous to-layer etching. In the fourth step, a cut feature is acted on *PS1* to control the beam width. Finally, wet etching solvent washes of the whole layer *SO*, then the component beam is released. The final 3D model is shown in figure 1.

This modeling procedure is totally different from that of the simulation-based modeling method [8]. It does not need mask as input. The modeling process is driven by the design intents. Designer can arbitrarily move their focus on the layers which they are interested with, then impose design feature on them. They need not worry about the manufacturability of the operations they make. The manufacturability has been checked during the process of feature mapping. This is a good design aid tool, rather than a design verification tool.

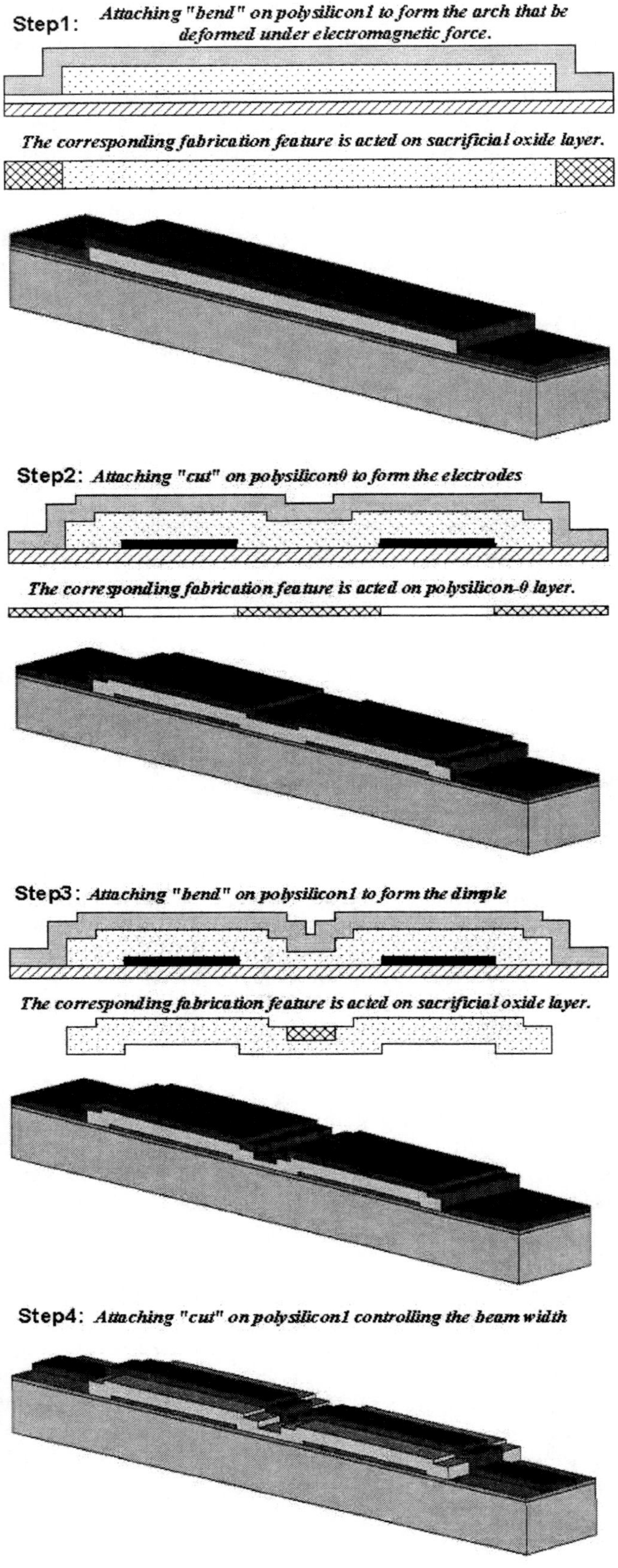

Step1: *Attaching "bend" on polysilicon1 to form the arch that be deformed under electromagnetic force.*

The corresponding fabrication feature is acted on sacrificial oxide layer.

Step2: *Attaching "cut" on polysilicon0 to form the electrodes*

The corresponding fabrication feature is acted on polysilicon-0 layer.

Step3: *Attaching "bend" on polysilicon1 to form the dimple*

The corresponding fabrication feature is acted on sacrificial oxide layer.

Step4: *Attaching "cut" on polysilicon1 controlling the beam width*

Figure 6: Design Process of RF Switch

5 CONCLUSION AND FUTURE WORK

In summary, the feature-based function-oriented geometric modeling approach greatly changes the picture of MEMS device geometric design. Under the support of function-oriented design tools, designers are allowed to concentrate on establishing suitable structure to implement desired functions. The fabrication data is automatically derived from design input. Finally, the mask set can be synthesized from derived. It is a practice of feature modeling technologies which are mature in macroworld on the domain of microdevice.

Work is continuing towards generalized feature representation, feature interference handling and process independent modeling.

REFERENCE

[1] Antonsson, E., "Structured Design Methods for MEMS, Final Report", *NSF MEMS Workshop, CalTech.*, 1995.

[2] Antonsson, E., "Microsystem Design Synthesis", Chapter 5 *Formal Engineering Design Synthesis*, Cambridge University Press, pp126-169, 2001.

[3] Fedder, G.K., "Structured Design of Integrated MEMS", *Technical Digest of the 12th IEEE International Conference MEMS'99*, Orlando, FL, January 1999.

[4] Senturia, S.D., "CAD Challenges for Microsensor, Microactuators, and Micro-systems", *Proc. of IEEE*, Vol.8 No.8, pp. 1611-1626, August 1998.

[5] de Silva, M.G., "Design For Manufacturability for MicroDevices", *NSF Workshop on 3D Nanomanu-facturing Partnering Industry*, Jan. 5-6, 2003, Birmingham, AL.

[6] Gao, F., Hong, Y. and Sarma, R., "Feature Model for Surface Micromachined MEMS", 2003 ASME *DETC/CIE-481*, September, 2003, Chicago, IL.

[7] Gao, F., and Sarma, R., "A Declarative Feature-Based Cross Sectional Design Tool for Surface Micromachined MEMS", 2001 ASME DETC2001/CIE-21775, Sept. 2001. Pittsburg, PA.

[8] Jorgensen, C.R. and Yarberry, V.R., "A 3D Geometry Model for the SUMMiT V MEMS Designer", *Proc. of MSM 2001*, pp.594-597. Feb. 2001.

New Accurate 3-D Finite Element Technology for Solving Geometrically Complex Coupled-Field Problems

I. Avdeev[*], M. Gyimesi[**], M. Lovell[*] and D. Ostergaard[**]

[*]University of Pittsburgh, Pittsburgh, PA, USA, mlovell@pitt.edu
[**]ANSYS, Inc., Canonsburg, PA, USA, miklos.gyimesi@ansys.com

ABSTRACT

A novel 3-D transducer formulation has been developed to solve geometrically complex coupled-field problems. The transducer is compatible with both structural and electrostatic solid elements, which allows for modeling complex devices. Through internal morphing capabilities and exact element integration the 3-D transducer element is one of the most powerful coupled field FE analysis tools available. To verify the accuracy and effectiveness of 3-D transducer a series of benchmark analyses were conducted.

Keywords: MEMS, FEA, coupled-field, 3-D transducer

1 INTRODUCTION

The increased functionality of MEMS fabrication and production techniques has lead to the ability of creating devices and components with complex geometrical configurations. These components require efficient FE modeling techniques to solve coupled electromechanical problems. The lumped models are no longer applicable for devices, such as combdrives or electrostatic motors, where fringing electrostatic fields are dominant [1]. There have been several numerical methods proposed for the treatment of electromechanical systems including: FE or boundary element methods using sequential physics coupling; strongly coupled but reduced order methods using fully lumped or mechanically distributed but electrically lumped 1-D, multi-dimensional or modal-space transducers. All of these methods need some extra meshing or morphing, introduce simplifying assumptions and may not be convenient to use [2].

In the present investigation, a novel strongly coupled 3-D tetrahedral transducer element is introduced for modeling analog electrostatic MEMS devices. This new transducer element, which can be utilized for a broad range of micro-system applications (combdrives, micromirrors, and electrostatic motors), is compatible with conventional electrostatic and structural 3-D finite elements. The element is capable of efficiently modeling interaction between deformable or rigid conductors that generate an electrostatic field. Strong coupling between the electrostatic and mechanical domains allows the static element formulation to be extended to transient and full harmonic analyses. Therefore, in many respects, the element is the most sophisticated FEA tool available for modeling MEMS problems where dominant fringing fields develop. The new technology is also very efficient in determining the pull-in parameters of complicated multi-electrode microdevices.

2 COUPLED-FIELD FE FORMULATION

The electrostatic 3-D domain of a coupled-field electro-mechanical problem is meshed with the new 3-D elements. The 3-D transducer element has a tetrahedral shape with the geometry fully defined by four nodes (see Figure 1).

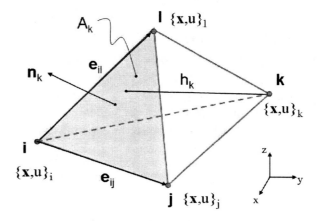

Figure 1: 3-D transducer element geometry and degrees of freedom (mechanical displacements and electrostatic potentials).

Each element node has four degrees of freedom: three components of a nodal displacement vector defined in a global Cartesian coordinate system, $\mathbf{x} = (u_x, u_y, u_z)$, and one potential of electrostatic field, u. There are three mechanical forces associated with each node: F_x, F_y, and F_z, and a nodal charge, Q. Transducer element potential energy is given by

$$W_e = \frac{\varepsilon}{2}\int_V \mathbf{E}^2 dv = \frac{\varepsilon}{2}\int_V (gradU)^2 dv \qquad (1)$$

In (1), ε is the element permittivity (assumed constant for the sake of simplicity), V is the element volume, $\mathbf{E}$ is the electrostatic field intensity vector, and U is the electrostatic potential. For a linear tetrahedral element the integral in (1) can be evaluated analytically:

$$W_e = \frac{V\varepsilon}{2}(gradU)^2 = \frac{V\varepsilon}{2}\sum_{i=1}^{4}\sum_{j=1}^{4}\frac{\mathbf{n}_i \cdot \mathbf{n}_j}{h_i h_j}u_i u_j \qquad (2)$$

In (2), h_i are the tetrahedral altitudes, $\mathbf{n}_i$ are the element's face normals, and u_i are the nodal electrostatic potentials. The energy is a function of the element geometry and nodal potentials. One of the important results is that (2) is an invariant formula for energy calculation. The mechanical forces and electrical charges in the element are calculated using virtual work principle [3]. The vectors of nodal mechanical forces and charges are calculated using the principle of virtual work by differentiating the element energy with respect to nodal mechanical displacements and electrostatic potentials [3]. The static non-linear equilibrium equations for the electrostatic domain meshed with transducer elements can be written in an incremental form suitable for the Newton-Raphson equation solvers

$$\begin{bmatrix} \mathbf{K}_{xx}(\mathbf{x},\mathbf{u}) & \mathbf{K}_{xu}(\mathbf{x},\mathbf{u}) \\ \mathbf{K}_{ux}(\mathbf{x},\mathbf{u}) & \mathbf{K}_{uu}(\mathbf{x},\mathbf{u}) \end{bmatrix}\begin{bmatrix} \Delta\mathbf{x} \\ \Delta\mathbf{u} \end{bmatrix} = \begin{bmatrix} \Delta\mathbf{f}(\mathbf{x},\mathbf{u}) \\ \Delta\mathbf{q}(\mathbf{x},\mathbf{u}) \end{bmatrix} \qquad (3)$$

In (3), $\Delta\mathbf{x}$ and $\Delta\mathbf{u}$ are the increments of vectors of the nodal mechanical displacements and electrostatic potentials, $\Delta\mathbf{f}(\mathbf{x},\mathbf{u})$ and $\Delta\mathbf{q}(\mathbf{x},\mathbf{u})$ are the increments of the out-of-balance nodal forces and charges respectively. The blocks of the coupled-field tangent stiffness matrix in (3) are computed by differentiating nodal forces and charges with respect to nodal mechanical displacements and electrostatic potentials [3]. To increase accuracy and to ensure a robust convergence of the non-linear solution, the integrals and derivatives of the element's formulation are calculated analytically.

3 MESH MORPHING

An internal mesh morphing capability is an important feature of the developed transducer element that separates it from lumped transducers and models based on a sequential coupling. Mesh morphing is a process of updating vectors of nodal displacements of the transducer elements during the solution of a non-linear problem. The number of elements remains constant so that element continuity is maintained during mesh morphing. In the element, there are out-of-balance mechanical forces acting upon each node. The interface nodes (nodes on the surface of electrodes) generate the electrostatic force that deforms the mechanical structure. Every inner node moves in a direction defined by the resulting out-of-balance force acting upon the node.

Structural stiffness of the transducer elements is inversely proportional to the element volume, i.e. the bigger elements are softer than the smaller ones. This is extremely important for modeling geometrical singularities such as sharp corners or edges. The mesh must be refined around the singularities in order to capture strong electrostatic fields and to accurately compute driving forces. High forces, however, can "invert" transducer elements if their structural stiffnesses are not big. For this reason, the structural stiffness of transducers is weighted based on their size (volume).

The convergence speed and solution accuracy of the element during a non-linear solution depends on many parameters, the most important of which are the convergence tolerance (CT) and the morphing acceleration factor (MAF). The first parameter is a convergence criterion used by an iterative Newton-Raphson solver. The second parameter is a factor used to stiffen or soften all of the transducer elements. Increasing the morphing acceleration factor produces a stiffer mesh, which could be necessary for strong singularities or small displacements.

4 NUMERICAL EXAMPLES

Several benchmark problems are solved using the new 3-D transducer element. The results are compared to experimental data available in the literature and to the solutions obtained using traditional techniques and commercial software (such as ANSYS\Multiphysics).

4.1 Parallel Plate Capacitive Transducer

As a first example to demonstrate the new element capabilities, we will compute a static equilibrium state of a parallel plate capacitive transducer [2]. The transducer consists of two electrodes separated by a gap that is a function of applied voltage and a stiffness of the suspending spring. This problem has an analytical solution for rigid electrodes and a lumped spring without accounting for fringing electrostatic fields [2]. The FE model consists of 100 transducers and one spring element (Figure 2).

CT	Time (s)	Iterations	Stroke	Error (%)
0.02000	2.84	7	0.16989	15.06
0.00500	3.71	12	0.19184	4.08
0.00050	5.34	21	0.19914	0.43
0.00005	6.97	30	0.19991	0.05

Table 1: Solution convergence speed and accuracy for various convergence tolerance (CT) values.

The electrostatic potential distribution is depicted in Figure 2. In Table 1, the solution accuracy and speed are compared for different values of convergence tolerance (morphing acceleration factor is set to one). At a smaller value of the tolerance, more iterations are required to converge, but the solution becomes more accurate. This is an intuitively simple, but very important result. In Table 2, the solution accuracy and speed are compared for different values of morphing acceleration factor (convergence tolerance is equal to 0.00005 N).

MAF	Time (s)	Iterations	Stroke	Error (%)
0.25	4.51	16	0.19991	0.05
1.00	6.97	30	0.19991	0.05
4.00	17.16	85	0.19990	0.05

Table 2: Table 4: Solution convergence speed and accuracy for various morphing acceleration factor (MAF) values

A smaller factor (softer mesh) leads to a faster convergence, while a larger factor leads to a slower convergence. Note, that the accuracy essentially remains independent of the acceleration factor.

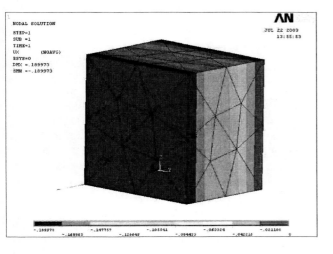

Figure 2: Parallel plate transducer: electrostatic potential distribution (3-D).

To achieve more accurate solution, one must therefore adjust the value of the convergence tolerance. The MAF can be further used to speed up the convergence for a given accuracy. It should be noted that there is a bottom MAF limit which cannot be reached without loosing the integrity of the transducer elements (transducer FE mesh becomes too soft mechanically). This limit varies for different problems and boundary conditions.

4.2 Electrostatic Torsion Microactuator

The second example problem that will be used to verify the 3-D transducer is an electrostatic torsion microactuator [4]. The finite element model of the microactuator is depicted in Figure 3. Structural brick elements (green) were used to model the microactuator. The 3-D transducer elements (purple) were used to model air gap between two driving electrodes and the actuator's plates. FE model contains 5,207 transducers and 2,096 structural ANSYS finite elements.

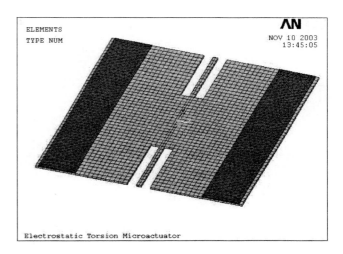

Figure 3: Electrostatic torsion microactuator: structural brick and transducer tetrahedral finite elements.

Alternating voltage between the two electrodes allows the micromirror to turn around the supporting beam axis. Increasing voltage leads to pull-in of the micromirror.

	Theory [4]	Experiment [4]	FEA (3-D)
θ_{pin}, (deg)	0.4042	0.385	0.397
V_{pin} (V)	11.59	11.50	11.55

Table 3: Comparison of theoretical, experimental, and FE pull-in parameters for torsion microactuator

The reported theoretical and experimental values of pull-in voltage are presented in Table 3. The results obtained using the 3-D transducer model are compared to the reported results. The FE results show very good agreement with the theoretical and experimental data [4].

4.3 Lateral Combdrive Transducer

The lateral combdrive electromechanical transducer [5] (Figure 4) is the final example problem to be solved with the new transducer element. FE coupled model contains 12,740 transducer elements. Structural domain is assumed to be rigid. Boundary conditions play an important role in the modeling and solution. The external nodes of the transducer mesh are free to move, while the boundary is fixed in certain directions. The outside boundary (far-field) is fixed with respect to all mechanical displacements. A voltage is applied between the rotor and the stator that produces strong fringing fields with potential distribution depicted in Figures 5 and 6. The mechanical displacement field along the combdrive axis is depicted in Figure 7 (surrounding air is hidden in the figure). The combdrive stroke as a function of the applied voltage was compared to semi-analytical solution and was found to be in a good agreement with it for a carefully chosen convergence tolerance parameter [2].

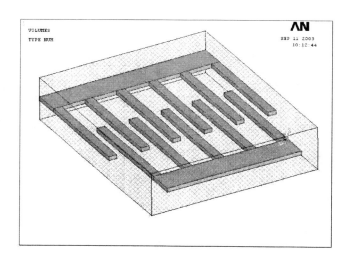

Figure 4: 3-D CAD model of the combdrive transducer.

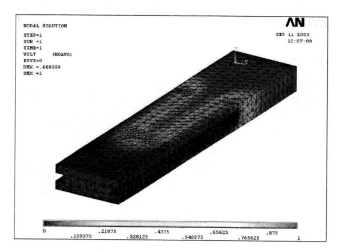

Figure 5: 3-D electrostatic potential distribution of a representative combdrive finger including surrounding air.

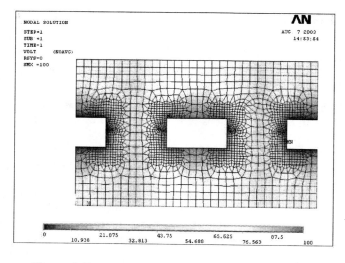

Figure 6: Potential distribution in the middle of the combdrive finger (2-D cross-section).

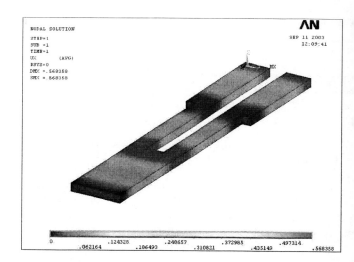

Figure 7: Mechanical displacement field along the combdrive finger axis (stroke).

5 CONCLUSIONS

A distributed 3-D transducer element formulation was developed for modeling a wide range of MEMS devices. The elements accounts for fringing electrostatic fields and the internal morphing capability of the element allows a designer to use the original mesh for solving large displacement non-linear problems. The computer FE code was developed using ANSYS as a platform and several numerical examples were presented which show good agreement with experimental data. The developed element is among the most sophisticated and effective techniques of solving 3-D coupled field problems presently available for designers and researchers working in the MEMS industry.

REFERENCES

[1] I. Avdeev, M. Lovell and D. Onipede, Jr., "Modeling In-Plane Misalignments in Lateral Combdrive Transducers," J. of Micromechanics and Microengineering, V.13, pp. 809-15 (2003).

[2] I. Avdeev, "New Formulation for Finite Element Modeling Electrostatically Driven MEMS," Ph.D. dissertation, University of Pittsburgh (2003).

[3] M. Gyimesi, D. Ostergaard and I. Avdeev, "Triangle Transducer for MEMS Simulation in ANSYS Finite Element Program," Proc. of the Fifth Int. Conf. on Model. Simulat. of Microsystems, San Juan, PR, April 22-25, pp. 380-83 (2002).

[4] O. Degani, E. Socher, A. Lipson, T. Leitner, D. Setter, S. Kaldor and Y. Nemirovsky, "Pull-in Study of an Electrostatic Torsion Microactuator," J. Microelectromech. Syst. V. 7, pp. 373–77, (1998).

[5] W. Tang, T. Nguyen and R. Howe, "Laterally driven polysilicon resonant microstructures," Sensors and Actuators A, V. 20, pp. 25–32 (1989).

Interdigitated Low-Loss Ohmic RF-MEMS Switches

R. Gaddi*, M. Bellei*, A. Gnudi**, B. Margesin*** and F. Giacomozzi***

* Centro di Ricerca "Ercole De Castro" (ARCES), University of Bologna
Via Toffano 2, 40125 Bologna, Italy, rgaddi@arces.unibo.it
** Dipartimento di Elettronica Informatica e Sistemistica (DEIS), University of Bologna
Viale Risorgimento 2, 40136 Bologna, Italy, agnudi@deis.unibo.it
*** ITC-irst, Via Sommarive 18, 38050 Povo (Trento), Italy, margesin@itc.it

ABSTRACT

An interdigitated design for MEMS RF-switches is applied to both a shunt and a series ohmic contact configuration. Interdigitated Al-Ti-TiN RF-signal paths and poly actuation electrodes are arranged underneath an electrodeposited gold plate, suspended by four thinner gold beam springs. Ohmic contact occurs at pull-in between the gold plate and the RF-signal elecrodes only. Measurements show insertion loss better than 0.8 dB and isolation better than 20 dB up to 13 GHz. Extracted lumped element equivalent circuits show intrinsic contact resistances of 1.6 Ω in the shunt and 4.5 Ω in the series switch. The interdigitated topology of RF-signal and actuation electrodes results in uniform contact pressure distribution and consistently low contact resistance.

Keywords: rf-mems, switch, ohmic, low-loss, model

1 INTRODUCTION

RF-MEMS switches have emerged in recent years as a potential alternative to solid state devices for improving reconfigurability in multistandard wireless systems [1]. Key specifications in this respect are isolation between the two RF-ports in open state and insertion loss of the closed switch [2]. Furthermore, compatibility with RF-CMOS circuitry for system-on-package integration would also require low actuation voltages, while simultaneous multistandard terminal operation would require specific transient response times. Compared to capacitive switches, ohmic contact based switches tend to be more broadband and have a better isolation vs. transient time trade-off. Besides, achieving a consistently low and repeatible contact resistance is still an issue to be addressed in terms of technology and topology design. Recently reported low loss ohmic switches were achieved through the use of a special technology including dimples on one of the electrodes during fabrication [3]. This paper presents an alternative solution for achieving low insertion loss while maintaining high isolation. An interdigitated topology for the signal and actuation electrodes is adopted with the aim of obtaining a uniform electrostatic pressure across the suspended

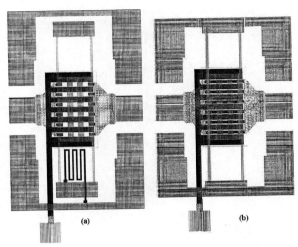

Figure 1: Layouts of the series (a) and shunt (b) interdigitated ohmic switches.

switch element, therefore resulting in consistently low contact resistances.

2 DEVICE DESCRIPTION AND CHARACTERISATION RESULTS

The technology process utilised for the fabrication of the ohmic switches is based on electrodeposited suspended gold membrane layer, one high resistivity poly layer for actuation electrodes and a Al-Ti-TiN multilayer for RF-signal path. Detailed process description is given in [4]. The shunt and series switch differ only for the topology of the RF-signal electrodes. The layouts of both topologies are shown in Figure 1. Input and output RF ports are physically connected by the signal fingers in the shunt switch. On the other hand, the series switch has two isolated RF ports by physically interrupting the signal fingers. The switch is therefore normally open. In both cases, direct contact is allowed only between the plate and the signal electrodes, by rising the signal metal above the level of the poly. This is achieved by placing poly dummy rectangular bricks underneath the part of the signal electrodes directly underlying the suspended plate. The actuated plate will rest on top of the signal electrodes and will not reach contact with the actuation pads.

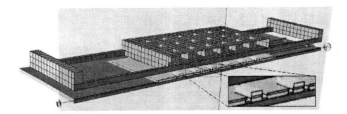

Figure 2: 3D model of the series-ohmic MEMS switch showing a section across the contact area.

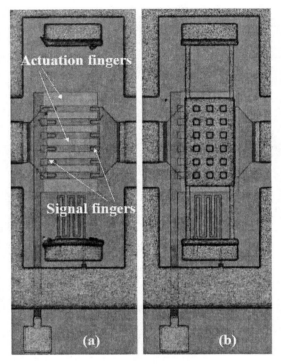

Figure 3: Series-ohmic MEMS switch with suspended plate removed (a) and present (b).

The top LTO layer is also removed from the whole area underneath the plate, through the contact via etch step, allowing direct Au-TiN ohmic contact. A cross section of the interdigitated electrodes topology is shown in the Coventor™ FEM 3D model of Figure 2. A $5\mu m$ thick electroplated gold layer is used for the plate to improve its rigidity, while thinner ($1.5\mu m$) gold implements the four beam springs. In the shunt configuration, the plate collapses and touches the signal fingers creating a low resistance path to ground that blocks the RF signal. On the other hand, in the series switch a high-resistance bridge-to-ground path is present, through a $40k\Omega$ meander poly resistor. This avoids short-circuiting the RF signal when the plate collapses. Therefore, the actuated plate realises a low resistance path between the two RF ports, closing the switch.

Figure 3 shows pictures of a series-ohmic switch, as fabricated and after the removal of the suspended gold plate for showing the underlying interdigitated electrodes.

The devices have been characterised through RF-probing using ground-signal-ground (GSG) configuration. S-parameters were measured from 0.5 to 13.5 GHz, and trough-reflect-match (TRM) calibration was carried out using trimmed gold standards on an alumina substrate. The applied actuation voltage was stepped through a double ramp in order to characterise both device's pull-in and pull-out behaviour. At each voltage step a complete frequency swept s-parameter measurement was taken.

Figure 4 shows s-parameter measurement results for the series switch, at the frequency of 2GHz, during pullin and pullout biasing. The sharp pull-in and pull-out electrical behaviour gives good hints on the correct functioning of the device, both in terms of the mechanical eletro-mechanical collapse and of the ohmic contact formation. Similar behaviour is observed for the shunt switch. Nonetheless, the latter shows generally better performances. Insertion loss better than 0.8dB up to 13.5 GHz, and better than 0.35dB between 2GHz and 4 GHz, is measured and repeatable. Isolation better that 20 dB up to 13.5GHz, and around 25dB below 4 GHz, is also measured and repeatable. On the other hand, both insertion loss and isolation of the series switch are fractionally worse. Pullin voltage is around 29V for both series and shunt switches. This value agrees with 3D FEM simulations provided a tensile residual stress in the suspended gold layer of 145 MPa is taken into account. Results from mechanical characterisation also confirm the presence of a residual tensile strain, which is believed to originate from the thermal annealing steps that follow gold electrodeposition. The design of lower actuation voltage devices with meander spring structures is currently under way, although a trade-off between actuation voltage and transient times is being considered.

Frequency domain s-parameters of the open and closed shunt switch are compared in Figure 5. From measured frequency domain s-parameters, lumped element equivalent circuits are directly extracted for both switches, in the actuated and non-actuated states. The circuit topology is assumed to be a succession of alternating Π and T shells, the outer one being associated to the shunt parasitic capacitance due to the GSG pads. The direct extraction procedure is based on subtracting Y- and Z-parameter matrixes from the measurements until all obtained lumped element values show a constant behaviour versus frequency [5]. Figure 6 shows the extracted equivalent circuits for both open and close shunt and series ohmic switches. Figure 7 compares measured and simulated s-parameters, in terms of isolation and insertion loss, for all extracted circuits. Simulations were performed using Spectre™ simulator in the Cadence™ environment. The shunt switch insertion loss is dominated by the series inductance and resistance of the parallel fingers connecting the two RF-ports. Their extracted values are 200pH and 0.85Ω re-

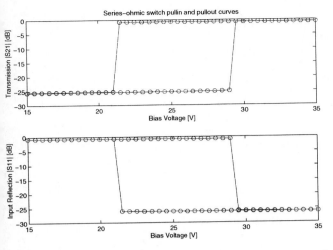

Figure 4: Measurement results at F=2GHz of reflection and transmission pullin and pullout of a series-ohmic switch.

spectively. When the switch is actuated, the intrinsic contact to the grounded plate has a 1.6Ω series resistance and 30pH inductance.

Isolation of the series switch is on the other hand given by the coupling between the two unconnected RF-ports. The extraction identifies one purely capacitive coupling of about 35fF. Improving isolation would therefore require increasing the distance between the two ports' electrodes. The down-state switch shows a series resistance between the two intrinsic RF-ports of 4.5Ω and an inductance of 280pH. This higher value is only partially due to having two contacts in series between the RF-ports, and partially to a reduced contact area compared to the shunt configuration ($\frac{1}{3}$ factor). Further studies are under way to better characterise contact resistance in terms of its dependence on pressure and contact area. From Figure 6 (c), it also emerges that the up-state series switch is not symmetric. This is due to a coupling RLC series network which is present only at the first RF-port. The overlap between the signal and actuation layers, which can be observed from the layout in Figure 1, is believed to be the origin of this. Only the series switch is sensitive to this coupling effect, which could be diminished through an increase in resistivity of the poly layer, together with a slightly modified layout that would reduce the overlap area.

3 CONCLUSION

The proposed interdigitated design of actuation and signal electrodes in ohmic RF-MEMS switches has proved to be a valid approach for achieving low-loss mechanical contacts with a standard surface micromachining technology. The measured insertion loss and isolation for the first prototype devices, in particular the shunt configuration, are compatible with multi-standard wireless RF-transceiver application specifications. A more

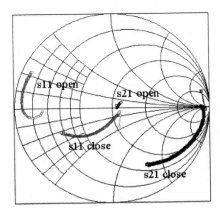

Figure 5: Open and close state frequency-domain s-parameters compared for the shunt switch (F=0.5-13.5 GHz).

even distribution of the electrostatic force across the plate is believed to result in uniform pressure for all contacts, yielding to lower contact resistance. Some improvements regarding both topology and technology have been identified for the next design cycle, especially regarding the series configuration. The extracted lumped element equivalent circuits have shown good agreement between measured and simulated s-parameters up to 13 GHz, and will provide the necessary tool for accurate RF circuit design.

4 ACKNOWLEDGEMENTS

This work was supported by the MIUR as part of the FIRB project "Enabling technologies for Wireless Reconfigurable Terminals" (RBNE01F582).

REFERENCES

[1] E.R. Brown, "RF-MEMS Switches for Reconfigurable Integrated Circuits," IEEE Trans. MTT, Vol.46, pp.1868-1880, Nov. 1998.

[2] J.B. Muldavin, G.M. Rebeiz, "High-Isolation CPW MEMS Shunt Switches - Part 1: Modeling," IEEE Trans. Microwave Theory Tech., Vol.48, pp.1045-1052, June 2000.

[3] A. Pothier, et Al., "Low Loss Ohmic Switches For RF Frequency Applications," 32nd Eu. Microwave Conf. Proc., Milan, IT, 23-27 Sept. 2002, pp.805-808.

[4] F. Giacomozzi, B. Margesin, L. Ferrario, V. Guarnieri, M. Zen, "Micromachined Switches: Realization and Test," Proc. 3rd MEMSWAVE Workshop, Heraklion, Greece, 26-28 June 2002.

[5] G. Dambrine, A. Cappy, F. Heliodore, E. Playez, "A New Method for Determining the FET Small-Signal Equivalent Circuit," IEEE Trans. Microwave Theory Tech., Vol.36, pp.1151-1159, July 1988.

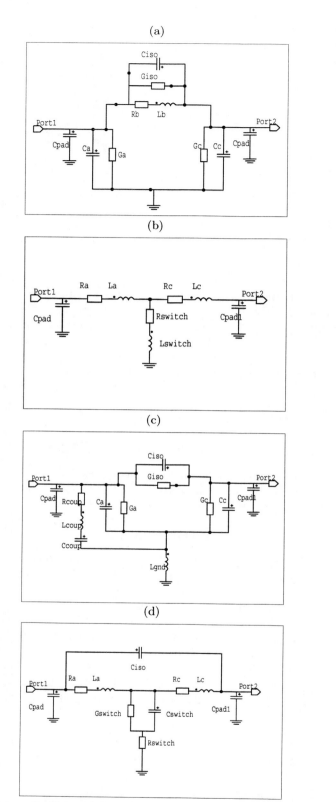

(a)

(b)

(c)

(d)

Figure 6: Extracted equivalent circuits: shunt switch up-state (a) and down-state (b); series switch up-state (c) and down-state (d).

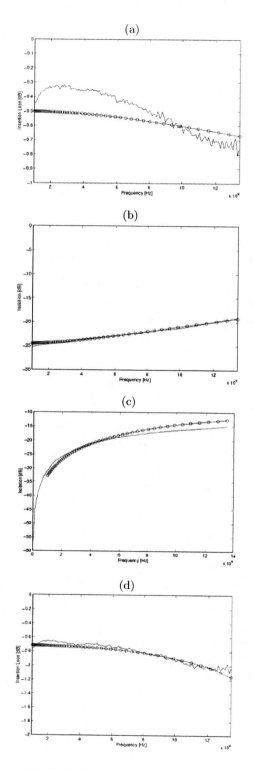

(a)

(b)

(c)

(d)

Figure 7: Measured (dots) and simulated (lines) s-parameters: shunt-switch insertion loss (a) and isolation (b); series-switch isolation (c) and insertion loss (d).

Compliant Force Amplifier Mechanisms for Surface Micromachined Resonant Accelerometers

C. B. W. Pedersen* and A. A. Seshia**

Department of Engineering, University of Cambridge,
Trumpington Street, Cambridge, CB2 1PZ, United Kingdom
*cbwp2@eng.cam.ac.uk and **aas41@eng.cam.ac.uk

ABSTRACT

The present work deals with the optimization of a compliant force amplifier mechanism in a surface micromachined resonant accelerometer. Figures of merit including noise floor and scale factor are critically dependent on the gain of the force amplifier mechanism and hence optimization of the force amplifier mechanism is necessary. The optimization is constrained by limitations imposed by the process and the device geometry.

The force amplifier mechanisms in this work are initially designed using continuum topology optimization. The results of topology optimization are seen to depend strongly on the size of the design domain, output and input stiffnesses and boundary conditions. The results of topology optimization are converted to beam element models that are used for a further shape and size optimization. Single-stage force amplification factors greater than 100 are obtained from the results of the optimization process.

Keywords: Topology, shape and size optimization, compliant amplifier, surface micromachining, resonant accelerometers.

1 Introduction

Resonant sensing principles have been applied to micromachined accelerometers [1] and gyroscopes [2]. These devices operate by a coupling a force acting on a proof mass (due to either an external acceleration or rotation rate) onto resonating force sensors. The output of these devices is then measured as a frequency shift of these resonant force sensors in response to the measured force. The scale factor of the devices is directly proportional to the force coupled onto the resonant sensors and hence amplification of the force to be measured before coupling onto the resonant sensors is desirable. This work examines the design of a compliant force leverage mechanism and the trade-offs imposed by technological and design constraints.

The accelerometer consists of a proof mass coupled to two resonant force sensors through a compliant leverage mechanism. A first generation surface micromachined

device was limited to a force amplification factor of approximately 9, a noise floor of 40 μg/$\sqrt{Hz}$ and a scale factor of 17 Hz/g [1]. This work aims to investigate the design of a force amplifier that would allow for an improvement upon these numbers by an increase in force amplification by a factor of at least 10x.

In recent years, several different optimization schemes combined with finite element modelling have been used to obtain optimized designs of compliant mechanisms [3]. Optimized compliant microleverage cantilever mechanisms in [4] have been demonstrated. The idea in the present work is to open up the design space for designing amplifier mechanisms in resonant accelerometers and not be simply restrict the designs to cantilever amplifier mechanisms.

2 The force amplifier

Figure 1 A shows a schematic drawing of the resonant accelerometer given by [1]. The resonant accelerometer has been implemented in a polysilicon surface micromachining process as described in [1].

Figure 1 B shows the design domain Ω for one of the four amplifiers in the resonant accelerometer. The

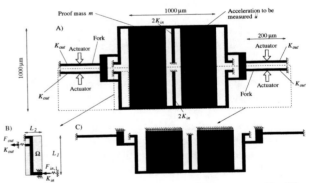

Figure 1: A; The resonant accelerometer. B; The subsystem containing the complaint amplifier mechanism. The spring stiffness K_{in} and K_{out} model the stiffness of the resonant accelerometer. The input force $F_{in,1} = \frac{1}{4} m \ddot{u}$ is one quarter of the inertial force for the proof mass m. The output force F_{out} is the axial force applied to one of the excited beams measuring the axial force F_{out}. C; Applied symmetry constraints.

size of the design domain is L_1 by L_2. The spring stiffness K_{out} substitutes the axial stiffness for one fork. These forks are actuated at resonance and variations in the natural frequency of the beam with applied axial force serve as the output of the sensor. The axial force, $F_{out} = K_{out} u_{out}$, can be determined where u_{out} is the axial displacement of the fork. The spring stiffness K_{in} is half the bending stiffness of the beams in the middle of the resonator.

Assuming that the proof mass m has the vertical acceleration $\ddot{u}$ the input force for one fork is $F_{in.1} = \frac{1}{4}m\ddot{u}$. It is assumed that it is valid to use a quasi-static model for the analysis and the input force is assumed to be $F_{in.1} = 0.1$ μN. The objective of the optimization is to maximize the amplification A of the amplifier mechanism. This corresponds to maximizing the ratio between the output force F_{out} and the input force $F_{in.1}$, given as

$$\max\left(A\right) = \max\left(\frac{F_{out}}{F_{in.1}}\right) = \frac{K_{out}}{F_{in.1}}\max\left(u_{out}\right) \quad (1)$$

where u_{out} is the displacement of the fork. From (1) it is observed that maximizing the displacement u_{out} corresponds to maximize amplification A.

Then the optimization for determining the mechanism with maximum amplification A in a given design domain is performed in the following way:
1) Topology optimization gives the topology of the amplifier mechanism in the design domain Ω with a size of L_1 by L_2, see figure 1 B.
2) Shape and size optimization of the amplifier where the topology is based upon the topology optimized design.
3) Inserting the shape and size optimized amplifiers into the entire resonant accelerometer and validate the solutions using [5].

2.1 Topology optimization, theory

The topology optimization formulation for (1) applied in the present work has been widely used in the literature [3]. The formulation was originally suggested by [6] and [7]. In the present finite element modelling geometrical nonlinearities are also included, see [8]. However, the examples presented in the literature use a ratio between the input spring stiffness and the output spring stiffness ($\frac{K_{out}}{K_{in}}$) close to one or a few magnitudes higher or lower. The real difference in this work is that the ratio ($\frac{K_{out}}{K_{in}}$) is 10^6 or higher.

2.2 Shape optimization, theory

The present section describes a combined shape and size optimization using rectangular 2D-beam elements of structures with given topologies. For varying the shape of the beams, the positions of the nodal points are varied. In the sizing optimization, the in-plane height h of

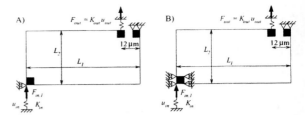

Figure 2: The rotated design domain containing the amplifier, see figure 1 B. The design domain is applied to two different sets of supports at the input port. In picture A one support at the input force is applied for avoiding any vertical displacement. In picture B more supports are added around the input force for avoiding rotation around the input port as well.

each beam is a design variable. For the modelling of the beams the finite element model in [9] is applied. This finite element model includes large rotations of the beam elements.

3 Topology optimization

The design domains are sketched in figure 2. The grey area denotes the design domain which is supported at the right top corner and the left bottom corner. Two set of supports are applied for the left bottom, support set A and support set B as shown in figure 2. The size of the design domain is L_1 by L_2 and the thickness is

L_2	Original cantilever beam design			
200 [μm]				
	$A_{wt} = 30$		$A_{ot} = 21$	
	Supports figure 2 A		Supports figure 2 B	
100 [μm]				
	$A_{wt} = 91$	$A_{ot} = 27$	$A_{wt} = 0$	$A_{ot} = 93$
200 [μm]				
	$A_{wt} = 115$	$A_{ot} = 6$	$A_{wt} = -3$	$A_{ot} = 98$
300 [μm]				
	$A_{wt} = 105$	$A_{ot} = 10$	$A_{wt} = 0$	$A_{ot} = 95$
400 [μm]				
	$A_{wt} = 104$	$A_{ot} = 0$	$A_{wt} = 0$	$A_{ot} = 45$

Table 1: Topology optimized amplifiers using the design domain in figure 2. The length L_2 of the design domain is varying from 100 μm to 400 μm. The number A_{wt} and A_{ot} indicate that A in (1) is determined using support set A and support set B in figure 2, respectively.

2.25 μm. For all the numerical examples the length L_1 is fixed at 500 μm whereas the length L_2 is varied. The amplifier is to be built in polysilicon with a nominal Young's modulus 160 GPa and the Poisson's ratio 0.25. An input force $F_{in,1} = 0.1$ μN is applied 4 μm from the left edge corner and the input spring ($K_{in} = 0.035$ N/m) is mounted at the point of application of the input force $F_{in,1}$. The output spring ($K_{out} = 3600$ N/m) is mounted 12 μm from the right corner. The volume of material is restricted to be 30% of the design domain to prevent parts of mechanisms to overlap during the deformation.

Here, we examine the impact on the topologies of the optimized amplifiers when the length L_2 of the design domain is changed and the set of supports around the input spring is changed from the support set A shown in figure 2 A to support set B shown in figure 2 B. In table 1 the length L_2 takes the values: 100 μm, 200 μm, 300 μm and 400 μm.

The reason for using two set of supports for the design domain in figure 2 is that it is not clear as to what kind of supports should be applied when considering the amplifier mechanism in isolation as opposed to the amplifier being integrated into the resonant accelerometer (see figure 1). Later by simulating half the resonant accelerometer structure in figure 1 C including the optimized amplifier mechanisms, we can determine a valid set of supports.

Several notable features are apparent by comparing the results of the optimization process (varying L_2). First, that the layouts of the optimized topologies are extremely sensitive to how the supports around the input port are simulated. Second, the response of the optimized structures are also extremely dependent upon which set of supports have been applied in the analysis. Third, that the highest amplification factor A is obtained for a length L_2 of 200 μm independent on the set of supports applied in the optimization. Fourth, that the optimized structures have higher amplification factors than the amplification factor of the original cantilever beam design. However, the optimized structures contain the so called hinges, see [3].

4 Shape and size optimization

From table 1 the two structures designed having a length L_2 of 200 μm are chosen as the topologies for the shape and size optimization. These two structures are then converted into 2D beam element models, see figure 3. Manufacturing constraints are now included in the optimization by setting the height of the beams to be greater than 2 μm.

We now examine the variation in the shape and size of the optimized amplifiers when the starting topologies are changed. Furthermore, the impact of having different supports are also examined. We note significant

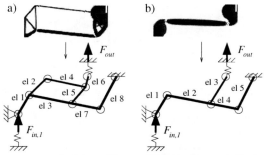

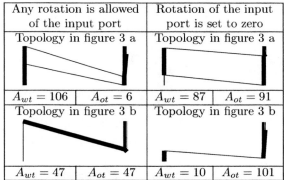

——— : heights of beams $\{h\}$ (design variables)

○ : positions of beams $\{x\}$ (design variables)

Figure 3: Two optimized topologies in table 1 are converted into two beam element models. These models are applied for the shape and size optimization.

Any rotation is allowed of the input port		Rotation of the input port is set to zero	
Topology in figure 3 a		Topology in figure 3 a	
$A_{wt} = 106$	$A_{ot} = 6$	$A_{wt} = 87$	$A_{ot} = 91$
Topology in figure 3 b		Topology in figure 3 b	
$A_{wt} = 47$	$A_{ot} = 47$	$A_{wt} = 10$	$A_{ot} = 101$

Table 2: Shape and size optimization of amplifiers using the different initial topologies given in figure 3. The number A_{wt} and A_{ot} indicate that there is no support for rotation and there is support for rotation in the beam model of the node where the input force $F_{in,1}$ is applied.

differences in amplification factor for the topologies simulated under the different support conditions.

5 The entire resonant accelerometer

In this section the optimized amplifier mechanisms obtained using beam element modelling in section 4 are converted into continuum models. Next, these optimized amplifier models are inserted into the symmetric resonant accelerometer shown in figure 1 C. The continuum modelling is done by [5] where geometrical nonlinearities are included to verify that the amplification is constant in the range of operation.

In table 3 the amplifications of the continuum structures are shown. Clearly, the results demonstrate that the support set B in figure 2, where supports are added to prevent the rotation around the input port leads to a reasonable modelling. It is clear that the results for the present optimization problem are extremely sensitive to how the supports for the design domain at the input port are modelled.

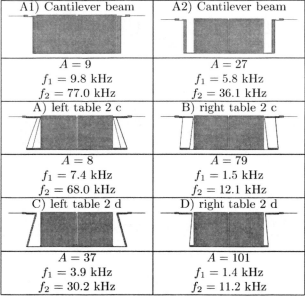

A1) Cantilever beam	A2) Cantilever beam
$A = 9$ $f_1 = 9.8$ kHz $f_2 = 77.0$ kHz	$A = 27$ $f_1 = 5.8$ kHz $f_2 = 36.1$ kHz
A) left table 2 c	B) right table 2 c
$A = 8$ $f_1 = 7.4$ kHz $f_2 = 68.0$ kHz	$A = 79$ $f_1 = 1.5$ kHz $f_2 = 12.1$ kHz
C) left table 2 d	D) right table 2 d
$A = 37$ $f_1 = 3.9$ kHz $f_2 = 30.2$ kHz	$A = 101$ $f_1 = 1.4$ kHz $f_2 = 11.2$ kHz

Table 3: Analysis of the continuum structures using 8-node biquadratic elements.

Figure 4: The eigenmodes for structure D in table 3.

The cantilever amplifier (described in [1], structure A1 and A2 of table 3) have an amplification of 9 and 27, respectively which is lower than the optimized.

The eigenfrequencies of the subcomponents for the resonant accelerometer should be higher than the sensor bandwidth. A value of 15 kHz works for many accelerometer applications. In the initial optimization this constraint is not applied. Therefore, an eigenfrequency analysis is performed to validate if this constraint is fulfilled for the optimized structures. Figure 4 show the eigenmodes for the three first eigenfrequencies of the structure D in table 3. From these numerical observations it can be concluded that the structures optimized using support set B in figure 2 almost fulfil the constraints on the eigenfrequencies even though they where not included in the analyses for the optimization.

6   Further results

In the paper [10] more results are shown. The theoretical bound for the amplification A is determined. Furthermore, it is shown that an increase in the stiffness K_{out} will lead to an increase in the amplification. The effect of adding more supports in design domain is studied. However, adding more supports seem not to improve the amplification.

7   Conclusion

A topology optimization based approach is used for the design of micromachined force amplifiers for inertial sensor applications. Technological constraints require that the results of the topology optimization mechanisms to be converted to beam element models and subjected to a further size and shape optimization. The dependence of geometrical and mechanical parameters on the optimization are studied. Single-stage force amplification factors greater than 100 are obtained from the results of the optimization process.

Amplifier mechanisms are currently being designed in the polysilicon surface micromachining process offered by Robert Bosch GmbH for experimental validation of the results described here.

REFERENCES

[1] A.A. Seshia, M. Palaniapan, T.A. Roessig, R.T. Howe, R.W. Gooch, T.R. Schimert, and S. Montague, "A Vacuum Packaged Surface Micromachined Resonant Accelerometer," Journal of Microelectromechanical Systems, 11, 784-793, 2002.

[2] A.A. Seshia, R.T. Howe, and S. Montague, "A Micromechanical Resonant Output Gyroscope," Proc. IEEE MEMS 2002, Las Vegas, 2002.

[3] M.P. Bendsøe and O.and Sigmund, "Topology Optimization - Theory, Method and Applications," Springer, 2003.

[4] X.P.S. Su and H.S.Yang,"Two-stage Compliant Microleverage Mechanism Optimization in a Resonant Accelerometer," Structural and Multidisciplinary Optimization, 22, 328-334.

[5] Hibbit, Karlson and Sorensen Inc., "ABAQUS/Standard User's Manual, Version 6.2," 2001.

[6] G.K. Ananthasuresh, S. Kota and Y. Gianchandani, "A Methodical Approach to the Design of Compliant Micromechanisms," Solid-State Sensor and Actuator Workshop, 189-192, 1994.

[7] O. Sigmund, "On the Design of Compliant Mechanisms using Topology Optimization", Mechanics of Structures and Machines, 25, 495-526, 1997.

[8] C.B.W. Pedersen, T. Buhl, and O. Sigmund, "Topology Synthesis of Large-Displacement Compliant Mechanisms", International Journal for Numerical Methods in Engineering, 50, 2683-2706, 2001.

[9] C.B.W. Pedersen, "Topology Optimization of 2D-Frame Structures with Path Dependent Response", International Journal for Numerical Methods in Engineering, 57, 1471-1501, 2003.

[10] C.B.W. Pedersen and A.A. Seshia, "On the Optimization of Compliant Force Amplifier Mechanisms for Surface Micromachined Resonant Accelerometers", Submitted, 2003.

Coupling of Resonant Modes in Micromechanical Vibratory Rate Gyroscopes

A. S. Phani†, A. A. Seshia†, M. Palaniapan*, R. T. Howe ‡, J. Yasaitis**

† Department of Engineering, University of Cambridge, Cambridge, CB2 1PZ, UK
* Department of ECE, National University of Singapore, Singapore 117576
‡ Department of EECS, University of California, Berkeley, CA 94720, USA
** Analog Devices Inc., Cambridge, MA 02139, USA
E-mail: **aas41@eng.cam.ac.uk**

ABSTRACT

Analytical models are presented to describe the resonant modal coupling behaviour of z-axis micromechanical vibratory rate gyroscopes fabricated in an integrated polysilicon surface micromachining process. The models are then applied to predict the extent of displacement and force coupling between the drive and sense axes of this device as a function of varying degrees of matching between the resonant frequencies associated with these modes. Two modelling approaches are presented. The first approach is based on linear vibration analysis. The second approach— a state-space based system identification method— is used to calculate the anisoelasticity parameters. It is shown that, as the resonant frequencies of the two modes are brought closer together, an improvement in overall resolution and scale factor of the device is obtained at the expense of an enhanced coupling of forces to displacements between the two axes and the onset of electrostatic instability for an open-loop sensing implementation.

Keywords: micromechanical gyroscope, resonant coupling, surface micromachining.

1 Introduction

Vibratory microgyroscopes operate on the principle of resonant modal coupling through the Coriolis force. A typical single-axis resonant microgyroscope consists minimally of two degrees of freedom, such as two masses coupled by springs. The first mass, called the drive mass, is set to resonance by driving it at its natural frequency using lateral comb drive mechanisms. When the coupled system is placed on a rotating platform, the Coriolis force induces displacements in the second mass, called the sense mass. This motion is detected using capacitive sensing structures (see figure 1), allowing estimation of the angular motion of the platform. The coupling of motions of the two masses is not only due to the Coriolis force but also due to the elastic forces (through the off-diagonal stiffness terms) and the off-diagonal damping terms (arising due to non-proportional damping). This introduces errors in the measured displacements of the sense mass. While the off-diagonal stiffness terms give rise to quadrature errors

i.e. errors in quadrature with the Coriolis component, the off-diagonal damping terms lead to in-phase errors. An approach to compensate for these errors is to design a control strategy to suppress the signals arising due to non-diagonal stiffness and damping terms [1] following the identification of these terms. The issue of identifying these matrices is discussed in a companion paper [2].

The focus of this paper is to develop analytical modelling techniques which can be used to predict the observed features in the sense output of a z-axis frame gyroscope described in [3]. Further details of the device are given in section 2. Experimental data, in the form of sense output spectrum, is presented and certain features of the data arising due to modal coupling are also described. Modelling of the device based on linear system theory and vibration analysis is described in section 3. A state-space based system identification approach is also presented to identify the angular mismatch between the drive-sense and principal stiffness axes. A comparison of the experimentally observed parameters, and those predicted by the modelling approach is given in section 4. Conclusions and further directions are summarised in section 5.

2 Device

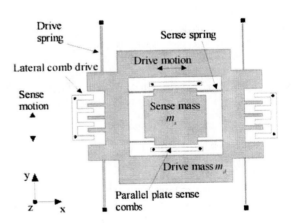

Figure 1: Schematic of a z-axis frame gyroscope with inner sensing and outer drive (ISOD) mechanism. See [3] for further details.

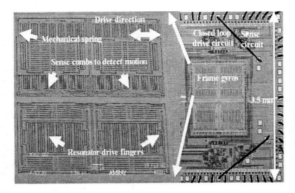

Figure 2: SEM and die photo of the frame microgyroscope fabricated in the Analog Devices Inc. Modular MEMS process.

A schematic of the device under consideration is shown in figure 1. It consists of two masses coupled via springs, the outer mass being attached to the substrate through a separate set of springs. The driven mode is excited by applying a force to the outer mass using an on-chip oscillator circuit and corresponds to the combined motion of the inner and outer mass along the x-axis. The sense mode corresponds to the case when the inner mass deflects relative to the outer mass along the y-axis. An SEM and chip micrograph of the gyroscope is shown in figure 2. The noise floor of the device is 0.01 deg/sec/$\sqrt{\text{Hz}}$. A detailed description of the gyroscope including rate data and functionality testing has been reported in [3].

2.1 Experiments

A limited set of experiments reveals the extent of modal coupling between the two axes. The driven mode is excited using an on-chip oscillator circuit described in [3]. A wideband electrostatic force input is applied to the inner mass along the y-axis using a separate set of electrodes. The sense mode frequency is tuned using a negative electrostatic spring effect and it is observed that the coupling from the force applied to the drive mass increases as a function of mode matching. The effect of mode matching under the described forcing conditions upon the output voltage spectrum of the gyroscope is shown in figure 3. It can be observed that the amplitude of the signal at the drive frequency increases when the sense mode is tuned to a frequency just below the drive resonance frequency. There is a characteristic anti-resonance behaviour at a frequency of about 16.327 kHz, where the output spectrum shows minimum response, both before and after the tuning of the sense mode. As the modes are brought closer together, the displacement of the structure along the sense direction increases due to the increased coupling between the two modes and a ceiling is reached governed

by the electrostatic pull-in associated with the operation of the capacitive sensing structures [4].

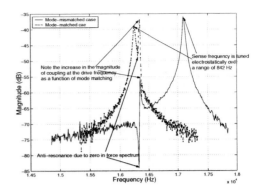

Figure 3: Experimentally observed output sense power spectrum corresponding to the displacement of the sense mass along the y-axis. Note the enhanced coupling of motion of the structure at the drive frequency when the modal frequencies are brought closer together. The measurements are made in the absence of an externally applied rotation rate.

3 Modelling

Two approaches to modelling the device dynamics are discussed in this section. The first approach is based on the frequency response analysis of linear vibrating systems. This technique can be used to predict the effect of modal coupling on the magnitude of the displacement of the inner mass along the y-axis as a function of mode matching. The second approach, based on state-space modelling, is used to identify the anisoelasticity parameters.

The governing equations of motion for a 2 DoF z-axis MEMS gyroscope can be written as:

$$\begin{bmatrix} m_d + m_s & 0 \\ 0 & m_s \end{bmatrix} \begin{bmatrix} \ddot{x} \\ \ddot{y} \end{bmatrix} + \begin{bmatrix} c_{dd} & c_{ds} \\ c_{sd} & c_{ss} \end{bmatrix} \begin{bmatrix} \dot{x} \\ \dot{y} \end{bmatrix} + \begin{bmatrix} k_{dd} & k_{ds} \\ k_{sd} & k_{ss} \end{bmatrix} \begin{bmatrix} x \\ y \end{bmatrix}$$
$$= \begin{bmatrix} F_d \\ F_s \end{bmatrix} + \begin{bmatrix} 0 & 2m\Omega_z \\ -2m\Omega_z & 0 \end{bmatrix} \begin{bmatrix} \dot{x} \\ \dot{y} \end{bmatrix} + \begin{bmatrix} E_d \\ E_s \end{bmatrix}$$
$$(1)$$

where: m's, k's and c's are respectively the mass, stiffness and damping coefficients. Coriolis forces appear in the above equations of motion as $2m\Omega_z$ terms. F_d and F_s are the external forces applied to the gyroscope by the control system. The extraneous effects such as external accelerations, centrifugal force and z-axis motion feedthrough are lumped in the terms E_d and E_s. Subscripts d and s denote the the drive and sense directions of the gyroscope respectively. Note that, in the drive mode, the sense and drive masses move together and hence the corresponding mass term is $m_d + m_s$. A linear vibration analysis based modelling approach is described first.

3.1 Frequency Response Analysis

The equations of motion can be rewritten in the frequency domain in the form of an input-output relationship for the 2 DoF gyroscope as follows:

$$\{O\} = [H]\{F\} \qquad (2)$$

where the output (displacement) vector $\{O\}$ and the input (force) vector $\{F\}$ are defined as:

$$\{O\} = \begin{bmatrix} O_d(\omega) \\ O_s(\omega) \end{bmatrix} \quad \{F\} = \begin{bmatrix} F_d(\omega) \\ F_s(\omega) \end{bmatrix} \qquad (3)$$

and $[H]$ is the Frequency Response Function (FRF) matrix given by:

$$[H] = \begin{bmatrix} H_{dd}(\omega) & H_{ds}(\omega) \\ H_{sd}(\omega) & H_{ss}(\omega) \end{bmatrix}. \qquad (4)$$

Using the equations of motion defined in equation 1 the expressions for the various FRFs can be written as:

$$H_{ss}(\omega) = \frac{D_{dd}(\omega)}{D_{ss}(\omega)D_{dd}(\omega) - (D_{sd}(\omega))^2} \qquad (5)$$

$$H_{sd}(\omega) = -\frac{D_{sd}(\omega)}{D_{ss}(\omega)D_{dd}(\omega) - (D_{sd}(\omega))^2} \qquad (6)$$

$$H_{dd}(\omega) = \frac{D_{ss}(\omega)}{D_{ss}(\omega)D_{dd}(\omega) - (D_{sd}(\omega))^2} \qquad (7)$$

where

$$\begin{aligned} D_{dd}(\omega) &= k_{dd} - (m_d + m_s)\omega^2 + ic_{dd}\omega, \\ D_{ss}(\omega) &= k_{ss} - m_s\omega^2 + ic_{ss}\omega, \\ D_{sd}(\omega) &= k_{sd} + ic_{sd}\omega. \end{aligned} \qquad (8)$$

Using equations 2, 3 and 4, the sense output of the gyroscope is given by:

$$O_s(\omega) = \underbrace{H_{ss}F_s(\omega)}_{\text{output due to force on the sense mass}} + \underbrace{H_{sd}F_d(\omega)}_{\text{output due to coupling}} \qquad (9)$$

At the drive resonance frequency $\omega = \omega_d$, the term $H_{ss}(\omega_d)F_s(\omega_d)$ is much smaller compared to the other term in equation 9, since $H_{ss}(\omega)$ is a high-Q FRF with a peak at at ω_s. Hence, the sense output can be simplified as:

$$O_s(\omega_d) \approx H_{sd}(\omega_d)F_d(\omega_d). \qquad (10)$$

Now the ratio of the sense output response under differing conditions of mode matching can be obtained as:

$$\frac{O_{s1}(\omega_d)}{O_{s2}(\omega_d)} \approx \frac{H_{sd1}(\omega_d)F_{d1}(\omega_d)}{H_{sd2}(\omega_d)F_{d2}(\omega_d)} \qquad (11)$$

where the subscripts 1 and 2 refer to the two differing conditions of mode matching shown in figure 3. Substituting $H_{sd}(\omega)$ from equation 6, the ratio of sense outputs can be written as:

$$\frac{O_{s1}(\omega_d)}{O_{s2}(\omega_d)} = \frac{D_{dd2}(\omega_d)D_{ss2}(\omega_d) - (D_{sd2}(\omega_d))^2}{D_{dd1}(\omega_d)D_{ss1}(\omega_d) - (D_{sd1}(\omega_d))^2}. \qquad (12)$$

Notice that the drive natural frequency (ω_d) remains unchanged before and after tuning as shown in figure 3. One can safely ignore terms like $(D_{sd}(\omega_d))^2$, provided the half power bandwidths of the two modes do not overlap, which is verified from the experimental data. With this simplification, the ratio of the sense outputs before and after tuning, using the definitions in equation 8, is given by:

$$\frac{O_{s1}(\omega_d)}{O_{s2}(\omega_d)} \approx \frac{c_{dd2}(k_{ss2} - m_s\omega^2 + ic_{ss2}\omega)}{c_{dd1}(k_{ss1} - m_s\omega^2 + ic_{ss1}\omega)}. \qquad (13)$$

At the sense resonance frequency $\omega = \omega_s$, $F_d(\omega_s) \cong 0$ and hence the ratio of sense outputs at the sense resonance frequency, using equation 9, is given by:

$$\frac{O_{s1}(\omega_{s1})}{O_{s2}(\omega_{s2})} = \frac{H_{ss2}(\omega_{s2})F_{s2}(\omega_{s2})}{H_{ss1}(\omega_{s1})F_{s1}(\omega_{s1})}. \qquad (14)$$

In this case also, the sense-sense FRF can be simplified, by ignoring terms containing $D_{sd}(\omega)$ as :

$$\begin{aligned} H_{ss}(\omega) &\approx \frac{D_{dd}(\omega)}{D_{dd}(\omega)D_{ss}(\omega)} \\ &= \frac{1}{D_{ss}(\omega)}. \end{aligned} \qquad (15)$$

By using the definition of $H_{ss}(\omega)$ from equation 8 in the above equation, the ratio of the gyroscope output at the sense resonance frequencies before and after tuning is given by:

$$\frac{O_{s1}(\omega_{s1})}{O_{s2}(\omega_{s2})} \approx \frac{c_{ss2}\omega_{s2}}{c_{ss1}\omega_{s1}}. \qquad (16)$$

By considering the magnitude of the cross FRF $H_{sd}(\omega)$ at the drive resonant frequency, and with the simplification described above, one obtains an approximate expression for the cross coupling stiffness as:

$$\begin{aligned} k_{sd} &\approx |H_{sd}(\omega_d)|c_{dd}\omega_d\sqrt{(k_{ss} - m_s\omega_d^2)^2 + \omega_d^2c_{ss}^2} \\ &= \left|\frac{O_s(\omega_d)}{F_d(\omega_d)}\right|c_{dd}\omega_d\sqrt{(k_{ss} - m_s\omega_d^2)^2 + \omega_d^2c_{ss}^2} \end{aligned} \qquad (17)$$

Estimating k_{sd} by using the above equation requires an accurate estimation of the diagonal damping terms, especially c_{dd}. An alternate method of measuring k_{sd} is to measure the output of the device in presence of a known applied rotation rate. The output voltage spectrum of the device in response to a sinusoidal rotation rate at 7 Hz is shown in figure 7 of [3]. The magnitude of the sense output displacement at the drive resonant frequency now differentiates between the two terms in equation 9. The DC term corresponds to the coupling of modes due to anisoelasticity while the AC term corresponds to coupling due to the Coriolis force. A ratio of these two terms gives an estimate for k_{sd}. An equation for k_{sd} can now be written as:

$$k_{sd} \approx 2(m_s + m_d)\Omega\omega_d\frac{O_s}{O_{sc}} \qquad (18)$$

Parameter	Theory	Experiment
Ratio of sense mode displacements before and after sense mode tuning at drive resonance frequency	0.121	0.126
Ratio of sense mode displacements before and after sense mode tuning at sense resonance frequencies	1.15	1.25
Instability point proof mass voltage (V)	9.9 V	9.6 V
Cross-coupling spring constant, k_{sd}		1.6×10^{-4} N/m

Table 1: Comparison of predicted and experimental parameters. Subscripts 1, 2 refer to mode-mismatched and mode-matched cases respectively. m, k, c and ω denote mass, stiffness, damping and resonant frequency.

where Ω is the applied rotation rate, O_s and O_{sc} are the sense outputs due to anisoelasticity (DC term) and applied rotation rate (AC term) respectively. An estimate for k_{sd} based on this analysis is shown in Table 1.

As the modes are tuned by varying the DC voltage on the proof mass, the sense displacement increases due to the increased coupling between the two modes and a ceiling is reached governed by electrostatic pull-in associated with the operation of the capacitive sensing structures. The expected point for the observation of electrostatic instability can be computed by following an analysis similar to that described in [4]. A good match is obtained between experiment and theory for the pull-in voltage as shown in Table 1. This instability limits the difference in the drive and sense frequencies.

3.2 State-Space Modelling

A state-space based modelling approach is used to identify the anisoelasticity parameters [2]. Figure 4 shows an 8th order model fitted to the measured sense output spectrum. This higher order model is reduced to a 4th order model (corresponding to the 2 physical modes of vibration) using the balanced reduction technique. The anisoelasticity *i.e.* the angular mismatch between the principal stiffness axes and the drive-sense directions is found to be 8° under the mode mismatched conditions.

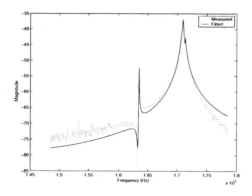

Figure 4: An 8th order fit to the measured output spectrum for the mode-mismatched case.

4 Comparison with experiment

The first row in Table 1 compares model and experiment for ratios of the displacement of the sense mass along the y-axis at the drive resonance frequency under two different conditions of mode matching. The second row in Table 1 compares model and experiment for ratios between the observed displacements of the inner mass along the y-axis at the sense resonance frequency under two different conditions of mode matching. The off-diagonal spring constant can be calculated as shown in equation 18. The predicted values for the ratios of displacements as a function of mode matching and the expected point for the observation of electrostatic instability are seen to closely match experiment.

5 Conclusion

Modal coupling in resonant z-axis frame gyroscopes has been investigated using the principles of linear vibration analysis and state-space modelling. These models are used to predict the features observed in the sense output spectrum of the device, as the drive and sense modal frequencies are brought closer through electrostatic tuning. Closed form expressions to predict the extent of observed modal coupling show a good match with experiments. A limit on the separation of the modal frequencies set by the electromechanical instability is also calculated and found to be in agreement with experiments. Future work will use these modelling principles for on-chip control algorithm implementation to suppress the errors due to modal coupling via off-diagonal stiffness and damping terms.

REFERENCES

[1] **Painter, C. C.** and **Shkel, A. M.**, SPIE Proceedings, March, 2001.

[2] **Phani, A. S.** and **Seshia, A. A.**, Nanotech 2004, 2004.

[3] **Palaniapan, M., Howe, R. T.** and **Yasaitis, J.**, IEEE-MEMS 2003, pp. 482–485, 2003.

[4] **Legtenberg, R., Groeneveld, A. W.** and **Elwenspoek, M.**, Journal of Micromechanics and Microengineering, Vol. 6, No. 3, pp. 320–329, 1996.

Numerical Modeling of a Piezoelectric Micropump

R. Schlipf[*], K. Haghighi[**] and R. Lange[***]

[*]Purdue University, Department of Agricultural and Biological Engineering, schlipf@purdue.edu
[**]Purdue University, Department of Agricultural and Biological Engineering,
1146 ABE Building, Purdue University, haghighi@purdue.edu
[***]ANSYS, Inc., rich.lange@ansys.com

ABSTRACT

An accurate description and understanding of any pumping method critical, especially on the micro scale. With the existence of a comprehensive and adaptable model, accurate preproduction predictions of performance are realized. A three-dimensional FEA approach for parametric design and optimization of a piezoelectrically actuated membrane micropump is presented. The model includes the piezoelectric material, membrane, pumping chamber, and cantilever valves. This numerical representation includes electro-mechanical coupling for piezoelectric actuation as well as consideration of fluid-structural interaction. Transient consideration of electrical, mechanical, and fluidic effects is included. The effects of independent factors such as component geometry, backpressure, and excitation voltage and frequency are each evaluated. Outputs include membrane deflection, flow pattern and velocities, and volumetric flow rate.

Keywords: finite element analysis, simulation, micropump, piezoelectric, modeling

1 INTRODUCTION

Research on micropumps began in 1980 [1] with significant advances in both pumping effectiveness and modeling capabilities resulting from this and subsequent work. Efforts continue to develop both analytical [2] and numerical models [3]. The valves on the pump in this work are based on those found in literature [4]. In general, efforts to characterize microscale pumps and their valve counterparts have been diverse and less than comprehensive. Complete understanding of the system is only possible when such factors as the component dimensions, backpressure, driving frequency, membrane deflection, and fluid flow are each included in and able to be analyzed by the model. The work described in this paper focuses on the development of a three dimensional numerical model using a commercially available finite element code (ANSYS 7.0) for the simulation of the transient behavior of a piezoelectrically actuated microscale membrane pump with cantilever valves. The development of this model is significant for several reasons. First, it allows the complete pump geometry to be contained in a single simulation environment, avoiding complex external iterative procedures and facilitating design optimization. This includes the electro-mechanical coupling required for piezoelectric simulations as well as the fluid-structure interactions (FSI) of a membrane pump. Next, numerical solutions allow for great design flexibility and analysis which is not constrained by known analytical equations. This model considers pump and valve geometry, ambient pressures, excitation voltage shape (sinusoidal, square, etc.), and magnitude, as well as material and fluid properties. Significant outputs include (but are not limited to) membrane deflection profile, operating pressure, and flow rate. In the sample geometry in Figure 1, the top (purple) region is the piezoelectric material, the next (dark blue) layer is the silicon membrane, and the remaining volumes (light blue) are the fluid. The fluid region includes the inlet (left) and outlet (right) valves. The overall goal of this project was to develop a comprehensive, self-contained, means of micropump evaluation.

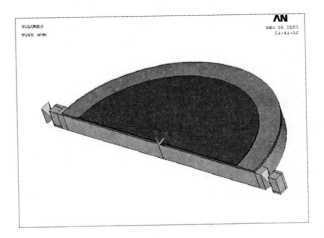

Figure 1: Isometric view of complete piezoelectrically actuated micropump with cantilever valves

2 THEORETICAL CONSIDERATIONS

Due to the extremely coupled nature of this device, care must be taken to consider all relevant physics. The applied voltage excites a structural deformation in the membrane, which causes a movement in the fluid resulting in a net fluid flow. The significant governing equations for the analysis are presented below, in order of input to output.

2.1 Electromechanical Coupling

First, the voltage is applied to the piezoelectric material resulting in a deformation. The electromechanical coupling is described by the following equations [5]:

$$\{\sigma\} = [c]\{\varepsilon_S\} - [e]\{E\} \tag{1}$$

$$\{D\} = [e]^T\{\varepsilon_S\} + [\varepsilon_{di}]\{E\} \tag{2}$$

where $\{\sigma\}$ is the stress vector, $[c]$ is the elasticity matrix, $\{\varepsilon_s\}$ is the strain vector, $[e]$ is the piezoelectric stress matrix, and $\{E\}$ is the electric field vector. Additionally in equation (2) we define the electric flux density vector $\{D\}$, and the dielectric matrix $[\varepsilon_{di}]$. Application of finite element procedures to equations (1) and (2) results in the global matrix equation of :

$$\begin{bmatrix} [M] & [0] \\ [0] & [0] \end{bmatrix}\begin{Bmatrix} \{\ddot{U}\} \\ \{\ddot{V}\} \end{Bmatrix} + \begin{bmatrix} [C] & [0] \\ [0] & [0] \end{bmatrix}\begin{Bmatrix} \{\dot{U}\} \\ \{\dot{V}\} \end{Bmatrix} + \tag{3}$$

$$\begin{bmatrix} [K] & [K^Z] \\ [K^Z]^T & [K^d] \end{bmatrix}\begin{Bmatrix} \{U\} \\ \{V\} \end{Bmatrix} = \begin{Bmatrix} \{F\} \\ \{L\} \end{Bmatrix}$$

where the $[M]$, $[C]$, $[K]$, $[K^d]$, and $[K^Z]$ are the structural mass, structural damping, structural stiffness, dielectric conductivity, and piezoelectric coupling matrix, respectively. Also, $\{U\}$ is the nodal displacement vector, $\{V\}$ the nodal electric potential vector, $\{F\}$ is the vector of nodal, surface, and body forces, and $\{L\}$ is the nodal charge vector. The single and double dot represent the first and second time derivatives, respectively.

2.2 Fluid-Structure Interaction

The boundary between the diaphragm and the fluid, as well as that between the flaps of the valves and the surrounding fluid is a significant factor in the analysis of the pump. The moving membrane (whose motion is described above) exerts a pressure on the fluid which moves in response to this stimulus. Similarly, the fluid exerts a force on the structure which affects its movement. The global system of finite element equations for this interaction is as follows:

$$\begin{bmatrix} [M_S] & [0] \\ \rho_f[R]^T & [M_f] \end{bmatrix}\begin{Bmatrix} \{\ddot{U}\} \\ \{\ddot{P}\} \end{Bmatrix} + \tag{4}$$

$$\begin{bmatrix} [K_S] & -[R] \\ [0] & [K_f] \end{bmatrix}\begin{Bmatrix} \{U\} \\ \{P\} \end{Bmatrix} = \begin{Bmatrix} \{F_S\} \\ \{F_f\} \end{Bmatrix}$$

Here the subscripts s and f denote solid and fluid, respectively. $\{P\}$ is the pressure vector. The matrix $[R]$ is a coupling matrix which accounts for the normal direction of each face. The density of the fluid is denoted ρ_f. $[M]$, $[K]$, $\{U\}$, and $\{F\}$ remain as described in section 2.1.

2.3 Fluidic

This model assumes an incompressible fluid. Also, because of the small length scales we can be assured of small Reynolds numbers, which supports the laminar flow assumption. The fluid flow is solved by a segregate sequential solver. The element matrix formulations are solved for each degree of freedom separately. In this problem, the velocities in the x, y, and z directions are solved respectively. From these results, the pressure is computed and the velocities are recalculated. If the solution has not converged, the algorithm proceeds to the next global iteration.

3 NUMERICAL MODELING

The model used the following elements to capture the various physical processes present in the pump. The entire fluid region of the model is meshed with the 3-D fluid element FLUID142. The pumping chamber and valves were meshed using 16500 swept hex elements. The piezoelectric material was meshed with 1620 tetrahedral SOLID98 coupled field solid elements; the pump membrane was meshed with 1720 tetrahedral SOLID92 elements. The flaps of the valves were meshed using 1350 structural SOLID45 elements. Additionally, the upstream face of the valves was meshed with the contact element CONTA174. This element (and its target element counterpart TARGE170), allows the modeling of contact between moving three dimensional bodies. In this case, the contact is to allow the flap of the valve to shut when a negative pressure gradient is applied across it. The entire mesh is shown in Figure 2 with an expanded view of the valve in Figure 3.

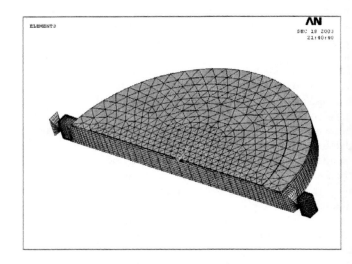

Figure 2: Isometric view of meshed pump

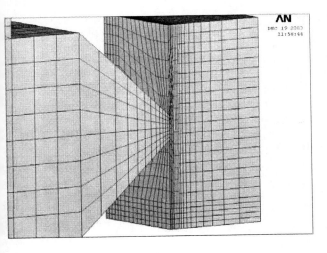

Figure 3: Expanded view of valve mesh

4 DESIGN PARAMETERS
4.1 Pump Geometry

The pumping chamber itself is circular, modeled as a semicircle due to symmetry considerations. The model consists of a fluid chamber and a piezoelectric actuator separated by a thin membrane. Common materials were used for the simulations presented here, namely, water, PZT-4, and silicon, respectively. The inlet and outlet ports are located on opposite sides of the chamber and each has a width of 800 μm. The pump radius (both the fluid and pump membrane) is 5 mm; the membrane has a thickness of 100 μm. The diameter was arbitrarily chosen to approximate existing pumps, the thickness is both a common membrane thickness and a value near the optimal value predicted by [2]. The PZT has a radius of 4.1 mm and a thickness of 30 μm. This size of actuator was chosen to reflect the results of optimization of the piezo/membrane system without fluidic effects. The radius and thickness were optimized simultaneously using both the ANSYS optimization algorithms and manual techniques.

The key pump parameters and the final dimensions are shown in Figure 4 and Table 1.

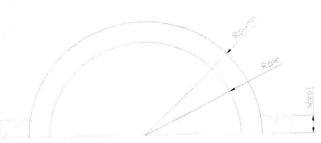

Figure 4: Pump parameters

R_{mem}	5000μm
T_{mem}	100μm
R_{pie}	4100μm
T_{pie}	30μm
W_{val}	800μm

Table 1: Pump dimensions

4.2 Valve Geometry

The valves incorporated into this model are simple cantilever-style valves. As in the valve demonstrated by [4] the channel area is reduced before the fluid encounters a movable flap. This style of valve allows flow in the forward direction while restricting it in the opposite. This causes a net forward flow as fluid is forced out of the pumping chamber. The valve parameters and the chosen dimensions for the analysis are shown in Figure 5 and Table 2 respectively.

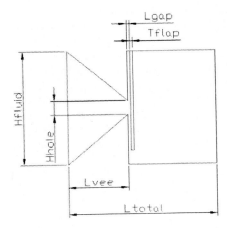

Figure 5: Valve parameters

H_{fluid}	800μm
T_{flap}	4μm
L_{gap}	1μm
L_{total}	1000μm
L_{vee}	400μm
H_{hole}	100μm

Table 2: Valve dimensions

4.3 Operating Parameters

In addition to the physical size and shape, several other factors have significant roles in determining the output characteristics of a pump. The magnitude, frequency, and shape (sinusoidal, square, etc.) of the actuation voltage are critical. This model uses a sinusoidal excitation of $300V_{0-P}$ with a frequency of 100Hz.

5 RESULTS

The results garnered from the complete model thus far are of varying degrees of success. The inherent problem with the model is that it does not solve using fluids of realistic compressibility. The model failed to converge for fluids of stiffness (defined by ANSYS as $\delta P/\delta \rho$) higher than 100 m^2/s^2. Pump operation using highly compressible fluids resulted in compression of the fluid rather than a mechanically forced fluid flow. Also, the model will solve when the geometry is altered to minimize the occurrence of thin, plate-like elements. This, too, however will not yield realistic results. Investigations into this alternative are continuing.

On a positive note, when solved using a lower compressibility, the trends and motion of the model certainly followed the predicted values and patterns. This was to be expected based on previous success in modeling the individual components of the pump. A sinusoidal input voltage generated a (slightly lagging) sinusoidal membrane displacement. A pressure difference across the valves was accompanied by both flap displacement and fluid flow. Additionally, the components interacted correctly. For example, the timing of the valve flap displacement corresponded well with the physical device though the value was several orders of magnitude lower than expected.

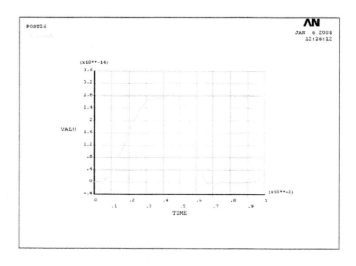

Figure 6: Inlet valve displacement (m) as pump draws fluid into the chamber (time 0-.005s) then expels it (time .005-.01s)

Note in Figure 6 that the flap opens (positive values of displacement) as the flow is drawn into the pump chamber; then, as the fluid is pressed out, the flap returns to its original closed position.

One means of understanding the compressibility effects is to view the velocity of the fluid in the x direction (from inlet to outlet). As can be seen from Figure 7 the fluid is not forced out of the valve but simply compressed; specifically, the flow has a high velocity in the pump chamber and a low velocity in (through) the valves.

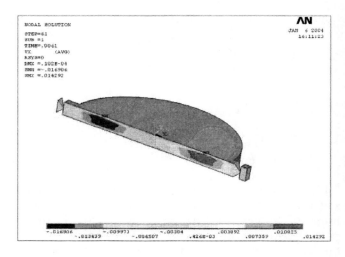

Figure 7: Velocity profile (m/s) in the x direction showing effects of compressibility

Work is continuing to be done in an effort to solve the current problems and provide a useful model for micropump optimization and characterization.

REFERENCES

[1] Gravesen, P., Jens Branebjerg, Ole Sondergard Jensen (1993). "Microfluidics - a review." Journal of Micromechanics and Microengineering 3(1993): 168-182.

[2] Li, S., and Shaochen Chen (2003). "Analytical analysis of a circular PZT actuator for valveless micropumps." Sensors and Actuators A 104: 151-161.

[3] Morris, C. J., Fred K Forster (2000). "Optimization of a circular piezoelectric bimorph for a micropump driver." Journal of Micromechanics and Microengineering 10: 459-465.

[4] Koch, M., A G R Evans and A Brunnschweiler (1996). "Coupled FEM simulation for the characterization of the fluid flow within a micromachined cantilever valve." Journal of Micromechanics and Microengineering 6(1996): 112-114.

[5] ANSYS Theory Reference v7.0

Identification of Anisoelasticity and Nonproportional Damping in MEMS Gyroscopes

A. Srikantha Phani, and Ashwin A Seshia

Department of Engineering, University of Cambridge
Trumpington Street, Cambridge, CB2 1PZ, UK,
E-mail: **skpa2@eng.cam.ac.uk**.

Abstract

A novel approach to identify nonproportional damping and structural anisoelasticities in vibratory MEMS gyroscopes is proposed in this paper. The proposed identification method is based on measured vibration data— in the form of frequency response functions (FRFs.) A new identification method, called the matrix perturbation method, is proposed to identify nonproportional damping from the measured FRFs. This method is based on a perturbation expansion of the FRF matrix of the nonproportionally damped system in terms of the FRF matrix of the proportionally damped system. A closed form expression for the off-diagonal modal damping matrix is obtained based on the perturbation theory. The mass and stiffness matrices required by this method are obtained using the instrumental variable (IV) method, proposed by Fritzen [1], which is also used to identify the structural anisoelasticities.

Keywords: MEMS gyroscope, nonproportional damping, anisoelasticity, modal coupling.

1 Introduction

Structural anisoelasticities and nonproportional damping lead to significant modal coupling between the sense and drive modes in a resonant gyroscope, thereby reducing the accuracy of the sense output. Accurate modelling and identification of these characteristics will significantly enhance the ability to compensate for these errors via feedback/feedforward control strategies thus leading to the next generation "smart" MEMS gyroscopes with self calibrating capabilities [2]. In this context, the present work is a novel approach to predict modal coupling between the drive and sense modes of a resonant gyroscope through the off-diagonal stiffness and damping terms.

Dynamics of a resonant gyroscope are reviewed in section 2. Two identification methods are studied to characterise the extent of anisoelasticities arising due to structural imperfections and nonproportional damping. An instrumental variable method is presented in section 3 to identify the mass and stiffness matrices and compute the anisoelasticity parameters from the identified stiffness matrix. Section 4 describes an algorithm to identify the non-proportional damping based on perturbation theory. In section 5, the identification methods are tested on a simulation example for various levels of random noise to assess their robustness. Conclusions drawn from the numerical study and future directions to improve the performance of resonant MEMS gyroscopes are outlined in section 6.

2 Gyroscope Dynamics

The governing equations of motion of a resonant gyroscope, measuring rotation about the z-axis, are given by:

$$\begin{bmatrix} m_x & 0 \\ 0 & m_y \end{bmatrix} \begin{bmatrix} \ddot{x} \\ \ddot{y} \end{bmatrix} + \begin{bmatrix} c_x & c_{xy} \\ c_{yx} & c_{yy} \end{bmatrix} \begin{bmatrix} \dot{x} \\ \dot{y} \end{bmatrix} + \begin{bmatrix} k_{xx} & k_{xy} \\ k_{yx} & k_{yy} \end{bmatrix} \begin{bmatrix} x \\ y \end{bmatrix}$$
$$= \begin{bmatrix} F_x^c \\ F_y^c \end{bmatrix} + \begin{bmatrix} 0 & 2m\Omega_z \\ -2m\Omega_z & 0 \end{bmatrix} \begin{bmatrix} \dot{x} \\ \dot{y} \end{bmatrix} + \begin{bmatrix} E_x \\ E_y \end{bmatrix} \quad (1)$$

where: the m's, k's and c's are respectively the mass, stiffness and damping coefficients. Coriolis forces appear in the above equations of motion as $2m\Omega_z$ terms. F_x^c and F_y^c are the external forces applied to the gyroscope by the control system. Extraneous effects such as external accelerations, centrifugal force and z-axis motion feedthrough are lumped in the terms E_x and E_y, where x and y denote the displacements of the gyroscope in the drive and sense direction respectively.

In an "ideal" gyroscope the principal stiffnesses are equal and the principal coordinates coincide with the measurement (X-Y) coordinates *i.e.* the drive and sense axes (see Figure 1). In the presence of structural imperfections the two principal stiffness values do not match. Hence there is an angular mismatch between the principal-axes and the measuring coordinate system. This "non-ideal" behavior results in a frequency mismatch (between the drive and sense resonant frequencies) and coupling of the modes leading to errors in the sense output [2]. Two types of errors are caused by the non-diagonal stiffness and damping matrices. While the non-diagonal stiffness matrix causes quadrature error *i.e.* error signal in quadrature with the measured Coriolis force, the non-diagonal damping leads to errors in-phase with the measured Coriolis force.

System identification, which forms the core of this study, is the first stage of any corrective procedure to compensate for the above errors.

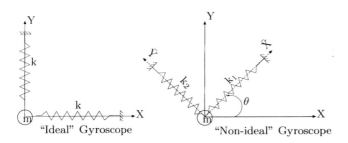

Figure 1: Structural imperfections shift the orientation of the principal axes, *i.e.* $\bar{X}$ and $\bar{Y}$ axes, relative to the measurement/excitation directions *i.e.* X and Y axes, in a non-ideal gyroscope. Note that the two principal stiffness values, k_2 and k_1, are different in a non-ideal gyroscope.

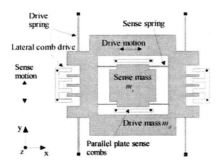

Figure 2: Schematic of a z-axis frame gyroscope with inner sensing and outer drive (ISOD) mechanism. Parameters obtained from a variant of the above device[3] were used for simulation.

3 Identification of Anisoelasticity

Anisoelasticity caused by fabrication tolerance limits and material inhomogeneity is quantified by the angular mismatch, denoted by θ, between the principal axes and X-Y coordinate system, and, the mismatch between the two principal stiffness values *i.e.* $2h = k_1 - k_2$ (see Figure 1). In order to measure these quantities, dynamic measurements can be made on the device by exciting along one degree of freedom (say drive axis), and measuring its response in the other degree of freedom (sense axis). It is assumed here that the 2 × 2 frequency response function (FRF) matrix governing the input-output relationships in the X and Y directions are measured. The FRF matrix, in the absence of rotation *i.e.* no Coriolis force, is given by:

$$H(\omega) = \left[-\omega^2 M + i\omega C + K \right]^{-1} \quad (2)$$

where the matrices M, K and C are as defined in equation 1. The real and imaginary parts of the dynamic stiffness matrix, obtained by inverting the FRF matrix H are given by:

$$\Re(D(\omega)) = K - \omega^2 M, \quad \Im(D(\omega)) = \omega C \quad (3)$$

respectively, where $D(\omega) = H^{-1}(\omega)$.

The above equation suggests an approach to identify the system matrices M, K and C. By considering the real and imaginary parts of the dynamic stiffness matrix D in equation (3) over a range of frequencies, a least-squares solution to the system matrices can be obtained. However, in practice, this simple approach is not possible and often leads to "biased" errors in estimates due to the presence of noise. In order to overcome this problem an instrumental variable technique [1] can be used.

3.1 Instrumental Variable Method

First developed by economists in the parameter estimation problems arising in the area of econometrics, this method was applied to problems in structural dynamics by Fritzen [1]. The development of the identification procedure is as follows.

Since $D(\omega) = H^{-1}(\omega)$, the error to be minimised in order to estimate the system matrices by least squares procedures is given by:

$$D(\omega)H(\omega) = I + E \quad (4)$$

where I is the identity matrix and E is the error matrix. By separating the real and imaginary parts and using the expression for FRF in equation (2), equation (4) can be written as:

$$\begin{bmatrix} \Re\left(D \begin{bmatrix} -\omega^2 I & i\omega I & I \end{bmatrix}\right) \\ \Im\left(D \begin{bmatrix} -\omega^2 I & i\omega I & I \end{bmatrix}\right) \end{bmatrix} \begin{bmatrix} M \\ C \\ K \end{bmatrix} = \begin{bmatrix} I \\ 0 \end{bmatrix} + \begin{bmatrix} \Re(E) \\ \Im(E) \end{bmatrix} \quad (5)$$

Stacking the above equation over a range of N discrete frequencies and rewriting:

$$A \begin{bmatrix} M \\ C \\ K \end{bmatrix} = \bar{I} + \bar{E} \quad (6)$$

where the matrix A is given by:

$$A = \begin{bmatrix} A_1 \\ \vdots \\ A_k \\ \vdots \\ A_N \end{bmatrix}, \quad A_k = \begin{bmatrix} \Re\left(D \begin{bmatrix} -\omega_k^2 I & i\omega_k I & I \end{bmatrix}\right) \\ \Im\left(D \begin{bmatrix} -\omega_k^2 I & i\omega_k I & I \end{bmatrix}\right) \end{bmatrix} \begin{bmatrix} M \\ C \\ K \end{bmatrix} \quad (7)$$

and

$$\bar{I} = \begin{bmatrix} \begin{bmatrix} I \\ 0 \end{bmatrix} \\ \vdots \\ \begin{bmatrix} I \\ 0 \end{bmatrix} \end{bmatrix} \quad \bar{E} = \begin{bmatrix} \begin{bmatrix} \Re(E_1) \\ \Im(E_1) \end{bmatrix} \\ \vdots \\ \begin{bmatrix} \Re(E_N) \\ \Im(E_N) \end{bmatrix} \end{bmatrix} \quad (8)$$

The instrumental variable method proposes an iterative solution of the form:

$$\begin{bmatrix} M \\ C \\ K \end{bmatrix}^{m+1} = \left[[W^m]^T A \right]^{-1} [W^m]^T \bar{I} \quad (9)$$

where W is the instrumental variable to be chosen by the user. In the present study W is chosen as the "analytical" dynamic stiffness matrix which can be obtained at each stage of iteration by the identified system matrices in the previous iteration $i.e.$ $W^{m+1} = \left[-\omega^2 M^m + i\omega C^m + K^m\right]^{-1}$. Here, M^m etc. are the matrices identified in the mth stage of iteration. The starting guess is provided by the linear least squares solution, obtained by using the real and imaginary parts of the "measured" dynamic stiffness $i.e.$ inverting the measured FRF matrix. Convergence can be checked based on the minimisation of the difference between the measured FRF matrix H and the reconstructed one H^{rm}, given by $H^{rm} = \left[K^m - \omega^2 M^m + i\omega C^m\right]^{-1}$. Thus the matrices M and K can be identified. The identified K matrix can be used to compute the anisoelasticity parameters.

3.2 Computation of Anisoelasticity Parameters

The elements of the stiffness matrix can be expressed in terms of the anisoelasticity parameters h, θ and the principal stiffness k_n as:

$$k_{xx} = k_n + h\cos(2\theta), \; k_{yy} = k_n - h\cos(2\theta), \; k_{xy} = k_{yx} = h\sin(2\theta) \tag{10}$$

The above relationships can be used to compute parameters h, θ and k_n given by:

$$\theta = \tan^{-1}\left(\frac{2k_{xy}}{k_{xx} - k_{yy}}\right), \; h = \frac{k_{xy}}{\sin(2\theta)}, \; k_n = \frac{k_{xx} + k_{yy}}{2} \tag{11}$$

Thus, a procedure to identify anisoelasticity in a MEMS device is:

- Measure the full FRF matrix H
- Identify the stiffness matrix K using the instrumental variable method
- From the identified stiffness matrix compute the anisoelasticity parameters using the equation (11)

The damping matrix identified by the IV method is not as accurate as the stiffness or mass matrices. Hence a more accurate method will be developed in the next section.

4 Identification of Nonproportional Damping

Using the procedure just outlined the matrices M and K can be identified from FRF measurements. These two matrices determine the analytical FRF of the undamped system given by:

$$H_u(\omega) = \left[-\omega^2 M + K\right]^{-1}. \tag{12}$$

However, the $measured$ FRF matrix is on the damped system and hence is given by the equation (2).

The natural question to ask at this sage is: how the measured FRFs of the damped system given by equation (2) can be related to the undamped FRFs given by equation (12). One can achieve this from a standard result of series expansion for the inverse of a sum of matrices from linear algebra. Firstly, for convenience, the expression for FRF in equation (2) can be written in modal coordinates determined by the undamped modes U as:

$$H(\omega) = \left[-\omega^2 I + i\omega C' + \Lambda\right]^{-1}, \tag{13}$$

where $C' = U^T C U$ is the damping matrix in modal coordinates, $\Lambda = U^T K U$ is a diagonal matrix with squared undamped natural frequencies as its diagonal entries and $I = U^T M U$ is the identity matrix. In experiments the modal damping factors determine the diagonal part of C', denoted by C'_d. The off-diagonal terms, denoted by C'_o, are identified by using the perturbation theory. Define $A(\omega) = \Lambda - \omega^2 I + C'_d$. Now the FRF expression in equation (13) can be written, to the first order in C'_o as:

$$H(\omega) = A(\omega) + i\omega C'_o \cong A^{-1}(\omega) - A^{-1}(\omega)C'_o A^{-1}(\omega). \tag{14}$$

The above relation can be used to identify C'_o, given $H(\omega)$ and $A(\omega)$. The convergence of the series in equation (14) is guaranteed if the matrix C' is diagonally dominant. This is only a sufficient condition [4] $i.e.$ nondiagonal dominance does not necessarily lead to nonconvergence of the series expansion in equation (14). With this background, a procedure for damping identification based on matrix perturbation expansion can be formulated as follows:

- Measure the full FRF matrix $H(\omega)$ on the test structure with a suitable choice of grid points.
- Identify the real modes U and damping factors (Q factors) and natural frequencies for each mode using modal identification procedures.
- The diagonal terms of the damping matrix are given by $C'_d = diag(2\zeta\omega)$
- Identify the off-diagonal terms of the damping matrix in modal coordinates using:

$$C'_o = \frac{A(\omega) - A(\omega)H(\omega)A(\omega)}{i\omega}. \tag{15}$$

- Obtain the modal damping matrix $C' = C'_d + C'_o$
- Transform into physical coordinates $C = U^{T^{-1}} C' U$

A least squares estimate of C can be obtained using the above procedure by considering a range of frequency points. The robustness of these methods will be studied next.

5 Numerical Example

The ISOD frame gyroscope parameters (see Table 1) are used to form the mass and stiffness matrices. Using the Q factors and natural frequencies the diagonal

Parameter	Value
Mechanical drive spring constant (k_x)	13.11 N/m
Mechanical sense spring constant (k_y)	28.45 N/m
Electrostatic spring constant (k_e)	-19.7 N/m
Drive mass (m_x)	1.1 μ g
Sense mass (m_y)	0.907 μ g
Q factor in sense mode (Q_x)	5000-1000
Q factor in drive mode (Q_y)	5000-1000

Table 1: MEMS gyroscope parameters used in simulation, obtained from [3].

Parameter	θ (rad/s)	h (N/m)	k_n (N/m)
Actual values	0.0873	2.18	10.930
0% noise	0.0873 ($\pm$0)	2.1800($\pm$0)	10.9300 ($\pm$0)
2% noise	0.0858 ($\pm$0.0031)	2.1809 ($\pm$0.0420)	10.9439 ($\pm$ 0.0305)
5% noise	0.0838 ($\pm$0.0077)	2.1826 ($\pm$0.1050)	10.9649 ($\pm$0.0758)
10% noise	0.0805 ($\pm$0.0153)	2.1866 ($\pm$0.2099)	11.0005 ($\pm$ 0.1500)
15% noise	0.0775 ($\pm$0.0229)	2.1920($\pm$0.3149)	11.0368($\pm$ 0.2229)

Table 2: Identification of anisoelasticity parameters.

part of the modal damping matrix is constructed. Small off-diagonal values are added, and the modal damping matrix is transformed back to physical coordinates using the undamped modes. From these three system matrices the frequency response function matrix is constructed using the equation (2). Random noise of 2%, 5%, 10% and 15% is added to the FRF matrix to simulate the measurement conditions. These "noisy" FRFs are treated as measurements input for the identification methods. A Typical noisy FRF matrix is shown in the figure 3(a).

The M, K matrices are identified using the IV method and the damping matrix C is identified using the matrix perturbation method. The identified K is used to compute the anisoelasticity parameters as described in section 3.2. Identified anisoelasticity parameters and their variance over 20 noise realisations is shown in the Table 2. It can be observed that even with a significant amount of added noise (15%) the errors in the parameters θ, h and K_n are respectively 4.04%, 18% and 0.3%. Note that the higher error on h is subjected to a large variation.

Finally the robustness of the non-proportional damping identification method to errors in the mass matrix and random noise is shown in figure 3(b). It can be seen that, while the sensitivity of the identification method to errors in the mass matrix is linear, the method is very robust to measurement noise. Given a good estimate of the mass matrix one can use this approach to identify the damping matrix.

6 Conclusion

Two algorithms are advanced to model and identify the structural imperfections and nonproportional damping from measured dynamic data on a MEMS device.

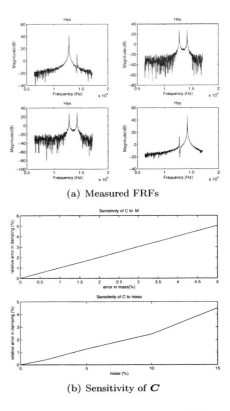

(a) Measured FRFs

(b) Sensitivity of C

Figure 3: Identification of nonproportional damping from vibration measurements: the subscripts x and y denote the drive and sense directions respectively.

These methods are validated on a numerical example whose parameters are based on an existing device. This study reveals that the proposed algorithms are robust to measurement errors. Future work will address the implementation of these algorithms on MEMS devices and on-chip control algorithm implementation.

REFERENCES

[1] **Fritzen, C.-P.**, *Identification of mass, damping, and stiffness matrices of mechanical systems*, Transactions of ASME, Journal of Vibration,Acoustics, Stress, and Reliability in Design, Vol. 108, No. 9, pp. 9–16, 1986.

[2] **Painter, C. C.** and **Shkel, A. M.**, *Identification of anisoelasticity for electrostatic "trimming" of rate integrating gyroscope*, SPIE Annual International Symposium on Smart Structures and Materials, New Port Beach, CA., Vol. March, 2001.

[3] **Palaniappan, M.**, IEEE-MEMS 2003, pp. 482–485, 2003.

[4] **Bhakar, A.**, *Estimates of errors in the frequency response of non-classically damped systems*, Journal of Sound and Vibration, Vol. 184, No. 1, pp. 59–72, 1995.

On the Air Damping of Micro-Resonators in the Free-Molecular Region

S. Hutcherson* and W. Ye**

Georgia Institute of Technology
771 Ferst Dr.,G.W.W. School of Mechanical Engineering
Georgia Institute of Technology, Atlanta, GA 30332 -0405, USA
*gtg552j@mail.gatech.edu **wenjing.ye@me.gatech.edu

ABSTRACT

Predicting air damping on micromachined mechanical resonators is crucial in the design of high-performance filters used in wireless communication systems. In the past, most of the work focuses on devices in which continuum theory can still be applied. In this work, we investigate damping on oscillating structures caused by air in the free-molecular region in which continuum theory is no longer valid. Such a study is important for devices operated at a very low pressure or for those whose characteristic length is on the order of nanometers. A careful examination of the previous work has been conducted. Mistakes and limitations have been found and reported in this paper. A molecular dynamics simulation has been developed and used in predicting quality factors of an oscillating micro beam operated at low pressures. Simulation results have shown an excellent agreement with experimental data.

Keywords: MEMS modeling, air damping, RF-MEMS

1 Introduction

There has been a lot of interest in using micromechanical resonator devices in wireless communication systems for various applications. One such example is to use them as high- frequency filters [1]. An electrostatically-driven, contour-mode disk resonator that oscillates at 733 MHz has been built [1]. Such designs have the potential to reach the gigahertz frequency range. A critical performance criterion of these resonators is the quality factor. A high quality factor is highly desirable in order to achieve high sensitivity. For this reason, there has been extensive research on investigating air damping on resonators [2,3]. Most of the studies focus on the devices with minimum feature size on the order of microns. At micro scale and standard pressure, continuum theory is still valid and was employed in the previous studies. In this work, we investigate air damping on resonators when the surrounding gas (air) is in the free-molecular region (Knudsen number > 10). This situation becomes important when resonators are operated at a very low pressure or the minimum feature size of the devices is on the order of nanometers. An application of this work could be the design of a gigahertz

disk resonator for which the gap between the resonator and the surrounding electrodes has to be in the submicron range (80 nm - 1 nm) in order to drive the stiff disk into resonance. With such a small gap, continuum theory breaks down and a molecular approach that accounts for interactions between the disk and individual molecules must be adopted.

Previous work on using molecular approaches to study air damping has been focused on micro-beam resonators operated at a very low pressure [4,7,8,9]. During the course of this study, we have found errors and limitations of these approaches. A molecular dynamic simulation has been developed to find the quality factor of the micro-beam resonator. With this approach, a much improved agreement with the experimental measurements has been achieved.

This paper is organized as follows. In the next section, a brief introduction of previous studies of air damping on a micro-beam resonator is given. It is followed by a careful examination of the previous approaches and corrections. In Section 3, the molecular dynamics simulation is described and the simulation results are presented.

2 The Theoretical Analysis of Micro-Beam Resonators

Previous work involving the analytical derivation of the damping effect on an oscillating beam was done by Christian [4], Kadar et al. [7], Li et al. [8], and Bao et al. [9]. Christian proposed a free molecular theory that was used to calculate the damping force acting on an oscillating vane in a vacuum by finding the pressure difference between the front and back of the vane using the Maxwell-Boltzmann speed distribution function for gases [4]. When this model was used to compute the quality factor of a micro-beam resonator, the quality factor was almost one order higher than the experimental results [6]. In an attempt to bring the theoretical values closer to the measurements, Kadar introduced a modification to Christian's method that involved a new molecule speed distribution function (the Maxwellian-Stream distribution), reducing Christian's quality factor results by a factor of π [7]. Li modified Kadar's Maxwellian-Stream (MS) distribution function to account for the velocity of the contact surface [8]. This further reduced Kadar's quality factor results, bringing the analytical solution even closer to the experimental results of Zook [6]. By pointing out the importance of the effects of damping caused by a nearby wall and the redundancy in Kadar's velocity distribution function, Bao introduced the energy transfer model as a way of directly calculating the energy loss [9]. He has shown that his results are closer to the experimental results than those based on Christian's model.

2.1 Errors in Kadar's Modification

Despite good agreements with experimental measurements, there are two fundamental mistakes in Kadar's approaches. As pointed out by Bao [9] (although without explanation), the Maxwellian-Stream distribution function is used redundantly. The redundancy comes from the way the number of molecules striking the micro-beam is computed. In Christian's model, it is calculated by constructing a control volume of $V = v_r dA dt$, where v_r is the relative velocity between the gas molecule (v) and the beam (u), dA and dt are the striking area and time respectively. Thus, the density of molecules that strike the beam within time dt and area dA is

$$ n \cdot \int_0^\infty V \cdot MB(v) dv $$

where $MB(v) = \left(\frac{m}{2\pi kT}\right)^{\frac{1}{2}} \exp{-\frac{mv^2}{2kT}}$ is the Maxwell-Boltzmann (MB) distribution function, m is the mass of the gas molecule, k is Boltzmann's constant, T is the absolute temperature of the gas, and n is the density of molecules. In Kadar's model, the Maxwellian-Stream (MS) distribution is used instead. Unlike the MB function that gives the velocity distribution of all the molecules in a general gas assembly, the MS function is the velocity distribution function of the molecules that strike the surface. Thus, when using MS to calculate the density of molecules that strike the micro-beam, the correct value should be $n \cdot \int_0^\infty MS(v) dv$. However, in Kadar's calculation, it is $n \cdot \int_0^\infty V \cdot MS(v) dv$. Kadar simply replaced the MB distribution function with the MS distribution function in Christian's calculation, which is not correct.

A second mistake comes from the coefficient in the MS distribution function used in Kadar's model. Directly adopted from [10], the coefficient in $C(c)$ is $\frac{1}{2} \left(\frac{m}{kT}\right)^2$ ([7], equation (12)). This coefficient, however, is obtained by considering only the molecules that strike the surface, i.e., integrating this distribution function in the velocity space gives only the total number of molecules that strike the surface, not the total number of gas molecules as a whole. The correct coefficient should be $\frac{m}{2kT}\sqrt{\frac{m}{2\pi kT}}$. With this coefficient and the correct way of computing the density of molecules that strike the surface using the MS distribution function, the net pressure

acting on the micro beam is

$$ p = 2m \int_0^\infty [n \cdot \sqrt{\frac{m}{2\pi kT}} e^{-\frac{mv^2}{2kT}} (v+u)^2 $$
$$ -n \cdot \sqrt{\frac{m}{2\pi kT}} e^{-\frac{mv^2}{2kT}} (v-u)^2] dv. \quad (1) $$

After taking the common factors out of the integral, the above equation corresponds exactly with the pressure equation derived from Christian's model (shown in equation (12) in [4]). This is not surprising since the fundamental principle used in Christian's and Kadar/Li's approaches is exactly the same.

2.2 Limitations of Bao's Energy Transfer Model

In [9], Bao et al. used a direct approach to determine the energy transfer between a resonating device and the equilibrium gas that surrounds it. By assuming elastic collisions between gas molecules and a resonating structure and ignoring the intermolecular collisions, Bao's model uses energy transfer mechanics to calculate the kinetic energy gain of molecules and therefore obtain the quality factor of the resonator structure. By applying this model to the oscillating micro beam, Bao was able to show that the calculated quality factor compared favorably with the measurements from Zook's microbeam experiment. Because it incorporates nearby walls and device geometry, Bao's energy transfer model provides a reasonable starting point for the analysis of resonating MEMS devices with various geometries. However there are three prevailing assumptions in the development of Bao's model. The first is that the velocity change of each particle is only considered after collisions with the resonating device have occurred; that is, the velocity of each particle is assumed to be constant for the entire period of interaction with the disk. This allows for a much simpler expression for the number of collisions ΔN, given by (2)

$$ \Delta N = \frac{\Delta t \times v_{xo}}{2(d_o - x)} = \frac{l v_{xo}}{2(d_o - x) v_{yzo}} = \frac{l}{2\sqrt{2}(d_o - x)} \quad (2) $$

where Δt is the time a gas molecule is within the region where it can interact with the device, l is the distance the molecule travels in the region, d_o is the initial gap between the device and the stationary wall, x is the displacement of the device as it oscillates, and v_{xo} and v_{yzo} are the initial velocities of the molecule in the x- and yz-directions, respectively. This implies that the number of collisions of a gas molecule whose velocity increases each time when it collides with the beam is the same as that of a molecule whose velocity decreases each time when it collides with the beam. Such an assumption could largely underestimate the energy gained by the gas molecules since the number of collisions of molecules who gain velocity is larger than the number of collisions of molecules who loss velocity. Figure 1 illustrates this

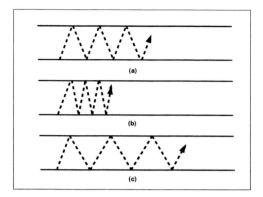

Figure 1: The Effects of the "Constant Velocity" Assumption

idea. In Figure 1(a), the path of a molecule is shown in time. This molecule's motion is based on the assumption that it continues with its initial velocity after every collision. Figures 1(b) and 1(c) show that if the velocity of the molecule increases or decreases after each collision, the time between two consecutive collisions will decrease or increase. For the same travel time (determined by l and v_{yzo}), the number of collisions will therefore be different. Since the final velocity of the molecule is based on the number of collisions ΔN, a change in this number will change the kinetic energy. As shown in the results section, this change is quite significant.

The second assumption is that the amplitude of the device's motion is much smaller than the gap between the moving device and any nearby walls. By this assumption, the device can be seen as stationary (distance-wise), thus further simplifying the calculation for the number of collisions that each molecule will have with the device to a constant. This assumption follows logically from the geometry and actuation mechanisms of the devices being studied.

The third major assumption is that the time for a gas molecule staying under the micro-beam (travel time) is much smaller than the oscillating cycle of the beam. Therefore, the velocity gained or lost by a molecule after each collision remains the same during its travel time. This is valid for the micro-beam being studied, as the oscillating period of the beam is almost one order smaller than the travel time of a molecule under the beam.

3 Molecular Dynamics Simulation

3.1 The MD Simulation

In order to develop a general approach to calculate the energy transferred between a resonating device and the gas surrounding it, a one-dimensional molecular dynamics simulation has been developed. The simulation tracks the position and velocity of one molecule as it moves within the gap between the device and nearby walls for a fixed period of time. Collisions between the molecule and the device or walls are assumed to be elastic and the energy transferred during the collision is calculated based on energy transfer mechanics, which incorporate the conservation of momentum and conservation of kinetic energy laws. The change of velocity of the molecule during each collision is recorded and the final velocity is obtained by adding the cumulative changes of velocity to the initial value. The final kinetic energy of the molecule was then calculated. To compute the total energy gained by the molecules during one oscillating cycle of the beam, the cycle is divided into many intervals (Δt). The changes of energy for molecules entering into the gap within each time interval are calculated and results are added to obtain the total energy change as shown in equation (3),

$$\Delta E_{cycle} = \frac{1}{4} n \bar{v} L \left(\sum \Delta e_k \right) \Delta t \qquad (3)$$

where $\bar{v}$ is the average velocity of the molecule, L is the peripheral length of the beam, and Δe_k is the change in kinetic energy of the molecule at each period division (Δt). The quality factor Q of the device was then calculated according to equation (4).

$$Q = \frac{2\pi E_B}{\Delta E_{cycle}} = \frac{2\pi \frac{1}{2} m_B A_0^2 \omega^2}{\Delta E_{cycle}} \qquad (4)$$

where m_B is the mass of the beam, A_0 is the amplitude of the beam displacement, and ω is the natural frequency.

3.2 Results - Quality Factor

To verify our MD simulation, the energy loss of the beam during one cycle was simulated first following exactly the same assumptions employed in Bao's model. The quality factors were calculated at different discretizations of the period (i.e., the different division of the period) and shown in Figure 2. As can be seen by Figure 2, the simulation results for Q_{sim} converge very closely to the value calculated from Bao's model (Q_{Bao}). The lower curve in Figure 2 (Q_{simV}) shows the resulting values when the molecular velocity is updated after each collision, instead of after the end of the travel time. It can be seen that the value converges to a little less than half of Bao's theoretical value. It is from this drop in quality factor that we can say that the "constant velocity" assumption made by Bao is incorrect. Figure 3 shows these results as compared to Zook's experimental and Christian's and Bao's theoretical results. Although

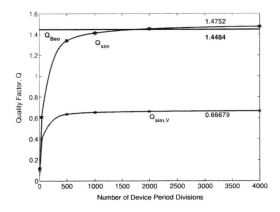

Figure 2: Convergence of Simulation Results

based on free-molecular assumption Bao's results most closely approximate the experimental values in the pressure range between 10^1 torr and 10^3 torr. At this range, the corresponding Knudsen number is between 4.4 and 0.044, indicating that the gas is in the transition or slip region in which the intermolecular collisions are important. The results from the simulation where the constant velocity assumption was relaxed show that the calculated values agree well with the experimental values in the pressure region lower than 10^1 torr (i.e., the Knudsen number is higher than 4.4) which is consistent with the assumption of free molecular region used in this study.

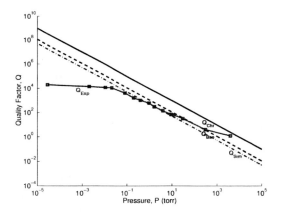

Figure 3: Simulation Results

4 Conclusion

In this paper, we report our studies on damping of oscillating microstructures caused by air in the free-molecular region. Previous work has been carefully examined and mistakes were found in Kadar/Li's approach. By correcting the mistakes, we demonstrate in this paper that Kadar/Li's approach leads to the same conclu-

sion as that of Christian's which is valid for a structure oscillating in a free space. For structures oscillating near a fixed wall such as the micro beam being studied, Bao's model is much more appropriate than Christian's. However, we also found that the "constant velocity" assumption used in Bao's model led to a much smaller energy loss prediction in the oscillating micro beam. To relax this assumption and to develop a general approach, a molecular dynamic simulation has been developed and used to obtain the quality factors of a oscillating micro beam operated at very low pressures. Results have shown an excellent agreement with the experimental data, particularly in the free-molecular region.

REFERENCES

[1] Nguyen, C., "MEMS Technologies for Communications", Technical Proceedings of the 2003 Nanotechnology Conference and Trade Show, Vol. 2, pp. 404-407, 2003.

[2] Cho, Y. H., A. P. Pisano and R. T. Howe, "Viscous damping model for laterally oscillating microstructures," Journal of Microelectromechanical Systems, Vol. 3, No. 2, pp. 81-87, 1994.

[3] Ye, W., X. Wang, W. Hemmert, D. Freeman and J. White, "Air Damping in Lateral Oscillating Micro Resonators: a Numerical and Experimental Study," Journal of Microelectromechanical Systems, Vol. 12, No. 5, pp. 557-566, 2003.

[4] Christian, R. G. "The theory of oscillating-vane vacuum gauges," Vacuum, Volume 16, Number 4, pp. 175–178, 1966.

[5] Halliday, et al. "The Kinetic Theory of Gases." Fundamentals of Physics: Extended, 5th Edition, pp. 484–501. New York: 1997, John Wiley and Sons.

[6] Zook, J. D., D. W. Burns, H. Guckel, J. J. Sniegowski, R. L. Engelstad, and Z. Feng. "Characteristics of polysilicon resonant microbeams," Sensors and Actuators A, Volume 35, pp. 51–59, 1992.

[7] Kadar, Z., W. Kindt, A. Bossche, and J. Mollinger. "Quality factor of torsional resonators in the low-pressure region," Sensors and Actuators A, Volume 53, pp. 299–303, 1996.

[8] Li, B., H. Wu, C. Zhu, and J. Liu. "The theoretical analysis on damping characteristics of resonant microbeam in vacuum," Sensors and Actuators, Volume 77, pp. 191–194, 1999.

[9] Bao, M., H. Yang, H. Yin and Y. Sun. "Energy transfer model for squeeze–film air damping in low vacuum," Journal of Micromechanics and Microengineering, Volume 12, pp. 341–346, 2002.

[10] Goodman, F. O., H. Y. Wachman. "Elementary Kinetic Theory of Gases at Interfaces," Dynamics of Gas-Surface Scattering, pp. 19–23. New York: 1976, Academic Press.

Simulation and Modeling of a Bridge-type Resonant Beam for a Coriolis True Mass Flow Sensor

S. H. Lee*[1], X. Wang*, W. Shin**, Z. Xiao*, K.K. Chin*, K.R. Farmer*

*New Jersey Institute of Technology, Microelectronics Research Center
121 Summit St. Rm. #206, Newark, NJ 07102, sl7@njit.edu
**Stevens Institute of Technology, wshin@stevens-tech.edu

ABSTRACT

This paper presents a simulation of a bridge-type resonating beam connected to the tube loop structure of a Coriolis true mass flow sensor[1]. The resonant beam technique, which is comparable to optical[2], piezoresistive[3], and capacitive[4] detection methods, is used to detect the vertical amplitude change of the tube-loop structure according to the Coriolis force of true mass flow by measuring the frequency shift of the resonating beam. The focus of this paper will be on the frequency shift caused by the amplitude of the bending mode of vibration, and the electrostatic pull-in effect as a measurement method for the frequency shift.

Keywords: resonating beam, Coriolis mass flow sensor, resonant frequency shift

1. INTRODUCTION

It is important to know the true mass flow in many flow applications. The Coriolis mass flow sensors measure the true mass flow. Various methods are used to measure true mass flow, including the optical[2] method, the capacitive[4] method, and the resonant beam detection method which tries to measure the bending or torsion displacement made in Coriolis sensors. Of the above methods, the micromachined resonant beam method is the most sensitive strain sensor. This method is able to provide high quality factor, so the resonant beam is used, not only in Coriolis sensors, but also in pressure sensors and accelerometers.

A resonant beam Coriolis mass flow sensor involves three vibrations. The 1st vibration is actuated externally with the lowest tube torsion resonant frequency depending on flow density. The 2nd vibration, introduced by the Coriolis force is an induced bending mode, and the Coriolis force is given by $F_c = -2mV \times \omega$ where mV is the mass flow momentum and ω is the angular frequency of the 1st torsion vibration. Angular frequency leads to a periodic Coriolis force imposed in the orthogonal direction of the 1st vibration, causing the elastic structure to undergo a 2nd vibration which, orthogonal to the 1st one, is operating in the bending mode of the tube structure. The measurement of mass flow now becomes the measurement of the peak amplitude of the 2nd vibration. The beam is periodically stretched or compressed. The resonant frequency of the beam under this periodic strain will have a periodic shift so that the detection of this frequency shift becomes the measurement of the 2nd vibration amplitude, thus a measurement of the mass flow. This paper presents the simulation of how the resonant beam is used and can be measured by pull-in effect to measure the 2nd vibration amplitude.

2. THEORY

It can be seen how the resonant beam works for measuring the bending displacement using the side view of the beam in figure 2. Assuming the Coriolis force in the z-direction on the top of the tube loop as the displacement, z_c, the bending moment $M(x)$, in the beam at location x is given by[5]

$$M(x) = M - Vx \tag{1}$$

The differential equation of the deflection of the beam is given by [6]

$$\frac{d^2 z_1(x)}{dx^2} = -\frac{M(x)}{EI} \tag{2}$$

Solving the differential equation with the boundary conditions (z(x) =0, z'(x) =0) gives the following [5]:

$$M(x) = \frac{6EIz_c}{L^2}\left(1 - \frac{2x}{L}\right) \tag{3}$$

The x-axis axial tensile stress is given by [5]

$$\sigma_x = c_{11}\varepsilon_x - vc_{12}\varepsilon_x \approx \frac{hz_c(c_{11} - vc_{12})}{L^2}\left(\frac{6x}{L} - 3\right) \tag{4}$$

where, c_{11} and c_{12} are stiffness coefficients, v the Poisson's ratio, σ_x the x component of stress, and $\varepsilon_1 = M(x)(h/2)/I$ the x component of strain. The stress is linearly proportional to the Coriolis induced displacement (z_c) of the bending tube loop structure. c_{11} and c_{12} are 16.6($\times 10^{10}$N/m^2) and 6.4, respectively, and v is 0.27 in single crystalline silicon. The natural frequency of a beam for the first mode is given by [7]

$$f_n = \frac{4.73^2}{2\pi L^2}\sqrt{\frac{EI}{\rho A}} \tag{5}$$

where f_n is a natural frequency of the resonant beam, E is Young's modulus, I bending moment of inertia, ρ density, A cross section of a beam, L length of a beam, and the coefficient for the first mode of natural frequency is 4.73[7].

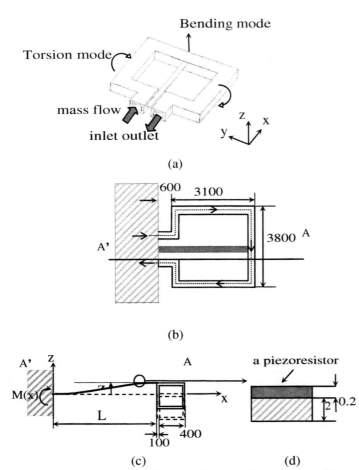

(a)

(b)

(c)   (d)

Figure 1: (a) Schematic of a Coriolis mass flow sensor with a resonant beam, (b) Top view of the structure, (c) Side view of the structure across the beam, (d) view of the beam including a piezoresistor. The unit of size is micrometers.

3. SIMULATION

The dimensions of the tube loop structure are shown in figure 1. The natural frequency of the tube loop structure is 2745Hz and 5110Hz in the first (bending) and third (torsion) modes, respectively. We assume the amplitude of the torsion excitation is fully transferred to that of the bending mode after the mass flow passes through the device. The bridge-type resonant beam is first investigated focusing on the relationship among the tensile stress, the resonant frequency shift and the damping effect, which can be controlled.

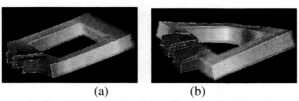

(a)   (b)

Figure 2: the 3D structure of bending mode (a) and torsion mode (b), which are the first and third mechanical vibration modes in the tube loop structure.

The size of a simulated resonant beam is 1000 (leng $\times$ 200 (width) x 2 (height) μm^3. Each end of a beam clamped and one end is displaced along the z-axis from 0 1µm. A travel range is considered to be 30 % (i.e. 0.74µ pull-in displacement of the 2µm gap due to the pull-in eff The 0.5um displacement for the Coriolis force of the m flow is applied to avoid the snap down of the beam by pu in displacement. The tensile stress increases linea according to the z-direction displacement at the edge of beam in figure 3. Even if the bending displacement of tube loop structure is along the z-axis, this displacement enough to make the x-axis tensile stress significant enou to affect the frequency shift. The frequency shift nonlinearly proportional to the square of the tensile stress figure 4. The tensile stress on the top and bottom of structure may cancel out each other in a beam . However, this case the beam can be regarded as a kind of a stri since the thickness of the beam is 2µm.

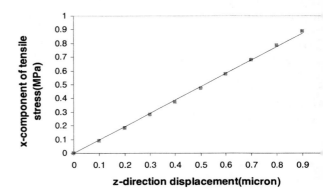

Figure 3: The relationship between the displacement and tensile stress.

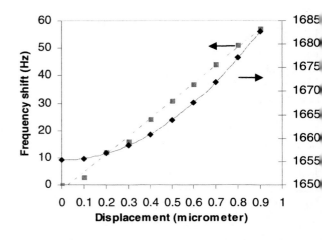

Figure 4: The frequency shift vs. the displacement by m flow

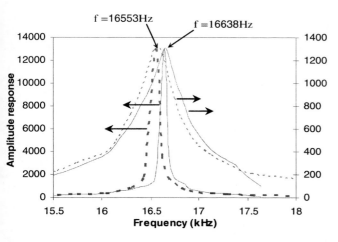

Figure 5: Effect of damping on the resonant frequency; 0.1% (thick lines) and 1% (thin lines) of the critical damping coefficient; dotted lines for no displacement, solid lines for 0.5μm. The resonant frequency shift of 86Hz by 0.5μm displacement is shown by a variation of 0.1% of the critical damping coefficient.

The resonant beam has been simulated with the electrostatic pull-in effect to make the resonance detect the frequency shift. As electrostatic actuation is applied to the resonant beam, the beam is deformed, tensile stress is caused at both ends and the extra frequency shift is produced. The range of the frequency shift caused by the pull-in effect is enough to measure the frequency shift caused by the Coriolis force, as shown in Table 1.

Voltage (V)	Pull-in displacement (μm)	Frequency (Hz)	Capacitance (pF)
0	0	16553	1.052472
2	0.055	16322	1.100262
4	0.245	15574	1.213922
6	0.743	15233	1.306456

Table 1. A comparison of the frequency and the capacitance of the pull-in effect on the resonant beam as the pull-in voltage is applied.

4. DISCUSSION

The resonant frequency shift becomes significant as the resonant frequency increases nonlinearly, in proportion with the square of the tensile stress. However, the ratio of the frequency shift to the tensile stress linearly increases. The electrostatic pull-in effect on the resonant beam has been applied to pick up the frequency shift, instead of using a piezoresistive or a capacitive method. The range of frequency shift produced by the Coriolis force is 86Hz from the 0 to 0.5μm displacement. The range of the frequency shift produced by the pull-in effect is 200Hz and 989Hz at 2V and 4V applied voltage, which means that the sensitivity

is 45% and 9%, respectively. The pull-in voltage of 2V applied to the beam facilitates the detection of frequency shift caused by mass flow. The smaller a pull-in voltage on the resonant beam, the greater the sensitivity. There can be a coupling problem between the frequencies of the resonant beam and the tube loop structure, however, the relative higher resonant frequency of the beam compared to that of the tube loop structure minimizes the effects of this problem.

As shown in figure 4 and table 1, the frequency shift increases with tip deflection on the double clamped beam whereas the frequency shift decreases with the pull-in effect. This can be explained by the relationship between the tensile stress and length of a beam. The frequency increases if the increase of length is so small that the tensile stress is the only factor, which affects resonant frequency when tensile force is applied to the axial direction of a beam. Otherwise, the frequency decreases as the length increases. [7] This sensitivity requires a high vacuum system that brings out the small damping coefficient to give a better quality factor, which affects the measurement of the frequency shift. A variation of 0.1% less than the critical damping coefficient is required to sense a significant resonant frequency shift, shown in figure 5.

As noted above, the resonant beam method can be compared with other detection methods for Coriolis mass flow sensors. Optical detection has a high sensitivity, but still needs a larger scale device. The capacitive method can sense a very small amount of capacitance change such as fF, and is comparable to the resonant beam method. The resonant beam method is the most sensitive relative to the other detection methods, but its disadvantage is that it requires very good vacuum sealing for high quality factor. All the above simulations were carried out with the COVERTORWARE software.

5. CONCLUSION

We have simulated the use of a resonant beam structure to measure flow in a Coriolis true mass flow sensor. To illustrate the technique, we have determined a resonant frequency shift (86Hz) for a 0.5μm displacement of the end of a double clamped beam attached to a tube loop structure. We have also studied the pull-in effect as a method to measure the resonant frequency shift induced by mass flow with high sensitivity (45%) at a pull-in voltage of 2V for our test structure.

REFERENCES

[1] A. M. Young, "Coriolis based mass flow measurement," Sensors, vol. 2, Dec. 1985, pp. 6-10
[2] Peter Enoksson, et al, "A silicon resonant sensor structure for Coriolis mass-flow measurements", Journal of microelectromechanical system, vol. 6, no. 2, June 1997, pp. 119-125
[3] Siebe Bouwstra, et al, "Resonating microbridge mass flow sensor", Sensors and Actuators, A21-A23 (1990)

332-335

[4] Yafan Zhang, et al, "A micro machined Coriolis-force-based mass flow meter for direct mass flow and fluid density measurement", The 11th International Conference on Solid-State Sensors and Actuators, Munich, Germany, Jun 10-14, 2001

[4] Measurement system, Earnest O. Doebelin, 1990, p 600-605

[5] Jyh–Cheng Yu, Chin-Bing Lan, "System modeling of microaccelerometer using piezoelectric thin films", Sensors and Actuators A 88 (2001) p 178-186

[6] Mechanics of materials, James E. Gere & Stephen P. Timosenko, 4th edition, p 602-603, 1997

[7] Vibration problems in engineering, S. Timosenko, 4th edition, 1974

[8] Microsystem design, Stephen D. Senturia, 2000, p 211-213

---

[1]Sang Hwui Lee, 121 Summit St. Microelectronics building Rm. #206, NJIT, Newark, 07102, Tel.: +1-973-596-6368; Fax: +1-973-596-6495; E-mail: sl7@njit.edu

Computational Prototyping of an RF MEMS Switch using Chatoyant

M. Bails*, J. A. Martinez*, S. P. Levitan*, I. Avdeev**, M. Lovell**, D. M. Chiarulli***

University of Pittsburgh Departments of
*Electrical, **Mechanical Engineering, and ***Computer Science
348 Benedum Hall, Pittsburgh, PA, 15261, USA, mikeb@ee.pitt.edu,
jmarti@ee.pitt.edu, steve@ee.pitt.edu, ivast@pitt.edu, mlovell@pitt.edu, don@cs.pitt.edu

ABSTRACT

In this paper we demonstrate the capabilities of our system-level CAD tool, Chatoyant, to model and simulate an RF MEMS switch. Chatoyant is a mixed signal, multi-domain CAD tool that can be used to design and analyze complete mixed-technology micro-systems. We perform a system level simulation of an RF MEMS switch. This is accomplished by coupling mechanical and electrical domains of this system. We verify our mechanical results using the commercial simulation packages, ANSYS, and CoventorWare.

Keywords: microelectromechanical systems, modeling, microsystems, finite element analysis, radio frequency switch, circuit simulation

1 BACKGROUND

Chatoyant is a multi-domain system level simulation tool. It is optimized for loosely coupled systems incorporating complex components, including electrical, optical, and mechanical devices which are found in multi-domain microsystems [1]. In Chatoyant, the mechanical behaviors of MEMS devices are modeled as a set of differential equations that define their dynamics as a reaction to external forces. Each mechanical element (beam, plate, etc.) is characterized by a template consisting of a combination of mass, damping, and stiffness matrices in a Modified Nodal Analysis (MNA) representation. This template is created by transforming the second order ordinary differential equation (ODE) motion equation into a first order ODE for a piecewise linear (PWL) solution (Figure 1).

General motion equation
$$F = [K][U] + [B][\dot{U}] + [M][\ddot{U}]$$

Standard ODE Transformation
$$\begin{bmatrix} 0 & M \\ M & B \end{bmatrix} \begin{bmatrix} \ddot{U} \\ \dot{U} \end{bmatrix} + \begin{bmatrix} -M & 0 \\ 0 & K \end{bmatrix} \begin{bmatrix} \dot{U} \\ U \end{bmatrix} = \begin{bmatrix} 0 \\ I \end{bmatrix} F$$

Templates for every basic element (e.g. beam)
$$X = \begin{bmatrix} \dot{U} \\ U \end{bmatrix}; \quad [Mb]\dot{X} + [Mk]X = [E]F$$

Figure 1: Mechanical Matrix Representation

By controlling the degrees of freedom for the components in the system and using a PWL solver, a trade-off between speed and accuracy is realized. Electrical components are modeled using a similar MNA technique. The interaction between electrical and mechanical components is accomplished by modeling the coupled energy between the domains.

2 RF MEMS DEVICE

The RF MEMS device we model was designed and fabricated at the University of Michigan [2, 3]. It is composed of electrostatic actuation plates and a capacitive plate suspended over a coplanar waveguide by spring meanders (Figure 2). This device works as an electrically switched shunt capacitor. With no voltage applied to the actuation pads, most of the RF signal can pass through the signal line. Applying a voltage to the actuation pads, results in an increase in the coupling capacitance between the signal line and the central capacitive plate. This lowers the impedance between the signal and ground, which effectively "shunts" the RF energy to ground and stops the RF propagation through the signal line [4]. The sensitivity of the device is directly related to the number of meanders in the spring assembly. Increasing the number of meanders will ideally lower the required voltage for switch operation. For this research, we considered a device having four meanders.

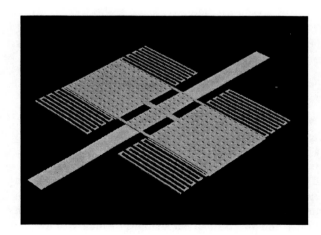

Figure 2: 3-D Rendering of RF MEMS Device

3 MODELING OF RF MEMS DEVICE

The general modeling techniques of Chatoyant are discussed in section 3.1. The remaining subsections are dedicated to the different aspects for prototyping the mechanical features of this device. These investigations include the stiffness properties of the spring meanders, the modal analysis of the spring assembly, and the dynamic analysis of the spring assembly.

The work is part of an ongoing analysis, which will include electrical models and system level analyses. The electrical models will be used to perform pull-in voltage analyses. Finally, a complete system simulation will be performed, including the RF signal, control voltages, and the MEMS switch verifying Chatoyant's mixed-domain capabilities.

3.1 Chatoyant Mechanical Modeling

The mechanical model for the device can be viewed as a set of ordinary differential equations that define its dynamics as a reaction to external forces. Each mechanical element (beam, plate, etc.) is composed of a set of characteristic matrices, K (stiffness), B (damping) and M (mass). Theses matrices are static and independent of the dynamics in the body. Figure 3 shows an example of the stiffness matrix used for a typical 2-D beam element [5]. These matrices can be extended to 3-D structures incorporating six degrees of freedom (x, y, z, rotation x, rotation y, rotation z).

$$K = \frac{EI}{L^3} \begin{bmatrix} 12 & 6L & -12 & 6L \\ 6L & 4L^2 & -6L & 2L^2 \\ -12 & -6L & 12 & -6L \\ 6L & 2L^2 & -6L & 4L^2 \end{bmatrix}$$

Figure 3: Sample 2-D Beam Element Matrix

The standard second order differential equation of motion, where U is position V is velocity and A is acceleration, is given by:

$$F = KU + BV + MA \qquad (1)$$

Knowing the velocity is the first derivative and acceleration is the second derivative, the above equation can be reduced to a standard first order form which gives a complete characterization of the mechanical system (see Figure 1).

The use of a PWL general solver decreases the computational task and allows for a trade-off between accuracy and speed. This technique can be used in both electrical and mechanical simulations, which merges the complex device interactions in these mixed domains.

3.2 Four Meander Spring Constant

The spring constant for a four meander spring is determined by applying a known force value (in this case, 1μN) to the free end of the structure and dividing by the resulting displacement. Chatoyant performs a dynamic analysis of the spring stiffness whereas ANSYS [6] and Coventor [7] use static analyses. For this device, the desired movement is in the z-direction, however, it is feasible that unwanted movement in x or y may exist. This unwanted movement could generate coupling variations to the incoming RF signal. Table 1 gives a comparison of the resulting spring constants in the x, y, and z directions for the different solvers, showing good agreement.

	Stiffness (N/m)		
	Kz	Kx	Ky
Chatoyant	0.0543	1.749	0.372
Ansys	0.0527	1.773	0.344
CoventorWare Architect	0.0546	1.863	0.396
CoventorWare Analyzer	0.0568	1.881	0.408

Table 1: Four Meander Spring Stiffness Constant

3.3 Spring Modal Analysis

The modal response of a four meander spring is necessary to determine the maximum frequency of operation for the switch. This is done to ensure that any unwanted resonance that may exist in the system is avoided.

The results of the first nine modal frequencies are shown in Figure 4. Table 2 provides a comparison of runtimes for the modal analysis for the different solvers. These analyses were performed on a P4 3.00 GHz processor with 2GB SDRAM.

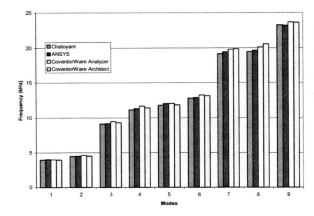

Figure 4: Modal Response of Four Meander Spring

Simulation Time - Spring Modal Analysis		
Solver	Simulation Time	Notes
Chatoyant	1.107 seconds	Two Nodes per Element with Six Degrees of Freedom per Node
CoventorWare Architect	2.330 seconds	Nonlinear - One Segment Beam with Six Degrees of Freedom at Each Beam End
CoventorWare Analyzer FEM*	134.000 seconds	Manhattan Bricks - 27-Node Parabolic Elements (1664 Elements)
ANSYS FEM*	30.000 seconds	3-D 20-Node Structural Solid Element, Solid95 (1664 Elements)

*The FEM Element size is 2.5um x 2.5um x 2um

Table 2: Simulation Times for Modal Analysis

3.4 Dynamic Behavior of RF Switch

The dynamic response of the switch was investigated in Chatoyant by applying a time-dependent force in the z-direction. A force of 1μN was applied to the plate structure over two different time sequences. The first dynamic response was based on a ramped input of 600ms (Figure 5). The second dynamic response was based on a ramped input of 600μs (Figure 6).

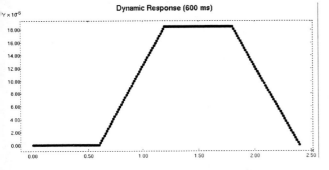

Figure 5: Slow Switch Response (600ms Rise Time)

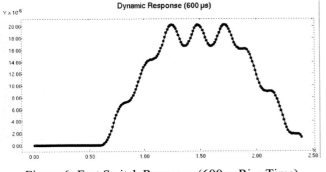

Figure 6: Fast Switch Response (600μs Rise Time)

With a slow input time of the force, the displacement of the switch displays a linear response. Increasing the input speed of the force by a magnitude of 100x, results in oscillations in the device. These oscillations could cause unwanted changes in the capacitance values during actuation. Considering this input force necessitates finding an input frequency which does not result in switch displacement resonance.

4 ELECTROSTATIC MODELING

For the ongoing analysis, the electrostatic models of the device will be investigated. The main factor in the electrical properties of the model stems from the pull-in voltage analysis. Analytically, pull-in voltage is related to the geometry of the parallel-plate capacitor and the stiffness coefficients of the spring meanders.

$$V_{pi} = \sqrt{\frac{8K_z g_o^3}{27\varepsilon_o A}} \qquad (2)$$

In the above equation, K_z is the total spring stiffness of the structure (equal to $4*k_z$, where k_z is the individual spring constant), g_o is the initial gap between the switch and the electrodes above the substrate, and ε_o is the free-space permittivity, and A is the area of the actuation plates. The pull-in equation is a result of equating the spring stiffness to the electrical spring softening k_e, where:

$$k_e = -\frac{dF_e}{dz} \qquad (3)$$

and,

$$F_e = \frac{1}{2}V^2\frac{dC}{dz} \qquad (4)$$

The electrical force is directly related to the parallel plate capacitance of the actuation pads and to the square of the input voltage.

The main concept of the RF switch of this type is to create low-voltage actuation. From the pull-in equation (2), it can be seen that the pull-in voltage can be altered by manipulating any of the geometrical aspects of the switch, namely, the initial gap, the area of the actuation pads, or by increasing the number of meanders in the spring (thereby lowering the overall spring constant).

5 FUTURE WORK: FULL SYSTEM LEVEL SIMULATION

The full system level analysis involves an end to end simulation of the device (Figure 7).

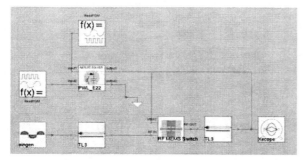

Figure 7: Chatoyant Schematic of Full System Simulation

The system, which uses a high frequency input signal (40GHz – 'singen' block), will be sent along the coplanar waveguide transmission lines (TL blocks). The device will be actuated by a pulsed input (PWL block and signal drivers). The speed of the switching device and the 'shunting' capabilities will be investigated to test the qualities of the device.

6 SUMMARY AND CONCLUSION

The ongoing analysis of the RF MEMS switch requires the capability to cross the mechanical and electrical domains in order to accurately model the device. The meander spring assembly determines the overall stiffness of the structure. The modal response helps to determine the frequency at which the device can be actuated. Dynamic analyses of the device were completed to investigate the switching capabilities of the device.

Other considerations for future analyses are to incorporate residual stresses associated with the fabrication of the device. This includes curvature of the device when released during the etching phase and differences in the geometry of the switch arising from fabrication and design.

REFERENCES

[1] S.P. Levitan, et al, "System Simulation of Mixed-Signal Multi-Domain Microsystems With Piecewise Linear Models." IEEE Trans. On CAD. Vol. 22 n. 2, Feb. 2003, pp. 139-154.

[2] S.P. Pacheco, et al, "Design of low actuation voltage RF MEMS switch," in IEEE MTT-S Int. Microwave Symp. Dig., vol 1, June 2000, pp. 165-168.

[3] Peroulis, D., et al, "Electromechanical Considerations in Developing Low-Voltage RF MEMS Switches," in IEEE Trans. On Micro. Theory and Techniques, vol 51, no. 1, Jan 2003, pp. 259-270.

[4] Yao, J. Jason, "RF MEMS from a device perspective," in J. Micromech. Microeng., vol. 10, 2000, pp. R9-R38.

[5] Przemieniecki, J.S. Theory of Matrix Structural Analysis., Dover Publications, Inc. New York, 1985.

[6] ANSYS Multiphysics/LS-DYNA, ANSYS Inc., Cannonsburg, PA. www.ansys.com.

[7] CoventorWare version2003.1, Coventor, Inc., Cary, NC. www.coventor.com.

Acknowledgements:
This work was supported, in part, by DARPA/AFOSR under grant number F49620-01-1-0536 and NSF under grant number C-CR9988319. The authors would like to thank Coventor, Inc. for their support, advice, and cooperation.

Effective Modelling and Simulation of Over-Heated Actuators

M. Zubert[*], M. Napieralska[*], A. Napieralski[*], J.L. Noullet[**]

[*]Technical University of Lodz, DMCS
Al. Politechniki 11, 93-590 Lodz, Poland, {napier,mnapier,mariuszz}@dmcs.p.lodz.pl
[**]INSA de Toulouse, AIME
135, avenue de Rangueil, 31077 Toulouse Cedex 4, France, noullet@aime.insa-tlse.fr

ABSTRACT

A mono-dimensional model of an over heated actuator is presented in this paper. The proposed distributed models are derived based on real physical phenomena described by multi-dimensional partial differential equations. The proposed model has been prepared taking into account real polycrystalline silicon parameters and their thermal dependencies. The considerations are based on the previous device analysis and FEM[1] simulation for the device operating in real working conditions in static and quasi-static mode.

Keywords: MEMS, over-heated actuator, polycrystalline silicon parameters, mono-dimensional model, RESCUER software

1 INTRODUCTION

The paper presents a continuation of work on accurate model of micro-machined over-heated actuator. The previous paper [20] only introduced to a distributed model laying special emphasis on the correct application of boundary conditions and the presentation of chosen simulation results obtained for the considered device performed in the ANSYS simulation environment.

The operating principle of this mechatronic structure (see Figure 1) is based on the coupling between electrical, thermal and mechanical phenomena. The actuator consists of suspended polycrystalline silicon beams. Two of them are anchored to the substrate at the two remaining ends $\partial\Omega_A$ and $\partial\Omega_B$ (see cross-sections S1 and S4, Figure 2).

Typical device dimensions: beams length: 30[μm], 115[μm], 145[μm]; beams width 5[μm], 20[μm]; vertical gap 5[μm]; polycrystalline silicon thickness 1.5[μm]. When electrical voltage is applied to these terminals, the Joule heat is dissipated in the beams. Since the beam consists of sections having different cross-section area, each section expands thermally differently and as the consequence the actuator bends (see Figure 2,3).

The presented device is relatively simple yet it involves many coupled phenomena having different nature thus it is an excellent example to illustrate multidomain simulation methodology. The mathematical description of the

---

[1] FEM – Finite Elements Method

phenomena occurring in the device and the gradual simplification of the description will be demonstrated in the following sections of the paper.

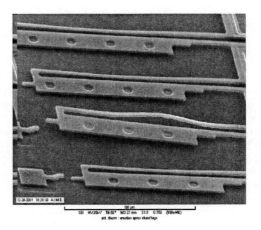

Figure 1 An example of fabricated devices manufactured in the AIME Laboratory.

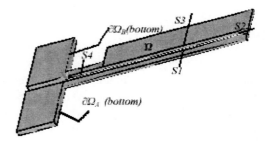

Figure 2 The shape of thermal linear motor [20].

2 ELECTRICAL DOMAIN

In the electrical domain, there can be distinguished two electrical terminals (**A**, **B**) and a polycrystalline silicon resistor with variable cross-section area. In this case, the electrical phenomena can be simplified and directly described mathematically using lumped network consisting of four resistors connected in series assuming idem potential material resistivity, simplified beam join resistances and independence of polycrystalline silicon resistivity on temperature [20]:

$$|j_i| = \frac{I}{S_i}; \quad I = (V_A - V_B)\left(\sum_{k=1}^{4} R_i\right)^{-1}; \quad R_i = \rho_{polySi}\frac{l_k}{S_k} \quad (1a)$$

$$p_{th,i} = \rho_{polySi} \cdot (I / S_i)^2 \quad (1b)$$

where I – device current, R_i - lumped beam resistance; S_i, l_k - cross-section area and length of the beam (for R_i) respectively; j_i – the current density in R_i, V_A, V_B - voltage applied to electrical terminals **A** and **B** respectively; I, k – the cross-section number (see S_i in Figure 2); $p_{th,i}$ - thermal power density of dissipated heat in R_i, ρ_{polySi} - polycrystalline silicon resistivity.

3 THERMAL DOMAIN

Fortunately, the coupling between electrical, thermal and mechanical phenomena allows the assumption of unidirectional power transfer from electrical to thermal domain and after that to mechanical one. Therefore a quasi-static thermal model can be adopted and consequently the thermo-dynamic coupling of the mechanical and thermal domain can be neglected in the heat transfer equation. Additionally, assuming homogeneous distribution of temperature in the device cross-section, the thermal phenomena can be modelled using the mono-dimensional form [20]:

$$\begin{cases} \partial_\xi(\lambda(\xi)\partial_\xi T(\xi)) + p_{th}(\xi) = h(\xi) \cdot (T(\xi) - T_{ext}) \\ \qquad \text{for} \quad \xi \equiv (x,y) \in \Omega; \\ T(\xi) = T_{ext} \\ \qquad \text{for } \xi \equiv (x,y) \in \partial\Omega_A \cup \partial\Omega_B \end{cases} \quad (2a)$$

where $T(\xi)$ - unknown temperature along particular beam in Kelvin degree (!); $h(\xi)$ - heat transfer coefficient; λ - thermal conductivity; Text -air temperature in Kelvin degree; ξ - "the natural coordinate" from the terminal **A** to **B**; $\partial\Omega_A$ and $\partial\Omega_B$ - the contacts of electrical terminals (**A** and **B**); $\partial\Omega = \partial\Omega_A \cup \partial\Omega_B \cup \partial\Omega_C$, $p_{th}(\xi) = p_{th,i(\xi)} \cdot S_{i(\xi)}$ etc

Then, the nonlinearity of equation (2) can be removed from the deferential operator by assuming that the thermal conductivity can be approximated by the following expression (for more see section 5):

$$\lambda(\xi) = \lambda_0(\xi)T(\xi)^{\lambda_1(\xi)} \Rightarrow$$

$$\hat{T}(\xi) = \frac{\lambda_0(\xi)}{\lambda_1(\xi)+1}T(\xi)^{\lambda_1(\xi)+1} + C_1 \quad (2b)$$

where $\hat{T}(\xi)$ - transformed temperature value; C_1 - constant parameter, we will assume that $C_1 = 0$.

$$\begin{cases} \partial_{\xi,\xi}\hat{T}(\xi) + p_{th}(\xi) = h(\xi) \cdot \left(\sqrt[\lambda_1(\xi)+1]{\hat{T}(\xi)} \cdot \varsigma - T_{ext}\right) \\ \qquad \text{for} \quad \xi \equiv (x,y) \in \Omega; \\ T(\xi) = T_{ext} \\ \qquad \text{for } \xi \equiv (x,y) \in \partial\Omega_A \cup \partial\Omega_B \end{cases} \quad (2c)$$

where $\varsigma = \sqrt[\lambda_1(\xi)+1]{\dfrac{\lambda_1(\xi)+1}{\lambda_0(\xi)}}$.

The detailed analysis shows that Raleigh number for the air flow in the air gap between two polycrystalline silicon walls (beams S1 - beam S3 and beam S1 - beamS4) is quite small $(Ra)_{l=5[\mu m]} \cong 85.73$ and consequently the heat transfer between walls can be approximated using the Fourier law[2] as follows:

$$q \cong \frac{\lambda_{air}}{\delta}\Delta T \quad (3)$$

where ΔT - wall temperature difference, λ_{air} - air thermal conductivity; δ - distance between walls; q - thermal flux. This phenomenon can be taken into account as an additional heat source increasing $p_{th}(\xi)$. In this case the heat transfer coefficients $h(\xi)$ have been estimated as the combination of Nusselt number for the whole beam [1] and vertical beam side (for more see [20]).

$$h(\xi) \cong \left(d^{*5/4}\alpha\, 0.372 \cdot \left(\frac{\nu/a}{0.7}\right)^{1/4} - \right.$$

$$\left. t^{5/4}(\alpha-1)0.6773\left(\frac{\nu/a}{(\nu/a)+0.952}\right)^{1/4}\right) \cdot$$

$$\cdot \frac{g\lambda_{air}(T_{ext})}{\nu^2 Text}(T(\xi) - Text)^{1/4}; \quad (4)$$

$$d^* = \frac{w(\xi)+t}{2}; \quad \alpha = \frac{t}{2w(\xi)+t}+1$$

where ν - coefficient of air kinematics viscosity, a – air thermal diffusivity; $w(\xi)$ - beam width; t – beam thickness.

---

[2] It takes into consideration only static phenomena. The values of air and polycrystalline silicon thermal compensation coefficients are bound by the relation $a_{polysilicon} \approx 2a_{air}$, where $a = \lambda/(c\delta)$), hence the thermal response in these materials are comparable. In other words "temperature covers twice as much distance in polycrystalline silicon as in air at the same time" ($x_{air} \approx 2x_{polysilicon}$). Fortunately this effect can be neglected due to the difference in air and polycrystalline silicon thermal conductivities ($\lambda_{polisilicon} \approx 350\lambda_{air}$).

4 MECHANICAL DOMAIN

The considered device consists of two non-equally heated beams. The device dimensions and thermal diffusivity allows the adoption of a non-isothermal large deflection model (form more see [2]-[5]) using modified Bernoulli-Euler equation [20].

$$\frac{d\varphi(\xi)}{ds} \cong \left(1-\beta\left(T(\xi)-T_0(\xi)\right)\right)^{-1}\frac{M(\xi)}{E\,J} \qquad (5a)$$

$$\frac{d\varphi(\xi)}{ds} = -\frac{d^2y(x)}{dx^2}\left(1+\left(\frac{dy(x)}{dx}\right)^2\right)^{-3/2} \qquad (5b)$$

$$S(\xi)\rho_{polySi}\frac{\partial^2 y(\xi,t)}{\partial t^2} = \frac{\partial^2 M(\xi,t)}{\partial \xi^2} - N\frac{\partial^2 y(\xi,t)}{\partial \xi^2} \qquad (5c)$$

where s – natural co-ordinate $s \in \langle 0, L \rangle$; L – beam length; $\varphi = \varphi(\xi) = \arctan \partial_x y(x)$ - the bending angle of the beam ($\partial_\xi \varphi(\xi) = R^{-1}$); J – beam inertial moment, $y(\xi)$ and $M=M(\xi)$ –beam bending and mechanical moment at point s respectively (see Figure 3); E - Young modulus; v - Poisson coefficient; β - thermal expansion coefficient; N – force associated with stress generation an axial force.

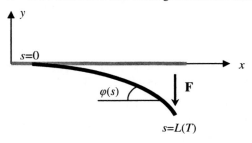

Figure 3 The schematic draft of large deflected beam

5 MATERIAL PROPERTIES

The most of the micro-machined elements similar to the presented micro-actuator are realized using polycrystalline silicon. Unfortunately, not all thermal and mechanical properties of this material have similar values to the bulk silicon properties. On the other hand, the thermal parameter dependence has to be taken into account for large temperature variation. Therefore, polycrystalline silicon parameters estimation will be presented in the next section.

The following investigation procedure and final approximation of polycrystalline silicon thermal properties in the temperature range between 250 K and 1200 K will be proposed for their lack in the professional literature.

In most cases, the silicon thermal conductivity is represented using only the first term of Taylor expansion. This approach is accurate inside a particular neighbourhood of an expansion point, but cannot be accurate for the large temperature variation range. A simple analysis of the experimental data [10]-[12] shows that the power approximation of the thermal conductance can be described

for the temperatures between 250 K and 1200 K by the below expressions proposed by the authors:

$$\lambda_{Si}(T) \cong 2533.25^{+740.79}_{-573.18}T^{-1.29697^{+0.03941}_{-0.03941}}\left[\frac{W}{cm \cdot K}\right] \qquad (6a)$$

$$D_{Si}(T) \cong 4269.39^{+3653.64}_{-1968.79}T^{-1.49425^{+0.09653}_{-0.09652}}\left[cm^2/s\right] \qquad (6b)$$

The thermal conductivity of polycrystalline silicon is reduced by the phonon grain boundary scattering effect [14]. This phenomenon is similar to the electrical resistance of segmented conductor or magnetic resistance of segmented ferromagnetic materials in an electro-magnetic field and similarly depends on the grain size. Its size and particular grain conductivity depends on grain localization, poly-silicon element thickness [13], material doping [14] and deposition pressure and temperature [16]. In the case of undoped polycrystalline silicon layers the thermal conductivity can be effectively approximated by the following expression (the experimental measurements for temperatures between 250 K and 300 K have been taken from [13]):

$$\lambda_{polySi}(T) \cong 15.0257^{+48.2547}_{-11.4579}T^{-0.738631^{+0.004381}_{-0.004382}}\left[\frac{W}{cm \cdot K}\right] \qquad (7a)$$

The proposed approximation can be also used for the extrapolation of polysilicon behaviour up to critical temperatures of 1150 K, (see also [15]). Additionally, assuming the same density and specific heat, the thermal diffusivity can be approximated by the following form:

$$D_{polySi}(T) \cong \frac{k_{polySi}(T)}{k_{Si}(T)}D_{Si}(T) \cong$$
$$\cong 25.3234 T^{-0.935911}\left[cm^2/s\right] \qquad (7b)$$

The temperature dependence of the linear expansion coefficient can be approximated by the following expression using silicon measurement data (the experimental measurements have been taken from [17]):

$$\beta_{polySi}(T) \approx \beta_{Si}(T) = -0.257366^{+0.41062}_{-0.41062} +$$
$$+13.0489^{+1.95453}_{-1.95453}\times10^{-3}\,T +$$
$$+12.8971^{+2.72858}_{-2.72858}\times10^{-6}\,T^2 +$$
$$+4.3542^{+1.1517}_{-1.1517}\times10^{-9}\,T^3\left[10^{-6}\,K^{-1}\right] \qquad (7c)$$

The approximation of electrical resistivity will be taken from the paper[3] [15]

$$\rho_{polySi}(T) \cong 8.(3)\cdot\left(1+9.1\times10^{-4}\left(T-300\right) + \right.$$
$$\left. +7.9\times10^{-7}\left(T-300\right)^2\right)\left[\mu\Omega \cdot m\right] \qquad (7d)$$

Other applied parameter values, e.g. material density, have been assumed to be equal to the bulk silicon ones.

$$\delta_{polySi} \approx \delta_{Si} = 2.329\left[g/cm^2\right] \qquad (8)$$

---

[3] The electrical conductivity depends on grain size.

As it was written before, the polycrystalline silicon parameters are dependent on the grain and element size and the deposition process parameters. Therefore, the research literature reports a Young's modulus ranging between 140 GPa and 210 GPa [19]. The value 165 GPa will be the most appropriate [16].

6 CONCLUSIONS

The paper presents an approach to the over heated actuator modelling taking into account its real behaviour. It requires the use of real values of polycrystalline silicon parameters and their thermal dependences. The modelling of such devices, despite their simplicity, is quite a difficult task because of the coupled multidomain phenomena occurring in the device. The proposed model can be implemented using the RESCUER language [6]-[8] and translated to the SPICE simulation environment. The final implementation of the complex model should additionally take into account both the dynamic model response and technological parameter distribution.

7 ACKNOWLEDGEMENT

The work reported in this paper was supported by the Internal University Grant K-25/1/2003-Dz.S.

REFERENCES

[1] R. Herman. Wärmeübergang bei freier Strömung am waagerechten Zylinder zweiatomigen Gasen. VDI – Forschungsheft nr 379, Berlin 1936

[2] L. D. Landau and E. M. Lifshitz, Theory of Elasticity 3rd Rev. Edit. Butterworth-Heinemann, February 1995 (Pergamon, 1975)

[3] S. Timoshenko Theory of Elasticity. 3rd edition McGraw Hill College, June 1970

[4] J. V. Clark, D. Bindel, W. Kao, E. Zhu, A. Kuo, N. Zhou, J. Nie, J. Demmel, Z. Bai, S. Govindjee, K. S. J. Pister, M. Gu, A. Agogino, "Addressing the Needs of Complex MEMS Design." To appear in MEMS 2002. L.V., Nevada, Jan, 2002.

[5] P. Allen, J. Howard, E. Kolesar, J. Wilken. "Design, Finite Element Analysis, and Experimental Evaluation of a Thermal-Actuated Beam Used to Achive Large In-Plane Mechanical Deflection". Tech Digest. Solid-state Sensor and actuator workshop, Hilton Head Island SC, pp.191-196, June 8-11, 1998

[6] Zubert M., Napieralski A.: "RESCUER – The revolution in multidomain simulations – Part I and II", Proc. 22nd International Conference on Microelectronics (MIEL2000), Niš, Serbia, 14-17 may 2000, vol.2 , pp. 549-559

[7] M. Zubert, M. Napieralska, A. Napieralski. "RESCUER – The new solution in multidomain simulations", Elsevier, Microelectronics Journal 31 (11-12) (2000) pp. 945-954

[8] M. Zubert, A. Napieralski, M. Napieralska. „The Application of RESCUER Software to Effective MEMS Design and Simulation", *Bulletin of the Polish Academy of Sciences – Technical Science,* No. 4, Warsaw, November 2001

[9] J. V. Clark, at all, "Sugar: Advancements in a 3D Multi-Domain Simulation Package for MEMS". *In Proc. of the Microscale Systems: Mechanics and Measurements Symposium.* Portland, OR, June 4, 2001

[10] B. Abeles, D.S. Beers, G.D. Cody, J.P. Dismukes. *Phys. Rev.* (USA) vol.125 (1962) p.44

[11] P. Turkes. *Phys. Status Solidi* A (Germany) vol.75 no.2 (1983) p.519-23

[12] C.J. Glassbrenner, G.A. Slack. *Phys. Rev.* (USA) vol.134 (1964) p.A1058

[13] S. Uma A.D. McConnell M. Asheghi K. Kurabayashi and K.E. Goodson. "Temperature dependent thermal conductivity of undoped polycrystalline silicon layers". *14th Symposium on Thermophysical Properties*, Boulder, Colorado, U.S.A., June 25-30, 2000

[14] M. Asheghi, K. Kurabayashi, R. Kasnavi, K. E. Goodson. "Thermal conduction in doped single-crystal silicon films". Journal of Applied Physics. Vol. 91, No 8, 15 April 2002, pp. 5079-5088

[15] S. Holzer, R. Minixhofer, C. Heitzinger, J. Fellner, T. Grasser, S. Selberherr "Extraction of Material Parameters Based on Inverse Modeling of Three-Dimensional Interconnect Structures". *9th THERMINIC Workshop*, 24-26 Sep. 2003, Aix-en-Provence, France, pp. 263-268

[16] "MEMS Reliability Assurance Guidelines for Space Applications", NASA, *Jet Propulsion Laboratory Publication 99-1*, California, January 1999

[17] Okada, Y. and Y. Tokumaru, *J. Appl. Phys.* **56**, 2 (1984) 314-320.

[18] R. Hull. "Properties of Crystalline Silicon", *INSPEC*, London, 1999

[19] D. W. Burns, Micromechanics of Integrated Sensors and the Planar Processed Pressure Transducer, Ph.D. Thesis, University of Wisconsin, Madison, 1988.

[20] M. Zubert., M. Napieralska., A. Napieralski, J.L. Noullet: "Generation of Reduced Models for Over-Heated Actuators. Multidomain Simulation Program". ", *Proceedings of the 9th International Conference "Mixed Design of Integrated Circuits and Systems" MIXDES'2002*, Wrocław, Poland 20-22 June 2002, pp. 435-440

Static and Dynamic Optical Metrology of Micro-Mirror Thermal Deformation

C.R. Forest*, P. Reynolds-Browne**, O. Blum-Spahn*, J. Harris**, E. Novak**, C.C. Wong*,
S. Mani*, F. Peter*, D. Adams*

* Sandia National Laboratories[1], Albuquerque, NM, cforest@mit.edu
** Veeco Instruments, Inc., Tucson, AZ, tbrowne@veeco.com

ABSTRACT

Metrology of MEMS can provide feedback for accurate modeling of thermal-mechanical behavior and understanding of thin film properties. Towards this end, measurements of the thermal-mechanical response of a micro-mirror design have been performed using interferometry. The thermal mechanical response of a 500 μm diameter mirror is induced with spot heating using a 00 μm diameter infrared laser at optical powers up to .5 W. The transition between elastic and plastic deformation is observed. Amplitudes of elastic and plastic deformation up to 120 nm and 3 μm are measured, respectively. Optical heating is then cycled in the elastic regime, permitting a novel application of stroboscopic interferometry. Results indicate that heating and cooling of the micro-mirror occurs over approximately 2.2 ms, as measured with a temporal resolution of 0.14 ms. Deformation amplitude is proportional to optical power, and depends on micro-mirror surface coating.

Keywords: stroboscopic interferometry, optical profilometry, MOEMS, optical switch, thermo-mechanical

1 INTRODUCTION

Accurate and repeatable metrology of micro-electro-mechanical systems (MEMS) is vital to their widespread application. Metrology provides feedback for design and manufacturing process improvement which leads to predictable, reliable performance. Since MEMS usually contain moving parts, the capability to measure dynamic behavior is also useful. Such applications include the characterization of vibration amplitudes, frequencies, and mode shapes, for example. The observation of mechanical responses at these length scales can lead to new understanding of thin film or micro-mechanical properties for better modeling of these phenomena.

.1 Integrated micro-switching system

A two-position micro-optical switch is being developed at Sandia National Laboratories for surety applications.

Figure 1: The optical shutter is a polysilicon gear, fabricated using micro-machining technology.

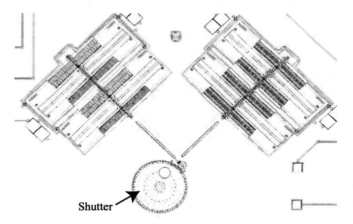

Figure 2: An electrostatic comb drive enables angular rotation of the shutter.

This system features an optical shutter, a vertical cavity surface emitting laser (VCSEL), and a photodetector. The optical shutter, shown in Figure 1, is a 500 μm diameter gear with an off-axis thru-hole. Its angular position is controlled by an electrostatic comb drive as shown in Figure 2. The device is fabricated by surface micro-machining technology (SUMMiT™ process).

In the original design [1], incident radiation from the laser passes directly through the off-axis hole and through a via in the underlying substrate to the detector. In the alternate switch state, the shutter is rotated to block the path of the incident radiation to the detector. Further design modification has led to utilization of the shutter as a micro-mirror as well. In this configura-

[1]Sandia is a multiprogram laboratory operated by Sandia Corporation, a Lockheed Martin Company, for the United States Department of Energy's National Nuclear Security Administration under Contract DE-AC04-94AL85000

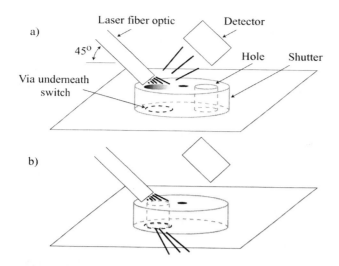

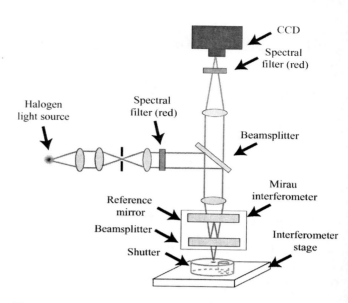

Figure 3: (a) Switch in the "on" position–incident laser radiation is reflected from the switch to the detector. (b) switch in the "off" position allows light to pass through.

tion, incident radiation reflects from the shutter's mirror surface to the detector as shown in Figure 3a. Figure 3b illustrates the "off" position, in which rotating the gear 180° allows the radiation to pass through the off-axis hole and through a via in the underlying substrate. The design modification is desirable since physical gear absence results in an "off" switch state.

Under these operating conditions, the VCSEL emits radiation for a period of tens of seconds before the shutter is rotated to the "off" position. For this duration, radiation incident on the mirror surface is partially absorbed, subjecting the shutter to localized spot heating. The resulting thermal expansion and deformation creates undesirable mechanical and optical performance, such as spurious reflections or binding between the shutter and its hub, which can prevent rotation. Understanding the heat transfer model for this system is crucial to mitigating these concerns. A thermo-mechanical heat transfer model in which absorbed radiation results in predictable deformation can be used to redesign the switch to prevent binding by minimizing deformation or moving deformation to an area where there is no performance impact [2]. Empirical data is required to generate this model. Therefore, we have pursued metrology of the deformed switch. This work, specifically, concerns the measurement of elastic and plastic thermally-induced deformation of the optical switch as a function of time and incident optical powers in order to create and validate a suitable heat transfer model.

2 OPTICAL SWITCH EXPERIMENTAL APPARATUS

For these experiments, a simple apparatus was devised to induce the optical shutter thermal response.

Figure 4: The Wyko optical profilometer uses a Mirau interferometer to measure the shutter at 10× magnification. The spectral filters reduce the bandwidth and protect the CCD array during the thermal measurements.

A semiconductor laser operating at 832 nm was connected to a multimode optical fiber. The bare fiber core was then positioned over the micro-mirror at 45° angle of incidence with a three axis translation stage, thus replicating the illumination scheme for the actual optical switch. The shutter actuator was not included since only the "on" position was of interest–the shutter angular position in which local spot heating occurs. This configuration resulted in an approximately 100 μm spot size on the 500 μm diameter micro-mirror. Experiments were performed with both "bare" and gold-coated polysilicon micro-mirrors. This 500 Å thick gold coating was applied to increase surface heat conduction. Using a variable power op-amp, the optical power incident on the micro-mirror could be varied from 0–1.5 W.

3 MEASUREMENT OF THERMAL DEFORMATION

To measure the thermal deformations, a Wyko optical profilometer (Veeco Instruments, Inc., NT3300 and NT 1100), utilizing a Mirau interferometer, was selected for its high spatial resolution, accuracy, and repeatability. The experimental apparatus was placed on the profilometer stage as shown in Figure 4. The instrument's CCD array was protected from the laser's intense infrared radiation by a narrowband red filter, matched to the instrument's source illumination.

3.1 Static measurements

Measurements of the switch were taken before heating and after the switch thermal response had reached

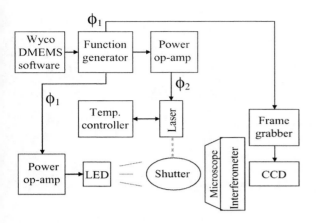

Figure 5: The function generator, controlled by Wyco software, sends drive square wave signals to the frame grabber and LED with phase ϕ_1 for image capture. A phase shifted square-wave (ϕ_2) of the same frequency is sent to the laser to induce thermal deformation.

steady state to determine the threshold optical power for plastic deformation and corresponding deformation amplitude. These static tests provided the boundary conditions for a purely elastic thermal response. The test procedure was as follows: a measurement of the room-temperature mirror surface was taken; optical power was set to 100 mW; a measurement of the heated, deformed mirror was taken while the laser was on. This cycle was repeated for increased optical powers for both the bare polysilicon and gold-coated mirror samples.

3.2 Dynamic measurements

The interferometer setup was modified to capture the mirror thermal deformations as they occurred. These dynamic measurements were performed using stroboscopic interferometry [3]. A schematic of the system is shown in Figure 5. As the name of the technique implies, pulse square-wave signals were used to drive the laser, CCD frame grabber and illumination. The phase of the laser heating, ϕ_1, was varied constantly with respect to the phase of the image capturing (frame grabber and illumination), ϕ_2. Both signals have the same frequency. This resulted in a set of data which fully recorded the mirror heating/cooling cycle. For these measurements, square wave drive signals at 20 Hz and 50% duty cycle were used. The phase shift, $\Delta\phi = \phi_1 - \phi_2$, was initially zero and was incremented by 1 degree per cycle. A higher signal frequency could have been used, up to a limit at which the laser would be cycled off before the mirror is heated to steady-state temperature and deformation. Smaller phase shift steps could also have been implemented if more temporal resolution of the transient behavior were desired.

For data integrity, the heating/cooling cycle in this stroboscopic interferometry measurement must be re-

Figure 6: Optical shutter physically deformed by locally heating with an infrared laser. Compare with Figure 1. Data absence at left edge of shutter caused by optic fiber obstruction of profilometer.

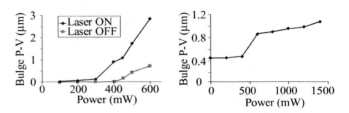

Figure 7: Static micro-mirror peak-valley (P-V) deformation for bare polysilicon (*left*) and gold-coated (*right*). Measurements taken with laser on, then off, for various optical powers.

peatable. Towards that end, results from the static measurement experiments were used to set the laser power below the threshold for plastic deformation. Thus, the purely elastic thermal response of the micro-mirror could be cycled thousands of times without residual mirror damage. Both optical power and mirror design (bare or gold-coated) were varied to study the dynamic thermal response.

4 Results and discussion

The static and dynamic experiments were performed to characterize the shutter thermal response. An interferometric image of a micro-mirror deformed elastically by spot heating is shown in Figure 6. A characteristic bulge on the surface is caused by local thermal expansion. The bulge amplitude for optical powers tested (0–1.5 W) ranged from 10's of nanometers to 3 μm.

Results from the static measurements are presented in Figure 7. As the bare polysilicon mirror is subjected to laser powers greater than 400 mW, the bulge present during heating does not entirely disappear after laser cessation. A micro-structural analysis of this plastically deformed region reveals recrystallization, indicating that melting has occured. Bulge P-V, measured with the laser on, was up to 150 nm for the elastic region and increased to over 3 μm for plastic deformation of the bare polysilicon mirror.

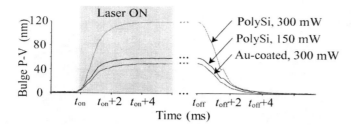

Figure 8: Experimental heating and cooling measurements, captured with stroboscopic interferometry. The bulge peak-valley (P-V) is proportional to incident power and depends on micro-mirror surface coating.

For the gold-coated micro-mirror, optical powers up to 1.5 W over an approximately 8000 μm^2 area failed to produce plastic deformation. Notably, however, the surface area of the bulged region was more than twice as large as for the bare polysilicon micro-mirror due to increased thermal conduction on the gold-coated surface. For this gold-coated mirror, elastic deformation increased nearly linearly with optical power, from 0.45 μm with no optical power to 1.1 μm at 1.5 W. Results from these static measurements indicate that the optical power on the bare polysilicon micro-mirror should be kept below 400 mW for the dynamic measurements to keep the deformation purely elastic.

The dynamic measurements were then performed. This technique permitted the observation of micro-mirror heating and cooling which occured over time scales of a few milliseconds. Results are shown in Figure 8. The heating and cooling cycles were measured at 150 mW and 300 mW for polysilicon, and at 300 mW for the gold-coated micro-mirrors. The bulge P-V is proportional to optical power, and agrees with the previous static measurements. Clearly, the gold coating distributes energy more and reduces the P-V deformation as compared with the polysilicon data. The time required to heat and cool the mirrors to steady-state is approximately 2 milliseconds.

Along with this quantitative analysis, qualitative images were captured for the heating and cooling of the devices. Figure 9 shows the transient cooling of a bare polysilicon mirror. The laser is completely off after 0.3 ms (frame 3), and thus energy is solely leaving the mirror, primarily by conduction to the underlying substrate via the hub [2]. Thus, stroboscopic interferometry has permitted the characterization of the dynamic thermal response of the micro-mirror.

5   CONCLUSIONS

For both the silicon and gold-coated mirrors, exposure to infrared radiation from a laser produced a steady-state deformation after approximately 2 ms. This thermal-mechanical behavior has been measured using

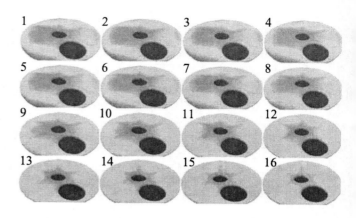

Figure 9: Sixteen images of a bare polysilicon micro-mirror over a span of 2.2 ms (resolution = 0.139 ms) illustrate transient cooling behavior following laser cessation.

interferometry, including a new application of stroboscopic interferometry. Results indicate that for the elastically deformed regime, deformation is proportional to optical power. Gold-coating reduced the P-V deformation for the same optical power and prevented plastic deformation for powers up to 1.5 W. In contrast, the bare polysilicon mirror was plastically deformed at powers greater than 400 mW. Amplitudes of deformation ranged from 10's of nanometers to microns for both the bare polysilicon and gold-coated mirrors.

These measurements provide a characterization of the micro-mirror thermal response that will be used to improve the design and analysis of this device. Issues such as how the heat can be better distributed, and how deformation can be reduced in critical locations such as the bearing hub can be better understood and solved.

These novel observations of repeatable, transient MEMS thermal response provide an essential contribution to understanding the heat transfer mechanisms in these thin films. Thermal-mechanical analysis software can be validated and improved by comparison with these experimental results. There are numerous other applications in which dynamic metrology (i.e., stroboscopic interferometry) of thermal-mechanical behavior could be useful, such as bolometers and temperature sensitive safety mechanisms.

REFERENCES

[1] A.D. Oliver, F.J. Peter, M.A. Polosky, "Microsystems based on surface Micromachined Mechanisms," Proceedings of IEEE, Sensors, (2002)

[2] C.C. Wong, S. Graham, "Investigating the Thermal Response of a Micro-Optical Shutter," IEEE/ASME Itherm2002 (2002)

[3] E. Novak, D. Wan, P. Unruh, M. Schurig, "MEMS metrology using a strobed interferometric system", ASPE Winter Topical Meeting, (2003)

Field-induced Dielectrophoresis and Phase Separation
for Manipulating Particles in Microfluidics

D.J. Bennett[*], B. Khusid[**], C.D. James[***], P.C. Galambos[****], M. Okandan[*****], D. Jacqmin[******],
A. Kumar[*******], Z. Qiu[********], and A. Acrivos[*********]

[*]New Jersey Institute of Technology, University Heights, Newark, NJ 07102 and Sandia National
Laboratories, POB 5800, Albuquerque, NM 87185, djbenne@sandia.gov
[**]New Jersey Institute of Technology, University Heights, Newark, NJ 07102, khusid@adm.njit.edu
[***]Sandia National Laboratories, POB 5800, Albuquerque, NM 87185, cdjame@sandia.gov
[****]Sandia National Laboratories, POB 5800, Albuquerque, NM 87185, pcgalam@sandia.gov
[*****]Sandia National Laboratories, POB 5800, Albuquerque, NM 87185, mokanda@sandia.gov
[******]NASA Glenn Research Center, Cleveland, OH 44135, fsdavid@tess.lerc.nasa.gov
[*******]City College of New York, 140th St., New York, NY 10031, anil@levdec.engr.ccny.cuny.edu
[********]City College of New York, 140th St., New York, NY 10031, qiu@lisgi6.engr.ccny.cuny.edu
[*********]City College of New York, 140th St., New York, NY 10031, acrivos@scisun.sci.ccny.cuny.edu

ABSTRACT

We report observations of a new electric field- and
shear-induced many-body phenomenon in the behavior of
suspensions. Its origin is dielectrophoresis accompanied by
the field-induced phase separation. As a result, a suspension
undergoes a field-driven phase separation leading to the
formation of a distinct boundary between regions enriched
with and depleted of particles. The theoretical predictions
are consistent with experimental data even though the
model contains no fitting parameters. It is demonstrated that
the field-induced dielectrophoresis accompanied by the
phase separation provides a new method for concentrating
particles in focused regions and for separating biological
and non-biological materials, a critical step in the
development of miniaturizing biological assays.

Keywords: dielectrophoresis, field-induced phase
separation, concentrating particles, separation of biological
and non-biological materials

Compared to other available methods, ac
dielectrophoresis is particularly well-suited for the
manipulation of minute particles in microfluidics [1]. The
ac dielectrophoretic force acting on a sphere immersed in a
liquid is [2]

$$\mathbf{F}_{dep} = \frac{3}{2}\varepsilon_0\varepsilon_f v_p \, Re[\beta(\omega)]\nabla\langle \mathbf{E}^2 \rangle \qquad (1)$$

where v_p is the particle volume, ε_0 is the vacuum
permittivity, ε_f is the dielectric constant of the liquid,
$Re(\beta)$ is the real part of the relative particle polarization

at the field frequency β, and $\langle \; \rangle$ denotes time averaging
over the field oscillation. Depending on the sign of $Re(\beta)$,
the particle moves toward the regions of high field strength
(positive dielectrophoresis) or low field strength (negative
dielectrophoresis). The concepts currently favored for the
design and operation of dielectrophoretic micro-devices
adopt the approach used for macro-scale electric filters [3].
This strategy considers the trend of the field-induced
particle motions by computing the spatial distribution of the
field strength over a channel as if it were filled only with a
liquid and then evaluating the direction of the
dielectrophoretic force, Eq. (1), exerted on a single particle
placed in the liquid. However, the exposure of suspended
particles to a field generates not only the dielectrophoretic
force acting on each of these particles, but also the dipolar
interactions of the particles due to their polarization.
Furthermore, the field-driven motion of the particles is
accompanied by their hydrodynamic interactions. These
long-range electrical and hydrodynamic interparticle
interactions are entirely neglected in analyzing the
performance of currently used micro-fluidic devices.

In [4-6], we demonstrated that a single-particle model
which only takes into account the dielectrophoretic force,
Eq. (1), the Stokes drag force, and the gravity force acting
on a particle predicts fairly well the rate of the field-driven
particle redistribution for dilute suspensions containing ~
0.1% (v/v) but does not predict the aggregation pattern
formed by these particles. Furthermore, the presence of the
interparticle dipole-dipole interactions was found to impose
a lower bound on the scale of microelectrode arrays for the
precise positioning of positively polarized particles in
selected locations of a dielectrophoretic micro-channel even
for ~0.1% (v/v) suspensions [7]. Now we report
observations of dielectrophoresis accompanied by a field-
induced phase separation whose origin is the interparticle
electric interactions. As a result, a suspension of negatively

polarized particles undergoes a field-driven phase separation leading to the formation of a distinct boundary between regions enriched with and depleted of particles.

Microfluidic devices for experiments were fabricated in Sandia National Laboratories Microelectronics Development LAB using Sandia's SwIFTTM (Surface micromachining with Integrated microFluidics Technology) process [8, 9] for producing highly-integrated monolithic multilayer structures. The fluidic channels, fabricated using single 6 inch Si wafers, are 40 μm wide, 6 μm high, and 570μm long and contain 6 μm wide p-doped polysilicon traces, with each trace being split into two electrically-connected electrodes located on the ceiling and the floor of the fluidic channel and arranged perpendicular to the flow (Fig. 1) [10]. Access ports to the channels were fabricated using a two step Bosch-etch process which consists of a counter-bore backside etch partly into the Si substrate, followed by a through-wafer etch to the channels on the front surface of the substrate. The suspension was delivered into the channel at a flow rate ranging from 0.24 pL/s to 9.6 pL/s. The electrodes were energized with a sinusoidal wave, 10 V peak-to-peak (ptp) and 15-30 MHz while the silicon substrate was grounded. These electrodes act like "dielectrophoretic gates" [according to Eq. (1)], that control the motion of polarized particles flowing through the channel. The suspensions were prepared by dispersing 1μm polystyrene spherical beads (1.05 g/cm^3, Duke Scientific Co.) in deionized (DI) water. Following [4, 5], the bead polarizability, $\beta \cong -0.45 - 0.27i$ for 10-30 MHz was calculated from the low-field measurements of the suspension complex dielectric permittivity. The photos presented in Fig. 1 show the time evolution of the particle distribution in a 0.1% (v/v)-suspension from which it is evident that the beads, experiencing negative dielectrophoresis, are repelled from the dielectrophoretic gate and accumulate in a region near the electrodes (Fig. 1). On account of compressive electric and shear stresses, the layer of these beads forms a round bolus with a distinct front between the regions enriched with and depleted of particles. The average particle concentration in the bolus was estimated [10] to run as high as 40-50% (v/v).

An electro-hydrodynamic model proposed for simulating the bolus formation [10] generalizes a thermodynamic theory [11-13] for a suspension subject to a spatially non-uniform ac field. This model encompasses the quasi-steady electrodynamic equations together with the momentum and continuity balance equations of the "mixture" model for a suspension, which are averaged over the field oscillations. The suspension is viewed as an effective Newtonian fluid the viscosity which varies with the particle concentration according to the Leighton-Acrivos expression. The bulk electric force exerted on a suspension, and the particle velocity relative to the suspending fluid are expressed in terms of the chemical potential of the particles [11-13]. Depicted in Fig. 2 is the simulation of the time evolution of the particle distribution using the data for water, for the particle properties, and for

the channel dimensions presented above. As can be seen, the particles, which accumulate in the region close to the microelectrodes, undergo a field-induced phase separation and form two concentrated layers near the channel top and bottom (Fig. 2). The theoretical predictions for the bolus growth, for the particle concentration in the bolus, and the fact that particles travel around the circumference of the bolus and are then drawn into the bolus side closest to the electrode, are consistent with the observations (Fig. 1). In line with the experiments, the simulations indicated that our experimental conditions lie in the transition region between the formation of one bolus located near the channel centerline and two boluses attached to the channel side walls. For example, a slight change (~8%) in the voltage distribution along the microelectrode leads to the appearance of two boluses [Fig. 2(d, h)].

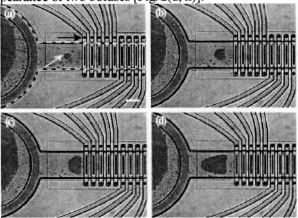

Figure 1. Dielectrophoresis (10V ptp, 30 MHz) of beads accompanied by a phase separation (white arrow). The dashed lines denote the outline of the fluidic channel, and the black arrow indicates the energized electrode. The dark regions on the polysilicon traces indicate the places where the ceiling electrodes are in direct contact with the fluid. The time is (a) 10s, (b) 70s, (c) 120s, and (d) 180s. The flow is from left to right. Scale bar = 20 μm [10].

The photo presented in Fig. 3 illustrates the dielectrophoretic separation of a heterogeneous mixture of polystyrene beads and heat-killed bacterial cells (Staphylococcus aureus; Molecular Probes) dispersed in DI water. The beads, experiencing negative dielectrophoresis, form two boluses adjacent to the channel walls [similar to the simulation results shown in Fig. 2 (d, h)]; whereas the cells, which are more polarizable than DI water at this frequency, experience positive dielectrophoresis and are collected on the electrode. The complete separation of the continuously flowing cell-bead mixture was achieved by cycling the activation of the electrode off and on. During the off phase, the cells and particles which were dielectrophoretically separated during the on phase, maintained their spatial separation and were carried along the channel to the outlet port, but, before the beads had reached the edge of the leading electrode, the field was turned back on, thereby repelling the beads again (Fig. 3) to distance ~ 20 μm from the electrode edge. In contrast, the cells which had not been collected on the electrode continued moving with the flow towards the outlet port of the device.

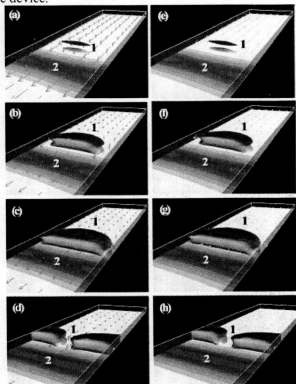

Figure 2. Numerical simulations of the concentration contours (1) for the 0.1 % (v/v)-suspension, 8.64 pL/s, 10 V ptp. In the panels, the fluid flow is from the upper right-hand corner to the lower left-hand corner. The color band (2) indicates the variation of E^2 (in kV^2/cm^2) along the channel bottom (white for up to 1.3, pink for 1.3-2.7, and blue for 10.6-11.9), whose maximum is located near the first electrode (Fig. 1). Time is (a, e) 10s, (b, f) 70s, (c, g) 180s, and (d, h) 120s. The computed values for c_{max} in the bolus [in % (v/v)] are (a) 8.73, (b) 45.4, and (c) 56.4, and (d) 54.1. In (a) to (d), the red arrows show the relative magnitude of the flow velocity [v_{max} (in μm/s) = (for t=0) 36, (a) 49.8, (b) 106.3, (c) 110.4, and (d) 139.9]. In (e) to (h), the green arrows show the relative magnitude of the particle velocity, $v_{p,max}$ (in μm/s) = (for t=0) 36, (e) 45.2, (f) 78.8, (g) 151.4, and (h) 171.6] [10].

To test theoretical predictions quantitatively, experiments were conducted [14] on 5%-15 %(v/v) suspensions of neutrally buoyant spherical (average diameter of 87 μm) polyalphaolefin particles (AVEKA, Woodbury, MN) suspended in corn oil (0.92g/cm^3, 59.7 cp at 23^{0}C) which exhibit negative dielectrophoresis. Following [4, 5], the particle polarizability was calculated from the low-field measurements of the suspension complex dielectric permittivity: $Re(\beta) \cong -0.15$ was found to equal, approximately, -0.15 for 0.1- 3.5 kHz,

whereas $Im(\beta)$ was found to decrease rapidly with frequency from $1.8 \cdot 10^{-3}$ for 0.1 kHz to less than $2 \cdot 10^{-4}$ for frequencies greater than 1 kHz. The experiments were performed in a horizontal parallel-plate chamber (6 cm wide, 12 cm long, and 3 mm high). The bottom of the chamber was equipped with electrodes (1.6 mm wide) which were embedded into groves at 2 mm intervals (Fig. 4) and were alternately connected to the high voltage and to the ground. The top of the chamber, consisting of a transparent glass coated with a conducting layer of indium tin oxide, was grounded. Under these conditions, the field configuration consists of two high-strength regions near the electrode edges and a wide low-strength region above the center of the grounded electrode near the channel midplane.

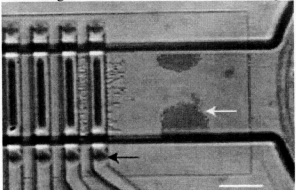

Figure 3. Dielectrophoretic (10V ptp, 15 MHz) separation of bacterial cells and beads. The cells adhere to the energized electrode (black arrow) while the beads experience negative dielectrophoresis accompanied by a phase separation (white arrow). The flow is from right to left. Scale bar = 20 μm [10].

The photos in Fig. 4 show the field-driven time evolution of the particles (seen as white spots in the photos) in the 10%(v/v)-suspension. After about 150 s a distinct front between the suspension and the suspending fluid containing very few particles [Fig. 4(d)] formed along the channel. As time progressed, the front slowly moved away from the high field regions and the particles became progressively confined to a thin column above the grounded electrode. Note that the same configuration of an electric field does not lead to the front formation in dilute suspensions, ~0.01%-0.2% (v/v) [4].

Simulations were conducted using the data of fluid and particle properties and the channel dimensions presented above. The theory predicts that the use of the characteristic time for dielectrophoresis τ_d, Eq. (2), [4, 5] makes it possible to combine the data on the time dependence of the front position for different voltages and frequencies (Fig. 5) into one band for $t/\tau_d \leq 20$. Here

$$\tau_d = \frac{3d^4\eta_f}{a^2\varepsilon_0\varepsilon_f|Re(\beta)|V_{rms}^2}, \qquad (2)$$

where η_f is the viscosity of the suspending fluid, d is the electrode width, a is the particle radius, and V_{rms} is the root mean square of the applied ac voltage. As can be seen from Fig. 5 (b), the simulation results are in a reasonable agreement with the experimental data.

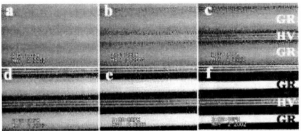

Figure 4. The particle distribution in a suspension with 10%(v/v) particle concentration (a) before and (b)-(f) following the application of a field 5kVrms, 100Hz at (b) 45s, (c) 90s, (d) 150s, (e) 300s, and (f) final state, ~39min. The electrode width is 1.6mm. HV and GR refer to high-voltage and grounded electrodes respectively [14].

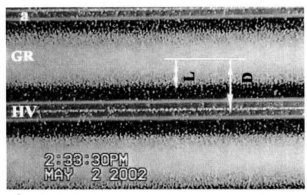

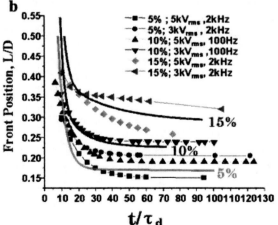

Figure 5. (a) The photo illustrates how the font location was measured; HV and GR refer to the high-voltage and grounded electrode respectively. (b) The experimental data (symbols) and computational results (solid lines) for 5%, 10%, and 15% (v/v) suspensions [14].

In conclusion, we demonstrated that dielectrophoresis accompanied by a field-induced phase transition provides a powerful method for strongly concentrating particles and for separating biological and non-biological particles. In particular, substances less polarizable than water (nearly all inorganic materials) can be removed from aqueous solutions. The predictions of the proposed electro-hydrodynamic model are consistent with the experiments even though the model contains no fitting parameters.

The work was supported, in part, by grants from the NASA, NAG3-2698 (B.K., Z.Q., and A.A), and from the DARPA through the Bioflips/Simbiosys Program, Mission Research Corporation/DARPA Contract # DAAH01-02-C-R083 (B.K.). D.B. thanks Sandia's MESA Institute and the NSF MAGNET/SEM program for support. The properties of polystyrene beads were measured by Len Duda (Sandia). The properties of polyalphaolefin particles were measured at the NJIT W.M. Keck Foundation Laboratory. Sandia is a multiprogram laboratory operated by Sandia Corporation, a Lockheed Martin Company, for the U.S. DOE under contract DE-ACO4-94-AL85000.

REFERENCES

[1] P.R.C. Gascoyne and J. Vykoukal, Electrophoresis 23, 1973 (2002).

[2] T.B. Jones, "Electromechanics of Particles," Cambridge University Press, 1995.

[3] "Handbook of Electrostatic Processes," Eds. J.-S. Chang, A.J. Kelly, and J.M. Crowley, Marcel Dekker, 1995.

[4] A. Dussaud, B. Khusid, and A. Acrivos, J. Appl. Phys., 88, 5463 (2000).

[5] Z. Qiu, N. Markarian, B. Khusid, and A. Acrivos, J. Appl. Phys., 92, 2829 (2002).

[6] N. Markarian, M. Yeksel, B. Khusid, K. Farmer, and A. Acrivos, J. Appl. Phys., 94, 4160 (2003).

[7] N. Markarian, M. Yeksel, B. Khusid, K. Farmer, and A. Acrivos, Appl. Phys. Lett., 82, 4839 (2003).

[8] M. Okandan, P. Galambos, S. Mani, and J. Jakubczak, "Proceedings of SPIE," 4560, 133 (2001).

[9] M. Okandan, P. Galambos, S. Mani, and J. Jakubczak, "Micro Total Analysis Systems 2001," 305-306.

[10] D.J. Bennett, B. Khusid, C.D. James, P.C. Galambos, M. Okandan, D. Jacqmin, and A. Acrivos, Appl. Phys. Lett., 83, 4866 (2003).

[11] B. Khusid and A. Acrivos, Phys. Rev. E, 52, 1669 (1995).

[12] B. Khusid and A. Acrivos, Phys. Rev. E 60, 3015 (1999).

[13] B. Khusid and A. Acrivos, Phys. Rev. E 54, 5428 (1996).

[14] A. Kumar, Z. Qiu, A. Acrivos, B. Khusid, and D. Jacqmin, Phys. Rev. E 69, (2004) (*accepted*)

Optimum Design of an Electrostatic Zipper Actuator

Michael P. Brenner*, Jeffrey H. Lang**, Jian Li***, and Alexander H. Slocum***

* Division of Engineering and Applied Sciences, Harvard University
Cambridge, MA 02138, brenner@deas.harvard.edu
Departments of Electrical** and Mechanical*** Engineering, Massachusetts Institute of Technology
Cambridge,MA 02139

ABSTRACT

We describe a method for predicting the optimal design of an electrostatic zipper actuator. The shape of the zipping beam and the electrode are optimized to maximize the applied force at a given stroke, while satisfying fabrication constraints, and the constraint that the pull-in instability occurs for a given applied voltage.

1 INTRODUCTION

This paper describes an algorithm for predicting the optimal design of an electrostatic zipper actuator[1], [2]. Such actuators operate [3] by charging up two electrodes (coated with thin dielectric layers) relative to each other. At least one of the electrodes is flexible, with a load attached to its end. When the charging voltage exceeds a critical value, the flexible electrode(s) "buckle" (the so-called "pull-in" instability), causing the two electrodes to come into contact (See Fig. 1). Further increase of the voltage causes the flexible electrode to "zip" along the other electrode, pushing against the applied load.

The force that an electrostatic actuator can exert depends on the applied voltage V, the thickness d of the dielectric layer (with dielectric constant ϵ_i), the beam thickness h, length L and extent b, and the elastic modulus E of the flexible electrode. In typical applications, the applied voltage is a modest ~ 100V; the force created by the actuation mechanism is used to deform silicon beams, with a bulk modulus $E \sim 10^{11}$Pa. Since the maximum energy density in the electric field $\epsilon_i V^2/d^2 \ll E$, such a small actuation voltage can only deform a silicon beam whose length L is much longer than its thickness. This constraint therefore sets the limit on the size of the device that can be successfully actuated.

Herein, we address the following optimization problem: what is the shape of the electrostatic actuator, undergoing the pull-in instability at a given voltage, and achieving a prespecified load, which has the smallest length L. The intuition that underlies our analysis is as follows: the force exerted by the zipper actuator clearly increases by increasing the thickness of the beam applying the force. However, increasing the thickness increases the threshold voltage for the pull-in instability. Therefore, the two requirements (achieving high force but remaining above the pull-in instability threshold) are competing constraints–designs which achieve pull-in at low applied voltages exert small forces; designs which exert large forces are hard to make pull-in. Our optimization strategy will aim to find the design which maximizes the force while satisfying a given pull-in voltage. Here we present the essential steps towards the solution of this problem for a flexible beam zipping on a straight electrode; a more complete presentation will include the generalization of the analysis to simultaneously compute the shapes of both the beam and the electrode.

2 THE MODEL

We model the electrostatic actuator as a laterally compliant beam (with moment of inertia $I(x) = 1/12bh^3$, where b is the breadth of the beam and h the thickness) above a fixed electrode, with a thin, uniform dielectric layer deposited on top of it. The beam and the electrode are held at a voltage V_0 relative to each other. If the deflection of the beam is $w(x)$, there is an electrostatic force pulling the beam and the electrode together of magnitude $0.5b\epsilon_0 V_0^2/(\epsilon_0 w + \epsilon_i d)^2$, where ϵ_0 (ϵ_i) is the dielectric constant of air (insulator), respectively. Hence the equation for the deflection of the beam is

$$(EIw'')'' = -\frac{b\epsilon_0}{2}\frac{V_0^2}{(\epsilon_0 w + \epsilon_i d)^2}. \qquad (1)$$

We nondimensionalize this equation by scaling w and S by the beam thickness h, and scaling all lateral length scales by the length of the electrode L. The spatial dependence of the beam thickness is accounted for by writing $I = bh^3/12\psi(x)$, so that

$$(\psi w'')'' = -\frac{\Gamma}{(w+\beta)^2}, \qquad (2)$$

where $\Gamma = \epsilon_0(V_0/h)^2(L/h)^4/E$ and $\beta = \epsilon_0 d(\epsilon_i h)^{-1}$. In these units, $w = w(y)$, where $0 \le y = x/L \le 1$.

We would now like to understand the solutions to (2) as the voltage is increased. We first assume that the end of the beam farthest from the electrode is held at fixed distance off the electrode so that $w(1) = w_0$; since there is no moment about the end we also have $w''(1) = 0$. When the electrostatic forces are small, the beam is essentially straight; above a critical Γ, there is a "pull-in" instability where the beam hugs the electrode.

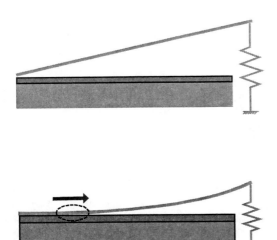

Figure 1: Schematic of the zipper actuator before (top
) and after (bottom) the pull-in instability. The top
electrode is charged to a voltage relative to the bottom
electrode, which is coated by a thin dielectric layer.

Increasing the voltage further causes the beam to zip up
along the electrode. In the first regime, before pull-in,
the boundary conditions at the beginning of the beam
depend on how the beam is held: i.e. $w(0) = w''(0) = 0$
if the end is free to rotate. In the second regime, after
pull-in, the boundary conditions are $w(0) = 0; w'(0) =
0; w''(0) = 0$, reflecting that the beam is tangent to the
electrode, and also the end is torque free. The last condition arises because if the end were not torque free then
the beam would zip further up the electrode. In the second regime, there are more boundary conditions than
allowed by the equation. The resolution to this is that
the 'zipping' point is arbitrary, so that the left boundary
condition is $w(s) = 0; w'(s) = 0; w''(s) = 0$, where s is
the zipping point.

Analysis of this equation [4] demonstrates that the
force-displacement relationship is given by

$$F_R \sim \left(\frac{\Gamma}{\beta}\right)^{3/4} \frac{1}{\sqrt{w_0}}, \qquad (3)$$

or in dimensional units

$$F \sim E^{1/4} \left(\frac{\epsilon_i V_0^2}{d^2}\right)^{3/4} \frac{d^{3/4} h^{3/4} b}{\sqrt{\Delta}}, \qquad (4)$$

where Δ is the the gap distance between the right hand
side of the zipper and the electrode. The force does
not scale as V_0^2, as one might expect from the force between two parallel capacitor plates. The physical reason
for the unusual dependence is that the force exerted on
the zipper also depends on the voltage-dependent *shape*
of the beam near the electrode. Numerical simulations
demonstrate that these formulae quantitatively predict

the force-displacement conditions over a wide parameter
range.

This formula implies that when the thickness of the
beam increases, the force increases like $h^{3/4}$; hence increasing the beam thickness increases the applied force.

3   PULL-IN INSTABILITY

The above analysis implies that increasing the force
exerted by the zipper requires increasing the beam thickness. However, there is a critical beam thickness above
which the pull-in instability does not occur for a given
voltage. Ultimately, our optimization analysis will use
a nonuniform beam shape ($\psi(x)$ not constant) to beat
these constraints as much as possible.

First we introduce a method computing the pull-in
voltage. Calculations of the pull-in voltage have been
previously carried out [3], [5], using energy methods; our
methodology for computing the pull-in voltage is easily
amenable to the optimization described below.

To calculate when pull-in occurs, we note that before
pull-in, the shape of the beam is determined by equation (2) with the boundary conditions $w(0) = w''(0) =
w''(1) = 0$ and $w(1) = w_0$. Before pull-in the deflection
from the straight beam shape is small, so $w = w_0 x + \zeta$,
where $\zeta \ll w_0 x$. Plugging into equation (2) and linearizing in ζ implies

$$(\psi\zeta'')'' = -\frac{\Gamma}{(w_0 x + \beta)^2} + \frac{2\Gamma}{(w_0 x + \zeta + \beta)^3}\zeta, \qquad (5)$$

with the boundary conditions $\zeta(0) = \zeta''(0) = \zeta(1) =
\zeta''(1) = 0$. When Γ is sufficiently small, the solution
can be expanded in powers of Γ: $\zeta = \bar\zeta = \sum_{n=0}^{\infty} \Gamma^n \zeta_n$.
At leading order we have

$$(\psi\zeta_1'')'' = -\frac{1}{(w_0 x + \beta)^2}.$$

In this regime, ζ depends continuously on Γ, so that
small changes in Γ lead to small changes in the shape of
the beam. At sufficiently large Γ corresponding to the
pull-in instability, this expansion breaks down. If we
write $\zeta = \bar\zeta + \xi$, this occurs when there is a nontrivial
solution to

$$(\psi\xi'')'' = \frac{2\Gamma}{(w_0 x + \bar\zeta + \beta)^3}\xi, \qquad (6)$$

satisfying the boundary conditions $\xi(0) = \xi''(0) = \xi(1) =
\xi''(1) = 0$. The existence of a nonzero ξ occurs at a
critical value of $\Gamma = \Gamma^*$, which signals a discontinuous
change in the solution, i.e. pull-in. Note that this is a
nonlinear eigenvalue problem, since $\bar\zeta = \bar\zeta(\Gamma)$.

A good approximation to Γ^* can be obtained by using a truncated expansion for $\bar\zeta$ in equation (6). For
example, a first approximation to Γ^* can be obtained

by solving the linear eigenvalue problem

$$(\psi\xi'')'' = \frac{2\Gamma}{(w_0 x + \beta)^3}\xi \qquad (7)$$

A next approximation can be obtained by solving the nonlinear eigenvalue problem

$$(\psi\xi'')'' = \frac{2\Gamma}{(w_0 x + \Gamma\zeta_1 + \beta)^3}\xi. \qquad (8)$$

This nonlinear eigenvalue problem (8) can be solved by iteration. Our numerical calculations show that the solution to the nonlinear problem differs from the solution to the linear problem by about a factor of two; In the limit that w_0 is large, the pullin voltage obeys the approximate law $\Gamma^* \approx \left(\frac{\pi}{4}\right)^4 w_0^3$. In dimensional units, this corresponds to the critical voltage

$$\epsilon_0 V^2 \approx E\frac{\Delta^3 h^3}{12L^4}\left(\frac{\pi}{4}\right)^4. \qquad (9)$$

The critical voltage increases with increasing gap thickness Δ and beam thickness, and decreases with increasing length of the beam.

4   UNIFORM ACTUATOR

The optimal actuator design is the device of smallest length L which achieves a given load. First we assume that that both the beam thickness does not vary along the actuator, so the optimization involves finding the the beam thickness h allowing the actuator to have smallest length.

To determine the optimum choice of (h, L), we require first that the pull-in instability occurs, and second that the actuator force $F_{actuator}$ is larger than the requisite applied load. The pull-in requirement is

$$L^4 > h^6\frac{E}{12\epsilon_0 V^2}\left(\frac{w_0}{h}\right)^3\left(\frac{\pi}{4}\right)^4. \qquad (10)$$

For the force requirement, the force applied by the actuator must exceed the applied load. The load typically requires deforming an elastic element, the size of which is comparable to the actuator itself: this is because the load force is minimized by making it as large as possible, and the size of the device is not compromised if the load is as large as an actuator. For definiteness, in the present analysis we assume the load is a relay switch [8], for which the force requirement implies

$$h > \frac{E}{12\epsilon_0 V^2}\left(800h_{switch}^3 d\right)^{4/3}\frac{w_0^{2/3}}{L^4}, \qquad (11)$$

where I_{switch} and h_{switch} are the moment of inertia and thickness of the switch. The two requirements are summarized in figure 2: Below the dashed line are the (h, L)

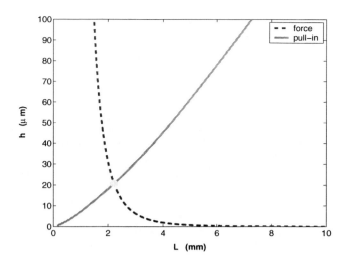

Figure 2: Phase diagram showing the allowed uniform zipper configurations. Below the dashed line are the (h, L) combinations that will undergo pullin; above the solid line are the combinations that satisfy the force constraint. Calculations assume $V = 100$Volts, $w_0 = 10\mu$m, and bulk modulus E corresponding to silicon.

combinations that will undergo pullin; above the solid line are the combinations that satisfy the force constraint.

The intersection point labelled in the figure shows the size of the smallest device satisfying the force and pull in constraints. This is the optimal device with a uniform thickness beam.

5   NONUNIFORM ACTUATOR

We now ask: can the length L of the device be made smaller by allowing the zipping beam to have nonuniform thickness? Intuitively, the reason this should be possible is as follows: the force can be increased by increasing the beam thickness at the pull in point; in contrast, the pull-in instability relies on an average beam thickness being thin. Therefore, perhaps it is possible to only thicken the beam in one region, which would increase the force but not affect the pull in instability.

In order to find the optimal device, we need several pieces of information: first, we must have a formula for how the force exerted by the actuator (at a given stroke w_0) changes upon changing the beam shape, $\psi \to \psi + \delta\psi$. Such a formula can be derived using the adjoint-method of optimal control theory[6]; the formula is of the form

$$\delta F = \int \delta\psi\lambda'' w'', \qquad (12)$$

where λ is a function that solves an adjoint equation $(\lambda''\psi)'' - \Gamma(2x/(w+\beta)^3 + 2\lambda/(w+\beta)^3) = 0$ with appropriate boundary conditions.

Second, we need a formula for how the pull-in voltage Γ changes on changing ψ. A straightforward analysis of the pullin equation (7) (analogous to [7], [8]) gives that

$$\delta\Gamma = C \int \delta\psi\zeta''^2, \qquad (13)$$

where $C = \int \zeta^2/(w_0 x + \beta)^3$.

Now, for any change in the beam shape $\delta\psi$, there will be a change in both the pullin voltage and the applied Force. Since the goal of the optimization is to produce a device with smaller extent L, we will use any increase in the applied force and decrease in the pull in voltage that we obtain to decrease L. In order to maximize our effort, we would like the decrease in L that is achievable from increasing the applied force to be *exactly* the same as that from decreasing the pull in voltage: Given the scalings $\Gamma \sim L^4$ and $F \sim L^{-3}$, this requirement translates into the constraint

$$-\frac{\delta F}{3F} + \frac{\delta\Gamma}{4\Gamma} = \int \delta\psi\left(-\frac{\lambda''w''}{3} + C\frac{\zeta'^2}{3}\right) = 0. \qquad (14)$$

The optimal device can now be found by carrying out an iterative procedure in which we choose $\delta\psi$ to increase the force, subject to the constraint equation (14). Figure 3 demonstrates the results of the iteration, assuming initial beam thickness $h = 20\mu m$, length $L = 2.2mm$; dielectric layer thickness $h_0 = 2\mu m$, stroke $w_0 = 65\mu m$, and voltage $V = 100$ volts. The top panel shows the change in the shape of the beam with iteration; the bottom shows the evolution of the force on the beam and the pull in voltage. The algorithm achieves an approximately 40 percent decrease in pull in voltage and a corresponding increase in the force: this is achieved by thickening the beam *near the zip point*, and thinning it throughout the rest of the beam. The average thickness of the beam therefore decreases (lowering the pull in voltage) , while the thickness of the beam at the touchdown point increases (raising the force). The optimal design for a given situation is chosen by the fabrication constraints.

Note that the above calculation shows the improved design allows an increase (decrease) in force (pull in voltage) for fixed L; in practice, we will keep the applied load and pull in voltage the same and decrease L.

To conclude, we have presented an algorithm for optimizing the force exerted by an electrostatic actuator while satisfying the pull in constraints, and shown that modest changes in the shape of an electrostatic actuator can lead to dramatic enhancement of the force. Larger enhancements can be obtained by simultaneously solving for the optimal beam and electrode shapes, a topic that we will present in detail elsewhere.

This research was supported by the NSF Division of Mathematical Sciences and a research grant from ABB Research Ltd, Zurich, Switzerland.

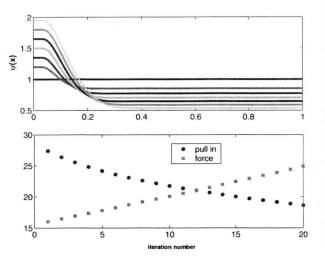

Figure 3: Evolution of the zipper profile upon optimization. Upper panel shows the evolution of the beam shape, and lower panel shows the corresponding decrease in pull in voltage and increase in applied force.

REFERENCES

[1] E. Thielicke and E. Obermeier, Mechatronics, 10, 431 455, 2000.

[2] G. Perregaux, S. Gonseth, P. Debergh, J Thiebaud and H. Vuillioenet, in Proc. IEEE 14th Ann. Int. Workshop on Micro Electro Mechanical Syst., Switzerland, 232-235 (2001) ; E. Thielicke, E. Obermeier, in Proc. IEEE/LEOS Optical MEMS 2002, 159-160 (2002); F. Sherman, C.-J. Kim and J. Woo and C.-M. Ho, in Proc. IEEE 11th Ann. Int. Workshop on Micro Electro Mechanical Sys., Germany, 454-459 (1998). ; G. Smith, J. Maloney, L. Fan and D. L. DeVoe, in Proc. of SPIE, 4559, pp. 138-147 (2001).

[3] R. Legtenberg, J. Gilbert and S. D. Senturia, J. Microelectromech. Syst., 6, 257-265 (1997).

[4] J. Li, M.P. Brenner, J. H. Lang and A. H. Slocum, submitted to J. Microelectromech. Syst.

[5] D. Bernstein, P. Guidotti and J.A. Pelesko, Proc. of Model. Simul. Micro 489-492 (2000); J.A. Pelesko, SIAM J. on Appl. Math, 62, 888-908 (2002).

[6] N V Banichuk, "Problems and Methods of Optimal Structural Design", Plenum Press, New York, 1983.

[7] J. B. Keller, Arch. Rat. Mech. Anal. 5, 275, 1960.

[8] M. P. Brenner, J. H. Lang, J. Li, J. Qiu, A. H. Slocum, Proc. Natl. Acad. Sci, 100, 9667-9773 (2003).